PRINCIPLES OF Botany

To native people of the Americas, the common sunflower was an important medicinal and food plant, a useful decoration, and a symbol used during worship; Incas carved images of the sunflower on their temple walls and jewelry. Today, baseball fans eat millions of seeds of sunflower plants during games throughout the season, and health-conscious chefs fry foods in the cholesterol-free sunflower oil that is high in polyunsaturated fats. Sunflowers are members of one of the largest flowering plant families, the *Asteraceae,* which includes over 20,000 different species. What appears as one large flower, such as the one you see wrapping around the cover, is actually a group of small flowers (not seen here); each of the small flowers may produce one seed that can be eaten or pressed for its oil.

The sunflower represents an economically valuable plant with interesting and important questions about its ecology and evolution. In this book, you'll see how the themes of ecology, evolution, and economic botany illustrate the relevance of plants as subjects of scientific study, unite the information about plants in general, and introduce the critical roles of plants in your life. You will learn how the activities and products of plants are essential to human existence and how scientists try to answer questions about plants in the fascinating study of botany.

PRINCIPLES OF
Botany

Gordon Uno
University of Oklahoma

Richard Storey
The Colorado College

Randy Moore
Texas A&M University—Kingsville

Boston   Burr Ridge, IL   Dubuque, IA   Madison, WI   New York   San Francisco   St. Louis
Bangkok   Bogotá   Caracas   Lisbon   London   Madrid
Mexico City   Milan   New Delhi   Seoul   Singapore   Sydney   Taipei   Toronto

McGraw-Hill Higher Education

A Division of The McGraw-Hill Companies

PRINCIPLES OF BOTANY

Published by McGraw-Hill, an imprint of The McGraw-Hill Companies, Inc., 1221 Avenue of the Americas, New York, NY 10020. Copyright © 2001 by The McGraw-Hill Companies, Inc. All rights reserved. No part of this publication may be reproduced or distributed in any form or by any means, or stored in a database or retrieval system, without the prior written consent of The McGraw-Hill Companies, Inc., including, but not limited to, in any network or other electronic storage or transmission, or broadcast for distance learning.

Some ancillaries, including electronic and print components, may not be available to customers outside the United States.

 This book is printed on recycled, acid-free paper containing 10% postconsumer waste.

3 4 5 6 7 8 9 0 QPD/QPD 0 9 8 7 6 5 4 3 2

ISBN 0-07-228592-3
ISBN 0-07-118087-7 (ISE)

Vice president and editor-in-chief: *Kevin T. Kane*
Publisher: *Michael D. Lange*
Senior sponsoring editor: *Margaret J. Kemp*
Senior developmental editor: *Kathleen R. Loewenberg*
Editorial assistant: *Dianne Berning*
Marketing managers: *Michelle Watnick/Heather K. Wagner*
Senior project manager: *Peggy J. Selle*
Senior production supervisor: *Mary E. Haas*
Design manager: *Stuart D. Paterson*
Cover/interior designer: *Jamie O'Neal*
Cover image: *Tony Stone Images*
Design features from: *Artville/Don Bishop*
Senior photo research coordinator: *Lori Hancock*
Photo research: *Shirley Lanners*
Senior supplement coordinator: *David A. Welsh*
Compositor: *Carlisle Communications, Ltd.*
Typeface: *10/12 Palatino*
Printer: *Quebecor Printing Book Group/Dubuque, IA*

The credits section for this book begins on page 531 and is considered an extension of the copyright page.

Library of Congress Cataloging-in-Publication Data

Uno, Gordon.
   Principles of botany / Gordon Uno, Richard Storey, Randy Moore.—1st ed.
      p.   cm.
   Includes index.
   ISBN 0-07-228592-3
   1. Botany.   I. Storey, Richard.   II. Moore, Randy.   III. Title.
QK47.U56 2001
580—dc21   00-024392
                                                        CIP

INTERNATIONAL EDITION ISBN 0-07-118087-7
Copyright © 2001. Exclusive rights by The McGraw-Hill Companies, Inc., for manufacture and export. This book cannot be re-exported from the country to which it is sold by McGraw-Hill. The International Edition is not available in North America.

www.mhhe.com

Dedication

This book is dedicated to my family and
friends who always thought more of me
than I actually was.—Gordon Uno

This book is dedicated to my family: including
Walter D., Frances, Martha, Justin, Elizabeth, and
Promise Storey for their undying support of my
passion for plants, nature, and teaching. I also
dedicate it to my long-time colleague, mentor,
and friend, Dr. Jack Carter, who taught me how to
teach botany but could never teach me how to
identify plants.—Richard Storey

I dedicate this book to Doyle, Tillie,
Janice, and Kris for their love, friendship, and
support. I also thank Lucy and Bumper for
doing their tricks.—Randy Moore

GORDON E. UNO was born and raised on the eastern plains of Colorado in the small farming town of Roggen. He received a B.A. in biology with education from the University of Colorado, Boulder, but decided that teaching in the pre-college classroom took more patience and resolve than he had. He continued with his formal education in the botany department at the University of California, Berkeley, where he received his Ph.D. in 1979. He immediately accepted a position in the botany and microbiology department at the University of Oklahoma (OU), where he vowed to stay only a year or two. Twenty-one years later, he is a David Ross Boyd Professor of Botany at OU and director of the Introductory Botany program. He is coauthor of three high-school biology texts and has written two handbooks for faculty members on teaching undergraduate science courses. In addition, he has served as President of the National Association of Biology Teachers, was a Program Officer at the National Science Foundation, and is a Fellow of the American Association for the Advancement of Science. He considers himself honored and lucky to have received three university, one state, and one national teaching award(s) and believes that longevity in the business really does pay off.

RICHARD STOREY was born in Roswell, New Mexico. He received a B.S. in Education (biology) from the University of New Mexico and, for a short time, taught biology at Manzano High School in Albuquerque. With support from the National Science Foundation, he received a Masters of Natural Science from the University of Oklahoma. In 1977, he was awarded a Ph.D. in botany from the University of Oklahoma and then conducted postdoctoral research at the C.F. Kettering Research Laboratory in Yellow Springs, Ohio. In 1978, he joined the biology faculty at the Colorado College in Colorado Springs, Colorado. During his time on the faculty, he taught botany, cell biology, and plant physiology and was awarded research grants to study protein metabolism, nitrogen fixation, and the nutritional qualities of potential crop plants. He has also received three faculty and student research and development grants from the Howard Hughes Medical Institute. In 1999, he served as President of the National Association of Biology Teachers and was named Dean of the College and Dean of the Faculty at Colorado College. He has been acknowledged with a Carnegie Association teaching award nomination from the Colorado College, has coauthored a high-school biology text, is a contributing author to three books on plant physiology, and has written over 50 scientific papers on botany. With his wife, Martha, he continues his poorly funded, but highly enthusiastic, pursuit to catch and release on a fly all of the trout inhabiting the beautiful streams of the Rocky Mountains.

RANDY MOORE received a B.S. in biology from Texas A&M University, an M.S. in botany from the University of Georgia, and a Ph.D. in biology from UCLA. Since then, he has published about 200 papers and a variety of books, including *Botany, Writing to Learn Biology,* and *Biology Laboratory Manual.* He has received "Best Professor" awards from two universities; local, state, and national teaching awards from organizations such as the National Science Teachers Association; research grants from agencies such as the National Science Foundation; and science writing awards from organizations such as the Education Press Association of America. He currently edits the popular journal *The American Biology Teacher* and serves on the editorial boards of a variety of other journals. Although he has dabbled in administration for more than a decade, his primary interests are in teaching and research.

BRIEF CONTENTS

Brief Contents

CONTENTS

CHAPTER 6

PLANT GROWTH AND DEVELOPMENT   133

CHAPTER 7

ROOT SYSTEMS AND PLANT MINERAL NUTRITION   153

CHAPTER 11

RESPIRATION 255

CHAPTER 12

FLOWERS AND FRUITS 271

CHAPTER 13

CHAPTER 14

CHAPTER 15

CHAPTER 16

BACTERIA, FUNGI, AND ALGAE   381

CHAPTER 17

BRYOPHYTES AND FERNS:
THE SEEDLESS PLANTS   409

CHAPTER 18

GYMNOSPERMS AND ANGIOSPERMS: THE SEED PLANTS 439

CHAPTER 19

ECOLOGY 469

PREFACE
Preface

WHY STUDY BOTANY?

The answer to the question *Why study botany?* differs, even among the three of us, and likely will be different from your answer; however, we feel that understanding plants is central to understanding life on earth and its evolution. Never before have so many people been concerned about the fragile, botanical wonders of this planet. With the exception of a few organisms at the bottom of the ocean and certain microorganisms, all other organisms rely on plants and algae directly or indirectly for their source of food and oxygen. Whether you think about plants or not, they are essential to your existence, and they form the foundation for all the interactions within terrestrial communities. It also may be surprising to learn that plants share many genes and characteristics with humans and other animals, and because of this, when we study plants, we learn about how other organisms function. Finally, plants are interesting organisms because of the diversity of habitats in which they live and the myriad of adaptations that have allowed them to survive.

THREE THEMES DEFINE THIS BOOK.

This textbook provides a basic background about the life of a plant and explains how plants affect the lives of humans. Within this framework, three themes emerged—evolution, ecology, and economic botany—illustrating how plants have evolved, what ecological roles they play in the world, and how species are economically important to humans. Evolution is a central theme of all biology, and the principles of natural selection operate on plants in similar ways to those of other organisms. In terms of ecology, plants and algae are the producers of most communities, and thus, to understand how animals and ecosystems function, a knowledge of plants is critical. Finally, plants are economically important because we eat, grow, wear, live in, and are healed and harmed by plants and their products on a daily basis. For these reasons and more, we find plants to be significant and interesting organisms, illustrative of the living world as a whole.

Focusing on these themes has allowed us to organize topics in the way we believe is most appropriate to understand plants, emphasizing form and function of plant structures. For instance, when we talk about the form of roots, we also discuss one of their important functions—absorbing minerals from the soil. Or, when we talk about the shape of leaves, we discuss how leaf form influences the function of transpi-

ration and movement of water in a plant. We begin this book with a description of natural selection, discuss evolution in detail near the end, and carry the discussion of both throughout the text in hopes you will ponder how structures and functions help plants survive and reproduce. We also begin with a discussion of important food and medicinal plants, conclude with appendices on biotechnology and chemistry, and include human uses of plants in every chapter and integrate ecological issues throughout the text. The use of themes is intended to help you organize information about plants, see that this information is connected, and understand that ideas build upon each other.

INQUIRY IS THE BASIS OF SCIENCE.

Throughout this book, you'll find questions—associated with photographs, opening the chapters, ending "Inquiry Summaries," and placed throughout the narrative. The questions are intended to pique interest and inspire further thought about the topic before continuing to read. Many of these questions have no single, correct answer; however, we hope there will be an attempt to answer them all. Science is not about correct answers and facts. Science is about making careful observations, asking good questions, and trying to answer the questions in a systematic and scientific way—science is a process of investigation. When all of us were younger, we asked many questions of our parents and teachers; children possess an innate curiosity about the natural world. Unfortunately, pre-college classes often inhibit student curiosity, and many students have become passive listeners and readers. Our goal is that our inquiry approach will re-spark an interest in the process of science. We hope that the information, ideas, and the many illustrations, references, and questions in this book will foster a new appreciation and respect for plants.

HELPFUL SUPPLEMENTS MAKE LEARNING EASIER.

1. An *Instructor's Manual* accompanies this text and provides chapter outlines, key terms, test and discussion questions, suggestions for class activities, and resources for teaching aids.

2. *Computerized Testing Software* offers the objective test questions that are in the Instructor's Manual in electronic format for ease in class testing and grading.

3. A set of *125 transparencies* is available to users of the text. The acetates duplicate key figures from the text that are important to understanding major concepts in botany.

4. A comprehensive *website* (http://www.mhhe.com/ botany) offers numerous resources for instructors and students alike via the *OnLine Learning Center*. The OLC includes live links to websites that provide more information on each chapter of the text, practice quizzing, key-term flashcards, animations, Internet activities, tips on writing term papers, career information, case studies, additional test questions, activities, sample syllabi, current global botanical issues, and much more.

5. *Botany Visual Resource Library CD-ROM.* This classroom presentation CD includes images from three botany textbooks and hundreds of photos and micrographs.

ACKNOWLEDGEMENTS

Production of this text required considerable time and work beyond our own, and we are indebted to Kathy Loewenberg, Peggy Selle, Marge Kemp, and Michael Lange for their continued effort and support in making this the best book possible. We thank them for their patience and expertise, understanding and sense of humor. All authors should work with such wonderful editors.

We are also indebted to our many colleagues who have devoted hours from their busy schedules to help us fine-tune our several drafts into what you now hold in your hand. We were humbled by, and truly appreciative of, their valuable feedback; however, all errors are, of course, our responsibility.

Gordon Uno
Richard Storey
Randy Moore

REVIEWERS

Paul W. Barnes	*Southwest Texas State University*
Robert W. Bauman, Jr.	*Amarillo College*
L. J. Davenport	*Samford University*
James T. Dawson	*Pittsburg State University*
Roger del Moral	*University of Washington–Seattle*
Robert Eddy	*Southeast Community College*
Frederick B. Essig	*University of South Florida*
Katharine B. Gregg	*West Virginia Wesleyan College*
Joel B. Hagen	*Radford University*
Laszlo Hanzely	*Northern Illinois University*
James D. Haynes	*Buffalo State College*
Jodie S. Holt	*University of California*
Elisabeth A. Hooper	*Truman State University*
John C. Hunter	*SUNY–Brockport*
William A. Jensen	*The Ohio State University*
Joanne M. Kilpatrick	*Auburn University–Montgomery*
Barbara E. Liedl	*Central College*
Bernard A. Marcus	*Genesee Community College*
Conley K. McMullen	*James Madison University*
Timothy D. Metz	*Campbell University*
Lillian W. Miller	*Florida Community College*
Dale M. J. Mueller	*Texas A&M University*
John Olsen	*Rhodes College*
Carolyn Peters	*Spoon River College*
Wayne C. Rosing	*Middle Tennessee State University*
Brian Shmaefsky	*Kingwood College*
James F. Smith	*Boise State University*
Anthea M. Stavroulakis	*CUNY–Kingsborough Community College*
Stephen L. Timme	*Pittsburg State University*
Andrea Wolfe	*Ohio State University*
Todd Christian Yetter	*Cumberland College*

Guided Tour

The features of this book are *Unique*

The organization and principle features of this book were planned with the students' holistic understanding of botany in mind:

LORE OF PLANTS

The Lore of Plants sections in each chapter provide interesting and intriguing highlights of the botanical world.

The Lore of Plants

Plant substances can be used to determine your blood type. Lectins are proteins from jack beans, lima beans, and lotus that bind to glycoproteins on the cell membranes of red blood cells. Because the cells of different blood types have different glycoproteins, cells of each blood type (A, B, O, AB) bind to a specific lectin. This is one example of the many clinical and research applications of plant lectins in human medicine. What is the role of lectins in a plant? Actually, no one knows, but it certainly is not to type blood. Perhaps they have a role in cell signaling.

INQUIRY SUMMARY

Molecules move across membranes by simple diffusion and by transport via selective membrane proteins. Facilitated diffusion via passive transport proteins does not involve metabolic energy and occurs when molecules move down a concentration gradient as they cross the membrane. Active transport requires metabolic energy and can move molecules against a concentration gradient using specific transport proteins. Active transport allows plants to accumulate high concentrations of molecules from dilute solutions in the soil and fosters movement of water into plants. Cells communicate with other plant cells by chemical signals. Membrane surfaces also recognize specific molecules from other organisms.

How might cell communication be important to the nitrogen economy of an ecosystem?

Plant Tissue Systems Help Us Understand the Evolution of Life on Land.

Plants evolved from algae 450 million years ago and then diversified in a way that forever changed our planet: they invaded land, thereby paving the way for the subsequent colonization of land by animals. Because the terrestrial habitats encountered by plants were more diverse than their ancestral aquatic environments, land plants faced many new environmental challenges and selective pressures. As a result, they evolved rapidly and became more complex and diverse than did their algal ancestors.

Foremost among these environmental challenges was life in air and soil instead of water. Maintaining an aquatic cellular environment (that is, plant cells filled with water) was imperative, since unprotected cells, whether of algae or land plants, desiccate and die within minutes after being exposed to dry air. Thus, the survival of plants on land depends on their ability to establish an aquatic cellular environment in the dry land environment. Establishing

FIGURE 4.20
Roots of pea plants with bacteria-containing nodules (×2). In these nodules, atmospheric nitrogen is converted to ammonia via a process called nitrogen fixation. The infection of roots by bacteria is enabled by root-hair molecules that bind to bacterial cell walls.

supply carbohydrates for bacterial growth, and the bacteria convert nitrogen from the air into ammonia, a metabolically useful form of nitrogen for plants. This conversion process is called **nitrogen fixation** (see chapter 7), and it is the main mechanism of nitrogen input into most ecosystems.

animals.) What are the ecological and evolutionary reasons for there being so many endemic species on Madagascar? (One answer is that plants and animals evolved in the unique environment of the island in isolation from other organisms on the continent of Africa and, over time, became species different from any others on earth.) What interactions between the rosy periwinkle and its environment may have promoted the production of vincristine and vinblastine? (One answer comes from the periwinkle's interaction with animals; the chemicals vincristine and vinblastine may have protected the plants from herbivores on the island.)

INQUIRY SUMMARY

Plants are a major source of medicines. Natural-products chemists and ethnobotanists are among the scientists who search the world for potentially new medicinal plants. One method used in this search includes studying the interaction between animals and the plants they avoid or that affect their lives. The evolution of economically important plants depends on their ecological relationships with other organisms in their native habitat.

Health food stores sell a variety of plant products that claim to have healthful effects. How can a person separate beneficial products from those with no effects?

What Is a Plant?

You already have some notion of what a plant is—an immobile, green organism that we eat, mow, and use for decoration and shade—but it's difficult to come up with a simple definition for the word *plant*. Not all plants are green, and some plants consume animals. Indeed, many plants do not look or seem like plants at all, but the most familiar plants share a few common characteristics.

Plants share many characteristics with other organisms.

Plants share the characteristics of life with all other organisms. All living organisms:

- take in and use energy, which allows them to conduct their daily activities;
- consist of a single cell or many cells—a large organism such as yourself or a small tree may have trillions of cells making up its body;
- reproduce;
- respond to the environment in which they live; and
- share parts of a common ancestry and have coevolved with other organisms, such as flowering plants and their pollinators (fig. 1.14).

FIGURE 1.14
Plants have coevolved with other organisms. In what ways might the relationship between this plant and animal be mutually beneficial?

A plant is an autotrophic organism with a unique set of characteristics, including cell walls made of cellulose.

Plants possess all of the characteristics previously listed as well as a combination of characteristics that is unique to them. In general, *plants are autotrophic*, or "self-feeding." This means they can produce their own sugars, proteins, vitamins, and amino acids. **Heterotrophic** organisms such as humans and other animals must obtain their food from other organisms and many of the necessary vitamins and amino acids from the food they eat. Most plants are autotrophic because they can convert light energy into the chemical energy of sugar in the process of **photosynthesis.** *Plants and algae possess special pigments, including chlorophyll, that absorb light.* Chlorophyll is not found in heterotrophic organisms. Once a sugar molecule is made, a plant can use the sugar as a source of energy, but it can also use the sugar to make cell parts, including a material called **cellulose.** *Plants have cell walls made of complex molecules such as cellulose.* This rigid cell wall surrounds each plant cell, provides some protection to the cell, and gives the cell shape. Animal cells lack a rigid, protective cell wall. *Plants are not motile*—they cannot move from one place to another once they start growing. This has immense consequences for the life of a plant. For instance, if the temperature is too high or the amount of precipitation too low in the place where a plant is rooted, it cannot move to an area with more suitable environmental conditions. However, even though they cannot move to a better location for growing, plants do have many movements, such as the head of some sunflowers turning toward the sun throughout the day or the leaves of a sensitive plant drooping when they are touched (see fig. 6.6). In addition, the offspring of plants may be carried great distances from the parent plant. *Plants have open,*

INQUIRY SUMMARIES

Inquiry Summaries at the end of each major section highlight key concepts, forming the building blocks of true comprehension.

QUESTIONS

Thought-provoking and open-ended questions are strategically placed throughout the chapters, challenging the student to stretch their analysis skills and really think about what they've just read.

Complete action statements or questions introduce major concepts within chapters, and also serve as a study aid for the student in a second reading of the material.

10.1 *Perspective* Perspective

The Evolution of the Light Reactions

The evolution of PS II forever changed life on earth. As PS II liberated O_2 into the atmosphere, the environment changed and allowed the evolution of aerobic respiration because the use of O_2 in respiration was a selective advantage. Why is aerobic respiration a selective advantage, and why is it important to you and other organisms? We answer these questions in chapter 11.

For nearly 35 years, botanists have been confident that they understand the major aspects of photosynthesis. Specifically, botanists have held that plants require two photosystems: PS I is essential for producing the reduced $NADP^+$ used to reduce CO_2, whereas PS II oxidizes water and releases O_2. But recently, a group of botanists at Oak Ridge National Laboratory in Tennessee questioned this longheld idea. These botanists studied mutants of *Chlamydomonas* (a single-celled alga) in which PS I does not operate. Nevertheless, the algae harvest light, fix CO_2, release O_2, and grow. These results suggest that the mutants can perform photosynthesis without PS I (i.e., with only PS II). While not all researchers are ready to accept this conclusion, the study may force botanists to consider alternatives to the model presented in figure 10.8. The study may also help us understand how photosynthesis evolved. The mutant without PS I grows better in the absence of oxygen, suggesting that PS II evolved first, under early anaerobic conditions that existed before photosynthesis filled the ancient atmosphere with O_2. That is, the earliest forms of photosynthesis used only PS II; PS I could have evolved later, as it may increase light harvest, stabilize PS II, and render electron flow more efficient.

FIGURE 10.8
The light-driven transport of electrons during photosynthesis creates a proton gradient across the thylakoid membrane. Th[...] dient, formed by the higher concentration of H^+ in the thylakoid space compared to the stroma, drives the formation of AT[...] thase. O_2 is released as H_2O is split in light.

BOXED READINGS

"Perspectives" boxed readings detail current discoveries and expand upon information introduced in the main text.

PHOTOGRAPHS & ILLUSTRATIONS

A richly diverse collection of photographs and diagrams reinforce and illustrate important concepts discussed in the accompanying text.

(a)

(b)

FIGURE 17.7
Gemmae cups of liverworts. (*a*) Gemmae cups ("splash cups") containing gemmae on the gametophytes of *Lunularia* (×1). Gemmae are splashed out of the cups by raindrops, after which the gemmae can grow into new gametophytes, each identical to the parent plant that produced it by mitosis. (*b*) Cross section of a thallus showing a gemmae cup (×10). How is reproduction via gemmae similar that via a stolon?

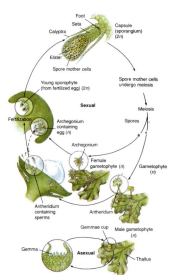

FIGURE 17.8
Life cycle of *Marchantia*, a thallose liverwort. During sexual reproduction, spores produced in the capsule germinate to form independent male and female gametophytes. The archegonium contains an egg; the antheridium produces many sperm. After fertilization, the sporophyte develops within the archegonium and produces a capsule containing spores. *Marchantia* also reproduces asexually by fragmentation and gemmae.

fig. 17.4). This is one of the ecological advantages of being small and nonvascular: bryophytes can live on impermeable substrates, thereby avoiding competition with vascular plants. Indeed, bryophytes grow in cracks of sidewalks, on moist soil, rooftops, the faces of cliffs, tombstones, and birds' nests; they carpet forest floors, dangle like drapery from branches, and sheathe the trunks of trees in rain forests. Examples of the extremes of bryophyte habitats include exposed rocks, volcanically heated soil (up to 55°C), and Antarctica, where summer temperatures seldom exceed minus 10°C. However, the most unusual habitat for bryophytes is reserved for *Splachnum*, the mammal dung moss (see fig. 17.4*d*). This moss produces a colored capsule and releases a putrid odor that attracts flies. Unlike the spores of other mosses, those of *Splachnum* are sticky and adhere to the visiting flies. The spores are disseminated when the flies move from the moss to piles of dung.

Bryophytes, which are often the first plants to invade an area after a fire, grow at elevations ranging from sea level to 5,500 meters. There are no marine bryophytes, but some, such as dune mosses, grow near the seashore. Bryophytes dominate the vegetation in peatlands. Mosses are especially abundant in the arctic and antarctic, where they far out-

SUMMARY

Flowers are the reproductive structures of angiosperms. A flower is a stem tip that bears some or all of the following kinds of appendages on a receptacle: sepals, petals, stamens, and pistils. The pistil consists of an ovary, which includes one or more ovule-bearing carpels, and a stigma that is usually borne on a style. During pollination, pollen is transferred from an anther to a stigma. A pollen tube grows into an ovule and carries sperm to the egg inside the ovule. After fertilization, ovules become seeds, and ovaries become fruits.

The angiosperms, which are the most dominant plants on earth, have flowers that include seeds in a carpel. Fossils of carpels and other parts of flowers are known from Cretaceous deposits that are at least 140 million years old. A 120-million-year-old fossil flower resembles several plants in different families of monocots and dicots, on the basis of simple features of the flower and its herbaceous type of body. Evidence suggests that flower parts evolved from leaves. The main force behind the rapid evolution of angiosperms may have been pollination by insects. The first flowers were probably pollinated by beetles; later angiosperms attracted butterflies and bees.

The great diversity in angiosperm species is based largely on variation in reproductive morphology. Angiosperms are divided into two large groups: the monocots and the dicots. Flowers may be complete or incomplete, perfect or imperfect, regular or irregular. They also may be solitary or arranged in an inflorescence with several other flowers. The diversity of angiosperms is so great that most of our knowledge about this group is divided into regional floras or taxonomic treatments of smaller groups of plants.

Most flowers are adapted for pollination by wind or by animals. Wind-pollinated flowers are usually incomplete and not showy, whereas insect-pollinated flowers and bird-pollinated flowers are often colorful. Night-blooming flowers attract nocturnal mammals or insects and are usually aromatic and white or cream colored.

Fruits may be either fleshy or dry. Some kinds of dry fruits remain sealed at maturity, whereas others split open at maturity. Fleshy fruits that are colorful and sweet tasting attract animals that eat them and disperse the seeds. Dry fruits, or the seeds from them, are also adapted for dispersal, often by the wind.

Writing to Learn Botany

Describe the role animals have played in the evolution of different kinds of flowers.

THOUGHT QUESTIONS

1. Dinosaurs and most other animals went extinct at the end of the Cretaceous period about 65 million years ago. Some scientists have proposed that such mass extinctions were caused by geological catastrophes. If this is true, how might angiosperms have escaped such catastrophes?

2. In addition to co-evolution with insect pollinators, what other kinds of co-evolutionary interactions may have occurred between flowering plants and animals?

3. Does the fact that angiosperms do not appear in the fossil record until the Cretaceous period mean that they were not around before then? Explain your answer.

4. Angiosperms usually have a shorter life cycle than do other plants. How might this have affected the evolution of angiosperms?

5. Why are strawberries, raspberries, and mulberries not berries? Since they are not berries, what are they?

6. Many seeds store their food reserves mostly as oil instead of carbohydrate. What are some examples of plants that produce oily seeds?

7. Why is it incorrect to refer to commercial sunflower "seeds" (those still in their shells) as seeds? Since they are not seeds, what are they?

8. Apples are fleshy, sweet-tasting fruits that attract and are eaten by fruit-eating animals that disperse their seeds. How could you explain the fact that the seeds of such an attractive fruit contain cyanide-producing chemicals?

9. What might the advantages be for self-compatibility versus self-incompatibility?

10. Recent research suggests that the production of more flowers decreases a plant's chances of mating with others by increasing the likelihood of self-pollination. How would you test this hypothesis?

11. Do plants compete to attract pollinators? Write a short paragraph to explain your answer.

SUGGESTED READINGS

Articles

Bakker, R. T. 1986. How dinosaurs invented flowers. *Natural History* 11:35–38.

Beattie, A. J. 1990. Seed dispersal by ants. *Scientific American* 263:76.

Burd, M. 1994. Bateman's principle and plant reproduction: The role of pollen limitation in fruit and seed set. *Botanical Review* 60:83–139.

SUMMARIES & THOUGHT QUESTIONS

The chapter "Summaries" and "Thought Questions" provide an opportunity for students to test their understanding of the material just covered.

WRITING TO LEARN BOTANY

"Writing to Learn Botany" segments suggest paper topics and stimulate creative thinking.

APPENDIX A
Appendix A

WHAT IS GENETIC ENGINEERING?

Genetic engineering, as discussed in chapter 5, is based on recombinant DNA technology. This technology relies on vectors (DNA carriers), such as viruses or bacterial plasmids (independent molecules of DNA apart from the primary bacterial genome; see fig. A.1). Viruses and plasmids are vectors because they carry inserted or foreign DNA as their own DNA, which is then replicated. In this way, DNA from plants or animals is inserted or foreign into bacteria or viruses. DNA cloning is an important tool in gene technology. It is used to find and study important genes and to make genetically engineered plants.

DNA CLONING AND RESTRICTION ENZYMES ARE FUNDAMENTAL TO GENETIC ENGINEERING

Before recombination, both the DNA molecules must be made receptive to each other. This job is done by bacterial restriction enzymes, which cut (i.e., restrict) DNA. The use of these elements in DNA cloning is discussed in the next few paragraphs. Different bacteria make hundreds of restriction enzymes, each of which recognizes a specific DNA sequence of four to eight nucleotides. In nature, these enzymes protect bacteria by destroying foreign DNA. To prevent self-destruction, bacterial DNA is chemically modified to be inert to its own restriction enzymes.

An example of a restriction enzyme is *EcoRI*, which is harvested for commercial sale from *Escherichia coli*. *EcoRI* recognizes the nucleotide sequence GAATTC and then cuts the DNA between the guanine (G) and the adenine (A) of the sequence. The six-nucleotide *EcoRI* site has identical sequences on both DNA strands because its complement is CTTAAG. This means that by cutting between guanine and adenine on each strand, the enzyme makes a zigzag cut that leaves short, single-stranded ends on each DNA double strand (fig. A.2). Such single-stranded ends are called sticky ends because they can be "glued" by hydrogen bonding to complementary sticky ends of other DNA molecules.

After cutting up a whole genome with a restriction enzyme, the next step in cloning DNA is to insert the DNA into a virus or plasmid vector. One type of vector is a bacterial plasmid, which is a small, circular molecule of DNA (see fig.

A.1 and chapter 16). After the foreign DNA is inserted into isolated plasmids, the plasmids are absorbed back into bacteria and reproduced (i.e., cloned) during normal DNA replication of the bacterial genes.

Foreign DNA and plasmid DNA are receptive to each other when they are cut by the same restriction enzyme. For example, an enzyme such as *EcoRI* makes the same sticky ends on both the foreign DNA and the plasmid DNA. When the two kinds of DNA are mixed, they anneal (come together and match) because their sticky ends complement each other.

At first, plasmid DNA is weakly held to foreign DNA by hydrogen bonds between complementary nucleotides. The linkage between them is strengthened when sugarphosphate bonds form between the two strands or molecules of DNA. Sugar-phosphate bonds are made by the enzyme DNA ligase. When this enzyme is added to the DNA mixture, it joins (i.e., ligates) phosphate groups to deoxyribose between adjacent strands of DNA.

Methods for cloning DNA in plasmids also work with viruses. Like plasmid DNA, viral DNA is cut by restriction enzymes, and foreign DNA is inserted into it. Since viruses are parasites, however, they must be reintroduced into their hosts before they can replicate the recombined DNA. For cloning DNA in viruses, the most convenient hosts are bacteria such as *E. coli* that can be easily cultured in the laboratory. The most commonly used viruses for cloning DNA are bacteriophages.

Libraries of Genes Can Be Made.

After a genome is cut by restriction enzymes, many of the resulting DNA fragments can be inserted into vectors. This has been done for many plants, animals, and fungi. When inserted into a vector, a fragment or gene of interest from any of these organisms can be retrieved from storage by culturing the appropriate vector. This storage and retrieval of genetic information is like having a library of genes, so scientists call the set of cloned fragments of a genome a *genomic library* (fig. A.3); such libraries may have several thousand to several million entries, each a recombined plasmid or virus that contains a different restriction fragment.

APPENDICES

The appendices, "What is Genetic Engineering?" and "Fundamentals of Chemistry for Botany Students" are helpful student resources covering two very important areas in botany.

INDEX

A detailed index includes both the common and scientific names of the plants mentioned in the text.

Cox, P. A. 1993. Water-pollinated plants. *Scientific American* 269:68–74.

Fleming, T. H. 1993. Plant-visiting bats. *American Scientist* 81:460–67.

Milius, S. 1999. The science of big, weird flowers. *Science News* 156:172–74.

Natural History. 1999. The Flower Issue. 108 (4).

Strauss, E. 1998. How plants pick their mates. *Science* 281:503.

Books

Barth, F. G. 1991. *Insects and Flowers: The Biology of a Partnership.* Princeton, NJ: Princeton University Press.

Bell, A. D. 1990. *Plant Forms: An Illustrated Guide to Flowering Plant Morphology.* New York: Oxford University Press.

Fenner, M., ed. 1992. *Seeds, the Ecology of Regeneration in Plant Communities.* Wallingford, England: CAB International.

Lloyd, D. G. and S. C. H. Barrett, eds. 1996. *Floral Biology.* New York: Chapman & Hall.

ON THE INTERNET

Visit the textbook's accompanying web site at http://www.mhhe.com/botany to find live Internet links for each of the topics listed below:

Characteristics of the flower
Fruit types
Flower evolution
Angiosperm life cycle
Monocots and dicots
Inflorescence types
Plant systematics
Floriculture
Flowers, pollinators and coevolution
Carl Linne

ON THE INTERNET

"On the Internet" lists key topics in the chapter that are repeated in the Online Learning Center as hyperlinks to web sites with additional information.

Supplements

INSTRUCTOR'S MANUAL

An *Instructor's Manual* accompanies this text and provides chapter outlines, key terms, test and discussion questions, suggestions for class activities, and resources for teaching aids.

Computerized Testing Software offers the objective test questions that are in the Instructor's Manual in electronic format for ease in class testing and grading.

WEB SITE & OLC

A comprehensive *Web site* (http://www.mhhe.com/botany) offers numerous resources for instructors and students alike via the *OnLine Learning Center*. The OLC includes live links to websites that provide more information on each chapter of the text, practice quizzing, key-term flashcards, animations, Internet activities, tips on writing term papers, career information, case studies, additional test questions, activities, sample syllabi, current global botanical issues, and much more.

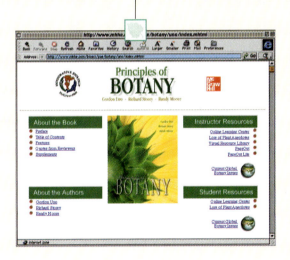

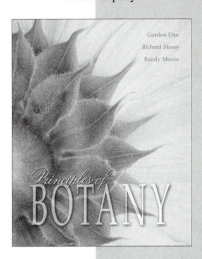

Instructor's Manual and Test Item File

to accompany

Gordon Uno

Richard Storey

Randy Moore

Principles of

BOTANY

Prepared by
Timothy Metz

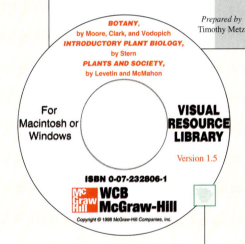

BOTANY,
by Moore, Clark, and Vodopich
INTRODUCTORY PLANT BIOLOGY,
by Stern
PLANTS AND SOCIETY,
by Levetin and McMahon

For Macintosh or Windows

VISUAL RESOURCE LIBRARY

Version 1.5

ISBN 0-07-232806-1

WCB McGraw-Hill

Copyright © 1998 McGraw-Hill Companies, Inc.

BOTANY VISUAL RESOURCE LIBRARY CD-ROM.

This classroom presentation CD includes images from three botany textbooks and hundreds of photos and micrographs.

TRANSPARENCIES

A set of *125 Transparencies* is available to users of the text. The acetates duplicate key figures from the text that are important to understanding major concepts in botany.

PageOut
Proven. Reliable. Class-tested.

More than 10,000 professors have chosen **PageOut** to create course websites. And for good reason: **PageOut** offers powerful features, yet is incredibly easy to use.

Now you can be the first to use an even better version of **PageOut**. Through class-testing and customer feedback, we have made key improvements to the grade book, as well as the quizzing and discussion areas. Best of all, **PageOut** is still free with every McGraw-Hill textbook. And students needn't bother with any special tokens or fees to access your **PageOut** website.

Customize the site to coincide with your lectures.

Complete the **PageOut** templates with your course information and you will have an interactive syllabus online. This feature lets you post content to coincide with your lectures. When students visit your **PageOut** website, your syllabus will direct them to components of McGraw-Hill web content germane to your text, or specific material of your own.

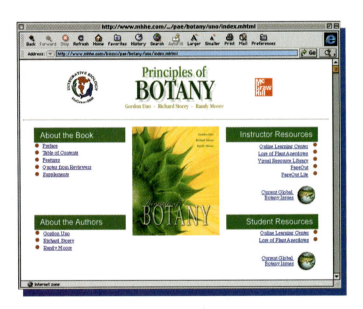

New Features based on customer feedback:

- Specific question selection for quizzes

- Ability to copy your course and share it with colleagues or use as a foundation for a new semester

- Enhanced grade book with reporting features

- Ability to use the **PageOut** discussion area, or add your own third party discussion tool

- Password protected courses

Short on time? Let us do the work.

Send your course materials to our McGraw-Hill service team. They will call you for a 30 minute consultation. A team member will then create your **PageOut** website and provide training to get you up and running. Contact your McGraw-Hill Representative for details.

Contact your local McGraw-Hill sales representative for more information or visit *www.mhhe.com*.

The Online Learning Center
Your Password to Success

www.mhhe.com/botany (click on cover)

This text-specific website allows students and instructors from all over the world to communicate. Instructors can create a more interactive course with the integration of this site, and students will find tools such as practice quizzing, key term flashcards, case studies, global botany issues/articles and chapter-related hyperlinks that will help them improve their grades and learn that botany can be fun.

Student Resources

Study questions
Quizzing
Hyperlinks to chapter-related websites
Case studies
Global botany articles
Key term flashcards

Instructor Resources

Instructor's Manual
Links to related websites to expand on
   particular topics
Lecture outlines
Case studies
Histology slides

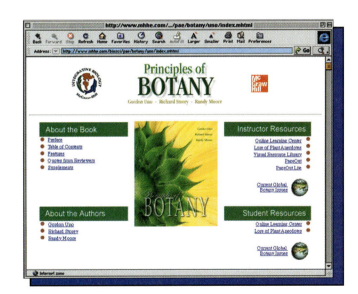

Imagine the advantages of having so many learning and teaching tools all in one place—all at your fingertips—FREE.

e-TEXT

E-TEXT is an exciting student resource that combines McGraw-Hill print, media, study, and web-based materials into one easy-to-use CD-ROM. This invaluable resource provides cutting-edge technology that accommodates all learning styles, and complements the printed text. The CD provides a truly non-linear experience by using video and art, as well as web-based and other course materials to help students organize their studies. Best of all, e-TEXT is free for students who purchase new copies of this McGraw-Hill title.

The following features illustrate, in depth, the benefits of e-TEXT:

- **Full textbook and study guide PDF files are interlinked.** This includes all narrative, art and photos, PLUS expertly crafted animations.

- Targeted web links encourage **focused web research.**

- A **Search feature** enables students to improve studying by locating targeted content quickly and easily.

- This hybrid CD is **compatible with both Macintosh and Windows** platforms.

- Required programs **Acrobat Reader and QuickTime are supplied** on the CD-ROM.

- **Bookmarks**—appearing on the left side of the screen—list all of the links available on that page.

- **Thumbnails** of the other pages within the chapter are shown for quick navigation.

- **Main menu links** are at the bottom of every screen as well as in the bookmark section.

- An explanation of features is provided on the **Help Page**.

- **Boldface terms** are linked to definitions in the glossary.

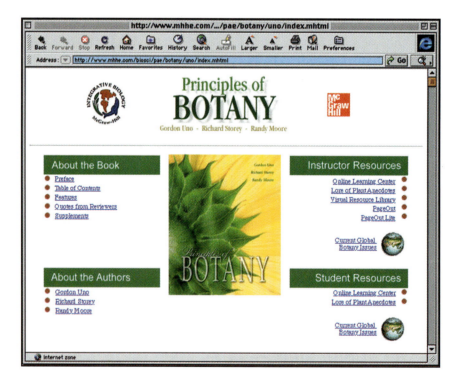

Contact your McGraw-Hill sales representative for more information or visit *www.mhhe.com.*

PRINCIPLES OF

Botany

Medicinal plants for sale in a market in Brazil.
Many plants are useful to humans, including
those used for medicinal purposes in many parts
of the world. For what reasons are plants
important in your life?

An Introduction to Plants and Their Study

LEARNING ONLINE

The Online Learing Center at www.mhhe.com/botany is full of helpful resources for you such as practice quizzes, key term flashcards, assistance with writing assignments, and hyperlinks that are specific to what you'll learn about in this chaper. For example, click on www.siu.edu/~ebl in the OLC for chapter 1 to see why plants are so important to humans.

Botanists are scientists who study plants. They have been heard to ask their students, "Have you thanked a green plant today?" This question actually has a serious basis because of the importance of plants in the lives of organisms on earth. Without plants, life on earth would certainly be different. Most notably, humans and almost all other organisms could not survive without their green cohabitants.

This chapter describes what botany is and why you will enjoy studying it. You will learn about the importance of plants to humans and to the world as a whole. You will also learn about a scientific method of investigation that underlies the process of scientific research. This information prepares you for the remaining chapters in this book, which integrate scientific knowledge about plants and discuss how we use this method of investigation to understand plants.

Plants Are Essential to Human Life.

On his bad days, Stephen felt tired and nauseated. He felt pain in his groin and one day noticed swelling in his left testicle. His doctor discovered that Stephen had cancer, and, at the age of 20, Stephen had surgery to remove the diseased testicle. Surprisingly, after the surgery Stephen was pain-free and comfortable. Unfortunately, the cancer had spread through his body, which meant more surgery followed by intense chemotherapy. This chemotherapy involved injections of a chemical soup consisting of high doses of vincristine, vinblastine, and other drugs. Vincristine and vinblastine are chemicals extracted from a tropical plant, the Madagascar rosy periwinkle (*Catharanthus roseus*) that is grown today in many American flower gardens (fig. 1.1). These extracts from the periwinkle slow the division of cancerous cells and thus slow the spread of the disease, while other parts of the chemical soup kill cancer cells. Today, Stephen is cancer-free, married, and is pursuing a career in medicine to learn how to fight cancer.

Stephen's battle against cancer was aided by extracts from a plant whose curative powers were discovered only recently. Modern scientists investigated the chemical extracts from the Madagascar rosy periwinkle because it had been used in certain folk remedies to treat diabetes. Although no treatment for diabetes was found in the plant, scientists found several chemicals, including vincristine and vinblastine, that are important in fighting cancers. Both of these drugs now are used to improve the quality of life for many patients and have helped save the lives of many more who suffer from Hodgkin's disease, childhood leukemia, and various forms of cancer.

The story about Stephen makes an important point: for many people and other animals on earth, plants are important sources of medicine. But for all animals, including humans, plants are essential for existence on earth as a source of food and oxygen. In fact, green plants and algae generate the oxygen and sugars that sustain almost all life on earth. In addition, the production of oxygen by these organisms enabled life to evolve into the familiar, large organisms that we see around us today.

FIGURE 1.1
Vincristine and vinblastine are cancer-fighting drugs extracted from the Madagascar rosy periwinkle (*Catharanthus roseus*).

Plants produce many medicines as well as the food and oxygen on which organisms depend.

Throughout this book, you will see many examples of how plants affect our lives. We eat fruits and vegetables produced by plants and are clothed by fibers from stems and leaves. Trees provide us with lumber, paper, string, rope, and welcome shade on hot days. In addition to medicines, many plants provide brilliant dyes, industrial chemicals, and useful oils. The colors and fragrances of flowers and foliage satisfy our aesthetic senses, while spices such as black pepper, nutmeg, ginger, cloves, vanilla, and cinnamon have enhanced our enjoyment of food since before the time of the Roman Empire (fig. 1.2). Spices also preserve foods, which made possible the colonization of the New World and food storage in areas without refrigeration. Recent evidence suggests many spices contain chemicals that kill microorganisms before they spoil food; thus, eating food laced with garlic, onion, allspice, and oregano may help reduce foodborne illnesses. These "spicy" chemicals may protect plants against harmful microorganisms; the evolution of the chemicals helped the spice plants survive.

(a)

(b)

FIGURE 1.2

(*a*) *Piper nigrum* is the source of black pepper, the most commonly used spice. The lure of pepper, the only spice that could make decaying or heavily salted meat edible, drew Columbus and medieval merchants to "discover" the rain-forested areas of earth. For Europeans, North America was a by-product of the maritime search for pepper. (*b*) Lavender Harvest. Romans used lavender (*Lavandula*) to scent and disinfect their baths. The genus name of lavender, *Lavandula* is from the Latin word *lavare,* meaning "to wash." (*c*) A woman in Madagascar cleaning cloves. Cloves are the dried flower buds of *Syzygium aromaticum.* (*d*) Peppers (*Capsicum* species) are favorites of people who like spicy food.
Source: (b) © Norfolk Lavender Ltd.

(c)

(d)

From the body paints of Amazon Indians to modern cosmetics, and from early Egyptian papyrus of more than 5,000 years ago to today's pulp mills, plants affect all aspects of our lives (fig. 1.3). We have even fought wars over plants. For example, the First Opium War (1839–42) was fought over an extract of the opium poppy (*Papaver somniferum*) (fig. 1.3*b*); one result of that war— Hong Kong was taken over by the British.

Today, almost everything we do is influenced, either directly or indirectly, by plants. However, the many uses of plants listed so far are only a small part of their importance to us. Consider the following examples:

ॐ We use spruce to make the 23,000 tons of newsprint needed to produce 65 million newspapers each day in the United States. A bumper issue of the *New York Times* clears about 400 hectares (nearly 1,000 football fields) of trees, and an average tree produces enough newsprint for about 400 copies of a 40-page tabloid. Each person in the United States uses an average of about 290 kilograms (640 pounds) of paper each year; that's roughly equivalent to 0.8 cubic meters (25 cubic feet) of wood.

ॐ Over 30 million people in the United States have home gardens. Thirty percent of these people use their gardens as a source of fresh fruits and vegetables, 25% as a source of better-tasting vegetables, and 15% as a source of income. Gardens provide $16 billion worth of food; that's about 23 kilograms (50 pounds) of food per gardener.

ॐ Although modern civilization would collapse without plants, not all plants benefit people. On the negative side, some plants cause allergies and others can poison us. Socrates was poisoned with poison hemlock, and Claudius, the father of Nero, was poisoned with monkshood juice and mushrooms (which are fungi). Many children get sick each year from eating poisonous plants. Plants such as castor bean, mistletoe, dumbcane, caladium, elephant's ear, philodendron, English ivy, rhubarb, and oleander contain potent toxins. In addition, thousands of people in the United States are addicted to illicit drugs derived from plants (fig. 1.4). Interestingly, some of these drugs are effective medicines at low dosages, and some hallucinogenic plants are important in religious ceremonies.

The Lore of Plants

Plants have been the "star witnesses" in several criminal trials, most notably in the 1932 trial of Bruno Hauptmann. In a trial based largely on botanical evidence (wood from Hauptmann's house was made into a ladder used in the crime), Hauptmann was convicted of kidnapping the baby boy of Charles and Anne Lindbergh and was sentenced to death. Botanical evidence has also been used to convict rapists, murderers, and thieves based on seeds and fragments of leaves and bark. In one murder case, a victim's stomach contained cabbage, beans, and green peppers—plant foods that are easily identified by botanists and trained investigators. Based on this evidence, the victim was traced to a local restaurant that served this combination of food, and investigators found witnesses there who were able to identify the victim and her dinner companion, who later murdered her.

(a)

(b)

FIGURE 1.3

(*a*) Papyrus (*Cyperus papyrus*) growing at the edges of a lake. Papyrus was once used to make paper, and to the Egyptians, the plant was once a symbol of youth and vigor. (*b*) Poppy (*Papaver somniferum*), the source of opium. Opium, the dried latex that oozes from cut fruits, is the source of alkaloids such as morphine and codeine.

- People have used fibers for more than 10,000 years. We use hemp to make rope; sisal to make brooms and brushes; ramie to make textiles; cotton to make our jeans; and flax to make linen, wrappings for Egyptian mummies, cigarette paper, and money.

- Plants supply many of our drinks. For example, tequila has been made from the century plant for hundreds of years, and Egyptians made beer from barley as early as 2500 B.C. Tea and coffee, the world's two most popular beverages (fig. 1.5), are also made from plants (see Perspective 1.1, "Coffee: Even If You Don't Know Beans"). Re-

cent studies indicate that drinking green tea may help slow the development of heart disease and certain cancers.

INQUIRY SUMMARY

Plants are essential to our lives because they produce the food and oxygen on which we, and all other animals, depend. Plants are also an important source of medicines, lumber, spices, paper, and fibers.

In what other ways are plants useful to you?

FIGURE 1.5
Tea is made from leaves of *Camellia sinensis*, a small evergreen shrub related to the garden camellia. Each year more than 2 million tons of tea are produced from about 25 countries. North Americans drink about 40 billion cups of tea per year.

FIGURE 1.4
The coca plant (*Erythroxylum coca*), the source of cocaine, a hallucinogenic chemical. What role might such chemicals play in the life of the plant?

Plants, Directly or Indirectly, Provide All the Food We Eat.

Our most important use of plants is as food. Today, as much as 90% of the total calories consumed by humans comes from crop plants in the following categories:

- Grains (cereals): wheat, rice, corn, sorghum, millet, barley, oats, rye (see Perspective 1.2, "Breakfast at the Sanitarium")
- Tuber and root crops: white potato, yam, sweet potato, cassava
- Sugar crops: sugarcane, sugar beet
- Protein seeds (legumes): beans, soybean, peas, lentils, chickpea, peanuts
- Oil seeds: olive, soybean, peanuts, coconut, sunflower, corn
- Fruits: citrus, mango, banana, apple
- Vegetables: cabbage, lettuce, onion

Most of the calories in our diet come from cereals such as wheat (fig. 1.6), rice, and corn, as well as potatoes, yams, bananas, coconuts, and cassava (the source of tapioca) (fig. 1.7). Even when you eat meat, you are eating plants indirectly because the meat came from animals that either ate plants or that ate animals that ate plants. Cereals provide much dietary carbohydrate, and the seeds of legumes such as beans are rich in protein. Cereals and legumes have different types of amino acids, the building blocks of proteins, so eating these two foods together provides a good balance of nutrients. The best plant-derived nutrition combines a cereal (a source of carbohydrate) with a legume (a source of protein, and often fat); a green, leafy vegetable (rich in vitamins and minerals); and perhaps small amounts of sunflower oil, avocado, or olives (which provide fats).

Although we eat many kinds of plants, cereals influence our lives more than any other plants. The importance of cereals cannot be understated. All of the world's great civilizations have been based on the cultivation of cereals: maize was the basis of the empires of the Incas, Aztecs, and Mayans; rice was the staple food in ancient China, Japan, and India; and the civilizations of Egypt, Rome, Greece, and Mesopotamia were based on wheat. Without an abundant and reliable source of these cereals, villages could not have grown into cities, nor cities into empires. Our lives today would be very different without cereals (see figs. 1.6 and 1.8).

(a)

FIGURE 1.6

(a) Wheat (*Triticum*) fields in Colorado. Grasses such as wheat provide a staple food for humans. (b) A few food products made from wheat.

(b)

FIGURE 1.7

Cassava (*Manihot esculenta*) is an important food crop and source of calories in the tropics, particularly in Africa. The large, tuberous roots are peeled, boiled, and mashed. Tapioca is made from cassava.

1.1 Perspective Perspective

Coffee: Even If You Don't Know Beans

Throughout this book, you'll see essays such as this that describe the excitement, lore, and recent discoveries of botany. Here, in the first of them, is "the rest of the story" about a plant extract that millions of people enjoy every day: coffee.

Coffee originated in the Middle East more than 1,000 years ago. Since then, it's been used as food, medicine, wine, and even as an aphrodisiac. Ever since the Boston Tea Party in 1773, coffee has been America's most popular beverage. Today, many people treasure quality coffees and pay more than $100 per pound for gourmet brands. Such value isn't a recent phenomenon—coffee beans are used as currency in some isolated parts of Africa, and the failure of a Turkish husband to provide his wife with coffee was once considered as "grounds" for divorce.

Coffee (*Coffea arabica*) is grown from seeds. After 2 years, the seedlings are about 0.5 meter tall and are transplanted to large plantations. Two to 3 years later, the plants produce white, jasmine-scented flowers that form fruit. When ripe, the fruits are red and sweet. Each fruit contains two green coffee beans that are harvested, dried, and sold.

Most grocery-store varieties of coffee come from arabica trees that produce about 5,000 fruits—1.4 kilograms (3 pounds) of coffee—per year. The coffee is about 2.5% caffeine.

Coffee berries mature in 7 to 9 months, turning from dark green to yellow, then red. Each fruit contains two green coffee beans that are harvested, dried, and sold.

Gourmet coffees comes from trees that are more susceptible to disease and more costly to harvest. These plants grow at elevations exceeding 1,800 meters (6,000 feet) and produce only about 2,500 fruits—0.7 kilogram (1.5 pounds) of coffee—per plant per year. Arabica coffee is about 1% caffeine. Espresso, the darkest roast, has even less because some caffeine burns off in the roasting process.

Given the potentially harmful effects of caffeine, several companies have marketed substitutes for coffee. The most popular of these was *Postum*, a nutritious beverage of wheat, molasses, and wheat bran sold by Charles W. Post in 1893. Post got the idea for this beverage while being treated for ulcers at the Western Health Reform Institute, operated by John Kellogg in Battle Creek, Michigan. Although Post left the institute uncured, he successfully marketed *Postum* with a clever advertising campaign that declared that *Postum* "makes your blood red" (to learn more about Post and Kellogg, see Perspective 1.2, "Breakfast at the Sanitarium"). Nevertheless, the public continued to want coffee. Sales increased greatly when Boston's Chase and Sanborn began selling roasted coffee in cans in 1878, and hoarding of coffee in the United States led to rationing in November 1942. Coffee's popularity has increased ever since.

Today, half of the people in the United States won't start the day without a cup of coffee; this amounts to an average consumption of about 13 pounds per person (in world-leading Finland, 38 pounds per year is the norm). Worldwide, the coffee industry generates $16 billion per year, and the more than 15 billion pounds of coffee produced annually in 50 countries provide more than 21 million jobs. Coffee shops such as Starbucks* are immensely popular throughout most parts of the world. Brazil produces more than one-third of the world's coffee, helping to make coffee the world's second-largest commodity—second only to oil—in international trade.

---

*Starbucks is named after the coffee-loving first mate in Herman Melville's classic adventure tale, *Moby-Dick*.

Agriculture provided a secure source of food for our early ancestors.

Obtaining food today is as simple as a trip to the supermarket or a visit to the garden or farm stand, but for our ancestors living 13,000 years ago, finding a meal was quite a challenge. Groups of people moved constantly in search of food (fig. 1.9). In what is now the United States, some people hunted the large plant-eating mammals such as bison, which roamed the Great Plains, while others collected seeds, roots, and fruits. These hunter-gatherers were at the mercy of the environment. If a drought or other natural disaster occurred, there might not be food where they were, so they had no choice but to move on or starve.

Over 12,000 years ago, people began to domesticate plants, consciously planting seeds, and then tending and harvesting the crop. (Some evidence indicates that rice may have been first domesticated in China several thousand years earlier.) In the Fertile Crescent, from the eastern Mediterranean to the Persian Gulf, the first plants to be grown agriculturally were rye, barley, and wheat, which are all cereals, followed soon thereafter by lentils and peas, which are both legumes (fig. 1.10). Several factors contributed to this first spark of civilization, including the close of the last ice age. As the glaciers receded, the land that was revealed became inhabited by a vast assemblage of animals and plants. Humans no longer had to follow herds of animals or gather berries to

1.2 Perspective Perspective

Breakfast at the Sanitarium: The Story of Breakfast Cereals

Cornflakes and several other breakfast cereals can be traced to Seventh-Day Adventists, an American religious sect. In 1866, the Adventists organized the Western Health Reform Institute in Battle Creek, Michigan. The institute was popular: patients such as Henry Ford, John D. Rockefeller, and Harvey Firestone flocked to the institute for recuperation and rejuvenation. The institute was later renamed the Battle Creek Sanitarium and, due to the sect's religious beliefs, served only vegetarian meals. Each morning, the flamboyant John Kellogg (a physician) and his quiet brother Will made breakfast for the sanitarium's patients by running a potful of cooked wheat through some rollers to form sheets of wheat. These sheets, when toasted and ground up, became a sort of cereal.

One night in 1895, the Kellogg brothers were called away on an emergency and left some wheat soaking for more hours than usual. When they tried to make their wheat-sheets the next day, each wheat grain formed its own flake. Patients at the sanitarium had a new breakfast food: wheat flakes. Four years later, the Kellogg brothers tried their new approach with corn, added some malt flavoring, and produced the first cornflakes. Patients at the sanitarium liked the flakes, especially ones made of corn, and thus was born Kellogg's Corn Flakes of the Kelloggs' cereal empire, which still exists in Battle Creek.* Although the Kellogg brothers later had a falling out (Will won use of the family name, which explains why his famous signature appears on the boxes of cereal), their product remained immensely popular. Indeed, within 15

years, more than 40 cereal companies were centered in or near Battle Creek, some of them operating out of tents.

Interestingly, one of the patients (a suspender salesman) at the sanitarium saw how the Kelloggs made their cornflakes. He later began a competing business that produced breakfast cereals, naming his first cornflakes "Elijah's Manna." However, the American clergy considered that name blasphemous, prompting the former patient in 1908 to change the name of the flakes. That patient was C. W. Post, who named his flakes Post Toasties.

---

*Interestingly, wheat flakes (e.g., Wheaties) weren't marketed until 26 years after cornflakes were sold. The process used by the Kelloggs to make flakes was soon modified to produce grains, shreds, and puffs. Puffs are made by heating the grains, causing pressure to build. When this pressure is released, the water vapor in the grain explodes the grain into a puff.

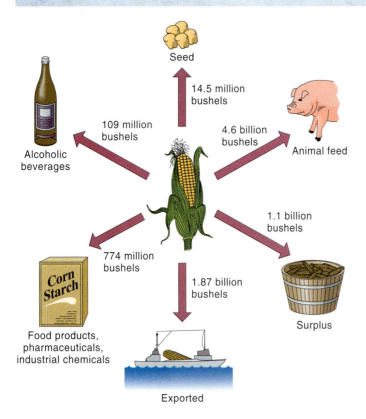

FIGURE 1.8

The fate of corn produced in the United States. Corn is also being used to produce ethanol, which is added to gasoline to make gasohol.

FIGURE 1.9

These hunter-gatherers in central Africa live much as our ancestors did. The woman carries the family dog; the man has caught a turtle. Hunter-gatherers such as these people do not build cities.

(a)

(b)

FIGURE 1.10

(*a*) and (*b*) Harvesting and winnowing of wheat in Tunisia, North Africa. Wheat has been cultivated for at least 12,000 years.

eat because food could be found in more areas. People could stay in one place, eating readily available small game, fish, and wild plants.

By about 10,000 years ago, humans learned to take care of animals such as wild sheep and goats, thus providing people with a continual supply of milk and meat. People also learned to manipulate the wild weeds that sustained many tribes, leaving seeds in the ground to ensure a future supply of food. Gradually, these early farmers discovered more about plant reproduction and used their knowledge to improve their food supply. They deduced by trial and error when and how to plant the correct seeds in the right soil and to see that the seeds got enough water to sprout. Thus, they had to learn about the **ecology** of the plants they were growing; that is, they had to learn how a plant was affected by its environment. They also learned to recognize when plants were ready to be harvested. People saved the seeds from one season's most useful plants to sow the next year's crop, thereby encouraging certain combinations of traits (a practice called **artificial selection**) that might not have predominated in the wild. Later, people learned to preserve their harvested food by drying it, thereby reducing even further their dependence on the unpredictable weather and climate. This domestication of animals and the intentional planting and cultivation of crops marked the birth of agriculture.

Farming likely arose independently in many places around the world and then spread from these centers (fig. 1.11). From the Fertile Crescent about 12,000 years ago, agriculture gradually spread to eastern Europe about 8,000 years ago and to the western Mediterranean

and central Europe about 7,000 years ago. Agriculture also spread throughout the Americas, based upon corn and other plants native to the region.

The distribution of crops across the planet today indicates that as humans colonized new lands, they often brought their native plants with them and likewise carried new plant varieties from the new land back to the old (e.g., corn was introduced to Europe from the Americas by Christopher Columbus in the late fifteenth century). Tracing the origins and evolution of crop plants is difficult, however, because archeological evidence is often destroyed over time.

In the New World, agriculture among Native Americans differed significantly from modern agriculture. Native Americans' agriculture began with the domesticaton of familiar, local plants. About 100 of the 120 known food crops were domesticated by Native Americans, including potatoes, avocados, beans, squash, pumpkins, bell peppers, tomatoes, peanuts, pecans, cashews, and many berries. They made chocolate (fig. 1.12), flavored their food and drinks with vanilla, grew pineapples (which later were sent to Hawaii), and even coated their popcorn with maple sugar, thus making a treat similar to Cracker Jack®. Native Americans used selective breeding to develop a wide variety of crops. For instance, they developed more than 300 varieties of corn from teosinte, a wild grass. Native Americans set up huge farms. Columbus reported cornfields nearly 30 kilometers (over 18 miles) long that included irrigation canals and the use of fertilizers. Recent evidence suggests that 1,000 years ago, Incas in Peru used

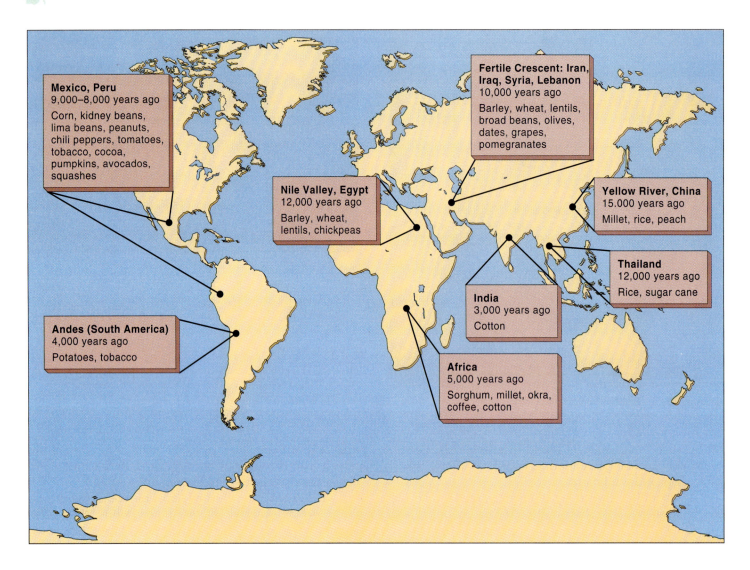

FIGURE 1.11

The domestication of agriculturally important plants. Farming arose independently in many places throughout the world.
Source: Data from Ricki Lewis, Life, 1992, McGraw-Hill Company, Inc.

conservation practices such as building canals and terraces along hillsides to control soil erosion and restore degraded farmland. Scientists base this suggestion on ecological evidence in the soil from the area; they have found high levels of pollen from *Ambrosia,* daisylike weeds that thrive in degraded soil, as well as sediments that indicate soil washed off the hillsides during heavy rains. On top of the older soil, scientists found soil with less ambrosia pollen but pollen and seeds from maize and other crops, suggesting that, over time, the Incas controlled soil erosion and were able to farm successfully.

Today, thanks to efficient transportation, many different kinds of plants are grown far from where they originated. Moreover, our relatively efficient farming methods free most people to pursue other activities. For example, only about 3% of the U.S. workforce are farmers; these people provide almost all of our food as well as more than 25% of our exports. This efficiency, however, comes at a high ex-

pense. Fossil fuels power the machinery that tills the soil, sows the crop seed, harvests the mature plants, and transports the crop to the factory where it is processed.

INQUIRY SUMMARY

Plants (and algae) provide all food for humans, either directly or indirectly. Most of the calories humans consume come from a few plants, the most important of which are cereals such as wheat, rice, and corn. Some agriculture began over 12,000 years ago in what is today's Middle East when the first farmers grew rye, wheat, and barley. Farming developed independently in several areas around the world and then spread from these centers of agriculture to other regions nearby.

Trace all of the food in your last lunch back to plants. From what plants did your lunch come, either directly or indirectly?

FIGURE 1.12

Fruits of the cacao (*Theobroma cacao*). Chocolate is made from seeds of cocoa. Because these seeds were believed to be of divine origin, the plant was named *Theobroma,* meaning "food of the gods." Most famous as the main ingredient of the drink given to Cortés by Montezuma in 1519, cocoa beans were also used as currency by the Aztecs, who paid their taxes with them until 1887. Chocolate has long been considered an aphrodisiac, and Casanova is said to have preferred chocolate to champagne as an inducement to romance. Today, chocolate remains a popular gift for romantics. On average, Americans eat about 11 pounds of chocolate per year.

The Plant Kingdom Is Nature's Medicine Cabinet.

Just as people learned to exploit plants for food, so they learned to use plants as medicine. Plants represent a huge storehouse of drugs: they produce more than 10,000 different known compounds, many of which help to protect the plants from hungry animals (herbivores). People who search for unknown compounds are natural-products chemists, such as Isao Kubo, a professor of natural-products chemistry at the University of California, Berkeley. The following excerpt is from his 1985 book, *From Medicine Men to Natural-Products Chemists:*

Several years ago, I was on a plant collecting trip near Serengeti National Park in East Africa. I was searching for the rare medicinal plant, *Kigeria africana.* After three days of combing the Serengeti plain, I came upon an outcrop of *K. africana* trees. Feeling very lucky indeed, I climbed up onto one of the tree's branches to collect the sausage-shaped fruit. But as I busily went about my collecting, I realized I was being watched. Two menacing eyes from an adjacent tree stared at me through the branches. Those eyes belonged to a rather large leopard. I tried with all my might to keep from falling from my perch as my knees knocked and sweat poured from my body. This brief encounter ended abruptly, however, when the leopard decided I was not a suitable meal and disappeared into the bushes. Well, I thought, another day in the life of a natural-products chemist.

Natural-products chemists such as Isao Kubo search through the bounty of chemicals made by organisms for substances that people can use (table 1.1). Here are a few of the more common drugs derived from plants:

- Steroids such as estrogen and testosterone are extracted from yams (*Dioscorea*) (fig. 1.13*a*). The progesterone used to make the first birth control pills was extracted from 10 tons of yams harvested from a Mexican jungle. Today, more than 60,000 tons of fresh yams are imported into the United States annually for the production of birth control pills, although much of the hormones today is produced synthetically.

- Cocaine comes from leaves of *Erythroxylum coca.* This shrub, the "divine plant" of the Inca civilization, is native to the eastern slopes of the Andes (fig. 1.13*b*). Cocaine, an ingredient of Coca-Cola until 1904, is grown mostly in Peru and Bolivia. It is a stimulant and hunger depressant, which explains why laborers can work for 2 to 3 days without food while chewing coca leaves. The laborers even measure distances in *cocadas*, the distance they can walk on one leaf that they chew. Today, cocaine is marketed illegally in highly addictive forms, with enormous costs to society.

- Reserpine is extracted from snakeroot (*Rauwolfia serpentina*), a low-growing evergreen shrub, and is used as a sedative and to decrease blood pressure. In India, snakeroot is used to make an antidote for snake bites. Many people with schizophrenia and other mental disorders can lead normal lives after being treated with reserpine. Annual prescriptions for reserpine in the United States exceed $100 million.

- Digitoxin, extracted from foxglove (*Digitalis purpurea*), a garden ornamental, is taken as a heart stimulant by more than 3 million Americans each day (fig. 1.13*c*).

- Tetrahydrocannabinol (THC) is derived from the hemp plant (marijuana; *Cannabis sativa*) and is used to make hashish, an illicit drug. However, smoking

TABLE 1.1   Some Important Medicinal Plants and Fungi

PLANT	PLANT PARTS USED	ACTIVE COMPOUNDS	USES IN MEDICINE
Atropa belladonna (belladonna)	leaves, roots	atropine, hyoscyamine	cardiac stimulant, pupil dilator, antidote for organophosphate poisoning
		scopolamine	motion sickness, antiemetic
Cannabis sativa (marijuana)	leaves, inflorescence	tetrahydrocannabinol (THC)	treatment of glaucoma, relief of nausea from chemotherapy
Catharanthus roseus (Madagascar rosy periwinkle)	leaves	vinblastine, vincristine	treatment of leukemia, Hodgkin's disease, and other cancers
Cinchona species (fever bark tree)	bark	quinine	treatment of malaria
Colchicum autumnale (autumn crocus)	corm	colchicine	treatment of gout
Digitalis purpurea (foxglove)	leaves	digitoxin, digoxin	cardiac stimulant, diuretic
Dioscorea species (yam)	tubers	steroids	production of cortisone, sex hormones, and oral contraceptives
Ephedra species (Mormon tea)	stems	ephedrine	decongestant, treatment of low blood pressure and asthma
Erythroxylum coca (coca)	leaves	cocaine	local anesthetic
Hydnocarpus kurzii (chaulmoogra tree)	seeds, fruits	ethyl esters of chaulmoogra oil	treatment of leprosy and related skin diseases
Hydrastis canadensis (goldenseal)	roots, rhizomes	hydrastine	treatment of inflamed mucous membranes
Nicotiana tabacum (tobacco)	leaves	nicotine	stimulant
Papaver somniferum (opium poppy)	latex from capsule	morphine, codeine	narcotic, analgesic, cough suppressant
Penicillium notatum (penicillin)*	hyphae	penicillin	antibiotic
Podophyllum peltatum (May apple)	roots, rhizomes	podophyllotoxin	treatment of venereal warts
Rauwolfia serpentina (rauwolfia)	roots	reserpine	treatment of high blood pressure and psychosis
Salix species (willow)	bark	salicin	analgesic, anti-inflammatory, treatment of rheumatoid arthritis and headaches
Taxomyces andreanae	hyphae	taxol	treatment of cancer
Taxus brevifolia	bark	taxol	treatment of cancer

*Fleming's discovery of penicillin involved *Penicillium notatum*, but most commercially produced penicillin today is derived from X-ray-induced mutants of *Penicillium chrysogenum*, which produce more than 1,000 times the penicillin originally derived from *P. notatum*.

(a)

(c)

(b)

FIGURE 1.13

(*a*) Yams (*Dioscorea* species) contain a variety of steroidal drugs, some of which are used to make birth control pills. (*b*) Woman gathering leaves of the coca plant (*Erythroxylum coca*) the source of cocaine. (*c*) Foxglove (*Digitalis purpurea*) contains cardiac glycosides used to treat congestive heart failure. These plants, along with *Digitalis lanata*, have saved the lives of many heart-attack victims and are helping millions of people with heart problems to lead normal lives.

marijuana has been effective in relieving the negative side effects of chemotherapy. Today, the former Soviet Union is the world's leading producer of hemp.

🌿 Pyrethrum, an extract from certain chrysanthemum flowers, is commonly used to treat head and body lice (and by organic gardeners as an insecticide) because it is a natural pesticide and non-toxic to warm-blooded animals.

Clues to possible medicines come from interactions between plants and animals.

Many chemicals in plants are potential drugs for humans. A chemical that oozes from a tree's bark or leaves and discourages hungry caterpillars from eating them may also be an effective drug for humans. An example comes from the Indian neem tree; a chemical in the bark of this plant keeps desert locusts off of it. The people of Serengeti National Park in East Africa chew the twigs of these trees to prevent tooth decay. Observing the behavior of native people around the world and of wild animals often provides a clue to possible medicinal plants (see Perspective 1.3, "Nature's Medicine Cabinet").

Today's natural-products chemists and ethnobotanists (scientists who study the use of plants by ethnic peoples) are

1.3 *Perspective* *Perspective*

Nature's Medicine Cabinet

To learn more about animal behavior, ecologist Holly Dublin spent most of 1975 doing something rather unusual: tracking a pregnant elephant in Kenya. Dublin noticed that the 60-year-old expectant mother seldom changed her routine of walking about 5 kilometers per day. However, one day the elephant changed her routine: she walked 28 kilometers to a riverbank and began eating leaves from a species of tree that Dublin had never seen an elephant eat before. Before leaving, the elephant ate the entire tree. Four days later, the elephant had her baby.

Dublin was puzzled by the elephant's abrupt change in behavior and unusual meal. Did the tree eaten by the elephant have anything to do with inducing birth? To her surprise, Dublin later learned that pregnant Kenyan women induced labor by drinking tea made from the tree's bark and leaves.

Dublin's observations are among a growing number of studies suggesting that animals use plants as drugstores:

- Chimps often eat leaves of the shrub *Vernonia amygdalin* when they're tired and sick. The plant is used by African tribes to treat the same symptoms. Similarly, chimps eat leaves of *Aspilia*, a member of the sunflower family. These leaves contain thiarurbine-A, a red, sulfur-containing oil that kills pathogenic bacteria and parasitic worms. Humans use extracts of the oil as anticancer drugs.

- Wild rhesus monkeys often eat soil with their food. That dirt contains much kaolin, a clay that detoxifies many poisons and is the active ingredient in Kaopectate, an antidiarrheal medicine.

Most biologists don't think that these and other examples are coincidence. Rather, they suspect that animals doctor themselves by using plants as preventive medicines.

modern versions of the medicine men and women who traditionally explored the healing power of plants. Herbal medical practices may have begun in prehistoric times when some individuals became botanical experts by sampling plants themselves. Clay tablets carved 4,000 years ago in Sumeria list several plant-based medicines, as do records from ancient Egypt and China. Roman philosopher Pliny the Elder wrote in the first century A.D., "If remedies were sought in the kitchen garden, none of the arts would become cheaper than the art of medicine." Modern-day medicine men, called *bwana mgana,* practice herbal medicine in the region of East Africa explored by Isao Kubo, quoted earlier.

A good example of medicines derived from plants are the group of chemicals called **alkaloids,** such as vincristine and vinblastine, which come from the periwinkle plant mentioned at the beginning of this chapter. Alkaloid narcotics derived from the opium poppy, including morphine, are excellent, but addictive, painkillers. Today, nearly half of all prescription drugs contain chemicals manufactured by plants, fungi, or bacteria, and many other drugs contain compounds that were synthesized in a laboratory but modeled after plant-derived substances. The many medicines we obtain from the plant kingdom are a compelling reason why we must stop destroying natural communities such as the world's tropical rain forests, where plant life is so abundant and diverse that all of the species have not even been identified, much less studied. In addition, much research is being conducted today on the role of plants in our diet and their relationship to our health. For instance, could eating more products from soybeans, such as tofu, help reduce the risk of cancer? Could a return to the menu of our ancient ancestors—to eat plants our ancestors evolved with—lower our risks of diseases such as diabetes, as it has for many Native Americans and Hawaiians? Today, plants dominate our lives and economy, just as they have in all civilizations.

The study of economically important plants includes investigation of their ecology and evolution.

Any plant that affects the lives of humans has economic importance. Weeds reduce crop productivity, poisonous plants may kill us, and plants that cause allergies cost us time and money. The lives of many people—including farmers, foresters, and grocers—revolve around plants, and medicinal and food plants are among thousands with commercial value. Understanding the life of economically important plants is essential to improving the quality of beneficial plants, to finding more helpful plants, and for treating the effects of harmful plants.

The study of plants having economic importance usually involves asking questions about their ecology and evolution. For instance, the rosy periwinkle, shown in figure 1.1, is native to Madagascar, a tropical island off the southeast coast of Africa. Most of the animals and nearly 85% of the island's plants are endemic to Madagascar, meaning that they grow nowhere else in the world. Unfortunately, the human population of Madagascar has risen dramatically, and native plants and animals have been driven to extinction because of agriculture, uncontrolled grazing, logging, and hunting. Is there a way to produce enough food and fuel for the people of the island and still save the habitats of the island's native species? (One answer could be to develop reserves on the island to protect native plants and animals.) What are the ecological and evolutionary reasons for there being so many endemic species on Madagascar? (One answer is that plants and animals evolved in the unique environment of the island in isolation from other organisms on the continent of Africa and, over time, became species different from any others on earth.) What interactions between the rosy periwinkle and its environment may have promoted the production of vincristine and vinblastine? (One answer comes from the periwinkle's interaction with animals; the chemicals vincristine and vinblastine may have protected the plants from herbivores on the island.)

INQUIRY SUMMARY

Plants are a major source of medicines. Natural-products chemists and ethnobotanists are among the scientists who search the world for potentially new medicinal plants. One method used in this search includes studying the interaction between animals and the plants they avoid or that affect their lives. The evolution of economically important plants depends on their ecological relationships with other organisms in their native habitat.

Health food stores sell a variety of plant products that claim to have healthful effects. How can a person separate beneficial products from those with no effects?

What Is a Plant?

You already have some notion of what a plant is—an immobile, green organism that we eat, mow, and use for decoration and shade—but it's difficult to come up with a simple definition for the word *plant.* Not all plants are green, and some plants consume animals. Indeed, many plants do not look or seem like plants at all, but the most familiar plants share a few common characteristics.

Plants share many characteristics with other organisms.

Plants share the characteristics of life with all other organisms. All living organisms:

- take in and use energy, which allows them to conduct their daily activities;
- consist of a single cell or many cells—a large organism such as yourself or a small tree may have trillions of cells making up its body;
- reproduce;
- respond to the environment in which they live; and
- share parts of a common ancestry and have coevolved with other organisms, such as flowering plants and their pollinators (fig. 1.14).

FIGURE 1.14

Plants have coevolved with other organisms. In what ways might the relationship between this plant and animal be mutually beneficial?

A plant is an autotrophic organism with a unique set of characteristics, including cell walls made of cellulose.

Plants possess all of the characteristics previously listed as well as a combination of characteristics that is unique to them. In general, *plants are **autotrophic,** or "self-feeding."* This means they can produce their own sugars, proteins, vitamins, and amino acids. **Heterotrophic** organisms such as humans and other animals must obtain their food from other organisms and many of the necessary vitamins and amino acids from the food they eat. Most plants are autotrophic because they can convert light energy into the chemical energy of sugar in the process of **photosynthesis.** *Plants and algae possess special **pigments**, including **chlorophyll**, that absorb light.* Chlorophyll is not found in heterotrophic organisms. Once a sugar molecule is made, a plant can use the sugar as a source of energy, but it can also use the sugar to make cell parts, including a material called **cellulose.** *Plants have **cell walls** made of complex molecules such as cellulose.* This rigid cell wall surrounds each plant cell, provides some protection to the cell, and gives the cell shape. Animal cells lack a rigid, protective cell wall. *Plants are not motile*—they cannot move from one place to another once they start growing. This has immense consequences for the life of a plant. For instance, if the temperature is too high or the amount of precipitation too low in the place where a plant is rooted, it cannot move to an area with more suitable environmental conditions. However, even though they cannot move to a better location for growing, plants do have many movements, such as the head of some sunflowers turning toward the sun throughout the day or the leaves of a sensitive plant drooping when they are touched (see fig. 6.6). In addition, the offspring of plants may be carried great distances from the parent plant. *Plants have open,* or indeterminate, growth, whereas most animals have closed growth. This means that the shape of the animal doesn't change much after it reaches maturity, but a plant grows indefinitely, producing more of its parts, such as roots, stems, and leaves, throughout its entire life. Finally, many *plants have the ability to reproduce asexually* (vegetatively) as well as sexually, whereas most animals practice only sexual reproduction. Asexual reproduction does not involve the fusion of a sperm and egg to produce an offspring; instead, a new plant may be produced directly from one of the parts of a plant.

INQUIRY SUMMARY

Plants share many characteristics with all other organisms on earth but possess a unique set of characteristics that includes being autotrophic, nonmotile, able to reproduce asexually, and having open growth and cell walls made of cellulose.

What does the fact that all organisms share many characteristics suggest about them having a common ancestor?

A Seed Develops into a Plant Having Roots, Stems, and Leaves.

When people picture a plant in their minds, the image conjured up usually has roots, stems, and leaves. While most people have an idea of what a "typical" plant looks like, many plants do not have roots, stems, leaves, or seeds. **Seeds** are produced by flowering plants, the **angiosperms,** and a group of plants often called evergreens, the **gymnosperms,** such as pines and firs. The inner part of each seed is an **embryo,** which is a young plant; a seed protects and nourishes this young plant until it can produce sugars on its own through photosynthesis.

A distinguishing feature of a seed is the **seed coat,** which is the outer, protective layer immediately surrounding the seed. The seed coat is often thin and papery, like that covering each peanut seed (the seed is the part you eat). At a baseball game, you may tear off a hard, brown shell to get to the peanuts inside. This shell is actually the fruit wall. The seeds of flowering plants are covered by a fruit; the seeds of evergreens such as pines are not completely covered but may be protected inside a cone.

Monocots and dicots have different kinds of seeds.

Flowering plants are divided into two large groups, distinguished in part by the characteristics of their seeds. **Monocots** (short for *monocotyledons*) are flowering plants that produce seeds each with only one **cotyledon,** or seed leaf. The cotyledon is a modified leaf involved in storing or supplying nutrients and energy for the embryo in a seed. All of the cereal plants, such as wheat and corn, are monocots and so

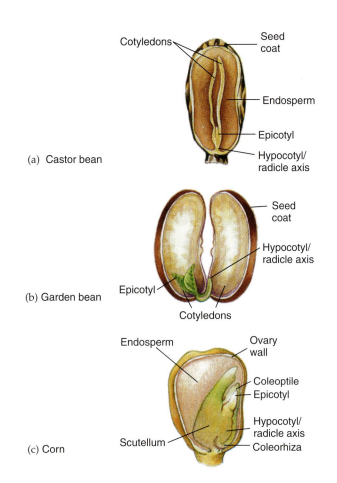

FIGURE 1.15

Seeds. (*a*) Seeds of castor bean (*Ricinus communis*) have abundant endosperm that surrounds two thin cotyledons. (*b*) The two cotyledons in each seed of garden bean (*Phaseolus vulgaris*) absorb the endosperm before germination. (*c*) Corn (*Zea mays*) has seeds in kernels; the single cotyledon is an endosperm-absorbing structure called a scutellum. Seedlings of bean and corn are shown in figure 1.16.

are the grasses that grow in your front lawn. The other large group of flowering plants includes the **dicots** (short for *dicotyledons*) (fig. 1.15), which produce seeds each having two cotyledons. An example of a dicot is a peanut; you can easily separate the peanut into two halves (each half is a cotyledon) with the embryo visible between them. Oaks, roses, and garden beans are other common dicots.

The region of the embryo above the attachment point of cotyledons is the **epicotyl,** which gives rise to the shoot, made up of stems and leaves. The region below the attachment point is the **hypocotyl.** An embryonic root, the **radicle,** is often distinguishable at the tip of the hypocotyl. The embryos of corn and other grass seeds are partially enclosed in protective sheaths.

The first root of the seedling supplies water and nutrients to the shoot before the shoot breaks through the surface of the soil. The shoot may or may not bear cotyledons when it emerges, depending on the type of plant. In the garden bean, for example, the fastest growth occurs in the hypocotyl.

As the new shoot pushes through the soil, the hypocotyl elongates and forces the embryonic **apical meristem,** or growing point at the apex of the shoot, along with the cotyledons, above the surface (fig. 1.16a). In contrast, cotyledons of the garden pea remain belowground because the fastest growth occurs at the embryonic shoot tip (fig. 1.16b). In both cases, the cotyledons supply energy from the **endosperm,** the nutritive material found in the seed for the growing seedling. Above the ground, cotyledons become green and photosynthetic in the garden bean. Cotyledons of the garden pea quickly shrivel as the nutritive material is used up by the seedling.

Grasses have the most complex type of seedling development. The seed, which remains enclosed in the ovary wall, germinates when the **coleoptile,** which sheathes the shoot, and the **coleorhiza,** which sheathes the radicle, both begin to grow (fig. 1.16c). The radicle then breaks through the tip of the coleorhiza and becomes the primary, or first, root. Similarly, when the coleoptile stops growing, the uppermost leaf pushes through it and becomes the first photosynthetic organ of the new seedling. During the germination of grass seeds, the cotyledon continues to absorb sugars from the endosperm until the new leaves make enough sugars for the plant to grow independently.

Roots anchor plants, store energy, and absorb and conduct water and minerals.

Mosses don't have roots, but mosses are small plants oftentimes restricted to moist places, such as along streams. What does this tell you about the advantage roots give to plants that possess them?

Plants with roots produce extensive root systems that tap into water and minerals dispersed in the soil. In only 6 weeks after germinating from its seed, a corn seedling produces more than 2,000 roots, and a 100-year-old Scotch pine (*Pinus sylvestris*) possesses more than 5 million roots. A root system may consist of a single large primary root or several large primary roots. In either case, smaller secondary roots grow from the primary root. The tip of all roots is covered by a protective **root cap;** the root cap produces mucilage, which helps the root as it moves through the soil. Thousands of single-celled **root hairs** cover each root beginning a few millimeters away from its tip (fig. 1.17).

Roots have four primary functions:

1. *Anchorage.* Roots permeate the soil, and in doing so, they anchor the plant in one place for its entire life (recall that plants are not motile).
2. *Storage.* Roots can store large amounts of energy reserves. In biennials (i.e., plants that complete their life cycle in 2 years), such as carrot and sugar beet, these reserves are concentrated in only one or a few primary roots. Humans, and other animals, harvest these roots

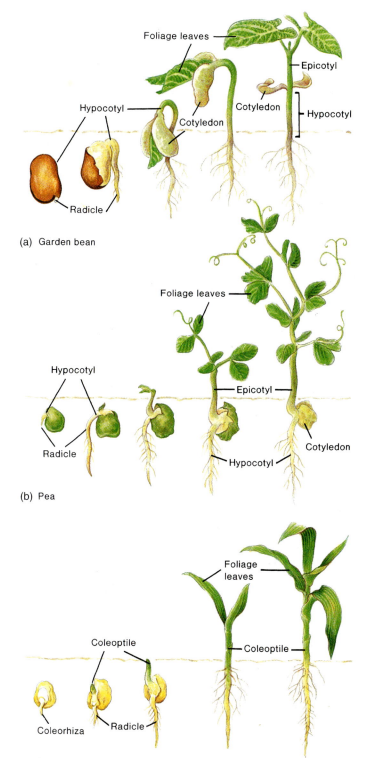

(a) Garden bean

(b) Pea

(c) Corn

FIGURE 1.16

Seed germination and seedling establishment. (*a*) Growth in the hypocotyl of garden bean forces the shoot apex and the cotyledons above the surface. (*b*) In garden pea, rapid growth at the shoot apex leaves the cotyledons belowground. (*c*) In corn and other grasses, the shoot apex grows out of the kernel through the tubelike coleoptile, and the root apex grows through the tubelike coleorhiza. Obtain seeds of some plant, germinate them, and observe the development of the seedlings.

Shoot

Primary root
with root hairs

Root cap

FIGURE 1.17
Corn seedling. Notice the many, fine root hairs extending from the primary root.

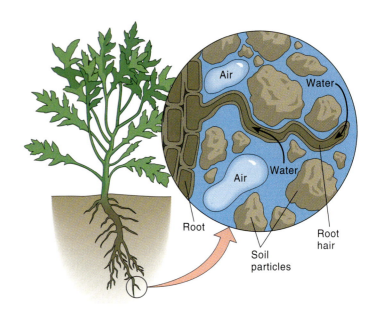

Air

Water

Air

Water

Root

Soil
particles

Root
hair

FIGURE 1.18
Root hairs absorb water and nutrients from the soil.

after the first year of growth, before the plant uses the stored energy for vegetative growth and reproduction.

3. *Conduction.* Roots transport water and dissolved nutrients to and from the shoot. The roots of plants such as quillwort (*Isoetes*) and shoreweed (*Littorella*) even transport carbon dioxide (CO_2) to leaves, where it is used in photosynthesis.

4. *Absorption.* Roots absorb large amounts of water and dissolved minerals from the soil. For example, the roots of a mature corn plant may absorb more than 2 liters of water per day. This is necessary because plants require huge amounts of water for optimal growth.

This water and its dissolved minerals are obtained from the soil by roots, which are adapted for absorbing water: their continuous growth extends into new territory, and their root hairs greatly increase the **surface area** for the absorption of water and nutrients. Surface area is the part of a plant in contact with its environment. A mature oak tree can have several million root tips distributed throughout the soil, each of which has thousands of root hairs absorbing water and dissolved minerals from different areas (fig. 1.18). Roots of a 4-month-old rye plant have a surface area exceeding 600 square meters (about the size of 1.5 basket-

ball courts), and the 150,000 root hairs in only 1 cubic centimeter of soil beneath a patch of bluegrass have a surface area equal to approximately two-thirds of this page. Exposing this much surface area to the environment greatly increases the absorptive capabilities of plants.

The root system may extend deeply into the soil.

Roots permeate the soil. Roots of plants such as salt cedar (*Tamarix*) spread tens of meters and burrow more than 30 meters deep. The primary root of a 12-meter oak tree goes down more than 5 meters, while its secondary roots span more than 40 meters. Most roots grow throughout the upper 3 meters of soil, where nutrients are most abundant. Grasses usually produce shallow root systems with many primary and secondary roots. This characteristic makes grasses favorite landscaping plants to control erosion in such places as steep embankments along roadsides. The extensive fibrous root systems of grasses help hold the soil in place and keep it from washing away.

Shoots consist of stems and leaves.

A stem is a collection of nodes, internodes, and axillary buds. **Nodes** are regions where leaves and axillary buds attach to stems, and **internodes** are the parts of stems between nodes (fig. 1.19). An **axillary bud** is found in the upper juncture between a leaf and the stem; each bud consists of an immature stem and immature leaves. Together, these parts make up the shoot system of a plant.

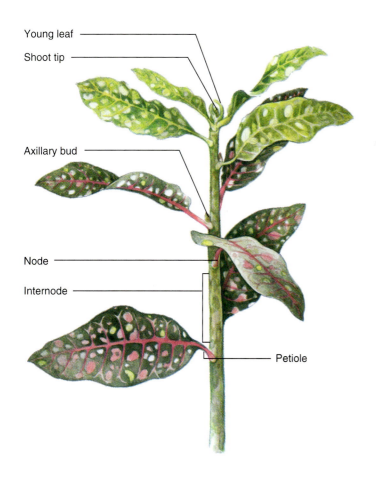

Young leaf

Shoot tip

Axillary bud

Node

Internode

Petiole

FIGURE 1.19

Stems consist of nodes and internodes. Nodes are regions where leaves and axillary buds attach to stems, whereas internodes are the parts of stems between nodes.

Stems perform three important functions:

1. *Support.* Stems support leaves, the solar collectors of plants, as well as flowers and fruits. Leaves must be held in a position where they can effectively absorb sunlight for photosynthesis. Flowers must be held in a position that makes them visible to potential pollinators, and fruits must be in a position to optimize their dispersal.

2. *Storage.* Stems can store large amounts of starch and water. For example, water accounts for as much as 98% of the weight of many cactus stems, and potatoes, which are modified underground stems, store much-valued starch.

3. *Conduction.* Stems transport water and minerals between roots and leaves. Stems link leaves with the water and dissolved nutrients of the soil. They also conduct the sugar produced in leaves to the roots. Stems of **herbaceous** (i.e., nonwoody) plants are green and photosynthetic. Although photosynthesis in stems is usually insignificant compared to that in leaves, in plants such as cacti it accounts for most of the plant's production of sugar.

Leaves are the main sites of photosynthesis.

Leaves are usually the most conspicuous organs of plants, and their most important functions are absorbing sunlight and producing sugar through photosynthesis. These functions are aided by the large surface area of the leaves exposed to the environment. For example, a maple tree (*Acer*) with a trunk 1 meter wide has approximately 100,000 leaves with a combined surface area exceeding 2,000 square meters (roughly the area of six basketball courts). On a global basis, photosynthesis in leaves produces more than 200 billion tons of sugars per year. Those sugars and the oxygen produced in photosynthesis sustain virtually all life on this planet.

INQUIRY SUMMARY

The body of seed plants consists of roots, stems, and leaves. Flowering plants are either monocots, possessing seeds with one cotyledon, or dicots, possessing seeds with two cotyledons. Seeds contain endosperm, which is an energy store for the developing embryo. Roots anchor the plant and absorb water and minerals; stems conduct these materials throughout the plant and support the leaves, which are the main sites of photosynthesis.

The root of a young plant develops more rapidly than does the shoot. What do you think is the significance of this?

What Is Botany?

Our need-driven uses of plants, along with our curiosity about the world, have spawned **botany,** the scientific study of plants. Today, botany is a thriving, exciting science that, directly or indirectly, deals with the largest part of our national and world economy. Botanists, other scientists, and nonscientists often share an almost insatiable curiosity about life and its diverse but related forms. We watch nature shows on television, visit zoos and botanical gardens, buy pets and houseplants, and tend our gardens. Botanists, however, also often watch life in more exotic places—for example, while perched in a treetop of a tropical rain forest or while in a greenhouse surrounded by a variety of plants being grown for experiments. Such intense observation has generated many questions about plants; most can be grouped as follows:

- How are plants constructed?
- How do plants work?
- How did plants get here; that is, how did they evolve?
- What roles do plants play in an ecosystem?
- In what ways are plants important to humans and other animals?

꙳ How do plants interact with other organisms and with their physical environment; that is, what is their ecology?

꙳ What kind of plants exist in the world, and how are they related to each other?

Botanists have answered these and other questions by using an experiment-based process—*a scientific method of investigation*—which is a systematic way to describe and explain the universe based on observing, comparing, reasoning, predicting, testing, concluding, and interpreting. All in all, it's not very different from the way that most nonscientists think about this fascinating thing we call life.

Observing and asking questions are the first steps in studying a problem.

Superficially, botany is a collection of information that describes and explains the workings of plants in the living world. Although this information is interesting, it is not the essence of botany. Rather, the excitement of botany lies in the intriguing observations and the carefully crafted experiments that botanists and others have devised to help us learn about plants. The important thing here is the *process* of botany—not just knowing the facts but appreciating how botanists discovered (and continue to discover) those facts. This process, a scientific method of investigation, is the distinguishing feature of science that you will read about throughout this book.

A scientific method of investigation usually begins with things with which we are all familiar: observation and curiosity. Such observations can happen just about anywhere—in a research laboratory, in a garden, or while you're reading a book or watching a television program. For instance, the observation that leaves of many plants follow the sun catches our attention and piques our curiosity. This may prompt us to ask, What enables the leaves of some plants to move throughout the day? Similarly, the observation that some plants in our garden are not attacked by whiteflies while others are infested may lead us to ask, What enables resistance to these whiteflies in some plants but not others? These kinds of questions are at the heart of any scientific method. Science is fundamentally about finding answers to such questions. To find these answers, botanists use past experiences, ideas, and observations to propose **hypotheses,** which lead to predictions. To determine whether predictions are accurate, botanists do experiments. If experimental results match the predictions of a hypothesis, the hypothesis is supported; if results do not match the predictions, the hypothesis is rejected. The effect of all this is to make scientific progress by revealing answers piece by piece. Experiments allow us to determine what causes a particular observation.

We all think like scientists at times.

The cycle of questioning and answering that defines a scientific method of investigation is second nature to a practicing scientist; it's more a philosophy than a set of rules. However, the use of a scientific method is not restricted to scientists: everyone observes, compares, predicts, plans, does experiments of some kind, and interprets and uses the results of these experiments. Even our ancestors used experimental trial and error to determine which plants were edible—sometimes with unfortunate results when they ate a previously unknown poisonous plant!

In an experiment in 1820, a New Jersey colonel named Robert Johnson bravely ascended the steps of a county courthouse and ate a basketful of tomatoes, much to the amazement of the 2,000 townspeople, who thought that tomatoes were poisonous. His bold experiment and its results (his survival) showed the townspeople that tomatoes are not poisonous. Parents do a similar experiment when they introduce food one item at a time to their children to check for allergic reactions. They also experiment by tasting the food in question to check its safety and taste.

Let's see how the process of scientific investigation might work in your life. Suppose that you wake up one night with a bothersome rash (*observation*). You want to know what caused this rash (*question*). You first try to figure out how this situation is similar to other situations with which you are familiar. Perhaps a classmate had a similar rash that was caused by an allergy to the grass at the neighborhood soccer field. You realize that your rash has appeared after all of the soccer matches that you've played at that field. The similarity between your friend's allergy and your rash leads to the explanation (*hypothesis*) that your rash is also caused by an allergy to the grass.

If your hypothesis is correct, you would not expect to get a rash from other kinds of plants (*prediction*). When you hike in the woods and are exposed to many different kinds of plants, but not the grass at your soccer field (*experiment*), you do not get a rash (*results*). Your results match the prediction of the grass-allergy hypothesis, from which you decide that your rash is caused by the grass you touch when playing soccer (*conclusion*). However, if this was your only hypothesis, you have not actually identified the cause. Perhaps your rash was caused by the paint used in the locker room or by the cologne a teammate uses. Or perhaps it was caused by fertilizers, pesticides, bacteria, molds, or other agents associated with the grass or soil at the soccer field. By testing a single hypothesis, you have not ruled out any of these other possibilities. To do so, you would have to devise alternative hypotheses, make predictions from them, and obtain experimental results to compare with the predictions. By this process, you may be able to reject all hypotheses, even the grass-allergy hypothesis. Either way, you make progress by testing several hypotheses, not just one. If you can reject all but the grass-allergy hypothesis, you can more confidently conclude the grass caused your rash. Furthermore, if the grass is in the soccer field but not in the baseball park, you may decide to become a baseball or softball player so that you can avoid the irritating plant (*use of new knowledge*). If all of this seems like common sense, it is. As Thomas Henry Huxley said, "Science is nothing but trained and organized common sense."

Although a conclusion marks an end to a scientific method of investigation for a particular experiment, it seldom ends the process of scientific inquiry. Any conclusion must be placed in perspective with existing knowledge. Moreover, an underlying characteristic of scientific thinking is that the sequence of observing, comparing, reasoning, predicting, testing, concluding, and interpreting is a cycle, with new ideas spawned at every step (fig. 1.20). In addition, results must be verifiable by other scientists using the same or a similar sequence of investigation. To the curious scientific mind, a conclusion is never the final answer. There is always something more to study, something new to learn. For instance, what part of the grass plant is causing your allergies? Are there other plants that give the same reaction? Will your children have the same allergies?

Our use of a scientific method of investigation has greatly advanced knowledge and improved our quality of life. For example, how would Gregor Mendel, a monk who described several principles of heredity more than a century ago, feel if he knew his observations of pea plants would one day help to explain how a perfectly healthy couple could have a child afflicted with a crippling genetic disease? Similarly, it took keen observation, experimentation, and reasoning to discover that the mild and supposedly harmless tranquilizer called thalidomide could cause 10,000 European children to be born without arms and legs and that a virus could disrupt the immune system and cause AIDS.

Biologists use a scientific method of investigation to answer questions about life.

In science, as well as life, it is important to keep an open mind to maintain the objectivity that a scientific method of investigation requires. An objective view is necessary when drawing conclusions and designing experiments so that biases or expectations do not cloud the interpretation of results. This is sometimes difficult because it is human nature to be cautious in accepting an observation that does not "fit" existing knowledge. For example, for centuries people assumed that life arose spontaneously from nonliving matter. This assumption was called *spontaneous generation*. Several simple experiments to test this assumption involved exposing decaying meat to the open air; the results of these experiments "proved" that life arises spontaneously from the nonliving, a conclusion that brilliant scientists such as Newton, Harvey, and Linnaeus never questioned. However, spontaneous generation finally was rejected by another set of different, more critical experiments done by Louis Pasteur. The

FIGURE 1.20

Scientific investigations begin with observations and questions. Describe what you see in the picture. What might be the cause of such a pattern of plants? How might you test your idea?

conclusion from those experiments, namely that life does not arise from decaying meat, surprised many who believed that mice were created by mud, that flies came from rotting beef, and that beetles sprouted from cow dung.

More recently, several studies have shown that animals that are fed large doses of vitamin E live longer than those receiving only average amounts of the vitamin. From those studies, many concluded that vitamin E retards aging in humans, thereby promoting all sorts of claims by vitamin and skin-cream manufacturers. However, the animals fed large amounts of vitamin E also lost weight, another factor that correlates positively with increased life expectancy. The original experiment did not distinguish between these two interpretations. As is usually the case, more experiments are necessary to eliminate the other possible interpretations.

Perhaps no one knew the conflicting feelings of joy and frustration that accompany a scientist's discovery of a

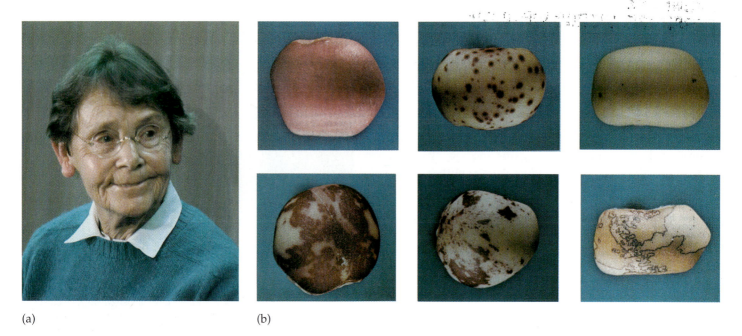

(a)   (b)

FIGURE 1.21

The work of botanist Barbara McClintock. (*a*) Barbara McClintock at the press conference announcing her Nobel Prize in October 1983 for her discovery of mobile genes ("transposable elements"). (*b*) Mutations caused by transposable elements. What should have been colored kernels lack pigments because a transposable element lies in or near the gene for pigment production. During development, the element sometimes leaves the gene, restoring pigmentation to that cell and all of its descendants. This produces the color variations of the kernels.

quirk of nature better than the late Barbara McClintock (fig. 1.21). In the 1940s, McClintock studied the inheritance of kernel color in corn by, as she put it, "Asking the maize (corn) plant to solve specific problems and then watching it respond." While watching her carefully bred plants respond, McClintock noticed that some kernels had a peculiar pattern of spotting. She concluded that the units of inheritance (genes) moved around in corn cells. This seemed preposterous at the time, for genes were thought to be immovable parts of chromosomes. Despite McClintock's evidence, the scientific community refused to accept her conclusion. However, after years of additional discoveries of mobile genes in other organisms, Barbara McClintock was finally recognized as the brilliant botanist that she had always been and was awarded the Nobel Prize. Her observation of transposable elements, or "jumping genes," in corn plants almost five decades ago may ultimately help explain how roving genes cause certain cancers in humans.

Biotechnology is an active area of scientific research as well as an industry that will affect your life.

Today, much of botany, and science in general, is driven by new technology. Just as microscopes revolutionized our understanding of cells, so too will new technology produce remarkable discoveries and new insights. That technology will overturn many of the "facts" that you read about in this or any other science book. This is why we concentrate so strongly in this book on the *process* of science; the facts and interpretations may change, but the process used to discover them does not. This process enables you to think critically, solve problems, and answer your own questions about health, the environment, and life on earth in the future.

Throughout this book, you also learn about a new, exciting, multibillion-dollar business: plant biotechnology. **Biotechnology** is a way of using organisms to make commercial products. Although we've used biotechnology for several decades, our ability to manipulate a plant's genome (its genetic makeup) began in 1983 when botanists transferred a gene from a bacterium into a plant. Presently, biotechnology and molecular biology are revolutionizing biology and biology-based industries. Here are a few of the ways we use biotechnology:

- *To make vitamins and drugs.* Just as pigs are now being used to make human hemoglobin, plants are being transformed into factories that make drugs and oils. A promising new vaccine for hepatitis B is made with baker's yeast, and plants are now used to make serum albumin, which is used for fluid replacement in burn victims. Genetically engineered tobacco plants now produce human antibodies useful for diagnosing and treating disease. Similarly, biologists have transferred a gene for an unusual protein—the one that keeps flounder from freezing in winter—into tomato and tobacco plants. This "antifreeze gene" could help extend the area in which these plants are grown. Botanists also have produced canola oil containing 40% laurate, an important ingredient of many soaps and shampoos.

- *To produce plants that resist drought, disease, and insects.* Botanists at Monsanto have used genetic engineering to produce disease-resistant plants. Others have transferred a gene from *Bacillus thuringiensis*, a common soil bacterium, into cotton plants (fig. 1.22). This gene produces a protein that, when eaten by pests such as the cotton bollworm, induces paralysis and death. Although most *B. thuringiensis* is used to fight caterpillars, other strains of the bacterium are being developed to fight beetles and mosquitoes. Botanists are searching for ways to help plants resist diseases that lay waste to, on average, 12% of crops worldwide.

INQUIRY SUMMARY

Botany is the study of plants, which uses a scientific method of investigation to discover how the world and plants work. A scientific method is a systematic way to describe and reveal the universe based on observing, comparing, reasoning, predicting, testing, and interpreting. Everyone uses a scientific method at times. Plant biotechnology is a multibillion-dollar botanical industry that is changing our lives and society. Biotechnology will affect our lives in many different ways in the future.

When was the last time you used a systematic method of investigation to answer a question or to solve a problem you had? What were the steps in your method?

You Can Be an Informed Agent for Change.

A major goal of this book is to help you appreciate living organisms. This is perhaps the most important goal of any biology book and any biologist's teaching, for such an appreciation of life is the first step toward respecting and conserving life. Appreciation and respect for life are critical, especially in light of our neglect of the environment for many decades, a neglect that is now costing us money and killing earth's organisms. Here are just a few examples:

- *Each year, businesses and local governments spend more than $100 billion trying to clean up messes in the environment. Businesses pass these costs on to consumers as higher prices. Pollutants such as lead, mercury, and DDT kill many organisms. Today, research is being conducted on the potential use of plants to

FIGURE 1.22

Monsanto's use of genetic engineering to improve crops. The improved cotton boll on the right with no insect damage is the result of combining traditional breeding techniques with the science of biotechnology. Insect control is achieved when a gene from a common soil bacterium, *Bacillus thuringiensis,* or *Bt,* is placed into plants, causing plants to produce a protein that acts as a natural insecticide against the tobacco budworm, pink bollworm, and cotton bollworm. An untreated boll on the left shows typical insect damage.

absorb and perhaps detoxify chemicals from the ground (see fig. 2.12).

- *About 5% of the 80,000 species of plants native to temperate regions are near extinction. This has been caused by habitat destruction by humans, overgrazing by domestic animals, the introduction of foreign plants, and the destruction of pollinators.

- *Each year, humans clear about 15 million hectares of tropical rain forest—an area approximately the size of Florida (fig. 1.23). By clearing these forests, we drive animals and plants to extinction and thereby lose many potentially valuable food sources and drugs that could save lives. Botanists estimate that as many as 50,000 species of tropical plants may be extinct within the next few decades because of the destruction of rain forests.

Such problems will not be solved by governmental panels, presidential commissions, blue-ribbon committees, or expert testimony before Congress. Such groups have been convened for decades, during which time our problems have only worsened. These problems will be solved only by a scientifically literate public. This book uses botany to teach such literacy; we hope it enables you to help solve the many problems we face.

This book helps you appreciate what botanists know, what they don't know, and what work they do. You learn not only "facts" but the *process of doing botany.* Your observations and ideas concerning life will be more meaningful the more experience and learning you have on which to

FIGURE 1.23
Fire in a Guatemalan rain forest to clear land for human uses. Such destruction helps drive plants and animals to extinction.

build. This book and your course will give you some of this valuable background, which should enable you to apply a scientific method of investigation to your own observations of the living world.

SUMMARY

Plants provide us with food, clothing, shelter, and medicine. Between 10,000–12,000 years ago, groups of people gradually changed from a hunter-gatherer lifestyle to an agricultural way of life, intentionally saving and planting seeds from the best individual plants of crops that could be used as food. Encouraging the propagation of certain individuals artificially selected particular traits; in this way, humans have influenced the evolution of domesticated plants.

Cereals such as corn, wheat, and rice are members of the grass family, whose edible seeds (grains) have been critical for civilizations around the world. For centuries, humans and other animals have used plants for their medicinal properties. Today, natural-products chemists use a combination of laboratory techniques and information from folklore and herbal medicine to make compounds with effects similar to those of plant-derived compounds in the search for new and more effective drugs.

A plant is an organism that shares many characteristics with all organisms on earth, including an energy requirement and the ability to reproduce and respond to the environment. Plants also possess characteristics that make them different from other organisms, including being au-

totrophic, possessing cell walls composed of cellulose, and having open growth.

Botany is the scientific study of plants. Plants have a fascinating history and lore, and our national and world economies are based on them. Virtually everything we do is influenced by plants. Botanists learn about plants and the world by using a scientific method of investigation, a systematic way to describe and explain the universe, based on observing, comparing, reasoning, predicting, testing, and interpreting. Everyone uses a scientific method of investigation at times. Plant biotechnology is a multibillion-dollar industry that is changing our lives and society.

Writing to Learn Botany

What are the strengths and weaknesses of a scientific method of investigation?

THOUGHT QUESTIONS

1. Why study plants?

2. Interferon is a protein that animals make to fight viral infections. Botanists recently inserted the gene for interferon into turnips, hoping the protein would make the plants more resistant to viral diseases. The experiment failed; engineered plants were no more resistant than were untreated plants. However, the interferon from the turnips appears to be active in animals, including humans. What does this tell you about the process of science?

3. How do you use a scientific method in your daily life?

4. Is this scientific method foolproof? Explain your answer.

5. How has our understanding of life been affected by technology?

6. Can a scientific method of investigation be used to make any judgments? Explain your answer.

7. How do plants influence your life?

8. Each year the National Cancer Institute tests extracts from about 5,000 plants for their effects on 60 kinds of human cancers and the AIDS virus. Fewer than 1% of these extracts are selectively toxic against these cancers. Indeed, these tests have produced only four marketable drugs (taxol is one of the program's successes). Given this low percentage and the program's high costs, should such tests be abandoned? Why or why not?

9. Report on recent attempts to preserve forests as areas of sustainable yield, where products are harvested indefinitely from plants without destroying the forest.

SUGGESTED READINGS

Articles

Darley, W.M. 1990. The essence of "plantness." *The American Biology Teacher* 52:354–57.

Gibbons, A. 1990. Biotechnology takes root in the Third World. *Science* 248:962–63.

Gibbons, B. 1990. The plant hunters: A portrait of the Missouri Botanical Garden. *National Geographic* (August):124–40.

Gibbs, A., and A.E. Lawson. 1992. The nature of scientific thinking as reflected by the work of biologists and by biology textbooks. *The American Biology Teacher* 54:137–52.

Lane, M.A., L.C. Anderson, et al. 1990. Forensic botany: Plants, perpetrators, pests, poisons, and pot. *BioScience* 40:34–39.

Moffat, A. 1998. Toting up the early harvest of transgenic plants. *Science* 282:2176–78.

Pringle, H. 1998. The slow birth of agriculture. *Science* 282:1446–50.

Sherman, P.W., and J. Billing. 1999. Darwinian gastronomy: Why we use spices. *BioScience* 49(6):453–63.

Stern, W.L. 1988. Wood in the courtroom. *World of Wood* 41:6–9.

Books

Imes, R. 1990. *The Practical Botanist.* New York: Simon and Schuster.

Morton, A.G. 1981. *History of Botanical Science. An Account of the Development of Botany from Ancient Times to the Present Day.* New York: Academic Press.

Overfield, R.A. 1993. *Science with Practice: Charles E. Bessey and the Maturing of American Botany.* Ames, IA: Iowa State University Press.

Simpson, B.B., and M. Conner-Ogorzaly. 1995. *Economic Botany: Plants in Our World,* 2d ed. New York: McGraw-Hill.

Talalaj, S., D. Talalaj, and J. Talalaj. 1991. *The Strangest Plants in the World.* Melbourne, Australia: Hill of Content Publishing.

ON THE INTERNET

Visit the textbook's accompanying web site at http://www.mhhe.com/botany to find live Internet links for each of the topics listed below:

Basic description of a plant

Plants and medicine

Plants and food

Agriculture and the human population

Plant chemicals

Angiosperms and Gymnosperms

The anatomy of a seed

Types of roots

A scientific method

Biotechnology and plants

Organisms produce more offspring than survive. Each winter squash (Cucurbita) fruit produces hundreds of seeds. What would happen if all offspring survived?

CHAPTER 2

The Ecology and Natural Selection of Plants

LEARNING ONLINE

Check out http://darwin.eeb.uconn.edu/evolution-sites.html in the Online Learning Center at www.mhhe.com/botany for chapter 2 to learn more about "living stones," or plants that resemble rocks. This, and other helpful information, such as practice quizzing, is available to you when you visit the OLC.

How can we explain that a plant species grows in one part of the world but not another or that plants of the same kind grow in a meadow but not in a forest next to the meadow? How can we explain the great diversity in size, shape, color, and function of plants in the world today? Why do plants such as the rosy periwinkle (see fig. 1.1) contain unique chemicals that are useful medicines while other plants do not? As we began to discuss in the first chapter, to understand the form and function of plants living today, we need to study their ecological and evolutionary histories.

In this chapter, we discuss the life of a typical plant, beginning from a seed. We investigate the fundamental ideas of ecology, including those factors in the environment that determine whether a plant survives in a particular place. In the final part of this chapter, we discuss natural selection and present the foundation for the detailed study of evolution found in later chapters. These major processes in biology—natural selection and evolution—help explain why plants of today grow where they do and are the way they are.

Ecology Is the Study of Interactions Between an Organism and Its Environment.

Near the edge of a forest in the eastern hills of Oklahoma, several bright yellow-brown fruits drop to the ground from a 20-year-old persimmon tree (*Diospyros virginiana*) (fig. 2.1). A wandering female raccoon, with three young raccoons shadowing her movements, finds the fruits, and the four begin an early evening snack. The raccoons eat all of the persimmons, swallowing the sweet pulp of the fruit and the seeds inside.

You might think the raccoons have destroyed the persimmon seeds, but this is not the case. The covering of the persimmon seed, the seed coat, is relatively resistant to the acidic juices of the raccoon's stomach, so the seeds remain unharmed. The flesh of the persimmon fruit, however, is broken down, providing energy and chemical building blocks for the raccoons. In a day or two, the undigested parts of the fruit are excreted by the animals along with the intact seeds of the persimmon (fig. 2.2). Thus, the raccoons disperse, or scatter, persimmon seeds far from the parent tree. If environmental conditions are suitable where the seeds are deposited, each seed will germinate, sprouting into a young seedling that develops roots, stems, and leaves. The young persimmons may eventually reach the height of the parent tree and produce flowers, fruits, and seeds of their own. However, not all of the persimmon seeds survive to become a tree. In fact, out of the thousands of seeds each parent tree produces, only one may live long enough to reproduce.

FIGURE 2.1
Fruits on a persimmon tree (*Diospyros virginiana*). These fleshy, colorful fruits are adapted to dispersal by animals.

FIGURE 2.2
Animals often eat fleshy fruits, and the seeds pass unharmed through the digestive tract, dispersing the seeds. In this scat (animal fecal droppings), light-colored, intact persimmon seeds are visible; they will germinate if environmental conditions are favorable.

Recall from chapter 1 that all living things respond to the environment in which they live. Investigating an organism's interaction with its environment is the study of **ecology.** In the next sections of this chapter, we explore the ecological reasons why so few seeds grow into adult plants and what parts of the environment control the growth and reproduction of plants. Also, recall from chapter 1 that organisms share parts of a common ancestry and have co-evolved with other organisms, as seen in the example of raccoons and persimmons. Ecology and evolution are intimately linked in the lives of organisms because those individuals that survive in their environment reproduce and contribute to the evolution of their species.

The environment of an organism consists of living organisms and nonliving factors.

What environmental factors must persimmon seeds and seedlings face and overcome if they are to survive? A **habitat** is the place where an organism lives and grows. Each habitat has an environment made up of several interacting components in two broad categories: (1) the **abiotic,** or nonliving, environment and (2) the **biotic,** or living, environment. The living environment consists of all those organisms that come into contact with other organisms—either directly or indirectly. For instance, our raccoon family is part of the persimmon's environment; however, so too are the coyote that kills one of the baby raccoons and the microscopic bacteria and fungi that destroy fruits and seeds that fall to the ground and rot. Also part of the biotic environment of the persimmon seedlings are the parent tree and all other plants growing near the seedlings.

The physical environment includes all the nonliving factors that affect the lives of organisms. These abiotic factors include precipitation, temperature, soil, nutrients, light, humidity, wind, and fire. Any one of these factors by itself can determine if an organism survives in a particular environment. For instance, if it gets too cold in the winter, all the young persimmon trees may die. In such an environment, the temperature is a **limiting factor,** which is an environmental factor that inhibits the growth, reproduction, or behavior of an organism.

Plants are indicators of the environment.

All organisms must be able to survive and grow in their environment before they can reproduce. Animals often can seek shelter in inclement weather, burrowing into the ground or flying away, but once a plant starts growing, it must remain in place and be able to endure all conditions in its habitat. If a plant can survive, it does so because its genes provide the potential for it to do so. An organism that

FIGURE 2.3

Plants are indicators of their environment. In what kind of environment would you expect this cactus (*Rathbunia alamoensis*) to grow?

can survive, grow, and reproduce is said to be *adapted* to its particular environment, and it has adaptations to do so. An **adaptation** is a heritable characteristic (one that can be passed on genetically to the next generation) that allows an organism to live in a particular environment. For example, your lungs are an adaptation that allows you to live on land and extract oxygen from the air. The fleshy fruit of the persimmon is an adaptation that attracts animals that may disperse its seeds.

Because plants have adaptations that allow them to live in their particular environment, you may be able to determine a plant's environment just by looking at its characteristics. This is why plants are often said to be "indicators of the environment." For instance, spines not only protect a cactus from herbivores but keep it from losing too much precious water in the dry, desert heat. Thus, finding a plant with spines may indicate that the environment in which this plant normally lives receives little precipitation (fig. 2.3).

INQUIRY SUMMARY

Ecology is the study of the interaction between an organism and its environment. A habitat is the place where an organism lives and grows, and each habitat has an environment made up of abiotic, or nonliving, and biotic, or living, factors. Those environmental factors that limit the growth, reproduction, or behavior of an organism are called limiting factors. Plants are indicators of their environment: their form and function often indicate physical and biotic factors with which they have evolved.

What are the major biotic and abiotic factors of the environment in which you live? What plants are indicators of the environment in which you live?

Dispersal of Fruits and Seeds Reduces Seed Predation and Competition.

The seeds of plants help us understand how many plants begin their lives and how these plants are adapted to their environment. If you have not yet done so, obtain seeds of some plant, germinate them, and make observations about the growth of the young plants. In later chapters, we study the reproduction of seed plants in detail, but for now, we will introduce some of the reproductive parts of a flowering plant.

The **ovary** of a plant's flower contains **ovules,** each of which contains an egg cell (fig. 2.4). The bright petals and smell of a flower often attract animal pollinators, which leave the flower carrying pollen grains and transport them to other flowers. Pollen grains produce sperm that fertilize the egg inside each ovule. A fertilized egg develops into an embryo, the ovule becomes a seed, and the ovary (and occasionally some of the associated tissues) becomes a fruit. Fruits usually contain from one to hundreds of seeds, although many edible fruits found in grocery stores are reproductive oddities and produce no seeds.

Plants such as persimmons, peas, and pines start their lives as a seed (recall the parts of a seed discussed in chapter 1), however, pines and other evergreens have no flowers and, thus, no fruit. Seeds must get from the parent plant to a favorable place for germination. While gardeners may plant pea seeds in well-tended soil, the best place for a new persimmon or pine may be where the parent plant is already growing, if the parent were not there. When a seed germinates next to a mature plant, the young seedling may not get enough resources it needs for continued growth, due to competition with the mature plant. **Competition** is an ecological interaction between two organisms to acquire a resource that both need and that is in limited supply—such as water, sunlight, or minerals (fig. 2.5).

If all seeds drop directly below the parent plant, a seed predator—an herbivore that eats seeds—may be able to find and eat most of them. The new generation of plants may have a greater chance of surviving if the seeds are moved away from the parent plant to a place that is favorable for seedling establishment, as well as away from herbivores that have found the mature plant. Thus, the first step in the survival of a seed plant is getting the seed to a favorable location for germination and growth.

Most seed-bearing plants have a means of dispersing their seeds or fruits with the seeds inside. Dispersal occurs either by physical or biological carriers. Wind and water are the physical carriers of fruits and seeds, whereas animals are the main biological carriers. Plants of a few species can forcefully eject their seeds away from themselves. For example, the seeds of touch-me-nots, witchhazel, dwarf mistletoe, and some legumes (e.g., African *Acacia*) may be flung several meters from the fruit by forceful ejection (fig. 2.6).

Some seeds and fruits are dispersed by wind and water.

The seeds of orchids are so small that they can be easily carried away from the fruit by light winds (e.g., *Goodyera repens* has seeds that each weigh 2×10^{-6} grams—about a half a million seeds per gram, which is half the weight of a dime!). The familiar plumes of dandelions, willows, milkweeds, poplars, and some buttercups are examples of further modifications for seed or fruit dispersal by wind (fig. 2.7). A maple fruit has a curved wing that causes the fruit to spin as it floats through the air. Even in a light wind, maple fruits spin away from the parent plant. In arid areas where strong winds are common, tumbleweeds (*Salsola* species) break off at the base of their main stems, and whole plants are blown around, releasing seeds as they tumble.

Coconuts are perhaps the most familiar example of an adaptation for dispersal by water (fig. 2.8). The buoyancy of a coconut comes from its fibrous husk, which allows the fruit to stay afloat for days or weeks before too much salty seawater seeps through the husk and kills the embryo. (You buy the seed of the coconut in the grocery store; most of the husk of the fruit has been stripped away before being sent

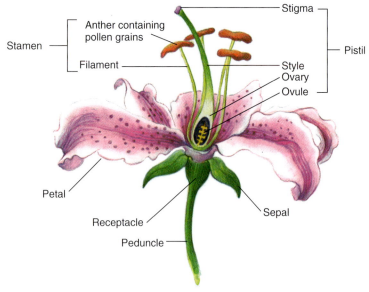

F I G U R E   2 . 4
The parts of a flower. This is a generalized flower.

FIGURE 2.5

Black spruce (*Picca mariana*) seedlings growing near adult trees. Competition may result between offspring and parents when seeds are dispersed short distances.

to market.) Many other plants that grow in or near water have fruits or seeds that float.

Animals disperse many seeds and fruits.

Fruits or seeds that are covered with a sticky substance (e.g., the seeds of mistletoes) or have barbs or hooks (e.g., the fruits of bur clover and puncture vine) cling to beaks, fur, and skin (fig. 2.7 and fig. 2.9). Such seeds or small fruits may be carried a few meters or many kilometers before they fall from or are brushed off the animal carrying them. The longest dispersal distances are probably achieved by seeds that get stuck to the muddy feet of migratory birds.

FIGURE 2.6

Self-propelled seed dispersal. Fruits of the touch-me-not (*Impatiens glandulifera*) explode open, hurling seeds as far as 2 meters.

The colorful and sweet-tasting fruits of many plants attract fruit-eating animals, such as in the case of the persimmons and raccoons described at the beginning of the chapter. Seeds from such fruits usually pass unharmed through the digestive tracts of birds and mammals, or the animals may regurgitate them. Passing through an animal's gut may actually promote seed germination. A few plants, such as the mescal bean (*Sophora secundiflora*), have brightly colored seed coats that attract birds (fig. 2.10). However, birds usually drop these seeds after a short distance because the seed coats are too hard for them to crack open.

In the Amazon Basin, the fruits of certain trees fall just as the trees' bases become seasonally flooded. Fish that swim under the trees during the flooding often catch the fruits in their mouths as they fall, and the seeds, which pass unharmed through their digestive tracts, are deposited elsewhere.

These examples illustrate how different organisms have coevolved; both organisms gain something in the relationship. This idea of coevolution between plants and other organisms permeates the study of botany.

The Lore of Plants

A single plant species, or two closely related ones, can be naturally separated by large distances (e.g., between North America and South America or Asia, between South America and Africa). While some of these separations can be explained by the drifting apart of continents, botanists think that many were caused by long-distance dispersal. The mechanisms of ancient long-distance dispersal, however, are harder to imagine. Possible mechanisms include such creative suggestions as seed transport by constipated, migratory birds, some of which may migrate over 16,000 kilometers—from the North Pole to the South Pole. Seed dispersal by far-ranging birds is also thought to account for the spread of several plant species found on tropical islands throughout the Pacific Ocean.

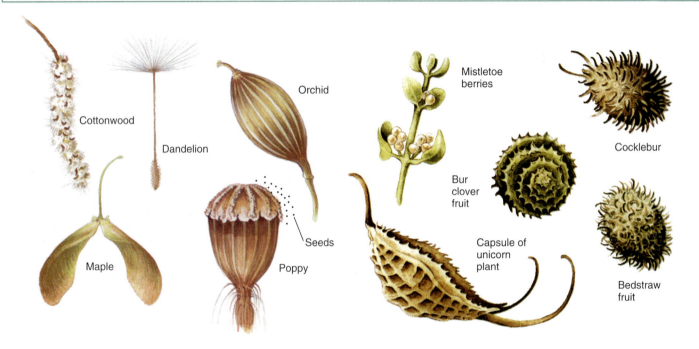

FIGURE 2.7

Wind-dispersed fruits and seeds are located on the left; five examples of animal dispersed fruits are on the right. Small seeds of orchids and poppies can be blown around by light breezes. Seeds or fruits of other plants have plumes or wings that catch the wind. Mistletoe seeds attach to animals by a sticky substance; other fruits attach to animals by hooks or barbs.

FIGURE 2.8

Fruits of the coconut palm (*Cocos nucifera*) may float in the ocean for weeks before reaching a beach where they can germinate.

INQUIRY SUMMARY

Successful reproduction involves transportation of seeds or fruits to favorable habitats, reducing predation and competition. Seeds and fruits are dispersed either by wind, water, or animals. Some seeds pass unharmed through the digestive tract of animals before they germinate. A few plants disperse their seeds by forceful ejection from the fruit. Animals may carry seeds or fruits on their fur or feet.

How might you test the hypothesis that the long-distance dispersal of a particular plant was due to migrating birds?

FIGURE 2.9
Seed dispersal by animals. This elk cow has seeds stuck to her fur.

*The germination of a seed and growth of a plant
depend on favorable environmental conditions.*

Within a seed, the embryo is **dormant,** meaning that it is alive but not actively growing. After a seed has been dispersed, it may begin to grow if environmental conditions are favorable. Different kinds of seeds require different internal or external stimuli to break dormancy and germinate. Common internal factors include changes in the kind of, amount of, or sensitivity to hormones (see chapter 6). The most common external factors are water, temperature, and light.

The most common stimulus for breaking dormancy is water. A dry seed absorbs water and expands, and the radicle bursts through the seed coat (see fig. 1.16). The seeds of many desert plants have a safeguard against premature germination after short rainfalls, which are common in deserts: the seeds contain a germination inhibitor that must be washed out completely before they will germinate. Even though the seeds absorb water, the rainfall may be insufficient to wash out the inhibitor. This adaptation to the desert ensures that seeds germinate only when there is enough water in the soil for the successful growth of the seedling. The seeds of rice and other semiaquatic plants (which spend part of their life in water) or aquatic plants germinate only in soil that is submerged in water.

Temperature is an important component of the environment. The optimum temperature for the germination of many seeds is close to room temperature (25°C), although the seeds of plants adapted to colder or warmer climates

FIGURE 2.10
Bright red seeds of the mescal bean (*Sophora secundiflora*) attract birds.

can germinate at lower or higher temperatures, respectively. Germination can occur at temperatures near freezing, 0°C, and as high as 45°C in different species. The seeds of woody plants in temperate climates often require a wet period that is followed by several weeks of cold before they germinate, a process called **stratification.** The seeds of apple trees and most pines can be artificially induced to germinate by wetting them and placing them in a refrigerator. How might stratification prevent a seed from germinating in the "wrong" season?

Although not all seeds are affected by light, the seeds of birch, certain grasses, and some varieties of lettuce require light for germination. These seeds will not germinate in the dark because they contain an inhibitor that is broken down only in the presence of light. Why do you think it would be an advantage for a seed *not* to need light for germination?

Seeds with thick or tough seed coats that do not split open may require special conditions for germination. A thick seed coat may reduce the uptake of water by the seed or prevent the embryo from expanding. For example, many legumes contain seeds with hard seed coats. Many such seeds germinate only after the seed coat is scratched or cracked or briefly soaked in concentrated acid. Such treatments are called **scarification.** In nature, seed coats may be scarified by bacterial action, repeated freezing and thawing, abrasive handling by rodents, or passing through the digestive tracts of animals, which may soak seeds in concentrated acid.

Many seeds lie dormant in a seed bank in the soil.

Many viable seeds do not germinate right after they are dispersed from the parent plant, nor do they germinate during the following growing season. Seeds can lie dormant for many years before conditions are suitable for germination. Everywhere that seed plants grow, the soil contains viable, ungerminated seeds in natural storage—that is, in a **seed bank.** Seeds in a seed bank may be dormant because of their own inhibitors, as in many desert plants. In other habitats, seed germination may be inhibited by chemicals released from nearby plants.

How long can seeds remain viable? Longevity varies among different kinds of plants and with environmental conditions. Some seeds, such as those of pumpkins, squashes, and other members of the melon family (Cucurbitaceae), can germinate after several years of storage. In some parts of the Mojave Desert, the seeds of weedy plants remain dormant for at least 15 years. The record for longevity, however, belongs to the arctic tundra lupine (*Lupinus arcticus*) (fig. 2.11). Seeds of this species were frozen in a lemming burrow that is estimated to be about 10,000 years old. Several of the seeds from this burrow germinated within 48 hours of planting, and one of the plants produced flowers within a year after planting.

FIGURE 2.11
Arctic tundra lupine (*Lupinus arcticus*). Seeds of this species germinated after being frozen in a lemming burrow for about 10,000 years.

Plants must deal with the environment both above- and belowground.

After germination, the seedling begins to grow and requires substantial amounts of water and nutrients. At this stage in its development, the young plant is sensitive to severe conditions in the environment, such as drought and extreme temperatures, and to parasites and herbivores. Any of these environmental factors can kill a seedling, which helps to explain why so many seedlings die. If a plant can survive this early stage in its development, then its chances are greatly increased that it will live long enough to reproduce.

Once the roots of the seedling begin to penetrate the soil, the plant is literally stuck there for life (unless a helpful gardener moves it). This means that the plant must be able to live in the environment in which it starts to grow. It also means that a plant must live in two different worlds at the same time: the relatively cool and moist underworld of the soil and the lighted, windy world aboveground that can fluctuate from below freezing to extremely hot temperatures. Living in two worlds requires that each plant have parts that are adapted to, or able to exploit, these vastly different environments. Most plants that you're familiar with have both a root system adapted to life in the soil and a shoot system adapted to life aboveground (fig. 2.12). Because of the environment in which they grow, what characteristics might a stem require that a root system would not?

Leaves provide an example of an ecological trade-off in a plant.

While roots absorb water and minerals for most plants, the shoots produce the sugar that provides the energy and

building materials needed for growth and reproduction. In general, the greater the number and size of leaves a plant possesses, the greater is its ability to catch sunlight and produce sugar through photosynthesis. However, there is an **ecological trade-off,** that is, a negative aspect of the characteristic for every positive aspect. In this case, the greater size of leaves, which gives the plant an advantage for photosynthesis, allows a greater loss of water through **transpiration.** Transpiration is water loss from a plant through its stem and, mostly, through its leaves. Almost all transpiration occurs through small pores called **stomata** (singular, stoma) in the surface of leaves. When stomata are open, carbon dioxide (CO_2) gas can enter the leaf and be used as a raw material to produce sugar during photosynthesis. However, as long as the stomata are open, water also is lost from the leaves through transpiration, and too much water loss can be deadly for a plant. Some plants, such as geraniums (*Pelargonium* species), possess hairs on their leaves that help reduce transpiration by decreasing the effect of drying winds. These hairs, however, do not prevent CO_2 from entering the leaf (fig. 2.13). How do you explain that plants with very large leaves often live in the water or in tropical rain forests?

Plants maintain a balance between their root system and shoot system.

The sugar produced during photosynthesis is the source of energy and chemical building blocks needed by a plant

FIGURE 2.12
A plant is adapted to life in the soil and aboveground. Consider this mulberry (*Morus*) tree that has been pulled from the ground, exposing most of its root system, to study the growth of plants in industrial wastes. Some plants help detoxify contaminated soils by absorbing harmful chemicals.

FIGURE 2.13
The hairs on some leaves, such as of this geranium (*Pelargonium*), help reduce transpiration, among other functions.

for its growth and reproduction. However, plants face a problem: all of their living cells require water and minerals that are absorbed by the root system, as well as sugar, which is produced by the leaves. Roots must obtain enough water and minerals for their own use as well as for the shoot. In addition, water lost during transpiration must be replaced. Similarly, shoots must produce enough sugar for their own growth and, because roots cannot produce any sugar of their own, for the root system as well. The minerals absorbed by roots may be used to help build the body of a plant; however, there is no energy in

these nutrients. Thus, the roots depend on the shoots for a source of energy and carbon-based building materials.

If shoot or root cells do not receive enough water, minerals, or sugar, then the entire plant is in jeopardy of dying. Thus, it is essential for each plant to have a balance between its aboveground and belowground parts—not necessarily in terms of size of parts but in terms of supplying necessary materials to all parts of the plant. For instance, if a plant produces too many roots, it may die because the shoot cannot produce enough sugar for the roots to survive. If a plant has not already stored much energy in its roots, the roots may "starve" because they lack the energy they need, followed shortly by the death of the entire plant. This happens when certain weed killers such as 2,4-D are sprayed on a young, unwanted plant. The plant does not immediately die after spraying because the herbicide contains a chemical that causes new roots to form without additional shoots. The leaves of the weed cannot produce enough sugar, so the root system, and then the plant, eventually dies.

If the shoot becomes too large, however, too much water may be lost through transpiration, and the roots may not be able to absorb enough water for the plant. The plant then loses so much water that its cells begin to shrink, causing the plant to **wilt,** a limp condition caused by insufficient water in the cells. Eventually, the plant may die if it does not absorb enough water. This often happens when a person tries to move a plant from one pot to another. Many fragile root hairs are stripped off the plant when it is first pulled out of the soil. When the plant is transplanted, not enough root hairs are left to absorb water, and the plant wilts and may die.

INQUIRY SUMMARY

Germination may be induced by several factors, including changes in temperature, water availability, or light. Some seeds require stratification or scarification by animals, bacteria, or abrasion during repeated freeze-thaw cycles. In nature, seeds may remain dormant for several years before environmental conditions are favorable enough to induce germination. Plants must simultaneously deal with the dark, cool environment of the soil and the dry and sometimes sunny environment aboveground. In addition, a plant must balance its roots with its stems; if either the shoot or root systems become too large, the plant may die. Ecological trade-offs are common in plants.

How might the balance between the root and shoot systems be considered an example of an ecological trade-off?

FIGURE 2.14
A population of California poppies (*Eschscholtzia californica*). Some populations of plants are composed of thousands of individuals.

All Individuals Are Part of a Population, Which Is Part of a Community of Organisms.

Recall the persimmon seeds and seedlings mentioned earlier in this chapter. Growing near the persimmon seedlings are older saplings and mature persimmon trees. Persimmons of all ages growing in the same area belong to a single **population,** which is a group of individuals of the same species sharing the same territory, or range, at the same time and interbreeding, or reproducing with each other (fig. 2.14). Members of the same species may be widely distributed and divided into many populations, but a population is specific to a given area, and members of a population must be close enough to interbreed and form offspring together.

Plants are the producers of a community.

Plants are also part of a **community,** which consists of populations of different species living and interacting in the same location (fig. 2.15). The Oklahoma forest community includes persimmons and all other plants and animals found there. In the community, plants absorb minerals, create shade, promote visits by pollinators, and produce substances that inhibit the growth of other plants. Plants also provide shelter for other organisms and interact with organisms in a mutually beneficial way. Plants are the **producers** in the community; that is, they are the organisms of the community that harvest the energy of sunlight and use that energy to assemble carbon dioxide from the air into sugar during photosynthesis. All organisms in a community depend either directly or indirectly on producers for a source of energy. For instance, a community may include

FIGURE 2.15
A community, such as this forest in Pennsylvania, consists of many different populations of plants and animals.

deer that eat persimmon leaves and birds that eat its fruits and seeds before they fall to the ground. These animals are called **herbivores** (plant eaters); they depend directly on plants for their energy and building-block nutrients. Other animals, **carnivores** (meat eaters), may eat herbivores to obtain energy and building blocks for their bodies. In this case then, persimmon trees are the base of a **food chain** that connects the producers of the community with the **consumers** of the community, the herbivores and carnivores. Another major group of organisms depends on dead organisms or their parts to make their living in the community. These **decomposers,** which include bacteria and fungi, obtain their energy and nutrients by breaking down the bodies of organisms or parts, such as leaves that have fallen off trees (fig. 2.16). Minerals from bodies and body parts that are not used by decomposers become part of the soil and can be reabsorbed by producers. Some materials such as plastic do not decompose. What does this tell you about the ability of fungi and bacteria to use these materials as a source of energy and nutrients?

Organisms in a community are linked by their use of energy and nutrients and are grouped into **trophic,** or nutritional, **levels,** such as the producer, herbivore, carnivore, and decomposer levels. Herbivores are referred to as primary (1°) consumers, and carnivores are secondary (2°) consumers. In any ecosystem, the producers and consumers form food chains or a **food web,** composed of many interlocking food chains, that determine the flow of energy through the different levels. Because most organisms have more than one source of food and are themselves often eaten by a variety of consumers, there are considerable differences in the length and complexity of food chains or webs (fig. 2.17).

A community and its abiotic factors constitute an ecosystem.

Where persimmon seeds are dropped by raccoons determines, in part, their fate. If they are dropped on clay soil, hard-baked by the sun, they may die. If some seeds fall into

FIGURE 2.16

Decomposers, such as fungi, use waste products, fallen leaves, and the bodies of organisms as sources of nutrients and energy.

a cool, moist crack in the ground, they may eventually germinate and start to grow. So, both biotic (living) and abiotic (nonliving) factors are important in the life of a plant. The populations interacting with one another in a community and their nonliving environment constitute an **ecosystem.** The distribution of a plant species is controlled by abiotic factors as well as the effects of other organisms in the ecosystem. For instance, the diversity of tree species in rain forests often varies with the levels of the minerals phosphorus and magnesium in the soil.

Plants have evolved with other organisms in their ecosystem.

All the plants in a particular habitat share and often compete for nutrients, light, water, and other resources needed for plant growth but found in limited supply. Of course, some plants are more efficient and therefore more compet-

---

 # The Lore of Plants

Cobra plants (*Darlingtonia californica*; fig. 2.18) occur in a few swampy places in the mountains of Oregon and northern California. These interesting plants, whose principal leaves form insect-trapping pitchers, have apparently survived because the swamps to which they are adapted have salt in concentrations high enough to kill most of their competitors. A pitcher leaf, which may be nearly 1 meter long, resembles a cobra head with its hood inflated; it even has "fangs" that insects follow right into the mouth of the trap. Once the insect is in the mouth, it encounters stiff, downward-pointing hairs that facilitate its descent and hinder its escape. Escape is made even more difficult by numerous small patches of transparent tissue on the back of the hood. The false windows mislead the victim in its efforts to find an escape route. Eventually the insect drowns in fluid at the base of the pitcher, where the insect's soft parts are digested by bacteria and small organisms, releasing nutrients useful to the plant.

Describe the food chain inside the cobra plant.

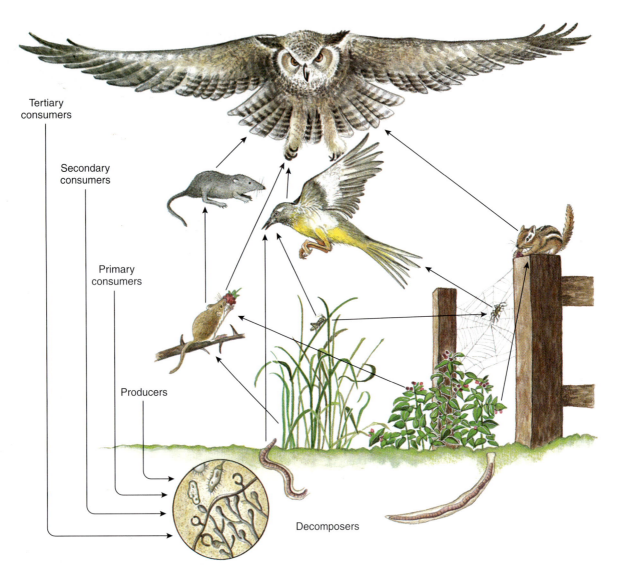

Tertiary
consumers

Secondary
consumers

Primary
consumers

Producers

Decomposers

FIGURE 2.17

A hypothetical food web. Although no real food web would be so simple, this diagram shows that plants are producers, animals are consumers, and microorganisms are common decomposers in an interlocking food web. Arrows show the path of energy. Tertiary consumers are carnivores that eat other carnivores.

itive in their use of resources. The result of this competition may be the elimination of one species by another, a drastic reduction in the number of one species versus its competitors, or the coexistence of competing species if the environment periodically changes. Under such pressures, plants slowly accumulate adaptations to their environment that increase their competitiveness and efficiency of growth. For instance, the competition for light among plants has resulted in the evolution of several mechanisms that allow species to adapt to different light intensities. Depending on the light that is available, leaf orientation may change throughout the day, plants may grow taller, the thickness of the leaf blades and number of chloroplasts

(the site of photosynthesis inside plant cells) in the leaves may change, and the size of leaves may be different. These mechanisms increase the maximum exposure of leaves to the light (fig. 2.19), thereby decreasing competition.

Plant-herbivore interactions are widespread in ecosystems, and a variety of plant defenses, such as chemical substances or structural modifications, such as spines, has evolved against herbivores (fig. 2.20). Many plants secrete substances that either offend insect consumers or function as insecticides. For example, the volatile oils produced by members of the mint family (Lamiaceae) deter insect larvae and several animals from eating the leaves. Some plant pests, however, possess enzymes

FIGURE 2.18

Cobra lily (*Darlingtonia californica*) lures insects by sight and smell into its pitcher trap. Light shining through the "windows" atop the dome fosters a false sense of escape.

FIGURE 2.19

Leaves of plants such as this ivy (*Parthenocissus tricuspidata*) form mosaics that help ensure that almost all leaves are exposed to light. Leaf mosaics often form in houseplants exposed to light from one direction.

that break down these chemicals, enabling the pests to thrive on plants that are toxic to other consumers. Members of the carrot family (Apiaceae), which includes carrots, parsley, and celery, produce aromatic substances that serve as insect repellents. However, larvae of the anise swallowtail butterfly feed exclusively on members of that family; the larvae have specific enzymes that degrade the repellent chemicals (fig. 2.21). In some cases, herbivores can actually stimulate the growth of plants such as grasses in the prairie. This stimulation is similar to when you mow the lawn: the mowing doesn't harm the grass but stimulates grass plants to produce more stems and leaves.

Another kind of interaction between organisms in an ecosystem involves **parasites,** which are organisms that obtain energy and nutrients from another living organism, the **host** (fig. 2.22). For instance, mistletoe

(*Phoradendron*) is a parasitic plant whose roots grow into another living plant, such as an oak tree (see fig. 7.18). The mistletoe extracts water and minerals from its host but makes its own sugar. While the parasite benefits from the relationship, the host is harmed. Many fungi are parasites on plants, including the fungus called wheat rust that attacks wheat plants. About 15% of the 3,000 members of the figwort family, Scrophulariaceae (which includes snapdragons and *Penstemon*), exhibit various degrees of parasitism. Species of *Harveya* and *Hyobanche* lack chlorophyll and depend entirely on their flowering-plant hosts for energy and other nutritional needs. *Gerardia aphylla* has pale leaves with about 10% of normal chlorophyll content; additional energy and nutrients come from its host plants. Parasitic species of *Pedicularis* and *Odontites rubra* have green leaves that furnish about half of their energy needs. Still other figworts

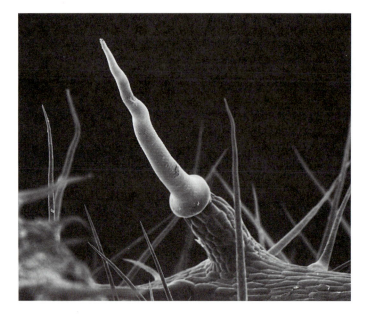

FIGURE 2.20
Stinging hairs of stinging nettle (*Urtica dioica*) are large cells filled with toxin (×275). The tip is pointed and glassy, which makes it brittle and easily broken. The small tip allows it to catch in an animal's skin and break off, thus releasing the toxin, much like a microscopic hypodermic needle.

FIGURE 2.22
Indian pipe (*Monotropa uniflora*) is a parasitic plant that does not conduct photosynthesis: it gets its nutrients indirectly from other plants, by way of a fungal bridge in the soil.

FIGURE 2.21
Anise swallowtail butterfly, whose larvae feed exclusively on members of the parsley family (Apiaceae). The larvae have specific enzymes that break down chemicals in the leaves that repel other insects.

(e.g., Indian paint brush, *Castilleja* species) may parasitize the roots of certain plants but can also live independently.

Many associations between plants and other organisms in an ecosystem are **mutualistic;** that is, the interaction is mutually beneficial to the organisms involved. For exam-

ple, an insect pollinator visiting a flower takes nectar and pollen for food while the plant benefits by having some of its pollen carried from one flower to another, thus helping the plant to reproduce sexually. Because both the insect and the plant benefit from their interaction, this is a mutualistic relationship. We learn more about pollinators and their flowers in chapter 12.

Several species of bleeding hearts (*Dicentra* species) have oil-bearing appendages on their seeds. Ants carry these seeds to their nests, strip off the appendages for food, and then deposit the seeds outside the nest, where the seeds germinate. Thus, ants help disperse the plants (fig. 2.23). Many species of violets (*Viola*) and spring beauty (*Claytonia virginiana*) are also dispersed by ants.

Interactions are a major influence in the lives of plants and in determining whether or not they live and reproduce. However, because plants are part of a community of organisms, what happens to one organism may affect many others. For instance, researchers recently found that when a beaver chews on a cottonwood tree, the tree releases a noxious chemical into its leaves—as a possible deterrent to other herbivores. However, one species of beetles eats these "noxious" leaves and stores the toxic chemical in their bodies as a defense against carnivores

FIGURE 2.23
Seeds of the Pacific bleeding heart (*Dicentra formosa*) have a mutualistic relationship with ants. The glistening white oil-bearing appendages are stripped from the seeds and eaten by ants.

that eat the beetles. Thus, the interaction between organisms of two different species affects a third species. Such interactions are a growing area of study in ecology.

INQUIRY SUMMARY

An environment in which an organism lives includes biotic (living) factors such as soil microbes, fungi, plants, and animals, as well as abiotic (nonliving) factors such as light, oxygen, climate, and physical disturbances. Producers and consumers defend themselves in ecosystems by such means as producing substances that make tissues unpalatable or toxic to consumers. Some consumers produce enzymes that enable them to break down toxic substances, thereby giving them an advantage over competitors. Individual organisms of the same species in one area belong to a population. All of the populations in an area are part of a community, of which plants are the producers. Producers provide the energy and minerals that are passed throughout the community's food web. Plants have many different interactions with other organisms of the community, being negatively affected by competition with other plants, herbivores that eat them, or parasites that are attached to and feed off them. Plants may also benefit from their interactions with other organisms, as in mutualistic relationships with pollinators. Interactions between two organisms often affect other species in the community.

Suppose you discover that two plants are interacting with each other in a community. How might you determine if this interaction is mutually beneficial to the plants?

Natural Selection Determines Which Individuals Survive and Reproduce.

As we said before, even if all persimmon seeds germinate in a habitat, few seedlings live to become trees. Dispersal has scattered the seeds, so the seedlings may be exposed to different environmental factors. Some seedlings may die from lack of rain, some are eaten by animals, and some may be washed away by a strong summer storm or killed by other factors in the environment. But even if the seedlings are placed in the same environment, not all grow to be adults because of their differences.

Just as you are slightly different from each of your biological brothers and sisters, so too is each persimmon seedling different from its siblings. These differences may seem slight, and perhaps a casual observer would be unable to see any variation; after all, the plants all look like persimmons. However, the seedlings may differ in how deep their roots grow, in their ability to use sunlight to form sugar in photosynthesis, in the size of their leaves, or in their tolerance to drought or insect pests. Those individual seedlings with the combination of characteristics that allows them to survive grow larger. Those individuals able to grow and reproduce in this habitat pass their genes to their offspring (each gene is part of a chromosome and controls a characteristic of an organism), and those unable to survive won't be able to pass on their genes. This simple idea describes the process of **natural selection** where individuals best adapted to their environment produce the most offspring and pass their desirable genes to the next generation. Natural selection affects persimmon seedlings, but it also affects every organism on earth. Now, when you study plants and other organisms, try to determine what characteristics have allowed the organisms to be adapted to their environment and to reproduce.

Within a population, different mature persimmon trees reproduce with each other, producing many offspring that are genetically different from each other. Suppose that approximately half of these persimmons have a gene that produces slightly thicker bark and that this bark protects trees from extremely low temperatures. If temperatures drop dramatically for a week during one winter, it may be that all the trees with thinner bark die. This means that thick-bark trees will produce more offspring than thin-bark trees the following year. Natural selection has occurred, and the environment has selected for trees with a thicker bark.

Artificial selection is controlled mating.

Natural selection is a fundamental process in nature and a cornerstone of biology. The process of natural selection was first proposed by Charles Darwin, an English naturalist who lived in the nineteenth century. Darwin observed that farmers often selected certain animals or plants for reproduction based on their desirable traits. Humans have practiced the selection and controlled the mating of plants and animals for centuries, a practice we now call **artificial selection.** Animal breeders know that to produce fatter hogs, they must allow only the fattest hogs to mate and must eliminate skinny hogs. Plant breeders know that to grow the largest tomatoes, they must allow cross-pollination only between plants that produce large tomatoes. These ideas are not new; most of our fruits and vegetables have been artificially selected for size, yield, pest resistance, and so on since agriculture began more than 12,000 years ago (see Perspective 2.1, "Selection for Monsters").

The practice of artificial selection led Darwin to conclude that a similar process occurs in nature, which he called *natural selection.* The main difference is that humans are the agents of artificial selection, whereas the biotic and abiotic forces of the environment are the agents of natural selection. Many of the principles of natural selection, therefore, can be deduced from the process of artificial selection.

Ample evidence from nature and the farm contributed to the development of the principles of natural selection.

The following four principles outline the main ideas of natural selection:

1. *Organisms have a tendency to produce more offspring than survive.* How many seeds does a tomato produce? How many does a tomato plant or a whole field of tomatoes produce? If the field of tomatoes is analogous to a population in nature, then many more seeds are produced than ever grow into new plants. If all seeds produced new plants, the earth would be knee-deep in tomato plants by the end of one growing season. The same is true for persimmons and other organisms on earth. The overabundance of tomato seeds is evidence that a population has the potential to increase. Imagine the potential for certain orchid populations to grow when each fruit may produce hundreds of thousands of seeds (fig. 2.24)!

2. *Only a fraction of the offspring in a population live to produce offspring.* In agriculture, most plants are harvested before they reproduce or before their offspring can be dispersed. Humans choose which individuals to harvest; that is, we are the agents of selection that limit reproduction. In nature, most plants die as seeds or seedlings. Conditions for germination or growth may not be good enough, pests may destroy the seeds or seedlings, or a freeze or drought may wipe them out. Saguaro cacti have been extensively studied in this regard (fig. 2.25). Fruits of the saguaro produce dozens of tiny black seeds, most of which germinate and grow under cultivation. A population of saguaros probably produces millions of seeds, yet in most populations, very few seedlings appear during any one year. Moreover, only a few new seedlings live more than a few years, and a saguaro must be 10 to 20 years old before it can form flowers, seeds, and fruit. Some populations have no seedlings for several years in a row.

FIGURE 2.24

An orchid fruit showing many tiny seeds. The fruit of some plants may contain hundreds of thousands of seeds.

3. *Individuals in a population vary, and these variations are inherited (at least in part) by their offspring.* Look at any group of organisms in a population and you see differences (fig. 2.26). Even in a field of corn, which may appear uniform, certain features vary from one plant to another. This was shown in a long-term experiment in artificial selection that began in 1896 at the Illinois Agricultural Experiment Station. At the beginning of the experiment, two types of corn kernels were selected for planting, one of slightly higher protein content than the other. Every year, the plants producing the highest amount of protein were mated with one another, and plants producing the lowest amount of protein were mated with one another. At the beginning of the experiment, the average protein content per kernel was 10.9%. After 50 generations, the low-protein line was down to 4.9%, and the high-protein line was up to 19.4%. In this case, heritable variation occurred in protein content in the kernel, and humans selected only certain individuals for breeding. The evidence for heritable variation in corn is that the protein content changed over time because of selection by humans.

2.1 *Perspective* Perspective

Selection for Monsters

Monstrous plants are plants that differ markedly from the norm. We have bred and selected for giant seeds, fruits, flowers, and vegetative organs. For example, we have selected and perpetuated cultivars of banana, grape, sugarcane, and other plants for increased bulk, sugar content, and visual appeal. Most of these varieties can no longer produce viable seeds, but our production of tasty tissue has been remarkable.

Nowhere is our selection for tasty tissue more intense than among pumpkin growers. These growers have spawned two groups—the World Pumpkin Confederation and the rival Great Pumpkin Commonwealth—that sponsor weigh-offs to crown world champion pumpkins. With tens of thousands of dollars of prize money at stake, the selection pressures are intense: farmers work in their patches as much as 5 hours per day on a pumpkin, caring for the soil, eliminating weeds, and covering the pumpkins with plastic and blankets at night. On a good day, a pumpkin can gain 30 pounds. How big will the winner be? Almost 1,000 pounds. Or, as one grower put it, "Envision a small Volkswagen; an orange one."

Of course, Volkswagen-size pumpkins won't spring from the kind of seeds that you pick up at the local grocery store; the winning pumpkins almost always come from winning parents. For example, one year, five of the top 20 pumpkins came from a pumpkin that weighed 600 pounds; seeds from that pumpkin produced seven pumpkins that each weighed over 800 pounds. These winners become stud pumpkins, each producing about 600 seeds. Thirty seeds are returned to the grower, while the rest are sold for as much as $25 *per seed*.

Our rampage through the genome of plants is also apparent in the genus *Brassica* of the mustard (Brassicaceae; also called Cruciferae) family. This genus contains perennial, biennial, and annual herbs, with 100 wild species found in north temperate parts of the world. You can easily recognize the family by its characteristic flower having four petals

Artificial selection has produced this variety of common vegetables from a single common ancestor. All of these vegetables are the same genus, Brassica *species, but various morphological features have been artificially selected to enhance different textures and flavors.*

that form a cross (hence the family name); six stamens, of which two are short; and its special fruit, the silique. Although the cultivated brassicas are probably derived from several wild species, a reasonable ancestor is colewort (*B. oleracea*), a scrubby perennial native to Europe and Asia. From colewort, we have selected and developed a wide array of common vegetables. One line produced kale, cauliflower, and broccoli.* A second line gave us kohlrabi, with a short stem enlarged into an aboveground tuberous vegetable, and rutabaga, whose tuberous storage stem develops belowground. The turnip's (*B. rapa*) storage organ is also a tuber, and its true root extends down from the swollen stem. Although all the leafy vegetable members of the genus

may not be derived from colewort, they are all closely related. Several species of mustard grow wild, and we exploit their edible leaves (*B. juncea*) and particularly their seeds (*B. nigra* and *B. alba*), which are used as mustard. Pakchoi (*B. chinensis*) is an important leafy vegetable in Asia, as are the closely related petsai (*B. pekinensis*) and false pakchoi (*B. parachinensis*). Rape, or colza (*B. napus*), is grown primarily for its seeds, which yield an edible oil.

Text on "Monstrous Plants" from *The Green World: An Introduction to Plants and People,* Second Edition by Richard M. Klein. Copyright © 1987 by Harper & Row, Publishers, Inc. Reprinted by permission of Addison Wesley Educational Publishers, Inc.

*Broccoli was created by Italian horticulturists. Among the descendants of these horticulturists was Albert Broccoli, who produced 17 James Bond movies.

Saguaro cacti (*Carnegiea gigantea*), some of which grow to be almost 20 meters tall.

4. *Individuals with favorable traits produce, on average, more offspring that survive to reproduce, than those with unfavorable traits.* High-protein corn, large hogs, and flavorful tomatoes are favorable traits for artificial selection. Organisms with these traits produce more offspring because people select them for breeding. In nature, if a plant is disease resistant, a good competitor for resources, and an efficient photosynthesizer, then we expect it to reproduce more frequently than plants that are not so disease resistant or competitive in the same environment. For example, a plant having a large leaf-surface area is adapted to a low-light environment and may have greater reproductive success than small-leaved plants in the same environment. Random events such as storms and floods may cause the unexpected, but under normal circumstances, the organisms that are most attuned to their environment are most successful.

F I G U R E   2 . 2 6
Natural populations, such as these maple (*Acer*) seedlings, are composed of genetically different individuals.

Together, these four principles describe the process of natural selection. A well-documented example of natural selection involves the work of Janis Antonovics with plants growing in soil containing waste products from mines, including high concentrations of copper, zinc, or lead. Most plants can't grow in soils contaminated with these metals; however, some species of grasses and weeds do grow around these mines. Surprisingly, these plants are the same species that grow in nearby pastures; apparently, some members of the species can tolerate toxic metals. In one experiment, Antonovics planted seeds from the pasture population of grasses in soil that was rich in copper. Only 1 in 7,000 survived. Thus, metal in the soil selected intensely against those individuals without the genetic makeup that allowed them to be tolerant of the copper: only 1 out of 7,000 plants was adapted to the mine soil (fig. 2.27). It is important to note that the environment did not cause the plant to become tolerant of copper; the genetic change occurred independently of the plant's exposure to copper.

Natural selection is meaningful only in the context of the environment.

Successful adaptations and natural selection are inseparable from the environment. For example, the succulent stem of a cactus that stores water is not a positive adaptation unless the cactus is living in a dry environment. Succulent stems may promote reproductive success in a desert but not in a rain forest. That is, the same characteristic that gives an organism an advantage over other plants of the population in one environment may not be advantageous in another environment. Thus, environmental conditions are the forces that determine the outcome of natural selection. These environmental conditions are called **selection pressures.** For example, fungal disease is a selection pressure. A dispersed seed may have a tough, protective seed coat that prevents

FIGURE 2.27

It is often difficult to get plants to grow on soil taken from mines, such as on this reclaimed mine in Oklahoma, because few plants are adapted to contaminated soils such as these.

fungal invasion. However, the seed coat may also prevent the penetration of water needed for germination. Thus, thicker seed coats may be advantageous in moist, fungus-rich environments, whereas thinner seed coats are selected in drier, less fungus-rich environments. If the embryo is disease resistant, however, then a thin seed coat may be best in moist as well as dry habitats. On average, individuals of a species that vary most strongly in ways favored by the environment survive and reproduce more offspring than others of the same species. Thus, favorable variations accumulate in populations as a result of natural selection.

Organisms must deal with various and often contradictory environmental factors. For example, warmth may increase metabolism for rapid growth, but it may also dry the soil rapidly. In addition, the environment of a habitat may change over time, thereby creating different selection pressures. A variety of adaptations have evolved in different species promoted by natural selection in the context of the environment.

You might think that the phrase *survival of the fittest* applies to the previous examples involving natural selection. It does; however, most people misinterpret this phrase to mean that *only the strong survive.* What **survival of the fittest** means is that those organisms best adapted to the environment produce the most offspring. More precisely, natural selection promotes characteristics that increase the passage of genes to the next generation. That is, it promotes successful reproduction. If only the strong survived, you might expect that trees of all kinds would have extremely thick bark, however, many do not. Can you think of a situation in which thick bark might be a *disadvantage* to a tree?

Natural selection is a driving force for evolution.

Evolution is measured by the change in the gene frequencies within a population over time. This means that evolu-

tion has occurred in a population when the genetic makeup of the population as a whole has changed. To understand evolution well, an introduction to genetics is necessary, and so we reserve a full discussion of evolution to later chapters. However, it is important to note that in the persimmon example described previously, evolution has occurred because the number of individuals in the population possessing the "thin bark gene" was dramatically reduced from one season to the next. Thus, evolution can happen in a very short period of time—one generation of an organism—but this usually results in a population of organisms that is only slightly different from the original population. Given enough time, however, changes within a population can accumulate and lead to a population that is substantially different from individuals of other populations—so much so that individuals of the two populations can no longer reproduce with each other. Once this happens, biologists may recognize that a new species has developed. Here too, evolution has occurred. Evolution, however, does not have to result in the formation of new species.

Evolution is not the same thing as natural selection; natural selection is one of the driving forces of evolution, causing changes in populations through time. You will learn much more about evolution throughout this book and particularly in chapter 14, but it is important for you to realize that on the earth today, the organisms and their parts are the result of the processes of natural selection and evolution. The size and shape of leaves, the depth and extension of roots, the number of seeds, and the kind of seed dispersal are all characteristics that have been selected for in different environments and shaped by evolution through time. In fact, all the parts of an organism can have selection pressures that shape them. Throughout the rest of this book, think about how different characteristics of an organism have been affected by the processes of natural selection and evolution: why are plants the way they are?

SUMMARY

Ecology is the study of the relationship between an organism and its environment. Seeds, and fruits containing seeds, may be dispersed away from the parent plant by wind, water, or animals, which reduce seed predation as well as competition between offspring. If a seed germinates in a favorable location, the developing seedling faces many selective pressures in its environment, both living and nonliving factors. Plants live in two worlds at the same time; their roots are in the soil and their shoots are in the air. Because they are anchored by their roots in one place, plants are considered indicators of the environment in which they grow. We can tell a lot about the environment of a plant simply by looking at the plant and its structures.

Individual plants of the same kind are part of a population, and different populations of organisms are part of a community. A community and its abiotic factors are called an ecosystem. Food chains interconnect to form a food web within a community, both of which describe who eats whom. Plants are the producers of a community, supplying the energy and minerals needed by other organisms in the ecosystem. Bacteria and fungi are decomposers, and animals are consumers, including herbivores that eat plants or carnivores that eat animals. Plants interact with many organisms in the community including herbivores, competitors, and parasites, all of which have negative effects on the plants. Plants also have mutualistic relationships with other organisms in which both organisms benefit from the interaction. Flowering plants and their pollinators, and seed plants and the animals that disperse them, are examples of mutualistic relationships.

The essence of natural selection is that individuals best adapted to their environment produce the most offspring and pass on their genes to the next generation. Organisms produce more offspring than needed to replace themselves and more than environmental resources can support. Most offspring die before maturity, and those with genes that confer the best adaptations to their environment reproduce the most. Natural selection functions at the population level, and a population, not an individual, is the smallest unit that can evolve. Populations evolve in the context of their environment, which confers success or failure to individuals with the best-adapted characteristics. Direct evidence for natural selection comes from experimentation and from comparisons with artificial selection. Natural selection is one of the main driving forces for evolution.

Direct evidence exists for several of the principles of natural selection: the overproduction of seeds indicates that populations can increase dramatically; a small fraction of seeds grows into reproductively mature plants; different individuals in a population have variable, genetically controlled features; plants that are better adapted to a habitat have more reproductive success in that habitat than plants that are not so well adapted to it.

Writing to Learn Botany

Discuss how the studies of artificial selection of farm animals and garden plants have contributed to the understanding of natural selection.

THOUGHT QUESTIONS

1. If a plant species is unnecessary for the functioning of an ecosystem, should we be concerned if it is eradicated? What might be some economic, moral, or aesthetic reasons for protecting an "unnecessary" plant?

2. Distinguish between an individual, a population, and a community.

3. Do you think that positive or negative interactions play a greater role in shaping a community? Explain your answer.

4. Describe an interaction between a local plant and an animal.

5. What would be the consequences if a population reproduced without the influence of natural selection?

6. Herbert Spencer was an eighteenth-century English sociologist who believed in applying Darwin's concepts to human society. He popularized the phrase *survival of the fittest* and suggested that the unemployed, the sick, and other "burdens on society" be allowed to die rather than be objects of public assistance and charity. What do you think of this idea and of the validity of his interpretation of the concept of natural selection?

7. Mimicry is a common biological phenomenon in which an individual from one species resembles some other organism. For example, the Notodontidae moth (*Antaea jaraguana*) from Brazil has a wing pattern that resembles a dead leaf and its venation. How would an insect benefit by resembling a plant?

8. If plants are indicators of the environment, what might plants look like that live in a windy habitat?

SUGGESTED READINGS

Articles

Batten, M. 1984. The ant and the acacia. *Science* 84:59–67.

Castello, J.D., D.J. Leopold, and P.J. Smallidge. 1995. Pathogens, patterns, and processes in forest ecosystems. *BioScience* 45:16–24.

Hartman, H. 1990. The evolution of natural selection: Darwin versus Wallace. *Perspectives in Biology and Medicine* 34:78–88.

Kearns, C.A., and D.W. Inouye. 1997. Pollinators, flowering plants, and conservation biology. *BioScience* 47:297–307.

Knapp, A.K., J.M. Blair, et al. 1999. The keystone role of bison in North American tallgrass prairie. *BioScience* 49:39–50.

Books

Crawley, M.J., ed. 1997. *Plant Ecology.* Cambridge, MA: Blackwell Science.

Darwin, C. 1967. *On the Origin of Species.* Facsimile 1st ed. of 1859. New York: Atheneum.

Endler, J.A. 1986. *Natural Selection in the Wild.* Princeton, NJ: Princeton University Press.

Molles, M.C., Jr. 1999. *Ecology: Concepts and Applications.* Dubuque, IA: McGraw-Hill.

ON THE INTERNET

Visit the textbook's accompanying web site at
http://www.mhhe.com/botany to find live Internet links for each
of the topics listed below:

Plant ecology

Abiotic factors that affect plants

Plant adaptive features

Fruits and seed dispersal

Plant competition

Plant as ecological producers

Plant/animal interactions

Natural selection

Charles Darwin

Like all organisms, plants convert energy from one form to another during metabolism. This corn (Zea mays) plant uses photosynthesis to convert light energy to chemical energy, which is then used to make the other molecules required during the plant's growth and development. What kinds of molecules do you and this plant share in common?

C H A P T E R 3

Energy and Cell Chemistry

CHAPTER OUTLINE

What is energy?

> There are two types of energy important to organisms: potential energy and kinetic energy.
>
> Perspective 3.1 Units of Energy
>
> The laws of thermodynamics govern the flow of energy in nature.

Organisms must obtain and convert energy to survive.

Metabolism converts the energy that powers life's work.

ATP is the main energy currency of cells.

Enzymes are required for energy conversions in cells.

The molecules of life are carbohydrates, proteins, nucleic acids, and lipids.

> The large organic molecules of life are polymers.
>
> Carbohydrates store energy and carbon and make up the wood of trees.
>
> Cellulose and starch are important polysaccharides in plants.
>
> Proteins make up membranes, function as enzymes, and store nitrogen in plants.

> Nucleic acids are the blueprints of life and the record of evolution.
>
> Lipids form membranes and also store energy and carbon in plants.
>
> Secondary metabolites produced in plants often have importance to plants and us.
>
> Perspective 3.2 What's Wrong with Tropical Oils?

LEARNING ONLINE

Did you know the Online Learning Center at www.mhhe.com/botany offers practice quizzes, key term flashcards to help you study for the upcoming quiz or exam? While you're there, visit www.hoflink.com/~house/index.html in the OLC for chapter 3 to learn how cellulose was important to the evolution of plants.

Organisms have evolved as a progression of increasingly complex levels of organization, starting with atoms and molecules, moving up to cells and their components, then tissues, organs, and, finally, the individual organisms. This increasing organization requires energy and chemical molecules as building blocks.

Energy is a word we all use daily but one that is not very well understood. In this chapter, you learn about energy and its importance to cells, organisms, and ecosystems. All living systems require energy to survive, grow, and reproduce. This is true for a rose or elm plant, your family pet, and you. What would happen to you if you didn't eat anything for several days? You would, of course, be very hungry. But soon you would lose weight, become very tired, and eventually die. What about plants? Can they starve? Plants don't "eat" like animals, but they do require energy. How would you starve a plant? If you answered "place it in the dark," you are correct and already know some important concepts about plants and energy.

Metabolism is another word we often use loosely. Metabolism includes all the chemical reactions in an organism. This includes the thousands of chemical reactions that are required for an organism to live and reproduce. Metabolism is central to energy and life: in organisms, some chemical reactions require energy, whereas others release energy. All organisms, including you and plants, use metabolism to harvest energy stored in chemical molecules such as sugars and fats.

In this chapter, we present the major kinds of chemical molecules found in organisms. These molecules are large organic molecules and include carbohydrates, proteins, nucleic acids, and lipids. The study of these molecules and how they are made provides the first insight into how plants live.

What is Energy?

Energy is the ability to do work or cause change—that is, to move matter against a force such as gravity or friction or to bring about change such as occurs during a chemical reaction. Because energy is the ability to do work, it is not always as obvious to us as matter, which has mass and occupies space. In turn, energy has no weight and does not take up space. We describe energy according to how it affects matter. Humans use energy for conspicuous activities such as dancing, mowing the lawn, playing baseball, and studying plants. However, plants may expend most of their energy in subtle, nearly unnoticeable ways via interesting adaptations that have evolved in them. For example, consider the *Philodendron* plant shown in figure 3.1. The plant's large leaves gather the energy available in sunlight and use it to fuel metabolism and growth. However, the calm existence of this plant changes when it reproduces. The clusters of flowers open for only a couple of days, but at night, when air temperatures often hover near freezing, the flowers can reach temperatures exceeding 46°C (by comparison, butter melts at 30°C). These furnacelike flowers maintain their high temperature for many hours in the cool night air. Plants such as skunk cabbage (*Symplocarpus foetidus*) and voodoo lily (*Sauromatum venosum*) also generate large amounts of heat, which can be used to help the plants reproduce. For example, heat produced by the voodoo lily helps to disperse compounds that smell like dung or rotting flesh. Insects attracted to the plants by these odors assist in pollination and, therefore, reproduction. This heat production requires much energy, which was previously stored in the plant.

You learned in chapter 2 that the flow of energy between producers, consumers, and decomposers in an ecosystem is necessary for life on earth. In chapter 6, you will read how plants use energy for growth and development. Clearly, understanding the energetics of plants helps us understand how they live and reproduce, how their adaptations evolved, and how ecosystems depend on energy flow.

All activities—ranging from cellular division and heat production to home runs and nuclear explosions—involve converting energy from one form to another. Humans convert the energy contained in coal and oil to electricity, and then convert the electricity into light energy to illuminate our homes, streets, and parks. Similarly, plants convert sunlight into chemical energy during photosynthesis. They then use this chemical energy to make DNA, build new cells, grow, and reproduce. Animals stay alive by eating other animals and/or plants and their stored energy. All aspects of the lives of organisms center on energy and energy conversions.

There are two types of energy important to organisms: potential energy and kinetic energy.

Potential energy is stored energy—that is, energy available to do work or cause change (fig. 3.2). Examples of potential energy include a teaspoon of sugar, an unlit firecracker, and a rock atop a hill. In organisms, potential energy is stored subtly in various molecules such as sugars and fats.

Kinetic energy is energy being used to do work (fig. 3.2). Examples of kinetic energy include burning sugar, an exploding firecracker, a rock rolling down a hill, light emitted

3.1 *Perspective* Perspective

Units of Energy

A calorie (note the small *c*) is the amount of energy required to raise the temperature of 1 gram of water by 1°C. The most common unit for measuring the energy content of food and the heat output of organisms is the **Calorie** (note the large C), which is the energy required to raise the temperature of 1 liter of water by 1°C. To help you put this into better perspective, consider that a slice of apple pie provides enough energy (365 Calories) for a woman to run for an hour. A Calorie is also called a kilocalorie, or large calorie.

FIGURE 3.1
The clublike spadix in the center of this *Philodendron* inflorescence heats up to temperatures as high as 46°C. This heat helps the plant disperse compounds that attract pollinators.

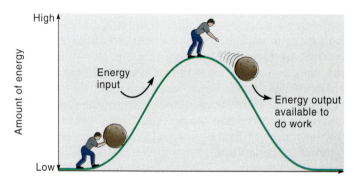

FIGURE 3.2
Pushing a boulder to the top of a hill requires an input of energy because it increases the potential energy of the boulder. The boulder atop the hill has potential energy because it can do work as it rolls down the hill. As the boulder rolls down the hill, potential energy is converted to kinetic energy.

Source: Postlethwait and Hopson, *Nature of Life,* 2d ed. Copyright © 1992 McGraw-Hill, Inc., New York, NY. Reproduced with permission of McGraw-Hill, Inc.

by a firefly, and a root forcing its way through soil. Kinetic energy affects matter by transferring motion to other matter, much as a moving pool ball transfers its kinetic energy to other pool balls. Similarly, flowing water can be used to turn a turbine, and a growing root can crack a concrete sidewalk. Kinetic energy moves objects, whether they be mountains, molehills, or molecules.

Heat is kinetic energy, but heat cannot be used to do work in a plant cell. Again, study the example shown in figure 3.2. The rock atop the hill contains much potential energy—the capacity to do work—because of its position, but since it is at rest, it has no kinetic energy. If the rock is given a nudge, it spontaneously rolls down the hill, transforming its potential energy into kinetic energy, which could be used to do work. Under most conditions, potential and kinetic energy are freely interconvertible, though not at 100% efficiency.

The laws of thermodynamics govern the flow of energy in nature.

Life depends on energy transformations, which happen when energy is converted from one form to another. For example, our bodies transform the chemical energy in food such as sugar and this enables us to grow, think, study, play, and dance. Our appliances convert electrical energy to light for reading and to heat for warming our homes.

Energy transformations are regulated by the **laws of thermodynamics.** These laws involve a **system** and its **surroundings.** The collection of matter or chemical reactions being studied is called the system, and the rest of the universe is referred to as the surroundings. A closed system, such as that illustrated by a perfect thermos bottle that loses no heat, is isolated from (i.e., does not exchange energy with) its surroundings. Conversely, an open system readily exchanges energy with its surroundings. Cells and organisms are open systems; they must be to exchange energy and matter with their surroundings. Don't let all of this intimidate you; the laws of thermodynamics are rather simple and are based on common sense. More important, they apply to all energy transformations, whether they be the combustion of gasoline in an engine, the breakdown of sugar in a cell, or the generation of heat by the *Philodendron* plant shown in figure 3.1. The laws of thermodynamics are important because they govern the existence, interactions, and diversity of all organisms.

The **first law of thermodynamics** is the law of conservation of energy. This law states that energy cannot be created or destroyed, but only converted to other forms. It can also be stated in other ways:

- In any process, the total amount of energy in a system and its surroundings remains constant.
- The amount of energy in the universe is constant.
- You can't get something for nothing.

For example, a power plant does not create energy; it merely transforms energy from one form (e.g., fossil fuel) to another (e.g., electricity). Likewise, green plants are not energy producers; they merely trap and convert the energy in sunlight into the energy stored in organic molecules. The first law of thermodynamics asserts that the energy in sunlight that warms our planet and drives photosynthesis must come from somewhere else in the system. In this case, it comes from the sun. Where does the sun get its energy?

Energy conversions often generate much heat. As you read this book, your body generates the heat of a 100 watt lightbulb. However, your body temperature is relatively stable because the heat generated by your body is radiated into your surroundings. According to the first law of thermodynamics, there is no change in the total amount of energy in the system (your body plus all the surroundings). That is, the energy radiated from your body can be traced to the energy contained in the food that you ate. This is why a crowded room can get so warm. The amount of energy released by your body cannot exceed the amount of energy contained in the food that you eat. If you stop eating, you eventually run out of energy, die, and stop releasing heat.

The **second law of thermodynamics** is the law of **entropy,** or disorder. This law states that all energy transformations are inefficient; that is, the amount of concentrated, useful energy decreases in all energy transformations. These statements express the second law in other ways:

- Systems tend toward increasing entropy (disorder).
- Any system spontaneously tends to become disorganized (like your dorm room or apartment!).
- In all energy transformations, some usable energy is lost as heat.
- A perpetual motion machine cannot exist; no real process can be 100% efficient.
- You can't break even.

These restatements of the second law are based on our world being overwhelmingly irreversible. This irreversibility is all around us, as shown by our aging, the futility of trying to unscramble an egg, and the impracticality of trying to "unpop" popcorn. Irreversibility results from the message of the second law of thermodynamics—namely, the loss of usable energy as heat during an energy transformation.

Consider a combustion engine in a car. Gasoline is a concentrated source of energy; its energy resides in the molecules of octane. However, when the chemical bonds break and new ones are formed in lower energy molecules (combustion with oxygen) and energy is released in the car's engine, less than one-fourth of the energy is used to move the automobile (i.e., most combustion engines are less than 25% efficient). The amount of energy used to heat the engine block (and the air around it), move the car, power the radio, and run the air conditioner equals that originally contained in the gasoline. However, applying heat to a car does not move the car. That is, the energy contained in the heated tires, pavement, air, and engine block cannot be recycled to run the car, because heat energy resides in randomly moving molecules and is therefore not in a concentrated, useful form. This heat represents the inefficiency inherent in any energy transformation and is the basis for the second law of thermodynamics. Because all energy transformations produce heat (i.e., an unusable form of energy), all things naturally become more disorganized.

The consequences of the second law of thermodynamics are important and familiar to everyone. For example, rocks tumble downhill rather than uphill, and pieces of a jigsaw puzzle never spontaneously fall into place when they're poured from the box. In short, the second law explains why disorder in the universe is increasing continually. All naturally occurring processes tend to reach maximum entropy (i.e., maximum disorder). Life exists at the expense of the universe, which is constantly running down and becoming more disordered. As soon as an organism dies, its body disintegrates because energy no longer is being taken in to keep the body organized.

INQUIRY SUMMARY

Energy is the ability to do work or cause change. Potential energy is energy available to do work, and kinetic energy is energy being used to do work. The first law of thermodynamics states that energy cannot be created or destroyed but only converted to other forms. The second law states that systems naturally tend toward increasing entropy (disorder).

How do the laws of thermodynamics relate to energy flow in an ecosystem? To your everyday life?

Organisms Must Obtain and Convert Energy to Survive.

Organisms obtain energy in different ways. Plants harvest energy from sunlight, and animals eat other organisms for energy. The cells of these organisms can obtain energy from the chemical reactions that break down large molecules such as starch and fat into smaller molecules such as water and CO_2. This cellular energy is used by organisms for activities such as growth, repair, maintenance, and reproduction. The chemical reactions that convert this energy are inefficient and release much heat. Indeed, the cells of most organisms extract less than half of their fuel's energy for useful work; the second law of thermodynamics explains this inefficiency and loss of energy. Thus, although organisms can channel the transformation of energy from one form to another—for example, they can hoard the energy in reserves (e.g., fat or starch) or use it for repair, movement, or reproduction—these diversions are only temporary. Eventually, all energy is transformed to heat, such as the heat that gives us a body temperature of 98.6°F.

The loss of useful energy as heat during energy transformations increases the entropy (disorder) in a system, such as a cell or an organism. Therefore, there is a natural tendency for things to become less organized as life-sustaining chemical reactions take place.

Because organisms are highly ordered and organized, it might seem that they are exceptions to the laws of thermodynamics. However, the second law describes closed (i.e., isolated) systems. Organisms remain organized because they are not closed systems: instead, they are open systems that exchange energy and matter with the environment. Plants and animals use inputs of energy, such as

sunlight and food, respectively, to overcome entropy, stay organized, and therefore stay alive.

The energy that keeps most organisms alive comes ultimately from the sun—that is, plants transform sunlight into the chemical structures of sugars, which humans and other organisms use as an energy source. As producers in an ecosystem, plants and other photosynthetic organisms (such as algae and some bacteria) are the first and most important part of the scheme (fig. 3.3).

The two primary energy transformations in plants are **photosynthesis** and **cellular respiration** (fig. 3.4). Photosynthesis uses light energy to convert carbon dioxide (CO_2) and water (H_2O)—both of which are energy-poor molecules—into sugars, which are relatively rich in energy. In the process, oxygen gas (O_2) is released as a byproduct. Cells extract energy from sugars via cellular respiration. Some of this energy is briefly conserved in molecules such as ATP (discussed later in this chapter). If all of the energy in the sugars were released at once (as in an explosion), the energy would be converted mostly to heat and produce lethally high temperatures. To counter this, energy from sugars and other molecules such as fats is extracted by slowly breaking apart the molecule in a series of chemical reactions (see chapter 11). During each reaction, there is a drop in the potential energy of the molecule. Much of this energy is lost as heat; however, some of it is trapped in the chemical structures of other molecules in the cell. The chemical reactions that transform energy and molecules in cells are collectively called **metabolism.**

INQUIRY SUMMARY

All of life's activities involve changing energy from one form to another. These reactions increase entropy (disorder). Organisms tend to resist entropy, but that requires a constant input of energy. Metabolism is the term for all the chemical reactions in an organism.

How do plants use energy to survive?

Metabolism Converts the Energy that Powers Life's Work.

Metabolism (from the Greek word meaning "change") is a fundamental property of life arising from the energy transformations in cells; it is the sum of the vast array of chemical reactions that occur in an organism. These reactions do not occur randomly; rather, they occur in step-by-step sequences called **metabolic pathways,** in which the product of one reaction becomes the starting point (i.e., the substrate) for another. The various metabolic pathways in a cell are much like the roads on a map (fig. 3.5).

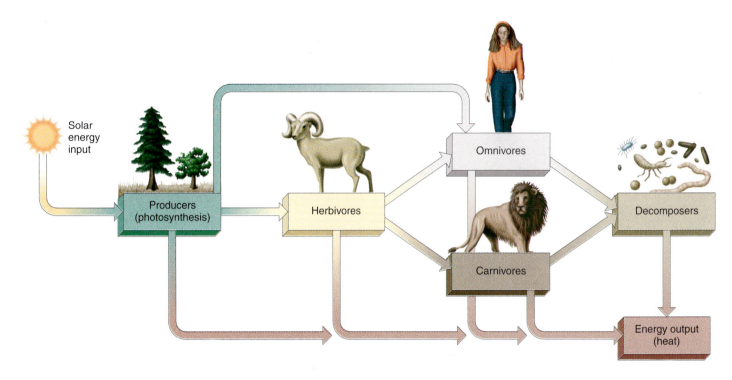

F I G U R E   3 . 3
Plants are producers; that is, they transform energy in sunlight into chemical energy. This energy then flows through and is used by other organisms. Some energy is converted to heat during each transformation. Decomposers also obtain energy from decaying producers and herbivores.

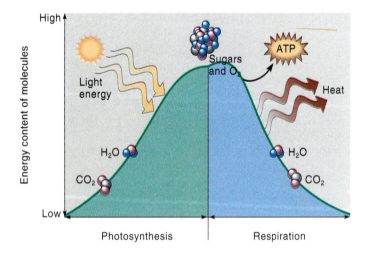

F I G U R E   3 . 4
Plants use photosynthesis, shown here as an uphill reaction, to make energy-rich sugars (and O_2) from energy-poor CO_2 and water. Plants and other organisms use respiration, shown here as a downhill reaction, to convert sugars to CO_2 and water. Some of the energy released during respiration is conserved in other molecules, especially ATP, which can be used to do work or cause change. The rest is lost as heat.

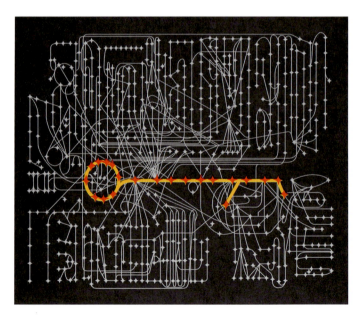

F I G U R E   3 . 5
Cellular metabolism. This diagram traces some of the metabolic reactions that occur in a cell. *Dots* represent molecules, and *lines* represent reactions—each catalyzed by a specific enzyme—that change the molecules. The stepwise sequences of reactions are called metabolic pathways. The pathway shown in *yellow* is central to most other pathways. Enzymes are explained later.

Each reaction in a metabolic pathway rearranges atoms in a molecule into new compounds as old chemical bonds break and new bonds form. The chemical reactions of metabolism either absorb or release energy. For example, the complete breakdown of sugar in a cell releases energy, some available to do work as ATP (discussed in the following), some unavailable to do work as heat:

$$sugar + O_2 \rightarrow CO_2 + H_2O + ATP + heat$$

During exercise, the heat released during this reaction is the heat that you feel as you perspire. The ATP can do work in a cell such as make a muscle contract as you lift a weight or provide energy for a plant root to grow.

In general, a chemical reaction in which a large molecule is broken down into smaller molecules and energy is made available to do work is called a **catabolic reaction (catabolism).** The breakdown of sugar, as just shown, is an example of a catabolic reaction. When larger molecules are assembled from smaller molecules (known as biosynthesis), energy is required for the "building-up" reaction, which is called an **anabolic reaction (anabolism).** The synthesis of starch from small sugar molecules is an example of anabolism. Humans use anabolic steroids to simulate anabolism—the "building-up" reactions in the body that produce muscle.

Most energy transformations in organisms involve chemical reactions called **oxidations** and **reductions.** Oxidation occurs when a substance combines with oxygen or when it loses electrons. Oxidative reactions, such as the breakdown of sugar to carbon dioxide and water, degrade molecules into simpler products and are therefore examples of catabolism. They are "downhill" or "breaking-down" reactions that release energy used to drive the work of a cell.

Reduction involves addition of electrons (e^-) to a molecule. These electrons can be added either alone or combined with protons as H_2 to a molecule (H_2, is two protons and two electrons). Electrons are negatively charged; therefore, the addition of electrons reduces the molecule. Reduction reactions such as the formation of sugars from CO_2 and H_2O often involve biosynthesis of complex molecules from simpler ones and are therefore examples of anabolism. They are "uphill" or "building-up" reactions that require a net input of energy. Electrons removed from one molecule during oxidation are used to reduce a second molecule. Thus, oxidation and reduction of reacting molecules always occur simultaneously; if one of the reacting molecules is reduced, another of the reacting molecules is oxidized (fig. 3.6). Oxidation and reduction reactions are important during metabolism. Oxidation usually means the release of energy during a reaction, whereas reduction implies a requirement for energy during a reaction. Many energy transformations in living systems involve oxidation and reduction of carbon compounds, such as sugars.

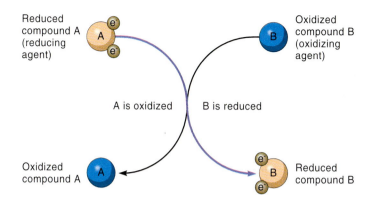

FIGURE 3.6
Oxidation and reduction occur simultaneously; as compound A is oxidized (i.e., donates electrons), compound B is reduced (i.e., accepts electrons). Oxidized compounds such as CO_2 contain less potential energy than do reduced compounds such as CH_4.

INQUIRY SUMMARY

Metabolism is the sum of the vast array of chemical reactions that occur in an organism. Metabolism occurs in stepwise sequences called metabolic pathways. Catabolism and anabolism are the types of metabolism. Many energy transformations in organisms involve chemical reactions called oxidations and reductions. Oxidation and reduction reactions occur simultaneously.

Overall, photosynthesis is anabolic, whereas respiration is catabolic. Can you explain why? Which process is oxidative and which is reductive?

ATP Is the Main Energy Currency of Cells.

Metabolism involves breaking-down reactions that release energy from energy-rich compounds such as sugars and fats. Organisms utilize this energy from the breakdown of energy-rich compounds via a set of catabolic reactions collectively called cellular respiration. Organisms, in turn, use this energy to drive anabolic reactions, which require energy to do work—for example, assembling proteins, building cell walls, and replicating genetic information (DNA synthesis). The most important molecule through which energy passes during cellular metabolism is a compound commonly known as **ATP (adenosine triphosphate).** Because the breakdown of ATP links energy exchanges in cells, ATP is the energy currency of the cell: when cells need energy to do something, they "spend" ATP.

ATP consists of adenine (a nitrogen-containing base), ribose (a five-carbon sugar), and three phosphate groups

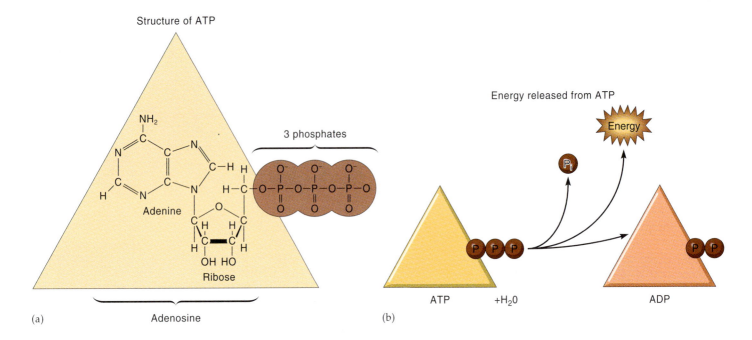

FIGURE 3.7

ATP (adenosine triphosphate), the energy currency of cells. (*a*) ATP consists of adenosine (adenine plus ribose) and three phosphate groups. (*b*) Transferring the terminal phosphate group from ATP to water releases energy, some of which can be used to do work.

(fig. 3.7). When cells require energy for a process, ATP is split enzymatically by reacting the ATP with water. Because water is involved in splitting ATP, the enzyme-catalyzed reaction is called **hydrolysis** (*hydro* = water; *lysis* = splitting). During the reaction, the energy contained in the ATP molecule is "released" (converted to do work) when the third phosphate group is broken away from the ATP molecule and new bonds are formed with water. The energy in ATP is converted in form and used by a cell to do work. This reaction with water forms ADP (adenosine diphosphate) plus P_i (P_i is the symbol for inorganic phosphate):

$$ATP + H_2O \rightarrow ADP + P_i + energy$$

ATP can be used to power many energy-requiring steps in cells because the bond to its third phosphate group is weak and can be broken relatively easily to power the formation of new bonds. This provides energy for the work of the cell.

Several properties of ATP make it ideally suited as the energy currency of cells. For example, the amount of energy released by converting ATP plus H_2O to ADP + P_1 is about twice as much as is needed to drive many cellular reactions, and it is about right for many others. Also, ATP does not cross cell membranes and is not considered a storage form of energy because it is unstable (reactive) and short-lived. Finally, the third phosphate bond of ATP is relatively weak and unstable; that's why it breaks so easily.

ATP is the most common energy currency in organisms, including you; all cells of all organisms use ATP for energy transformations. Like all of the different appliances that can plug into an electrical outlet and do different things, so too can different chemical reactions use a cell's ATP to do different kinds of work. Organisms use ATP for virtually all of their work, including making new cells and macromolecules, pumping materials, and moving materials through cells and throughout the organism. Accomplishing all of this work requires huge amounts of ATP. A typical human adult uses the equivalent of about 200 kilograms of ATP per day, but has only a few grams of ATP on hand at any one time. Even the hungriest of us doesn't eat 200 kilograms of food in a week, much less in a day. Nevertheless, organisms that run out of ATP—that is, those that go "bankrupt"—immediately die. Where does all this ATP come from? Obviously, we do not make 200 kilograms of ATP from scratch each day. Rather, we recycle our ATP at a furious pace, turning over (making and breaking) the entire supply every minute or so. In general, cells make ATP according to the following reaction:

$$ADP + P_i + energy \rightarrow ATP + H_2O$$

This is an anabolic reaction that requires energy to proceed. The required energy can come from the food you eat or, in plants, from the sugars produced during photosynthesis. ATP is used to drive two kinds of reactions in cells:

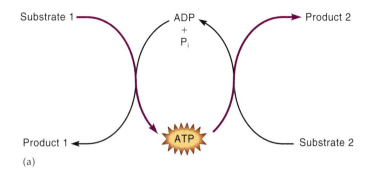

Substrate 1 → ADP + P$_i$ → Product 2
Product 1 ← ATP ← Substrate 2

(a)

ATP + C–C–C–C–C–C ⟶ C–C–C–C–C–C–℗ + ADP

Glucose   Enzyme   Glucose – 6 – phosphate

(b)

F I G U R E   3 . 8

(*a*) Coupled reactions. Energy released by a chemical reaction can be used to make ATP from ADP and P$_i$. This ATP can then be used to drive a synthesis reaction. (*b*) Phosphorylation of free glucose (six-carbon sugar) to glucose-6-phosphate, which becomes energized and reactive compared to free glucose. ℗ equals phosphate attached to a carbon skeleton, here glucose. P$_i$ is inorganic phosphate, phosphoric acid. –C–C– represent the carbon skeleton.

1. **Coupled reactions.** Just as cells couple or connect the breakdown of organic molecules to the production of ATP, so too do they couple the breakdown (hydrolysis) of ATP to power other reactions that occur at the same time and place in a cell. Energy from ATP is required to proceed. These coupled reactions often manufacture larger molecules at the expense of energy released during ATP hydrolysis (fig. 3.8*a*).

2. **Phosphorylation.** ATP also accomplishes much of its work by transferring its phosphate group to another molecule in a process called phosphorylation. Phosphorylations energize the organic molecules receiving the phosphate group, which are then used in later reactions (fig. 3.8*b*).

   Other energy compounds are involved in metabolism. **NAD$^+$**, like ATP, is made of adenine, ribose, and phosphate groups. However, the active part of NAD$^+$ is a nitrogen-containing ring, called nicotinamide, which is a derivative of nicotinic acid (niacin, a B vitamin). Niacin is one of the compounds added to food products, such as cornflakes, to make them "vitamin fortified." NAD is nicotinamide adenine dinucleotide. In all cells, NAD$^+$ is reduced to **NADH** when it accepts electrons (e$^-$) and a proton (H$^+$) during a chemical reaction; this is an example of a reduction reaction. NADH can be used to make ATP and to reduce other compounds in cells during reactions. You'll learn how pathways "cash in" the energy contained in NADH for ATP in chapter 11 on respiration.

INQUIRY SUMMARY

ATP is the energy currency of cells. ATP hydrolysis releases about the right amount of energy needed to drive most cellular reactions. Energy transformations involve coupled reactions in which the energy released by ATP drives other coupled reactions or phosphorylations where ATP energizes another molecule. NADH is also involved in metabolism.

How are coupled reactions important in metabolism?

Enzymes Are Required for Energy Conversions in Cells.

Chemical reactions require a bit of energy to get activated, like a charcoal fire requires a bit of lighter fluid to get started. **Enzymes** speed reactions by lowering the energy required for activation of the reaction (activation energy; fig. 3.9). Enzymes are organic catalysts. The breakdown of sugar can serve as an example:

$$\text{sugar} + O_2 \xrightarrow{\text{enzymes}} CO_2 + H_2O + ATP + \text{heat}$$

Without enzymes, this reaction would occur, but it might take decades. Organisms cannot wait that long. With enzymes, only a second or two is required. Enzymes accomplish this rather amazing feat by specifically binding to the reactant called a substrate (here, sugar or O_2) so that the reaction can occur quickly and its products (CO_2, H_2O and ATP) are formed. This example, in fact, summarizes the several chemical reactions of respiration as you will learn about in chapter 11.

Enzymes help control energy transformations by regulating the rate of metabolic reactions. If metabolism is likened to a series of interconnected roads (see fig. 3.5), then enzymes would function as traffic lights that control the flow of energy in cells. Many herbicides that kill plants target an enzyme, inhibit it, and thus kill the unwanted plant. Interestingly, several human diseases, such as lactose intolerance, arthritis, and phenylketonuria (PKU) are due to enzyme malfunctions, and this serves to remind us of the importance of enzymes.

Many enzymes are inhibited by other molecules. Drugs such as sulfanilamide and toxins such as lead and nerve gas are potent inhibitors of specific enzymes. Penicillin is an antibiotic that binds irreversibly to an enzyme that makes cell walls in bacteria. This blockage of cell-wall synthesis ultimately kills the bacteria, thus accounting for the antibiotic effect of penicillin.

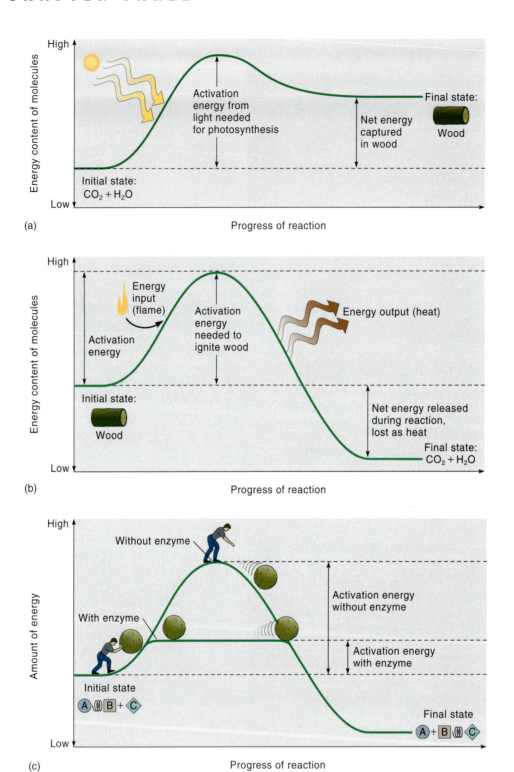

FIGURE 3.9

Energy of activation. (*a*) Plants use photosynthesis to make sugars that are, in turn, used to make products such as wood. The energy to do this—that is, the activation energy for photosynthesis—comes from sunlight. (*b*) The energy captured in wood is released when wood burns. To start this reaction, activation energy must be provided to ignite the wood. (*c*) Enzymes speed the rate of spontaneous reactions by lowering the energy of activation of the reaction, ⚭ illustrates bonds.

INQUIRY SUMMARY

Enzymes catalyze biological reactions by lowering the energy of activation. Enzymes are critical to life because they speed reactions to a biologically useful rate. Enzymes control metabolism and can be inhibited by other molecules.

Could life as we know it have evolved without enzymes?

The Molecules of Life Are Carbohydrates, Proteins, Nucleic Acids, and Lipids.

Except for water, the bulk of plant cells and tissues consists of four types of large organic compounds—carbohydrates, proteins, nucleic acids, and lipids. These large compounds, or biomolecules, are often called the molecules of life.

The process that begins the manufacturing of all these molecules from CO_2 is photosynthesis. During photosynthesis, plants manufacture all of these large organic molecules from CO_2 in the air and from water and minerals in the soil. Thus, through photosynthesis, plants make themselves out of carbon from the air, and water and minerals from the soil. Stated another way, the chemical elements of life are organized into complex molecules that make up the plant cell, and this organization begins with photosynthesis. (You will learn about photosynthesis in depth when you study chapter 10.)

What is the source of energy that drives photosynthesis? It is sunlight, the initial source of energy for the vast majority of ecosystems on earth. Plants convert light energy to chemical energy, and then use that energy to make organic compounds. This is why plants are called producers in an ecosystem. Can you name an ecosystem that does not depend on sunlight?

The large organic molecules of life are polymers.

Carbohydrates, proteins, nucleic acids, and lipids are **polymers,** which means they consist of many identical or similar single units. Each unit, or **monomer,** is a smaller molecule that can be combined with other monomers, forming a larger polymer (mono means one; poly means many). Examples of polymers include cellulose, starches, enzymes, DNA, waxes, lignin, and tannins. Glucose, a simple sugar, is the monomer that is linked to form starch. In contrast, most plant proteins consist of 20 different kinds of monomers known as **amino acids.** There are several hundred different amino acids in nature, but in the evolution of organisms, only 20 of them became essential in the assembly of pro-

teins. In plants, there are probably more than 50,000 unique proteins made from these 20 slightly different amino acids. Glutenins, proteins in wheat that help make wet flour sticky and hold the flour together as bread rises, are an example of plant proteins.

Four different kinds of monomers, known as **nucleotides,** make a DNA molecule. Every organism has the four different nucleotides sequenced uniquely into different sets of DNA molecules that make up a special configuration of genes in an individual. These DNA molecules and genes comprise chromosomes (see chapter 5). Thus, DNA is the polymer that makes up genes, the hereditary units of life. Variations in the amount and types of polymers account for the main differences we see between plants and animals that exist now or have existed in the past.

In addition to the four major classes of large compounds common to all organisms—carbohydrates, proteins, nucleic acids, and lipids—plants may also contain other complex polymers. For example, a polymer known as **lignin** helps form the tough plant cell walls that form wood. Plants make many other kinds of large compounds, usually in lesser amounts, including chemicals having such exotic names as phenolics, alkaloids, terpenoids, sterols, and flavonoids. Many of these strange-sounding molecules have particular importance to humans, such as the alkaloids vincristine, vinblastine (chapter 1), caffeine and nicotine and others we will explore throughout this book.

Carbohydrates store energy and carbon and make up the wood of trees.

Glucose, fructose, and other simple sugars, as well as their polymers, are called **carbohydrates.** Carbohydrates generally contain one oxygen and two hydrogens for every carbon. For example, the simple sugars glucose and fructose each consist of 6 carbons, 12 hydrogens, and 6 oxygens and have the formula $C_6H_{12}O_6$. The structure of glucose is shown as part of figure 3.10. Other simple sugars may have this same formula, differing only in the arrangement of the elements on the carbon backbone; this is why the compounds are different yet have the same chemical formula. Structural differences between these simple sugars give the molecules significantly different properties. For example, glucose has little flavor, whereas fructose (also known as grape or fruit sugar) is the sweetest natural sugar known. While it is important to understand the structures of carbohydrates, as it is for the other polymers, this understanding means little without thinking about their functions.

Carbohydrate monomers in plants are usually bound to other kinds of molecules or are linked together into larger carbohydrates. To distinguish different sizes of carbohydrates, monomers are called single sugars (**monosaccharides**), whereas dimers (molecules with two monomers) are called double sugars (**disaccharides**). In turn, long carbohydrate

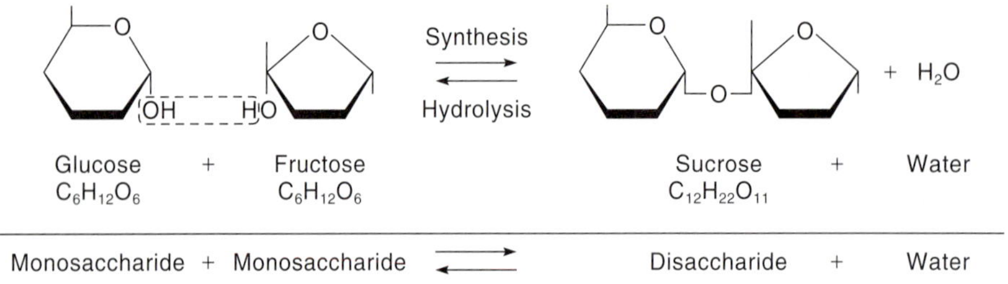

Glucose   +   Fructose   Sucrose   +   Water
$C_6H_{12}O_6$ $C_6H_{12}O_6$ $C_{12}H_{22}O_{11}$

Monosaccharide + Monosaccharide ⇄ Disaccharide + Water

FIGURE 3.10

The synthesis and hydrolysis of sucrose. During synthesis, a bond forms between glucose and fructose, and a molecule of water is removed. Hydrolysis occurs as a molecule of water is added and the bond between glucose and fructose is broken. The enzyme that drives hydrolysis is sucrase; the synthesis of sucrose is actually a multistep reaction that involves several enzymes and requires energy.

polymers are called **polysaccharides.** Single and double sugars are known as **simple sugars,** while the polysaccharides are **complex carbohydrates.** You may have heard of complex carbohydrates in relation to human nutrition.

The most common simple sugar in plants is **sucrose,** which is a disaccharide consisting of glucose and fructose (fig. 3.10). Sucrose is common table sugar (also called cane sugar or beet sugar) and is the major form of carbohydrate that circulates in plants. Recall that sugars contain chemical energy and carbon, which are essential for life. Thus, sucrose moves carbon and energy from the leaves, where sucrose is a product of photosynthesis, to other parts of a plant such as roots. This transport of sucrose around the plant "feeds" energy and carbon to the nonphotosynthetic parts of the plant.

People exploit such transport when they collect sap from sugar maples (*Acer saccharum*) to make maple syrup. Of course, table sugar (made from sugar beets or sugarcane) is very important in our diet. We consume more than 150 pounds of table sugar per person in the United States each year. Do you know what plants all this table sugar comes from?

When insects visit a flower, they are usually after the sucrose in nectar. Bees harvest the sucrose and use it as an energy source. In the hive, bees break the sucrose down into glucose and fructose, which becomes part of the honey that bears and humans consume. Honey tastes sweeter than table sugar (sucrose) because it contains fructose. Its flavor comes from the oils in the nectar, and the color comes from yellow pigments in the pollen grains collected at the same time.

Cellulose and starch are important polysaccharides in plants.

Cellulose is the most abundant structural polysaccharide in plants and the most abundant polymer on earth. A structural polysaccharide is a portion of the body of a plant; it forms part of the organization and architectural arrangement of a plant. Cellulose is found in the stiff wall that surrounds all plant cells. Between 40% and 60% of a plant cell wall is cellulose, and it, along with other polymers, makes the walls of some plants hard and strong enough to form wood. Can you identify several uses of wood in which strength is important? Cellulose occurs in almost pure form in some plants—for example, in the fibers surrounding cotton seeds (*Gossypium hirsutum*). These tough fibers are then used to make cloth that ends up in your cotton clothes and towels; for example, fibers make your jeans strong and durable.

The strength of cellulose comes from its organization (fig. 3.11). Cellulose is a linear polymer consisting of 100 to 15,000 glucose units. A higher level of organization occurs in cellulose when a thousand or more of these linear polymers twist together into **microfibrils.** Microfibrils are like strong, tiny cables.

Several microfibrils intertwine to make larger **cellulose fibrils.** Layers of fibrils are then cemented into a strong, three-dimensional grid by other kinds of structural polysaccharides (fig. 3.12). For example, the gluey polysaccharides that cement cellulose fibrils together are called **pectins** and **hemicelluloses.** Structurally, you can think of plant cell walls as being similar to reinforced concrete—that is, made of cement and steel rods. The cement is like the pectins and polysaccharides, and the rods are like cellulose fibrils. The plant cell wall surrounds and holds the plant cell together like reinforced concrete surrounds and holds the rooms of a concrete building together except, unlike the concrete, cell walls allow free passage of water and dissolved material through the wall. The evolution of cellulose structure and of plant cell walls is extremely important to life as we know it. Can you identify some reasons why this is true? We will learn more about cell walls in chapter 4.

Some of the most interesting hemicelluloses may not be structural polysaccharides at all. These are usually exuded from stems, roots, leaves, or fruits in a sticky mixture called gum. Gums are complex, branched polysaccharides consisting of several kinds of monomers. For example, a gum called gum arabic (from *Acacia senegal*) is almost everywhere in our daily lives: it is used to stabilize postage-stamp glue, beer suds, hand lotions, and liquid soaps. Look for this polymer on the label of some household products. This is not, however, the gum you chew in class.

Two commercially important polysaccharides of some algae are agar and carrageenan. These polymers are

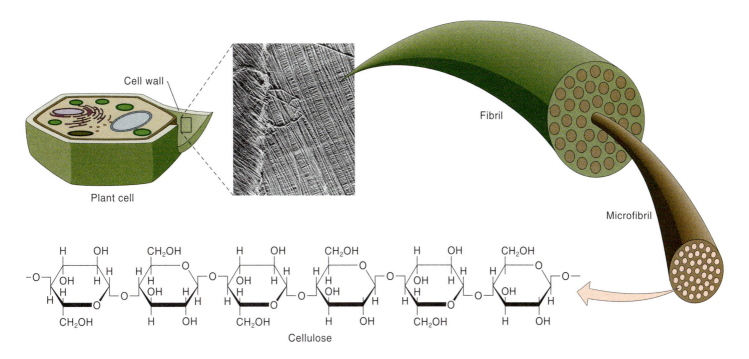

FIGURE 3.11

Model for the arrangement of fibrils, microfibrils, and cellulose in cell walls. The scanning electron micrograph shows the fibrils in a cell wall of the green alga *Chaetomorpha* (×30,000).

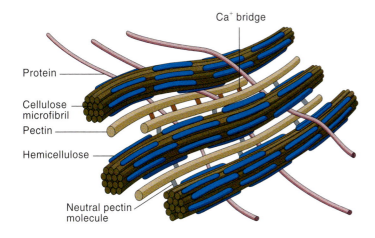

FIGURE 3.12

Model of the interconnections among major components of primary cell walls. Hemicellulose molecules bind to the surface of cellulose microfibrils and to pectin molecules by crosslinks with other pectin molecules. Calcium (Ca$^+$) holds pectin molecules to one another. Glycoproteins are tightly woven into the cell-wall matrix.

the slippery substances that surround the cell walls of certain red algae. Each polymer is a different mix of monomers. Agar is used to make drug capsules, cosmetics, gelatin desserts, and temporary preservatives; it is also used in scientific research to solidify culture medium for growing microorganisms and plant tissue cultures. Car-

rageenan is used primarily as a stabilizer in paints and cosmetics, in foods such as salad dressings, in processed dairy products, and as a smoothing agent in salad dressings.

The most common storage polysaccharide in plants is **starch**—a long polymer made of repeating glucose monomers. Starch in corn, wheat, and rice grain serves as a stored carbon and energy supply during germination and growth of the young seedling until the plant begins photosynthesis on its own—hence the name storage polysaccharide. Animals, such as us, take advantage of this and use the stored starch in grains and other foods such as pea and bean seeds and potatoes for food. Pasta, made from wheat grain (flour), derives most of its taste, texture, and calories from this complex carbohydrate, starch.

Animals contain digestive enzymes that break down starch in the diet. Thus, unlike cellulose and dietary fiber, animals get the calories (and carbon) from starchy foods. In fact, starch is important in the diet of many animals that feed directly on plants in an ecosystem. What other foods in our diet are rich in starch?

Photosynthesis is the important process in plants that uses the energy of the sun to produce carbohydrates, primarily sucrose and starch. In a plant, starch can accumulate in leaf cells during very active photosynthesis in bright light. During the night, after photosynthesis stops in a leaf, the accumulated starch can be broken down and used to make sucrose, which is exported from the leaf and distributed throughout a plant. As sucrose is imported by storage organs, such as a potato tuber or a lima bean, it is converted

The Lore of Plants

Cellulose and other related structural polysaccharides are important in the diet of animals, including us, because they provide bulk or roughage. Bulk helps the large intestine retain water and helps move undigested foods quickly through it. This is the source of the old saying "An apple a day keeps the doctor away."

Dietary fiber is the name nutritionists have given to cellulose and its structural relatives. Most animals lack the digestive enzymes to break down dietary fiber, so it is not absorbed, and the calories are not taken in by the body. Thus, dietary fiber is indigestible and no energy is derived from it.

What foods in your diet are rich in fiber?

---

to starch, the storage polysaccharide. Thus, in evolving plants, starch has become a molecule that stores carbon and energy for later use. Sucrose is the transported form of the carbon and energy, whether it comes from starch breakdown or from photosynthesis. We will focus on how sucrose and starch are made following photosynthesis in chapter 10.

Although starch is the most common storage polysaccharide in plants, it is not the only one. Many plants make inulin instead of (or in addition to) starch. Inulin is a polymer of fructose that functions as a storage polysaccharide in dahlias (*Dahlia* species), Jerusalem artichoke (*Helianthus tuberosum*), globe artichoke (*Cynara scolymus*), chicory (*Cichorium intybus*), and sweet corn. The storage organs of these plants (e.g., tubers of Jerusalem artichoke, kernels of sweet corn) taste sweet because inulin releases fructose.

INQUIRY SUMMARY

Carbohydrates include simple sugars, such as glucose and sucrose, and complex polysaccharides, such as cellulose and starch. Cellulose is an important component of plant cell walls. Sucrose and starch from photosynthesis transport and store carbon and energy in plants, and both are important food sources for animals.

What makes starch and cellulose so different even though they are made of glucose?

Proteins make up membranes, function as enzymes, and store nitrogen in plants.

After cellulose, **proteins** make up most of the remaining biomass of many plant tissues. A protein is a polymer consisting of one or more long chains of monomers called amino acids, which are linked by chemical bonds between adjacent carbon and nitrogen atoms. Thus, amino acids contain nitrogen and cannot be manufactured in a plant without an external supply of nitrogen. This is a major reason why plants require nitrogen; it is why people supply nitrogen-containing fertilizer to their houseplants, garden plants, lawn, and crops.

The smallest proteins have fewer than 100 amino acids, whereas the largest have several thousand. There are thousands of different kinds of proteins each playing different roles in plants. The proteins are made different by the different sequence and numbers of the 20 different amino acids that are known to occur in protein polymers. Proteins may also include sugars or other kinds of small molecules attached to the amino acid chain.

Like polysaccharides, proteins are polymers and are important in cell structure and as storage reserves. Unlike carbohydrates, most proteins are enzymes that catalyze (speed up) biochemical reactions. Also, unlike polysaccharide polymers such as starch and cellulose, proteins do not occur as chains of the same monomer. Instead, each protein contains a unique, prescribed order of the 20 different kinds of amino acids. Thus, each amino acid is like a letter in an alphabet, and each protein is like a long word that usually contains the entire alphabet, with some letters repeated many times. The huge diversity of proteins in nature results from these different sequences of amino acids in each different protein.

The sequence of amino acids in a protein is the protein's **primary structure** (fig. 3.13). As we will see, the primary structure is governed by a plant's genes. However, proteins are not simply long polymers that drift randomly in the cell; rather, they fold and twist into unique three-dimensional shapes determined by their primary structure. It is the distinctive three-dimensional structure that determines the specialized function of each protein.

Proteins are difficult to classify because of their great diversity. For simplicity, we will discuss proteins relative to their three major functions: structural proteins, storage proteins, and enzymes.

Structural proteins occur in cell walls. In addition to containing carbohydrates, cell walls include from 2% to 10%

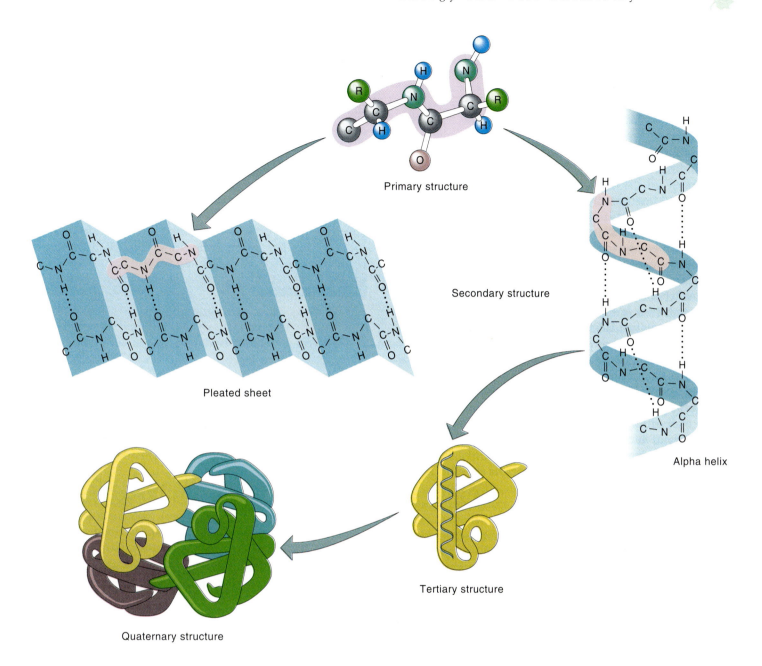

Primary structure

Secondary structure

Alpha helix

Pleated sheet

Tertiary structure

Quaternary structure

FIGURE 3.13

Models of the primary, secondary, tertiary, and quaternary structures of a protein. The primary structure is the sequence of amino acids. Two linked amino acids are shown here. Each *R* group is different for each amino acid. The secondary structure may consist of coils (alpha helices) or folds (pleated sheets), depending on which amino acids form bonds with each other. (For simplicity, R groups are omitted from the pleated sheet and the alpha helix.) Tertiary structure is maintained by different kinds of chemical interactions within the protein. Quaternary structure occurs between different tertiary structure subunits.

protein. Some of these proteins help cell walls enlarge by allowing slippage between the cellulose fibers that form the wall, thereby allowing additional water to flow into the relaxed plant cell. This can result in expansion and growth of a cell and therefore a plant. Several of the structural proteins in cell walls help protect or repair damaged cells after a plant is attacked by insects, fungi, or overzealous gardeners.

Structural proteins also occur in the assorted membranes of plant cells. Each kind of membrane has a different protein composition. For example, the internal membranes of the subcellular structures called mitochondria and chloroplasts are about 75% protein, whereas the membrane that surrounds the cell, just inside the cell wall, is usually about 50% protein. Lipids comprise most of the nonprotein portion of membranes. Membranes are discussed in more detail later in this chapter.

Storage proteins accumulate in seeds as they develop in the fruit of a flowering plant. These proteins are made from

The Lore of Plants

There are no gelatin recipes that include pineapple juice, perhaps because fresh pineapples contain a protein-degrading enzyme (bromelain) that hydrolyzes gelatin, a protein. Hydrolyzed gelatin cannot form a gel, so gelatin desserts containing fresh pineapple juice would be too runny to appeal to the typical consumer. Gelatin desserts made with processed, canned pineapple will gel. Can you explain why?

amino acids imported by the developing seed. Accumulated storage proteins—which are simply polymers of amino acids—are broken down to the constituent amino acids during seed germination and subsequently used as a source of nitrogen for the early development of seedlings. This gives seed-bearing plants, such as elm, sunflower, and grasses, an evolutionary advantage because they package the nitrogen-containing amino acids needed for early growth of a new plant. After germination and seedling establishment, young plants can manufacture their own amino acids and then use these to make proteins and other nitrogen-containing compounds such as chlorophyll and DNA. Of course, manufacturing these nitrogen-containing compounds depends on the availability of nitrogen in the soil.

The composition of seed storage proteins varies in different plants. For example, corn grains produce a storage protein called zein, named for the genus of corn, *Zea.* As mentioned earlier, some of the complex storage proteins of wheat grains are called glutenins. The sticky glutenins hold bread dough together as it rises and contribute to the taste of bread. If you have ever kneaded bread, you know how sticky these storage proteins can be and how good fresh bread tastes. Storage proteins, called globulins in legume seeds (such as pinto beans and peas) and called glutenins in cereal grains (such as rice, corn and wheat), are especially important because they are a major source of amino acids for humans, cattle, and other animals. Soy proteins, from modern varieties of soybean, are important in our diet because they are a good source of amino acids required in our diet. Read the label on prepared food containers and see how many contain soy proteins. Many people eat tofu, a product made from soy protein. Soybeans have been produced in the United States for many years but only recently have they been raised for their protein. Do you know the original, primary use of soybeans? They were originally fermented into soy sauce.

Nutritionally, corn and barley are low in the essential amino acid lysine. Beans, peas, and other legume seeds such as peanuts are deficient in the essential amino acids cysteine and methionine, but their lysine content is adequate. People who depend on these plants for protein in their diet must be certain to combine them daily to get all the required amino acids. This mixing of foods to ensure the intake of all the essential amino acids is called protein complementing. Interestingly, soybeans have become a major crop in the United States because the seeds of newer varieties produce complete proteins—that is, those with enough of all the amino acids required in the human diet. There is considerable research in genetic engineering to develop tasty varieties of corn and wheat that contain complete proteins in the seeds. Of course, seed proteins in nature serve as a major source of amino acids consumed by animals that eat plants (herbivores and omnivores). This is another reason plants are called producers in an ecosystem.

Seeds of some plants also contain proteins that have undesirable nutritional effects on animals that eat them. For example, up to 10% of the protein in many cereal grains inhibit certain digestive enzymes of animals. Similarly, the seed proteins of a few plants are toxic; these include proteins such as ricin from the castor bean (*Ricinus communis*) and abrin from the rosary pea (*Abrus precatorius*). Cooking usually renders these digestion-inhibiting and toxic proteins harmless. Can you propose a selective advantage for the production of these proteins by the plant?

Almost all enzymes are proteins (some RNA molecules are catalysts). In fact, cells have been called "bags of enzymes" to underscore the importance of enzymes in life. Enzymes are often named by adding the ending *-ase* to the name of a molecule involved in the reaction that the enzyme catalyzes. One example is sucrase, the enzyme previously mentioned that breaks sucrose into glucose and fructose. Enzymes catalyze reactions that make or digest all compounds in plants.

Enzymes often remain active even after they are removed from the cell. Pure enzymes that maintain their activity are commercially important. These enzymes include the protein-degrading enzymes, called proteases, from papaya (*Carica papaya*), which digest the muscle tissue of animals. This is why papaya extract is a major ingredient in commercial meat-tenderizers. It is also useful in neutralizing wasp stings because the injected toxin is a protein. Another plant protease is a drug used to treat the slippage of disks in the spinal column. The enzyme is injected directly into the problem area, where it dissolves the proteinaceous cartilage of which the problem disk is made.

INQUIRY SUMMARY

Proteins consist of 20 different kinds of amino acids bonded together in a specific sequence called the primary structure. The subsequent three-dimensional shape, which depends on the primary structure, determines the specific function of the protein. The major roles of proteins in plants are structural proteins, storage proteins, and enzymes.

What would be the ecological importance of storage proteins in a grassland ecosystem?

Nucleic acids are the blueprints of life and the record of evolution.

Perhaps the most complex biological polymers are the two types of **nucleic acids—DNA** (deoxyribonucleic acid) and **RNA** (ribonucleic acid). DNA makes up genes that are found in chromosomes. Genes are units of heredity that pass along traits to offspring and direct the expression of the traits in the individual. This is one of the major roles of DNA. RNA is involved in carrying out the expression of genes and assisting in the assembly of amino acids into the specific sequence of a protein. In addition, RNA is known to have catalytic capabilities in cells, which suggests ancient life could have used RNA to speed chemical reactions until proteins evolved as more efficient enzymes.

The complexity of nucleic acids results from their primary structure (sequence of nucleotides) of five different monomers, each of which is called a nucleotide. Each nucleotide consists of a phosphate group, one of two types of a simple sugar, and one of five nitrogen-containing bases (fig. 3.14). A nitrogen base is a molecule made of carbon, hydrogen, oxygen, and nitrogen. The five nitrogen bases are guanine (G), adenine (A), cytosine (C), thymine (T), and uracil (U). Only four of the five occur in DNA or RNA. The simple sugar is either ribose, which occurs in ribonucleic acid (RNA), or deoxyribose, which is the sugar in deoxyribonucleic acid (DNA).

A DNA molecule consists of four different kinds of nucleotides containing either guanine (G), adenine (A), cytosine (C), or thymine (T). The base composition of RNA is the same as that of DNA, except that RNA includes ribose sugar and the nucleotide uracil (U) instead of thymine. DNA occurs as a double-stranded spiral, like a spiral staircase, called a **double helix** (fig. 3.14). In protein synthesis, DNA is the blueprint for RNA, which in turn directs the assembly of amino acids into proteins within the cell.

Nucleic acid molecules vary in size from about 30 nucleotides to several million nucleotides. The smaller nucleic acids are primarily RNA. Even the smallest nucleic acids, however, have a seemingly infinite number of nucleotide sequences. For example, four different kinds of nucleotides can be arranged in a 10-monomer sequence in 4^{10} (1,048,576) different ways. This is analogous to using an alphabet with four letters to write more than a million 10-letter words. This analogy should help you understand the diversity of life on earth. This point is worth emphasizing: although DNA consists of the same four nucleotides in all organisms, the amounts and sequences of different nucleotides vary immeasurably. Thus, there are an enormous number of different genes in nature. These differences account for the immense variety of organisms on earth and act as a target upon which natural selection can act, as described in chapter 2.

Nucleic acids are unique because they can replicate themselves and guide protein synthesis. Exact copies or replicas of DNA (and therefore of the genes) can be made in a plant cell as it ages. In turn, the replicated DNA—which serves as copies of the genes in the parent cell—is equally distributed to offspring cells during cell division. This is the major factor in inheritance, as genes, and the traits they carry, are passed from parent to offspring.

DNA guides the synthesis of RNA, which in turn guides the assembly of proteins. Stated another way, the sequence of nucleotides in DNA acts as a blueprint for the sequence of nucleotides in RNA, which then acts as the code for the specific sequence of amino acids that becomes the assembled protein. This sequence of DNA → RNA → protein is the basis of protein synthesis in all organisms and, therefore, of the expression of genetic traits that are carried in the genes made of DNA. We return to this important concept in chapter 13.

Nucleic acids form the molecular and evolutionary foundation for every living organism. The roles of nucleic acids and genes are so central to life that several chapters in this book consider their biology, chemistry, and evolutionary importance.

INQUIRY SUMMARY

DNA and RNA molecules are made of nucleotides that contain nitrogen. The sequence of these nucleotides forms the genetic foundation of life. DNA has two major roles in plant cells. First, DNA makes up the genes carrying the traits that are passed from parent to offspring. Second, DNA directs the expression of those traits by acting as a genetic blueprint for protein synthesis.

Some biologists suggest RNA evolved before DNA. What evidence supports this hypothesis?

Lipids form membranes and also store energy and carbon in plants.

Unlike other biological polymers, **lipids** are not formed from specific, repeating monomeric subunits. Rather, they are characterized by their water-repellent property and inability to dissolve in water. The major plant lipids include

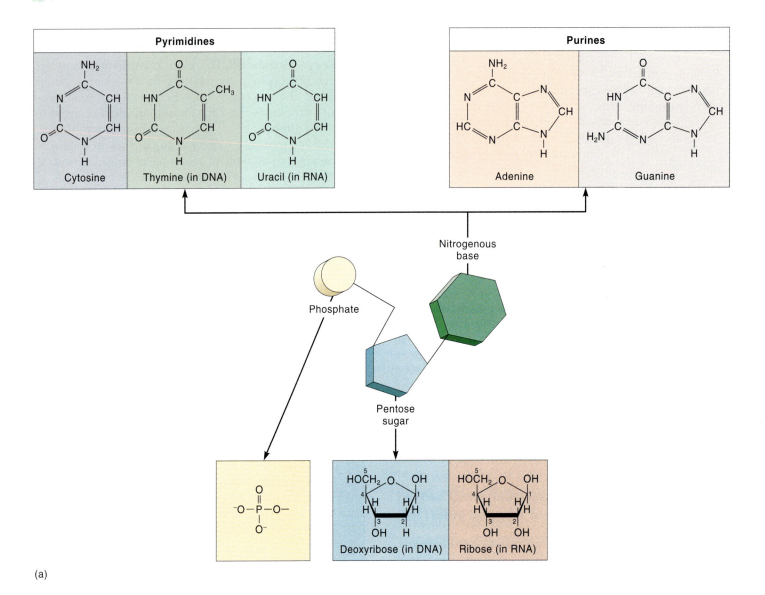

(a)

FIGURE 3.14

The structure of DNA. (*a*) Each nucleotide monomer consists of three smaller building blocks: a nitrogenous base, a five-carbon pentose sugar, and a phosphate group. (*b*) Nucleotide monomers are bonded to each other by covalent bonds between the phosphate of one nucleotide and the sugar of the next nucleotide. (*c*) DNA is usually a double strand held together by hydrogen bonds between nitrogenous bases: adenine (A) pairs only with thymine (T), and cytosine (C) pairs only with guanine (G). The double strand is twisted spirally into a double helix. (*d*) A space-filling model shows the close stacking between nitrogenous bases. Color code for atoms: *yellow* = P, *dark blue* = C, *red* = O, *turquoise* = N, and *white* = H.

oils, phospholipids (phosphate attached to a lipid), and waxes. The term lipid includes fats and similar molecules, but we use lipid and fat interchangeably for convenience. Lipids help form membranes and serve as long-term storage forms of carbon and energy.

Oils are fats that are liquid at room temperature. A fat or oil molecule is made of glycerol with three chains of organic acids, called **fatty acids,** linked to it (fig. 3.15). The linkage between a fatty acid and glycerol is called an acylglyceride linkage. Thus, a single fat molecule is a triacylglyceride, or, more commonly, a **triglyceride.** Most of the fat in our diet is triglycerides, and when we speak of

fat or oil in food, technically we mean a mixture of several kinds of triglycerides; each triglyceride may have distinctive fatty acids attached to glycerol. Accordingly, each of the three fatty acids that make up a triglyceride may differ in two fundamental ways: (1) the length of their carbon-based chain, and (2) the number and placement of double bonds between the carbon atoms forming the fatty acid chain.

In a fatty acid, carbon atoms are bonded to other carbon atoms. A fatty acid having no carbon-to-carbon double bonds is saturated (that is, saturated with hydrogen and all bonds between adjacent carbon atoms are single bonds; see

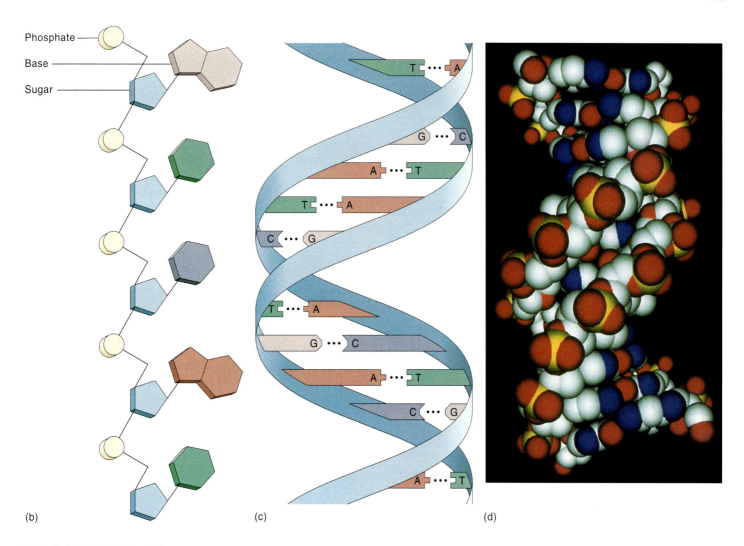

Phosphate

Base

Sugar

(b)

(c)

(d)

FIGURE 3.14
(continued)

The Lore of Plants

Many people believe a vegetarian diet is more healthy than the traditional American diet. Is this true? Fatty acids in plant oils are usually rich in either mono-unsaturated (they have one double bond between two carbon atoms) or polyunsaturated (two or more double bonds) fatty acids. You probably have heard these terms relative to human nutrition. Common fatty acids include oleic acid (one double bond), linoleic acid (two double bonds), and linolenic acid (three double bonds). Linoleic acid and linolenic acid are essential fatty acids; this means they are necessary for human growth and development, but our bodies cannot synthesize them. Yes, plants are a good source of essential fatty acids and generally provide them in a more healthy form than animal foods (but there are new health concerns about polyunsaturated fatty acids). Corn, olive, and sunflower oil are good examples of plant lipids. What is another advantage of eating vegetables? No cholesterol! Yet, you should remember even plant oils, like other fats, contain 9 kilocalories per gram: they can lead to obesity if consumed in excess. In addition, before you decide about a vegetarian diet, see Perspective 3.2, "What's Wrong with Tropical Oils?"

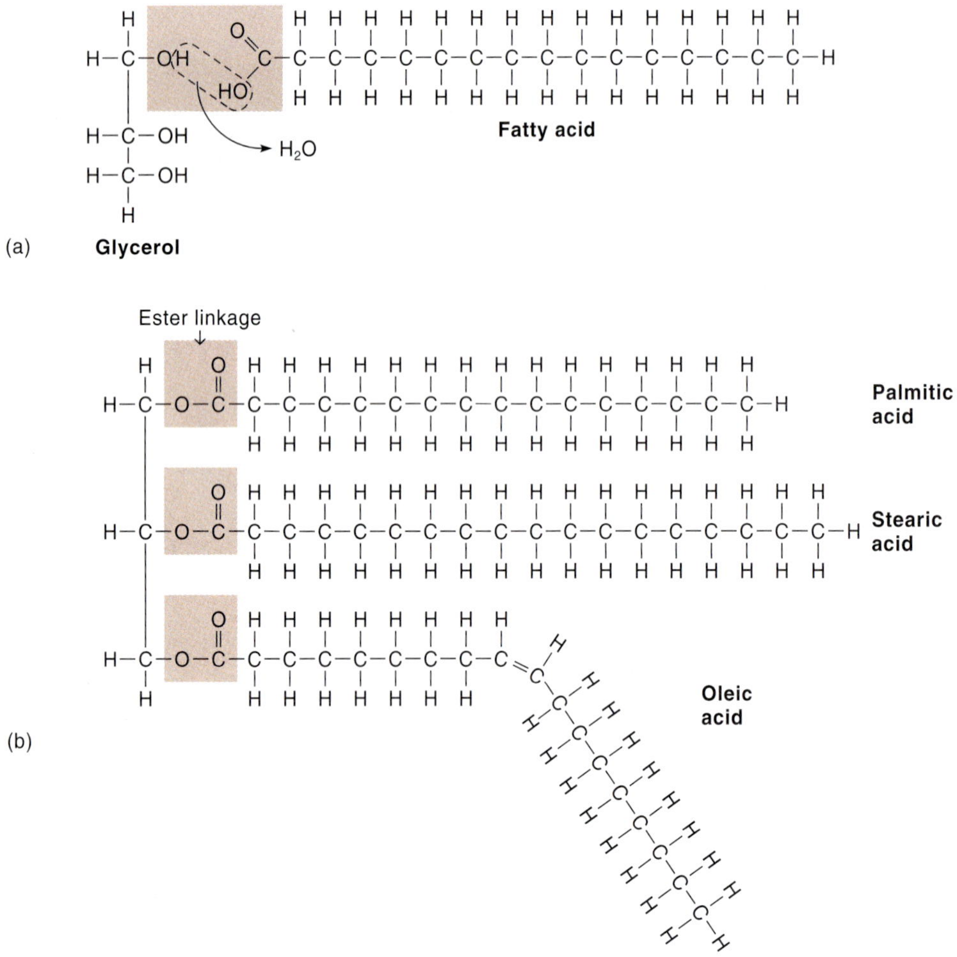

(a)   **Glycerol**

(b)

FIGURE 3.15

The structure of a fat. (*a*) An acylglyceride bond forms when a fatty acid links to a glycerol, with the removal of a water molecule. (*b*) Fats are triacyglycerides whose fatty acids vary in length (number of carbon atoms) and also in the presence and location of carbon–carbon double bonds. Oleic is a mono-unsaturated fatty acid due to the single carbon–carbon double bond.

ergy as oil is that more energy can be packaged in fat than in starch (9 Calories or kilocalories per gram for fat compared to 4 Calories for starch).

As mentioned, some seeds and fruits contain enough oil to be commercially valuable. The best known of these are cotton, sesame, safflower, sunflower, olive, coconut, peanut, corn, castor bean, and soybean. You can buy most of these oils at grocery stores. Indeed, more than 90% of the oils that we harvest from these plants are used for products such as margarines, shortenings, salad oils, and frying oils; the rest are used for nonfood products such as lubricants, fuels, coatings, and soaps. Some plant fats, such as palm and coconut oil, are saturated but tend to be liquid at room temperature because they contain short-chain fatty acids. These oils are common in foods and are often used to pop the popcorn at athletic events and in theaters.

We have explained that lipids accumulate in developing seeds, where they serve to store carbon and energy. Remember that all the carbon in a plant, whether in a carbohydrate, protein, nucleic acid, or lipid, came originally into the plant as CO_2 and was converted to sugar by photosynthesis. The lipids, and other organic molecules, are then made from the carbon in the sugar during metabolism.

Membranes contain triglyceride molecules with a phosphate group substituted for one of the three fatty acids (fig. 3.16). Such triglycerides are called **phospholipids.** The charged or polar phosphate group on the molecule dissolves in water or forms a bond with a membrane protein. This property causes phospholipids to have a dual solubility on the same molecule, since the phosphate end is water soluble (hydrophilic; Greek for "water loving") and the fatty end is uncharged, nonpolar, and water insoluble (hydrophobic; Greek for "water hating"). Consequently, in membranes, phospholipids can interact with the polar groups of proteins or water at one end and form an oily matrix at the other. This versatility is important because it helps organize and stabilize membranes as they control the passage of substances.

Waxes are complex mixtures of fatty acids and other lipid molecules and often comprise the outermost layer (called the waxy cuticle) of leaves, fruits, and herbaceous

palmitic and stearic acid in fig. 3.15). Oils are liquid at room temperature because (1) the fatty acid chains in their constituent triglycerides are shorter and (2) because many of the chains are unsaturated; that is, they have double bonds between one or more pairs of carbon atoms in a fatty acid (see oleic acid in fig. 3.15). Double bonds provide molecular rigidity at sharp angles, which prevents the molecules from packing tightly into a solid. This is why unsaturated lipids such as corn oil and sunflower oil are liquids at room temperature. Saturated fats, such as those found in animals, tend to be solid at room temperature. While we call the liquid form an oil, chemically both are a fat made of triglycerides.

In plants, triglycerides (usually oils) are most abundant in seeds. Like starch, seed oils are a form of chemical energy and stored carbon that are used when the seed germinates. A selective advantage to storing carbon and en-

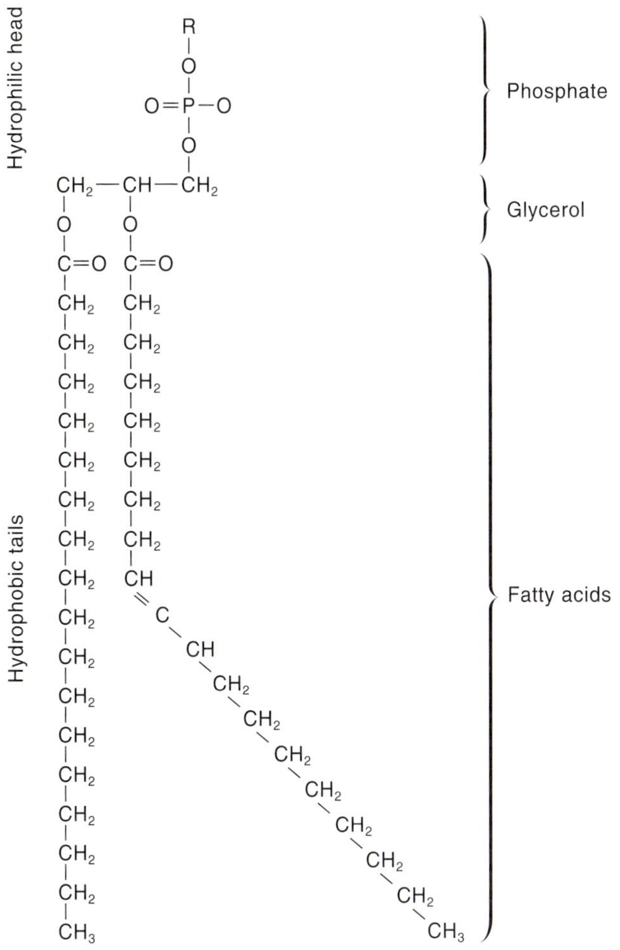

FIGURE 3.16

The structure of a phospholipid. A phospholipid consists of glycerol that is bonded to two fatty acids, which are hydrophobic ("water-hating"), and to one phosphate group, which is hydrophilic ("water-loving"). Phospholipids vary by their fatty acids and by side chains (R) that are attached to the phosphate. The side chains include glycerol, sugars, and nitrogen containing carbon chains.

stems. Because of genetic variability, the chemical structures of different waxes vary depending on the plants that produce them, but the role is similar in waterproofing the plant.

Waxes are usually harder and more water-repellent than other fats, which is why you wax but don't "fat" your car or kitchen floor. Wax from plants such as the carnauba palm (*Copernicia cerifera*), the source of carnauba wax, is especially hard. Other commercial waxes come from bayberry (*Myrica pensylvanica*), which is used to make novelty candles, and from candellila (*Euphorbia antisyphilitica*). Candellila wax, which is often substituted for carnauba wax, was once the main component of the hard "melts in your mouth, not in your hands" coating of M&M chocolate candies. Unfortunately, candellila is near extinction because the plants have been overcollected for their wax. Collecting this plant is forbidden in the United States, but it is still heavily collected in Mexico.

Cutin consists of a variety of fatty acids linked together and, as mentioned, makes up most of the cuticle covering many plant parts. A similar substance, called **suberin,** occurs, for example, in the cork cells of bark. Cutin and suberin differ mainly in the kinds of fatty acids they contain, but both function as barriers to water movement. In part, the old saying "floats like a cork" comes from the water-repellent properties of the suberin produced by cork cells and air trapped in the cork. What are some uses of cork?

INQUIRY SUMMARY

Lipids are water-repellent compounds made mostly of carbon and hydrogen. The major lipids of plants are triglycerides, phospholipids, and waxes, which all contain fatty acids. Triglycerides function mainly as energy and carbon storage in oilseeds. Phospholipids are important components of membranes. Waxes and wax-like substances prevent water loss from plants.

How might cork have been important in the evolution of land plants?

Secondary metabolites produced in plants often have importance to us.

Plants make a variety of less widely distributed compounds, such as morphine, caffeine, nicotine, menthol, and rubber. These compounds are **secondary metabolites,** which are chemicals that occur irregularly or rarely among plants and that initially had no known role in plant cells. Most plants have not yet been examined for their secondary products, but new ones are discovered almost daily. Many of these secondary compounds were first discussed in chapter 1.

Most secondary metabolites can be grouped into classes based on structural similarities, the way they are manufactured in a plant, or the kinds of plants that make them. The largest such classes are the alkaloids, the terpenoids, and the phenolics. Examples of several kinds of secondary metabolites are presented in table 3.1.

Secondary metabolites often occur in plants in combination with one or more sugars. Such combination molecules are called glycosides. For example, digitoxose (used to produce digitalis for heart patients) is found in purple foxglove (*Digitalis purpurea*; fig. 3.17c), and apiose is unique to parsley (*Petroselinum*) and its relatives. Cyanogenic glycosides are sugar-containing compounds that release cyanide gas when broken down. These compounds occur in many plant families, but they are especially common in the pea and rose families. Cyanide from such plants is not usually fatal, but two cups of well-chewed apple seeds would probably kill an adult human. There is recent evidence lycopene from tomato (fig. 3.17d) may reduce the risk of certain cancers.

TABLE 3.1		**Examples of Well-Known Secondary Metabolites**

COMPOUND	EXAMPLE SOURCE	COMMENT
Alkaloids		
coniine	poison hemlock	nerve toxin; killer of Socrates
strychine	strychnine tree	potent nerve stimulant and convulsant
tubocurarine	curare tree	component of arrow poisons; used as muscle relaxant during surgery
thomasine	tomato leaves	inhibits feeding by Colorado potato beetles, which are pests of tomatoes
morphine	opium poppy	main painkiller used worldwide
codeine	opium poppy	cough suppressant
atropine	belladonna	used in eye exams to dilate pupils and as an antidote to nerve gas
vincristine	Madagascar rosy periwinkle	main treatment for certain kinds of leukemia
quinine	quinine tree	bitter flavor of tonic in gin and tonic: used to prevent malaria
Other Nitrogen-Containing Compounds		
canavanine	jack beans	toxic nonprotein amino acid
buteonine	roots of garden beet	red pigment
sinigrin	horseradish	"burning" spice of mustards
amygdalin	apricot seeds	releases cyanide; may be active ingredient of unapproved cancer drug called Laetrile
Terpenoids		
menthol	mints and eucalyptus	strong aroma; used in cough medicines
camphor	camphor tree	component of disinfectants and plasticizers
nepatalactone	catnip	very attractive to cats
smilagenin	sarsaparilla	steroidal glycoside
digitalin	purple foxglove	cardiotonic used to stimulate heart action
oleandrin	oleander	heart poison (similar to digitalin)
lycopene	tomatoes	red/orange pigment
rubber	rubber trees	component of rubber tires
taxol	Pacific yew	drug that inhibits cancerous tumors, especially of ovarian cancer
Phenolics		
salicin	willow trees	folk medicine against headaches and fever; precursor of aspirin*
gallic acid	walnut husks	main component of some tannins
myristicin	nutmeg	main flavor of the spice
rutin	buckwheat	common "bioflavonoid" sold in nutrition stores
cyanidin glucoside	chrysanthemums	deep red pigment
limonin	grapefruit	bitter flavor

*The medicinal reputation of willow (*Salix alba*) stems from salicin, a chemical first isolated from willow bark in 1827. Although salicin was medically useless because of its severe side effects, related compounds were isolated later from other plants (e.g., salicylic acid from meadowsweet, *Spiraea ulmaria*) Finally, in 1899, Felix Hoffman produced acetylsalicylic acid to help his father deal with arthritis. Acetylsalicylic acid, which produced no bad side effects, was mass-produced later by Fredrich Bayer & Co. Bayer named the compound *aspirin*—"a" for acetyl and "spirin" for *Spiraea*—which, although catchy, is misleading, because *Salix* was there first. Today, aspirin is the world's most frequently taken medicine; Americans swallow almost 40 tons of it every day.

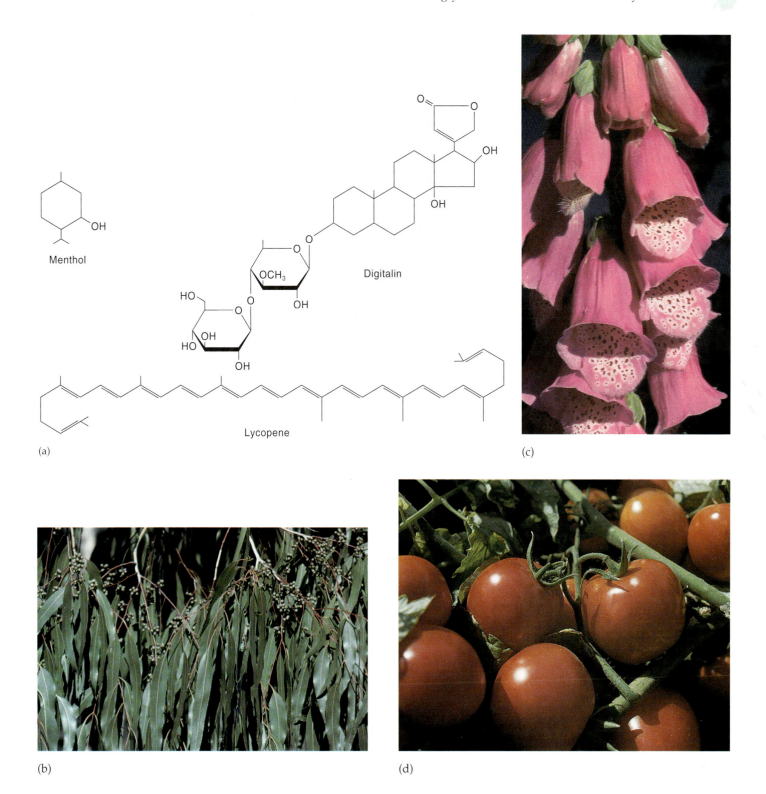

Menthol

Digitalin

Lycopene

(a)

(b)

(c)

(d)

FIGURE 3.17

Terpenoids. (*a*) Structures of three terpenoids. (*b*) *Eucalyptus* species produce menthol. (*c*) The purple foxglove (*Digitalis purpurea*) is a source of digitalin. (*d*) Lycopene is the main red pigment of tomatoes.

3.2 *Perspective* Perspective

What's Wrong with Tropical Oils?

In most people, saturated fats in the diet raise blood-cholesterol levels more than do mono-unsaturated and probably polyunsaturated fats. In fact, blood-cholesterol levels are raised more by saturated fats than by cholesterol in the diet. High blood cholesterol has been linked with accumulation of cholesterol deposits in the arteries and an increased risk of cardiovascular disease (mostly heart attack). Current dietary guidelines recommend that your total fat intake should be less than 30% of your total energy intake (Calories). Of this total, saturated fat intake should be less than 10% of energy intake.* Thus, it is clearly important to minimize saturated fats in our diets.

How does this often-repeated advice relate to tropical oils? Fat from seeds of palms, coconuts, and other tropical plants contain mostly palmitic acid, which is saturated. Thus, tropical oils are a hidden source of saturated fats: They have long been used to make a variety of processed foods, including breakfast cereals, crackers, cookies, coffee creamers, baked goods, and microwave popcorn. Palm, palm kernel, and coconut oils are used in cocoa mixes, diet desserts, instant soup, and chocolate, but their use is declining.

It seems like a simple task to avoid tropical oils, because all you have to do is read the ingredients list on a food label. Some manufacturers are already making this easier by printing "No Tropical Oils" in bold, colorful letters on the fronts of packages. However, many such products still contain saturated fats in the form of "hydrogenated vegetable oil." When hydrogen is added to the double bonds of a polyunsaturated fat, the fat becomes saturated. This means that linolenic acid, with three double bonds, becomes palmitic acid when it is hydrogenated. Since palmitic acid from linolenic acid does not come from palms or coconuts, the "no tropical oils" claim is legitimate; however, the saturated fat content may still be high, and the increased risk of cardiovascular disease remains.

For comparison, canola, olive, and peanut oils contain 50% to 80% of total fat as mono-unsaturated fatty acids, while corn, sunflower, soybean, and safflower oils contain 55% to 75% of total fat as polyunsaturated fatty acids. All of these are relatively low in saturated fatty acids. Animal fats contain 40% to 60% of the total fat as saturated fatty acids.

What can you do? Become a label reader and try to include mono-unsaturated fat in your diet (strive for 10% of your total Calories) instead of saturated, and probably even polyunsaturated, fat.

*The same 10% target is true for polyunsaturated fat because there is evidence linking it to increased accumulation of cholesterol in arteries, but the evidence against saturated fat is most compelling. Carbohydrate intake should constitute 50% or more of energy intake.

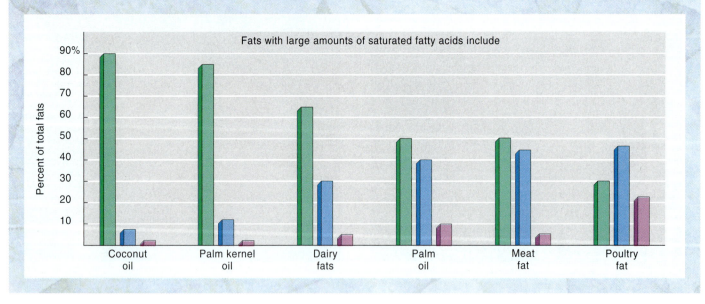

Alkaloids are a diverse group of secondary products that contain nitrogen. They often produce dramatic physiological effects in humans and other animals. For example, the alkaloids coniine, strychnine, and tubocurarine are infamous toxins, whereas morphine, codeine, atropine, and vincristine (chapter 1) are important therapeutic drugs. Alkaloids are often bitter; one of the most bitter substances known is the alkaloid quinine. Terpenes such as geraniol and menthol, from the leaves of mints and eucalyptus, are volatile and usually have strong odors. Pines and other resinous plants synthesize terpenoids that we use to make terpentine. The largest terpenoids are known as rubber. Although about 2,000 plant species make rubber, most of them make this substance in amounts too small for com-

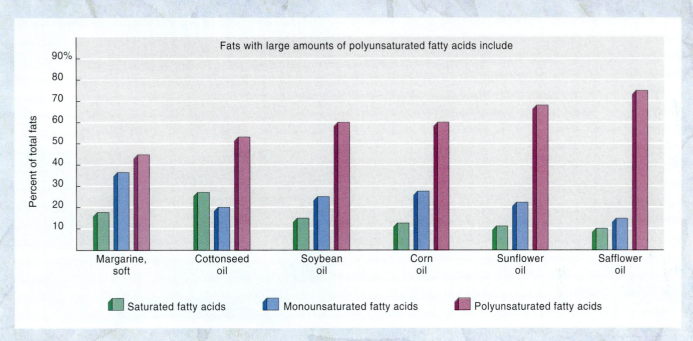

Fats with large amounts of polyunsaturated fatty acids include

Dietary fat and fatty acid proportions in different foods.
Reprinted with permission from /it/Diabetes Forecast./xit/ Copyright © 1991 American Diabetes Association. For information on joining ADA and receiving /it/Diabetes Forecast./xit/ call 1-800-806-7801.

The Lore of Plants

Tabasco (meaning "land where soil is humid") brand pepper sauce is made at Avery Island, Louisiana. Peppers that are handpicked before 3:30 P.M. each day are mashed and mixed with Avery Island salt and fermented in white oak barrels for 3 years. The fermented peppers are then mixed with white vinegar, the seeds are strained out, and the sauce is put into bottles. (The sauce was originally sold in cast-off cologne bottles.) Peppers are the fruit of the pepper plant (*Capsicum* species), and the "heat" is from the alkaloids (capsaicin) stored in the parenchyma of the pepper wall and seeds. Tabasco is often a secret ingredient in buffalo-wing sauces.

The Avery Island factory produces almost 80 million 2-ounce bottles of Tabasco sauce per year, generating more than $50 million in sales throughout the world. Labels for Tabasco sauce have been printed in 15 different languages, and seeds extracted from the original lineage of pepper plants are kept in a bank vault.

If you're ever near Avery Island, stop by the factory for a free tour, and a taste.

mercial use. A simple phenolic, salicylic acid, was originally extracted from willow plants and used to make aspirin.

Flavonoids are sold in nutrition centers and health food stores, usually as supplements with vitamin C. The most commonly available flavonoid is rutin, which is easily obtained from eucalyptus or buckwheat leaves. Myristicin, the main flavor of nutmeg (table 3.1), is an example of a familiar secondary compound (fig. 3.18). Flavonoids called tannins can tan leather and are astringent to the taste; they are the compounds that impart the "dryness" (astringency) to dry wines. The most significant phenolic polymer is lignin (fig. 3.19), a major structural component of wood, which has enormous economic and ecological importance. Wood has many well-known uses so the economic importance should be obvious to you. However, the ecological roles are less well known. For example, lignin is highly resistant to decay by fungi and bacteria. Accordingly, it has been suggested that lignin may contribute to the long life of trees and the slowness of wood to rot.

There are many other secondary metabolites in plants. Many are familiar to us as the unique flavors and odors of common edible plants or the colors of garden

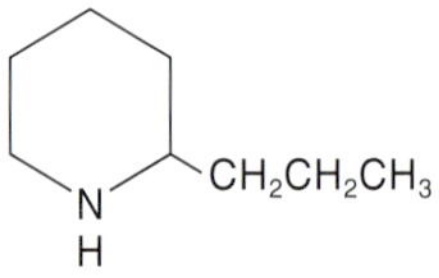

Coniine

(a)

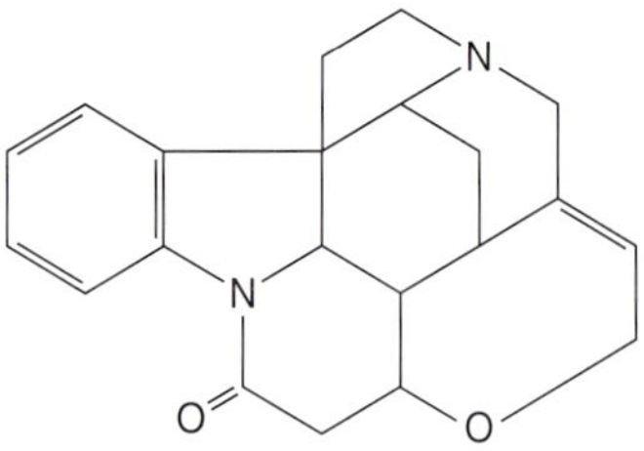

Strychnine

(b)

Tomatine

R = galactose + glucose + glucose + xylose

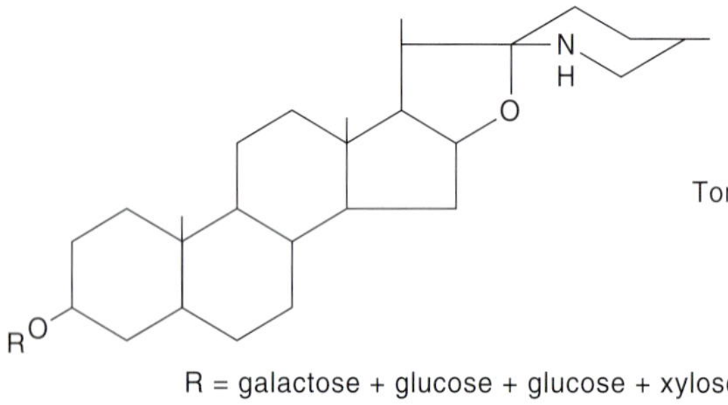

(c)

FIGURE 3.18
Secondary metabolites in plants. Structures of three alkaloids. Note they all contain nitrogen and carbon.

FIGURE 3.19
Lignin is the strengthening polymer that makes wood valuable commercially. These baseball bats are made of the wood of white ash (*Fraxinus americana*). To learn how these bats are made, see Perspective 8.2, "The Bats of Summer: Botany and Our National Pastime."

flowers. The tremendous diversity of secondary metabolites shows that plants are amazing biochemical factories. It is well worth preserving the diversity of plants on earth. What other plant products can you name that are of economic importance to us? Why is it important to identify and preserve endangered plant species and their habitats?

Several secondary metabolites influence interactions between plants and other organisms. For example, many secondary products are brightly colored pigments that attract insects and other animals for pollination or dispersal of fruits and seeds. Conversely, nicotine and other toxic chemicals may protect plants against attacks by hungry insects or disease-causing microbes. This defensive role seems to be true in some cases but not in others. For example, phenolics in soybeans provide resistance to fungal infection, and nicotine from tobacco is toxic to many insects. However, tobacco hornworms have little difficulty eating tobacco leaves that contain up to 30% nicotine, and monarch butterflies are unaffected by the cardiac glycosides of milkweeds. Instead of protecting the plants from being eaten by butterfly caterpillars, these toxins remain in the adult butterflies and protect them from insect-eating birds. What does this imply about the evolution of these plants and animals? Although there are many examples of the ecological functions of these kinds of chemicals, many secondary metabolites have not been examined for their potential ecological roles in plants.

SUMMARY

Energy is the ability to do work or cause change. Potential energy is energy available to do work, whereas kinetic energy is energy being used to do work.

All of life's activities require that energy be transformed from one form to another. These energy transformations are governed by the laws of thermodynamics. The first law of thermodynamics is the law of conservation of energy; it states that in all chemical and physical changes, matter and energy cannot be created or destroyed, only converted to other forms. The second law of thermodynamics states that all energy transformations are inefficient. This inefficiency results from the loss of usable energy in the form of heat. The measure of the disorder created by any spontaneous reaction is entropy. All spontaneous reactions increase entropy.

Metabolism is the sum of the vast array of energy and matter transformations in cells, which occur in step-by-step sequences called metabolic pathways. Each reaction of a metabolic pathway rearranges atoms into new compounds and changes the amount of energy available for work. This change in energy is the amount of energy available to form other bonds. Energy conversions occur as old bonds are broken and new ones are formed during chemical reactions.

Cells use ATP to do work. ATP is suited for this function because (1) most of its energy is available immediately to cells and (2) it has about the right energy needed to drive most cellular reactions.

The elements of living cells occur in thousands of combinations, forming thousands of different molecules. Molecules containing carbon are called organic molecules. Polymers are larger organic molecules made up of smaller, single-unit molecules called monomers. The four polymers of life are carbohydrates, proteins, nucleic acids, and lipids. Carbohydrates are simple sugars and complex polysaccharides. They serve to store energy and carbon from photosynthesis and make up plant cell walls. Proteins, which are made of amino acids, serve many roles in plants, including enzymes and membrane structure. The nucleic acids are DNA and RNA. DNA, the material of which genes are made, directs the activities of the cell. RNA, which is synthesized on a DNA blueprint, is involved in protein synthesis. Lipids form membranes and store energy and carbon in seeds.

Writing to Learn Botany

Creationists frequently claim that the evolution of increasingly complex organisms during the history of life contradicts the second law of thermodynamics, which is an unbreakable rule. Therefore, they claim, biological evolution is invalid. How would you respond to this argument?

THOUGHT QUESTIONS

1. Why are the laws of thermodynamics important in the study of ecosystems?

2. What's the difference between potential and kinetic energy?

3. What is entropy? How does the second law of thermodynamics affect plants?

4. What is energy? How is it used to "fight the entropy battle"?

5. What is metabolism? What two major components does it include?

6. What forms the primary structure of a protein?

7. What ultimately determines the primary structure of a protein?

8. How do the chemical and physical properties of saturated fat and unsaturated fat differ? How are they alike?

9. One of the earliest explanations for the function of secondary metabolites was that they are waste products of plant metabolism. Suggest reasons why this might be a good explanation; also suggest reasons why it might not be a good one. How could you test each idea?

10. How might a spontaneous change in the sequence of nucleotides in the DNA sequence for a protein enzyme alter the course of evolution of a species?

SUGGESTED READINGS

Articles

Davis, G. R. 1990. Energy for planet earth. *Scientific American* 263:54–62.

Duchesne, L. C., and D. W. Larson. 1989. Cellulose and the evolution of plant life. *BioScience* 39:238–41.

Seymour, R. S. 1997. Plants that warm themselves. *Scientific American* 276:104–9.

Books

Dey, P. M., and J. B. Harbourne, eds. 1998. *Plant Biochemistry.* New York: Academic Press.

Hopkins, W. G. 1999. *Introduction to Plant Physiology,* 2d ed. New York: John Wiley and Sons.

Sackheim, G. 1996. *Introduction to Chemistry for Biology Students,* 5th ed. Redwood City, CA: Benjamin/Cummings.

Snyder, C. 1995. *The Extraordinary Chemistry of Ordinary Things,* 2d ed. New York: John Wiley and Sons.

ON THE INTERNET

Visit the textbook's accompanying web site at http://www.mhhe.com/botany to find live Internet links for each of the topics listed below:

Forms of energy

The laws of thermodynamics

Photosynthesis and respiration

ATP

Enzymes

Plant carbohydrates

Cellulose

Proteins, DNA and RNA

Fats and calories

Waxes and oils

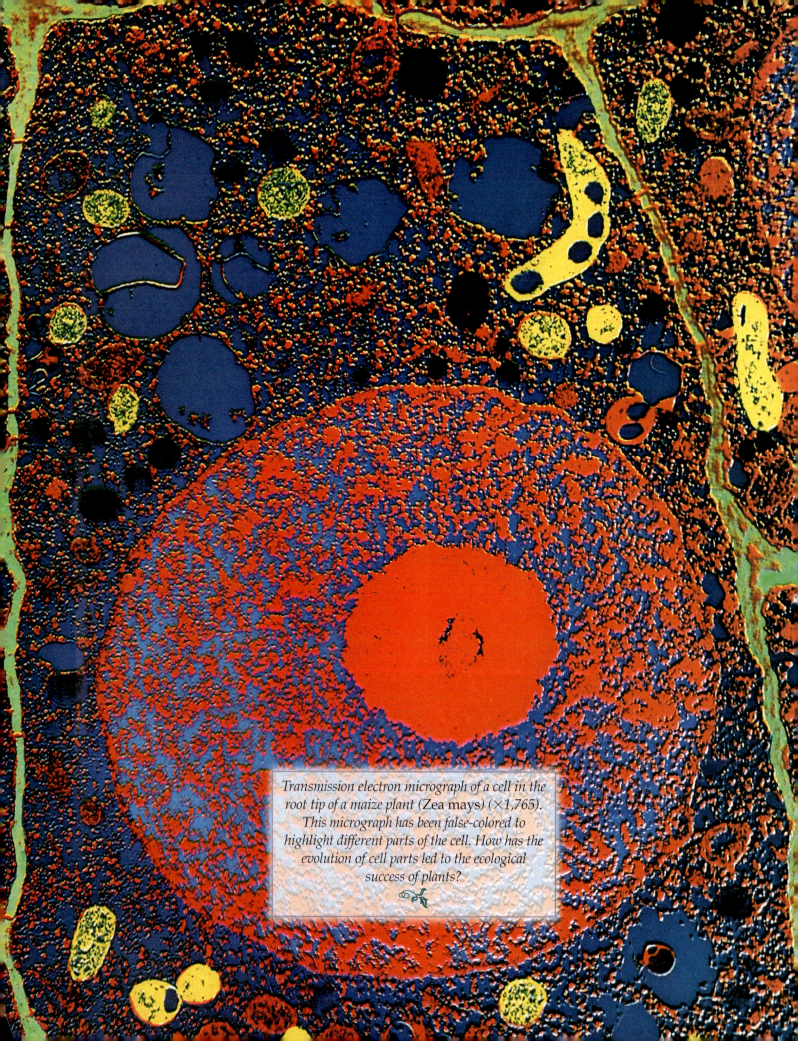

Transmission electron micrograph of a cell in the root tip of a maize plant (Zea mays) (×1,765). This micrograph has been false-colored to highlight different parts of the cell. How has the evolution of cell parts led to the ecological success of plants?

CHAPTER 4

Plant Cells and Tissues

LEARNING ONLINE

One of the key features of the Online Learning Center at www.mhhe.com/botany are the chapter web links. These hyperlinks are correlated to the listing of important topics at the end of every chapter, and provide additional information about each topic. For example, to learn more about plant cells, visit the interactive web site www.hcs.ohio-state.edu/hcs/TMI/HORT300/index.htm in the OLC for chapter 4.

We learned previously that plants consist of organized parts beginning with atoms to molecules, then cells, tissue systems, organs, and, finally, whole plants. Perhaps, the most fundamental level of organization includes cells and their internal compartments, called organelles. Chloroplasts and nuclei are examples of organelles in plant cells. Some organelles occur in all cells, while others occur only in the cells of leaves, roots, or other parts of plants.

Largely depending on their organelles, cells can divide, grow, transport sugar and water, photosynthesize, secrete nectar or resin, or harvest energy from organic molecules. Thus, the traits that distinguish a leaf from a root, or even a giant redwood from a dandelion, emerge from the types and arrangements of their molecules into cells and tissue systems. The tree you see out the window consists of thousands of different kinds of molecules, and dozens of different kinds of cells, some of which are shared by all life, and some of which are unique to that kind of tree.

As we learn in chapter 5, plant cells divide by completion of the cell cycle. Following this, most new cells do not divide again but instead become specialized as they grow and develop into cells of the four types of tissue systems found in plants: (1) meristems, (2) ground tissue, (3) epidermal tissue, and (4) vascular tissues. Cells and the tissues they ultimately form are the topics of this chapter.

Plants Consist of Cells and Products of Cells.

All plants consist of cells and the products of cells that together combine into tissue systems, then organs, and finally organisms; that is, the entire plant (fig. 4.1). To understand the evolution and importance of plant cells, we need to think about basic cell structure and learn some descriptive terms. As mentioned, plant cells contain internal compartments called **organelles,** which are discrete structures that have specific functions. Thus, organelles provide a "division of labor" that makes a plant cell more efficient. The nucleus, vacuole, mitochondria, and chloroplasts, as discussed in this chapter, are examples of organelles.

Beside organelles, there are some important terms botanists use to describe cell anatomy (figs. 4.1 and 4.2). For example, all of the material inside the cell membrane but outside the nucleus has traditionally been termed **cytoplasm.** In turn, the **cytosol** is the semifluid matrix that surrounds and bathes the organelles outside the nucleus. So, cytoplasm is made of cytosol and all the organelles outside the nucleus. The entire plant cell, from the cell membrane inward is called a **protoplast.** The **cell wall,** which is assembled from cellulose and other polymers produced by the plant cell (chapter 3), is actually considered to be extracellular—outside the cell, or the protoplast. This is why we say the plant body is made of cells and products of cells, because the plant cell produces the cell wall and other material outside the protoplast. As we discuss the structures and functions of plant cells, think about the evolutionary advantages of compartmentalizing a cell into structured, working parts called organelles.

INQUIRY SUMMARY

Cells, and the product of cells, form tissues and then organs that make up a plant. Plant cells are compartmentalized internally into structures called organelles. The plant cell, or protoplast, is surrounded by a cell wall.

What is the selective advantage of organs, such as leaves, divided into small cells?

Highly Structured Organelles Have Evolved Inside Plant Cells.

Figure 4.2 is a diagram of a typical plant cell from a leaf. As you study the structures found in plant cells, be certain you think about the functions of each part and how these are related; that is, how structure supports function. Also think about possible selective advantages to each structure and function.

The cell membrane is a remarkably changeable, multipurpose membrane.

The **cell membrane** (also called the plasma membrane) surrounds and defines the plant cell protoplast and regulates the flux of molecules into and out of the cell interior. Some substances can pass through the membrane; others cannot. Thus, a cell membrane acts as a selectively permeable barrier to substances inside and outside a cell. Partly because of its outer position, the cell membrane receives chemical and environmental signals from outside the cell. Signals received by the cell membrane can change

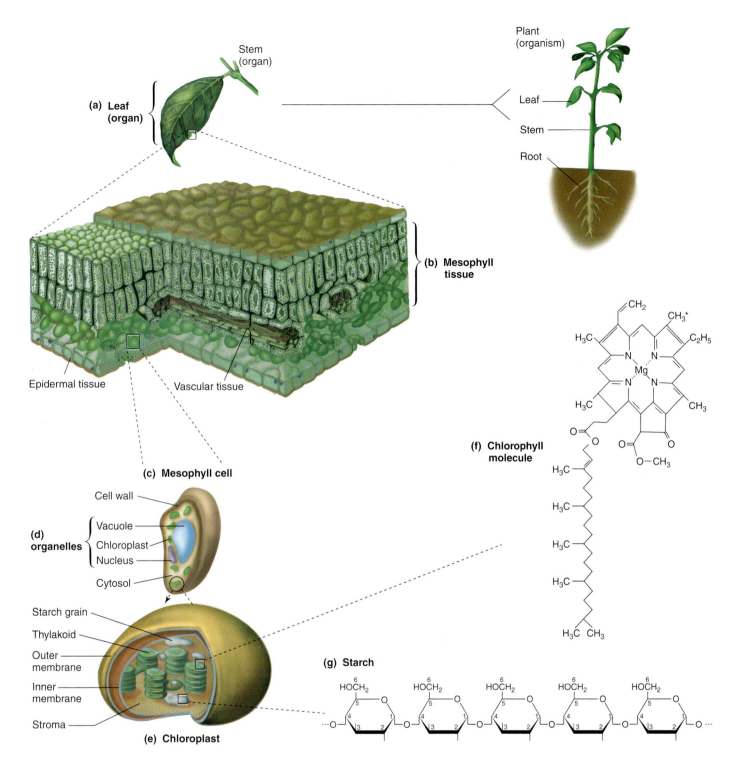

(a) Leaf (organ)

Stem (organ)

Plant (organism)

Leaf

Stem

Root

(b) Mesophyll tissue

Epidermal tissue

Vascular tissue

(c) Mesophyll cell

(d) organelles

Cell wall

Vacuole

Chloroplast

Nucleus

Cytosol

Starch grain

Thylakoid

Outer membrane

Inner membrane

Stroma

(e) Chloroplast

(f) Chlorophyll molecule

(g) Starch

FIGURE 4.1

This model shows the levels of structural and functional organization in a plant. Plants are organisms made of organs including roots, stems, and leaves (*a*) Leaves, like other organs, are composed of tissues; epidermal, vascular, and mesophyll tissue are the examples shown here. Mesophyll tissue (*b*) is found in the middle (*meso*) of a leaf (*phyll*) and has evolved to carry on much of the photosynthesis of the plant. The cells (*c*) that make up the mesophyll tissue contain organelles (*d*) such as a vacuole, nucleus, and chloroplasts. Chloroplasts (*e*) are green because thylakoid membranes inside the organelle are made with chlorophyll molecules (*f*) that help trap light and fuel photosynthesis. Starch molecules (*g*) form inside the illuminated chloroplast as a product of photosynthesis.

The levels of organization show a division of labor, a specialization of tasks that was selected for during evolution because of increased efficiency of operation. In this example, chlorophyll molecules absorb light, starting photosynthesis inside chloroplasts. Chloroplasts are abundant in mesophyll cells, which make up the photosynthetic mesophyll tissue of leaves. Overall, a major function of leaves is photosynthesis, but it is accomplished by increasing levels of organization from chlorophyll molecules to chloroplast organelles, to mesophyll cells, and then mesophyll tissue and leaves as organs making up an organism—the plant. We learn the particulars of these structures and functions here and in subsequent chapters.

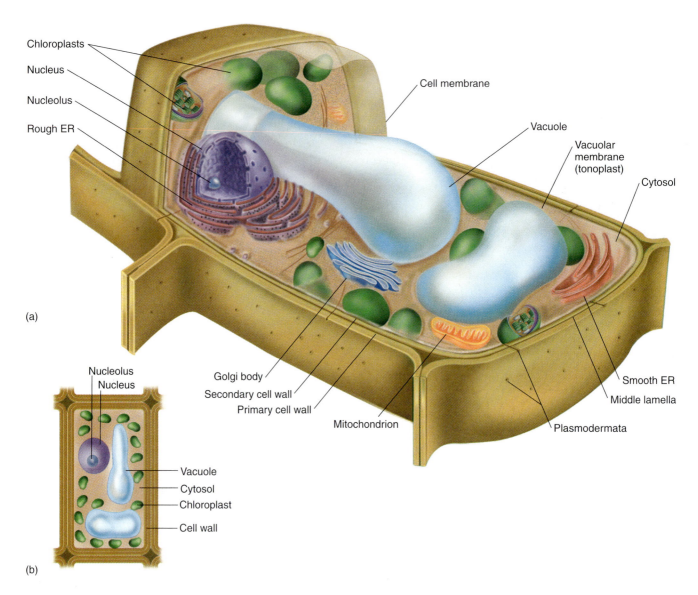

Chloroplasts
Nucleus
Nucleolus
Rough ER
(a)

Cell membrane
Vacuole
Vacuolar membrane (tonoplast)
Cytosol
Smooth ER
Middle lamella
Plasmodermata
Mitochondrion
Primary cell wall
Secondary cell wall
Golgi body

Nucleolus
Nucleus
(b)

Vacuole
Cytosol
Chloroplast
Cell wall

F I G U R E   4 . 2

(a) A leaf cell enlarged to show microscopic features. The nucleus of the enlarged cell would be about 10 micrometers in diameter, and the cell itself would be about 60 micrometers long. (b) A leaf cell diagrammed with the aid of a light microscope.

cellular activities. For example, hormones (see chapter 6), as chemical messages, received by the cell membrane can initiate a series of reactions that cause the cell to enlarge. The cell membrane accepts packets of raw materials from other membranes inside the cell and directs the assembly of these materials into cell walls, outside the protoplast. The detailed structure and function of membranes are discussed later in this chapter.

Membranes also surround and form the exterior of organelles. Thus, these membranes form the compartments—that is, the organelles—and maintain specific environments inside the membrane (see fig. 4.1). Different organelles often contain enzymes and chemicals that are necessary for the specific functions of the organelle. A chloroplast containing the enzymes specialized for photosynthesis is an example. Compartmentation provided by organelles results in increased efficiency of the metabolic reactions and prevents interference from enzymes and chemicals outside the organelle. For example, an enzyme inside an organelle might be sensitive to, even inhibited by, some chemical in another part of a cell. Inside an organelle, the sensitive enzyme is sequestered from the potential inhibitor, and the cellular activities remain efficient. Compartmentation allows many different kinds of reactions to occur in a cell simultaneously without interfering with each other. In this sense, a eukaryotic cell functions

4.1 *Perspective* Perspective
Perspective

Cell Evolution

Biologists currently separate all organisms in nature into two structurally distinct groups based on their cell structure—**prokaryotes** and **eukaryotes.** Prokaryotes are the bacteria, including the photosynthetic cyanobacteria, and they lack membrane-bounded organelles such as a nucleus. Eukaryotes include plants, fungi, and animals, and their cells contain membrane-bounded organelles, including a nucleus. This grouping is a fundamental concept in biology and provides a framework for modern classification and, therefore, an understanding of the evolution of organisms.

Some 3.8 billion years ago, the original prokaryotes evolved as the first known life on earth. They were small, single-celled bacteria that lacked a nucleus or other membrane-bounded organelles. Today, because of their ability to adapt and evolve in almost all environments, prokaryotes are ubiquitous and very successful; that is, bacteria live just about everywhere and are doing quite well as a group. Is it the age of bacteria, not mammals? Think about it. There are more bacteria on and in your body than there are people on earth.

About 2.7 billion years ago, eukaryotic cells evolved from their ancestor prokaryotes. Considerable evidence suggests smaller bacteria were engulfed by larger ones and, instead of being broken apart, formed the organelles, such as chloroplasts and mitochondria, characteristic of eukaryotes. In fact, the term eukaryote can be explained as meaning "true nucleus." The cells of eukaryotes contain a nucleus and other organelles, while cells of prokaryotes do not have a nucleus or other organelles.

Bacteria, which some now divided into *eubacteria* (bacteria) and the older *archaebacteria* (archaea), are the known prokaryotes. Many human diseases, such as bubonic plague, cholera, ulcers, typhoid, and strep throat, can be caused by bacteria. Eukaryotic cells are those of fungi, algae, protozoa, animals, and plants—the so-called "higher organisms." In fact, all true multicellular organisms are eukaryotes and so are many single-celled ones. The evolution of eukaryotes allowed the divergence of organisms into the vast diversity of the plant world and the animal world, including you.

The Lore of Plants

Every cookbook describes how to prevent browning of fruits and vegetables before they are canned or frozen. When fruit and vegetable tissues are exposed to air, browning is caused by the enzyme-catalyzed oxidation of phenolics in damaged cells. The enzymes (phenolic oxidases) and the phenolics are kept in separate organelles inside intact cells. When damaged, the organelles leak the enzymes and phenolics, and browning occurs. The chemical reaction that causes browning can be stopped by covering the freshly cut tissue with lemon juice or vinegar, which contain organic acids. These organic acids inhibit the browning reactions catalyzed by phenolic oxidase enzyme.

like a highly organized factory, with a division of labor that promotes efficiency and productivity.

The nucleus is surrounded by a double-membrane envelope and contains genes that control the activities of the plant cell.

The nucleus is surrounded by a double membrane called the nuclear envelope, which has pores to allow passage of materials in and out of the nucleus (fig. 4.3). A double-membrane envelope is two membranes that are separate but very close together like the two separate paper sheets of an envelope. The nucleus contains most of a cell's DNA (chloroplasts and mitochondria contain some DNA), which occurs with proteins in threadlike structures called **chromosomes.** Chromosomes contain genes. **Genes,** which are sequences of DNA, direct most of the activities of the cell via the synthesis of RNA. All types of RNA are manufactured in the nucleus using DNA as a pattern or template. The template serves as a molecular blueprint for the assembly of the nucleotides into the RNA molecule. After initial assembly, all RNA molecules are structurally modified inside the nucleus and then exported to the cytosol, where they direct protein synthesis in the cell. Chapter 5 is devoted to DNA and its functions.

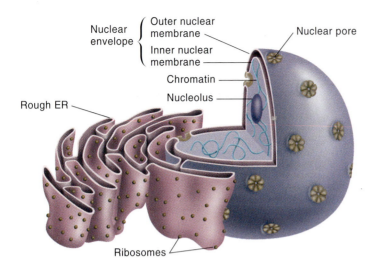

FIGURE 4.3

The nuclear envelope has pores that link the cytoplasm with the inside of the nucleus. The outer nuclear membrane is continuous with the ER. Chromatin is DNA plus protein that together constitute chromosomes.

The internal membrane system is a continuum of organelles inside the cell.

Many membranes of a cell are connected either by direct contact with each other or by the exchange of membrane segments. These interconnected membranes function as a membrane system that includes the cell membrane and the internal membranes of the **endoplasmic reticulum (ER), dictyosomes (Golgi),** nucleus and perhaps the **vacuole** (see fig. 4.2). Many biochemical processes occur in or on membranes, including synthesis of polymers to make more membranes or cell walls.

Much of the membrane surface area inside cells occurs in the ER, which is an important site of polymer biosynthesis. As discussed in chapter 3, biosynthesis is chemical assembly of larger molecules (often polymers) from smaller ones inside a cell. When shown in three dimensions, the ER is a system of flattened tubes and sacs (fig. 4.4) that are continuous between the cell membrane and the outer membrane of the nuclear envelope. Internal compartments of the ER, where biosynthesis occurs, are loaded with enzymes.

Two regions of the ER can be distinguished in electron micrographs. One region is called the **rough ER** because of many attached **ribosomes** that give it a rough appearance. Ribosomes have been called the workbench of protein synthesis because amino acids are assembled into proteins there. Rough ER is a major region of protein synthesis in a cell. In contrast, the second region is called **smooth ER** because it has no ribosomes attached to it. Smooth ER helps make phospholipids and assemble new membranes. Both types of ER form small, membrane-bounded organelles called vesicles that break away and fuse with other membranes. Vesicles are packages containing proteins and other molecules that are products of the particular type of ER. ER probably was a selective advantage in the evolution of cells because it provides a concentration of enzymes within a

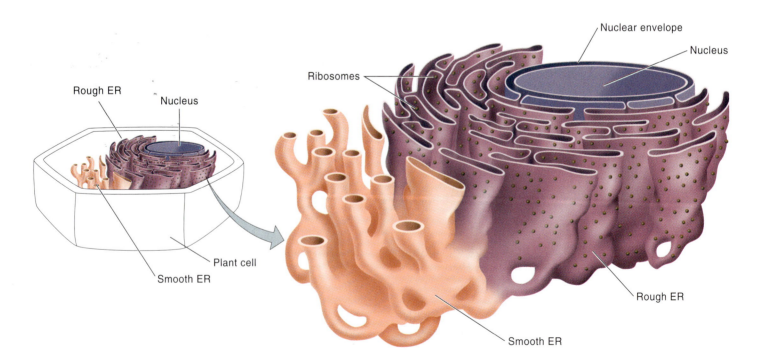

FIGURE 4.4

The three-dimensional system of flattened tubes and sacs of the ER. Ribosomes are the site of protein synthesis.

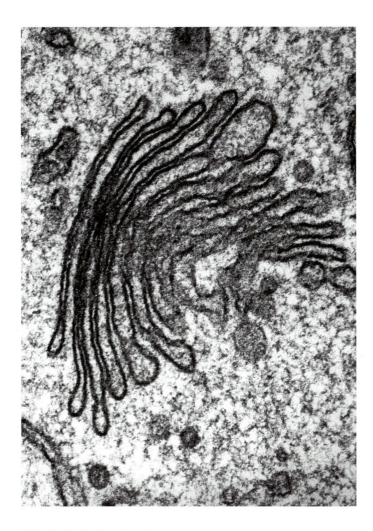

FIGURE 4.5
Electron micrograph of a dictyosome (×75,000). Note the vesicles near the dictyosome.

relatively small compartment resulting in efficient, organized, cell-activities and metabolism. In general, compartmentation is a selective advantage to eukaryotic cells.

Stacks of flattened vesicles are called **dictyosomes,** or sometimes Golgi bodies (fig. 4.5). Dictyosomes receive proteins, lipids, and other material from the ER via vesicles. In the dictyosomes, the materials may be chemically modified and then sorted into separate export packets again called vesicles.

Vesicles that move to the cell membrane (secretory vesicles) fuse with it and secrete their contents to the exterior of the cell. For example, secreted polysaccharides, such as cellulose, become part of the cell wall. Other substances are secreted from specialized cells and tissues via dictyosomes. These substances include nectar that flowers secrete as an attraction for animals and the oils or other chemicals secreted from the glands on leaves and stems of mints and many other fragrant plants.

Plasmodesmata, which are membrane-lined channels through the walls of adjacent cells, also contain ER. These plasmodesmata channels form a continuous internal membrane between plant cells (fig. 4.6). Materials can move from cell to cell through plasmodesmata.

Vacuoles help regulate the water content of plant cells and contain numerous enzymes and pigments.

Vesicles from the ER and dictyosomes often fuse to form a larger membrane-bound sac called a vacuole (fig. 4.7). A mature plant cell typically has one large vacuole that can occupy up to 95% of the cell's volume. The membrane of the vacuole has its own name, the tonoplast. As plant cells grow, most of their enlargement results from the absorption of water by vacuoles. This absorption of water by the vacuole expands and pushes the rest of the cell's contents into a thin layer against the cell wall. Water-filled vacuoles create pressure, called **turgor pressure,** on the cell walls, which contributes to rigid structure of the cell and, therefore, the plant. When a plant receives too little water, turgor pressure decreases and the plant wilts. An analogy would be a car tire. When fully inflated, it is hard and stiff. When deflated, the tire is soft and flexible.

We can see the effects of turgor pressure by letting carrots or celery dry out and become soft and flaccid; we can make them crisp again by putting them in water. This reacquired crispness (turgor) is caused by vacuoles that have refilled with water. The crispness is revealed when you bite into carrots or celery. The snap you hear is the release of turgor pressure.

Vacuoles are versatile organelles, as indicated by the diversity of substances that occur in them in addition to water. Certain vacuole enzymes can digest storage materials and components from other organelles for recycling into the cytosol. Pigments in vacuoles, especially red and blue anthocyanins, impart bright colors to flowers, fruits, and other plant parts. Some plants harbor toxic chemicals in their vacuoles. These chemicals may deter insects and other animals from eating the plants that contain them.

Ribosomes are the workbench of protein synthesis.

Ribosomes are very small and consist of approximately equal amounts of protein and RNA. The RNA in ribosomes is called **ribosomal RNA (rRNA).** Ribosomes are essential in cells because they are the site of protein synthesis. It is in ribosomes where amino acids are brought together in just the right order and assembled into various proteins.

Ribosomes may be attached to the ER (making it rough ER as explained above) or remain unattached (free ribosomes) in the cytosol. During protein synthesis, ribosomes occur in clusters attached to a single molecule of messenger RNA that directs protein synthesis (discussed in

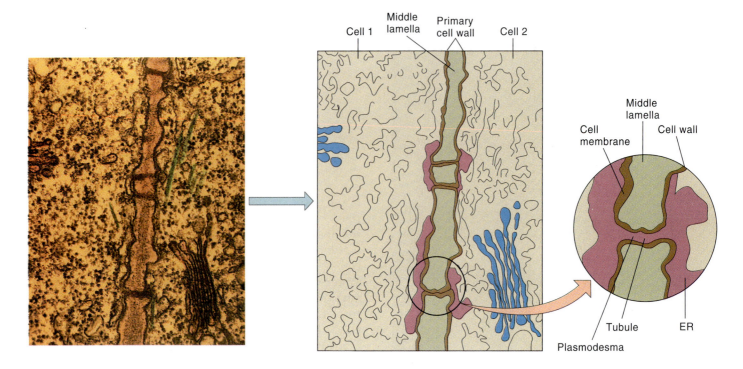

FIGURE 4.6

This transmission electron micrograph (×17,000) and accompanying drawing of a primary cell wall show the plasmodesmata (which is lined by the cell membrane) and a channel that connects the ER between adjacent cells.

chapter 5). Unlike the nucleus and other organelles, ribosomes are not surrounded by membranes, and, therefore, some biologists do not consider them organelles. All organisms contain ribosomes in their cells. What does this suggest about their importance in the evolution of cells?

Mitochondria and chloroplasts are organelles of energy conversion.

Two kinds of plant organelles, **chloroplasts** and **mitochondria** (see fig. 4.2), convert most of the energy needed for cellular metabolism. Chloroplasts are the site of photosynthesis in plant cells. A general equation for photosynthesis is

$$CO_2 + H_2O + light\ energy \rightarrow sugars + O_2$$

In chloroplasts, the green pigment chlorophyll traps the light energy that is used to fix (capture and incorporate) carbon dioxide (CO_2) into sugars, primarily sucrose and starch. Chloroplasts also use light energy to make amino acids and fatty acids from carbon fixed via photosynthesis.

Mitochondria have been called the powerhouses of the cell because these are the organelles where the energy stored in the sugars produced during photosynthesis is converted to ATP, which powers much of the work of the cell. This release of energy from sugars to ATP is called cellular respiration, shown by the general equation:

$$sugars + O_2 \rightarrow CO_2 + H_2O + ATP$$

Cells thrive on the energy of ATP. Recall from chapter 3 that ATP is important inside cells because it can power many individual steps in the work of the cell. ATP is used by cells to grow and divide, and to synthesize proteins, lipids, carbohydrates and nucleic acids, to move molecules between cells and to power energy-requiring chemical reactions.

Chloroplasts and mitochondria contain a small amount of their own DNA, RNA, and ribosomes that synthesize some of the enzymes specific to each organelle (nuclear DNA provides instructions to make additional enzymes). Finally, chloroplasts and mitochondria grow and divide in the cell on their own (see Perspective 4.2, "Cellular Invasion"). We shall focus on these two important organelles in chapters 10 and 11.

The cytoskeleton forms a network within the cell.

The **cytoskeleton** is a network of three kinds of very small protein filaments that form a mechanical support system in a cell (fig. 4.8). These filaments run throughout the plant cell like a web or network and are so tiny they can be seen only with the most powerful electron microscopes. We have not shown the cytoskeleton in our model of the cell (see fig. 4.2) because the diagram would become too cluttered. The cytoskeleton filaments probably have many

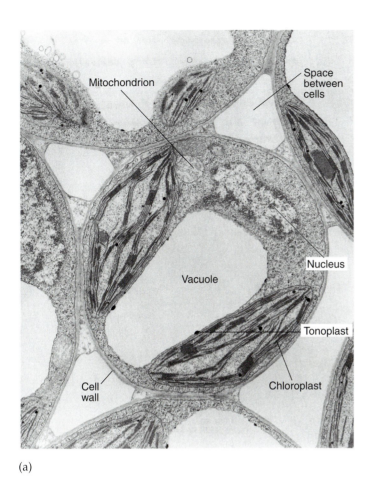

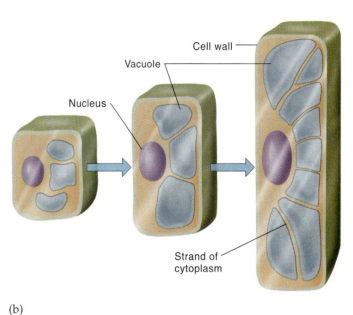

(a)

(b)

FIGURE 4.7

Vacuoles. (*a*) Transmission electron micrograph from a coleus leaf (*Coleus blumei*). The expanding central vacuole compresses the rest of the cell contents into a small space against the cell wall (×20,000). (*b*) Schematic diagram of cell growth, showing how an increase in cell volume can occur without a large increase in the amount of cytoplasm. The peripheral layer of cytoplasm may be interconnected through the vacuole by cytoplasmic strands that radiate from the region of the nucleus.

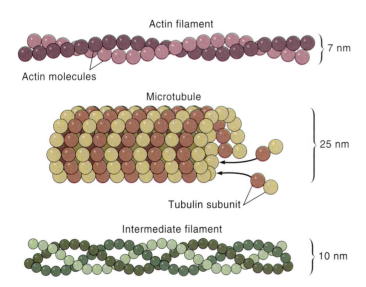

FIGURE 4.8

Structural models of the three kinds of cytoskeletal filaments.

roles in cells. They help maintain organelle position and organization within the cell, direct cell expansion, and control the movement of chromosomes during nuclear division. Groups of filaments may form channels for transporting large molecules within the cell. Much remains to be learned about functions of the cytoskeleton in cells. Like ribosomes, no membrane forms the cytoskeleton and some botanists do not consider it an organelle.

INQUIRY SUMMARY

Plant cells are bordered by a cell membrane that regulates the flux of materials into and out of the cell interior. Inside the cell membrane is a complex group of highly structured, functional organelles that give the cell an efficient division of labor. These organelles include the nucleus, vacuoles, chloroplasts, and mitochondria The cytoskeleton forms a slender network inside a cell.

How were organelles important in the evolution of plant cells?

4.2 *Perspective**Perspective*

Cellular Invasion: Evolution of Chloroplasts and Mitochondria

Chloroplasts and mitochondria are similar to bacteria: they are about the same size, they contain DNA arranged in circular form, they have ribosomes that are smaller than those made in nuclei and found attached to ER, and they have a common gene structure that differs from that of nuclear genes. Furthermore, the inner membranes of chloroplasts and mitochondria are similar to the cell membranes of bacteria. Protein synthesis in chloroplasts and mitochondria is inhibited by antibiotics that have similar effects on ribosomes in bacteria but not those made in nuclei. Such similarities between these organelles and bacteria are evidence for the **endosymbiotic hypothesis** of the origin of chloroplasts and mitochondria. According to this hypothesis, small bacteria invaded or were engulfed by larger bacteria. The smaller bacteria, now represented by chloroplasts and mitochondria, developed symbiotic relationships with their larger hosts. The mitochondrial precursors may have provided aerobic (oxygen using) metabolism for a host that was probably unable to use oxygen. The pho-

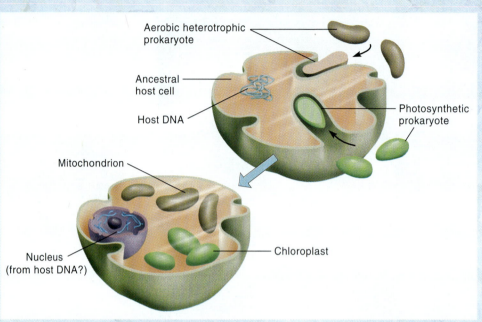

Model of cellular invasion to illustrate the endosymbiotic hypothesis of the origin of chloroplasts and mitochondria. The origin of the nucleus is not explained by this hypothesis.

tosynthetic precursors of chloroplasts provided the ability to make carbohydrates.

Modern organisms provide good clues about the bacterial relatives of chloroplasts and mitochondria. For example, one extant photosynthetic bacterium, *Prochlorothrix*, contains chloro-

phyll *b*, which is otherwise known only in plants and some algae. Other photosynthetic bacteria lack this pigment. While there is considerable evidence for the evolution of chloroplasts. The origin of the nucleus has not been explained by this hypothesis.

Cell Walls Surround and Shape the Plant Cell.

Under a microscope, often the most easily observed part of a plant tissue is the cell wall (see fig. 4.2). In some tissues, such as the cork in bark, the protoplast is gone and only a relatively thick cell wall remains. Cell walls are dynamic products of cells that can grow and change in shape and composition. Without cell walls, plant evolution, if it occurred at all, would have been very different.

Young cells and cells in actively growing areas of plants have primary cell walls that are relatively thin and flexible (see figs. 4.2 and 4.10). Examples of such cells include the dividing cells at tips of roots and shoots. The primary wall, which is produced early in the life of a new plant cell, is usually less than 25% cellulose, the remainder being hemicelluloses, pectins, and glycoproteins (proteins with a sugar molecule attached). Many cells surrounded only by a primary cell wall can change shape, divide, or differentiate into other kinds of cells. Some primary cell walls also contain small amounts of the polymer lignin.

Certain kinds of plant cells stop growing when they reach maturity. When this occurs, these cells form a secondary cell wall inside a primary cell wall (fig. 4.10). The secondary cell wall is more rigid than the primary cell wall and therefore functions as a strong support structure. Although cellulose is one of their main compo-

FIGURE 4.9
Scanning electron micrograph of a water-conducting vessel element, showing secondary wall with pitting (×200). Note the protoplast is absent.

nents, the secondary walls surrounding cells in woody stems are up to 25% lignin, which adds hardness and resists decay. Largely because of its lignin content, wood is one of the strongest materials known. Secondary cell walls in other plant tissues and organs may become lignified also, adding strength to this part of the plant. Unlike primary cell walls, secondary cell walls are rigid; they do not expand and grow. Many types of cells that produce secondary cell walls die when they reach functional maturity, and the wall is left empty of the protoplast (see fig. 4.9).

Some cell walls, such as those of cork cells, are coated with suberin—the waxy substance described in chapter 3. Recall that suberized tissues block water loss through bark, which is why cork from the cork oak (*Quercus suber*) is useful in making stoppers for wine bottles (see Perspective 8.1, "Cork"). Suberized tissue is also why tree trunks do not dry out.

Cells that adjoin one another are probably held together by pectins, which form the pectic layer between cells, and is called the middle lamella (fig. 4.10). Some tissues, such as the flesh of apples, are so rich in pectins that

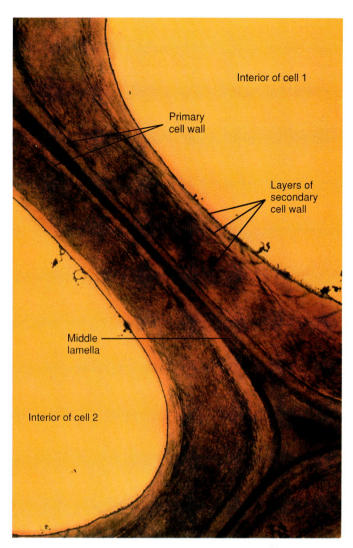

Interior of cell 1

Primary cell wall

Layers of secondary cell wall

Middle lamella

Interior of cell 2

FIGURE 4.10
Transmission electron micrograph of cell walls. The primary wall is constructed when the cell is young. Thicker secondary walls are added later when the cell stops growing (×3,000). What is the selective advantage of lignin in the secondary walls?

these polysaccharides are extracted for use as thickening agents in making jams and jellies.

INQUIRY SUMMARY

Plant cells produce a rigid wall that surrounds and protects the cell protoplast, which consists of the cell membrane and its contents. Young cells produce a relatively thin primary cell wall, while older cells may produce a thicker, lignin-containing, secondary cell wall.

What is the evolutionary advantage of secondary cell walls?

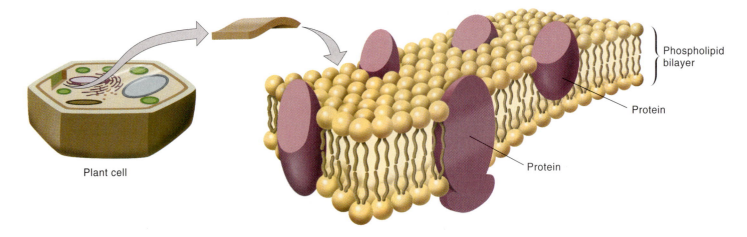

FIGURE 4.11

The fluid mosaic model of membrane structure. According to this model, proteins are dispersed in the phospholipid bilayer.

Membranes Regulate the Flux of Materials into and out of the Cell and Organelles.

As discussed, membranes surround cells, connect cells to one another, form complex internal networks, send and receive chemical signals, and divide cells into distinct compartments. The abundance of membranes in cells underscores their importance, but it does not begin to show the diversity of their functions. Membranes are active and changeable participants in cellular metabolism: virtually everything that happens in a cell is directly or indirectly associated with a membrane.

Membranes consist primarily of proteins and phospholipids as represented in the **fluid mosaic model** shown in figure 4.11. The backbone of membranes is the double layer of the phospholipids. Proteins are involved in the many functions of membranes. Membranes also contain other lipids such as sterols that influence the fluidity of the membrane.

Proteins control most membrane functions, but the lipid layer is important also.

There may be 50 or more different kinds of proteins in a cell membrane and perhaps as many in the membranes that surround organelles. The proteins may be on the surface or they may extend through the lipid bilayer made primarily of phospholipids, as explained previously. This diversity of proteins is reflected in the enormous range of activities associated with membranes:

§ The cell membrane generally allows the unrestricted movement of water and certain small, dissolved substances into or out of the cell. Water balance is crucial for maintaining turgor pressure (which makes the cell

rigid) as it drives cellular expansion during growth, and it is also important in metabolism.

§ Membranes can control or block the passage of some kinds of molecules; such membranes are referred to as selectively permeable membranes. Certain ions, such as potassium ions (K^+) and hydrogen ions (protons, H^+), are actively pumped through membranes. Ion pumps in the cell membrane are proteins that use energy from ATP to move or pump ions from the cell, while ion pumps in mitochondrial and chloroplast membranes are important for making ATP. Enzymes that cooperate in multistep processes, such as ATP synthesis or the absorption of light energy, often occur together in a particular spot on a membrane.

§ The cell membrane contains proteins that bind molecules, such as hormones, released from other cells in the process of **cell signaling.** Once bound to an external molecule, these proteins can activate other proteins in the membrane that cause metabolic changes in the cell. For example, some hormones act by binding to the cell membrane proteins, often causing dramatic changes in the metabolism and structure of the cell, including cell division.

Phospholipids form the thin bilayer of membranes.

A lipid bilayer, which is two sheets of phospholipids sandwiched together, serves as the basic membrane structure (fig. 4.11) and is a permeability barrier that blocks movement of most molecules. Because of the nature of the lipids present, the bilayer can be thought of as a fatty, fluid film with proteins attached. Fluidity, or movement of molecules within the membrane, is important for membrane function. Fluidity allows proteins to move about in the bilayer and interact with one another. This is critical for cell signaling and allows membranes to fuse. Membranes also have carbohydrates associated with them. For example, sugars are often attached to proteins (glycoproteins; *glyco* means "sugar") on the exterior surface of the membrane and often facilitate cell signaling.

Diffusion, convection and bulk flow account for much of the movement of molecules in plants.

All molecules display random thermal motion. Thus, a solute molecule has a tendency to move around in a solution. **Solute** is a molecule dissolved in solution; examples of solute include protons (H^+), mineral ions such as potassium (K^+) and magnesium (Mg^{2+}), or small organic compounds such as some sugars and amino acids. Salt water is a solution of salts dissolved in water which is the solvent. One result of the random movement of dissolved molecules (solute) is that they diffuse outward from regions of high concentration to regions of lower concentration. **Diffusion** is the net movement of molecules from an area of greater concentration to an area of lesser concentration. Diffusion by random movement continues until the distribution of molecules becomes even throughout the solution (fig. 4.12). The rate of diffusion depends on (1) the size of the molecules (larger molecules move slower), (2) the temperature of the solution (higher temperatures cause faster movement), and (3) the solubility of the molecules in the solvent (molecules that do not dissolve do not diffuse). However, diffusion, with no other forces acting on dispersal, is very slow.

Molecules in solution and water also spread via convection or bulk flow when pressure (e.g., gravity, mixing) or temperature is involved. This phenomenon is easily illustrated by placing a drop of a dye into a beaker of water (fig. 4.12), or by opening a bottle of perfume, or by running water through a garden hose.

Dispersal results when molecules move down a concentration gradient, from greater to lesser concentration. We can think of solutes as marbles and the gradient as a hill; a bunch of marbles rolling down a hill would be analogous to solutes moving down a concentration gradient. Furthermore, a steeper hill would be analogous to the steeper gradient caused by a higher concentration of solutes. Thus, a steeper gradient causes a higher net rate of solute movement. Just as marbles can move up the hill if there is a force to push them, so too can solutes move from lower to higher concentrations if there is energy to push them. As you will see later in this chapter, cellular energy is often used for moving solutes "up" their concentration gradients.

Water potential can be used to predict the flow of water through a plant.

Like solutes, water also has potential energy to flow to where it is less concentrated. The potential energy of water has a special name: **water potential.** Water tends to move down a water-potential gradient—that is, from a region of high water potential to a region of low water potential. By general agreement, the water potential of pure water is zero. This means

FIGURE 4.12

Beakers of water before and after the addition of the dye bromthymol blue. Convection and random movements of water and dye molecules cause bulk flow and diffusion, eventually resulting in a uniform concentration of the dye. Note: If the dye could be placed in the water without any convection, dispersal would be quite slow compared to this demonstration.

that the water potential of a solution has a negative value because the water is less concentrated than in pure water. Also by general agreement, water potential is expressed in units of pressure. Thus, water potential may be expressed in bars or, more recently, megapascals (MPa). One bar equals atmospheric pressure at sea level and room temperature (25°C), and 0.1 MPa is 1 bar. The potential of seawater, because it contains so much salt, is about −25 bars, the water potential in the cells of a freshly watered plant may be more than −1 bar (meaning closer to zero), and the water potential in dry seeds is about −200 bars. Thus, water moves into dry seeds when they are soaked. For comparison, the recommended air pressure for most car tires is about 2 bars.

Three key points: (1) the water potential of pure water is zero; (2) water flows down a water potential gradient, toward the more negative water potential; and (3) the addition of solutes to a solution lowers the water potential.

Osmosis is the diffusion of water across a membrane.

The diffusion of water through a selectively permeable membrane has a special name: **osmosis.** To appreciate the importance of osmosis, consider a membrane that is permeable to water but impermeable to glucose. When such a membrane separates two halves of a container, each having a different concentration of glucose, water diffuses by osmosis into the side having the higher glucose concentration and the lower water concentration (fig. 4.13). The net movement of water stops either when both sides of the container have the same concentration of glucose and water or

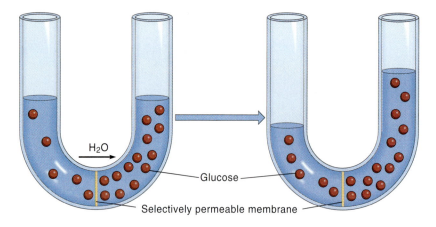

F I G U R E   4 . 1 3

Osmosis is demonstrated by the movement of water through a selectively permeable membrane in a U-tube. Glucose, which cannot pass through the membrane, is more concentrated in the right-hand part of the tube than in the left-hand part. Net movement of water into the more concentrated glucose solution causes the volume of the right-hand solution to increase. In this model, glucose is the solute and water is the solvent. Because glucose dissolves in water, we say this is a glucose solution.

when the force of gravity equals the force of water movement, whichever occurs first. Note that the side that began with a higher concentration of glucose increases in volume. This relative increase in volume can contribute to the movement of water and solute through a plant.

This example of osmosis hints at another feature of the process. Immediately before osmosis begins, the dilute solution in the tube on the left side of the membrane has a higher water and lower glucose concentration compared to the lower water but higher glucose concentration on the right side of the membrane. The concentration difference means water on the left side (fig. 4.13) has a greater potential to move across the membrane. Its movement, however, can be prevented if a piston is placed on the right side with just enough pressure to keep the volume constant (fig. 4.14). A pressure gauge on the piston measures the force required to maintain a constant volume. This pressure, which is called **osmotic pressure,** is a measure of the ability of osmosis to do work in a system, such as moving water through a plant.

Most plants and indeed plant cells are surrounded by an environment with a lower concentration of solute and higher concentration of water. This difference in solute concentrations results in water entering a plant, which usually has a lower water potential than the surroundings. The cells and the plant absorb as much water as they can hold. Remember that the outward pressure of the cell membrane against the cell walls is called turgor pressure because it keeps the cells turgid (fig. 4.15).

Turgor pressure is vital to plants in many ways. During growth, cell expansion is caused by turgor pressure on cell walls that have become relaxed. Turgor pressure also keeps herbaceous (nonwoody) plants upright, supports the fleshy stalks and leaves of trees and shrubs, and keeps supermarket

vegetables crisp when they are sprayed with water. Changes in turgor also cause movements in plants, such as the opening and closing of stomata and the curling of grass leaves. Some movements caused by changes in turgor are dramatic, such as leaf movement in the sensitive plant (see chapter 6).

Cells lose turgor when they are placed in a dry environment or high-salt solution. The continued loss of turgor causes the cell membrane and protoplast to shrink away from the cell wall (fig. 4.15). Dryness causes most loss of turgor in plants.

Osmotically induced shrinkage of the cytoplasm is called plasmolysis. This phenomenon occurs in crop and garden plants when salt accumulates in the soil from extensive use of hard (i.e., mineral-rich) water. It also occurs when people apply too much fertilizer, causing a high concentration of salt outside the plant. As a result, water exits the plant via osmosis and diffuses to a region of lesser water concentration (i.e. water potential) compared to inside the plant. The loss of turgor in these plants causes their leaves and stems to wilt. Key to turgor pressure is the cellular import and export of molecules other than water. This occurs via membrane transport.

Selectively permeable membranes use transport proteins to regulate the flux of molecules.

As mentioned, biological membranes are selectively permeable. This is one of their most important properties, because it keeps metabolically important substances inside the cell or

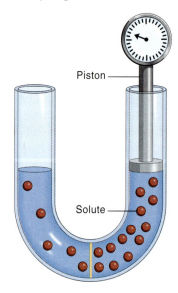

F I G U R E   4 . 1 4

Higher water potential in the left-hand solution causes water to move into the right-hand solution. The amount of pressure that is necessary to maintain constant volume on the right equals the force of water movement across the membrane.

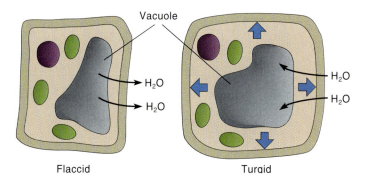

FIGURE 4.15

Flaccid and turgid cells. The plumpness of the turgid cell is maintained by turgor pressure. Note the difference in the size of the vacuole. The arrows show water movement.

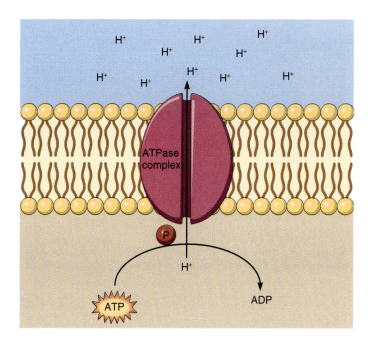

FIGURE 4.16

Active transport uses energy released by splitting ATP via ATPases in the membrane. This energy is used to transport ions (H^+) against their concentration gradient. Phosphate (P) from ATP binds to ATPases during the process.

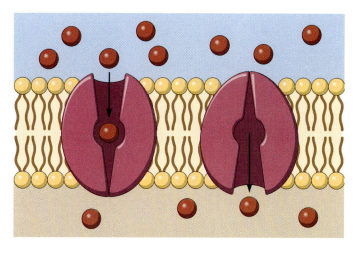

FIGURE 4.17

A possible mechanism for facilitated diffusion by passive transport. The transport protein (*purple*) accepts specific solute molecules (*red spheres*) on one side of the membrane and releases them on the other side. The transport protein alternates between two forms, depending on whether it is "open" to one side of the membrane or the other. Transport proteins may also form channels through the bilayer that allows certain ions to flow down a gradient.

organelle and prevents inappropriate or toxic substances from entering. Membranes also enable ions and larger polar molecules, such as sugars, to pass into the cell through specific membrane proteins called **transport proteins.**

There are two main types of transport proteins in membranes. One type includes proteins that work by **active transport,** which requires energy from ATP to move solutes up a concentration gradient from lower to higher concentrations (fig. 4.16). The other type includes facilitated diffusion via **passive transport** proteins, which do not require metabolic energy (fig. 4.17). During passive transport, proteins merely act as carriers or channels for the dif-

fusion of certain molecules down their concentration gradients. Each passive transport channel is specific for one or two solute molecules. In contrast to simple diffusion through the phospholipid bilayer, passive transport via membrane proteins is called facilitated diffusion because movement of the diffusing molecule is assisted (facilitated) by a membrane protein. Sugars typically move by facilitated diffusion through channels in a membrane formed by transport proteins.

Active transport allows plants to accumulate molecules in higher concentrations than would be available through diffusion or passive transport alone. In turn, this allows plants to take in more water from the soil in response to the higher concentration of solute inside the plant than outside in the soil. Thus, active transport allows plant roots to accumulate high concentrations of molecules from dilute solutions in the soil and to acquire water from the soil.

Although water moves freely across biological membranes in response to concentration differences, control of osmosis occurs by regulating the concentrations of solutes inside cells. For example, loss of turgor (from loss of water) can stimulate cells to actively transport potassium ions (K^+) to the inside. As the concentration of K^+ increases inside the cell, the relative concentration of water decreases, and more water enters the cell via osmosis. Active transport is extremely important in the physiology of plants and other organisms, even you, because it allows cells to accumulate ions in high concentrations as needed.

Simple diffusion, passive transport (facilitated diffusion), and active transport are all direct movement of substances through membranes. Large molecules also can

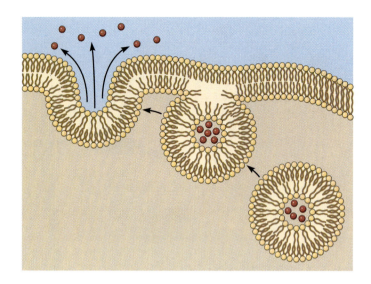

FIGURE 4.18
Exocytosis transports large molecules out of cells. This kind of transport involves membrane-bounded vesicles that fuse with the cell membrane. The vesicles usually come from the dictyosomes. Cell walls may be manufactured in this manner

move across membranes by a process called **exocytosis** (fig. 4.18). For example, via exocytosis, plant cells can export polysaccharides and proteins across the cell membrane for assembly into cell walls. Moreover, cells of root tips secrete a slimy polysaccharide that lubricates their passage through soil as they grow, and cells covering leaves exude waxy substances onto their surfaces to inhibit water loss. Leaves of the Venus's flytrap and other insectivorous plants secrete enzymes that digest insects (see fig. 6.7).

Membranes are important in cell-to-cell signaling in plants.

Cells in a complex organism interact with their environment, with one another, and with the cells of other organisms. Cell-to-cell interactions occur when chemical or electrical signals released from one cell are received by another, and therefore change some aspect of metabolism. Auxin, the plant hormone, is an example of an internal chemical signal—that is, a signal that moves from cell to cell in the same plant (fig. 4.19). For example, auxin may signal cells to elongate or differentiate. External signals are those that pass between different organisms—for example, between plants and bacteria or fungi that may trigger growth responses.

Each plant is surrounded by other organisms, including animals, bacteria, fungi, and other plants. Interactions are common among plants and many of the organisms in their environment. Most of our knowledge about these kinds of interactions concerns the effects of microbes that cause plant diseases. Can you develop an explanation of the selective advantage of mutual signaling between plant roots and soil bacteria?

Auxin (an internal signal from shoot tip to stem cells below)

Polysaccharides (external signals from soil microbes to roots)

FIGURE 4.19
Auxin is an example of an internal signal between cells. Microbial polysaccharides may function as a signal from bacteria to roots.

To fight infections caused by microbes, the cell membranes of many plant cells can send signals into their own cytosol that induce a defensive response. For example, polysaccharides secreted by the fungus *Colletotrichum lindemuthianum* cause cells in kidney-bean leaves to make several different kinds of chemicals that inhibit the growth and spread of the fungus. Many companies that manufacture pesticides are conducting research to better understand these chemical and cellular mechanisms in plants.

Cellular recognition is perhaps best understood among plants in the legume family and bacterial species of the genus *Rhizobium*. For example, specific molecules on the surface of the roots of white clover (*Trifolium repens*) bind only to the cell walls of the bacterium *Rhizobium trifolii*. Once bound to the root, these bacteria infect the roots and cause them to swell into nodules (fig. 4.20). The interaction between root-nodule bacteria and plants is mutually beneficial because the organisms trade useful molecules: the plants

The Lore of Plants

Plant substances can be used to determine your blood type. Lectins are proteins from jack beans, lima beans, and lotus that bind to glycoproteins on the cell membranes of red blood cells. Because the cells of different blood types have different glycoproteins, cells of each blood type (A, B, O, AB) bind to a specific lectin. This is one example of the many clinical and research applications of plant lectins in human medicine. What is the role of lectins in a plant? Actually, no one knows, but it certainly is not to type blood. Perhaps they have a role in cell signaling.

FIGURE 4.20
Roots of pea plants with bacteria-containing nodules (×2). In these nodules, atmospheric nitrogen is converted to ammonia via a process called nitrogen fixation. The infection of roots by bacteria is enabled by root-hair molecules that bind to bacterial cell walls.

supply carbohydrates for bacterial growth, and the bacteria convert nitrogen from the air into ammonia, a metabolically useful form of nitrogen for plants. This conversion process is called **nitrogen fixation** (see chapter 7), and it is the main mechanism of nitrogen input into most ecosystems.

INQUIRY SUMMARY

Molecules move across membranes by simple diffusion and by transport via selective membrane proteins. Facilitated diffusion via passive transport proteins does not involve metabolic energy and occurs when molecules move down a concentration gradient as they cross the membrane. Active transport requires metabolic energy and can move molecules against a concentration gradient using specific transport proteins. Active transport allows plants to accumulate high concentrations of molecules from dilute solutions in the soil and fosters movement of water into plants. Cells communicate with other plant cells by chemical signals. Membrane surfaces also recognize specific molecules from other organisms.

How might cell communication be important to the nitrogen economy of an ecosystem?

Plant Tissue Systems Help Us Understand the Evolution of Life on Land.

Plants evolved from algae 450 million years ago and then diversified in a way that forever changed our planet: they invaded land, thereby paving the way for the subsequent colonization of land by animals. Because the terrestrial habitats encountered by plants were more diverse than their ancestral aquatic environments, land plants faced many new environmental challenges and selective pressures. As a result, they evolved rapidly and became more complex and diverse than did their algal ancestors.

Foremost among these environmental challenges was life in air and soil instead of water. Maintaining an aquatic cellular environment (that is, plant cells filled with water) was imperative, since unprotected cells, whether of algae or land plants, desiccate and die within minutes after being exposed to dry air. Thus, the survival of plants on land depends on their ability to establish an aquatic cellular environment in the dry land environment. Establishing

4.3 *Perspective* Perspective

The Estimated Age of Eukaryotic Cells

Research scientists from the University of Sydney in Australia recently discovered that eukaryotic cells—those with nuclei—probably lived on earth some 2.7 billion years ago. This is about 1 billion to 500 million years earlier than previously thought. The scientists actually examined rocks from an ancient, underground seabed about 2.7 billion years old. By employing new approaches, they found special biological molecules in the rocks. These molecules, called steranes, are produced by eukaryotic cells with nuclei.

The discovery of eukaryotic life so early in the earth's history has many implications for evolution. For example, one possibility is that prokaryotic cells may have actually evolved earlier than previously thought. The current "old-age record" for bacteria is from rocks in Greenland that are about 3.85 billion years old. The question remains however: just when did the first life evolve on earth?

---

this environment is a complex problem and requires specializations for gas exchange, ventilation, support, water retention, and protection. Plant tissues are the evolutionary "solutions" to these problems.

Plants Consist of Four Basic Kinds of Tissue Systems: Meristematic, Ground, Dermal, and Vascular.

To understand the ecology and evolution of higher plants, you must learn about the structure and function of their main vegetative organs—roots, stems, and leaves (fig. 4.21)—introduced in chapter 2. Along with this, you should understand the structure and function of the reproductive organs. In flowering plants, the organs for sexual reproduction are flowers, fruits, and seeds; in later chapters, we focus on the relationships between structures and functions of the tissues and cells in these organs. But first, we must know something about the tissues that make up the organs to understand how the entire plant works and fits into an ecosystem.

In chapter 2, we discussed germination and development of a persimmon seedling as an example of the development of the primary body of a flowering plant. In chapter 5, we will stress a basic feature of cells: cellular division during mitosis. Following division, most of the new cells become structurally and functionally specialized. Taken together, these processes produce the primary body of a plant, which is an axis consisting of a root and shoot. Shoots consist of stems and leaves (fig. 4.21). You may want to review the parts of the primary plant body in chapter 2.

The primary body of plants consists of four tissue systems: meristems, ground tissue, dermal tissue, and vascular tissues. Tissue systems can be described as a group of common cells performing a common function or functions. Meristems are localized regions of cellular division that form a

plant's cells. Like meristems, the ground, dermal, and vascular tissues consist of distinctive types of similar cells and usually have more than one function. For example, the common cells of ground tissue may perform photosynthesis in a leaf, store starch in a root, or strengthen a stem. Dermal tissue is made of cells that absorb nutrients and protect plants from desiccation, whereas vascular tissue provides support and moves water and solutes throughout the plant. The functions of each of these tissue systems represent evolutionary adaptations to life on land. Indeed, plants are designed much like buildings: they have supporting structures, ventilating and plumbing systems, storage areas, and protective coverings. These life-support activities in plants are performed by the ground, vascular, and dermal tissues.

Meristems are the site of cell division in plants.

Growth means to permanently increase in size by a natural process. However, not all parts of a plant grow at the same time or rate. Rather, plant growth is largely restricted to specialized regions of active cell division called **meristems.** Without meristems, normal plants would not grow. Special cells called **initial cells,** which complete the cell cycle (i.e., they divide), are found in meristems.

Plants have four types of meristems (figs. 4.21 and 4.22):

1. **Apical meristems** occur near the tips of roots and shoots and produce primary tissues. These meristems account for **primary growth,** which is elongation of roots and shoots. Primary growth is important because it enables a plant to extend into new environments that may provide light, water, and nutrients. Primary growth can be powerful enough for roots to break through the concrete of sidewalks or invade pipes in the ground.

2. **Axillary buds** are "left behind" during primary growth as a stem elongates. They occur in the axil of a leaf (the angle between the leaf and stem) and typically

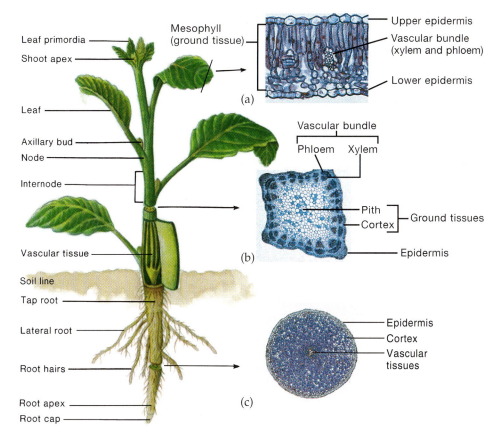

Leaf primordia

Shoot apex

Mesophyll
(ground tissue)

Upper epidermis

Vascular bundle
(xylem and phloem)

Lower epidermis

(a)

Leaf

Axillary bud

Node

Internode

Vascular bundle

Phloem   Xylem

Pith

Cortex — Ground tissues

Epidermis

(b)

Vascular tissue

Soil line

Tap root

Lateral root

Epidermis

Cortex

Vascular
tissues

Root hairs

Root apex

Root cap

(c)

FIGURE 4.21

The tissues and organs of a herbaceous plant. Cross sections of a (*a*) leaf, (*b*) stem, and (*c*) root.

that plants are a mixture of young dividing cells, maturing cells, and mature cells, all of which are derived from meristems. As a result, plants function as mature organisms while still growing. Plants are especially sensitive to damaged meristems, however, because such damage often stops growth. In shoots, meristems are protected with young leaves and by forming dormant reserve meristems (i.e., buds) that can take over if the apical meristem is damaged. Similarly, roots protect their apical meristems with a root cap. If the primary root is damaged, its function is often assumed by a lateral root that grows from inside of the main root.

Ground tissue constitutes most of the primary body of a plant.

Ground tissue occurs throughout a plant; it often is mixed with other tissue systems such as vascular tissue. The cortex and pith of stems and roots consist almost entirely of ground tissue (fig. 4.22). Ground tissue has several functions, including storage, metabolism, and support. These functions are performed by its three kinds of cells in the tissue: parenchyma, collenchyma, and sclerenchyma.

Parenchyma (from the Greek words *para,* meaning "beside," and *en + chein,* meaning "to pour in") cells are the most abundant and versatile cells in plants. They are identified by their relatively inconspicuous structure; parenchyma cells have few distinctive structural characteristics. In practice, botanists classify parenchyma as any cell not assignable to another structural or functional class. The shortcomings of this classification scheme become apparent when we consider some of the diverse functions of parenchyma cells:

§ *Storage.* Plants store nutrients in parenchyma cells. Most nutrients in plants such as corn and potatoes are contained in starch-laden parenchyma cells.

§ *Metabolism.* Parenchyma cells are often the site of metabolism in plants, including photosynthesis, respiration, and protein synthesis. Clearly, all parenchyma cells are not alike.

Parenchyma cells are made by all of a plant's meristems and occur throughout the plant body. They comprise the photosynthetic tissue of a leaf, the flesh of fruit, and the

undergo a dormant period. Axillary buds are important because they are a shoot's insurance policy: they contain inactive cells that can form a branch, a leaf, or a flower when the bud breaks dormancy and resumes growth and development.

3. **Lateral meristems** are cylindrical meristems that form in mature regions of the roots and shoots of many plants, particularly those that produce woody stems. Lateral meristems produce **secondary growth,** which increases the girth or width of the plant. Secondary growth is important because it makes a plant sturdier. In turn, the plant may grow taller and intercept more light. Wood, such as a baseball bat or the lumber used to build walls of a house, comes from secondary growth in trees.

4. **Intercalary meristems** occur between mature tissues. These meristems are most common in grasses, where they occur throughout slender leaves. Intercalary meristems are important because they help regenerate parts removed by grazing herbivores (or lawn mowers).

Localized growth in meristems has important implications for plant growth and development: it means

The Lore of Plants

In 1825, a retired brewer named Richard Cox plucked an apple from a tree while walking across an abandoned farm. He was so enamored with the apple's taste that he grafted buds of the apple tree onto other "normal" apple trees to propagate his prized discovery. Subsequent growth of these grafted buds formed stems that produced apples identical to the one he'd discovered during his walk. Similar grafts have been made thousands of times. Today, Cox's orange pippin apples can be traced to the neglected tree he discovered more than 160 years ago. Similarly, the seedless navel orange of California can be traced to a tree that today grows in Riverside, California. This tree was sent there as a bud-grafted tree discovered by a missionary in Bahia, Brazil, in 1870.

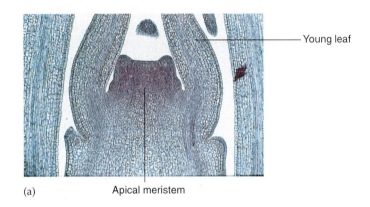

(a)

Young leaf

Apical meristem

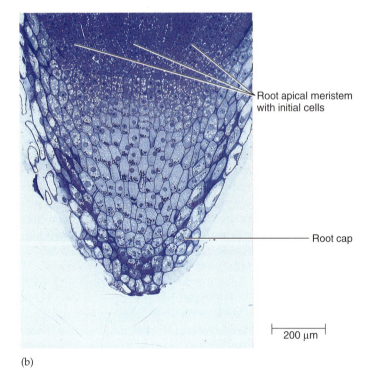

(b)

Root apical meristem with initial cells

Root cap

200 µm

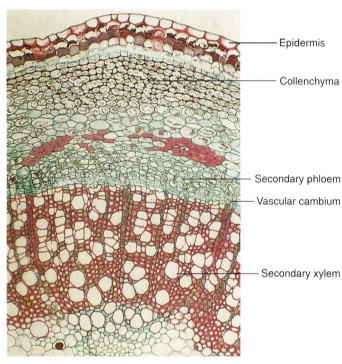

(c)

Epidermis

Collenchyma

Secondary phloem

Vascular cambium

Secondary xylem

FIGURE 4.22

Meristems in plants. Apical meristems in shoots (×380) (*a*) and roots (×76) (*b*) produce *primary growth*, which elongates the organs. The shoot apical meristem (*a*, shown in longitudinal section) produces young leaves and axillary buds, which usually overarch the meristem. The root apical meristem (*b*, also shown in longitudinal section) is covered by a thimble-shaped root cap that protects the meristem as the root grows through the soil. The vascular cambium (*c*, shown in cross section) is a lateral meristem that produces *secondary growth*, or thickening, of the root or shoot of a woody plant. The vascular cambium produces wood (secondary xylem, which conducts water and dissolved minerals) to the inside and secondary phloem (which conducts dissolved organic compounds) to the outside. The young stem shown in (*c*) is covered by a protective epidermis, under which is collenchyma tissue that helps support the stem (×100).

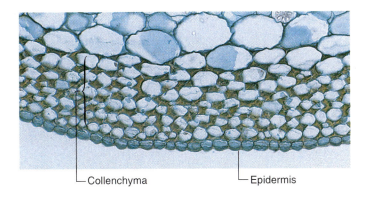

Collenchyma   Epidermis

FIGURE 4.23

Light micrograph of a transverse section of collenchyma tissue from a petiole of rhubarb (*Rheum rhaponticum*) (×185). In fresh tissue like this, the unevenly thickened collenchyma cell walls have a glistening appearance. Collenchyma supports growing regions of plants.

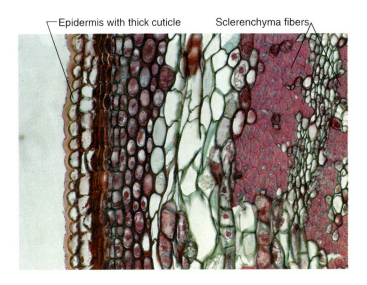

Epidermis with thick cuticle   Sclerenchyma fibers

FIGURE 4.24

Sclerenchyma fibers from a stem of basswood (*Tilia americana*) seen in cross-sectional view (×2.50). Sclerenchyma fibers have thick walls and usually occur in groups, as shown here.

storage tissue of roots and seeds. Parenchyma cells are alive at maturity and usually are surrounded only by a primary cell wall. Finally, parenchyma cells often have large vacuoles. When full of water, these vacuoles make the cells turgid. This turgidity is what makes the lettuce or carrots in your salad crisp. Chlorenchyma cells are chloroplast-containing parenchyma specialized for photosynthesis, usually in leaves.

Collenchyma (from the Greek word *kolla,* meaning "glue") cells are relatively long (up to 2 millimeters long) cells having unevenly thickened primary cell walls (fig. 4.23). They support growing regions of shoots and are therefore common in expanding leaves, petioles and elongating stems. Collenchyma cells are exquisitely adapted for support: their cell walls can stretch because they lack lignin. Thus, the cells can elongate as the tissue they are a part of elongates. Furthermore, collenchyma cells often form in strands or as a cylinder just beneath the epidermis. This arrangement maximizes support, because a cylinder provides more support than does a rod located in the center of a stem or petiole. You're already familiar with strands of collenchyma they are the tough strings in celery (*Apium graveolens*) stalks (which are actually petioles, not stems).

Sclerenchyma (from the Greek word *skleros,* meaning "hard") cells are rigid and produce thick, nonstretchable secondary cell walls (fig. 4.24). They support and strengthen nongrowing, nonextending regions of plants such as mature stems. Sclerenchyma cells are usually dead at maturity; this means the protoplast dissolves, leaving behind the cell wall. Thus, the support provided by sclerenchyma is attributable to its cell-wall "skeleton," which is produced before the cell dies. Sclerenchyma cells occur in all mature parts of plants, including leaves, stems, roots, and bark. There are two types of sclerenchyma cells: *sclereids* and *fibers,* both of which differentiate from parenchyma. Sclereids are relatively short, have variable

shapes, and usually occur singly or in small groups, while fibers are long, slender cells typically occurring in strands or bundles.

Sclereids occur throughout plants, including in roots, leaves, stems, seed coats, and even in the shells of peanuts (*Arachis*). Sclereids often form hard layers; for example, the tough core of an apple consists mostly of sclereids, as does the shell of a walnut. Sclereids also produce the gritty texture of pears. Fibers are often associated with vascular tissue. They differentiate from parenchyma or the vascular cambium and vary in length; for example, fibers in sisal (*Agave sisalana*) are 18 millimeters long, while those of ramie (*Boehmeria nivea*) may be over half a meter long.

Humans have used fibers for more than 10,000 years. People living in the northwestern United States extracted and bound fibers into cords as early as 8000 B.C., and a complete bag made of fibers dates to 5000 B.C. Flax (*Linum usitatissimum*) and hemp (*Cannabis sativa*) have been cultivated for fibers for more than 5,000 years; in fact, hemp was one of the largest crops grown by George Washington on his Virginia plantation (see Perspective 4.4, "Hemp"). Today, humans cultivate plants from more than 40 families for fibers, many of which you encounter every day. For example, Manila hemp (*Musa textilis*) is used to make ropes and cords (fig. 4.25). Century plant (*Agave sisalana*) is used to make coarse ropes and twines, and Mauritius hemp (*Furcraea gigantea*) is used to make ropes, cords, and coarse fabrics. Hemp is used to make twine and rope, while flax is used to make linen. Linen is flexible, soft, and lustrous but not as elastic and flexible as cotton. Ramie is used to make fine Oriental textiles and, more recently, Western clothing. Jute (*Corchorus capsularis*) is used to make coarse fabrics,

4.4 *Perspective* Perspective

Hemp

Hemp, a fabric made from fibers of *Cannabis sativa*, was once as common as cotton. Thomas Jefferson farmed hemp, and George Washington proclaimed, "Sow it everywhere!" His soldiers wore hemp uniforms, and the first two drafts of the Declaration of Independence were written on hemp paper. Even Betsy Ross's American flag was red, white, and blue hemp.

Hemp is a low-maintenance crop: it requires little water and no fertilizers or pesticides. Its deep roots prevent erosion. Hemp looks and feels like linen; only an expert can distinguish the two fibers. Despite these advantages, hemp was eventually replaced by cheap cotton and wood pulp. In 1937, a federal tax—which recognized the marijuana connection—raised prices of hemp to prohibitive levels, thereby ending production and making the fabric obsolete.

Today, the prospects of growing hemp in the United States are unclear. In 1971, Congress repealed the 1937 tax, so it is now up to the states to decide whether to permit production. But because the cultivation of hemp involves growing marijuana, the federal Drug Enforcement Agency can still intervene. To avoid drug-related problems, hemp merchants import hemp from countries where cultivation is legal, primarily China and Hungary. France has developed cultivars of *Cannabis sativa* that lack the psychoactive agent tetrahydrocannabinol (THC).

The demand for hemp is great; global hemp sales rose from $3 million in 1993 to more than $100 million in 1999. Sales in the United States alone now exceed $60 million per year. Hemp is an expensive fabric; on average, it costs about six times as much as cotton, and a pair of hemp jeans cost about $80. Yet despite its cost, hemp is becoming increasingly popular: products made of hemp (e.g., cosmetics, carpets, salad oil, snacks, and biodegradable auto parts) are sold throughout the United States.

FIGURE 4.25

Harvesting *Musa textilis,* commonly known as abaca, or Manila hemp. The trunk, which consists of leaf stalks, is used for its fibers; these fibers make high-grade cordage.

bags, burlap, and sacks. After cotton, jute is the most important textile fiber in world commerce.

Dermal tissue forms the epidermis that covers the plant body.

The **epidermis** has several functions, including the absorption and retention of water and minerals, protection against herbivores, and control of gas exchange. Each of these functions is attributable to one or more of the unique features of the epidermis, such as the presence of a cuticle, few intercellular spaces, and multifunctional outgrowths called trichomes.

The outer walls of most epidermal cells are covered with a waterproof cuticle made of a fatty material called **cutin** (fig. 4.26). The cuticle protects the plant from desiccation by helping to maintain a watery, aquatic environment inside the plant. The cuticle also protects a plant from microbes because it resists microbial infection and degradation. This resilience accounts for why cuticles have been recovered from fossils millions of years old. On the undersides of leaves of wax palm (*Copernicia cerifera,* from the Latin word *cerifera,* meaning "wax bearer"), the whitish layer of wax is often more than 5 millimeters thick. Wax from leaves of this plant is called carnauba wax and is used to make polishes, car wax, candles, and lipstick. It takes about 300 large leaves to produce a kilogram of wax.

Most epidermal cells are flat and packed together like bricks in a brick wall (fig. 4.27). These cells typically lack chloroplasts and are transparent; it is the underlying

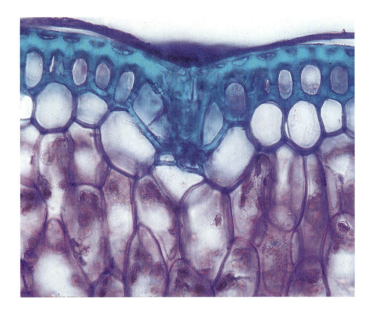

FIGURE 4.26
Epidermal cells cover the primary body of a plant (×390). These cells are covered by a water-resistant cuticle that protects the underlying cells and tissues.

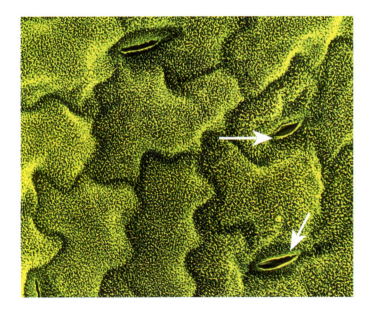

FIGURE 4.27
Scanning electron micrograph of epidermal cells on the upper surface of a leaf of garden pea (*Pisum sativum*). Epidermal cells are packed closely together; the only spaces occur between specialized cells that form stomata (arrowheads). Stomata allow for gas exchange for photosynthesis (×100).

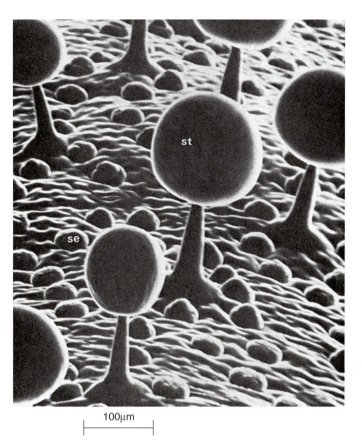

100μm

FIGURE 4.28
Scanning electron micrograph of part of a leaf of butterwort (*Pinguicula grandiflora*), a carnivorous plant. The stalked (*st*) and smaller sessile (*se*) trichomes secrete enzymes and other compounds that help trap and digest animals.

chlorenchyma cells that give leaves their green color. However, the vacuoles of epidermal cells occasionally contain pigments. For example, red cabbage is colored by anthocyanins in epidermal cells, as are flowers and colored parts of variegated leaves of plants such as *Coleus*. The epidermis of leaves of *Peperomia* and the rubber plant (*Ficus elastica*) is typically several cells thick; the additional layers of cells store water.

As we learned in chapter 2, **stomata** (singular, **stoma**) are specialized structures in the epidermis (see fig. 4.27). Stomata may form part of the epidermis of stems, leaves, flowers, and fruits but tend to be most abundant on the undersides of leaves of many land plants. Stomata allow CO_2 to enter the leaf and fuel photosynthesis.

Trichomes are single-celled or multicellular outgrowths of epidermal cells. The most economically important trichomes are cotton fibers, which are up to 6 centimeters long and are made by the epidermis of cotton seeds. The contents of trichomes of some plants are used by humans. For example, menthol is a volatile oil collected from trichomes of peppermint (*Mentha piperita*), and trichomes of *Cannabis sativa* produce the tetrahydrocannabinols (THC) responsible for marijuana's effects. Trichomes of carnivorous plants such as sundew (*Drosera*) and butterwort (*Pinguicula*) secrete enzymes that help the plant digest its trapped prey (fig. 4.28). Trichomes absorb nutrients released as the prey are dissolved in traps.

Some trichomes protect plants from larger prey, including humans. For example, the leaves of stinging nettle (*Urtica*) have brittle, vase-shaped trichomes (see fig. 2.20). When touched, the tip of the trichome breaks off, leaving behind a sharp, syringe-shaped cell that readily penetrates your skin. Pressure from the contact also injects the trichome's irritating chemicals into your skin. Ouch!

As you learned in chapter 1, root hairs are outgrowths of epidermal cells specialized for absorbing water and minerals from soil (see fig. 1.17). They occur near the tips of roots, where they are abundant (e.g., many plants have as many as 40,000 root hairs per square centimeter). Recall that root hairs increase the absorptive surface area of roots several thousandfold, thereby enabling the plant to extract water and dissolved minerals more effectively from nooks and crannies in the soil.

Vascular tissue is the plant's plumbing system.

Vascular tissue is specialized for long-distance transport of water and dissolved solutes such as sucrose from photosynthesis or minerals absorbed from the soil. Thus, vascular tissue is like the plumbing system in a building. In fact, vascular tissue is like long pipes that spread throughout the plant. They are easily seen as veins in leaves like the plumbing in the walls of a building. *Xylem* and *phloem* are the two kinds of vascular tissues in plants. In this section, we will concentrate on their differing structures and functions and on the ways in which these properties adapt plants to life on land.

Xylem transports water and minerals, mostly from roots to shoots. Xylem (from the Greek word *xylos,* meaning "wood") transports water and dissolved nutrients taken in from the soil in an unbroken stream from the roots to all parts of a plant. The water transported in xylem replaces that lost via evaporation through open stomata in the leaf epidermis.

There are two types of xylem:

1. **Primary xylem,** which occurs throughout the primary body of a plant, differentiates from cells formed by apical meristems. During primary growth, as the plant elongates and develops, young xylem cells are stretched as they mature. The secondary cell walls of mature xylem become elaborately sculptured into hoops, bands, or springlike helices that allow elongation and prevent the cells from being crushed by adjacent cells or by internal tensions generated by water transport (fig. 4.29). Primary xylem is formed as organs, such as roots and stems, elongate. In fact, the botanical term *primary* refers to formation by an apical meristem, which is concerned with the business of forming cells as plants lengthen.

2. **Secondary xylem** is produced by a lateral meristem called *vascular cambium* and is commonly called

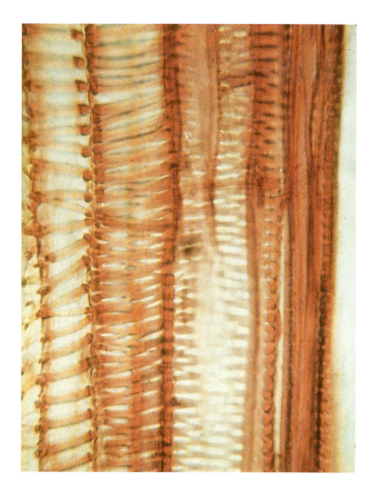

FIGURE 4.29

The secondary walls of some of the first-formed xylem cells are often deposited in rings or spirals (×280). These thickenings strengthen the cells while enabling them to elongate during growth.

wood, although some plants produce secondary xylem and are not "woody." In woody plants, such as elms or pine trees, the large, woody trunk of a tree is mostly secondary xylem, or wood (chapter 8).

There are two kinds of conducting cells in xylem tissue: *tracheids* and *vessels* (fig. 4.30). Both are relatively long, dead at maturity (the plant protoplast has disintegrated, leaving the extracellular wall behind), and have thick, lignified secondary cell walls. Because water is typically pulled through xylem at a negative pressure (i.e., tension), these thick walls help prevent the cells from collapsing. The thick walls of xylem also help support the plant. This is a good example of structure supporting function in plant tissues.

Tracheids are the least specialized xylem cells. Therefore, they are considered the most primitive or earliest xylem to evolve. Tracheids are the only water-conducting cells in most woody, nonflowering plants such as pine or spruce. Tracheids are long, slender cells

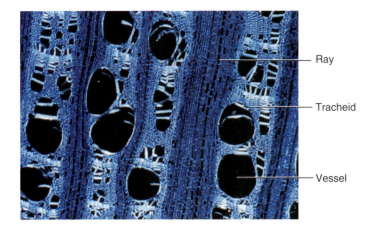

FIGURE 4.30

Light micrograph of a cross section of wood of a flowering plant, showing tracheids, vessels, and rays (×85). The wood of most flowering plants (e.g., oak) has tracheids and vessels to conduct water and dissolved materials from the roots to leaves. You'll learn more about wood in chapter 8.

with tapered, overlapping ends. Water moves upward in roots and stems from tracheid to tracheid through thin areas in cell walls called pits.

Vessel cells are more evolutionary advanced than tracheids and occur in several groups of plants, including angiosperms. With only a few exceptions, all angiosperms contain vessels and tracheids. Vessel cells, which are shorter and wider than tracheids, are arranged end to end. Their end walls are partially or wholly dissolved, forming long, hollow vessels (like straws) through which water moves. This feature of vessels, along with their relatively large diameter, enables them to transport water more rapidly than can smaller, thinner tracheids. The evolution of vessel cells in angiosperms with their greater water-transport capacity is a major reason why angiosperms—the flowering plants—dominate today's landscapes.

Phloem transports organic materials and ions throughout a plant. Phloem (from the Greek word *phloios*, meaning "bark") transports dissolved organic materials (especially sucrose) throughout the plant. Unlike xylem, which primarily transports water upward, phloem transports solutes (dissolved molecules) upward and downward.

Primary phloem, like primary xylem, is produced by the apical meristem and extends throughout the primary body of a plant. Thus, primary phloem comes from apical meristems. *Procambium* is a type of undifferentiated plant tissue that gives rise to vascular tissue.

The conducting cells of phloem are called *sieve elements*. Sieve elements lack a nucleus (it disintegrates after division) but have otherwise intact protoplasts at maturity. They also have thin areas along their cell walls called sieve areas that are perforated by sieve pores (fig. 4.31). Solutes

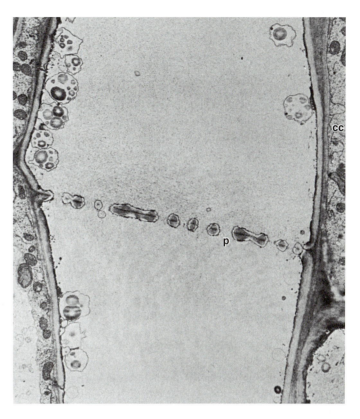

FIGURE 4.31

Sieve elements of tobacco (*Nicotiana tabacum*) (×4,400). Note that the pores (*p*) of the sieve plate are open. Sieve elements conduct water and dissolved organic matter (especially sugars) in all directions in the plant (*cc* = companion cell, a specialized parenchyma cell closely associated with sieve elements).

move from sieve element to sieve element through these pores. Sieve plates produce thin, primary cell walls and are the most delicate cells in plants. They usually live for less than a year, although in palms they have been reported to live for more than a century.

INQUIRY SUMMARY

The four types of meristems are apical meristems, axillary buds, lateral meristems, and intercalary meristems. Parenchyma cells are the primary components of ground tissue and are the most abundant and least structurally specialized cells in plants. The epidermis covers and waterproofs the primary body of a plant. Outgrowths of epidermal cells called trichomes have many functions, including absorption and protection. Xylem conducts water and dissolved minerals from roots to leaves. Phloem conducts water and organic solutes (especially sucrose) in all directions in plants.

How might terrestrial ecosystems be different without the vascular tissues of plants?

Secretory Structures Are Found in Many Plants.

Plants secrete a variety of substances from structures called **secretory structures.** Secretory structures are seldom classified as a separate type of plant tissue because they often intergrade with other tissues. For example, some secretory structures are epidermal trichomes, whereas others are parts of the ground and vascular tissues.

Nectaries are structures that secrete nectar, a sugary exudate that attracts insects, birds, or other animals. Most nectaries are associated with flowers and are called floral nectaries. Their *nectar* is 10% to 50% sugar (especially sucrose, glucose, and fructose) and contains amino acids. Nectar typically is pushed or diffuses through the walls of secretory cells, but it may ooze from stomata. Nectar in floral nectaries attracts insects and other animals that visit a flower and then transfer pollen between plants (see chapter 12). Plants usually secrete relatively small amounts of nectar, which forces foraging animals to visit several flowers and transfer a large amount of pollen, before getting a full meal. Thus, a single animal can pollinate tens or hundreds of plants. Nectaries also can occur on vegetative parts of plants. These nectaries often attract animals that defend the plant. For example, the nectaries of plants such as trumpet creeper (*Campsis radicans*) and *Costus* attract ants that eat the nectar and, in return, defend the plant from leaf-eating insects. These tiny defenders are surprisingly effective: *Costus* plants deprived of ants are quickly devastated by insects and produce only one-third as many seeds as plants protected by ants.

Internal secretory cells are large cells containing substances such as oils, tannins, resins, and crystals. These cells often occur in groups and have several functions, including the storage and production of chemical deterrents to foraging animals. Secretory cells are a source of many oils used by humans. For example, peanut oil comes from secretory cells in peanut seeds (*Arachis hypogaea*), safflower oil from seeds of safflower (*Carthamus tinctorius*), clove oil from flowers of clove (*Eugenia caryophyllata*), palm oil from fruits of palms (such as *Elaeis guineensis*), and cinnamon oil from the secondary phloem of *Cinnamomum zeylanicum.*

Many plants secrete oils and resins into internal canals, ducts, and cavities. For example, myrrh, frankincense, *Citrus* oils, *Eucalyptus* oil, and pine (*Pinus*) resin are extracted from internal cavities, ducts, and canals; these oils deter grazing animals, and resin rapidly seals wounds. As already mentioned, resins and oils help plants defend themselves. However, some animals have evolved traits for not only tolerating these resins but also using them for their own defense. Sawfly larvae, for example, eat pine and store its resin in their bodies. When disturbed by another animal, the larvae rear up and secrete a drop of resin onto the intruder. This typically troubles the rude visitor and ensures privacy for the larvae.

Laticifers are secretory structures that contain latex, which is made of a hodgepodge of carbohydrates, organic acids, alkaloids, terpenes, oils, resins, enzymes, and rubber. Latex oozes from wounded surfaces of plants such as *Euphorbia* and dandelion (*Taraxacum*). It may be white (as in *Euphorbia*), colorless (as in some species of mulberry, *Morus*), or orange-yellow (as in *Cannabis*). Latex is important because it deters grazing animals with its bitter taste and because it helps seal wounds in a plant. Humans have many uses for latex. For example, opium, morphine, and codeine are derived from latex of *Papaver somniferum*, the opium poppy. Latex from the rubber tree (*Hevea brasiliensis*) is 30% to 50% rubber; Amazon Indians used this latex to make rubber balls and containers more than 450 years ago. Today, *Hevea* remains the primary source of natural rubber. The latex of several other plants contains rubberlike particles: gutta percha used to make dentures and golf balls is derived from the latex of *Palaquium,* and chicle (as in Chiclets) gum comes from *Achras sapota.*

SUMMARY

The plant body is composed of cells and their products. All plant cells are surrounded by a rigid cell wall that is produced by the cell. Inside the wall is the cell membrane, which selectively regulates the movement of materials into and out of the cells. Plant cells are compartmentalized into organelles that are structurally organized to support their function. The nucleus, chloroplast, ribosomes, and vacuole are examples of plant cell organelles.

Molecules flow into and out of the cells by simple diffusion, facilitated diffusion via passive transport, and active transport. Osmosis is the diffusion of water across a membrane. Osmosis occurs in response to a water potential gradient produced by differences in solute concentration on each side of a membrane.

Several adaptations to land were required for the evolution of terrestrial plants. These adaptations include the four kinds of tissue systems in plants: meristems, ground tissue, dermal tissue, and vascular tissue. Meristems are localized regions of cell division that form cells. Ground tissue, which usually constitutes most of the plant body, may perform photosynthesis in a leaf, store starch in a root, or add strength to a stem. Dermal tissue forms the epidermis, which surrounds the plant body and serves to absorb nutrients and protect a plant from desiccation. Vascular tissues, the xylem and phloem, provide support and move water and solutes throughout the plant. The activity associated with apical meristems, at the tips of shoots and roots, results in growth in length and forms the primary body of the plant—roots, stems, and leaves.

Writing to Learn Botany

Enzymes called beta-glucosidases hydrolyze bonds between glucose molecules that make up cellulose. What kinds of animals have this enzyme? What selective advantage might the presence of the enzyme give the animal? What nutritional and health problems would humans have if we had this kind of enzyme?

THOUGHT QUESTIONS

1. What is the selective advantage of the small size of cells compared to the large size of organisms?

2. How does the function of rough ER differ from that of smooth ER?

3. What are the structural components of the cell membrane, and what are their roles?

4. What are the major roles of mitochondria, vacuoles, and chloroplasts in plant cells?

5. Can you explain why sucrose, but not starch, concentration affects the relative concentration of water?

6. Certain brands of water purifiers for the home clean water by "reverse osmosis." How do you think this process might relate to osmosis?

7. What adaptations to land were involved in the evolutionary transitions from an aquatic to terrestrial life for plants?

8. What are meristems, and why are they important?

9. In the 1800s, German plant anatomist Anton de Bary said, "The plant forms cells, not cells the plant." What does this mean, and how is it related to plant evolution?

10. Leaves of many grasses grow from meristems located at the base of the leaf. What is the ecological importance of this adaptation in grassland ecosystems?

11. Vessel cells evolved from tracheids in several groups of flowering plants. What do you conclude from this observation?

SUGGESTED READINGS

Articles

de Duve, C. 1996. The birth of complex cells. *Scientific American* 274: 50–57.

Feldman, L. 1998. Not so quiet quiescent centers. *Trends in Plant Science* 3:80–81.

Jergens, G. 1992. Genes to greens: Embryonic pattern formation in plants. *Science* 256:487–88.

Niklas, K. J. 1989. The cellular mechanisms of plants. *American Scientist* 77:344–49.

Reiter, W. 1998. The molecular analysis of cell wall components. *Trends in Plant Science*. 3:27–32.

Storey, R. D. 1990. Textbook errors and misconceptions in biology: Cell structure. *The American Biology Teacher* 52:213–18.

Books

Becker, W. L. Kleinsmith, and J. Hardin. 2000. *The World of the Cell*, 4th ed. San Francisco, CA. Benjamin/Cummings Publishing Co.

Hopkins, W. 1999. *Introduction to Plant Physiology*, 2d ed. New York: John Wiley & Sons.

Lyndon, R. F. 1990. *Plant Development: The Cellular Basis.* London: Unwin Hyman.

Mauseth, J. D. 1988. *Plant Anatomy.* Menlo Park, CA: Benjamin/Cummings.

Sachs, T. 1991. *Pattern Formation in Plant Tissues.* Cambridge: Cambridge University Press.

Taiz, L., and E. Zeiger. 1998. *Plant Physiology*, 2d ed. Sunderland, MA: Sinauer Assoc.

ON THE INTERNET

Visit the textbook's accompanying web site at http://www.mhhe.com/botany to find live Internet links for each of the topics listed below:

Cellular organelles

Chloroplasts and other plastids

Evolution of the cell

The nucleus

The vacuole

Cell membranes

Cell walls

Diffusion and osmosis

Plant tissues

The cell theory

A molecule of DNA. A double strand of DNA is about 2 nanometers thick, but may reach lengths of many centimeters. This is an electron micrograph of the DNA from a virus that infects bacteria. ($\times$ 20,000) What might the amount of DNA in a cell indicate about the evolutionary development of an organism?

CHAPTER 5

DNA, Genes, and Cell Division

LEARNING ONLINE

You've heard of DNA and genes, but do you know how are they passed on to new plant cells? To learn more about how cells make new cells, visit http://gened.emc.maricopa.edu/bio181/BIOBK/BioBookTOC.html in chapter 5 of the Online Learning Center at www.mhhe.com/botany. The OLC provides lots of useful information like this weblink to help you better understand the material in your textbook.

The discovery of DNA structure by James Watson and Francis Crick was a major breakthrough in science. They proposed a model of a double helix that has become the most famous and intensely studied molecule in science. The double helix has appeared on the cover of major magazines and is studied by students from kindergarten to graduate school. Everyone has heard of DNA and genes, but few people really know their importance or what they actually do in a cell.

In this chapter, you will learn about DNA, genes, and chromosomes and how they can control the activities of a cell. Also, you will learn about cell division and the life of a dividing cell during the stages of the cell cycle. A nucleus contains DNA associated with proteins that make up the chromosomes. Plant genes are found in the chromosomes, and DNA is the genetic material.

Plant genes serve two critical functions: (1) transmission of the genetic information and (2) expression of the genetic information. What does all this mean, and how does it happen in plants? Is it different in plants than in you? What makes plants different from animals? We consider these interesting questions in this and other chapters.

The methods and discoveries of molecular genetics have greatly improved crop productivity, gene therapy, and other aspects of applied biology, which rely on our knowing how to change existing genes, or transfer genes from one organism to another. This work is the basis for a highly sophisticated, multibillion-dollar industry called gene technology, which is examined in this chapter and in appendix A. This work by biologists has touched, and will continue to touch, your life in many ways.

Mendel's Work Began Modern Genetics.

Gregor Mendel's classic studies on peas, conducted in his Augustinian monastery garden from 1856 to 1862, began modern genetics and led to the first idea of traits and genes (called elemente by Mendel). In fact, Mendel was the first to describe quantitatively heredity and a genetic trait in plants when he studied the round versus wrinkled shapes of mature pea seeds. The genetic basis for this trait was not described until 1990, however, by Madan Bhattacharyya and several colleagues at the John Innes Institute in Norwich, England. This group identified the enzyme that controls the chemical reaction leading to seed roundness. Roundness is apparently controlled by one form of the enzyme that works on the structure of starch. This form of the enzyme is active in the early development of round seeds but is absent in wrinkled seeds. Round seeds have large amounts of starch and less sucrose than wrinkled seeds, which have less starch and proportionately more sucrose. Wrinkles occur because sucrose causes the seeds to absorb water, then swell and stretch as they develop on the plant. When wrinkled seeds dry as they mature, the loss of water makes them shrivel. In contrast, starch does not cause water absorption; therefore, round seeds do not stretch as much as wrinkled ones. Hence, when round seeds mature, they remain full and round. Mendel noticed this difference over 130 years ago and reported this trait in his papers.

Bhattacharyya's study is an example of how genes are found and how they influence **phenotypes**—that is, the appearance of an organism as controlled by the genes. An explanation of the genetics of wrinkled seeds, as described by Mendel, and of other characteristics of organisms, requires an understanding of the structure and mechanisms of genes. These are the subjects of this chapter.

Many questions about genes remain to be answered, mainly through discoveries relying on the powerful techniques of molecular biology. One fundamental question that intrigues botanists is how can roots be roots and leaves be leaves when their cells have the same genes and the same DNA? Similar questions can be asked about the genes of other organisms, including you.

INQUIRY SUMMARY

Gregor Mendel's classic studies of peas began modern genetics and led to the first idea of traits and genes. Genes influence the phenotype, which is the appearance of an organism.

What might be a selective advantage to wild pea plants producing round seeds?

How Does the Nucleus Control the Activities of the Cell?

In chapter 4, you learned about the various compartments, called organelles, that are part of the plant cell. You also learned that one of these organelles, the nucleus, controls the bulk of activities of the plant cell. How is this accomplished? To begin to answer this question, we must first understand the structure and function of the components of the nucleus—that is, DNA, genes, and chromosomes.

As you read this chapter, keep in mind that cells, and parts of cells including the nucleus and its components, have undergone natural selection and evolved. In fact, the evolution of eukaryotic cells depended on the DNA forming genes in the nucleus and becoming specialized to direct the varied activities of different cells in a plant or animal.

What are genes, and why are they important?

Previously you learned that DNA makes up genes and that genes are organized into chromosomes. We know a great deal about DNA and chromosomes, but what exactly is a gene? A gene is not so easy to define because there are exceptions to most definitions and because we are constantly learning more about these basic hereditary units. Early biologists described a single gene as a section of DNA that codes for one protein—the long-held "one-gene, one-protein" definition. A more current, but fairly standard, definition says that a **gene** is a sequence of DNA that codes for a gene product, usually a protein (genes also code for RNA as a product). Most recently, it has been shown that a gene can actually code for several proteins, so our definition might state "single or several proteins", or RNA. Also, some segments of DNA have a regulatory role in transcription and form no known gene product. As you can see defining a gene can be confusing, but most biologists agree that a gene is the basic unit of heredity. The next time a friend states that he or she has "the gene," for something, ask them what they mean by a gene and see what kind of an answer you get.

Keep in mind two roles of genes that give them importance in nature:

1. *Genes are the genetic information passed to the next generation of offspring cells following division of the parental cell.* Genes are instructions passed to offspring cells as parental cells divide. This role is really that of heredity, as genes, which carry the genetic information, are transmitted from parent to offspring. Genes are the units of inheritance.

2. *In cells, genes are the molecular blueprints for the production of proteins; these proteins control the structure, development, and activities of a cell as it grows and develops* (chapter 3). DNA directs the synthesis of RNA, which in turn directs the assembly of amino acids into proteins. The sequence of DNA → RNA → protein has been called the dogma of protein synthesis since it occurs in all organisms including plants and animals. Thus, expression of genetic information is a second role of genes.

   Recall that most proteins are enzymes, and enzymes control the chemical reactions that determine the genetic characteristics of an organism.

The nucleus contains DNA associated with proteins that make up the chromosomes.

The importance of nuclei was first suggested decades ago after biologists noticed that nuclei contained thin "strings" that would stain into "colored bodies." These unique colored bodies were called **chromosomes,** and the early researchers could see them only in dividing cells, such as those at the tip of an onion root. These biologists were fascinated by the colored bodies but really had little notion of their critical role in the cells.

We now know that chromosomes are made of genes that control the cell. In turn, genes are made of DNA. It is important to know basic chromosome and DNA structure to understand the critical functions of genes. Remember that structure-function relationships are of major importance in botany. Chromosomes and DNA provide a good example of this relationship at the molecular level.

Figure 5.1 is a model of a stained plant chromosome during cell division as viewed through a light microscope. Chromosomes examined by electron microscopy appear as small beads on a thin string. The beads are actually called **nucleosomes** and are the basic structural unit of all eukaryotic chromosomes. Nucleosomes probably help keep the long DNA molecules untangled and organized inside the nucleus. Also, nucleosomes may help control which genes are expressed, and which are not, in different organs of a plant.

Nucleosomes consist of: (1) DNA and (2) five unique kinds of chromosomal proteins called **histones.** DNA appears to wrap around four kinds of histones, forming a nucleosome that looks like a small bead (fig. 5.1) and most certainly contains the DNA of numerous genes. Nucleosomes are separated by spacer DNA, which is associated with a fifth histone that apparently binds nucleosomes together in a regular, repeating array. As mentioned previously, you can think of the nucleosome as a way to package a very long thread—the DNA—in a very small compartment—the nucleus. The entire DNA plus histone association, including the nucleosomes, is called **chromatin** (figs. 5.1 and 5.2).

Chromatin is folded into loops (see fig. 5.1) that extend from the main axis of the chromosome, but the mechanisms of loop formation and maintenance are unknown. Some loops coil and fold even further, until stained chromatin becomes visible with a light microscope. This is why stained chromosomes were called "colored bodies" by early researchers. The visible, more condensed form of chromatin may mean the genes that make up a chromosome are packed too tightly to be functional; This is one explanation for how some genes can be inactive at different times or in different organs during the life of the cell. When stained during cell division, chromosomes are readily visible with a good light microscope, but the nucleosomes can be seen only with an electron microscope.

The discovery of the structure of DNA was a major breakthrough in science.

Progress in understanding the gene was accelerated in the 1950s by a new theory of the structure of DNA. In the years

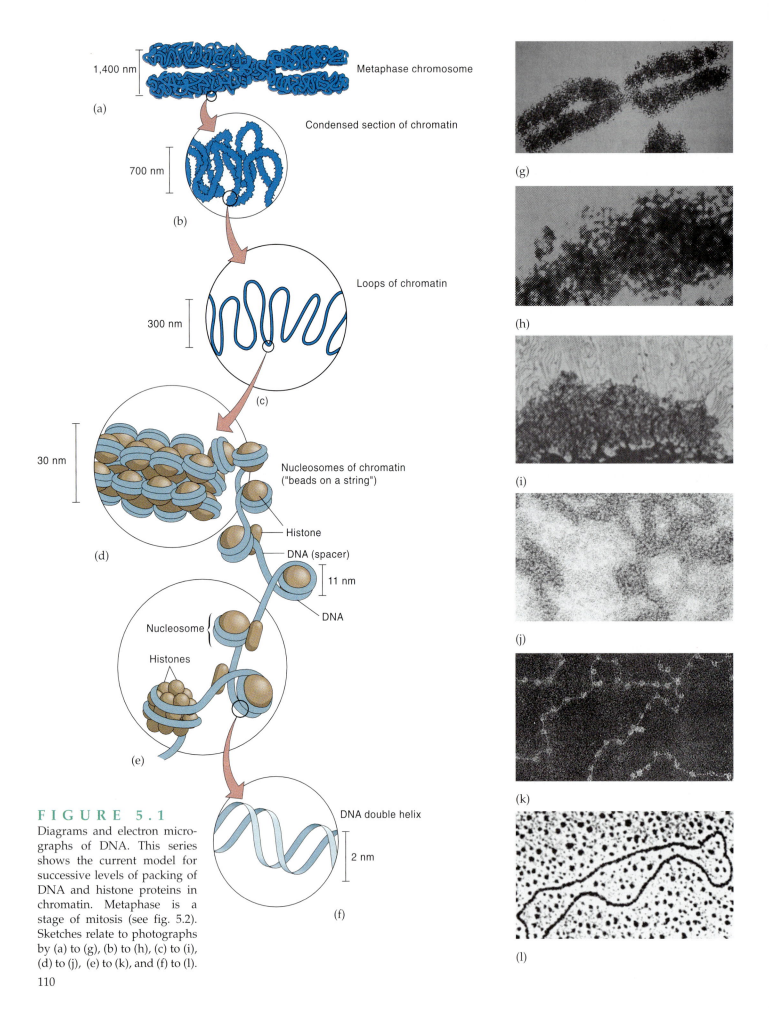

(a) 1,400 nm — Metaphase chromosome

(b) 700 nm — Condensed section of chromatin

(c) 300 nm — Loops of chromatin

(d) 30 nm — Nucleosomes of chromatin ("beads on a string")

Histone
DNA (spacer)
11 nm
DNA

(e) Nucleosome
Histones

(f) DNA double helix
2 nm

(g)

(h)

(i)

(j)

(k)

(l)

FIGURE 5.1

Diagrams and electron micrographs of DNA. This series shows the current model for successive levels of packing of DNA and histone proteins in chromatin. Metaphase is a stage of mitosis (see fig. 5.2). Sketches relate to photographs by (a) to (g), (b) to (h), (c) to (i), (d) to (j), (e) to (k), and (f) to (l).

110

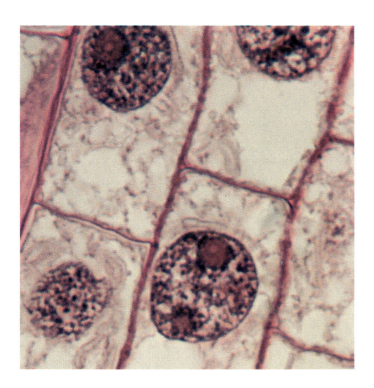

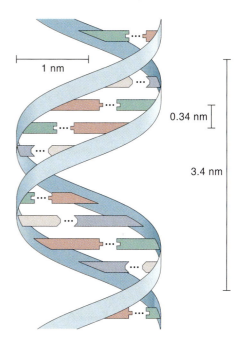

FIGURE 5.3
Diagrammatic representation of the double helix of DNA. Ribbons represent the deoxyribose-phosphate backbones of each strand. The bars between ribbons represent paired nitrogenous bases.

FIGURE 5.2
Interphase nucleus from a cell of an onion root tip (*Allium cepa*), stained to show chromatin (×500). DNA synthesis and cell growth occur during interphase, which takes place before mitosis.

prior to this, scientists had studied chromosomes intensively and even debated whether it was DNA or protein that carried the genetic information. As you will see, our knowledge of inheritance has come far since then.

In the early 1950s, interest in the structure of DNA was at its peak. Evidence indicated that DNA was the substance of inheritance—the genes—and several different kinds of studies began to reveal the complexity of the DNA molecule. These studies were like the pieces of a puzzle, the structure of DNA being the full picture. The relevance of various kinds of information to understanding the structure of DNA was first correctly recognized by James Watson and Francis Crick of Cambridge University. They are therefore credited with first describing the structure of DNA, the most significant discovery in biology of the twentieth century. In 1962, Watson and Crick, with associate Maurice Wilkins, received a Nobel Prize for their landmark effort.

Watson and Crick's model of DNA was a double helix—that is, a two-stranded spiral (figs. 3.14 and 5.3). The use of the term model in this case is not abstract. Watson and Crick actually built a model of a DNA molecule by putting together pieces of metal to represent the chemical components of DNA. Note the model is not a twisted ladder but is more like a spiral staircase or ribbon wrapped around a barber pole. The nitrogen-containing bases of each nucleotide

show specific pairing; that is, thymine pairs only with adenine, and cytosine pairs only with guanine (fig. 5.4). This pairing joins the two strands made of the deoxyribose sugar and phosphate that form the backbone of a DNA molecule.

Three postulates made by Watson and Crick make up the theory of DNA structure and duplication:

1. DNA consists of a double helix. The two strands of the molecule are made of alternate deoxyribose and phosphate components.

2. Pairs of bases form links between opposite deoxyribose molecules in the two strands. The base-pairs are adenine to thymine (A–T) and cytosine to guanine (C–G) (fig. 5.4).

3. Base-pairs may be in any sequence along a double strand, but the pairing rule in postulate 2 always holds.

Figure 5.4 shows the nucleotide with cytosine (C) as the end base sketched on the lower right side of the DNA molecule. If the sketch continued downward, what would be the next base in the sequence? This is really a trick question because the next base could be any of the four choices of possible nucleotides—A, T, C, or G. This variability of the base sequence in DNA explains the enormous diversity of genes and organisms in nature, for it is the sequence of these bases that determines genetic variability. Remember though, the next nucleotide, regardless of which one it might be, would then be base-paired according to postulate 2.

F I G U R E   5 . 4
Components of a portion of a DNA molecule shown in fig. 5.3. The two strands are held together by hydrogen bonds between pairs of nucleotides, adenine pairing with thymine and cytosine pairing with guanine. A nucleotide is a nitrogen-containing base bonded to a molecule of deoxyribose with phosphate attached.

The impact of the Watson-Crick model was immediate. It implied how DNA could exist in many different sequences of nucleotide bases, thereby being sufficiently complex to produce the variability of genes found in nature. The model hinted that different genes could exist as base sequences of different lengths. The structure of DNA also explains its role in directing RNA synthesis inside the nucleus as DNA serves as a template for the assembly of RNA. In short, the Watson and Crick model provided a structural explanation for the actions of genes.

somes that organize DNA and help control gene activity. DNA plus histone proteins are called chromatin, which makes up a eukaryotic chromosome. Chromosomes are chains of genes made of DNA.

DNA is a double helix characterized by base-pairing of the nucleotides depicted as A to T and G to C. The Watson-Crick model for the structure of DNA explains how DNA can be complex enough to accommodate a large number of genes in a multitude of nucleotide sequences.

How does the structure of DNA explain its functions?

INQUIRY SUMMARY

Genes are responsible for the transmission and expression of genetic information. Chromosomes are made of DNA and histone proteins arranged, through the course evolution, in structural units called nucleo-

How Do Genes Work?

Imagine a nucleus as an enormous dictionary containing the words that define a particular organism but in which the words are written in a microscopic code. As recently as the

The Lore of Plants

To appreciate the ability of DNA to store information, consider a species such as humans whose genome is about 3 billion base-pairs (A to T and G to C). If this DNA were stretched out, there would be about 2 meters of DNA per cell. However, all of this DNA is stored in a nucleus, which is only about 0.005 millimeters (0.002 inches) in diameter, or about 1/500 the thickness of a dime. These 3 billion base-pairs—that is, 6 billion bits of information—store 21 times more information than is found in the *Encyclopedia Britannica*. Almost all known plant genomes are much larger than the human genome. Imagine the information stored and the packing problem in plant nuclei.

1950s, this code seemed so complicated that scientists thought it might be beyond human comprehension. Furthermore, the code was so large for any organism that if its letters could be printed in the same size as this type, the dictionary would fill several hundred volumes the size of this book. Now we know that the code is made of DNA and that DNA is organized into "words" called genes. But how is the code translated into organisms? There is still no complete answer for this intriguing question, although pieces of the answer have been discovered by finding out how genes work at the cellular level.

In cells, proteins are made in the cytosol in ribosomes, not in the nucleus, where most of the plant's genes (DNA) occur. This observation was one of the first clues that genes do not make proteins directly and that RNA is involved in the process. This observation is a start to answer our question about how genes work.

RNA is transcribed in the nucleus and translated in the cytosol.

The directed assembly of amino acids into a protein is a multistep process that begins when the DNA of a gene in the nucleus is used as a template for making RNA. This process of RNA synthesis is called **transcription.** RNA fresh off the DNA template is often referred to as the primary transcript, or pre-RNA, because, while still in the nucleus, RNA is edited (that is, chemically modified, then spliced and reassembled) before export as mature RNA into the cytosol.

There are three main kinds of RNA transcribed in the cell nucleus on a DNA template.

1. **Messenger RNA (mRNA)** is the actual message coded from DNA that directs the synthesis of a protein in a ribosome, and carries the DNA message from gene to ribosome.

2. **Transfer RNA (tRNA)** binds to amino acids and brings (transfers) them to a ribosome as specified by mRNA.

3. **Ribosomal RNA (rRNA)** combines with ribosomal proteins, forming a **ribosome,** the workbench of protein synthesis.

In every cell—plant, animal, fungus, bacterium—all three kinds of RNA are essential for protein synthesis. The process of linking amino acids into a protein in a ribosome is called **translation,** or protein synthesis. The directions for translation are contained in mRNA. Once translation is complete, proteins are often assembled into multichain units or otherwise modified before becoming functional proteins. Overall, the way genes work (gene expression) to transfer genetic information from a DNA template into the structure of a protein can be summarized as follows:

Nuclear DNA (genes)

↓ *Transcription*

Pre-mRNA

↓ *Editing, then export to cytosol*

Mature mRNA

↓ *Translation*

Protein

↓ *Assembly, modification*

Functional protein (here an enzyme)

↓ *Catalysis*

Chemical reaction that determines a trait

To appreciate this process, consider the red pigment in a rose petal. Synthesis of the pigment requires several specific, functional enzymes to catalyze the chemical reactions necessary to form the pigment. These enzymes are assembled according to the cascade of events just shown. Thus, DNA ultimately directs the synthesis of the red pigment, which serves as an example of the expression of a genetic trait—red color in the petal. Can you think of other examples of traits in plants? In animals?

DNA directs the synthesis of RNA in the nucleus.

During transcription of a gene, DNA serves as a template for making RNA. The entire gene is transcribed into molecules of pre-RNA that can be about the same length as the DNA template (fig. 5.5). Recall from chapter 3 that, unlike DNA, RNA contains ribose instead of deoxyribose, with uracil (U) substituted for thymine (T). Thus, the DNA of a gene sequence such as ATGCCTGGA will be transcribed into an RNA sequence of UACGGACCU. RNA nucleotides, like DNA nucleotides, are manufactured in the plant cell.

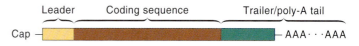

FIGURE 5.6

Model of mRNA that is exported from the nucleus. The edited or processed mRNA consists of a cap, a leader sequence, a coding sequence for a polypeptide, and a trailer. The coding sequence is made of RNA codons.

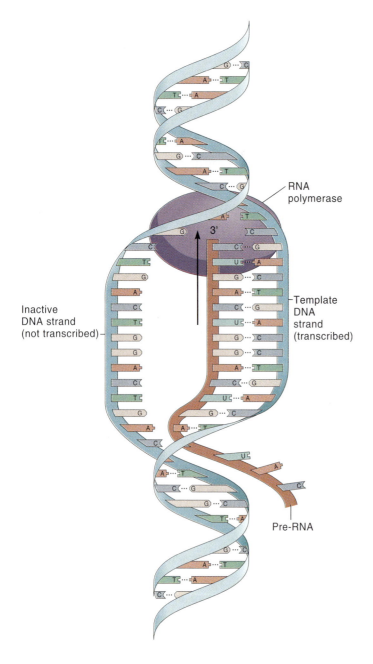

FIGURE 5.5

Diagrammatic representation of transcription. RNA polymerase assembles RNA nucleotides in a sequence that is complementary to the DNA template. Pre-RNA is therefore identical to the non-transcribed strand of DNA, except for the substitution of uracil (U) for thymine (T) in RNA.

Transcription of all three types of RNA occurs in three continuous steps: initiation, elongation, and termination. Initiation begins when special enzymes, including RNA polymerase, attach to the nucleotide sequence of DNA at the start of the gene (fig. 5.5). Next, during the elongation step, the polymerase enzyme moves along the strand of DNA, sequentially adding RNA nucleotides to the growing (elongating) chain of RNA (fig. 5.5). Finally, RNA polymerase reaches the end of the gene. This is the point where tran-

scription stops and the new RNA molecule is released from its DNA template. The termination section of the DNA is like a period at the end of a sentence, and the gene is the sentence.

In this chapter, we focus on the forms of RNA that help make proteins in the cytosol (mRNA, tRNA, rRNA). You should understand that certain kinds of RNA molecules called **ribozymes** can also function as enzymes (see Perspective 5.1, "Ribozymes and the Origin of Life"). The evolution of the varied functions of RNA was likely a pivotal event in the molecular origin of life.

All three types of RNA are required for protein synthesis in the cytosol.

Messenger RNA (mRNA) gets it name because it contains the "message" (i.e., coding sequence) for a protein. A mature mRNA molecule, after editing, has four main parts: the cap, the leader sequence, the coding sequence, and the trailer sequence (fig. 5.6), all of which are required for the mRNA molecule to function. The working unit of the mRNA coding sequence is the **codon.** Each codon is three mRNA nucleotides long and represents the central unit of genetic information passed from DNA to mRNA to protein.

The "transfer" in transfer RNA (tRNA) refers to its role in bringing amino acids together during translation. The probable three-dimensional structure of a tRNA molecule is shown by computer simulation in figure 5.7. Note that the molecule is twisted and folded back on itself, so that four double-stranded regions and three loops are formed.

Loop 2 of tRNA has three nucleotides called the **anticodon.** The tRNA anticodon sequence is complementary to the mRNA codon sequence, each with three nucleotides. All tRNA molecules also contain an amino acid acceptor site, which specifically binds to one of the 20 kinds of amino acids that will later make up proteins. The amino acids are synthesized in the plant cell before attachment to tRNA.

Ribosomes are the sites of protein synthesis. Each ribosome consists of a complex of rRNA and ribosomal proteins, which are organized into a large subunit and a small subunit (fig. 5.8). During translation, the two subunits join together and form a structural and functional workbench for protein synthesis. Several ribosomes are attached to a single mRNA molecule during synthesis.

5.1 PerspectivePerspective

Ribozymes and the Origin of Life

In 1981, Thomas Cech, of the University of Colorado, discovered that primary RNA transcripts fresh off the DNA template (pre-RNA) in *Tetrahymena*, a single-celled protist, can be processed in the absence of protein enzymes. Processing occurs because one of the **introns** (see fig. 5.14) is self-splicing, which means that RNA acts as an enzyme to catalyze a chemical reaction—in this case, splicing out the introns of the pre-RNA. Unlike a protein enzyme, however, the intron works on itself. To acknowledge this difference, Cech named this intron a ribozyme. His work on ribozymes earned Cech, who is now head of the Howard Hughes Medical Institute, a Nobel Prize in 1986.

Cech's discovery sparked a search for ribozymes in other organisms. So far, they have been found in the nuclei and organelles of plants, animals, and fungi and in the genomes of bacteria. Additional functions also have been discovered for ribozymes. For example, they work like nucleases, which are enzymes that digest nucleic acids. Ribozymes also mimic polymerases by joining nucleotides in short chains and by joining short RNA sequences into longer ones. They also have catalytic functions inside ribosomes.

What do ribozymes have to do with the origin of life? They may be the answer to the "chicken versus egg" paradox between nucleic acids and proteins. The paradox is that if proteins came first, how could they be encoded without DNA or RNA? Conversely, if nucleic acids came first, how could they be replicated without enzymes? This paradox rests on the now debunked dogma that only proteins can be enzymes and only nucleic acids can contain genetic information. However, we now know that RNA can play both roles; many viruses have an RNA, not DNA, genome, including the virus that causes AIDS. Thus, the first organisms may have contained RNA before DNA or proteins evolved.

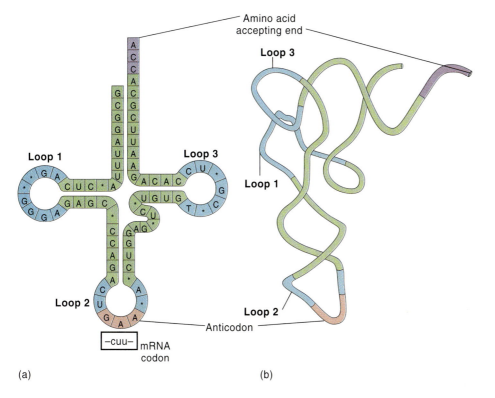

(a)

(b)

FIGURE 5.7

The structure of transfer RNA. (*a*) Two-dimensional representation, showing three loops and four regions of internally complementary nucleotides that form double strands. At one end of the molecule is the amino acid attachment site, and at the middle loop (loop 2) is the mRNA recognition site of GAA (anticodon). The tRNA anticodon attaches to the mRNA codon shown as CUU. Asterisks denote altered nucleotides. (*b*) Three-dimensional representation.

The mRNA is translated in the cytosol.

Protein synthesis (translation) is like RNA synthesis (transcription) because it uses enzymes to make a polymer. Translation differs, however, because it links amino acids into proteins, not nucleotides into RNA as in transcription. Remember that a gene (DNA) expresses the genetic traits via transcription of DNA into RNA and then translation of RNA into a specific protein. Thus, genes control the structure and function of an organism.

Molecules of tRNA must bind their appropriate amino acids before translation can start (fig. 5.9). Each tRNA molecule is attached to a specific amino acid. For example, the amino acid methionine is attached only to molecules of tRNA-Met (the tRNA molecule that is specific for methionine), and the amino acid valine is attached only to molecules of tRNA-Val. This means the correct amino acid is attached to the corresponding tRNA molecule. Each amino acid–tRNA complex (amino acid bound to tRNA) is made in a

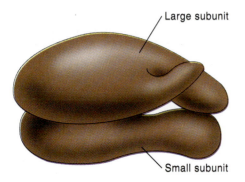

Large subunit

Small subunit

FIGURE 5.8
Structure of a eukaryotic ribosome. Each ribosome consists of a large subunit and a small subunit. The two subunits join together during protein synthesis.

process that utilizes energy from ATP. Every amino acid–tRNA molecule has two main functions: (1) to carry the right amino acid to the ribosome as specified by the mRNA code and (2) to activate the amino acid so that it reacts with another amino acid, linking them together and ultimately forming a protein.

Translation occurs in three continuous phases in a ribosome: initiation, elongation, and termination. Translation starts when mRNA and an initiator tRNA–amino acid complex bind to a small ribosomal subunit (fig. 5.10). Remember the tRNA anticodon and the mRNA codon recognize each other as the codon "pops" up from the subunit. After the initiator tRNA–amino acid complex, the mRNA, and the small subunit are in place, the large ribosomal subunit binds to the small subunit. This phase of initiation completes the assembly of a functional ribosome.

The next phase, elongation, begins when a second tRNA molecule is attached next to the initiator tRNA on the large ribosomal subunit (fig. 5.11). The two amino acids are joined by a transferase enzyme. A third tRNA anticodon then recognizes the exposed, corresponding mRNA codon bringing another amino acid with it. This process continues to link amino acids, one by one, until the chain of amino acids is complete. Translation stops when the ribosome arrives at a stop codon on the mRNA molecule (fig. 5.12). The result is a protein with an amino acid sequence arranged according to the DNA blueprint sent to the ribosome as a mRNA message.

How fast is translation? The simple answer is, impressively fast! It takes less than a minute to translate mRNA into a polypeptide of 400 amino acids. Simultaneous translation by many ribosomes attached to a single mRNA molecule (a polyribosome) allows each mRNA

FIGURE 5.9
Synthesis of an activated amino acid-tRNA complex. First, the enzyme aminoacyl-tRNA synthetase binds to a molecule of ATP and to a particular amino acid. ATP provides the energy for the synthesis reactions. AMP ($\boxed{A}$-P) remains attached until the appropriate molecule of tRNA binds to the enzyme and a chemical bond is formed between the tRNA molecule and the amino acid. The product is an activated amino acid-tRNA complex that can move to a ribosome during protein synthesis.

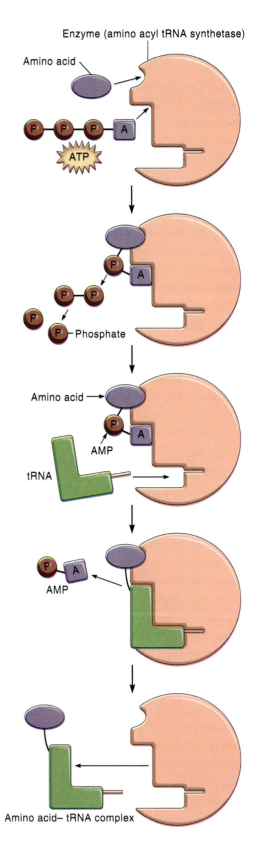

Enzyme (amino acyl tRNA synthetase)

Amino acid

ATP

Phosphate

Amino acid

AMP

tRNA

AMP

Amino acid– tRNA complex

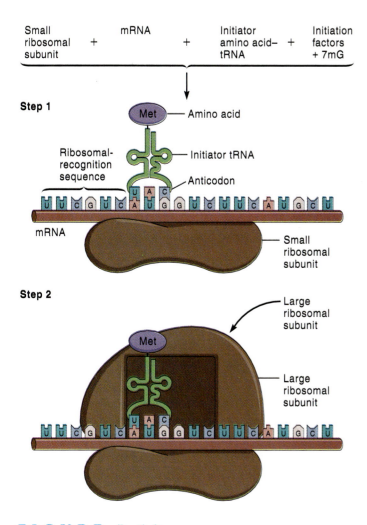

FIGURE 5.10

Initiation of protein synthesis. First, the ribosomal recognition sequence attaches to the small ribosomal subunit. Then, an initiator amino acid–tRNA binds to the start codon on mRNA, in a complex together with the small ribosomal subunit. Protein synthesis begins after the large ribosomal subunit is attached to this complex. The codon for Met is AUG. The anticodon is UAC in the tRNA initiator. This is a simplified explanation and ribosome structure is more complex than shown.

molecule to make as many as 10 proteins per minute. By using several thousand mRNA molecules at the same time, a cell can translate tens of thousands of proteins per minute. Continued translation at this rate produces millions of proteins during the life of a single cell.

The genetic code is a key to protein synthesis.

While the gene is in the DNA blueprint, the **genetic code** can be considered a sequence of mRNA codons that specify amino acids during translation. The code links the nucleotide sequence of genes through mRNA to the sequence of amino acids in a protein. Accordingly, the genetic code is the code of life, for it works in organisms, including you.

As mentioned, the smallest unit of the code is a codon. Each codon is three consecutive mRNA nucleotides.

By employing three nucleotides, the genetic code has 64 possible codons (4^3 possible combinations of the four kinds of nucleotides). The presence of 64 possible codons means there are several more codons than would be needed for 20 amino acids plus a stop signal codon. What do the other codons do? It took a long time to answer this question, but eventually all 64 combinations were found to be useful in the genetic code as each codes for an amino acid. Thus, several amino acids have more than one matching tRNA. There are, as you might expect, also 64 different tRNA anticodons consisting of three tRNA nucleotides that each complement a mRNA codon. The 64 codons are listed in figure 5.13.

How is gene expression regulated?

With an understanding of how genes are expressed, we should ask how they might be regulated. While a detailed answer is beyond the scope of this book, regulation can occur at each step we have discussed, including the points of splicing pre-RNA. For example, during development the initiation of transcription may be turned on or off, depending on the growth stage of the cell or tissue.

Plants have several kinds of organs and tissues, each with some specialized traits, such as photosynthesis in a green leaf cell compared to starch accumulation in storage cells of a white potato tuber. This high degree of specialization requires finely tuned regulation of gene expression in specialized cells, which all contain an identical set of genes. The growth and development of a plant depends on when particular genes are turned on or off.

Eukaryotic chromosomes contain interrupted genes.

A recent, surprising discovery in biology showed that the DNA of most eukaryotic genes contain noncoding regions that interrupt the coding regions within a gene. Here, "noncoding" means the DNA regions do not end up in mature mRNA. The interruptions were named *intervening sequences,* and they are transcribed into introns of pre-mRNA. Coding regions are transcribed into exons of pre-mRNA. Thus, in pre-mRNA, exons are separated by introns (fig. 5.14; also see Perspective 5.1). During the editing of pre-mRNA, introns are spliced out and exons are joined together, like a film editor removes unwanted sections of a film and splices together the desired sections. Ultimately then, exons of pre-mRNA are spliced together and form mature mRNA, which codes for the protein. Introns were named because they came from intervening sequences of a gene. In most genes, introns actually make up the bulk of pre-mRNA. Both tRNA and rRNA also are edited in plant cells, but more is known about processing mRNA.

The functions of introns remain largely unclear. Some biologists have suggested they are just "junk DNA," evolutionary baggage accumulated over time like old, undiscarded

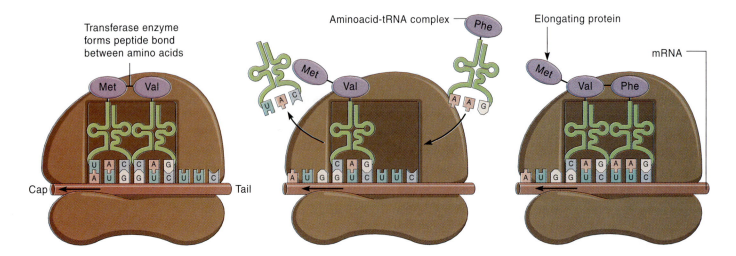

FIGURE 5.11

Steps in elongation. After initiation, a second amino acid–tRNA complex (here Val-tRNA) binds to the ribosome next to the initiator tRNA, upon which the two amino acids, Met and Val, are joined by a peptide bond. The initiator tRNA is then released, the mRNA advances one codon through the ribosome, and a third amino acid–tRNA (Phe-tRNA) binds to the spot on the ribosome that was vacated by the second amino acid–tRNA. A peptide bond forms between the second and third amino acids. These steps are repeated for each additional amino acid in the elongating protein. This is a simplified representation of the most important steps.

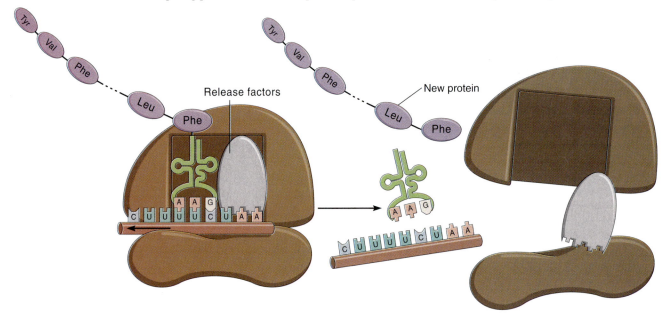

FIGURE 5.12

Translation stops when the ribosome reaches a stop codon (UAA). The stop codon binds to several protein release factors, thereby interrupting translation. As the complex falls apart, the new protein is released. This is a simplified sketch.

clothes in the back of a closet. Other scientists speculate introns could regulate plant development.

INQUIRY SUMMARY

Transcription is RNA synthesis that entails three steps: initiation, elongation, and termination. Three main kinds of RNA work together to make proteins. Mes-

senger RNA carries the DNA message from a gene to a ribosome. Transfer RNA is directed by mRNA to bring amino acids to a ribosome in the proper order. Ribosomes, which contain protein and ribosomal RNA, are where mRNA and tRNA come together, making proteins in the cytosol. During translation, each codon in mRNA binds to a specific molecule of tRNA, which carries a certain amino acid. These amino acids join

Second Base

First Base	U	C	A	G	Third Base
U	UUU \} Phe F UUC / UUA \} Leu UUG /	UCU \} UCC | Ser UCA / S UCG /	UAU \} Tyr Y UAC / UAA Stop UAG Stop	UGU \} Cys C UGC / UGA Stop UGG Trp W	U C A G
C	CUU \} CUC | Leu CUA | L CUG /	CCU \} CCC | Pro CCA | P CCG /	CAU \} His H CAC / CAA \} Gln Q CAG /	CGU \} CGC | Arg CGA | R CGG /	U C A G
A	AUU \} AUC | Ile I AUA / AUG Met or M Start	ACU \} ACC | Thr ACA | T ACG /	AAU \} Asn N AAC / AAA \} Lys K AAG /	AGU \} Ser S AGC / AGA \} Arg R AGG /	U C A G
G	GUU \} GUC | Val GUA | V GUG /	GCU \} GCC | Ala GCA | A GCG /	GAU \} Asp D GAC / GAA \} Glu E GAG /	GGU \} GGC | Gly GGA | G GGG /	U C A G

FIGURE 5.13

The genetic code for mRNA. Both the three-letter and one-letter abbreviations are given for each amino acid. For example, Leu, L are for leucine; Ser, S for serine. Three of the 64 codons are stop signals; they do not translate into amino acids.

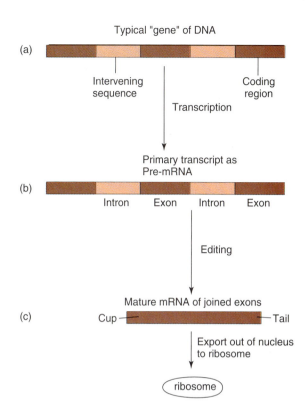

FIGURE 5.14

Transcription and editing of mRNA. A typical plant gene (*a*) has intervening sequences (introns) and coding regions (exons) that are transcribed into pre-mRNA (*b*). After editing (*c*), only the exons remain as the exons are spliced out.

into a chain forming the protein. In this way, genes, which are composed of DNA, control the synthesis of proteins that, in turn, control most activities of the cell and the plant.

The genetic code consists of 64 codons in mRNA. Codons are three mRNA nucleotides long. Each codon specifies a particular amino acid or acts as a start or stop sign for protein synthesis.

Eukaryotic genes contain intervening sequences of DNA, which become introns of pre-mRNA.

How does the recent discovery of intervening sequences make a definition of a gene even more difficult to pin down?

Genetic Engineering Is a Major Focus of Modern Biology.

Can you imagine a blue rose, a potato that makes plastic, a tobacco plant that glows in the dark, or a truly tasty tomato from the supermarket? All of these products already have been made by genetic engineering, which is the artificial manipulation of genes or the transfer of genes from one organism to another (fig. 5.15; see appendix B on the techniques of genetic engineering). Such artificial (i.e., human-

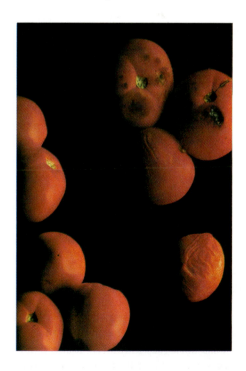

FIGURE 5.15

Genetically engineered tomatoes: spoilage-resistant (*left*) and normal tomatoes.

5.2 *Perspective* Perspective

Public Issues in Plant Genetic Engineering

Regardless of the seemingly unlimited potential for human benefit from transgenic plants, the scientific and commercial aspects of plant genetic engineering bring forth some emotional, ethical, and legal issues for the public. One of the main developments giving rise to these issues is that the natural barriers to transferring genes from one organism to another are rapidly disappearing. Unlike breeding experiments, where hybridization transfers genes between closely related species, genetic engineering enables the transfer of genes between organisms from different kingdoms.

Organizations already have been formed by people who object to genetic engineering. Some groups believe that it is too dangerous to make transgenic organisms because we might accidentally make and release superresistant disease bacteria or uncontrollable weeds into the environment. Other groups believe that it is morally or ethically wrong to manipulate genes artificially. These are just some of the social issues that have been explored in conferences, books, articles, and government reports over the past several years. These kinds of issues are certain to remain at the forefront of public consciousness for years to come.

---

directed) exploitation of genes is also called recombinant DNA technology. In the examples above, the rose has genes from the petunia that control the synthesis of blue floral pigments. The potato has bacterial genes that make polymers that can be used to make biodegradable plastics. The luciferase gene, which causes fireflies to glow, has been transferred to tobacco plants as an easily identifiable marker gene for the detection of successful gene transfers; if the plants glow, then the transfer of the luciferase gene and other genes linked to it was successful (thus the luciferase gene marks the transfer). These plants are all **transgenic,** which means they contain genes from other organisms. Conversely, tomatoes were genetically engineered when the tomato gene for a specific carbohydrate-degrading enzyme was altered in the laboratory and reinserted into the tomato plant. The vine-ripened fruits from these plants, which have all their natural flavor, stay firm during shipping and storage (fig. 5.15), but marketing problems have removed them from the market.

Genetic engineering has had, and will continue to have, an impact on society (see Perspective 5.2, "Public Issues in Plant Genetic Engineering"). Newspapers announce some breakthrough or controversial outcome of genetic engineering almost daily. The cloning of Dolly the sheep is one headline example. Do you know what cloning means or how genes are combined in different organisms?

Botanists have discovered a variety of genes involved in disease resistance by plants. For example, the RPS2 gene in *Arabidopsis thaliana* (see Perspective 5.3, "An Important Little Weed") confers resistance to pathogenic bacteria such as *Pseudomonas syringae.* Similarly, tobacco mosaic virus (TMV) causes a mosaic pattern of infection on tobacco leaves (fig. 5.16). When the gene for the protein coat of TMV is transferred to tobacco, the plants become resistant to the virus. The TMV responds to the genetically engineered to-

FIGURE 5.16

This tobacco leaf is infected with tobacco mosaic virus (TMV). Resistance to this virus can be engineered into tobacco by using a gene from the virus itself.

bacco cells as if they were already infected, and since the virus cannot infect cells that are already infected, the plant becomes immune to infection.

Crops often can be protected against insects by being sprayed with bacteria that contain the toxin gene from *Bacillus thuringiensis.* This gene also has been inserted into tomato, potato, and tobacco plants. Caterpillars that normally eat the leaves of these species do not eat plants that contain the toxin made by this gene. Perhaps the most important crop for genetically engineering resistance to insects is cotton (*Gossypium hirsutum*); indeed, most of the chemical pesticides used in the United States are used on cotton. Experimental cotton plants that contain the toxin

5.3 *Perspective* Perspective

An Important Little Weed

More than a half century ago, German botanist Friedrich Laibach claimed that he had found the perfect plant for studying genetics: *Arabidopsis thaliana,* a commercially useless member of the mustard family. Although Laibach's advice was initially ignored, today *Arabidopsis* is one of the most prized organisms for studying how genes control development. Here's why:

- *Arabidopsis* is small and easy to grow. Thousands of plants can be grown on a lab bench under fluorescent lights. No tractors or overalls are required.

- *Arabidopsis* has a short generation time; its entire life cycle takes only about a month. Moreover, each plant produces thousands of seeds.

- *Arabidopsis* has the smallest known genome of any flowering plant. Indeed, its 30,000 genes are on only five pairs of chromosomes. For comparison, the genome of corn is 30 to 40 times larger than that of *Arabidopsis.* The genome of *Arabidopsis* is only about 20 times larger than that of the common bacterium *Escherichia coli.* Unlike most other plants, *Arabidopsis* has almost no repetitive DNA; this is why plants such as wheat and tomato—which have about the same number of genes as *Arabidopsis*—have so much more DNA than does *Arabidopsis.* The relatively simple genome of *Arabidopsis* makes it easier for botanists to associate specific genetic functions with certain regions of DNA.

Today, botanists are using *Arabidopsis* to answer a variety of questions about plant growth and development, evolution, and genetics. Studies of *Arabidopsis* have uncovered genes that regulate processes ranging from photosynthesis to disease resistance and flowering; one group of botanists has inserted bacterial genes into *Arabidopsis* to create plants that produce biodegradable plastics. There's even an international program to sequence the genome of *Arabidopsis.*

gene from *B. thuringiensis* typically have better resistance to insects and require fewer chemical pesticides.

Grains from cereal grasses are deficient in lysine, which is an essential amino acid in our diet. However, induced mutations in cultured rice cells have increased their production of lysine and protein. Plants from these mutant cell cultures produce grains containing more lysine and protein than do normal rice plants. Botanists hope to transfer the gene for the high-lysine mutation to other cereal grasses.

In countries such as Brazil, farmers often lose 20% to 40% of their crops to storage pests such as weevils. To address this, plants are being engineered whose fruits are pest resistant. For example, botanists have produced bean plants with fruit that produces an inhibitor of α-amylase (a universal digestive enzyme in animals). This inhibitor blocks the α-amylase secreted by attacking weevils, thereby reducing the harvest's susceptibility to the pest.

Genes from desert petunias have been transferred into cultivated petunias. The transgenic petunias require much less water than the normal plants. If crop plants likewise can be altered genetically to require less water, there will be less need for irrigation. Such crops will become more important as the demand for water increases in semiarid agricultural regions of the southwestern United States.

Sales of nonfood products from genetically transformed plants will grow from about $21 million per year in 1997 to $320 million or more by 2005. The first nonfood commercial product of plant genetic engineering was lauric acid, a 12-carbon fatty acid that is used in making soaps and detergents. The transgenic plant that makes this chemical is rapeseed (*Brassica napus*). Normally, fatty acids in this plant are 18 carbons long, but scientists at a company named Calgene have inserted into rapeseed a gene from the California bay tree (*Umbellularia californica*) that shuts off fatty acid synthesis after 12 carbons. Oil from a few thousand acres of transgenic rapeseed, which is easy to grow and a good oil producer, was marketed in 1995. Other transgenic rapeseed lines, in different stages of development, will produce lubricants, nylon, and fatty acids for making margarine and shortening. Besides fatty acids, other polymers from transgenic plants include a biodegradable plastic that is made by bacterial genes inserted into *Arabidopsis thaliana* and polyesterlike fibers from similar genes inserted into cotton. Neither of these polymers has yet reached commercial production.

INQUIRY SUMMARY

Genetic engineering, which is the artificial manipulation of genes or the transfer of genes from one organism to another, has produced many important products. Such artificial (i.e., human-directed) exploitation of genes is also called recombinant DNA technology.

There is much public concern over perceived dangers of genetic engineering. Why is this so? What do you think?

How Do Plant Cells Divide?

How do plants become larger? If you have ever raised a plant in your house or in a garden, you may have wondered how it grew from a seed or small seedling into a larger, different-looking adult plant. Before a plant can grow and its parts can become specialized, new cells must be produced. How does this occur? Here we begin to answer this important question. Chapter 6 will further help you solve this mystery.

Every plant begins as a single cell. It grows and develops as cells repeatedly expand and divide and then become specialized as some genes are turned on and others turned off. As the plant matures, fewer cells continue to divide. Most cells instead have become specialized for a limited set of functions, such as photosynthesis, storage, and transport. Cells that continue to divide usually occur only in specific meristematic regions of the plant, such as stem tips, root tips, and buds; in these regions, cellular expansion and division can theoretically continue indefinitely. The uninterrupted repetition of cell growth and division, together called the **cell cycle,** gives plants the potential for unlimited growth. Some plants, such as giant redwoods, come closer to this potential than others. An overview of the important events of the cell cycle in plants allows us to understand how plant cells divide and, therefore, how plants grow and develop.

Most of the attention devoted to the cell cycle involves the mechanisms of DNA synthesis and the behavior of chromosomes during mitosis. These subjects are the focus of the remainder of this chapter. When coordination of the events of the cycle is lost, cells can divide out of control. In humans, this is known as cancer. Do plants get cancer?

The cell cycle is completed in cells that divide.

The first dividing cell in the life history of a plant is the **zygote,** which forms by fusion of a sperm and an egg. The zygote grows into an **embryo** by cellular division and expansion as groups of cells become specialized into tissues. The embryo has specialized regions, called **meristems,** that undergo cell division; that is, certain cells in meristem tissue can complete the cell cycle. In flowering plants, as a seed germinates and the seedling develops, meristematic cells at the growing tips of developing roots and shoots never stop dividing. In contrast, cells in leaves and certain tissues of stems and roots stop dividing and become specialized; the cell cycle in these cells is arrested, and they do not normally produce new cells.

Cells also divide and grow in several other parts of flowering plants. For example, flowers, fruits, and branches arise by cell division in buds, which contain meristems that are active only during specific periods of plant development. Such buds are usually attached to stems; you may have seen buds on trees and bushes around campus or else-where. Roots also contain meristematic regions that form branch roots. Many kinds of nonmeristematic cells in stems and roots and, in the leaves of some plants, can be stimulated to divide when they are damaged. Cell division in damaged tissues continues until the wound is sealed with a layer of protective cork. This is how a tree trunk can repair itself, at least partially, after someone foolishly carves his or her initials in it. Thus, the cell cycle can be resumed, when needed, for the formation of reproductive organs, stem and root branches, and wound-repair tissue. Botanists say that plant cells are **totipotent** because they can become meristematic, resume cell division, and develop new tissue. In fact, because each individual plant cell contains a complete and identical set of genes, single cells can develop into entire new plants (see Perspective 5.4, "The Many Uses of Plant Tissue Culture").

The cell cycle involves growth followed by cell division.

In 1858, nearly 200 years after cells were first seen through a microscope, the German physiologist Rudolf Virchow proposed that all cells must come from preexisting cells, setting the foundation of the cell theory. This now obvious statement was a breakthrough in the thinking of biologists at the time. Since then, we have discovered the formation of new cells includes processes of the nucleus and cytoplasm. During the past century, however, much more attention has been given to events that occur in the nucleus. Biologists focused their research on the nucleus because chromosomes in dividing cells were relatively easy to see with light microscopy.

The life of a dividing plant cell has two main chapters: cell growth and cell division. Traditionally, the stages during cell growth have been collectively called interphase, and the stages of cell division are called mitosis and cytokinesis. Mitosis is the division of nuclei, and cytokinesis is the division of cytoplasm into two identical cells. More recently, these two stages have been separated into four continuous phases called the cell cycle. The first is the G_1 **phase** (meaning the first gap or growth phase), which occurs between the end of mitosis and the onset of DNA synthesis. This is followed by the **S phase,** which is when DNA is synthesized. The third part of interphase is the G_2 **phase** (second gap or growth phase), which begins at the end of the S phase (i.e., when DNA synthesis is complete) and lasts until the beginning of mitosis, which has been named the **M phase.** Mitosis is a continuous process that is, for our convenience, divided into four phases: **prophase, metaphase, anaphase,** and **telophase.** Cytokinesis usually follows telophase and completes the formation of the two offspring cells. Figure 5.17 is a model of a plant cell cycle.

The cell cycle is completed only in dividing cells. Most cells do not divide after they mature. Because all living cells are totipotent, however, they may revert to division under certain conditions. Knowledge of the cell cycle

5.4 *Perspective*Perspective

The Many Uses of Plant Tissue Culture

Virtually all cells in a plant have identical sets of chromosomes in their nuclei. This observation leads to the prediction that each cell is totipotent—that is, each cell has the same genes and therefore the same genetic potential to make all other cell types. The first confirmation of this prediction came from experiments in the 1950s. Small pieces of tissue from carrots were grown or cultured in a nutrient broth. Eventually, cell division and development produced new plants.

Whole plants of many species are now routinely regenerated from cell cultures. This process is used, for example, to grow clones of orchids and other ornamental plants inexpensively and to make disease-free clones of potatoes and other crop plants. Your house fern probably came from tissue culture. Cloning by cell culture is also important in genetic engineering of plants, as discussed earlier.

The main steps involved in culturing vegetative cells and regenerating a new plant from cell culture. Unspecialized cells divide and differentiate into all of the cell types needed to grow into another plant. The results of this experiment are evidence that each cell in a carrot has the genetic potential to make all other cell types; that is, plant cells are totipotent.

allows us to understand how plants grow and develop specialized organs.

Interphase includes G$_1$, S, and G$_2$ of the cell cycle.

Most meristematic cells (see chapter 4) spend about 90% of their time in interphase (fig. 5.17). However, throughout the first half of this century, little was known about what happens during interphase, since distinct chromosomes could not be seen by light microscopy for this part of the cell cycle. In fact, early biologists described interphase as the "resting stage" because they could not see chromosomes or any other activities. We now know better than to describe interphase as a time of rest.

Growth begins during the G$_1$ phase. Cellular activities are reactivated during the G$_1$ phase following cell division. For example, cytoskeletal microtubules reassemble to support the growing cell. Also, the cell enlarges, organelles multiply, mitochondrial and plastid DNA increase severalfold, and many enzymes and structural proteins are made during the G$_1$ phase. One of the main tasks of G$_1$ in plants is the synthesis of nucleotides that will be used to make DNA in the S phase to follow.

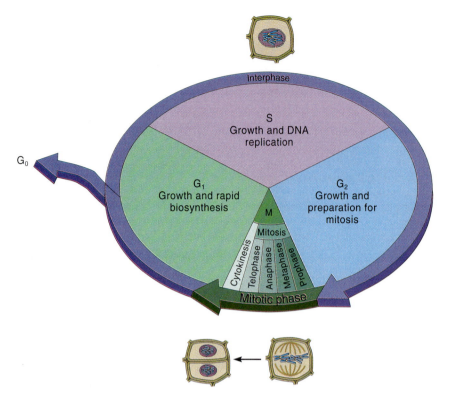

FIGURE 5.17

Periods of the cell cycle. Most meristematic cells spend about 90% of their time in interphase and about 10% of their time in mitosis and cytokinesis. The majority of plant cells do not complete the cycle and enter G_0; that is, most plant cells become specialized and never divide, remaining in G_0.

G_1 is the most variable phase in the cell cycle; it can be almost absent in rapidly dividing cells, or it can last for hours or days in slowly dividing cells. After division, most new plant cells become specialized for a certain task, such as leaf cells for photosynthesis, and never leave the G_1 phase; this is renamed the G_0 stage for cells departed from the cell cycle. Often under the influence of plant hormones (see chapter 6), some meristem cells (and even some non-meristem cells) proceed beyond G_1 and begin the S phase as they continue the cell cycle.

DNA is synthesized during the S phase by semiconservative replication. As mentioned previously, when Watson and Crick proposed the double-helix model for the structure of DNA, they immediately saw that this structure revealed a potential mechanism for self-replication. They envisioned that the two strands would "unzip" between base-pairs by breaking the hydrogen bonds between them. Each single strand could then act as a template for guiding the formation of a new chain onto the old one (fig. 5.18). According to this postulate, two double-stranded molecules would be produced, each made of one parent strand and one new strand. This is referred to as **semiconservative replication** because half of each offspring molecule is conserved from one strand of the parent double helix. Replica-

tion is DNA synthesis on a DNA template; it also has been called DNA duplication.

DNA replication occurs when DNA polymerases catalyze the bonding of newly made nucleotides into a new strand (fig. 5.19). The order of nucleotides in the new strand is directed by the sequence of the single-stranded DNA template in the parent strand. Thus, only bases complementary to the template can be added to the growing chain, which keeps adenines (A) opposite thymines (T) and cytosines (C) opposite guanines (G) in the developing double strand. The significance of the S phase is not in doubling the DNA content but in making an exact copy of the DNA and, therefore, the genes of the cell.

The G_2 phase is relatively short in plant cells and is the end of interphase. The G_2 phase begins after chromatin synthesis is complete, when newly replicated chromatin gradually begins to coil and condense. During G_2, the cell finishes preparing for mitosis. This preparation includes making tubulins for mitotic microtubules, making proteins for processing chromosomes, and breaking down the nuclear envelope. During G_2, as well as G_1, the amino acids required for protein synthesis are manufactured as are the nucleotides for RNA synthesis. The cell also makes all three types of RNA and thousands of proteins.

The average length of a G_2 phase is 3 to 5 hours, and it continues until mitosis begins. The end of G_2 is the end of interphase, but do not think it is a clear stop and start; some events may overlap, and there may be no clear, absolute boundary in these human-made designations.

Mitosis and cytokinesis form new, identical cells during the M phase of the cell cycle.

In one of the earliest descriptions of dividing nuclei, German cytologist Walther Flemming referred to the various stages of chromatin as "mitosen" (threads), giving rise to the name *mitosis* for the process of nuclear division. Flemming's work in the 1870s inspired many studies of mitosis. By the start of the twentieth century, biologists knew that this process was the foundation for nuclear division in all plants and animals.

Mitosis refers to the separation of chromosomes and the formation of two genetically identical offspring nuclei. Mitosis is usually followed by cytokinesis, and together these events comprise the **M phase** of the cell cycle. Mitosis can be studied in several types of meristematic cells, either in whole plants, in cell cultures, or in wound or tumor

The Lore of Plants

As often happens in biology, "normal" processes, such as those of the cell cycle, have many exceptions. One common exception to the cell cycle is that, in many organisms, the S phase can occur many times before mitosis. The result is that the nuclei of some cells can have much more than the expected amount of DNA. For example, hair cells and glandular cells may have from 16 to more than 4,000 times the "normal" amount of DNA. The record for plants is probably in certain cells of the seeds of the lords-and-ladies *Arum* (*Arum maculatum*), which have more than 24,000 times the "normal" amount of DNA. But this number pales compared to the giant neurons of a certain species of mollusk, which have at least 75,000 times the amount of DNA.

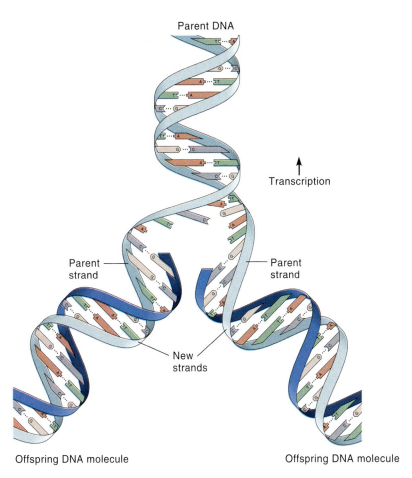

FIGURE 5.18

Semiconservative replication means that half of each parent molecule of DNA is conserved in each offspring molecule; the other half is newly synthesized. After replication, the two offspring molecules are identical to the parental double helix, and will become the two chromatids.

tissues. For example, mitosis can be readily observed in onion root tips, as shown in figure 5.20.

Chromosome condensation starts prophase of mitosis. Chromatin of the G_2 nucleus condenses rapidly during early prophase (fig. 5.21a). As condensation proceeds, chromatin appears as a mass of elongated threads. By late prophase, individual, replicated chromosomes are visible, appearing as two parallel threads attached at a constriction point called the **centromere** (figs. 5.21a and 5.22). In this double-threaded form of the replicated chromosome, each thread is called a **chromatid;** however, together the two attached chromatids are still considered to be one chromosome because there is just one centromere. Each chromatid is one of the two double strands of semiconservatively replicated DNA and associated histones from the S phase. Nonetheless, since the two chromatids are attached at a centromere, together the two are considered one replicated chromosome (the chromatids remained attached following replication in the S phase).

The end of prophase is marked by the disappearance of nucleoli and the nuclear envelope. Without a nuclear boundary, chromosomes protrude into the cytoplasm, and their behavior appears to be chaotic and uncontrolled. As we will see, this could not be further from the truth.

Chromosome alignment occurs during metaphase, and separation occurs during anaphase. After the nuclear envelope dissolves, several nearby parallel fibers, called spindle fibers, accumulate near the chromosomes in early metaphase. Spindle fibers, which are made of microtubules, are collectively called the spindle apparatus. The mature spindle apparatus is elongated, with its ends pointing to opposite poles of the cell (fig. 5.22).

As viewed by light microscopy, the metaphase chromosomes line up independently of one another in a circle around the circumference of the spindle apparatus. By the end of metaphase, the chromosomes all have been forced into a tight plane at a specific position in the cell. "Disorder" has become order. Each chromosome can now separate at its centromere with subsequent separation of the two chromatids.

During anaphase, which is the shortest stage of mitosis, chromatids attached to the same centromere separate

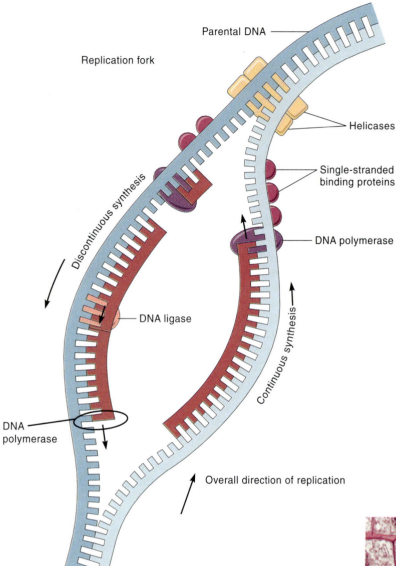

Parental DNA

Replication fork

Discontinuous synthesis

DNA ligase

DNA polymerase

Helicases

Single-stranded binding proteins

DNA polymerase

Continuous synthesis

Overall direction of replication

FIGURE 5.19

DNA replication requires enzymes called helicases to unwind the double helix, single-stranded binding proteins to stabilize the unzipped double strand, DNA polymerase to assemble nucleotides into a new strand, and DNA ligase to bind fragments together from discontinuous synthesis, which goes in steps. Note the overall direction of synthesis. Continuous synthesis occurs one DNA nucleotide at a time.

Formation of new nuclei occurs during telophase, while two identical offspring cells form during cytokinesis to complete the cell cycle. During telophase, the formation of two new offspring nuclei seems to mimic prophase in reverse. The spindle apparatus disappears, a nuclear envelope forms around each of the two sets of chromosomes, and nucleoli appear in each new nucleus (see fig. 5.21d). The chromosomes steadily elongate and decondense once again into diffuse chromatin.

During late anaphase or early telophase, the cell begins cytokinesis. As part of this, a cylindrical system of short microtubules forms in a plane between the offspring nuclei, which will be the division plane of the cytoplasm. As cytokinesis continues, the new cell walls of the offspring cells begin to form. This structure is called the cell plate, which grows outward, forming the new cell wall (see fig. 5.21d and e).

and move to opposite poles of the spindle (see fig. 5.21c). Upon separation, the term chromatid no longer applies; each chromatid is considered a separate chromosome because it now has its own centromere. Because of the accuracy of replication of the two chromatids during the S phase, each of the two new separated chromosomes is a copy of the other chromosome from the pair. Thus, each new offspring cell ends up with the same set of duplicate chromosomes; using modern terms, they are clones of each other because they are identical.

In most cells, the spindle elongates during anaphase, thereby increasing the distance between the poles. This movement further separates the two sets of chromosomes as they move along the spindle. By the end of anaphase, the two genetically identical sets of chromosomes have been separated and are ready to be packaged in the two new nuclei.

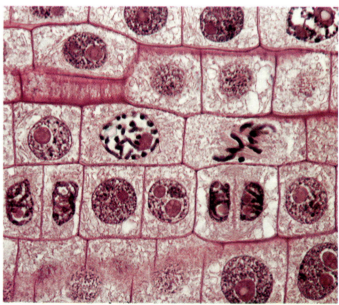

FIGURE 5.20

Mitosis can be seen easily among the nuclei of actively dividing cells, such as these of onion root tip (*Allium cepa*). At any one time, most nuclei are in interphase (×280).

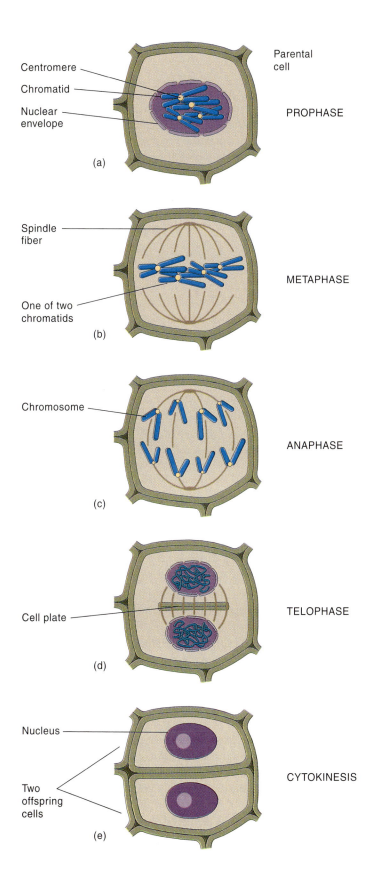

Centromere
Chromatid
Nuclear envelope

Parental cell

PROPHASE

(a)

Spindle fiber

METAPHASE

One of two chromatids

(b)

Chromosome

ANAPHASE

(c)

Cell plate

TELOPHASE

(d)

Nucleus

CYTOKINESIS

Two offspring cells

(e)

FIGURE 5.21

Mitosis and cytokinesis following G_1, S, and G_2 phases. (a) Chromosome condensation: prophase. (b) Chromosome alignment: metaphase. (c) Chromosome migration: anaphase. (d) Chromosomes decondense and new nuclei are formed: telophase. The cell plate forms at this point if telophase is to be followed by cytokinesis. (e) Cells divide: cytokinesis.

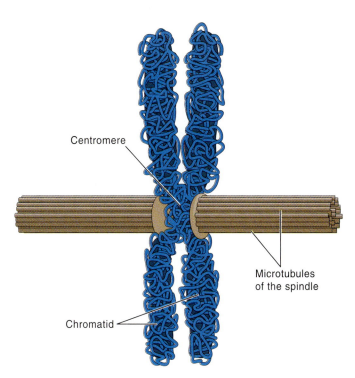

Centromere

Microtubules of the spindle

Chromatid

FIGURE 5.22

Diagram of a chromosome in metaphase with spindle fibers from opposite ends of the cell attached at the centromere. A spindle is made of microtubules. Centromeres appear as a constricted region of a replicated chromosome (fig. 5.21a) where the two chromatids are held together.

INQUIRY SUMMARY

Plant meristems are regions where cells can grow and then divide in two continuous stages, originally designated as interphase and as mitosis. The cell cycle model further separates these two stages into four phases: G_1 S, G_2 (interphase), and M (mitosis and cytokinesis) as a set of processes that foster cell growth and division. G_1 involves growth and metabolism. S is the phase of DNA synthesis or replication in which exact copies of the genes are synthesized. G_2 is the growth and metabolism phase after DNA synthesis but before cell division (the M phase). Mitosis is a continuous process that produces two nuclei containing identical sets of chromosomes. Cytokinesis splits the cytoplasm of a dividing cell into two genetically identical offspring cells. These meristematic cells have competed the cell cycle and may continue to divide or become specialized as part of a plant tissue. Most cells do not exit the G_1 phase and remain undivided in G_0.

The importance of replication and mitosis is not so much in doubling the number of cells but in forming two genetically identical offspring cells. The two cells are a clone of the parent cell, so plants have been cloning themselves long before humans got into the game of cloning.

What is the biological significance of the S phase? The M phase?

How Is Cell Division Controlled?

After the cell cycle was described, botanists began to ask, What controls the cycle? Why do some cells divide rapidly while others divide more slowly? For example, under rather meager growing conditions, buffalo gourd (*Cucurbita foetidissima*) can grow 25 to 35 centimeters a day, requiring rapid production of new cells via cell division. Jade plants (*Crassula ovata*) may not grow that much in a year, even in their normal habitat, so production of new cells in these plants is much slower than buffalo gourd. What controls the different rates of cell division in species?

During the past 10 years, research on the cell cycle has focused on the properties of **cyclins,** which are proteins that activate cell cycle enzymes and **cell division cycle (cdc) genes.** Most of the cdc genes that have been discovered so far regulate transitions from the G_1 phase to the S phase or from the G_2 phase to mitosis. There is evidence cdc genes are influenced by plant hormones and by external stimuli (chapter 6).

Studies of one of the yeast genes that controls the timing of the cell cycle, called a mitotic inducer gene (cdc25), have revealed how plant cdc genes might function. This discovery was made when cdc25 was transferred to tobacco. (Appendix B describes how genes can be transferred between organisms.) The cells of tobacco plants containing cdc25 underwent faster division in some cells of their flower buds. This resulted in smaller than normal offspring cells. Such plants also produced flowers faster but without petals. This experiment shows how a gene that controls the timing of cell division affects cell size and floral development. Plant scientists believe that the action of cdc25 inserted into tobacco provides clues about how the apparently similar cdc genes work in plants. This is the first step toward understanding the molecular regulation that links the cell cycle with cell size and plant development. Can you state some reasons why botanists, particularly those working in agriculture, might be interested in control of the cell cycle?

As we shall learn in chapter 6, plant growth and development are complicated processes, subject to various environmental conditions and genetic controls, including plant hormones. Nonetheless, the final target of these controls is often at the cell cycle, which can influence whether or not cells divide.

SUMMARY

Mendel's classic studies on peas began the field of modern genetics. Genes serve two crucial functions: (1) transmission of the genetic information to offspring cells following cell division, and (2) expression of the genetic information in a cell.

The nucleus contains DNA associated with histone proteins that make up the chromosomes. Plant genes are found in the chromosomes. DNA is the genetic material. The small beads of chromosomes, revealed by electron microscopy, are called nucleosomes and are the basic structural unit for packaging the DNA of the chromosome. The entire DNA plus protein association, including the nucleosomes and the DNA/protein between them, is called chromatin.

The discovery of the structure of DNA was a major breakthrough in science. Watson and Crick reconciled all of the available data and proposed a model of a double helix—that is, a two-stranded spiral.

Gene expression requires RNA synthesis (transcription) followed by protein synthesis (translation). While still in the nucleus, RNA is edited (cut, spliced, or otherwise modified) and then exported into the cytoplasm. There are three main kinds of RNA made in the nucleus on a DNA template, edited, and exported to the cytosol:

- Messenger RNA (mRNA) is the actual message for assembly of a protein from amino acids.
- Transfer RNA (tRNA) binds to amino acids, so that they can be brought together during protein synthesis in a ribosome.
- Ribosomal RNA (rRNA) along with ribosomal proteins comprise a ribosome, the workbench of protein synthesis.

The near-universal genetic code is a set of RNA nucleotide messages that specify amino acids during translation in a ribosome. The code is made of sets of three nucleotides called codons. Overall, genes work by transferring information from DNA to protein, and by directing the sequence of nucleotides in mRNA, which determines the sequence of amino acids in the protein. The specific amino acid sequence determines the exact structure and therefore the distinctive function of a protein.

The cell cycle helps us understand cell growth and division. The two main stages of the cell cycle are interphase, which includes DNA synthesis, and cell division. Interphase is further recognized to have three phases: (1) the growth phase between the end of the previous mitosis and DNA synthesis (G_1 phase), (2) the phase involving DNA synthesis (S phase), and (3) the growth phase between the end of DNA synthesis and the beginning of mitosis (G_2 phase). Finally, cell division involves mitosis followed by cytokinesis, which constitutes the M phase of the cell cycle. Most plant cells do not divide.

DNA synthesis is referred to as semiconservative replication because each new double helix contains one strand from the previous double helix. Just one strand is newly synthesized. Semiconservative replication uses a single strand of parent DNA as a template for guiding the synthesis of the new strand. The sequence of nucleotides in the new strand complements that of the parent strand. Histones are also synthesized during the S phase, resulting in duplicate chromosomes formed as two identical chromatids joined at a centromere.

The main phases of mitosis are easily observable by light microscopy. Chromosomes condense and become visible in prophase, align along a division plane in metaphase, split into two identical sets of chromosomes and move to opposite poles of the cell in anaphase, and organize into separate offspring nuclei in telophase. Mitosis is usually followed by cytokinesis, which begins as early as late anaphase as new cell walls start to form between offspring cells. Completion of the cell cycle results in two genetically identical offspring cells, which can then expand and contribute to the growth of the plant. The cell cycle is, at least in part, thought to be under control of cyclin proteins and cdc genes.

Writing to Learn Botany

Should we be concerned about releasing artificially produced genes into nature? Why or why not?

THOUGHT QUESTIONS

1. Define a gene.

2. Mitosis is said to be "observable" with a light microscope. However, meristematic cells are killed before they are mounted onto a glass microscope slide for study. How, then, do we observe this living process by examining dead cells?

3. DNA polymerases catalyze the assembly of DNA chains. Where do the nucleotides come from?

4. Organellar and nuclear genomes must be passed to offspring cells during cell reproduction. During which period(s) of the cell cycle are plastids and mitochondria most likely to be reproduced? Why?

5. Onion and wall cress have a 63-fold difference in their respective genome sizes but less than a 4-fold difference in S-phase duration in their root tip cells. How can we account for the comparatively much faster duplication of the onion genome?

6. What properties of histones do you think are especially important for the function of these proteins in maintaining chromatin structure? Why?

7. The genomes of organisms contain gigantic amounts of information. For example, each human parent contributes to an embryo a set of DNA consisting of almost 3.3 billion nucleotides arranged in 23 chromosomes. If such a DNA sequence were typed in a 10-point font on a continuous ribbon, the ribbon would stretch from San Francisco to Chicago, then to Baltimore, Houston, Los Angeles, and finally back to San Francisco. Even individual genes consist of huge amounts of information. How would you organize that much information so that you could compare it with the findings of other biologists?

8. How may introns regulate genes?

9. What is the major biological significance of the cell cycle?

10. Starting with the gene, what are the fastest and slowest steps in protein synthesis? What makes the fast steps fast and the slow steps slow?

11. Develop an explanation of why RNA synthesis is called transcription.

12. Develop an explanation of why protein synthesis is called translation.

SUGGESTED READINGS

Articles

Barton, J.H. 1991. Patenting life. *Scientific American* 264:40–46.

Boulter, D. 1995. Plant biotechnology: Facts and public perceptions. *Phytochemistry* 40:1–9.

Chang, F., and P. Nurse. 1993. Finishing the cell cycle: Control of mitosis and cytokinesis in fission yeast. *Trends in Genetics* 9:333–35.

Francis, D., and N.G. Halford. 1995. The plant cell cycle. *Physiologia Plantarum* 93:365–74.

Gasser, C.S., and R.T. Fraley. 1992. Transgenic crops. *Scientific American* 266:62–69.

Grasser, K. 1998. HMG1 and HU proteins: Architectural elements in plant chromatin. *Trends in Plant Science* 3(7):260–65.

Jacobs, T.W. 1995. Cell cycle control. *Annual Review of Plant Physiology and Plant Molecular Biology* 46:317–30.

Kappeli, O., and L. Auberson. 1998. How safe is safe enough in plant genetic engineering? *Trends in Plant Science* 3(7):276–81.

Moffat, A.S. 1995. Exploring transgenic plants as a new vaccine source. *Science* 268:658–60.

Murray, A.W. 1993. Cell-cycle control: Turning on mitosis. *Current Biology* 3:291–93.

Powledge, F. 1995. Who owns rice and beans? *BioScience* 45:440–44.

Books

Alberts, B., D. Bray, J. Lewis, M. Raff, K. Roberts, and J.D. Watson. 1994. *Molecular Biology of the Cell*, 3d ed. New York: Garland.

Anderson, J.W., and J. Beardall. 1991. *Molecular Activities of Plant Cells*. Oxford: Blackwell Scientific Publications.

Becker, W. L. Kleinsmith, and J. Hardin. 2000. *The World of the Cell* 4th ed. San Francisco, CA: Benjamin/Cummings Publishing Co.

Lehninger, A.L., D.L. Nelson, and M.M. Cox. 1992. *Principles of Biochemistry,* 2d ed. New York: Worth.

Taiz, L., and E. Zeiger. 1998. *Plant Physiology,* 2d ed. Sunderland, MA: Seigneur Associates.

ON THE INTERNET

Visit the textbook's accompanying web site at
http://www.mhhe.com/botany to find live Internet links for each
of the topics listed below:

Gregor Mendel

Cell division: Mitosis and meiosis

Structure of DNA

The gene and its function

Chromosomes

Fundamentals of protein synthesis

The genetic code

Gene regulation and expression

Genetic engineering in plants

The cell cycle

The Cape Sundew (Drosera capensis) *on the left and the Boston Fern* (Nephrolepis exaltata) *on the right were both grown from cultured meristematic cells. How is it possible for one cell to develop into an entire organism with different structures?*

CHAPTER 6

Plant Growth and Development

LEARNING ONLINE

The Online Learning Center at www.mhhe.com/botany is full of useful resources for you such as chapter specific hyperlinks that send you right to Internet sites with information on exactly what you're stydying. For example, visit www.plant-hormones.bbsrc.ac.uk/education/kenhp.htm in chapter 6 of the OLC to learn more about how genes, hormones, and the environment shape plant growth.

Plants have a long and interesting evolutionary history. They evolved from algae around 430 million years ago and then diversified in a way that forever changed our planet: they invaded land, thereby paving the way for the subsequent colonization of land by animals. The terrestrial habitats encountered by plants were more diverse than their ancestral aquatic environments. Land plants faced many new environmental challenges and selective pressures. As a result, they evolved rapidly and became more complex and diverse than did their algal ancestors.

Previously, we learned that life on land presented numerous environmental challenges (a.k.a. selective pressures) to evolving plant populations. For example, sustaining a moist internal environment was critical since unprotected plant cells exposed to dry air quickly dessicate and die. Therefore, the successful establishment of flora in dry air and soil depended on adaptations that ensured continued dampness inside the plant. Waxy coverings, extensive root systems, stomata, structural support mechanisms and even modified leaves evolved at different times in different environments. The growth

and development of plant tissue systems and organs was the adaptative "solution" to the environmental pressures of life evolving on land.

In chapter 5, we learned plant cells grow and divide by completing the cell cycle. Mitosis and cytokinesis form two identical offspring cells. Most of these cells do not divide again but instead become specialized as they differentiate into the cells of tissue systems found in plants (see chapter 4). When cells differentiate, they undergo a progressive, developmental change to a more specialized form or function.

How do cells differentiate (i.e., become specialized) and form the organs of the primary body of a plant as described in chapter 2? This is a fundamental question in botany that we answer here as we discuss plant hormones and environmental cues. If plants did not form tissues and then organs, all of their parts would look and function alike; there would be no carrots (root), asparagus (stem and leaf), artichokes (bud), beans (seed), or apples (fruit) to eat. All of the vegetables in our diet might all look and taste the same.

Plant Growth and Development Are Controlled by Many Internal and External Factors.

We have learned about the kinds and importance of the four plant tissue systems, and in later chapters, we see how these tissues are formed into vegetative organs—roots, stems, and leaves, and into reproductive organs—cones, flowers, fruits, and seeds. Yet, a fundamental question in botany remains to be answered: What controls the growth and development of these tissues? How do cells specialize into different tissues, and how do these tissues form specialized organs? Why are root cells root cells, and why are leaf cells leaf cells? Why does a carrot look and taste like it does, and how is it so different from broccoli or an ear of corn? How do genes control what a plant cell becomes? Does the environment affect plant development? While we have some of the answers to these and similar questions, we certainly do not have them all. In fact, today, as you read this book, botanists are conducting research to fill in the blanks and advance our understanding of the green world.

To begin to answer these questions, there are some basic concepts we need to introduce. Plant growth and development result from interactions of cellular division, cellular enlargement, and the patterned differentiation of cells into tissues and tissues into organs. Roots and shoots grow from

meristems, which are embryonic regions of cellular division and expansion.

Remember, growth enlarges a plant, but it doesn't necessarily result in a plant adapted to its environment. Rather, suitable adaptation results from differentiation, whereby new cells are organized into tissues and organs specialized for different functions. Because organs such as leaves, roots, and stems are produced continually, each can foster acclimation and adaptation to changing environmental conditions. For example, leaves of plants grown in dim light are often larger and less green than leaves on plants grown in bright light. Growth and development are limited by a plant's genome (its complement of genes) and are integrated by environmental signals such as light, moisture, and gravity.

What is growth? This seemingly simple question can be tricky, and we have postponed a detailed discussion until now. Growth can mean several different things, such as an increase in size, volume, weight, or number of cells. For example, a piece of dry wood swells (a.k.a. "grows") when placed in water, and a child who is "growing up" becomes taller. We shall define **growth** as any permanent increase in the size of an organism or its parts. Growth is accompanied by metabolic processes that occur at the expense of metabolic energy. Thus, the production of a leaf is growth, whereas the water-induced swelling of wood is not. Cell division usually accompanies cell enlargement, which usually results in growth.

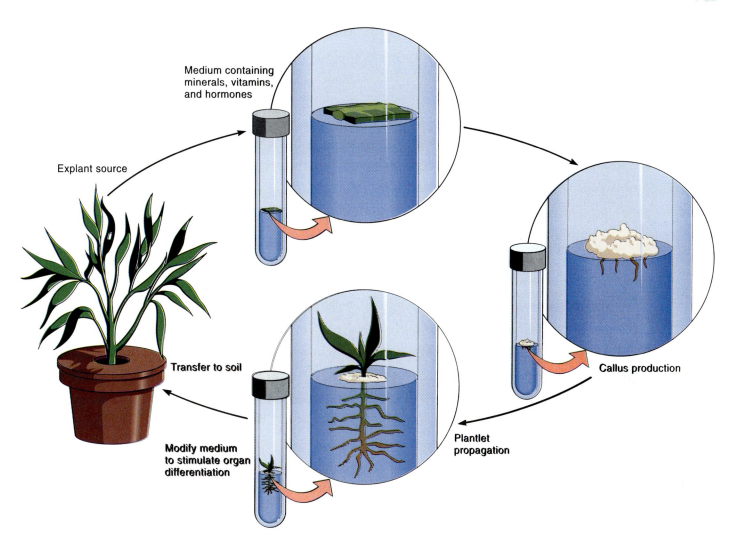

FIGURE 6.1

Plant development and propagation by tissue culture. Cells taken from a plant can be transferred to a suitable growth medium and allowed to grow and develop (called *regeneration* by tissue culture botanists) into a new plant under proper conditions. The medium is modified with nutrients and hormones to stimulate tissue and organ development. Callus is a mass of similar, undeveloped plant cells grown in culture.

Two events lead to cell enlargement:

1. *The cell wall loosens and therefore can stretch.* One means by which walls are loosened is by secretion of acids into the cell walls. These acids loosen the cell wall by activating pH-dependent enzymes that break bonds between cellulose molecules in the wall. Cell-wall acidification associated with cell elongation is strongly influenced by auxin, a plant hormone. You'll learn more about auxin later.

2. *Positive turgor pressure occurs.* Turgor pressure is pressure in a cell resulting from the uptake of water that occurs primarily in response to increasing concentrations of solutes in the vacuole. Thus, filling the vacuole with water is what "fills up" an expanding cell; for example, cell expansion in onion roots corresponds to a 30- to 150-fold increase in the size of the vacuole. This means of enlargement is more efficient in terms of en-

ergy use than is the synthesis of an equal volume of protein, enzymes, and organelles. It can also be fast; for example, petioles of a tropical water lily (*Victoria regia*) can lengthen more than 2 centimeters per hour.

Plants can't grow without cell enlargement. Since enlargement requires turgor, which, in turn, requires water, plant growth and development are intimately linked with a plant's water status. Fluctuations in water availability also influence metabolism, which controls events such as seed germination, bud growth, and the cell elongation that causes roots and shoots to respond to environmental signals such as light and gravity.

Cell division and enlargement are important for growth but do not by themselves constitute development. To understand this, consider the model (fig. 6.1) of how plants are grown in tissue culture—a laboratory technique that has been called "gardening in test tubes." Much of

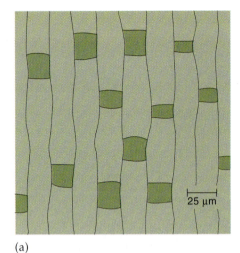

(a)

(b)

FIGURE 6.2

Stomatal differentiation in leaves. (*a*) Early in development, epidermal cells divide asymmetrically. The small cells (*shaded*) divide to form stomatal mother cells, while the larger cells become ordinary epidermal cells. (*b*) This type of cell lineage produces regularly spaced stomata. Stomata regulate gas exchange.

what we know about differentiation in plants comes from studies using tissue culture, a technique for growing fragments of plants in an artificial medium. Typically, a section of a plant is placed on a Jell-O-like medium, and the cells begin to divide. This division forms the callus masses growing in vitro shown in figure 6.1. The cells divide and form the callus of unspecialized parenchyma tissue. These callus cells divide and expand rapidly but do not resemble a plant because they lack specialized tissue that forms roots, stems, and leaves. Producing these characteristic parts of a plant is referred to as plant development and requires cellular specialization; that is, it requires that cells differentiate as shown in figure 6.2.

Development is a genetically programmed progression from a simpler to a more advanced or complex form.

Differentiation, as we stated earlier, involves progressive, developmental change to a more specialized form or function. Botanists often use development and differentiation interchangeably.

Researchers have learned the "genetic tricks" and environmental cues needed to regenerate a plant from some types of callus tissue (see fig. 6.1). This is how modern house ferns are mass-produced, and this knowledge is also important in crop improvement and genetic engineering (see appendix B). Recall that early tissue-culture experiments were important because they showed plant cells are totipotent and, therefore, given the proper environment, a single totipotent cell can grow and develop into an entire plant. This has significance in the study of plant development because it means a plant cell contains all the genes to grow and develop into a new, entire plant. It also suggests that specific genes are turned on during differentiation. Finally, tissue culture is an important means for studying factors involved in cell differentiation because culturing is relatively simple and gives a researcher considerable experimental control. When plants are grown in culture, the effects of compounds such as minerals, sugars, and hormones can be tested by observing what happens to the callus tissue when these compounds are added or removed from the growth medium.

INQUIRY SUMMARY

Growth is an irreversible increase in size of an organism or its parts and results from increases in the number and sizes of cells. Development via differentiation is an orderly process by which the structure and function of genetically identical cells become different. Differentiation produces cells, tissues, and organs specialized for different functions. Much of what we know about differentiation in plants comes from studies using tissue culture, a technique for growing regenerated plants in an artificial medium.

How might differentiation of dermal tissue important in plant evolution on land?

Various Signals Regulate Plant Growth and Development.

Early botanists suggested that cell differentiation was controlled by unique expression of a cell's genetic information. They also suspected that internal signals (i.e., chemicals inside the plant cells and tissues) were involved. These suggestions, like that of totipotency, have been supported by many experiments and today form a foundation of molecular biology and genetics. But if all cells in a plant have the same genes and the same genetic potential and if specialization results from differential expression of this potential, what controls which part of a cell's potential is expressed? Stated another way, what determines which genes are turned on and, therefore, the developmental fate of a cell? This question has intrigued botanists for centuries. Even Charles Darwin conducted experiments and wrote papers about plant development and plant responses to the environment.

We now know that internal signals are involved and that they can control plant growth and development by influencing the expression of the genes. These signals are complex and involve hormonal, positional, biophysical, and genetic controls discussed throughout this chapter.

Plant hormones serve as chemical signals between cells and tissue systems.

Internal chemical signals were suspected when early botanists discovered that one part of a plant affects other parts of the plant. For example, removing the shoot apex usually stimulates bud growth, and seeds often germinate faster when they're removed from fruit. These effects have often been attributed to chemical signals called **plant hormones,** which are small organic molecules produced in plants that serve as highly specific chemical signals between cells. Hormones seem to be involved in just about every aspect of plant growth and development. They usually elicit a physiological response in the target cells by binding to specific cellular proteins called receptors. Target cells are those cells that respond to the hormone.

Currently, there are five customarily recognized groups of plant hormones: *auxin, cytokinins, gibberellins, abscisic acid,* and *ethylene* (table 6.1).

Several aspects of growth and development are influenced by application of plant hormones. For example, auxin strongly affects the differentiation of procambium into young leaves. Here's how we think it may happen: (1) young leaves produce large amounts of auxin; (2) this auxin moves down the stem; (3) auxin moving down the stem induces differentiation of vascular tissues from the target procambium cells.

As you might predict, replacing leaves of a shoot apex with synthetic auxin stimulates differentiation of procambium. Furthermore, applying auxin to a callus mass in tissue culture (fig. 6.1) induces differentiation of procambium into vascular tissue at the site of application. These observations suggest that auxin formed in leaves and transported down the stem stimulates the differentiation of vascular tissue in the leaf. Thus, in young leaves, auxin controls the development of vascular tissues.

Auxin is probably the most thoroughly studied plant hormone. While much remains to be learned about how auxin works, it is clear that applied auxin can rapidly influence the expression of genes in various tissues and cause an array of physiological and morphological changes in a plant such as described earlier. Several of these genes have been identified in plants, but the mechanisms of how auxin is recognized as a plant hormone and how such a simple molecule (table 6.1) can influence numerous responses in various tissues remains to be resolved. Indole acetic acid (IAA) is the principal natural auxin produced by plants. Dichlorophenoxy acid (2, 4-D) is a synthetic auxin used as a herbicide.

Cytokinins were first discovered as a breakdown product of DNA during studies on plant tissue culture. Cytokinins are associated with an activation of cell division and regulation of the plant cell cycle. For example, when a soil bacterium, *Agrobacterium tumefaciens,* infects some plants, a tumorlike growth called a crown gall is formed as cell division is stimulated in the infected region of the plant. The gall is often like a big lump on the lower stem of the plant, and it contains substantial amounts of cytokinins and auxin. (See appendix B for more information on crown gall and genetic engineering.)

Gibberellins (GA) are often sold to home gardeners because applications of GA can cause considerable elongation of the internodes of various plants. The largest responses usually occur in genetically dwarfed plants, which can be made to grow tall with GA. However, normally tall plants often show no response to GA applications, probably because GA turns on genes only in the dwarf plant. Commercially, GA can be applied to the flowers of apple trees to induce fruit formation.

Abscisic acid (ABA), when first discovered, was thought to be specific for abscission of plant parts as they age (abscission means the detachment of leaves, flowers, or fruits). We now think ABA has several roles in plants (table 6.1). Commercially, ABA can be applied to landscape plants during drought to induce closure of the stomata and reduce water loss. A major drawback is that ABA is very expensive and its use is not practical on a large scale. What other reason might gardeners choose not to artificially close a plant's stomata?

Ethylene, the only gaseous plant hormone, regulates plant development, including aging and responses to pathogen attack. Specific ethylene receptors in target cells have been found and are critical in plant responses to the hormone. In forestry, dwarf mistletoe is a major parasite on certain evergreens, causing, for example, the weakening and possible death of ponderosa pine in large areas of the Rocky Mountains. Ethylene, applied in a reactive liquid compound that releases the gas around a plant, is currently the only known control of the dwarf mistletoe. Like ABA though, its application is expensive and not practical for large-scale management of the problem.

Each of the five types of known hormones may have many different and redundant effects in different tissues; that is, they may influence an array of physiological and developmental events. For example, auxin can stimulate or inhibit cellular enlargement, depending on the concentration applied to the experimental plant and whether it is applied to roots or shoots. Cytokinin, as well as auxin, often stimulates cell division. ABA is involved with seed dormancy, leaf water loss (through stomata), and adaptation to environmental stresses such as drought, cold, and high

TABLE 6.1   Functions of Plant Hormones

HORMONE	MAJOR FUNCTIONS	WHERE PRODUCED OR FOUND IN PLANT
Auxin (such as IAA)	stimulates stem elongation, root growth, differentiation and branching, apical dominance, development of fruit; instrumental in phototropism and gravitropism	endosperm and embryo of seed; meristems of apical buds and young leaves
Cytokinins (such as kinetin)	affect root growth and differentiation; stimulate cell division and growth, germination, and flowering; delay senescence	synthesized in roots and transported to other organs
Gibberellins (such as GA$_1$)	promote seed and bud germination, stem elongation, leaf growth; stimulate flowering and development of fruit; affect root growth and differentiation	meristems of apical buds, roots, and young leaves; embryo
Abscisic acid	inhibits growth; closes stomata during water stress; counteracts breaking of dormancy	leaves, stems, green fruit
Ethylene	promotes fruit ripening; opposes or reduces some effects of auxin; promotes or inhibits growth and development of roots, leaves, flowers, depending on species	tissues of ripening fruits, nodes of stems, senescent leaves

salinity. Ethylene usually inhibits cell expansion but may promote it in some submerged plants. Ethylene and auxin may interact in the control of abscission of leaves and fruit, while auxin probably stimulates ethylene production in some cases. Thus, plant hormones are not as specific as animal hormones (see Perspective 6.1, "Are Plant Hormones Like Animal Hormones?"), and probably no phase of plant growth and development is controlled exclusively by one hormone. Rather, hormones are probably integrating agents that are necessary for, but do not completely control,

a particular response. For example, cytokinins are necessary to break the dormancy of buds, but they do not control the subsequent growth of the bud.

In addition, hormones are often nonspecific and are influenced by other hormones and other substances, such as calcium ions (Ca^{2+}). Calcium can change the way a plant tissue responds to a hormone and can influence developmental events ranging from cellular elongation to responses to light and gravity (see Perspective 6.2, "Calcium and Cell Signaling").

The Lore of Plants

The ancient Chinese knew that fruit would ripen faster if placed in rooms containing burning incense. We now know the factor responsible for this hastened ripening was not heat but ethylene released as the incense burned. In 1910, an annual report to the Japanese Department of Agriculture recommended that oranges not be stored with bananas, because oranges released something that caused premature ripening of the bananas.

This "something" was not identified until 1934, when R. Gane showed that ethylene is made by plants and that it causes faster ripening of many fruits, including bananas. Because most ethylene-induced effects result from ethylene in the air, the effects of ethylene can be "contagious." One bad apple can indeed spoil the barrel. Ethylene made by one overripe apple can induce rapid ripening of an entire barrel of apples.

6.1 *Perspective* Perspective

Are Plant Hormones Like Animal Hormones?

Probably not. Most animal hormones are synthesized in one part of the animal and exported, often via the bloodstream, to the target cells and tissues. Insulin, for example, is produced in the pancreas and affects muscle and other cells. Original models and definitions of plant hormones were developed using research and knowledge of animal hormones, but we now know the plant hormones often can be quite different from those of animals. For example, unlike animal hormones, each group of plant hormones has shown a lack of specificity; that is, their actions in laboratory experiments often overlap, and they seldom, if ever, have unique effects. Also, plant hormones do not seem to be made in specialized hormone-producing tissue as in animals. To emphasize these differences, some botanists refer to plant hormones as *plant growth regulators*.

Plant hormones affect growth and differentiation of cells that are genetically programmed to differentiate in a certain way. This programming can be a function of the cell's position in a plant or its age and is controlled by genes. For decades, botanists thought the concentration of the hormone was the basis for it causing a response. Now we know the target cells must also be responsive to the hormone for the hormonal-induced change to occur. Responsiveness involves receptor proteins for the hormone in the target. Receptor proteins, which are produced and influenced by genes, seem to change with age and external stress such as drought, wounding, or cold. For example, if a porcupine eats a section of the tender bark of an Aspen tree (*Populus tremuloides*), the plant responds initially to the wounding (an environmental cue) by a change in internal hormone levels and cellular sensitivity to these hormones. The result of this cell signaling is a sealing of the wounded bark as new cells and tissues form. Without the wounding, the bark cells are not responsive to the hormones even if they are present, because the receptor proteins are not active and the cells are insensitive to the hormones involved. Wounding can cause receptors as well as hormone levels to change.

Although botanists have long assumed that genetic controls underlie the specialization of cells in meristems, no such developmental genes had been discovered. In 1989, however, a team of scientists led by Sarah Hake discovered a gene that causes odd growths (knots) on the leaves of maize (*Zea mays*). To Hake's surprise, the DNA sequence of this gene, called *knotted-l*, or *knl*, included a segment that was already known in the developmental genes of fruit flies. By early 1994, Hake and her colleagues had found several other maize genes containing such segments. We will return to genetics in chapter 13.

Just how many groups of plant hormones are there?

There is now considerable evidence for at least one other group of plant hormones from the steroid family of organic molecules. This group was discovered in plants from the mustard family (*Brassica*) and have been called **brassinosteroids.** These plant steroids are important internal chemical signals involved in structural changes that often occur after exposing a plant to light. For example, stem elongation and leaf abscission are affected by brassinosteroids.

In addition, several other signal molecules have been reported, including salicylic acid (see table 3.1), jasmonic acid (or jasmonate), cell-wall fragments, and systemin (a small protein) that have roles in plant resistance to fungal

6.2 *Perspective* Perspective

Calcium and Cell Signaling

Plant cells have evolved ways to perceive different signals from their environment, both external to the plant (such as cold, light, heat, low oxygen, drought, and high salt) and internal (such as hormones). Following perception of the signal, plant cells respond by changing various biochemical functions and anatomical developments using a signal transduction pathway that has been difficult to elucidate. Recently, it was shown that Ca^{2+} acts as a secondary messenger in plant signal transduction, as was first shown in animals. Ca^{2+} triggers numerous cell and tissue responses that influence growth and development of a plant. The calcium-binding protein, calmodulin, is involved in executing Ca^{2+} signals by affecting transcription of genes and regulating target proteins involved in the response.

---

disease and defense against predator animals. Ongoing research with plants continues to add to, and even subtract from, our list of chemical signals important to growth and development and to protection in the botanical world. For example, cytokinins, long recognized as a group of hormones able to stimulate cell division in plants, are found throughout plants but are most plentiful in growing tissue. Recently, botanists have asked if cytokinins are actually produced by plants. There is considerable evidence cytokinins are produced by microorganisms associated with plants, but it has never been proven they are synthesized by plant cells. Does this eliminate cytokinins as a group of plant hormones? Only further research can answer this question.

INQUIRY SUMMARY

Plant hormones are internal chemicals involved in the control of growth and development of plant cells. There are five recognized major groups of plant hormones: auxins, gibberellins, cytokinins, abscisic acid, and ethylene. The competence of a target cell to respond to hormones is influenced by receptor proteins for the hormone and can be affected by age, other cells, and the timing of the signal. It may be true that plant hormones are necessary but do not control the response in question. New groups of plant hormones have been reported.

How would you conduct experiments to determine the influence of ethylene on leaf aging?

The Environment Is also Important in Plant Growth and Development.

Plant growth and development are strongly influenced by the environment. Some responses of plants—such as stomatal opening or roots growing downward—are relatively rapid; other responses, such as flowering, are relatively slow and are associated with changes of seasons. Regardless of their duration, most plant responses to environmental signals are due to growth and often to differentiation and are influenced, at least in part, by hormones and the sensitivity of cells to the hormones. The relationships of plant growth and development and environmental signals determine the seasonality of growth, reproduction, and dormancy, ultimately fostering survival of the plant and the species in a given environment.

Plant growth toward or away from an external stimulus such as light or gravity is called a **tropism.** There are several kinds of tropisms, each of which is named for the stimulus that causes the response (see Perspective 6.3, "Tracking the Sun"). For example, **phototropism** is a growth response to light (fig. 6.3), and **gravitropism** is a growth response to gravity (fig. 6.4). Tropisms result from one side of the responding organ elongating faster than the other side. This differential rate of tissue elongation causes the curvature characteristic of a tropism. Auxin and perhaps other hormones are involved in this tropism.

Thigmotropism is a growth response of plants to touch. The most common example is the coiling of tendrils or of entire stems of plants such as morning glory (*Ipomoea*) and bindweed (*Convolvulus*) (fig. 6.5). Before touching an object, tendrils and twining stems often grow in a spiral pattern that increases their chances of contacting an object to which they can cling. Thigmotropism is an adaptation that allows plants to climb objects, thereby increasing the possibilities of light absorption and photosynthesis. Contact with an object is detected by specialized epidermal cells, which induce differential growth in the tendril. Such growth can be extremely rapid: a tendril can encircle an object within 5 to 10 minutes. Thigmotropism may be influenced by auxin and ethylene; when applied to a receptive plant in a laboratory, these hormones can induce thigmotropiclike curvature of tendrils even in the absence of touch.

Unlike tropisms, **nastic movements** are independent of the direction of the stimulus: they occur in an anatomi-

The Lore of Plants

Recent research has shown tobacco plants infected by a virus release a vaporized chemical signal that warns neighboring plants of the attack. Discovery of this airborne alert system comes long after botanists found plants could emit such signals about insects. When the vapor alerts healthy plants to the viral infection, the plants initiate their own defenses against the virus. Infected plants turn an internal chemical signal called salicylic acid (a cousin of aspirin) into the fragrant compound we call oil of wintergreen, which evaporates and becomes the vapor signal between the plants. Uninfected plants turn the oil of wintergreen back into salicylic acid, which then signals genes to produce defense proteins against the virus. Other crops may also communicate via this or another vapor signal system.

How might this research lead to new, natural pesticides for agriculture?

6.3 Perspective

Tracking the Sun

One of the most unusual and adaptive movements in plants is solar tracking, also known as heliotropism (from the Greek word *helios,* meaning "sun"). Solar tracking is how the organs of some plants track the sun across the sky, much like a radio telescope tracks a satellite. Like several other plant movements, solar tracking is caused by turgor changes in motor cells located at the base of leaves and leaflets.

The tracking movements of leaves depend on environmental and physiological conditions. For example, the so-called compass plant (*Silphium laciniatum*) orients its leaves parallel to the sun's rays, thereby decreasing leaf temperature and minimizing desiccation. Other plants often orient their leaves perpendicular to the sun's rays, thereby increasing the amount of light intercepted by the leaf for photosynthesis. Finally, some plants orient their leaves more obliquely to the sun when it is hot than when it is cooler. This variation in solar tracking helps keep the leaf temperature near the optimal temperature for photosynthesis.

Sunflowers (Helianthus annuus) *oriented in the same direction, toward the sun.* Helianthus *means "sunflower."*

Solar tracking occurs in many plants (e.g., cotton, alfalfa, beans, and soybeans) and is not restricted to leaves. What happens on cloudy, overcast days? On these days, leaves are oriented horizontally in their "resting" position. If the sun appears from behind the clouds late in the day, leaves rapidly reorient themselves: they can move up to 60 degrees per hour, which is four times faster than the movement of the sun across the sky.

Solar tracking is controlled not only by the sun's position; leaves begin orienting themselves toward the direction of sunrise several hours before sunrise. How plants "remember" the position of the last sunrise is a mystery, but the plant auxin IAA may be involved because IAA from leaves can alter turgor of cells. Whatever the mechanism, plants respond quickly: only four sunrises are needed to entrain solar tracking.

cally predetermined direction, rather than toward or away from the stimulus. Nastic movements also occur in response to environmental stimuli and include some of the most unusual as well as spectacular responses in all of the plant kingdom. For example, there is a nastic movement resulting from contact or mechanical disturbances such as shaking. Movements are based on a plant's ability to rapidly transmit a stimulus from touch-sensitive cells in one part of the plant to responding cells located elsewhere. Among the most dramatic of these responses are those exhibited by the sensitive plant (*Mimosa pudica*): touching a leaf causes the leaflets to fold and the stems to droop (fig. 6.6).

The most spectacular and macabre of plants' responses to environmental stimuli is the nastic response of Venus's-flytrap (*Dionaea muscipula*), a popular novelty plant. Leaves (i.e., traps) of Venus's-flytrap have two

FIGURE 6.3

Phototrophic curvature of corn shoots. Light from one direction causes plant hormones to move to the shaded side of the stem. The hormones stimulate growth on the shaded side, causing the shoot to curve toward the light as cells lengthen on the darker side of the stem. Auxin is involved in the process.

lobes, each of which has several sensitive "trigger" hairs overlying the motor cells. For the trap to close, at least two trap hairs must be touched in sequence (the trap won't close if you merely touch the same hair twice). When a meandering animal touches these hairs, the flytrap snaps shut (fig. 6.7).

Similar nastic movements occur in a few other plants. These movements are interesting, but what is their adaptive significance? Nastic movements may scare insects off of plants, thereby decreasing a leaf's chances of being eaten. Presumably, folded leaves are more difficult to see,

which also may divert a herbivore's attention to some other plant.

INQUIRY SUMMARY

Plant growth toward or away from an environmental stimulus is called a tropism. Phototropism is growth toward light. Gravitropism occurs in response to gravity. Nastic movements occur in response to various stimuli. Thigmotropism is a growth response to touch. Tropisms may be influenced by auxin and ethylene.

What might be a selective advantage of thigmotropism and phototropism to rain forest plants?

Flowering Is Influenced by Genes and the Environment.

One of the most striking and predictable responses of many plants to changes of season is flowering, which is the reproductive growth and development in angiosperms (flowering plants). Farmers and gardeners have long known that many plants flower only during certain times of the year. For example, clover develops flowers during summer, and most asters bloom during the short days of fall.

Botanists in the early 1900s discovered that an environmental signal for flowering is the **photoperiod,** which is the ratio of the length of day to the length of night over a 24-hour period. Subsequent research led botanists to classify plants into four groups based on the plant responses to photoperiod in greenhouse studies: day-neutral plants, short-day plants, long-day plants, and intermediate-day plants (fig. 6.8).

♪ **Day-neutral plants** flower without regard to photoperiod; that is, day length has no effect on their flowering, and they form flowers when they reach

(a)　　　　　　　　　　(b)　　　　　　　　　　(c)

FIGURE 6.4

Gravitropism by horizontally oriented roots of corn. (*a, b*) Downward curvature begins within 30 minutes and is completed (*c*) within a few hours. Curvature results from faster elongation of the upper side of the root than of the lower side. Again, auxin is involved.

FIGURE 6.5
Thigmotropic coiling of a stem of bindweed (*Convolvulus*).

maturity. Examples of day-neutral plants include roses, snapdragons, cotton, carnations, dandelions, sunflowers, tomatoes, cucumbers, and many weeds. Most plants seem to be day-neutral, especially as they age.

🌢 **Short-day plants** flower only if daylight is shorter than some critical length (conversely, there is a critical length of night required for flowering in short-day plants; in fact, the length of night is probably most critical). For example, ragweed plants flower only when exposed to 14 hours or less of light (i.e., 10 or more hours of uninterrupted dark) per day. Because of their light requirement, short-day plants usually flower in late summer or fall as the days become shorter, but a few flower in the short days of early spring. Asters, strawberries, dahlias, poinsettias, potatoes, soybeans, goldenrods, Christmas cactus, and some chrysanthemums are short-day plants.

🌢 **Long-day plants** usually flower in the spring or early summer; they flower only if light periods are longer than a critical length, which is usually between 9 to 16 hours of light (or 15 to 8 hours of uninterrupted dark). For example, wheat plants flower only when light periods exceed 14 hours (and the night is less than 10 hours long). Lettuce, spinach, radish, beet, clover,

(a)

(b)

(c)

FIGURE 6.6
Nastic movement of leaves and leaflets of the sensitive plant (*Mimosa pudica*). In undisturbed plants (*a*), leaves are erect. Touching a leaf (*b*) causes leaflets to fold and the petiole to droop (*c*).

FIGURE 6.7

A Venus's-flytrap captures its prey. (*a*) The unsuspecting fly shown here was attracted to the trap by compounds secreted on the leaf's surface. (*b*) When this fly touched two of the trap's trigger hairs in sequence, it was snared.

gladiolus, and iris are long-day plants and can be expected to flower in the spring or early summer.

  🌶 **Intermediate-day plants** flower only when exposed to days of intermediate length; they grow vegetatively if exposed to days that are either too long or too short. Examples of intermediate-day plants are sugarcane and purple nutsedge.

The measuring system in many plants can be remarkably sensitive: henbane (*Hyoscyamus niger*) will flower when exposed to dark periods of 13.7 hours (10.3 hours of light) but will not flower when exposed to dark periods of 14.0 hours (10 hours of light). Plants stripped of their leaves are not responsive to changes in photoperiod, suggesting leaves detect photoperiod.

How strict is the requirement for a flower-inducing photoperiod? For some varieties of soybean (*Glycine max*), the requirement for a photoperiod to induce flowering is absolute: these plants are short-day plants and will not flower unless exposed to long nights. Conversely, other plants such as Christmas cactus (*Schlumbergera bridgesii*) will eventually flower without an inductive photoperiod. In these plants, the inductive photoperiod merely causes flowering to occur sooner. Flowering of many plants is also influenced by moisture, soil conditions, nearby plants, and temperature.

Ecologically, the control of flowering by photoperiod influences the distribution of plants. For example, few short-day plants grow in the tropics because day length is

FIGURE 6.8

Kinds of flowering responses to day length. Flowering by day-neutral plants is not affected by day length; short-day plants require a light period that is shorter than some critical length; long-day plants flower only if light periods are longer than some critical period; intermediate-day plants do not flower if the light period is too long or too short. Botanists originally thought that flowering was controlled by day length. Subsequent studies showed that the length of the *dark* period is what controls flowering in most plants.

always too long to induce flowering. Similarly, short-day plants such as ragweed do not grow far into northern areas because they do not have time to form seeds. How might flowering have affected the evolution of animals in the tropics and northern areas?

The Lore of Plants

Horticulturists have long exploited our knowledge of photoperiodism to produce flowers for sale. For example, chrysanthemums are short-day plants that usually flower only in the fall. However, mums (i.e., flowers of chrysanthemums) can be made available year-round by using lightproof shades to create inductive, long-night photoperiods. Similarly, extended dark periods or supplemental light can be used to induce out-of-season flowering in such plants such as irises, Christmas cactus and poinsettias.

Flowering can occur only if a plant has passed from its juvenile stage and becomes reproductively mature, or ready to flower. For example, ragweed is a short-day plant that flowers only in the short days of autumn. Why doesn't it also flower during the short days (long nights) of spring? The answer is that ragweed is not reproductively mature in the spring. The time required for reproductive maturation from the juvenile stage ranges from only a few days (as in garden peas or in the Japanese morning glory, *Ipomoea nil*) or weeks (as in annuals) to several years. Indeed, some species of *Agave* require decades before being ready to flower—hence their name, century plant. Similarly, the giant bamboos of Asia flower only every 33 or 66 years, even when transplanted to new environments. The delay in flowering caused by reproductive maturation may help ensure that the plant has stored enough food reserves to form and maintain flowers. Plant hormones are thought to be involved in reaching reproductive maturity. How might the environment also affect this process?

INQUIRY SUMMARY

Photoperiod is the ratio of the length of day to the length of night over a 24-hour period. Botanists classify plants into four groups based on the plants' responses to photoperiod: day-neutral plants, short-day plants, long-day plants, and intermediate-day plants.

Flowering tends to spread out over time among different plant species within an ecosystem. What might be the selective advantage of this variation in flowering date?

Phytochrome Is an Hourglass for Measuring Important Environmental Cues.

Early experiments on flowering prompted botanists to search for a pigment that controls photoperiod. Their search was aided by two clues observed in a laboratory:

1. Flowering is inhibited most effectively by interrupting the dark period with red light.

2. An interruption of the dark period with red light can be reversed if it is followed immediately by exposure to far-red light (the shortest wavelengths of infrared light).

These clues suggested that the pigment involved with flowering in some plants was sensitive to day length and existed in two forms, one that absorbed red light and another that absorbed far-red light. The bluish protein pigment was soon isolated and identified as **phytochrome,** which can exist in two interconvertible forms, **Pr** and **Pfr.** More recently, botanists have discovered there is more than one phytochrome in plants and possibly as many as four or five. The genes that code for the synthesis of the enzymes that make the four forms of phytochrome have been identified. These four forms of phytochrome have been designated types A, B, C, and E in angiosperms. Type A is the most common, and we focus on it next.

Phytochrome is synthesized in the dark as Pr, which is the inactive form of the pigment. Laboratory experiments showed Pr absorbs red light (fig. 6.9; or white light, the entire visible spectrum of light) and is converted to Pfr, the active form of phytochrome in cells. In turn, active Pfr absorbs far-red light (730 nanometers) and is converted to Pr (fig. 6.9). In nature, Pr is synthesized at night and converted to Pfr, the active form, during the day (in white light). Then, the following night, the reverse can occur, and Pfr can be converted back to Pr. This conversion of the active form (Pfr) to the inactive form (Pr) in the dark is a critical factor in determining the ratio of Pfr to Pr (active to inactive form) and whether or not a long-day or short-day plant flowers. Phytochrome enables plants to measure and adapt to temporal changes in light conditions. It is nature's hourglass.

As you might expect, plant hormones link the light-measuring and timekeeping functions of phytochrome and the resulting changes in gene expression that lead to altered plant form and activity, such as flowering. Pfr can activate or terminate gene expression; that is, *active* means it can inhibit as well as activate changes in the plant.

What is the significance of the interconvertibility of Pr and Pfr? Ecologically, the answer appears rather simple: because Pr is converted to active Pfr during the day, the presence of a large amount of Pfr could chemically signal the plant that it is in light and allow a gauge of the day length (and therefore the season of the year). When the plant is

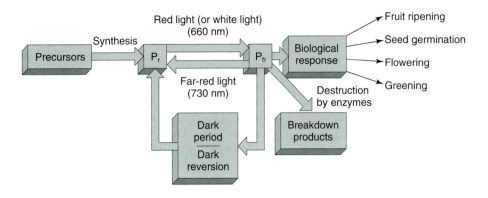

FIGURE 6.9

Phytochrome is synthesized as P_r, which absorbs red light and is converted to P_{fr}. P_{fr} has several possible fates: it can absorb far-red light and be converted to P_r, initiate a biological response, be destroyed, or revert to P_r if maintained in darkness. Far red is light in the 730 nanometer spectrum. Red light is 660 nanometers.

placed in darkness, Pfr is slowly converted to inactive Pr, and this seems to be the critical process. These conversions of Pfr and Pr were originally thought to start a set of reactions that enabled the plant to measure the length of darkness relative to the length of light and would signal the plant to flower or not to flower, depending on the species (remember, the active form can induce or inhibit flower formation, depending on the type of plant). Indeed, this reasoning would explain why an uninterrupted period of dark controls plant responses to day length, or the photoperiod (remember, the length of night is critical). However, research has shown the dark-reversion of Pr to Pfr requires only 3 to 4 hours and therefore cannot fully account for the light-dark sensing mechanism that controls flowering. Rather, flowering is probably controlled by the combined actions of phytochrome and an internal "clock," but relatively little is known about this system. A pigment that is a receptor of blue light may send a signal to the internal clock. Nonetheless, phytochrome is an important timekeeping mechanism that enables some plants to respond to day length (relative to night length) and flower upon environmental cues.

Another aspect of plant development influenced by phytochrome is seed germination. In certain types of lettuce and weedy species, laboratory experiments have shown red light stimulates germination and far-red light inhibits germination in some plants. Alternating exposure of seeds to red and far-red light can be repeated indefinitely, but seed germination is affected only by the last exposure (fig. 6.10). In nature, seeds of many species germinate when there is enough light to stimulate the conversion of inactive Pr to active Pfr. Most seeds with this sensitivity contain small amounts of stored food. It is unlikely that these seeds could survive germination from deep within the soil. Thus, the absence of active Pfr (due to no sunlight) may account for the lack of germination by seeds buried deep in soil. This is highly adaptive, since it promotes the persistence of a "seed bank" in the soil. The seeds may germinate in the sunlight when environmental forces, such as

flooding or animals digging in the soil, bring them to the surface. Plowing does the same thing.

Phytochrome also influences the early growth of seedlings. When a seed germinates underground or in darkness, the seedling has abnormally elongated stems, small roots and leaves, a pale color, and it appears spindly. This condition is termed **etiolation,** and it is most easily detected by yellowing of the normally green tissue. The rapid elongation of etiolated plants helps expose the leaves to light before exhausting their stored food reserves. Once a plant is in light, etiolated tissue is replaced by production of normal, green (photosynthetic) tissue that contains chlorophyll. The light-controlled transformation from etiolated to normal growth is controlled by phytochrome. The synthesis of chlorophyll is called greening.

Phytochrome and photoperiod also affect several other responses, including shoot gravitropism, stomatal formation, leaf abscission, chlorophyll synthesis, chloroplast development, leaf expansion, spore germination, and nastic movements. In each of these responses, phytochrome is a light receptor. Phytochrome triggers light-responsive elements of DNA and may thus alter membrane permeability, enzyme activity, gene expression, or the movement of hormones into and out of cells, thereby regulating cellular activities. Indeed, concentrations of most growth-promoting plant hormones increase immediately after exposure to red light. Although plants detect light intensity and duration (photoperiod) with other photoreceptors, phytochrome is the most widely studied and well described. But the specific signal pathways among light perception, hormones, light responsive elements, and gene expression, although the subjects of intense research, remain unclear.

Plants also contain other light-sensing pigments. For example, pigments that react to ultraviolet (UV) light, blue light, and red light have been reported in angiosperms and lower plants. Many plant responses have been assigned to these receptors, including phototropism and fruit ripening.

INQUIRY SUMMARY

Photoperiodism is a response to changes in the relative lengths of night and day. It is an important adaptation in plant evolution because it enables plants to "measure and respond" to seasonal changes in climate. Phytochromes absorb red and far-red light and influence many aspects of plant growth and development, including flowering. Phytochromes monitor the light

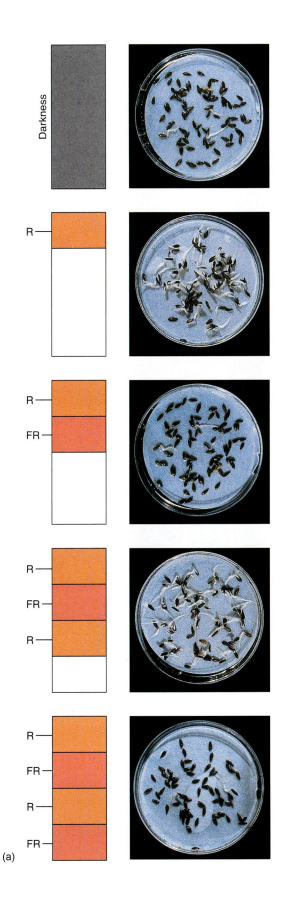

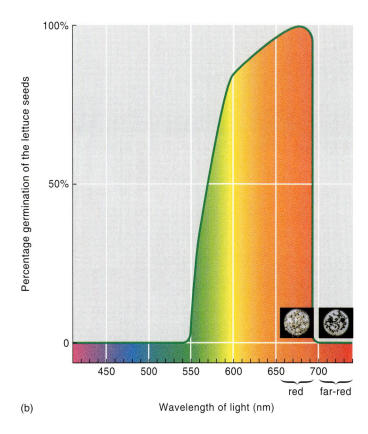

(b)   Wavelength of light (nm)

FIGURE 6.10

Control of lettuce seed germination by red (R) and far-red (FR) light. (*a*) If the last exposure is to red light, most of the seeds germinate. The seeds remain dormant if the last exposure is to far-red light. This sensitivity to red and far-red light is controlled by phytochrome.

(*b*) Lettuce seeds germinate in red light, but not in far-red light.

Source: Data from M. Wilkins, *Plantwatching: How Plants Remember, Tell Time, Form Partnerships, and More,* 1998, Macmillan Publishing Company.

environment and influence patterns of gene expression that enable plants to optimize growth and development in accordance with prevailing conditions.

How would you determine if a response in a plant to an environmental cue was due to phytochrome?

Senescence and Dormancy Are Genetically Programmed Events in Plants.

Earlier in this chapter, we stressed that plants usually respond to environmental stimuli by altering their patterns of growth and development. However, some stimuli may induce **senescence,** which is a collective term for the processes accompanying aging that lead to the death of a plant or plant part. If you have ever seen a leaf yellow and fall off a houseplant, you have witnessed senescence. This

FIGURE 6.11

Leaf senescence is a complex, energy-requiring process that is often controlled by photoperiod. The spectacular colors of leaves in the fall result from the destruction of chlorophyll, which reveals the presence of other pigments such as carotenoids and anthocyanins.

senescence may be caused by environmental stress, such as drought or lack of light, or simply by natural aging of the leaf.

In some instances, senescence is rapid; for example, flowers of plants such as wood sorrel (*Oxalis europaea*) and heron's bill (*Erodium cicutarium*) shrivel and die only a few hours after being formed. Other examples of senescence, such as the colorful "turning" of leaves in autumn, last longer and are triggered by changes in photoperiod and temperature (fig. 6.11). Whatever its duration, senescence is

not merely a gradual cessation of growth; it is an energy-requiring process associated with metabolic changes in the plant. For example, leaf senescence in an aspen (*Populus tremuloides*) tree begins during the shortening days of late summer and is first noticed by us as the leaves turn from green to yellow. Leaf senescence involves mobilization of nutrients out of the aging leaf. This includes a breakdown of leaf proteins by enzymes called proteases, which release the constituent amino acids from the protein. These amino acids are transported to storage tissue for later assembly into new proteins when the plant resumes growth. By the time a senescent leaf is shed from a plant, the leaf contains little more than cell walls and remnants of nutrient-depleted protoplasts. Most of its nutrients have long since been moved to the roots or other areas for storage. These nutrients are used in the spring when the plant resumes growth.

What controls leaf senescence? Senescence is strongly influenced by plant hormones; for example, fruits of plants such as soybeans are thought to produce a "senescence factor" that is transported to leaves and induces senescence. In the laboratory, leaf senescence can be delayed by applying cytokinins, gibberellins, and/or auxin and can be promoted by applying abscisic acid or ethylene.

Leaf senescence is only one aspect of a plant's preparation for cold or drought. The shortening days of autumn induce **dormancy,** which is a period of decreased metabolism in the plant body. Like senescence, dormancy involves structural and chemical changes within the plant. Cells make sugars and amino acids that function as antifreeze (i.e., they lower the freezing point in cells); these changes, as well as dehydration, prevent or minimize damage from cold. Also, inhibitors accumulate in buds, transforming them into winter buds covered by thick, protective scales. These changes in preparation for new environmental conditions (such as winter) are called **acclimation.** As a result of acclimation, a plant can better survive a cold, dry winter and is said to be cold-hardy. Cold-hardiness plays a major role in determining the distribution of a species: the poor cold-hardiness of some plants restricts their growth to tropical or warm temperate regions.

Resumption of active growth in spring is usually influenced by photoperiod and/or temperature. The lengthening days of spring can break dormancy in birch and red oak, whereas fruit trees such as apple and cherry resume growth only after exposure to winter's cold. This cold requirement for breaking dormancy presents problems for apple and cherry trees planted in warm climates. Since these climates usually do not have a cold or long winter, spring growth either does not occur or is greatly delayed.

In other plants, dormancy is released by factors unrelated to photoperiod or temperature. For example, rainfall alone releases dormancy in many desert plants, while plants such as potato require a dry period before renewing growth. The mechanisms by which photoperiod or cold break dormancy are unknown but probably involve changes in amounts of hormones or the sensitivity of tissues to hormones.

differentiation involve the progressive specialization of cells into tissues and then plant organs. Plant hormones constitute most of the internal chemical signals in plants. The five recognized groups of plant hormones are auxins, gibberellins, cytokinins, abscisic acid, and ethylene. Plant hormones are not specific in their action; probably no phase of plant growth and development is controlled exclusively by one hormone. Among other factors, responses to hormones depend on the plant's age and environmental conditions. In addition, the ability of a target tissue to respond to plant hormones is influenced by receptor proteins for the hormone and the timing of the signal. Plant hormones are probably necessary for, but do not control, the response in question. Signal transduction pathways remain to be understood.

Flowering is an example of environmental responses during plant development. Phytochrome, a photoreversible protein pigment, is involved in light-sensing responses in plants and is associated with several responses including flowering, gravitropism, leaf abscission, chlorophyll synthesis, and nastic movements.

Senescence and dormancy are controlled by plant genes and subject to environmental stimuli. Senescence, a genetically programmed death of a plant or its parts, may be caused by drought, lack of light, or simply by natural aging of the plant. Dormancy, a time of inactivity, involves chemical changes within the plant.

Writing to Learn Botany

- Discuss the commercial use of plant growth regulators in agriculture and horticulture.
- What is the selective advantage to a plant of shedding its leaves? Its fruit?

THOUGHT QUESTIONS

1. What hormones might have been involved in the evolutionary transitions of plants from an aquatic to terrestrial life?

2. What are plant hormones? How are they similar and how are they different from animal hormones?

3. What criteria would be necessary for a molecule to be considered a plant hormone?

4. Calcium influences several effects of plant hormones. Why, then, isn't calcium considered a plant hormone?

5. What are the differences and similarities between a plant hormone and an enzyme?

6. Define and give examples of environmentally influenced growth and development.

7. The same concentration of auxin that stimulates formation of adventitious roots will subsequently stop

SUMMARY

Plant growth and development are controlled by several environmental and genetic factors. Growth is any irreversible increase in size of an organism or its parts. Development and

the growth of these roots unless it is significantly reduced. Explain the significance of this observation.

8. What is the selective advantage of plants anticipating
   seasonal changes by measuring the photoperiod
   rather than rainfall, temperature, or some other
   weather factor?

9. What would be the consequences of transplanting a
   short-day plant to the tropics? Explain your answer.

10. Hoeing a garden or tilling a field often triggers the
    germination of many seeds buried in the soil. Why?

11. Is it necessary to orient a seed in a particular way
    when you plant it? Explain.

12. What might be a selective advantage of solar track-
    ing? Explain.

SUGGESTED READINGS

Articles

Bennett, M. J., et al. 1996. Arabidopsis *Aukl* gene: A permease-
like regulator of root gravitropism. *Science* 273:
948–50.

Fluhr, R. 1998. Ethylene perception: From two-component
signal transducers to gene induction. *Trends in Plant
Science* 3:141–46.

Guilfoyle, T., G. Hagen, and J. Murfett. 1998. How does
auxin turn on genes? *Plant Physiology* 118: 341–47.

Holland, M. 1997. Occom's razor applied to hormonology:
Are cytokinins produced by plants? *Plant Physiology*
115:865–68.

Reed, J. 1998. Phytochrome phosphorylation—no longer a
red/far-red herring. *Trends in Plant Science* 3:43–44.

Salisbury, F. B. 1993. Gravitropism: Changing ideas. *Horti-
cultural Reviews* 15:233–78.

Snedden, W., and H. Fromm, 1998. Calmodulin, calmodulin-
related proteins and plant responses to the environment. *Trends in Plant Science* 3:299–304.

Trewavas, A. 1991. How do plant growth substances work?
Plant, Cell and Environment 14:1–2.

Books

Hopkins, W. 1999. *Introduction to Plant Physiology,* 2d ed.
New York: John Wiley & Sons.

Lyndon, R. F. 1990. *Plant Development: The Cellular Basis.*
London: Unwin Hyman.

Sachs, T. 1991. *Pattern Formation in Plant Tissues.* Cambridge: Cambridge University Press.

Smith, R. 1992. *Plant Tissue Culture: Techniques and Experiments.* San Diego: Academic Press.

Taiz, L., and E. Zeiger. 1998. *Plant Physiology,* 2d ed. Sunderland, MA: Sinauer Assoc.

ON THE INTERNET

Visit the textbook's accompanying web site at
http://www.mhhe.com/botany to find live Internet links for each
of the topics listed below:

Plant hormones

Plants and phototropism

Plant movement types

Photoperiod

Plant phytochromes

Plant senescence

Plant dormancy and acclimation

*Notice how the roots of these grasses hold
the soil in place. Does a plant get any
of its mass from the soil?*

CHAPTER 7

Root Systems and Plant Mineral Nutrition

CHAPTER OUTLINE

Roots are a plant's link to the underground environment.

> Taproot systems have one large primary root and many smaller branch roots.
> Fibrous root systems consist of a mass of similarly sized roots.
> The root cap protects the growing parts of a root as the root grows through the soil.
> The subapical region of the root includes the zones of cell division, cell elongation, and cell maturation.
> The root consists of layers of different tissues.
> The rhizosphere surrounds each root.

Several factors control the growth and distribution of roots.

> Depending on their environment and age, plants shift resources to roots or shoots.

Roots often possess special adaptations to the environment.

> Many epiphytes are adapted to life entirely above the ground.

Roots play key roles in the ecology and evolution of plants.

> Mycorrhizae are mutualistic relationships between a plant and a fungus.
> Mutualistic relationships may evolve between two organisms that require different resources.
> Roots of legumes often establish mutualistic relationships with nitrogen-fixing bacteria.

> Because of their root systems, some plants affect the ecology of an entire community.
> Perspective 7.1 Putting Things Back

Many roots are economically important.

Soils are the source of minerals for plants and other organisms in an ecosystem.

> The physical environment of the soil is a major factor that controls plant growth.

Plants obtain essential elements from the soil.

> Essential elements are required by plants for normal growth and reproduction.
> Most essential elements have several functions in plants.
> Deficiencies of essential elements disrupt plant growth and development.
> Some plants accumulate large amounts of various elements.
> Soils in rain forests are deficient in most nutrients.

LEARNING ONLINE

Visit http://koning.ecsu.ctstateu.edu/plant_physiology/roots.html in chapter 7 of the Online Learning Center at www.mhhe.com/botany for additional information on the structure and function of roots. The OLC offers key hyperlinks like this one to help with your assignments.

Most people don't think about roots of plants because roots are "out of sight, out of mind." Hundreds of years ago, however, people relied on plants as their most important source of medicines, and roots were a main constituent of their herbal concoctions. Some of the roots used long ago have been found to have medicinal uses today. Also hundreds of years ago, people believed the mass of a plant came from the soil absorbed by the plant's roots. Although this belief was disproved, plants do absorb important materials from the soil—water and minerals that are used as building blocks for plants and, eventually, for animals in a community. Roots keep soil from eroding and interact with thousands of different species of organisms that dwell in the ground. So, next time you look at a plant, think about the parts you don't see, for they play important roles in the ecosystem and in the lives of plants and animals.

In this chapter, you learn about the structure of roots and about some of the major functions of typical roots, including the absorption of water and minerals. You are introduced to the study of plant mineral nutrition, which includes the kinds and uses of minerals required by plants. Minerals are chemical building blocks; minerals are added to sugar molecules and to molecules made from sugars to construct the bodies of plants. In some environments, the major functions of roots are something other than absorption of water and minerals; thus, in this chapter, you learn about adaptations of roots to different environments, some special functions of roots, and how these functions might have evolved.

Roots Are a Plant's Link to the Underground Environment.

Life is not possible without water, and most plants get water by absorbing it through their roots. (In chapter 9, you learn how water gets into and up a plant.) Some plants live their entire existence aboveground—roots and all—but recall from chapter 2 that most plants live in two worlds at the same time, aboveground and belowground. Most roots are adapted to the dark, relatively moist, and mineral-rich underground world. Roots, however, represent only part of the whole plant, and the aboveground shoot depends on roots for its supply of water and minerals. The shoot uses these supplies to make sugar and other organic molecules that are sent to the root, which then uses these molecules to grow and collect more water and minerals.

Taproot systems have one large primary root and many smaller branch roots.

Recall that the first structure to emerge from a germinating seed is the radicle, or primary root (see fig. 1.16). In most dicots, the radicle enlarges and forms a prominent **taproot** that persists throughout the life of the plant. Many progressively smaller **branch roots** (also called secondary or lateral roots) grow from the taproot. This type of root system consisting of a large taproot and smaller branch roots is called a **taproot system** and is common in cone-bearing trees and dicots (fig. 7.1). In plants such as sugar beet and carrot, fleshy taproots are the plant's "food" pantry; they store large reserves of carbohydrates. (The use of the word *food* here means a substance that provides both energy and chemical building blocks for a plant. A plant gets no energy from the soil, thus, a plant does not absorb food from the soil. Although a plant absorbs minerals through its roots, a plant's "food" is produced through photosynthesis in its leaves.)

Roots conform to the general biological principle of **form and function,** which is the idea that the form of a structure often reveals, and is directly related to, its function. For instance, the large, fleshy taproot of a dandelion or carrot has a form that is adapted to store many food reserves, one of its main functions. Not all taproots evolved a function for storage. For example, the long taproots of poison ivy (*Rhus toxicodendron*) and mesquite (*Prosopis*) absorb water from deep in the ground. As you read this book, try to determine how the form of a plant part could be advantageous to the part itself and to the plant as a whole.

Because taproots of most dicots grow faster than branch roots throughout the life of the plant, many plants have long taproots. Indeed, engineers digging a mine in the southwestern United States uncovered a mesquite root 53 meters down (as deep as a 15-story building is tall). If you'd like to see firsthand how big roots can get, visit the Roto-Rooter Monster Root Hall of Fame in Des Moines, Iowa. There you can see "Moby Root," a 31-meter specimen pulled out of a drainage pipe of a parking garage in 1994.

Fibrous root systems consist of a mass of similarly sized roots.

Most monocots (including grasses) have a **fibrous root system** consisting of an extensive mass of similarly sized roots (fig. 7.2). In these plants, the radicle is short-lived and is replaced by a mass of **adventitious roots** (from the Latin word *adventicius*, meaning "not belonging to"), which are roots that form on organs other than roots. Fibrous roots of a few plants are edible; for example, sweet potatoes are fleshy parts of fibrous root systems of *Ipomoea batatas*. Some plants have two

FIGURE 7.1

Taproot system of dandelion (*Taraxacum*), consisting of a prominent taproot and smaller branch roots.

FIGURE 7.2

Fibrous root system of a grass. Fibrous root systems consist of many similarly sized roots that form extensive networks in the soil.

kinds of root systems; for example, clover (*Trifolium*) has a taproot and an extensive fibrous system of roots.

Corn plants have fibrous roots belowground and aboveground adventitious roots that help to support the tall upright stem (fig. 7.3). These adventitious roots, called **prop roots,** are common in tropical trees in areas where windstorms often topple unsupported trunks. Adventitious roots also grow from underground stems, called **rhizomes,** of ferns, club mosses (*Lycopodium*), and horsetails (*Equisetum*). In many plants, adventitious roots are a pri-

mary means of vegetative, or asexual, reproduction; that is, the formation of offspring without the fusion of sperm and egg but by growth and development of a vegetative organ such as a root, stem, or leaf. For instance, what appear to be many individual trees in a forest of quaking aspen (*Populus tremuloides*) are often a single large plant with many stems (a clone) spread by adventitious roots. We use adventitious roots on cuttings to propagate many plants (fig. 7.4), including raspberries, apples, cabbage, brussels sprouts, and many kinds of houseplants.

Adventitious roots form in all sorts of places on plants, including leaves, petioles (the thin stalk that attaches a leaf to a stem), and stems. The formation of adventitious roots is controlled by hormones such as auxin (see chapter 6), a form of which is the active ingredient in commercial rooting compounds, such as Rootone, that promote the propagation of cuttings. Although they arise at

FIGURE 7.3

Prop roots on corn (*Zea mays*) are adventitious roots that arise from the stem. As their name suggests, prop roots help support the plant.

FIGURE 7.4

Adventitious roots at the base of a plant cutting. Wounding induces parenchyma cells near the cut surface to become meristematic cells. These meristematic cells then divide and produce roots on the cutting.

FIGURE 7.5

Root tip (*upper right*) with its thimble-shaped root cap (*lower left*) removed (×10).

different locations, primary roots and adventitious roots may have similar structures and functions.

The root cap protects the growing parts of a root as the root grows through the soil.

The tips of most roots are covered and protected by a thimble-shaped **root cap** (fig. 7.5). Cells of root caps can secrete large amounts of **mucigel,** a slimy substance (fig. 7.6) containing sugars, enzymes, and amino acids. A root weighing only 1 gram can secrete as much as 100 milligrams of mucigel per day. Although this rate of secretion may seem rather insignificant, the total amount of mucigel secreted by an actively growing group of plants can reach impressive proportions. For example, the roots of 1 hectare (about 2.5 acres) of corn secrete more than 1,000 cubic meters of mucigel during a growing season; that's enough mucigel to fill a typical two-story, four-bedroom house.

Mucigel has four major functions:

1. *Protection.* Mucigel protects roots from drying out. It also may contain compounds that diffuse into the soil and inhibit growth of other roots. For example, the mucigel of giant foxtail can decrease the growth of nearby corn roots by about 33%. What advantage might this be for the giant foxtail?

2. *Lubrication.* Mucigel lubricates roots as they push between soil particles.

3. *Water absorption.* Soil particles cling to mucigel, thereby increasing the root's contact with the soil. The water-absorbing properties of mucigel help maintain the continuity between roots and soil water.

4. *Nutrient absorption.* Mucigel makes it possible for minerals, along with water in the soil, to be better absorbed by roots.

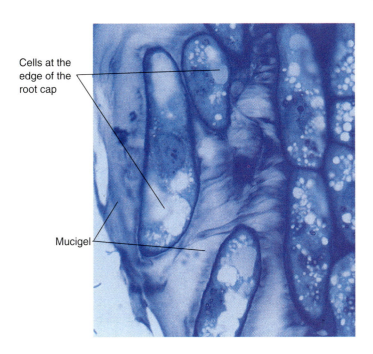

Cells at the edge of the root cap

Mucigel

FIGURE 7.6
The outermost cells of root caps secrete mucigel, a polysaccharide that lubricates the tip of the root as it moves through the soil (×760). In this micrograph, mucigel surrounds the root cap of corn (*Zea mays*).

There is no structure in shoots that corresponds to a mucigel-producing root cap. What might be the evolutionary explanation for this difference?

The subapical region of the root includes the zones of cell division, cell elongation, and cell maturation.

The part of the root just behind the root cap is called the subapical region. This region is traditionally divided into three regions: the zones of cell division, cell elongation, and cell maturation (fig. 7.7). These zones intergrade with one another and are not sharply defined. In general, however, the **zone of cell division** consists mainly of the apical meristem of the root, where new root cells are produced. These cells divide every 12 to 36 hours; in some plants, the meristem produces almost 20,000 new cells each day. In the **zone of cell elongation,** the newly formed cells elongate by as much as 150-fold, primarily by filling their vacuoles with water. Cell elongation in this zone shoves the root cap and apical meristem through the soil at rates as high as 4 centimeters per day. Cells behind the elongating zone do not elongate. In the **zone of cell maturation,** the immature, elongated root cells begin to take on a specific function, and root hairs develop on the outside of the root (see figs. 1.18 and 7.7*b*). Root hairs form only in the maturing, non-elongating region of the root; thus, roots have no lateral appendages at their tips. What would be the disadvantage of producing root hairs at the tip of the root?

The root consists of layers of different tissues.

A cross section through a mature dicot root gives us a look at its primary structure (fig. 7.8*a*). A typical dicot root is surrounded by a layer of epidermis cells. The root epidermis usually either lacks a cuticle—a thin coating of wax—or has a thin cuticle that does not significantly affect water absorption. (On leaves and stems, the waxy cuticle helps keep water from evaporating from the plant.) Xylem and phloem cells are packed together in the center of the root. Cortex cells constitute the majority of cells in the root and often store most of the starch (fig. 7.8*b*). In many plants, the outermost layer(s) of the cortex is a protective layer called the **hypodermis,** with cells lined with suberin, a waxy substance that is impervious to water. The hypodermis is usually most prominent in roots growing in arid soil or near the soil's surface. Suberin in the hypodermis slows the outward movement of water and dissolved nutrients by as much as 200-fold, thereby helping roots retain water and minerals they've absorbed. How might the hypodermis help a plant live in a dry climate?

Most of the cortex consists of thin-walled, starch-storing parenchyma cells separated by large intercellular spaces that can occupy as much as 30% of the root's volume. The innermost layer of the cortex is the **endodermis** (see fig. 7.8*a*). Cells of the endodermis are packed tightly together and lack intercellular spaces. Furthermore, four of the six sides of each endodermis cell are impregnated with a **Casparian strip** made of lignin and suberin and arranged similarly to a rubber band around a rectangular box (fig. 7.9). If endodermis cells are likened to bricks in a brick wall, then the Casparian strip is analogous to the mortar surrounding each brick. The Casparian strip redirects the inward movement of water and nutrients as they flow through the endodermis; water and dissolved minerals must pass through the cell membranes of endodermis cells to reach the vascular tissues in the center of the root (fig. 7.10). As a result, the endodermis functions somewhat as a valve that regulates the movement of minerals into the vascular tissue. This arrangement is critical for helping to eliminate leaks, for conserving certain minerals in the vascular tissue, and to establish a negative water potential (chapter 4) inside the center of the root, the stele.

The **stele** of a root includes all of the tissues in the middle of the root and consists of the pericycle, vascular tissues (xylem and phloem), and sometimes a pith. The outermost layer of the stele is the **pericycle,** a layer of thin-walled meristem cells one to several cells thick. The pericycle is important because it produces branch roots. This means that branch roots form near the center of the root and then grow outward (fig. 7.11). The earliest sign of branch-root formation is cell divisions in the pericycle. Soon thereafter, a root cap and primary tissues form. The branch root then forces its way through the cortex and epidermis of the parent root, much as it will later force its way through the soil. The vascular tissues of branch roots link with those of the parent root.

Inside the pericycle is the root's vascular tissue. Roots of most dicots and some monocots (e.g., wheat and barley)

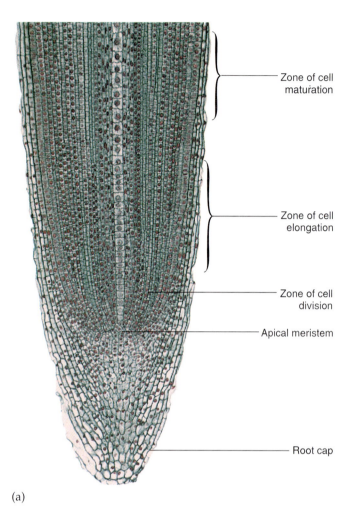

Zone of cell maturation

Zone of cell elongation

Zone of cell division

Apical meristem

Root cap

(a)

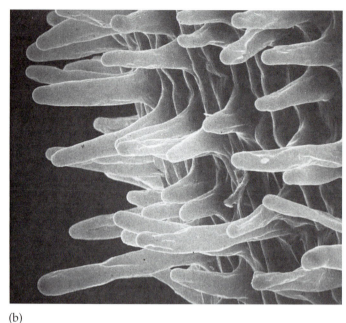

(b)

FIGURE 7.7

(*a*) The subapical region of roots includes the zone of cell division, zone of cell elongation, and zone of cell maturation. (*b*) Scanning electron micrograph of root hairs in the zone of maturation on a primary root of a radish (*Raphanus sativus*) (×25). Root hairs, which are extensions of epidermal cells, greatly increase the absorptive surface area of the root.

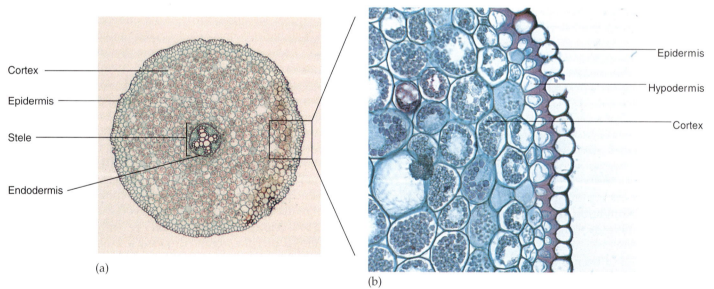

Cortex

Epidermis

Stele

Endodermis

(a)

Epidermis

Hypodermis

Cortex

(b)

FIGURE 7.8

Cross sections of the root of a buttercup (*Ranunculus*). (*a*) Overall view of a mature root. The stele includes xylem and phloem tissues specialized for long-distance transport of water and solutes, while the epidermis forms a protective outer layer of the root (×10). (*b*) Cortex cells contain many starch-laden amyloplasts (stained darkly in this micrograph), which are starch bodies, and are usually separated by intercellular spaces (×220). In many plants, the outermost layer of the cortex is specialized as a hypodermis, which retards the loss of water and dissolved nutrients. Surrounding the root is the epidermis.

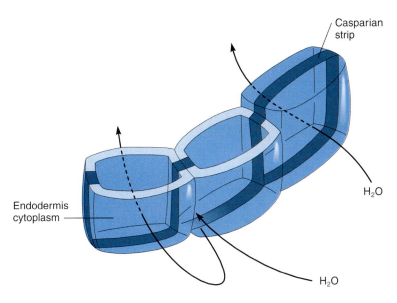

FIGURE 7.9

The Casparian strip of the endodermis blocks the movement of water and dissolved minerals in the cell walls and between cells. As a result, water and dissolved minerals are directed through the cell membrane and protoplast of endodermal cells, where subsequent use and movement are more controlled.

have a lobed, solid core of xylem in the center (see figs. 7.8*a* and 7.11). Phloem is found between the lobes of xylem. The roots of many monocots and a few dicots have a ring of vascular tissue surrounding a pith made of parenchyma cells (fig. 7.12).

The rhizosphere surrounds each root.

The narrow zone of soil surrounding a root is called the **rhizosphere** (fig. 7.13). The rhizosphere extends up to 5 millimeters from the root's surface and is a complex and ever-changing environment. Growth and metabolism of roots modify the rhizosphere in several ways. First, roots force their way through crevices and between soil particles. Later, when the roots die and decay, they leave open channels that help aerate the soil; this aeration may improve the root growth of other plants. Roots also alter the chemical composition of the soil. For instance, respiration in roots decreases the concentration of oxygen and increases the concentration of carbon dioxide in the rhizosphere. This lowers the pH of the soil, increasing its acidity, which may change the ability of roots to absorb minerals. This is because minerals are absorbed while dissolved in water, and their ability to dissolve is affected by pH—the availability of some minerals such as iron may increase and others such as potassium may decrease in acidic soils.

Roots enrich the soil with organic matter. Some plants transport as much as 60% of the sugar not used by

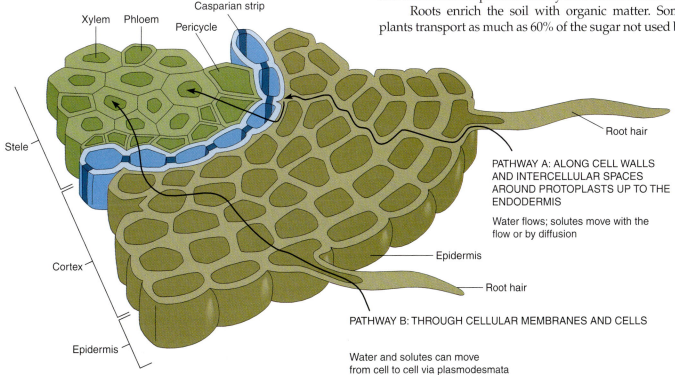

FIGURE 7.10

Pathways show routes water may take as it moves from the soil, through the root, into the stele. Casparian strips in cell walls of the endodermis block off the stele of the root. As a result, water and dissolved minerals enter the stele through the cell membranes of endodermis cells ($\times 100$).

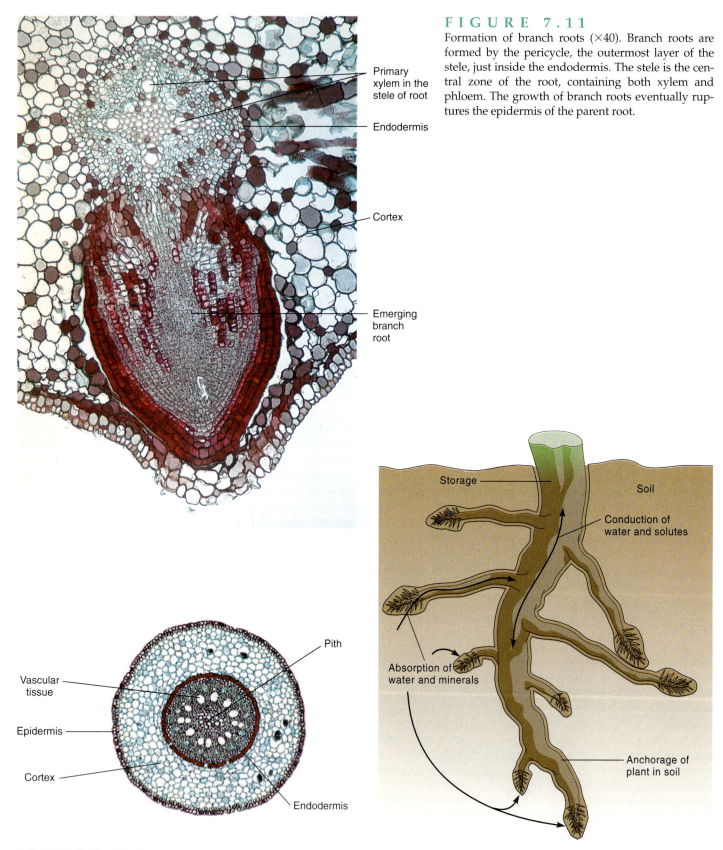

Primary
xylem in the
stele of root

Endodermis

Cortex

Emerging
branch
root

FIGURE 7.11
Formation of branch roots (×40). Branch roots are
formed by the pericycle, the outermost layer of the
stele, just inside the endodermis. The stele is the central zone of the root, containing both xylem and
phloem. The growth of branch roots eventually ruptures the epidermis of the parent root.

Pith

Vascular
tissue

Epidermis

Cortex

Endodermis

Storage

Soil

Conduction of
water and solutes

Absorption of
water and minerals

Anchorage of
plant in soil

FIGURE 7.12
Cross section of a root of greenbrier (*Smilax*), a monocot (×50).
The most obvious tissues are the epidermis, cortex, endodermis
and the vascular tissues of the vascular cylinder.

FIGURE 7.13
Diagram summarizing the major functions of roots. The rhizosphere is the narrow zone surrounding the entire root.

the shoots to roots, and much of this sugar is deposited in the soil as mucigel and other compounds. In plants such as wheat, the amount of carbohydrate lost to the soil often exceeds that stored in the plant's fruits. Roots leave massive amounts of organic matter in the soil when they die and decay. This organic material enriches the soil by returning minerals to the soil that the living root absorbed. In addition to sugar, some plants secrete proteins from their roots that may help protect the roots and the plant against disease-causing bacteria and other organisms in the soil. Scientists are trying to make economic use of root protein secretions by growing plants **hydroponically;** that is, in a nutrient-rich solution rather than in soil. Investigators genetically manipulate the kind of protein the plant produces and then collect the protein secreted by the roots. In the future, this might become an economically important way of "milking" plants that serve as protein factories.

As a result of roots, the rhizosphere usually contains large amounts of energy-rich organic molecules. These molecules are often used as food for more than 10^{10} microorganisms (microbes) in each *cubic centimeter* of soil—populations that are 10 to 100 times more dense than those in the rest of the soil. Most microbes in soil live near roots. These microbes, in turn, secrete compounds that affect the growth and distribution of roots. Equally important, microbes often increase the uptake of minerals from the soil; this is why plants growing in sterile soil absorb fewer minerals than do those growing in rhizospheres that include microbes.

INQUIRY SUMMARY

Most dicots have a taproot system consisting of a large taproot and smaller branch roots that maximize storage. Monocots have fibrous root systems consisting of similarly sized roots that maximize absorption. Adventitious roots form on organs other than roots. Roots, as well as other plant structures, conform to the principle of form and function. Ends of roots are covered by a root cap, which produces mucigel, a substance that protects a root, lubricates it, and helps it absorb materials from the soil. The subapical region of the root includes the zones of cell division, cell elongation, and cell maturation. The epidermis surrounds the cortex, which has three layers: hypodermis, storage parenchyma, and endodermis. The hypodermis protects roots, and storage parenchyma tissue stores reserves for subsequent use. The endodermis is lined with the Casparian strip, which diverts water and dissolved minerals into the cytoplasm of endodermal cells. The stele includes all of the tissues inside the cortex, including the pericycle and vascular tissues. The pericycle produces branch roots. The narrow zone of soil surrounding a root is called the rhizosphere. Many microbes live in and affect the rhizosphere.

Do you think plants with fibrous roots or with taproots would be better adapted to sandy, unstable soils such as those at beaches? Why do you think so?

FIGURE 7.14

The interior of the rhizotron at the University of Michigan Biological Laboratory. The underground laboratory's windows contain removable glass panes that provide researchers with direct access to the soil. Roots that press against the glass panes can be studied relatively easily.

Several Factors Control the Growth and Distribution of Roots.

Much of what we know about roots comes from studies of potted plants or seeds germinated in artificial environments such as sterile dishes or moist paper towels. However, there is little evidence that roots grow similarly in the ground. Understanding how roots grow in their natural environment requires some ingenious as well as tedious work and involves underground cameras and painstaking excavations (see fig. 2.12). The most elaborate methods for studying roots involve laboratories called *rhizotrons*, which are underground walkways with glass walls (fig. 7.14). As they grow, many roots press against these glass walls and can be easily studied.

Observations in rhizotrons have been supplemented with other studies to reveal that the growth and distribution of roots are primarily controlled by four factors:

1. *Gravity.* Different roots respond differently to gravity. For example, primary roots grow down; that is, they are **positively gravitropic** (see figs. 6.4 and 7.15). In plants such as castor bean (*Ricinus communis*), branch roots grow laterally for several centimeters, after which they become responsive to gravity and grow downward. As a result of this differential responsiveness to gravity, roots permeate the soil at various angles and efficiently absorb water and minerals.

2. *Genetic differences.* The growth and distribution of roots are also controlled genetically. For instance, plants such as locoweed (*Astragalus mollissimus*) grow deep

Gravity

FIGURE 7.15

Primary roots of corn (*Zea mays*) are positively gravitropic; that is, they grow downward. They do so whether the grain germinates in a vertical, horizontal, or inverted position in this petri dish standing up on its edge. What do you predict would happen if a light was placed directly above the petri dish? What if the petri dish were placed in the dark?

taproots, whereas most grasses produce shallow, fibrous root systems. Similarly, the root systems of corn are denser near the soil surface than those of soybeans, regardless of the type of soil in which they grow.

3. *Stage of plant development.* Just before and during fruit formation, most of a plant's resources are used for shoot rather than root growth. Thus, root growth typically slows during flowering and fruit formation. This shift in the allocation of resources is most obvious in some monocot crops such as corn, in which much of the root system dies before harvest.

4. *Soil properties.* Roots of many species grow best in loosely packed soil. For example, roots of wheat grow four times faster in loose sand than in tightly packed clay. Roots that grow in tightly packed soil are usually shorter and thicker than those that grow in loosely packed soil. Similarly, roots usually grow deeper in moist, aerated soil than in soaked, poorly aerated soil.

If you water your lawn shallowly each day, the root system will be restricted to the upper levels of the soil. This means that if you go on vacation during the summer, your lawn grasses may be threatened because the top layers of soil dry out first and, because your lawn's roots are limited to this dry region, the entire lawn is in jeopardy of dying. However, if you water infrequently, but deeply, the roots of your lawn grasses can grow deeply into the soil. Thus, even if the top layers of soil dry out, the grass plants have roots deep enough in the soil to absorb the water necessary to keep them alive.

Competition with other plants is minimized by growing into different areas of the soil. Consider the root systems of mesquite (*Prosopis*) and a saguaro cactus (*Carnegiea gigantea*), two plants that grow in dry environments. Mesquite produces long taproots that obtain water from deep underground. Saguaro cacti (see fig. 2.25) survive in the same environment by producing an extensive mass of shallow roots that spreads as far as 30 meters and that maximizes water absorption after infrequent rains in the desert.

Depending on their environment and age, plants shift resources to roots or shoots.

One of the major management problems faced by a plant is the allocation of resources between roots and shoots. Plants produce shoots that absorb light for photosynthesis, but efficient photosynthesis requires an equally efficient root system for gathering water and minerals from soil. A plant, however, has only a limited amount of energy that can be used for growth. Thus, plant growth requires a mechanism that effectively allocates energy for the growth of roots versus shoots. Botanists study this problem in a variety of ways, such as by determining the ratio of the weights of roots and shoots in plants living in different environments. This root:shoot ratio tells us how the environment affects the growth of the plant below- and aboveground. The root:shoot ratio is relatively large for seedlings and decreases gradually as a plant ages. What might be the selective advantage of relatively large roots for a seedling?

Decreasing the amount of minerals available to roots decreases the root:shoot ratio, as does diminishing light. This tells us that roots are affected more strongly by minerals and light than are shoots. If stressed by extreme environmental conditions, such as a drought, most plants route proportionally more energy and materials to their shoots.

INQUIRY SUMMARY

The study of roots is aided by laboratories such as rhizotrons. The growth and distribution of roots are controlled by gravity, genetic differences, the stage of plant development, and properties of the soil. Plants compete for the soil's nutrients and water by producing many roots, permeating the soil, and growing roots in places different from those of other plants. The growth of roots is affected by the environment. Compared to older plants, a seedling allocates more resources to its root than its shoot.

How do you think an annual plant (one that lives less than a year) might allocate resources to its roots and shoots?

FIGURE 7.16
This stand of aspen is probably a clone of asexually reproducing organisms. This clone is adapted to its high-mountain, summer-drought environment. It produces leaves early in spring and apparently reproduces only asexually.

FIGURE 7.17
In mangroves, gases diffuse to submerged roots through modified roots that stick up in the air.

Roots Often Possess Special Adaptations to the Environment.

Although most roots absorb water and minerals and anchor the plant in place, the following are a few of the most common functions of modified roots; that is, roots for which other major functions have evolved:

- *Storage.* In plants such as beets, turnips, radish, dandelion, and cassava, roots store large amounts of starch. Sugar produced in leaves during photosynthesis moves throughout the plant via the phloem and is transformed into starch in the roots. Roots of other plants store carbohydrates as sugars; for example, the roots of sweet potato (*Ipomoea batatas*) contain 15% to 20% sucrose. Roots also can store large amounts of water; the taproots of some desert plants store more than 70 kilograms of water. In the desert of South Africa, a plant called Hottentot bread (*Dioscorea elephantipes*) is named after the people of the region who use the huge (350 kilograms; over 770 pounds) underground root as a source of starch during famines. This member of the yam family, however, also stores water for its own use during long periods between rainfalls.

- *Vegetative reproduction.* The roots of plants such as cherry, pear, apple, and teak produce shoots called **suckers** that emerge from the soil. When separated from the parent plant, suckers become new individuals. The production of suckers is a relatively common form of vegetative reproduction in plants. For example, most groups of creosote bushes (*Larrea tridentata*)

are clones (genetically identical individuals) derived from a single plant. Some of these clones are more than 12,000 years old; this means the seeds that started these clones germinated approximately 4,000 years before humans began to communicate through writing. The most massive plant in the world is believed to be a quaking aspen (*Populus tremuloides*) that has grown thousands of suckers from the same root system (fig. 7.16). This plant, which grows in the Wasatch Mountains of Utah, consists of more than 47,000 tree trunks, each with the usual complement of leaves and branches. It covers almost 43 hectares (over 100 acres), and its mass has been estimated to be about 6 million metric tons. The discoverers of this aspen named it *Pando,* a Latin word meaning "I spread." What evidence would you need to determine that this is actually only one plant?

- *Aeration.* Many plants grow in stagnant water and mud that contains much less oxygen than does the air. In this environment, the roots of most plants would suffer a similar fate to that of any person without oxygen—although roots can tolerate anaerobic conditions for much longer before dying. Plants such as black mangrove (*Avicennia germinans*) that grow in low-oxygen environments avoid suffocation because specialized roots have evolved that import oxygen from the atmosphere (fig. 7.17). These specialized roots contain as much as 80% **aerenchyma** (tissue containing large air spaces) and grow up into the air. These roots function like snorkels through which oxygen diffuses to submerged roots.

FIGURE 7.18
Mistletoe (*Phoradendron flavescens*), a parasitic plant. This oak tree is heavily infected by mistletoe.

FIGURE 7.19
Epiphytic bromeliad growing on a tree.

🍃 *Movement.* Contractile roots evolved on plants such as lily, gladiolus, ginseng, and dandelion. These roots contract as cells shrink in the cortex; this shrinkage twists the xylem cells into a corkscrew and wrinkles the surface of the root. As a result, contractile roots can shrink more than 50% in only a few weeks. Contractile roots are important because they pull a plant into the more stable environment of deeper soil.

🍃 *Nutrition.* Several plants are **parasites,** meaning that they are attached to a host plant by specialized root structures and harm these plants by siphoning nutrients from them (fig. 7.18). For example, cancer root (*Orobanche*) and mistletoe (*Phoradendron*) parasitize hardwood trees, whereas trees such as sandalwood (*Santalum*) obtain their nutrients from nearby grasses. Many parasitic plants lack chlorophyll and therefore depend entirely on their host for their sugar as well as water and minerals. However, the presence of chloro-phyll in a parasite does not guarantee that the plant is photosynthetic; witchweed (*Striga*) is green yet grows only as a parasite.

Many epiphytes are adapted to life entirely above the ground.

Rain forests contain many **epiphytes** (from the Greek words *epi,* meaning "upon," and *phyton,* meaning "plant"), which are plants that grow independently on other plants (fig. 7.19). Common epiphytes include orchids, staghorn ferns (*Platycerium bifurcatum*), and even some cacti. These plants grow slowly and must absorb their nutrients through aerial roots from sources other than the soil or the plant on which they grow. Aerial roots are adventitious roots formed by aboveground structures such as stems.

Aerial roots have different functions in different kinds of plants:

⑤ *Water retention.* The epidermis of roots of some epiphytic orchids is almost impermeable to water. The primary functions of the epidermis are mechanical protection of the internal tissues and the retention of water in the root.

⑤ *Photosynthesis.* In plants such as philodendron, vanilla orchid, and several aquatic plants, the roots are photosynthetic. In orchids such as *Microcoelia smithii,* aerial roots are flat and green and look remarkably like shoots.

⑤ *Support.* Prop roots are supportive aerial roots that grow into the soil; they are common in plants that grow in mudflats, such as banyan trees (*Ficus benghalensis*), as well as in plants such as corn (see fig. 7.3). Banyan trees can produce thousands of prop roots that grow down from horizontal stems and form pillarlike supports. Roots of the flowering plants called ball moss and English ivy anchor the plant's shoot to the tree on which it grows. Shallow-rooted tropical trees such as *Khaya* and fig (*Ficus*) produce remarkable *buttress roots* at the base of their trunks (fig. 7.20). These planklike roots are often more than 4 meters high and are specialized for support. They contain large amounts of fibers.

Bird's-nest ferns (*Asplenium nidus*) accumulate rainfall and litter between their closely packed leaves; the ferns get their nutrients by growing roots among these leaves. One of the most specialized epiphytes is Spanish moss (*Tillandsia usneoides*), which grows throughout the southeastern United States. These flowering plants lack roots; they absorb water and nutrients through hairs that coat their stems and leaves.

INQUIRY SUMMARY

Roots of some plants are modified for functions other than absorption and anchorage. These modified roots include those that store energy reserves for a plant as well as those that are involved in vegetative reproduction and obtaining oxygen. Parasitic plants obtain nutrients directly from other plants; many parasites have specialized root systems that tap into the tissues of their host. Aerial roots help adapt epiphytes to life aboveground.

Suppose you find one plant (plant A) attached to another (plant B). How could you tell if plant A was a parasite or just an epiphyte on plant B?

Roots Play Key Roles in the Ecology and Evolution of Plants.

Soil teems with life: on average, about 0.1% of the weight of soil is living organisms. Although this percentage may not impress you, consider that nearly 170,000 species of soil organisms have been identified. One kilogram of fertile soil contains about 2 trillion bacteria, 400 million fungi, 50 mil-

FIGURE 7.20
Modified roots. Some tropical trees produce planklike buttress roots at the base of their trunk. Buttress roots help stabilize and support the tree.

lion algae, 30 million protozoa, nematodes, other worms, and insects. Together, these organisms exert a tremendous influence on soil and the plants it supports.

Organisms mix and refine the soil. For instance, one earthworm can digest more than a ton of soil in 1 year. This is why Aristotle called earthworms "the intestines of the earth": they aerate and refine the soil by processing it through their guts. Organisms also add **humus** (decaying organic material) to the soil. Plants, especially grasses, are the primary source of organic matter in soil. A 4-month-old rye plant produces almost 11,000 kilometers (a length greater than the round-trip distance between Miami and Seattle) of roots and root hairs, most of which eventually die and become humus. Because humus contains large amounts of nutrients, the biological activity of soil is usually an excellent indicator of its fertility.

Organisms can affect the availability of nutrients to plants. Roots of lupine (*Lupinus albus*) growing in

FIGURE 7.21

Scanning electron micrograph of a root of a lodgepole pine (*Pinus contorta*) associated with filaments of mycorrhizal fungus.

FIGURE 7.22

Mycorrhizae stimulate the growth of plants. This photo shows how mycorrhizae affect the growth of lemon trees (*Citrus limon*): the plants on the *right* were grown with mycorrhizae, whereas those on the *left* were grown without mycorrhizae. These trees are all 4.5 months old.

phosphorus-deficient soil can decrease the pH of the soil and thus increase the availability of other nutrients to plants. Similarly, plants and microbes secrete enzymes into the soil that liberate phosphate from mineral particles, making more phosphate available to plants.

Some organisms make the soil inhospitable for plants. For instance, many plant pathogens, such as nematodes, live in the soil. Furthermore, some plants produce chemicals that inhibit the growth of other plants; as an example, the soft-leaved purple sage (*Salvia leucophylla*) produces chemicals that inhibit the growth of nearby plants. As a result, these sage plants are often surrounded by bare zones in which no other plants grow. Some scientists refer to this interaction as a form of "chemical warfare" called allelopathy.

Mycorrhizae are mutualistic relationships between a plant and a fungus.

Many relationships have evolved between roots and beneficial fungi. Such associations between a root and a fungus are called **mycorrhizae.** In mycorrhizal roots of trees such as pine and oak, the fungus produces a mass of filaments on the surface of the root. These filaments invade the root and form an extensive netlike structure between the cells of the cortex. A mature tree may have thousands of mycorrhizae.

Mycorrhizae are a type of mutualism, meaning that both the plant and the fungus benefit from the association. The fungus absorbs nutrients from the soil that the tree uses; in return, the host plant provides the fungus with sugar, amino acids, and other organic substances. Mycorrhizal roots often lack root hairs, suggesting that the fungi replace the absorptive functions of root hairs. Mycorrhizae increase mineral absorption, especially for phosphorus. Roots with mycorrhizae absorb as much as four times more phosphorus than do roots that lack them. Two factors contribute to this efficient absorption: (1) the fungal body,

made up of hairlike filaments called **hyphae,** has a much greater surface area than roots, and (2) the fungal hyphae permeate a greater soil volume than do the roots (fig. 7.21).

As a result of this increased absorption, mycorrhizae often dramatically increase plant growth; they can increase the growth of wheat by more than 200%, corn by 100%, and onions by more than 3,000%. Mycorrhizae also improve a plant's endurance of drought, disease, extreme temperatures, and high salinity and are important for reclaiming strip-mined land and establishing nursery stock plants. Many plants *require* mycorrhizae for vigorous growth; orchid seeds germinate only in the presence of a mycorrhizal fungus, and many citrus trees and gymnosperms are difficult to grow unless mycorrhizae are present (fig. 7.22). Mycorrhizae occur in more than 80% of all plants studied.

Mycorrhizae play critical roles in determining whether tree species can invade a disturbed area and are especially important for plants growing in nutrient-poor soils. **Pioneer plants**—plants such as alder (*Alnus*) that colonize bare or disturbed soils—growing on nutrient-deficient soil invariably have mycorrhizae, as did the first plants that invaded the land millions of years ago.

Mutualistic relationships may evolve between two organisms that require different resources.

When plants invaded the land between 400 and 500 million years ago, one of the biggest problems they faced was obtaining enough nutrients. Their aquatic habitat contained

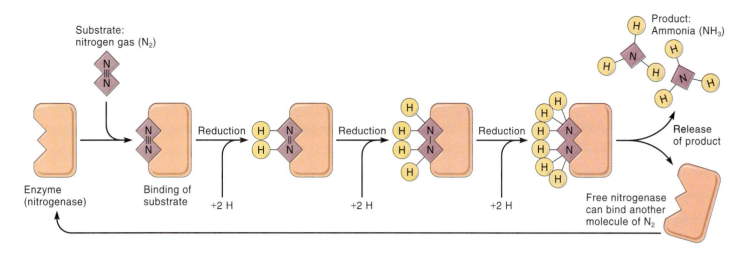

FIGURE 7.23

Nitrogenase is a bacterial enzyme complex found in nitrogen-fixing bacteria that converts atmospheric nitrogen gas (N_2) to ammonia (NH_3). Plants can use ammonia to make animo acids and other nitrogen-containing compounds for growth.

many decomposers that degraded large molecules, thereby increasing the availability of nutrients. As a result, aquatic organisms were literally bathed in nutrients. On land, however, the situation was different: most soils contained few fungi and therefore did not possess enough nutrients to sustain plant growth. One of the ways that plants adapted to life on land was by forming mutually beneficial associations between their roots and fungi. In effect, plants and fungi *co-evolved* for a life on land that helped ensure the continued availability of nutrients for their growth.

How might such a mutualistic relationship between two organisms evolve? A current hypothesis about the evolution of such beneficial interactions is that when one organism (species A) is better than another (species B) at acquiring a needed resource, species A benefits by specializing on the resource and trading any excess to species B. In return, species B specializes at obtaining a different resource and trades the excess to species A. This trading of resources is what many human businesses do, but scientists are now applying the model to living organisms. A mycorrhizal fungus and a plant both get more of the resources they need by specializing at obtaining different resources and trading their excesses with each other. Plants benefit from the ability of fungi to absorb and make available nutrients from the environment; fungi benefit because the plants provide them with sugar for growth. This relationship allows both organisms to benefit.

Roots of legumes often establish mutualistic relationships with nitrogen-fixing bacteria.

Nitrogen is often deficient in soils; this deficiency limits the growth of plants and, as a consequence, animals. The deficiency of nitrogen in soil is somewhat paradoxical, because

the atmosphere—the air you breathe—is almost 80% nitrogen gas (N_2). The chemical bond holding the two atoms of nitrogen gas together makes the N_2 molecule extremely stable and therefore unusable by most organisms, including humans. However, a few species of soil bacteria contain a complex enzyme called nitrogenase that can convert, or *fix*, atmospheric N_2. These microorganisms convert the N_2 to ammonia (NH_3), which can be used by plants to make nitrogen-containing compounds, such as amino acids needed for growth (fig. 7.23). These **nitrogen-fixing bacteria** live alone as well as symbiotically with plants.

Some plants have established mutualistic relationships with nitrogen-fixing bacteria; these plants obtain nitrogen from compounds produced by the bacteria. The roots of legumes such as beans and alfalfa have a mutualistic relationship with *Rhizobium* bacteria. These bacteria secrete carbohydratelike molecules that trigger the formation of **nodules** in the roots of the legume plants (fig. 7.24). *Rhizobium* invades the roots through root hairs, and legume roots form localized swellings, the nodules, in response to the bacteria. Unlike free-living nitrogen-fixing bacteria, *Rhizobium* fixes almost 10 times more nitrogen than it uses. The excess nitrogen is released into the legume, where it is used for growth. In return, the bacteria get sugars from the host plant.

Nitrogen-fixing bacteria form mutualistic relationships with such diverse plants as mosses (*Sphagnum*), liverworts (*Blasia*), and tropical trees (*Parasponia*). These associations, like those with legumes, often have important agricultural implications. For example, the cyanobacterium *Anabaena* grows symbiotically in leaves of water fern (*Azolla*) (fig. 7.25a). When infected *Azolla* are grown in rice paddies, some of the nitrogen fixed by *Anabaena* dissolves in the water of the paddy and becomes available to rice plants (fig. 7.25b). As a result, rice crops in China and southeast Asia can be grown in

FIGURE 7.24

Nodules on the roots of these soybeans (*Glycine max*) consist of cortex cells infected with nitrogen-fixing bacteria.

nitrogen-deficient soils. Botanists are now searching for new strains of *Anabaena* to increase rice production.

Harvesting crops removes more than 21 million metric tons of nitrogen per year from the soil. Nitrogen fixation by bacteria replenishes almost two-thirds of this loss. The remaining nitrogen is resupplied to the soil in fertilizers and other sources. Nitrogen fixation helps many farmers. For example, much of the nitrogen fixed in nod-

ules of legumes is released into the soil. Farmers exploit this by practicing crop rotation, by which legume crops are alternated with nonlegumes as a means of maintaining relatively high levels of nitrogen in the soil. Plowing under a legume crop such as alfalfa adds organic material and up to 300 kilograms of nitrogen per hectare—more than 45 times that contributed by free-living bacteria. Nitrogen fixation also reduces the need to apply expensive fertilizers (see Perspective 7.1, "Putting Things Back"). In fact, addition of nitrogen fertilizer inhibits formation of root nodules on legumes and nitrogen fixation. Molecular biologists are trying to incorporate the genes involved in nitrogen fixation into other crop plants such as corn and wheat that do not have associations with nitrogen-fixing bacteria.

Because of their root systems, some plants affect the ecology of an entire community.

As we saw in chapter 2, ecological relationships between organisms in a community are often complex—as in the food web of a community. An example of a complex ecological relationship involves the tamarisk, or saltcedar (*Tamarix* species), which thrives along the stream banks of many western states. Originally introduced to America from its native habitats in Asia and the Mediterranean, the tamarisk was planted as an ornamental tree and as a way to stabilize

(b)

FIGURE 7.25

Anabaena and rice production. (*a*) Cultivation of rice is aided by nitrogen fixation by the cyanobacterium *Anabaena azollae*. *Anabaena* lives and fixes nitrogen in cavities of this water fern, *Azolla*. (*b*) *Azolla* is grown in rice paddies of the warmer parts of Asia. Later, the paddies are drained, and the *Azolla* is plowed into the soil, thereby enriching the soil with nitrogen for the rice crop. Nitrogen fertilizers do not have to be used in such paddies.

(a)

7.1 *Perspective* *Perspective*

Putting Things Back

Plants extract large amounts of nutrients from soil. For example, during one growing season, a wheat crop on 1 hectare of land removes 85 kilograms of nitrogen, 47 kilograms of potassium, and 17 kilograms of phosphorus from the soil. Some of these nutrients are replenished by decaying humus and by plowing under the remaining parts of the crop. However, such replenishment does not match what is lost when the crop is harvested if crops are grown repeatedly on the site. Consequently, these lost nutrients must be replenished with fertilizers. For example, the yield of an unfertilized soil that initially produced 100 bushels of corn per acre diminished to only 23 bushels per acre in 70 years. When this soil was fertilized, the yield increased to more than 130 bushels per acre. Although fertilization increases plant growth and crop yield, it rapidly reaches a point of diminishing returns: doubling the yield of already fertile soil often requires adding as much as five times more fertilizer.

Chemical Fertilizers

Most chemical fertilizers have a rating that consists of three numbers, such as 12–6–6. These numbers refer to the amounts of nitrogen, phosphorus, and potassium, which are the three elements most likely to be deficient in soil. Thus, a 12–6–6 fertilizer contains 12% nitrogen (usually as ammonium salts), 6% phosphorus (as phosphoric acid), and 6% potassium (as potash).

Nitrogen, which is the most expensive of these elements to produce, is incorporated into fertilizer via the *Haber-Bosch* process:

$$3\,H_2 + N_2$$

500°C,
300 atmospheres pressure

$$2\,NH_3$$

The application of fertilizer increases plant growth by increasing the availability of nutrients.

Nitrogen can either be added directly to the fertilizer as an ammonium salt or be converted to nitrate and then added as a nitrate salt (e.g., $NaNO_3$).

More than 40 million metric tons of nitrogen produced by the Haber-Bosch process are added to soil each year. However, this represents only about one-fifth the amount of nitrogen added to the world's soil by nitrogen-fixing bacteria (nitrogen is also added by thunderstorms and atmospheric deposition). Furthermore, the Haber-Bosch process is expensive in terms of energy; producing 2.5 kilograms of ammonia via the Haber-Bosch process requires the energy equivalent of 1,000 kilograms of coal. The costs of producing nitrogen account for about half of our $16-billion fertilizer bill. Consequently, manufacturing nitrogen-containing fertilizer requires more energy than any other aspect of crop production in the United States. To compound this problem, applications of nitrogen-containing fertilizers are inefficient, because crops absorb only about half of the nitrogen that is applied. The rest is absorbed by other organisms, leached from the soil in rainfall, or reconverted to gaseous nitrogen (N_2) by denitrifying bacteria such as *Micrococcus denitrificans*.

Chemical fertilizers are concentrated, easy to apply, and allow a grower to apply specific amounts of various nutrients. However, these fertilizers do not replenish humus in the soil. To maintain humus, growers usually plow under either the unharvested plants or a subsequent cover crop of barley or rye. The latter process is called green manuring and provides an excellent example of another kind of fertilizer: organic fertilizer.

Organic Fertilizers

Organic fertilizers are essentially the same thing as humus. Although hardly new (planting a fish with corn seed is proverbial), the increased costs of chemical fertilizers have prompted a growing number of gardeners and farmers to rediscover organic fertilizers, which increase both the water retention and fertility of soil. Organic fertilizers include manure, dead animals and plants, fish scraps, and cottonseed meal. On a smaller scale, backyard gardeners often use compost, fish meal, lawn clippings, garbage, and a concoction called manure tea as organic fertilizers. We do not recommend fertilizing your houseplants with manure tea if guests are coming.

Foliar Fertilization

Despite the presence of a thick cuticle, many plants can absorb nutrients through their leaves and stems. For example, iron is sprayed on azaleas and pineapples, and copper and zinc are sprayed on citrus to prevent mineral deficiencies. This type of fertilization is called foliar fertilization and is restricted primarily to micronutrients (i.e., nutrients of which plants need small amounts).

The Lore of Plants

Although it often seems that bacteria are the aggressors, nodule formation is an example of plants taking the first step in controlling the metabolism of another organism. Roots of alfalfa and other legumes secrete chemicals called flavonoids that bind to and activate a bacterial nodulation gene called *nodD*. The *nodD* gene product activates other *nod* genes, whose products then turn on plant genes that control the growth and function of nodules. Nitrogen-fixing bacteria live inside these nodules in a mutually beneficial relationship with the legumes.

stream banks. Tamarisk trees have spread rapidly, now covering almost 500,000 hectares (over 1,900 square miles) in 15 states. Why has it been so successful? A single tree can produce more than half a million seeds a year, and it has taproots that grow deeper into the dry soils of the west than do the roots of native species. Because of their root system, the trees can outcompete, and replace, most native plants along streams. Thus, most people consider tamarisks to be pests; however, at the bottom of the Grand Canyon, the trees provide habitat for a variety of animals, including the southwestern willow flycatcher, a bird that was on the endangered species list. In one part of Colorado where tamarisk is considered a pest, beavers are helping control the spread of these trees. Beavers use tamarisk trees to build their lodges and dams. The dams create pools of water that provide willow trees with enough water to survive and outcompete the tamarisk. In addition, in their native habitats in Asia, the saltcedar is not a pest because it has natural herbivores that keep the population in check. In the United States, the trees have no natural enemies, which also helps their spread. One plan is to import a Chinese leaf-eating beetle that lives on saltcedar in Asia in the hope that the beetle will reduce the number of saltcedars. However, some ecologists are worried that the beetle may switch to native plants in the community and keep them from returning to their former population size. Thus, the tamarisk is viewed as both a positive and negative member of different communities, and questions remain on how best to control the plants because of complex interactions with other organisms.

INQUIRY SUMMARY

Living organisms in soil refine and mix the soil, add organic matter, increase the amount of nutrients in the soil, and often make soil uninhabitable for other organisms. Mycorrhizae are mutually beneficial associations between roots and fungi. Mycorrhizal fungi increase the absorption of nutrients and tolerance of environmental stresses.

Mutualistic relationships may have evolved as organisms traded resources. Many plants, especially legumes, establish mutually beneficial relationships with nitrogen-fixing bacteria. The bacteria convert atmospheric N_2 to ammonia, which the legume uses for growth. These relationships have great benefits for farmers around the world.

Many fungi are parasites of crop plants and forest trees. Why might it be a bad idea to spray a fungicide (which kills all fungi) on a forest to control a parasitic fungus?

Many Roots Are Economically Important.

Roots have long been among our favorite foods. For example, we've cultivated carrots (*Daucus carota*) for more than 2,000 years, and sugar beets (*Beta vulgaris*) are a common source of table sugar. Similarly, radish, horseradish, sweet potatoes, and turnips appear regularly on our tables. Of these roots, sweet potato is the most nutritious: it is about 5% protein and contains large amounts of calcium, iron, and other minerals. Conversely, cassava, which is used to make tapioca, contains almost no protein, yet it provides more starch per hectare than any other cultivated crop. Cassava also can grow in nutrient-poor soil and is a staple for millions of people who live in the tropics. Spices such as licorice, sassafras, and sarsaparilla (the flavoring used to make root beer) are derived from roots.

Roots also provide drugs such as ipecac (used to cause vomiting in case of poisoning), ginseng, the tranquilizer reserpine, and the heart relaxant protoveratrine. Members of the coffee family provide several dyes, as do carrots: their carotene is sometimes used to color butter. Finally, a woodland shrub called the wahoo plant (*Euonymus*) is sold in some novelty shops as a cure to "uncross" victims of witches' spells. Folklore has it that the victim is saved from the curse by holding a piece of the plant's root overhead and screaming "wahoo" seven times.

INQUIRY SUMMARY

Plants are an important source of food for humans. In addition, they provide drugs and dyes.

Do you get more calories in your diet from roots, stems, or leaves of plants?

Surface litter

Topsoil

Subsoil

Weathering bedrock

FIGURE 7.26

Soil layers. The layers can sometimes be seen in road cuts such as this one in Australia. The upper layers developed from bedrock. The dark upper layer is home to most of the organisms that live in the soil.

Soils Are the Source of Minerals for Plants and Other Organisms in an Ecosystem.

Soil is like a layer cake. It has several layers, each of which has distinguishing characteristics (fig. 7.26) and which can vary enormously in thickness from one habitat to another. The "frosting" of this cake is the **surface litter** covering the soil, typically only a few centimeters thick and consisting of fallen leaves. Just beneath the surface litter, the uppermost layer of soil usually extends 10 to 30 centimeters below the surface. In most fertile soils, this **topsoil** is not acidic and contains 10% to 15% organic matter, which gives this layer a dark color. The layer below the topsoil consists of larger soil particles than those in the topsoil and extends 30 to 60 centimeters below the soil surface. This layer, called **subsoil,** usually contains relatively little organic matter and is therefore lighter in color than topsoil. In many regions, this layer contains large amounts of minerals washed by rainfall from the topsoil. Mature roots commonly extend into this layer,

where minerals accumulate. The next lower layer often is referred to as the **weathering bedrock,** because it consists primarily of rock fragments. This layer usually lacks organic matter. In chapter 19, you learn about the formation of soils and how nutrients cycle through soils.

The physical environment of the soil is a major factor that controls plant growth.

Regardless of the different properties of their layers, all soils contain the same five components: mineral particles, decaying organic matter (humus), air, water, and living organisms. Differing amounts of these materials define the soil's properties and, therefore, the plants it can support. Just as there is no such thing as a typical cell, there is no typical soil. Indeed, there are more than 70,000 kinds of soils in the United States alone based on bedrock, organic material, and other characteristics.

Weathering breaks rocks into progressively smaller pieces, the smallest of which are called **soil particles.** These particles consist of **minerals,** which are naturally occurring inorganic compounds that are used by plants as building blocks. All soils contain three kinds of soil particles: sand, silt, and clay. Sand particles are the largest (0.02–2 millimeters), clay particles are the final product of weathering and the smallest of these particles (< 0.002 millimeters), and silt particles are between sand and clay particles in size. Different kinds of soils contain different proportions of sand, silt, and clay. Soils having a mixture of sand, silt, and clay—the soils in which most plants grow best—are called **loams.** Sandy soils are usually nutrient deficient and poorly suited for growing crops.

The amount of water held in a soil is proportional to the surface area of its particles: the larger the surface area, the greater the retention of water. Clay particles are smaller than sand and therefore have a larger surface area per unit of soil volume than does sand (because there are so many more clay particles per unit volume). Indeed, the surface of clay particles in the upper few centimeters of soil in a 2-hectare cornfield equals the surface area of North America. As a result, clay soils retain much more water than do sandy soils. This property of clays would seem to make them ideal for plant growth. However, this is not entirely true because the small size of clay particles results in their packing tightly together—so tightly that the clay has low amounts of oxygen, due to small air spaces. This tight packing also retards the penetration of water into the soil (e.g., water penetrates clay about 20 times slower than it penetrates sand). As a result, much of the water that falls on clay soil runs off and is unavailable for plant growth. This runoff water also contributes to erosion. Tightly packed clay also can impede root growth.

Humus is the decomposing organic matter in soil. The amount of humus in soil varies, usually from 1% to 30%. The most extreme examples are the soils of swamps and bogs, which may contain more than 90% humus. These soils are usually so acidic that decomposers can hardly grow in them. As a result, humus accumulates faster than it is broken down in these watery habitats.

The amount of humus in soil affects the soil and its plants in several ways. The lightweight and spongy texture of humus increases the water-retention capacity of the soil. Water absorption by humus decreases runoff, thereby slowing erosion. Humus swells and shrinks as it absorbs water and later dries. This periodic swelling and shrinking aerates the soil. Humus is also a reservoir of nutrients for plants. Like time-release vitamins, humus gradually releases nutrients as it is degraded by decomposers. Humus also provides habitats for many organisms that mix and concentrate nutrients. Most plants grow best in soil containing 10% to 20% humus.

Because humus is constantly being decomposed by bacteria and fungi, it must be replenished. In natural ecosystems, humus is replenished by leaves that fall off plants and the addition of animal wastes. Many gardeners replenish humus in soil by **composting,** a method of converting organic wastes to fertilizer. The process is simple: organic wastes (e.g., lawn clippings) are gathered into a compost pile, where they are degraded by fungi and bacteria. During winter, this decomposition may take several months, while during summer, it may be completed in as little as 3 weeks. The resulting compost costs less than commercial fertilizer, provides an efficient means of disposing of waste products, and contains nearly all of the nutrients needed by plants. Moreover, composting releases nutrients gradually as the materials decompose, thereby providing plants with a continuous balance of nutrients. On a larger scale, farmers often plow under the stubble of plants remaining after a harvest to maintain the humus in soil used to grow crops.

About 25% to 50% of the volume of most soils is air. This "empty space" has a critical role: it is the conduit by which oxygen reaches the roots. When water replaces a soil's air for a long period, such as in flooded soil or overwatered houseplants, roots can become anaerobic and die. Clay soils, with their tightly packed clay particles, contain less air than do sandy soils. Thus, aeration of a clay soil can be improved by adding silt, sand, or organic material, which increases the penetration of water into the soil, thereby stimulating root growth.

INQUIRY SUMMARY

Soil forms from rocks by a process called weathering. Soils are often arranged in distinctive layers, each of which has unique properties. There are three kinds of soil particles: sand, silt, and clay. Plants usually grow well in soils called loams that consist of nearly equal amounts of all three particles. Clay particles are the smallest of these particles and form soils that are often hard-packed, promoting erosion and inhibiting root growth. Humus is the decomposing organic matter in soil. Humus retains water and is a source of nutrients

for plants. Composting is one way to produce fertilizers from lawn and garden plant wastes.

Garden supply stores sell bags of peat moss (Sphagnum). What might be the advantage of adding peat moss to clay soils before planting your garden, lawns, or shade trees?

Plants Obtain Essential Elements from the Soil.

Over 300 years ago, Jan-Baptista van Helmont, a Dutch physician, studied how a plant gains weight. Up to the time of his experiments, people believed that plants obtained their weight from the soil itself. Van Helmont measured the growth of a willow tree for several years. That tree gained more than 70 kilograms, yet reduced the weight of the soil by less than 60 grams. Van Helmont thus disproved the idea that plants get their weight from soil, but he attributed the growth of the willow tree to the water he added to the soil. This conclusion was partially correct because most of a plant *is* water. However, only later did botanists discover that plant growth also depends on carbon dioxide from the air and several other minerals that, like water, are absorbed from the soil and are essential for growth.

Essential elements are required by plants for normal growth and reproduction.

More than 60 elements have been found in plant tissues. These elements range from those as common as carbon and hydrogen to those as exotic as platinum, uranium, and gold. Are all of these elements essential for growth? If so, what functions do they perform? Answering these questions has proven to be a formidable task. We must first define what we mean by "essential." An element is **essential** if (1) it is required for normal growth and reproduction, (2) no other element can replace it and correct the deficiency, and (3) it has a direct or indirect action in plant metabolism; that is, in the normal plant functioning. Thus, the mere presence of an element in a plant does not necessarily mean it is essential.

Armed with this definition, botanists began studying the 92 naturally occurring elements to determine how each might influence plant growth. Their approach was simple: an element would be deemed essential only if a plant could not grow and reproduce without it (fig. 7.27). Many of the early experiments to determine essential elements involved hydroponics, a technique for growing plants without soil in a solution of water and specific chemicals (fig. 7.28). By the middle of the 1800s, these experiments had shown that plants require at least nine elements: carbon, hydrogen, oxygen, phosphorus, potassium, nitrogen, sulfur, calcium, and magnesium. Because these elements are required in relatively large amounts (i.e., usually more than

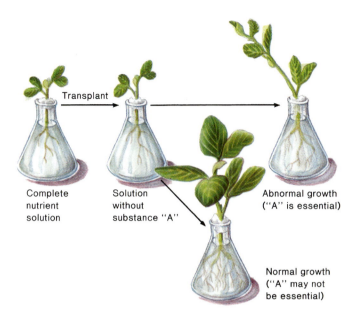

FIGURE 7.27

Discovering essential plant nutrients. Plant physiologists transplant a seedling to a solution lacking only one of the ingredients thought to be essential for growth, substance "A" in this example. If the plant grows and reproduces normally after being transplanted, the missing ingredient is assumed to be nonessential.

FIGURE 7.28

Hydroponic farming. In this apparatus, a solution of nutrients flows over the roots of lettuce growing in water.

0.5% of the dry weight of the plant), they became known as **macronutrients** (fig. 7.29).

Subsequent studies of the mineral nutrients encountered several problems, including contamination. For example, "pure" water and chemicals used to prepare nutrient solutions often contained enough mineral impurities to

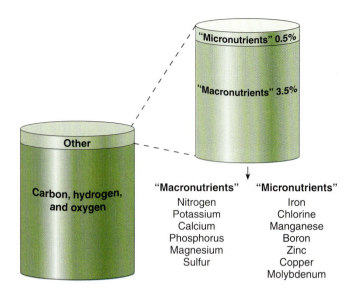

FIGURE 7.29

The proportional weights of various elements in plants. Macronutrients and micronutrients together total only 4% of the total weight of the plant, but they are essential to the plant's life and growth.

satisfy plants' requirements for several elements. Furthermore, the growth containers used in hydroponics experiments were often made of glass that contained large amounts of boron. When these containers were filled with nutrient solutions, enough boron dissolved out of the glass to satisfy the plant's requirement for boron. Such impurities, combined with insensitive methods for detecting many elements, made it impossible to determine the essentiality of many elements.

New techniques and more sensitive instruments helped botanists discover seven other essential elements: iron, chlorine, copper, manganese, zinc, molybdenum, and boron. Because these nutrients are required in relatively small amounts (i.e., usually only a few parts per million), they are called **micronutrients,** or trace elements. For example, the micronutrient boron is required, but as little as five parts per million are needed by some plants. This means that if you were to count all the atoms of a plant, out of every million atoms, only five of them would be boron.

Although required in smaller amounts than macronutrients, micronutrients are equally essential for growth. The 4 metric tons of carbon, hydrogen, and oxygen are no more important to an Idaho potato farmer's crop than are the 2.5 grams (less than the weight of a dime) of copper. The study of how a plant uses these essential elements is **plant mineral nutrition.**

*Most essential elements have
several functions in plants.*

The functions of the 16 elements essential for plant growth are summarized in table 7.1. A quick survey of this

		Some Functions and Deficiency Symptoms of Essential Elements		
ESSENTIAL ELEMENTS	PERCENTAGE OF DRY TISSUE	NUMBER OF ATOMS RELATIVE TO MOLYBDENUM	FUNCTIONS	DEFICIENCY SYMPTOMS
Micronutrients				
molybdenum (Mo)	0.00001	1	part of nitrate reductase; essential for N fixation	chlorosis or twisting of young leaves
copper (Cu)	0.0006	100	component of plastocyanin, a plastid pigment; present in lignin of xylem elements; activates enzymes	young leaves dark green, twisted, and wilted; tips of roots and shoots remain alive; rarely deficient
zinc (Zn)	0.002	300	necessary for formation of pollen; involved in auxin synthesis; maintenance of ribosome structure; activates enzymes	chlorosis, smaller leaves, reduced internodes; distorted leaf margins; older leaves most affected
manganese (Mn)	0.005	1,000	photosynthetic O_2 evolution; enzyme activator; electron transfers	interveinal chlorosis; appears first on older leaves; necrosis common; disorganization of lamellar membranes
boron (B)	0.002	2,000	essential for growth of pollen tubes; regulation of enzyme function; possible role in sugar transport	death of apical meristems; leaves twisted and pale at base; swollen, discolored root tips; young tissue most affected
iron (Fe)	0.01	2,000	required for synthesis of chlorophyll; component of cytochromes and ferredoxin; cofactor of peroxidase and some other enzymes	interveinal chlorosis; short and slender stems; buds remain alive; affects young leaves first
chlorine (Cl)	0.01	3,000	activates photosynthetic elements; functions in water balance	wilted leaves; chlorosis; necrosis; stunted, thickened roots
Macronutrients				
sulfur (S)	0.1	30,000	part of coenzyme A and the amino acids and cysteine and methionine; can be absorbed through stomata as gaseous SO_2	interveinal chlorosis; usually no necrosis; affects young leaves; rarely deficient

ESSENTIAL ELEMENTS	PERCENTAGE OF DRY TISSUE	NUMBER OF ATOMS RELATIVE TO MOLYBDENUM	FUNCTIONS	DEFICIENCY SYMPTOMS
phosphorus (P)	0.2	60,000	part of nucleic acids, sugar phosphates, and ATP; component of phospholipids of membranes; coenzymes	stunted growth, dark green pigmentation; accumulation of anthocyanin pigments, delayed maturity; affects entire plant; second to N, P is element most likely to be deficient
magnesium (Mg)	0.2	80,000	part of chlorophyll; enzyme activator; protein synthesis	chlorosis and reddening of leaves; leaf tips turn upward; older leaves most affected
calcium (Ca)	0.5	125,000	membrane integrity; in middle lamella; functions as "second messenger" to coordinate plant's responses to many environmental stimuli; reversibly binds with calmodulin, which activates many enzymes	death of root and shoot tips; young leaves and shoots most affected
potassium (K)	1.0	250,000	regulates osmotic pressure of guard cells, thereby controlling opening and closing of stomata; activates more than 60 enzymes; necessary for starch formation	chlorosis and necrosis, weak stems and roots; roots more susceptible to disease; older leaves most affected
nitrogen (N)	1.5	1,000,000	part of nucleic acids, chlorophyll, amino acids, protein, nucleotides, and coenzymes	general chlorosis, stunted growth; purplish coloration due to accumulation of anthocyanin pigments; N is element most likely to be deficient
oxygen (O)	45	30,000,000	major component of plant's organic compounds	rarely limiting enough as nutrient to cause specific symptoms
carbon (C)	45	35,000,000	major component of plant's organic compounds	rarely limiting enough as nutrient to cause specific symptoms
hydrogen (H)	6	60,000,000	major component of plant's organic compounds	rarely limiting enough as nutrient to cause specific symptoms

table leads to two important generalizations. First, plants require different amounts of different elements. For example, most plants require about 35 million times more carbon than molybdenum. These differing requirements reflect the various uses of these elements: carbon is in almost all compounds in plants, whereas molybdenum occurs in only a few. Secondly, most elements have several functions. Potassium is involved in starch synthesis, affects protein shape, and activates enzymes. Interestingly, the functions of most micronutrients in plants are similar to those in animals.

Although the functions of essential elements are diverse, they can be grouped into four general categories.

1. Essential elements can be parts of cell structures. Carbon, hydrogen, and oxygen are part of all the biological molecules, such as carbohydrates. Similarly, nitrogen is an integral part of proteins and nucleic acids.

2. Essential elements can be parts of compounds involved in energy-related chemical reactions in a plant. Magnesium is part of chlorophyll, and phosphorus is part of ATP and nucleic acids.

3. Essential elements can activate or inhibit enzymes. Magnesium stimulates several enzymes involved in cellular respiration, whereas calcium inhibits several enzymes. In some cases, these enzymes may be those responsible for synthesizing plant hormones.

4. Essential elements can alter the osmotic potential of a cell. For example, the movement of potassium into and out of guard cells causes them to open and close stomata, the tiny openings in the epidermis of leaves.

Deficiencies of essential elements disrupt plant growth and development.

Obtaining some of the essential elements is not a problem for plants. For example, carbon dioxide is always present in the air, and water is usually abundant enough to support at least minimal growth. As a result, deficiencies of carbon, hydrogen, and oxygen rarely occur in plants. Plants also require large amounts of nitrogen, phosphorus, and potassium; low levels of these elements in soils almost always limit plant growth, and, as a result, these are the elements most likely to be deficient in soil. Thus, plant growth and development largely depend on a plant's ability to absorb these minerals. Overcoming deficiencies of these elements in crop fields and gardens often requires adding large amounts of fertilizer.

Deficiencies of micronutrients are corrected more easily. For example, molybdenum-deficient soil can be replenished by annually adding less than 4 grams (less than the weight of a nickel) of molybdenum to a hectare of land, and deficiencies of zinc and iron have been corrected by driving zinc-coated (galvanized) tacks or nails into the trunks of trees. Examples of deficiency symptoms of essential elements are found in table 7.1 and shown in figure 7.30.

FIGURE 7.30

Nutrient deficiencies produce distinctive symptoms. (*a*) Potassium deficiency causes leaves to curl at their edges. (*b*) Iron deficiency causes chlorotic leaves (veins remain green). A deficiency of manganese produces similar symptoms.

(a)

(b)

Some plants accumulate large amounts of various elements.

Plants such as alpine pennycress (*Thlaspi caerulescens*) are **hyperaccumulators,** meaning that they concentrate certain elements in their bodies at levels 100 times (or more) greater than normal. The hyperaccumulation of metals also benefits the plants by helping them to evade "weak-stomached" predators, including caterpillars, fungi, bacteria, and humans; high-metal plants can poison unwelcome guests. The saltcedar gets its common name because it accumulates salt and then gets rid of it by pumping the salt out onto its leaves. The salt helps keep some herbivores from eating saltcedar leaves. An example of a metal concentrated by hyperaccumulators is lead (Pb), which was once an antiknock component of most gasolines and was deposited along highways from automobile exhaust. Plants growing near these highways often absorbed this lead, as did some grasses growing near lead mines. Radioactive strontium (^{90}Sr) is abundant in plants growing near nuclear test sites and sites of nuclear accidents such as Chernobyl, Russia.

Gold (Au) often occurs in plants such as *Phacelia sericea* that grow near gold mines. These plants have been used by geologists to locate deposits of gold in soil. Some researchers are experimenting with different species such as *Brassica juncea,* a member of the mustard family that grows very quickly on mine wastes and is a hyperaccumulator of gold. By growing these plants in soil containing minute quantities of gold, investigators were able to obtain almost 1 milligram of gold per gram of dry plant tissue. Investigators hope to increase this gold yield and thereby use such plants to "mine" for gold.

Selenium may account for more than 1% of the dry weight of locoweed (*Astragalus*). If locoweed is eaten by livestock, the large amounts of selenium in these plants affect the animals' nervous systems, causing them to stagger about as if intoxicated (fig. 7.31). This illness, called "blind staggers," often kills livestock and has even caused ranchers to abandon their ranches. Similarly, bees that harvest pollen from locoweed often produce honey containing large amounts of selenium.

The use of hyperaccumulators to remove toxic substances from soil is part of a growing practice called **phytoremediation** (see fig. 2.12). For example, at one site, a field of *Brassica juncea* decreased the concentration of selenium by more than half, down to a depth of 1 meter. Phytoremediation is expected to save some of the billions of dollars the United States has budgeted for soil cleanups over the next few years (costs range from $190,000 to $2,800,000 per hectare). Phytoremediation might be an economically important way of dealing with an environmental problem in such areas as military sites laced with the heavy metals lead and cadmium and municipal waste dumps, where copper and mercury are problems. What might be an evolutionary *disadvantage* for a plant to become a hyperaccumulator?

Soils in rain forests are deficient in most nutrients.

In Amazon rain forests, soils are notoriously deficient in most nutrients. The frequent deluges of rain in these forests leach large amounts of calcium, phosphorus, and magne-

FIGURE 7.31

The prince's plume (*Stanleya elata*), shown here growing in Death Valley, California, requires selenium. Its selenium level is so high that this plant is toxic to browsing mammals.

sium from the leaves of the trees. Consequently, plants in a forest depend largely on their ability to gather the relatively nutrient-rich rainwater that trickles down stems. Many plants do this by forming extensive, shallow root systems, whereas other plants intercept nutrients before they reach the soil. For example, the roots of trees such as *Eperua purpurea* grow upward in the fissures of the bark of nearby trees. These fissures are predictable pathways for flowing nutrients, which the roots absorb from the fissures without letting them enter the soil. Like their counterparts in the soil, these roots are extensively branched and occur throughout the canopy; they may extend more than 13 meters into the air. Interestingly, the roots of *Eperua* grow up along its own stems as well as those of nearby trees, but roots of nearby trees do not grow onto *Eperua* stems. As a result, *Eperua* can recycle its own nutrients as well as those of its neighbors.

Why do the roots of these tropical trees grow up, while most others do not? Do these roots respond to nutrients coming down the tree trunks, or do they fortuitously climb any vertical support that they happen to contact? To answer these questions, Robert Sanford of Stanford University sank drainpipes into the forest floor, attaching a reservoir atop each pipe. One-third of the reservoirs contained nutrient-rich cattle manure, one-third contained leaf litter, and the remaining third were empty. Within 4 months, climbing roots appeared on all of the "trees" containing manure, on 25% of those containing litter, and on none of the empty ones. Thus, the growth pattern of the roots of these plants is an adaptation to the low soil-nutrient availability of the Amazon rain forest. The roots are responding to nutrients coming down the tree trunks.

SUMMARY

Three different types of roots are found in plants: taproots, fibrous roots, and adventitious roots. These roots differ in their form and function. Taproots are single, relatively large roots, whereas fibrous root systems consist of many similarly sized roots. Although most roots absorb water and minerals from the soil, modified roots have many other functions, including asexual reproduction, storage of energy, aeration, and protection. Some roots are even photosynthetic. Roots are composed of layers of tissues including the epidermis, cortex, vascular tissue, and endodermis, which helps filter water and minerals that are entering the stele of the root. Just behind the tip of a root are the zones of cell division, cell elongation, and cell maturation.

The roots of many plants interact with organisms that dwell in the soil, including bacteria, fungi, and nematodes. Plants establish several kinds of associations with other organisms to obtain nutrients. The roots of almost all plants have mutually beneficial interactions with fungi. These mutualisms, which are called mycorrhizae, increase the absorption of water and nutrients. Many plants, especially legumes, establish mutually beneficial symbioses with nitrogen-fixing bacteria. The bacteria convert atmospheric N_2 to ammonia, which the plant uses for growth. In return, the bacteria receive sugars from the host plant. Parasitic plants get their nutrients directly from other plants.

Soil is made up of sand, silt, and clay particles. Soil also contains air, water, organisms, and the decomposing bodies of these organisms, which makes up humus. The contents of the soil determine how well plants grow.

Plants require at least 16 essential elements that are necessary for normal growth and reproduction, cannot be substituted for, and directly or indirectly affect metabolism. Carbon, hydrogen, oxygen, phosphorus, potassium, nitrogen, sulfur, calcium, and magnesium are required in relatively large amounts and are called macronutrients. Iron, chlorine, copper, manganese, zinc, boron, and molybdenum are required in relatively small amounts and are called micronutrients. Macronutrients and micronutrients function as parts of structural units, parts of compounds involved in metabolism, and activators and inhibitors of enzymes. Most essential elements are absorbed by roots. This absorption is facilitated by the large surface area of roots.

Some plants are hyperaccumulators of minerals, which may provide protection from some herbivores. Humans are making use of these hyperaccumulators when cleaning up toxic materials in soil. This practice of phytoremediation may save millions of dollars.

Rain in tropical rain forests contains many nutrients leached from the forest's dense canopy. Some trees in the rain forest have roots that grow up along the trunks of adjacent trees and absorb the nutrients from rainfall trickling down the trees.

Writing to Learn Botany

Crop plants absorb only about half of the nitrogen supplied in fertilizer. What happens to the rest? What are the consequences of this?

THOUGHT QUESTIONS

1. Of what selective advantage are the compounds that inhibit the growth of other plants' roots?

2. What's the difference between a small root and a root hair?

3. What are the advantages and disadvantages of a fibrous root system? A taproot system?

4. Why are roots important to plants? Why are they important to animals?

5. Describe how roots modify the soil.

6. What is the difference between an essential element and a beneficial element?

7. Criteria for establishing the essentiality of an element for plants have been around for several decades. Why, then, are essential elements still being discovered?

8. How are root nodules different from mycorrhizae?

9. Explain why removing living organisms from a soil can modify the properties of a soil.

10. Discuss the adaptations of plants for obtaining sufficient amounts of nutrients.

SUGGESTED READINGS

Articles

Adler, T. 1996. Botanical cleanup crews: Using plants to tackle polluted waters. *Science News* 150:42–43.

del Rio, C. M. 1996. Murder by mistletoe. *Natural History* (February):64–68.

Meeks, J. C. 1998. Symbiosis between nitrogen-fixing cyanobacteria and plants. *BioScience* 48:266–76.

Richter, D. D., and D. Markewitz. 1995. How deep is soil? *BioScience* 45:600–609.

Stewart, G. R., and M. C. Press. 1990. The physiology and biochemistry of parasitic angiosperms. *Annual Review of Plant Physiology and Plant Molecular Biology* 41:127–51.

Vijn, I., et al. 1993. *Nod* factors and nodulation in plants. *Science* 260:1764–65.

Wall, D. H., and J. C. Moore. 1999. Interactions underground. *BioScience* 49:109–17.

Books

Bowes, B. G. 1996. *A Color Atlas of Plant Structure.* Ames, IA: Iowa State University Press.

Hopkins, W. 1999. *Introduction to Plant Physiology,* 2d ed. New York: John Wiley & Sons.

Taiz, L., and E. Zeiger. 1998. *Plant Physiology*, 2d ed. Sunderland, MA: Sinauer Assoc.

ON THE INTERNET

Visit the textbook's accompanying web site at http://www.mhhe.com/botany to find live Internet links for each of the topics listed below:

Types of roots

Root growth, soil, water, and air

Root anatomy

Lateral roots

Agriculture and hydroponics

Roots and epiphytism

Mycorrhizae

Roots as food

Plant mineral nutrition

Rainforest soils

Majestic trees, such as these sequoias, are produced in part by secondary growth of the stem. How do plants stand upright without a skeleton?

Stems and Secondary Growth

LEARNING ONLINE

To learn more about stem structure and function, and how stems are important to humans, visit http://koning.ecsu.ctstateu.edu/plant_physiology/stemslec.html in chapter 8 of the Online Learning Center at www.mhhe.com/botany.

The autotrophic life of plants requires that large amounts of surface area be exposed to the aboveground, sunlit environment. The most conspicuous surface area is the plant's shoot, which is an integrated system of stems and leaves (leaves are discussed in chapter 9). Stems support leaves—that is, stems are the "hat racks" on which plants hang their solar-collecting leaves. The most important functions of leaves and stems are photosynthesis and support, respectively. The structure and function of stems and leaves are sensitive to changes in light and moisture, which allows plants to adapt to different and changing environments. Through evolution, stems of some plants have become modified for climbing, storage, and reproduction.

In the second part of this chapter, you learn about the formation of an extremely durable and strong tissue—wood—that is produced by secondary growth. Wood is one of the most important and useful plant products to humans.

Stems Are Composed of the Four Kinds of Tissues Found in All Organs of a Plant.

Stems consist of four basic tissues—epidermal, vascular, ground, and meristematic tissues (fig. 8.1), which were first introduced in chapter 4. These tissues also occur in roots and leaves, and although they have some similar functions in all parts of the plant, stem tissues possess distinctly different characteristics.

Young stems are surrounded by a transparent epidermis that is usually one cell thick and often bears hairlike structures called **trichomes.** Some trichomes have a protective function (see fig. 2.20); for instance, the trichomes of tomato plants secrete an irritating juice that deters hungry insects, while hook-shaped trichomes often entangle insects and prevent them from feeding as they struggle to free themselves. In older stems, the epidermis is replaced by bark, which is discussed later.

Xylem and phloem are vascular tissues—the "plumbing system" of a plant. In stems, xylem and phloem are located in **vascular bundles** (figs. 8.1 and 8.2). Phloem forms on the outside of the bundle, whereas xylem forms on the inside (i.e., toward the center of the stem). A group of **fibers** (sclerenchyma, see chap. 4) is often found next to each vascular bundle, with each fiber surrounded by a thick cell wall. Because of their thick cell walls, fibers are strong, rigid structures. As mentioned in chapters 1 and 4, humans have made use of durable plant fibers for many years, weaving them into textiles and ropes.

A layer of meristematic cells is sandwiched between the xylem and phloem within each vascular bundle of a dicot stem (fig. 8.2). This meristem is called the **vascular cambium** because it produces new vascular tissue—xylem and phloem. These vascular tissues are produced during **secondary growth,** as the plant's stem grows thicker, and therefore are called secondary xylem and secondary phloem. Monocot stems have no vascular cambium. What does this suggest about how thick the stems of monocots can grow?

Vascular bundles are arranged differently in stems of monocots and dicots (see fig. 8.1). Monocots, such as corn, have vascular bundles scattered throughout the stem. Most dicots, such as alfalfa (*Medicago sativa*) and sunflower (*Helianthus annuus*), have a single ring of vascular bundles near the edge of the stem.

Grasses, such as bamboo and Bermuda grass, elongate at meristems between mature tissues at the bases of their nodes and leaf sheaths (fig. 8.3). These meristems are called **intercalary meristems** and continue to produce new cells long after tissues in the rest of the internode are fully mature. Intercalary meristems are important because they re-form the stem or leaf of a grass when its tip is torn off by a grazing animal. Because grasses are a primary food for grazing animals such as cows and sheep, and because humans and other predators often depend on grazers for food, intercalary meristems are largely responsible for keeping the meat counter open at your local supermarket.

Intercalary meristems can produce phenomenal rates of growth in many plants. For example, bamboo stems can elongate almost a meter per day (that's more than an average child grows between birth and its tenth birthday) and can grow to be over 35 meters tall. Intercalary meristems also have a direct impact on us: no matter how many afternoons we spend vibrating behind our lawn mowers, we can be sure the grass blades will soon grow back, thanks to intercalary meristems.

Between the epidermis and ring of vascular tissue in dicots is the **cortex** (see fig. 8.1b), which is ground tissue consisting mostly of parenchyma cells. The cortex is photosynthetic in some plants such as geraniums and often stores starch. In dicots, the parenchyma in the center of the stem is specialized for storage and is called **pith.** Because monocots have vascular bundles scattered throughout the stem, their stems do not have a readily discernable pith (fig. 8.1a).

The evolution of taller plants was aided by the mechanical support of rigid cells.

As plants evolved on land, stems became longer and supported more leaves. Problems arose for these taller plants, including falling over under their own weight or being

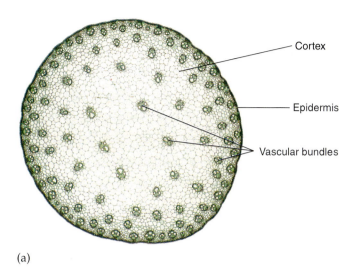

(a)

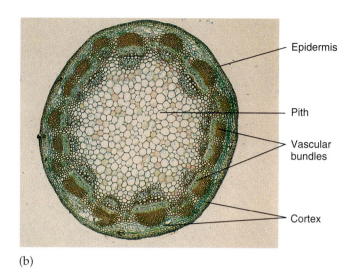

(b)

FIGURE 8.1

Cross section of a stem of (*a*) corn (*Zea mays*), a monocot (×5), and (*b*) sunflower (*Helianthus annuus*), a dicot (×10). In monocots, vascular bundles occur throughout the cortex. In dicots, vascular bundles typically occur in a ring surrounding a pith composed of parenchyma cells. Meristematic tissue is found in the vascular bundles of the dicot stem (but not in the monocot stem).

blown over by the wind. Different plant species have overcome this problem in a similar way through the evolution of stiff building material that tends to develop just beneath or near the external surface of vertical stems. As engineers have long known, the best location for placing strong structural materials to support a tall building is the external surface or just beneath it. Similarly, tall plant stems are supported by the collection of rigid fibers possessing thick cell walls. These fibers are packed closely together near the external part of the stem where they function like steel girders in a building to help keep the stem upright. These fibers help hold up the stem; support leaves, flowers, and fruits that hang from the stem; and help keep the plant upright in strong winds. Other plants use different parts that provide the same mechanical support; for instance, banana trees possess stiff, persistent leaf bases that help support their stems.

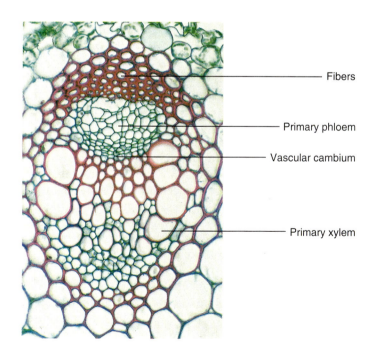

FIGURE 8.2

Vascular bundle from a stem of *Ranunculus*, a dicot (×400). A bundle consists of primary phloem, primary xylem, and vascular cambium. Vascular bundles are often associated with bundles of fibers. The primary phloem is made up of sieve tube members and companion cells (smaller cells).

INQUIRY SUMMARY

Stems consist of epidermal, ground, vascular, and meristematic tissues. The vascular tissue is located in vascular bundles embedded in ground tissue, which often produces and stores food. Young stems are covered by a thin, protective epidermis. The meristematic vascular cambium produces secondary growth on dicots. Fibers and other rigid parts of a plant help keep stems upright.

Do you think intercalary meristems would have evolved if there had been no herbivores? Explain your answer.

Sugar Is Transported Through Phloem in a Stem.

Recent studies with radioactive tracers show that sugars and other organic substances move almost exclusively

FIGURE 8.3

Bamboo growing in Indonesia. These plants elongate rapidly from intercalary meristems at the bases of their nodes.

through stems in sieve tubes of the phloem. Sieve tube members are cells arranged end to end to form sieve tubes. Each sieve tube member is associated with a parenchymalike cell called a **companion cell** (see fig. 4.31). Companion cells and sieve tube members function as a single unit. Sieve tubes in most plants are short-lived; they usually function only during the season in which they are formed. In these plants, sieve tubes are eventually replaced by cells produced by the vascular cambium. Sieve tube members are cylinders connected by sievelike areas called **sieve plates,** each of which has numerous **sieve pores** 1 to 5 micrometers in diameter (see fig. 4.31). Sieve pores may occupy more than 50% of the area of a sieve plate, and sugar traverses as many as 12,000 sieve plates per meter.

Sugars move in the phloem from sources to sinks.

Sugar moves fast in the phloem: peak rates of transport may exceed 2 meters per hour in some plants. The consequences of this rate of transport are impressive; as much as 20 liters (about 5 gallons) of sweet sap can be collected per day from the severed stems of sugar palm, a tropical plant that is used as a source of sugar. At the cellular level, solute transport is turbulent. A phloem cell 0.5 millimeter long empties and fills every 2 seconds, and a gram of sugar passes through a square centimeter of sieve tube area every 20 seconds.

Diffusion is much too slow to account for phloem transport in plants (chapter 4). In 1926, the pressure-flow model was proposed to explain how sugar moves in phloem. This model suggests that sugar moves through sieve tubes along a pressure gradient in a manner similar to the movement of water through a garden hose. A pressure gradient occurs in plants between sources and sinks. **Sources** are the sites where sucrose is made by photosynthesis or the breakdown of starch. Sources contain large amounts of sugar, and their solute concentration is therefore very high. Chlorenchyma cells in leaves are examples of sources because photosynthesis occurs there. **Sinks** are sites where sugars are used during metabolism or stored as starch. Roots, active meristems, and developing flowers and fruits are examples of sinks.

The pressure-flow model is attractive because it explains source-to-sink movement in plants (fig. 8.4). In this model, sucrose produced at a source is loaded into a sieve tube by companion cells (see fig. 8.2). This loading causes water to enter the sieve tube by osmosis. The influx of water into sieve tubes creates a pressure, which carries sucrose to a sink, where it is unloaded and used or stored. Removing sucrose at the sink causes water to move out of the sieve tube at the sink, which reduces the pressure in the cells. Thus, for instance, sugar moves from high pressure in the leaf (source) to low pressure in the root (sink). Water exiting the sieve tube at the sink returns to the xylem and is

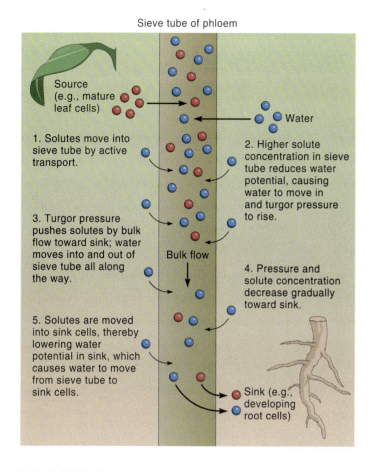

Sieve tube of phloem

Source (e.g., mature leaf cells)

1. Solutes move into sieve tube by active transport.

Water

2. Higher solute concentration in sieve tube reduces water potential, causing water to move in and turgor pressure to rise.

3. Turgor pressure pushes solutes by bulk flow toward sink; water moves into and out of sieve tube all along the way.

Bulk flow

4. Pressure and solute concentration decrease gradually toward sink.

5. Solutes are moved into sink cells, thereby lowering water potential in sink, which causes water to move from sieve tube to sink cells.

Sink (e.g., developing root cells)

FIGURE 8.4

Proposed mechanism for flow of sugar in the phloem of flowering plants. Solutes, such as sugar, move from sources to sinks.

recirculated. The ability of the pressure-flow model to account for and accurately predict the characteristics of phloem transport makes it the most widely accepted model for phloem transport.

Apparently, sugars can move in all directions; these directions are determined by the presence of sources and sinks. When sucrose is required simultaneously by several sinks, larger sinks receive more sugars than other sinks. Developing fruits and flowers are large sinks that often take priority over other sinks, thereby explaining why vegetative growth often slows or stops until fruit formation is complete. Sources contribute primarily to the nearest sink rather than equally to all sinks. For example, leaves near roots usually transport their sugars to roots, while leaves near tips of stems transport sugars to the shoot apex. Leaves midway between roots and shoot tips transport their sugars to both of these sinks. Sieve tubes themselves are not the sources or sinks; that is, sugars are neither produced, needed, nor stored in sieve tubes. Rather, sieve tubes *link* sources and sinks.

FIGURE 8.5

Aphids are phloem-feeders. An aphid feeding on a plant stem; excess phloem sap (droplet of honeydew) passes through the aphid.

Investigations of phloem contents have been aided by insects.

One seemingly easy way to determine the contents of a sieve tube would be to collect phloem sap as it oozes from a wounded stem. However, there are problems with this approach because wounds rupture and kill other cells and their contents mix with and contaminate the sap from phloem. Ideally, a botanist could delicately insert a microneedle into a single sieve tube without a sudden release of pressure. Such a microneedle was discovered in 1953. It was provided by insect physiologists who were studying the feeding habits of aphids, which are insects that obtain food from phloem (fig. 8.5). The scientists noted that aphids have natural phloem probes that tap into individual sieve tubes: each insect has a needlelike mouthpart called a stylet. Because the contents of sieve tubes are under a positive pressure, their contents surge into the aphid as it enjoys its sugary meal. This rush of phloem sap is often overwhelming; when this occurs, the aphid secretes excess sap as a drop of sugary honeydew, which eventually drops onto underlying plants, sidewalks, and cars. But how can we be sure the aphid doesn't alter the contents of the sap before secreting it as honeydew? Some clever microsurgery solved this problem. An aphid was allowed to insert its stylus into a sieve tube member. When the feeding started, the aphid was anesthetized with a gentle stream of carbon dioxide. The stylet was then severed from the aphid's body, leaving only the open-ended stylet inserted into a sieve tube member. Pressure in the sieve tube pushed sugary sap through the severed stylet for several days: collection rates averaged about 1 cubic millimeter of phloem exudate per hour.

Using this technique, it was discovered that the most abundant compound in a sieve tube is water. More than 90% of the solutes in sieve tubes are carbohydrates, largely or entirely as sucrose. The concentration of sucrose may be as high as 30%, thereby giving phloem sap a syrupy thickness. However, not all plants transport only sucrose; some plants transport sugar alcohols in their phloem. For example, apple and cherry trees transport sorbitol, and mannitol moves in the phloem of ash (*Fraxinus*); both sorbitol and mannitol are used to sweeten sugarless gum. While the biblical manna that was miraculously supplied to the Israelites may have come from various sources, commercial manna (the source of mannitol) is the dried phloem exudate of manna ash (*Fraxinus ornus*) and related plants. Sieve tubes also transport amino acids, hormones, alkaloids, viruses, and inorganic ions.

INQUIRY SUMMARY

Sugars and other organic compounds move throughout plants in the phloem. These solutes move through cells under a positive pressure. Transport of sugar in phloem is best explained by the pressure-flow model. This model states that sugar is carried by water along a gradient of pressure from sources to sinks.

Can you think of any situation in which a leaf could be a sink and not a source?

Stems Play Key Roles in the Ecology of Plants.

Most plants respond to mechanical stimulation such as wind and shaking. In one experiment to test the effects of mechanical stimulation, plants that were rubbed had stems that were thicker and 45% shorter than plants that were not rubbed. This growth response in response to touch is called thigmomorphogenesis (*thigmo,* meaning touch, and *morphogenesis,* meaning development). In windy habitats, plants often possess short, thick stems that resist bending and breaking—an adaptive response to the constant buffeting by the winds.

Most stems support leaves and other parts of plants, as well as conduct water, minerals, and sugar between roots and leaves. Plants whose stems do not elongate are called **rosette plants** (fig. 8.6); these plants have very short internodes with tightly packed leaves. The stems of rosette plants are short and consist almost entirely of overlapping leaf bases. These structures can reach impressive proportions. For example, although the large, overlapping leaf bases of a banana plant form a "trunk" that can raise leaves more than 10 meters, its pyramid-shaped stem barely reaches above ground (fig. 8.7). Rosettes fail to elongate because they may not make enough of the hormone gibberellin (see chapter 6), or they may not be sensitive to it. Nonetheless, treating rosettes with gibberellin typically causes internodes to lengthen.

Giant rosette plants (*Espeletia, Lobelia,* and *Senecio* species) grow on the sides of tropical mountains in the An-

FIGURE 8.6

A rosette plant, *Echeveria gilva,* with leaves attached to a very short stem. Note the flowering stalks growing from the axils of leaves.

des and in Africa at altitudes where the temperatures can drop below freezing during the night (fig. 8.8). Different species of these tree-sized plants, with tightly packed leaves clustered in a rosette at the top of an elongated stem, possess a variety of adaptations that allow them to resist freezing, including retaining dead leaves or having dense hairs that cover the stem, both of which help insulate the stem. Other species collect several liters of water in their stems that warm up during the day and store heat that can be released during the night, thereby keeping the stem from freezing.

Modified stems adapt plants to different factors in the environment.

Although most stems support a plant and conduct water and minerals, modified stems have unusual, and even bizarre, shapes and functions. **Stolons,** or **runners,** are

FIGURE 8.7

The "stem" of the banana (*Musa*) plant consists of a series of large, overlapping leaf bases.

FIGURE 8.9

A modified stem that grows aboveground. Stolons of beach strawberry (*Fragaria chilensis*). What advantage might these plants have in very loose, unstable soils?

FIGURE 8.8

Giant rosettes (*Espeletia*) growing in the Andes Mountains of Ecuador possess clusters of leaves at the top of an elongated stem.

FIGURE 8.10

The tips of the tendrils of the Virginia creeper (*Parthenocissus quinquefolia*) have adhesive pads that stick to walls and allow the stem and leaves to grow up the side of a building.

horizontal stems that grow along the soil surface and function in vegetative, or asexual, reproduction. For example, buds at nodes of strawberry runners produce shoots and roots that eventually form new plants (fig. 8.9). Other plants having stolons include common houseplants such as the Boston fern and spider plant and common lawn plants such as Bermuda grass and crabgrass. Plants along sandy beaches often possess runners, which help stabilize the dunes.

Tendrils and twining shoots of plants such as morning glory and sweet potato coil around objects and help support the plant. Tips of tendrils of plants such as Virginia creeper (*Parthenocissus*) have adhesive pads that stick to walls, trees, or other upright objects as the creeper stem creeps up them (fig. 8.10). Tendrils and twining shoots are

FIGURE 8.11
Tendrils are whiplike structures that coil and help a plant cling to other objects.

FIGURE 8.12
The flattened stem, the cladophyll, of this plant (*Homalocladium platycladum*) looks like a long leaf and produces sugar through photosynthesis, however, unlike a leaf, flowers develop along its sides at nodes.

important because they help support leaves, flowers, and fruits in the air and sunlight; without tendrils, these relatively weak-stemmed plants would lie on the ground and, perhaps, in the shade and out of sight of pollinators and dispersers (fig. 8.11). Twining shoots are stems with long internodes that move in circles through the air, thereby increasing the probability they will contact a supportive structure (see chapter 6 on plant movements). Twining shoots may be more than 3 meters long and often die if they do not find a structure on which to attach. Examples of plants with twining shoots include honeysuckle and *Wisteria*.

Cladophylls are flat, leaflike stems that conduct photosynthesis (fig. 8.12). Examples of plants with cladophylls include butcher's broom, greenbrier, asparagus, and orchids (e.g., *Epidendrum*). Some **thorns** are sharp-pointed, modified stems that are adapted to poke the sensitive noses of would-be herbivores. Examples of plants with stem thorns are honey locust, firethorn, hawthorn, and bougainvillea. **Succulent stems** of plants such as cacti have a low surface-to-volume ratio. Succulent stems store large amounts of water and are common in plants growing in deserts. The form of cladophylls, thorns, and succulent stems are all related to their function. What significance might a low surface-to-volume ratio have for a plant that lives in a desert?

Modified belowground stems often help plants survive cold winters.

Bulbs are rosette stems surrounded by fleshy leaves that store energy reserves and nutrients. When these supplies are used by the plant (such as during flowering), the fleshy leaves collapse into a papery scale leaf surrounding the stem. Onion, lily, hyacinth, tulip, narcissus, and daffodil are examples of plants that produce bulbs (fig. 8.13). Bulbs are important because they enable many plants to survive stressful periods, such as winter, when no photosynthesis occurs.

Rhizomes are underground stems that grow horizontally near the soil surface. They typically have short internodes and scale leaves and produce adventitious roots along their lower surface. Rhizomes store energy reserves for renewing growth of the shoot after periods of stress, such as cold winters. This explains why plants such as Johnson grass (*Sorghum halepense*) are so difficult to eradicate; although their aboveground shoots die back during winter, their dormant rhizomes renew growth the following spring. Other examples of plants having rhizomes include quack grass, *Iris, Canna, Begonia,* and ginger.

Tubers are swollen underground stems that store energy reserves for subsequent growth. Irish potatoes (*Solanum tuberosum*) are tubers produced on rhizomes or stolons that burrow into the soil (fig. 8.14). The "eyes" of potatoes are buds in the axils of small, scalelike leaves located at nodes; the areas between a potato's eyes are internodes.

Corms are stubby, vertically oriented stems that grow underground. Corms, which have only a few thin leaves, store energy and nutrients. Like rhizomes, corms enable many plants to survive winter. Examples of plants having corms are *Gladiolus, Cyclamen,* and *Crocus.* Plants growing in cold climates often have tubers, corms, bulbs, or rhizomes.

Leaves

Stem

FIGURE 8.13

A modified stem that grows belowground. Onions are bulbs consisting of layered, fleshy leaf bases attached to a short stem.

INQUIRY SUMMARY

Stems of different plants have evolved for reproduction, climbing, photosynthesis, protection, and storage. These modified stems adapt plants to different environments; plants with bulbs, rhizomes, tubers, or corms are often adapted to cold climates; succulent stems are adapted to dry climates; and stolons are adapted to unstable, sandy soils.

Beside helping a plant survive cold winters, can you think of any other selective advantages structures such as corms, bulbs, tubers, and rhizomes might give a plant?

Wood Is a Secondary Tissue Produced by the Vascular Cambium.

Recall from chapter 4 that primary growth is due to the apical meristem found at the tip of roots and shoots. A plant grows longer through primary growth from the tip of the roots downward and from the tip of the shoots upward. Secondary growth results from a lateral meristem, the vascular cambium, that produces secondary phloem and secondary

FIGURE 8.14

The Irish potato (*Solanum tuberosum*) is a tuber. Sprouts growing from the "eyes" of this potato will produce new plants.

xylem. Bark, secondary phloem, and **wood,** which is secondary xylem, increase the plant's girth, thereby making the plant sturdier and able to grow taller. This increased height decreases the chances that the plant's leaves will be shaded by other plants.

Wood is formed in the following manner. As a dicot stem matures, cortex cells between the vascular bundles develop into vascular cambium cells. In this way, these new vascular cambium cells connect with those in the vascular bundles, forming a thin band of cells that appears as a circle in cross section of a stem. As the vascular cambium cells divide, they produce rings of secondary xylem (wood) to the inside and secondary phloem to the outside (fig. 8.15). On average, the vascular cambium produces four to ten times more xylem than phloem. Occasional divisions of vascular cambium cells perpendicular to the divisions producing xylem and phloem expand the vascular cambium to accommodate the increasing girth of the stem.

What controls the formation of xylem cells to the inside of the stem and phloem to the outside? Sensitivity to hormones, such as auxin, is important as is the genetic makeup of cambial cells. To understand this, consider the results of an elegant experiment. Plugs of tissue were removed from the region of the stem between vascular bundles and reoriented 180 degrees prior to formation of the ring of vascular cambium (i.e., so the end of the plug that originally was nearest the epidermis was now toward the center of the stem). When this was done, vascular cambium formed as it normally

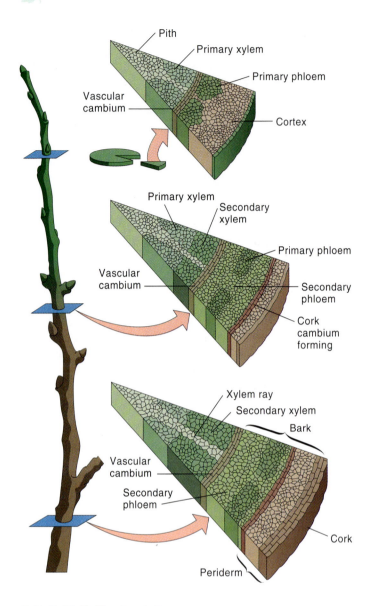

FIGURE 8.15

Secondary growth of dicot stems. The vascular cambium differentiates between the primary xylem and primary phloem and forms an open-ended cylinder. The vascular cambium produces secondary phloem to the outside and secondary xylem (wood) to the inside. Derivatives of the secondary phloem form a protective layer called the *periderm.*

would; however, it produced secondary tissues that were 180 degrees off; that is, it produced secondary xylem to the *outside* and secondary phloem to the *inside.* These and other experiments suggest the cortex cells that ultimately develop into vascular cambium are genetically predetermined.

The activity of the vascular cambium is influenced by hormones.

During autumn and winter, the vascular cambium is inactive. The moist, warmer, and longer days of spring reacti-

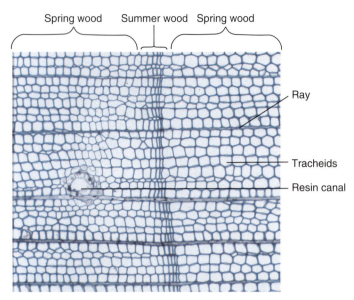

FIGURE 8.16

Cross section of softwood of eastern white pine (*Pinus strobus*), a gymnosperm (×95). Softwood is produced by non-flowering trees and is relatively homogeneous because it lacks vessels. Note the larger spring wood cells and the smaller summer wood cells that mark the end of the growing season.

vate the vascular cambium, inducing cell division. What controls this reactivation? Evidence shows the cambium is inactivated when the shoot apex is removed; this, in turn, removes the source of the hormone auxin. The cambium, however, cannot be reactivated by adding auxin to the cut surface. Moreover, the concentration of auxin in the vascular cambium remains high even when the cambium is dormant, suggesting that the activity of the vascular cambium is not due to a deficiency of auxin. Rather, the dormancy of the vascular cambium is probably due to a decreased *sensitivity* to auxin: auxin affects the vascular cambium cells in the spring only after these cells regain their ability to respond to auxin. Other hormones also may be involved in this activity.

The environment affects the production of different kinds of wood.

Wood—secondary xylem—formed by the vascular cambium during the moist days of spring (early in the growing season) is called **spring wood** and consists of large cells (fig. 8.16). Later in the season, the drier days of summer gradually slow the activity of the vascular cambium and cause it to produce **summer wood,** made of smaller cells.

Differences between spring and summer wood are abrupt in trees such as pine and are visible in most trees as **growth rings,** or annual rings (fig. 8.17). In temperate regions, the predictable seasons usually produce one growth ring per year in a woody plant, thus accounting for the term annual ring. However, trees such as ebony (*Diospyros ebenum*) and jacaranda (e.g., *Jacaranda arborea*) that grow in the seasonless tropics lack growth rings, and plants growing in arid and semiarid regions often produce more than one growth ring per year in response to sporadic rains.

Growth rings appear as concentric circles in cross section and are usually 1 to 10 millimeters wide. Each growth ring increases the diameter of the tree trunk; once formed, each ring remains unchanged in size or position during the life of the tree. Because climate (especially the availability of water) strongly influences the formation of growth rings, a cross section of wood is a diary of the climatic history of a particular region. The science of interpreting history by studying growth rings is called **dendrochronology** and is an important tool for meteorologists, anthropologists, and ecologists. For example, the White Mountains of eastern California are home of bristlecone pines (*Pinus longaeva;*

fig. 8.18), whose slow growth packs almost 1,000 growth rings into only 13 centimeters or so of wood. By comparing the growth rings of living and dead trees, dendrochronologists have reconstructed the area's climate for the past 10,000 years. They've used this information to date Native American cliff dwellings and to document droughts in the years 840, 1067, 1379, and 1632—each 200 to 300 years apart. Dendrochronologists studying annual rings of bald cypress trees (*Taxodium distichum*) have also suggested that the "Lost Colony" of Roanoke, Virginia, had problems because the settlers arrived during the middle of a severe drought that lasted several years.

The vascular cambium is reactivated each year and is virtually immortal—at least as long as the woody plant is alive. Many plants are more than 1,000 years old, and a bristlecone pine named "Methuselah" growing in California is almost 5,000 years old; this means the seed that produced this plant germinated more than 500 years before the Egyptian pyramids were completed. The perpetual reactivation of the vascular cambium for thousands of years produces impressive trees—the largest and most conspicuous plants ever to live. The fattest tree is a chestnut (*Castanea*) named the "Tree of One Hundred Horses," which lives on Sicily's Mount Etna: this tree is 58 meters around (it would take about 30 students with arms outstretched and touching fingertips to encircle this tree). A

(a)

FIGURE 8.17

(*a*) Growth rings are seen in this cross section of a locust tree showing the dark heartwood (inner 12–13 rings) and the lighter sapwood. The transport of water and dissolved minerals occurs only through the sapwood. The dark color of heartwood results from the presence of resins, gums, and other chemicals. (*b*) This diagram of a tree trunk shows a hypothetical situation that would result from growth during average conditions (no floods or drought). In general, growth begins at the top of the tree and progresses toward the base. Although the width of each growth ring decreases as you move toward the edge of the trunk, the total cross-sectional area remains the same for each year.

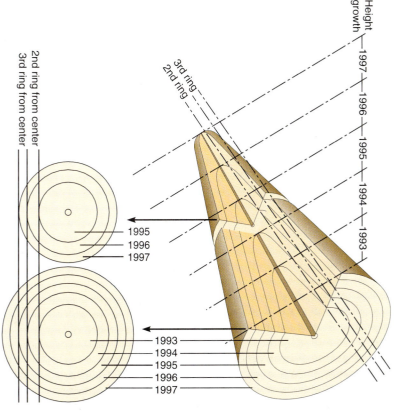

(b)

FIGURE 8.18
Bristlecone pines (*Pinus longaeva*) are the oldest known plants that are not clones.

FIGURE 8.19
The sequoias are among the most impressive members of the plant kingdom. The volume of a mature sequoia's trunk is almost twice that of a typical two-story house.

2,000-year-old tule tree (*Taxodium mucronatum*) near Oaxaca, Mexico, is almost 45 meters in circumference, and at last measurement, a California redwood (*Sequoia sempervirens*) named the "Mendocino Tree" stands 112 meters (367.5 feet) (fig. 8.19). That redwood weighs approximately 1,600 tons, which is roughly equivalent to the weight of 10 blue whales. Yet, these giant trees are supported by roots that seldom go deeper than 1.8 meters into the soil. Mature redwood trees would tower above our nation's Capitol, dwarf the Saturn V rockets that flew astronauts to the moon, and reach from home plate to the outfield bleachers of any baseball park in the country. A 20-year-old redwood sapling, a mere sprig, is often more than 15 meters high.

The oldest giant sequoias are about 3,200 years old, meaning that the trees growing today were already 200 years old when King David ruled the Israelites. In its seventies, the sunset years for humans, a sequoia is still a "teenager" and bears its first seeds. A mature tree produces about 2,000 cones per year, each of which contains about 200 seeds. Each of these seeds weighs less than 0.01 gram (about eight-thousandths of an ounce), a mere one-hundred-billionth of the weight of a full-size tree. The average sequoia needs more than 1,100 liters of water per day. The energy required to lift this water to the tree's leaves is enough to launch a can of soda into a low orbit of earth.

INQUIRY SUMMARY

Wood is made up of secondary xylem and helps support trees. The vascular cambium is a meristem that produces secondary xylem toward the inside, and secondary phloem toward the outside, of mature dicot stems. Spring wood, made during the moist days of spring, contains large xylem cells. Summer wood, made during the drier days of summer, contains smaller xylem cells. Differences between spring and summer wood are visible as growth rings. The activity of the vascular cambium produces enormous organisms.

What would the growth ring look like in a tree whose leaves had been stripped by leaf-eating caterpillars?

FIGURE 8.20
The thin, paperlike bark of paper birch (*Betula papyifera*).

Trees Have Evolved Chemical and Physical Characteristics that Help Fend Off Infections and Pests.

As we have just seen, some trees attain gigantic sizes and old ages despite their lack of an immune system and their inability to flee enemies. Trees attacked by pathogens (disease-causing organisms) possess a defense mechanism called **compartmentalization,** which involves walling off infections, fortifying cell walls near wounds. Some trees also produce antimicrobial compounds such as phenols. This evolved strategy is effective but has important consequences for a tree. Phenols are toxic chemicals produced by plants that can kill invading microbes but that also damage the tree by poisoning its tissues. Trees survive these sacrifices of their parts by continually producing new tissue. However, the walled-off compartments of a tree are inaccessible, which forces the tree to grow a "new" tree over the sealed compartments. This new tree is visible as new growth rings, which are the new tissues that perform the tree's activities.

A tree's survival depends on how fast and effectively it can react to and wall off its invaders; American chestnut (*Castanea dentata*) is nearly extinct, and American elm (*Ulmus americana*) faces possible extinction because these trees cannot react fast enough to their pathogens. Too many walled-off compartments (i.e., too many infections) can stop the movement of water within a tree and thereby kill the tree.

The bark is a tree's external, protective barrier.

All of the tissues outside the vascular cambium constitute **bark** (see fig. 8.15). Bark is important to trees because it pro-

vides a protective barrier to the living tissues just beneath it. In mature trees, bark consists of (1) secondary phloem, which transports sugars between leaves and roots, and (2) **periderm,** which is an outer layer that protects and insulates underlying tissues. Periderm consists of cells impregnated with suberin, a substance that makes the bark impermeable to water.

Secondary phloem is the tissue produced by the vascular cambium toward the outside of the stem and just inside of the primary phloem. Only the inner centimeter or so of secondary phloem contains functional cells that conduct sugar. Secondary growth eventually ruptures the epidermis of young stems and roots. The ruptured epidermis is replaced by periderm, which protects the underlying tissues. Periderm consists of three tissues: cork cambium, cork, and secondary cortex cells. The **cork cambium** is a meristem that produces the periderm.

As trees grow wider through secondary growth, the bark often splits. Trees have adapted to this problem by continually forming new layers of cork cambium. In trees such as paper birch (*Betula*), the replacement cork cambium is cylindrical; thus, these trees shed their bark in strips or large, hollow cylinders (fig. 8.20). Trees that continually form overlapping plates of cork produce scales of bark rather than large rings of bark and include sycamore, elm, maple, and pine.

Cork cells are dead at maturity, leaving cell walls that are suberized (i.e., impregnated with suberin), and lacking intercellular spaces. In one season, the cork cambium produces as many as 40 layers of cork cells that waterproof, insulate, and protect the plant. In some trees, such as the cork oak (*Quercus suber*), these layers are very thick and are harvested to make bottle stoppers and insulation materials. Cork also forms over wounds, at the spots where leaves drop from stems, and on many fruits and potatoes. Thus, when you eat the "skin" of a baked potato, you are eating parts of bark!

Bark prevents oxygen gas from readily diffusing into the tree to the living cells of the cork cambium, secondary phloem, and vascular cambium. Without oxygen, these tissues will die. Gas exchange, involving oxygen gas and carbon dioxide, occurs across the bark through **lenticels,** which are raised, localized areas of loosely packed cells (fig. 8.21). Lenticels are formed by the cork cambium and range in shape from round to elliptical. They are easily visible on pears and apples and are the dark spots and streaks on corks (see Perspective 8.1, "Cork").

Trees can be killed by girdling.

A tree is **girdled** when a strip of its bark is removed around the circumference of its stem (fig. 8.22). Girdling usually kills a tree. This can happen accidentally when you move your "weed-eater" too close to a young tree

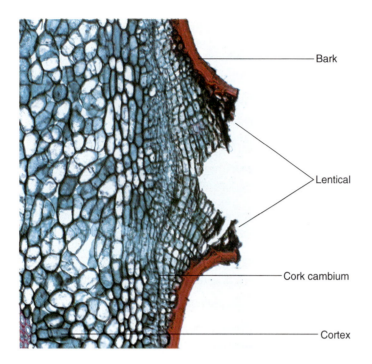

Bark

Lentical

Cork cambium

Cortex

FIGURE 8.21
Cross section of part of a young stem of elderberry (*Sambucus*) (×80). Gas exchange across the bark occurs through lenticels.

FIGURE 8.22
This girdle on a black cherry tree (*Prunus serotina*) blocked the flow of sugars through phloem from above, stopping growth below. What happens to roots in girdled trees?

and strip its bark off near ground level. On older trees with tougher, thicker bark, girdling takes more effort but happens, for instance, when male deer rub their antlers against a tree or when beavers chew the bark off a tree.

Girdling removes the bark, including phloem, and vascular cambium. Because the phloem has been removed, sugar can no longer move from the leaves of the tree to its roots. In addition, no new phloem can be produced to replace that which was removed because the vascular cambium also has been stripped off in the girdle. Thus, the roots of a girdled tree eventually die because they cannot receive the sugar necessary to sustain them. After the roots die, the tree also dies.

Imported wood-boring insects may upset the natural ecology of American forests.

The wood of trees represents a tremendous source of energy in a community, but few organisms use it because chemicals in the wood and bark are difficult to digest. However, it should not be surprising to find that some insect herbivores make their living in wood. One wood-boring insect has been accidentally introduced to the United States, and some scientists feel that this Asian long-horned beetle (*Anoplophora glabripennis*) could cause more than $100 billion in damage to trees across the United States. It is a pest in China, Korea, and Japan but was brought to the United States as a hitchhiker on wooden pallets or packing material used for imports from the Far East.

The beetle spends most of its life as a grub inside wood, first eating the vascular cambium, the phloem with the sugar it is carrying, and other nearby tissues. It then burrows into the secondary xylem and spends the winter in the tree's wood. The grub matures into an adult inside the tree and then burrows its way out to mate. Their chewing abilities make the beetles lethal to sugar maples, horse chestnuts, willows, elms, black locusts, poplars, and several other species of trees. Their activities could hurt the maple sugar industry, fall-foliage tourism throughout much of the Northeast, and communities of trees in parks and cities. Currently, the battle against these insects includes finding infected trees, cutting them down, and destroying the wood and beetle grubs inside before they mature. The problem is how to locate infected trees. Ecologists are studying the possibility of using high-tech sound detectors to pick up grubs that are chewing on wood; they are noisy eaters!

INQUIRY SUMMARY

Bark includes secondary phloem, which transports sugars, and periderm, which protects and insulates underlying tissues. Gas exchange across the periderm occurs through lenticels. Girdling kills the roots of a tree first; then the entire tree dies. Bark, toxic chemicals, and compartmentalization all help protect a tree from

8.1 *Perspective* Perspective

Cork

Each year we use almost 200,000 metric tons of commercial cork for things as diverse as stoppers for wine bottles, insulation for the space shuttle, and grips on symphony conductors' batons. Commercial cork is the outer bark of cork oak (*Quercus suber*), an evergreen oak grown in cork plantations in the western Mediterranean.

The first periderm, which forms from the epidermis, is commercially useless. Consequently, it is stripped and discarded when the tree is approximately 10 years old. A usable cork layer 3 to 10 centimeters thick can be harvested when a tree is 20 to 25 years old (i.e., when the tree is about 40 centimeters in diameter) and thereafter once per decade until the tree is approximately 150 years old. Since cork oaks grow crookedly, cork is harvested by hand rather than machine; workers use hatchets to peel it off in lumberlike slabs. The cork is then dried and boiled before being marketed. In contrast to that of most other trees, the cork of a cork oak breaks away at the cork cambium and can be peeled without harming the tree (see photos).

Commercial cork has several properties that make it ideal for a variety of uses. It consists of densely packed, suberized cells (about 1 million cells cm^{-3}) that make it impermeable to liquids and

The bark of this cork oak (Quercus suber) has just been harvested for its cork. The inner bark remains, so that the tree does not die when the cork is harvested.

gases. Half of a cork's volume is trapped air; thus, it is four times lighter than water. It is also compressible—cork can tolerate 400 kg cm^{-3} and expand to its original shape within one day. Cork is virtually indestructible, fire-resistant, and durable; it resists friction; and it absorbs vibration and sound.

---

pathogens; however, herbivores still may attack the wood of trees.

Can a tree have both bark and epidermis tissue at the same time? Explain your answer.

Secondary Growth Is Produced by Different Types of Plants and Their Parts.

We classify wood on the basis of the kind of plant that produces it (i.e., softwoods versus hardwoods) as well as the location and function of the wood in the plant. The following section describes the differences between softwoods and hardwoods.

Softwoods are nonflowering, seed-producing trees, most of which are evergreen and native to temperate zones. Pine, spruce, larch, fir, and redwood are examples of softwoods. Their wood is relatively homogeneous because it lacks vessels and is about 90% tracheids (see fig. 8.16 and chapter 4). Many softwoods, such as pine and spruce, also contain **resin canals** filled with resin secreted by living parenchyma cells that line the canals. Resin hardens when exposed to air and is an effective means of sealing plant wounds. Fossilized resin is amber, a substance described in some mineralogy books as a gem and that has trapped unwary insects for millions of years.

Most softwoods are relatively light (i.e., soft) and are easily penetrated by nails, which explains their widespread use in home construction. Softwoods contain a large amount of lignin, a chemical that impregnates and stiffens

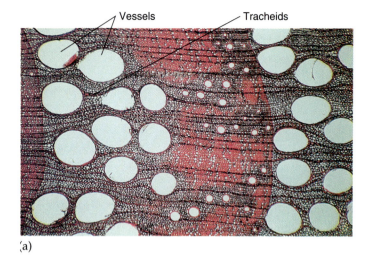

(a)

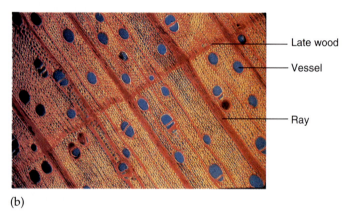

(b)

FIGURE 8.23

Cross sections of wood from hardwoods, which contain tracheids and vessels. Tracheids and vessels transport water and dissolved minerals from roots to shoots. (*a*) Because the largest vessels in oak (*Quercus* species) occur only in early wood, oak wood is referred to as ring-porous wood (×46). (*b*) Because the largest vessels in sugar maple (*Acer saccharum*) are distributed throughout the wood, maple wood is referred to as diffuse-porous wood (×120).

cell walls. In wood such as white pine, lignin makes the wood ideal for lumber because it makes wood stable and unlikely to warp. Most commercial plantations grow pines or firs because these trees grow faster and are more profitable than dicot trees. Pulp (ground-up wood used to make paper) and plywood come primarily from softwoods.

Hardwoods are flowering, dicot trees native to temperate and tropical regions. Oak, maple, ash, walnut, and hickory are examples of hardwoods. Unlike softwoods, which contain mostly tracheid cells, hardwoods contain tracheids, fibers, and vessels (fig. 8.23). Fibers make most hardwoods stronger and denser (i.e., harder) than softwoods, which explains why woods like oak, walnut, maple, and hickory are hard to nail.

In hardwoods, water and minerals are conducted through tracheids and vessels. Vessels are often visible to the naked eye and are frequently called pores by woodworkers. If the largest vessels occur only in spring wood, the wood is called **ring-porous wood** (fig. 8.23*a*). Because of their targetlike arrangement of vessels, ring-porous woods such as elm, chestnut, oak, and ash have easily discernible growth rings. In most trees, there is little contrast in the size or pattern of vessels across growth rings; that is, vessels in these trees are distributed uniformly (fig. 8.23*b*). Such wood (e.g., maple and birch) is called **diffuse-porous wood.**

Because of their many fibers, hardwoods are usually denser than softwoods. However, there are many exceptions to this generalization. For example, hardwoods such as poplar and basswood are softer than softwoods such as hemlock and yellow pine. Similarly, balsa (Spanish for "raft"; *Ochroma lagopus*) and the even lighter pith-plant (*Aeschynomene aspera*) are hardwoods whose wood is lighter than cork.

Wood is classified by its function and location.

We also classify wood as either sapwood or heartwood. This distinction involves position, function, and appearance of a tree's wood and is based on different parts of a woody stem specialized for different functions.

Water and dissolved nutrients, the sap, move in the outer few centimeters of a woody plant. The wood that transports sap is called **sapwood** and is usually light, pale, and relatively weak (see fig. 8.17*a*). The dry wood in the heart (i.e., center) of a tree is called **heartwood** and is the dump site for some products of the tree's chemical reactions, such as resins, gums, oils, and tannins, which are gradually deposited in heartwood. These products fill the xylem cells and eventually clog the wood. Heartwood, which can be more than a meter wide in large trees, is darker, denser, and more aromatic than sapwood.

Secondary growth also occurs in roots.

The pericycle of roots is initially lobed and surrounds the spokes of primary xylem (see figs. 7.8*a* and 7.11). The vascular cambium in roots is produced from the pericycle and becomes roughly circular after it begins to divide, producing secondary growth. In roots, however, secondary growth is usually less extensive than in shoots. Nevertheless, secondary growth in roots has similar consequences as in shoots: it increases the girth of roots by producing secondary xylem, secondary phloem, and periderm. These tissues—especially the periderm—stop the absorption of water. The portions of roots that have undergone secondary growth help anchor the plant.

Many plants have unusual secondary growth.

Not all dicots have the typical kind of secondary growth described earlier in this chapter.

The unusual secondary growth of some dicots results from one of three anomalies:

1. *Unusual behavior of a typical vascular cambium.* In plants such as carrot (*Daucus*) and some cacti, the vascular cambium produces large amounts of storage parenchyma. In cacti, these parenchyma cells store water, whereas in carrots, they store carbohydrates.

2. *Presence of more than one cambium.* In plants such as beets (*Beta vulgaris*), several cambia are arranged in concentric circles, producing concentric rings of phloem toward the outside and xylem toward the inside, along with large amounts of storage parenchyma.

3. *Differential activity of a vascular cambium.* In plants such as Dutchman's pipe (*Aristolochia*), parts of the cambium produce parenchyma, whereas other parts produce vascular tissue. As a result, the stems of these plants often possess ridges. In some species of *Bauhinia*, only two opposite areas of the vascular cambium are active, which produces a flattened stem.

Typically, monocots do not have secondary growth.

Most monocots have only primary growth. However, some arborescent (i.e., treelike) monocots such as *Dracaena* and *Cordyline* form a vascular cambium just outside their vascular bundles. Unlike that of dicots, the monocot vascular cambium does not produce secondary xylem, or wood, but secondary cortex cells.

Secondary growth of some monocots can be impressive; for example, some species of *Dracaena* can be more than 12 meters around. However, not all treelike monocots exhibit secondary growth. For example, palms are overgrown herbs that become treelike by **primary thickening,** in which intense expansion of cells occurs just behind the apical meristem (fig. 8.24), causing the stem to expand in girth. Periderms of monocots also differ from those of dicots. All cells of a monocot's periderm are suberized, and there are no distinguishable layers of cells.

INQUIRY SUMMARY

Softwoods are nonflowering seed plants native to temperate regions. Their wood is homogeneous and contains mostly tracheids. Hardwoods are dicots whose wood contains tracheids, vessels specialized for transport, and fibers specialized for support. Sapwood transports water and dissolved nutrients and comprises the outer few centimeters of wood. Heartwood is in the center of a tree, where many chemicals are stored. Heartwood does not transport water and minerals. Secondary growth occurs in roots of some plants and in a few monocots.

Do you think that hardwood or softwood trees grow faster? How might this be related to the wood they produce?

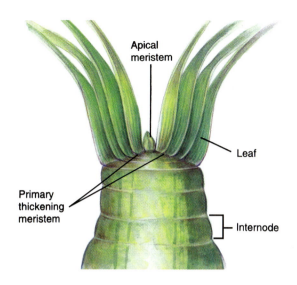

FIGURE 8.24

Primary thickening in monocots. In palms, primary thickening occurs just below the apical meristem, causing the stem to grow in girth.

Many Stems Are Economically Important.

Humans have many uses for stems. Sugar and molasses from sugarcane stems (*Saccharum officinalis*) go into our soft drinks and over our pancakes. Stem fibers are used to make fabrics, such as linen from the fibers of the flax plant. We also eat a few stems, such as young bamboo shoots and tubers of potatoes. In some parts of the world, potato tubers are the most important part of the diet. Andean tribes mash and dry tubers to form *chuño,* which is added to just about everything they eat. Indeed, members of the tribes say that "stew without *chuño* is like life without love." Although we don't expect you to take your potatoes quite that seriously, we do hope you'll remember what you're eating next time you order some fries. (Read about the history of the Irish potato famine in the mid-1800s to see how important potatoes were to these Europeans.) Charcoal, cork, insulation, life preservers, quinine—an antimalarial medicine (fig. 8.25)—cinnamon, and even a diesel-like fuel also are derived from stems.

Wood is one of the world's most economically important plant products.

Secondary xylem, or wood, comprises about 90% of a typical tree. About one-third of the United States is covered by trees, of which approximately 65% are farmed commercially. We have all sorts of uses for the 1.7 million cubic meters of wood produced by these forests each day. Each person in the United States uses an average of 2.25 cubic meters of wood per year. About 50% of this wood is used for lumber, 25% for pulp and paper products, 10% for

FIGURE 8.25
Cinchona bark being dried in preparation of the extraction of quinine.

plywood and veneer, and the rest for fuel and miscellaneous products such as telephone poles, oars, toothpicks, bowling pins, chopsticks, baseball bats, baskets, acoustic instruments such as violins and guitars, croquet balls, and cue sticks (see Perspective 8.2, "The Bats of Summer"). Americans use 250,000 tons of napkins and 2 million tons of newsprint and writing paper per year. These and other uses of wood make forestry a $40 billion-per-year industry in the United States alone (see Perspective 8.3, "The Lore of Trees").

We make more than 5,000 products from the secondary xylem of the species of trees that grow in the United States. Each year, people in the United States cut down about 3 billion trees. This harvest exceeds that of all other countries, including Brazil. Here are some of the most important products from trees:

§ *Lumber.* In the United States, most of the harvested wood is used for lumber and comes from about 35 species of trees. We use this lumber for fences, houses (white pine), furniture, and paneling (black walnut), as well as barrels and flooring (white oak). Because wood conducts heat poorly, wood floors are a good way to insulate a house. We also use lumber to make veneers, plywood, and particleboard.

Veneers are thin (about 1 millimeter thick) sheets of wood peeled from large logs of oak, maple, walnut, hickory, and other hardwood trees. They are often glued over cheaper and less attractive wood to provide a relatively tough, attractive surface for furniture. Plywood is made by gluing several veneers together at right angles. These layered veneers make plywood strong, hard to split, and well-suited for construction. Particleboard is made by gluing and pressing together small particles (i.e., chips) of wood into flat sheets. Particleboard is used as insulation and to make boxes.

§ *Pulp and paper.* Paper is made from wood pulp, which is pulverized wood. To make pulp, paper mills first strip the bark from harvested trees. The trunks are then either pulverized on a grindstone or cooked in a batch of chemicals. Both of these processes degrade the wood into pulp, which is about 99% water and 1% fibers. The untreated pulp is fed into a machine that spreads and dries it into newsprint, an inexpensive type of paper containing only about 5% water. In about 1 minute, a papermaking machine can produce a sheet of paper covering more than 22,000 square meters, an area bigger than five football fields. An average tree harvested by a lumber company produces enough pulp to make about 400 copies of a 40-page newspaper.

Because newsprint is made from untreated pulp, it contains much lignin; this lignin contributes to the paper becoming yellow and brittle after a few days. To make higher-quality paper, paper mills first remove the lignin from the pulp. They then add chemicals to whiten the paper and, if necessary, fillers such as china clay, latex, and alginates (compounds extracted from kelps) to produce smooth, glossy paper. All of these processes increase costs and decrease yield by 20% to 50%, which is why fine stationery costs more than newsprint and why magazines cost more than newspapers. Although more than 90% of paper is derived from wood, special types of paper are derived from a few other plants. For example, money and cigarette paper are partially made from flax (*Linum*).

Most wood pulp is used to make paper, for which there is an ever-increasing demand. Each person in the United States uses an average of 680 grams of paper per day for things ranging from botany texts and advertisements to greeting cards, clothes, and copy paper. This year, almost 2 trillion pieces of copy paper will move through offices.

Wood pulp is also used to make products such as cardboard, fiberboard, nitrocellulose explosives, synthetic cattle food, fillers in ice cream and bread, linoleum, and plastic films such as cellophane. In fact, you may be wearing a derivative of wood pulp: rayon (artificial silk) and several other synthetic fibers are made from the dissolved cellulose of wood pulp. At one time, wood pulp even was added to some bread to increase its fiber—talk about roughage!

8.2 *Perspective* Perspective

The Bats of Summer: Botany and Our National Pastime

Baseball is a sport influenced heavily by plants. Pitchers use rosin to improve their grip, hitters coat their bat handles with pine tar, and an occasional cheater "corks" his bat by filling its barrel with cork to improve bat speed and drive. No part of baseball conjures up as much superstition and irrational behavior than the game's icon—the baseball bat. Players have prayed for them, coated them with manure, and even slept with them to improve their hitting. One aspiring minor leaguer in the Philadelphia Phillies' farm system even painted eyes and glasses on his bat after going 0 for 16, to help his bat see the ball better. He must have nearly blinded his bat that night when he hit a pair of home runs.

Baseball bats are made from white ash (*Fraxinus americana*) growing in the northeastern United States, especially New York and Pennsylvania. Ash is ideal for bats because it is tough, light, and will drive a ball. Ash trees harvested to make bats are each about 75 years old (40 centimeters in diameter) and produce about 60 bats per tree. Technicians use hand-turned lathes to produce one major league bat every 10 to 20 minutes. The wooden bats on sale at your local sporting-goods store receive somewhat less attention—machine lathes produce a retail bat in only 8 seconds.

Following lathing, a bat's trademark is branded on the wood's tangential surface and against the grain. Thus, a hitter who follows the adage of "keeping the trademark up" will strike a ball with the grain and where the bat is strongest. When Yogi Berra dismissed this advice by saying he went to the plate "to hit, not to read," his bat makers moved the trademark of his bats a quarter of a turn, thus saving the lives of countless bats.

Although ash bats are durable and strong, they frequently break. Each major leaguer uses six to seven dozen bats per season—approximately one bat per base hit. Hillerich and Bradsby, the makers of Louisville Sluggers, cut down about 200,000 trees per year to meet major leaguers' demands for bats. Today, bats made of aluminum—"aerospace alloy," as their manufacturers say—have replaced wooden bats in amateur baseball. As a result, Hillerich and Bradsby produce fewer than 1 million bats per year, down from 6 million several years ago. They sell these bats to major league baseball teams for $24 to $30 each.

---

§ *Fuel.* More than half of all wood used worldwide is used for fuel: nearly 2 billion people get at least 90% of their energy for heating and cooking by burning wood, while another 1 billion use wood for at least half of their heating and cooking needs. In developing countries, almost 90% of wood is used for fuel, whereas only 10% of the wood consumed in the United States is for fuel. In some parts of the world, growing populations overharvest local trees to cook their food and warm themselves, causing long-term environmental damage. When people chop down trees for fuel or to clear land for farming, the root systems die, and soil erosion increases because the once-extensive root systems no longer can hold soil in place (fig. 8.26).

§ *Charcoal and related products.* Charcoal is made by burning blocks of hardwood in the presence of relatively little air. Subsequent distillation of the charcoal briquettes produces other economically important products, including methanol (wood alcohol), methane, acetic acid, and acetone. Distilling blocks of softwoods produces turpentine and a sticky rosin used to make paint, ink, lacquer, soap, and polishes. Rosin is also rubbed onto the bows of violins, the shoes of boxers and ballerinas, and the hands of baseball pitchers to increase friction.

§ *Fabrics and rope.* Fibers from the inner bark of baobab (*Adansonia digitata*) trees are woven into fabric and

FIGURE 8.26

Madagascar's Betsiboka River. After farmers cleared land of trees, the soil was eroded by the region's heavy monsoon rains. In a few short years, the red soil washed into the river. Photo taken from the space shuttle *Discovery* in 1984.

rope. Baobab ropes are strong; as natives of tropical Africa say, "As secure as an elephant bound with baobab rope." Baobab trees are often more than 10 meters wide, and their hollowed trunks have been used for everything from morgues to city jails.

8.3 *Perspective* Perspective

The Lore of Trees

Trees, like all plants, have a fascinating history and lore:

 § There are fewer than 1,000 species of trees in the United States but more species of trees in the Appalachian Mountains than in all of Europe. There are more species of trees in 1 hectare (10,000 square meters) of Malaysian rain forest than in all of the United States.

 § The average tree planted today along a street in New York City lives only about 7 years.

 § A fully grown deciduous tree can pull more than a ton of water from the soil each day.

 § Planting three or four trees around every American house would save 10% to 50% on air-conditioning bills.

 § The most isolated tree in the world is a Norwegian spruce growing in the wasteland of Campbell Island, Antarctica. Its nearest arboreal neighbor is over 190 kilometers away in the Auckland Islands.

 § The fastest-growing tree is *Albizia falcata*, a member of the pea family. One tree in Malaysia grew more than 10 meters (in height) in only 13 months and more than 30 meters (in height) in just over 5 years.

 § The California Department of Forestry and Fire Protection estimates that a single tree that lives for 50 years provides about $200,000 worth of services to its community. These services include providing oxygen ($32,000), recycling water and regulating humidity ($37,000), producing protein ($2,500), controlling pollution ($62,500), providing shelter for animals ($31,000), and reducing erosion ($31,000).

 § Each year, the average tree takes in about 12 kilograms of carbon dioxide, an amount equivalent to that emitted by a car on a 7,000 kilometer trip. The tree also releases enough oxygen to keep a family of four breathing for a year.

 § Last year, people in the United States planted more than 4 billion trees. However, if you live in an urban area, you might not have noticed: in some of these areas, as many as four trees die for each one that is planted.

 § Several cities in the United States impose severe penalties for killing a tree. In New York City, anyone who cuts down one of the city's 2 million park trees or 700,000 street trees can be fined $1,000 and put in jail for 90 days. In New Jersey, anyone found guilty of cutting down a shade tree can be fined up to $1,500 and pay a "replacement fee" of up to $27 per square inch of tree. This means that someone convicted of cutting down a white oak 75 centimeters in diameter and 1.5 meters tall would have to pay—in addition to the fine—a replacement fee of $19,085.

 § *Sugar and spice.* Cinnamon is the outer bark of a tropical shrub, *Cinnamomum zeylanicum*; it is sold either as a pulverized powder or as "sticks," which are the curled outer-bark segments split from twigs. Maple syrup originates from the sugary sap taken from the stem of sugar maples.

 § *Dyes.* Dyes extracted from the inner bark of alder (*Alnus*) are used to color wool and other fabrics. Logwood dye is obtained from the heartwood of a small, thorny tree of tropical America (*Hematoxylon campechianum*); the dye is used to stain laboratory microscope slides.

 § *Drugs.* The bark of *Cinchona* trees contains quinine, a drug that kills the protozoan responsible for malaria, the world's leading killer. Although quinine has been a prized drug since the seventeenth century, the synthetic drug chloroquine is also used to treat malaria. Quinine, however, still flavors tonic water, producing a somewhat bitter aftertaste in gin-and-tonics. A more recent discovery is that the bark of the Pacific yew contains a promising drug (taxol) that arrests the growth of tumors. To extract only 1 kilogram of the drug requires that we cut down about 12,000 trees and strip from them more than 27,000 kilograms of bark. Botanists are now producing taxol in cells grown via tissue culture (see Perspective 18.1, "A Fungus that May Save the Yews").

 § *Tannins and other chemicals.* Tannins from chestnut and tan oak are used to prepare animal skins for sale, creating leather from cowhides. Natural chewing gum is extracted from sapodillo (*Manilkara zapota*) trees (most chewing gums sold today are synthetic polymers). *Hevea brasiliensis* produces natural rubber, which is still used to make airplane tires, and turpentine is distilled from the stems of longleaf and slash pines. Gum arabic from *Acacia* trees is used to glaze doughnuts and to make the adhesive on stamps and envelopes. Balsa wood is used to make model airplanes and life jackets.

 § *Symbolic and commemorative uses.* Each December, we chop down about 50 million Christmas trees. Also, trees are commemorated each year on Arbor Day, a largely ignored holiday that falls on the last Friday in April. Although Arbor Day was established in 1872

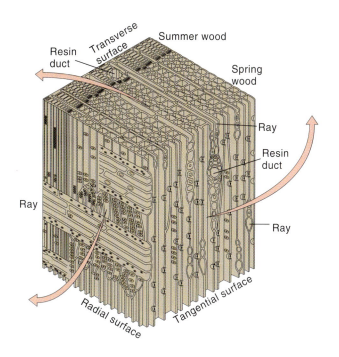

FIGURE 8.27

The different kinds of sections of wood of eastern white pine (*Pinus strobus*): transverse (i.e., cross) section, radial section (longitudinal section through the center of the stem), and tangential section (longitudinal section not through the center of the stem).

FIGURE 8.28

Knots in wood. Each knot is a cross section of a branch that was covered by lateral growth of a tree's main stem.

by the Nebraska state legislature, it didn't go into effect until several years later because overeager settlers had already cut down most trees in the state. The first year, repentant Nebraskans planted a million trees; within 16 years, citizens had planted 600 million trees. The rest of the country was apparently impressed and gradually acknowledged Arbor Day. Although Arbor Day is a feature of the federal calendar, no one gets the day off.

Different characteristics are revealed in wood depending on how it is cut.

Interpreting the features of wood requires an understanding of how the wood you are examining was cut. Figure 8.27 shows the three ways that wood is cut.

Growth rings of tree trunks cut in cross section are arranged in concentric circles, much like the circles on a target. **Ray cells** (parenchyma cells that transport water and dissolved minerals radially in the stem) radiate from the center of the section, like spokes from the center of a wheel. Cross sections of trees are seldom used in construction because they often split after they're cut. A radial section is a longitudinal section that goes through the center of the stem. In boards made from radial sections, growth rings appear as parallel lines oriented perpendicular to rays. A tangential section is a longitudinal section that does not go through the

center of the stem. In boards made from tangential sections, growth rings are arranged in large, irregular patterns of concentric *V*s.

Knots are bases of branches that have been covered by the lateral growth of the main stem. Some knots fit tightly in wood, while others fit loosely and usually fall out (i.e., producing a "knothole" in the wood); the "fit" of a knot in wood depends on whether the branch was dead or alive when the trunk grew around it (fig. 8.28). Knots produced by dead limbs have no continuity between the xylem of the limb and the main stem and therefore fall out when the wood dries. Conversely, knots from living branches do not fall out because their xylem is continuous with that of the main axis. Knots usually weaken lumber and decrease its value, but many people find wood with knots attractive for paneling and furniture.

Grain is the appearance of wood due to the arrangement of cells. Grain may have regular patterns or, as in trees such as English elm (*Ulmus procera*), be irregular and lack patterns. **Texture** refers to the arrangement and size of the pores in wood. Texture depends primarily on the size and distribution of vessels, tracheids, and, to a lesser extent, rays. Trees such as oak have large vessels and a coarse texture, whereas softwoods lack vessels and have a fine texture. **Density** of wood is expressed as specific gravity, which is the ratio of the density of the wood to that of water. In general, fast-growing conifers, such as pine, and diffuse, porous hardwoods are less dense (i.e., have a lower specific gravity) than slow-growing hardwoods. Less dense wood floats in water. Among the densest woods are lignum vitae (*Guaiacum officinale*) and black ironwood (*Krugiodendron ferreum*). Blocks of these and a few other woods sink when placed in water.

Durability of wood refers to its ability to resist weathering and decay. Woods such as redwood, teak, incense cedar, and black locust contain large amounts of natural preservatives, such as tannins, and are extremely durable, while those such as cottonwood, willow, and fir contain relatively few preservatives and decay rapidly. These natural preservatives make redwood a favorite lumber for building outdoor patio furniture that can weather the change in climate, which is one reason redwood trees are being harvested from forests in the western United States. One of the most durable woods is lignum vitae, a hard and resinous wood that has been found in near-perfect condition in 400-year-old sunken Spanish galleons. Another durable wood is sequoia, which is rich in tannins. A well-preserved sequoia log was carbon-dated in 1964 and found to be 2,100 years old; this means the log has been intact since the beginning of the Roman empire.

Water content of wood accounts for as much as 75% of the weight of a living tree. Once harvested and cut, wood is dried (i.e., cured or seasoned) in large ovens that reduce the water content of the wood to between 5% and 20%. Dry wood is about 65% cellulose, 20% lignin, and 15% secondary compounds such as resins, gums, oils, and tannins. Drying wood is important because wood with a water content less than 20% is essentially resistant to attack by most fungi, which can cause wood to rot. Why do you think that logs are the last things to decay in a forest?

SUMMARY

Stems support leaves, produce and store food, and transport water, minerals, and sugar between roots and leaves. Internodal elongation of stems helps plants intercept light. Many grasses elongate via the activities of intercalary meristems, which are meristems between mature tissues at the bases of grass leaves. Plants whose stems do not elongate are called rosette plants. Stems are made of epidermal, ground, vascular, and meristematic tissues. Vascular tissues are arranged in vascular bundles: in monocots these bundles are scattered throughout the stem's ground tissue, whereas in most dicots they are arranged in a single ring. Phloem cells conduct sugar from sources (such as leaves) to sinks (such as roots).

Stolons, tendrils, twining shoots, cladophylls, thorns, succulent stems, bulbs, rhizomes, tubers, and corms are stems modified for reproduction, climbing, photosynthesis, protection, and storage.

Secondary growth is growth in girth produced by two lateral meristems: the vascular cambium, which produces secondary xylem (wood) and secondary phloem, and the cork cambium, which produces bark (including the periderm). Secondary growth is important because it enables plants to grow taller and increases their chances of intercepting light for photosynthesis. The evolution of taller plants required support from stiff, structural cells developing near the external surface of stems.

The vascular cambium produces spring wood during moist days of spring and summer wood during the drier days of summer, with differences visible as growth rings. All of the tissues outside of the vascular cambium constitute bark, a protective layer. Gas exchange across the periderm occurs through lenticels. Trees fight pathogens by compartmentalization, which involves walling off infections, fortifying cell walls near wounds, and producing antimicrobial compounds such as phenols.

Softwoods are nonflowering plants native to temperate regions; their wood is homogeneous and contains mostly tracheids. Hardwoods are dicots whose wood contains tracheids, vessels, and fibers. Sapwood transports water and dissolved nutrients and comprises the outer few centimeters of wood. Heartwood is located in the center of a tree and does not transport water and nutrients. Knots, grain, texture, durability, density, and water content are distinguishing features of wood. We use wood and bark to make lumber, pulp, fuel, paper, charcoal, fabrics, rope, spice, dyes, and drugs.

Writing to Learn Botany

Many trees have been considered sacred in some cultures. For example, people in the Valley of Mexico prayed to willow trees when they were threatened by storms. What other trees have been viewed as sacred throughout history? Why?

THOUGHT QUESTIONS

1. Are there growth rings in the secondary phloem? Would the number of growth rings in roots always equal that in shoots? Explain your answers.

2. Which would produce more uniform paper—pulp from a softwood or from a hardwood? Why?

3. What has happened to the epidermis, cortex, and primary phloem of a 200-year-old oak tree?

4. Why would extensive secondary growth in stems of annual plants be a "poor investment"?

5. Beached blue whales are crushed by the weight of their bodies. Why aren't trees crushed by their weight?

6. Discuss the contribution of each of the following to secondary growth: vascular cambium, rays, periderm, secondary phloem, heartwood, sapwood, and compartmentalization of damage.

7. How would wood produced during a wet year differ from wood produced during a dry year?

8. How could thick bark benefit a tree?

9. The structural diversity of plants results from different arrangements and proportions of tissues rather than from the presence of different types of cells. Use specific examples to support this statement, and indicate why it is important for understanding the structure of plants.

10. Two major functions of a tree trunk are conduction and support. Describe the anatomical basis for each of these functions.

SUGGESTED READINGS

Articles

Duchesne, L. C. and D. W. Larson. 1989. Cellulose and the evolution of plant life. *BioScience* 39:238–41.

Milius, S. 1999. Son of long-horned beetles. *Science News* 155:380–82.

Stewart, D. 1990. Green giants. *Discover* (April):61–64.

Stokstad, E. 1999. How Aztecs played their rubber matches. *Science* 284:1898–99

Books

Altman, N. 1994. *Sacred Trees.* San Francisco: Sierra Club Books.

Menninger, E. A. 1995. *Fantastic Trees.* Portland, OR: Timber Press.

Niklas, K. J. 1997. *The Evolutionary Biology of Plants.* Chicago: University of Chicago Press.

Prance, G. T., and A. E. Prance. 1993. *Bark.* Portland, OR: Timber Press.

Zuckerman, L. 1999. *The Potato: How the Humble Spud Rescued the Western World.* North Point Press.

ON THE INTERNET

Visit the textbook's accompanying web site at http://www.mhhe.com/botany to find live Internet links for each of the topics listed below:

External features of the stem

Herbaceous stem

Woody stem

Modified stems

Economic importance of bark

Stems as food

Wood to paper

Wood to clothing

Wood as fuel

Lumber

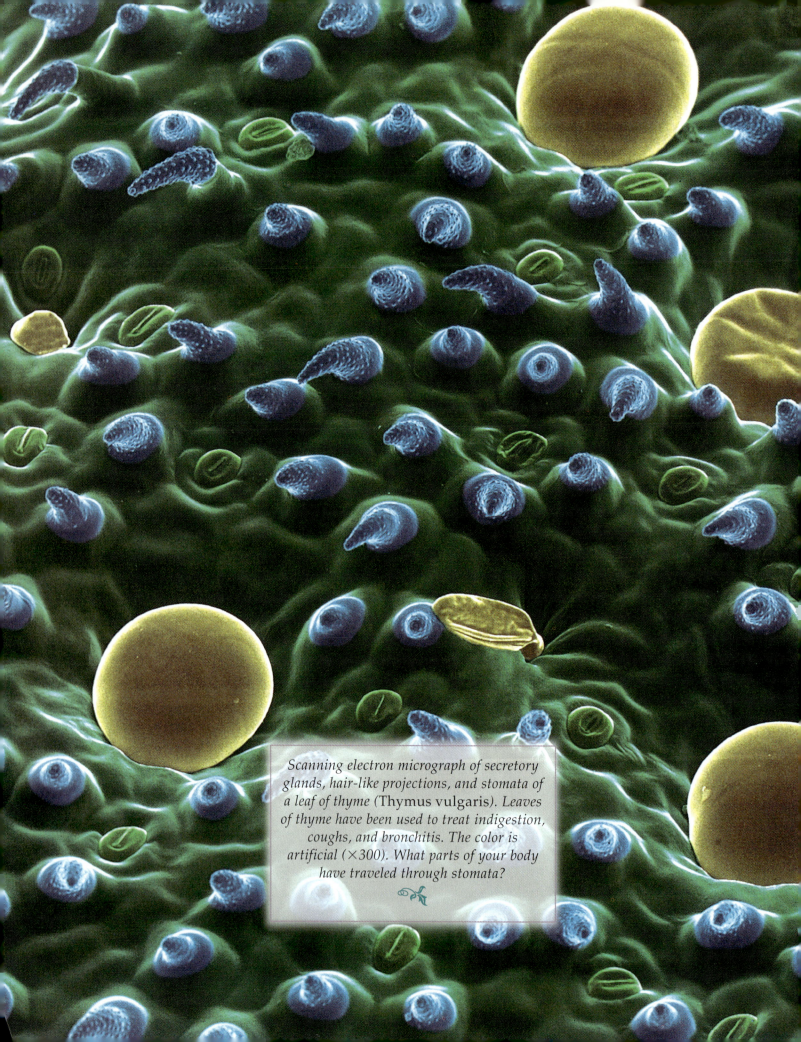

Scanning electron micrograph of secretory glands, hair-like projections, and stomata of a leaf of thyme (Thymus vulgaris). Leaves of thyme have been used to treat indigestion, coughs, and bronchitis. The color is artificial (×300). What parts of your body have traveled through stomata?

CHAPTER 9

Leaves and the Movement of Water

CHAPTER OUTLINE

The form of leaves is important to their function.

> *The arrangement of leaves maximizes the absorption of sunlight for photosynthesis.*
> *Two main types of leaves are simple and compound.*

A leaf's tissues support its primary function—photosynthesis.

> *The organization of the mesophyll promotes gas exchange in a leaf.*

Most water absorbed by a plant's roots is lost from the shoot.

> *The arrangement of cells inside a leaf increases transpiration.*
> *Several environmental factors affect transpiration.*
> *Adaptations have evolved that reduce transpiration.*
> *Deciduous plants drop all their leaves each year.*

Water movement in a plant is aided by the form of xylem cells.

> *Multicellularity and colonization of land promoted the evolution of vascular tissue.*
> *Perspective 9.1 The Risks of Having Vessels*
> *Water can be pushed up a plant a short distance by root pressure.*
> *Water is pulled up plants by evaporation from leaves.*

Leaves are affected by environmental factors.

> *Leaves can move.*

The evolution of plants has produced a wide variety of modified leaves.

> *Some carnivorous plants possess modified leaves that contain a community of organisms.*

Many defense mechanisms have evolved against herbivores.

> *Some plants poison their attackers.*
> *Plants can change the lives of their attackers.*
> *Some plants are difficult for herbivores to digest.*

Many leaves are economically important.

LEARNING ONLINE

The Online Learning Center at www.mhhe.com/botany is full of helpful resources for you such as hyperlinks that are specific to what you'll read about in this chapter. To learn more about leaves and water movement in plants, visit http://koning.ecsu.ctstateu.edu/Plants_Human/leaves.html in chapter 9 of the OLC.

205

The biblical phrase "all flesh is grass" has a biological meaning that is important to us; that is, our bodies are reconstituted plant material. Because most of the visible parts of grass plants are leaves, your body may consist mostly of reorganized leaf material. Recall from chapter 1 that plants supply all of our food either directly or indirectly, so whether we eat leaves or eat animals that eat leaves, the food we take in becomes the flesh of our bodies. Tracing food to our flesh leads us to the question, Have you ever been through a stoma? A stoma is one of the many tiny pores in leaves through which carbon dioxide (CO_2) passes, which then becomes part of sugar molecules during photosynthesis. These sugar molecules become part of the plant body and, later, part of our body. Thus, much of our flesh once began as molecules of CO_2 that passed through the stomata of leaves and became part of a plant. In this chapter, we begin the magical trek taken by molecules that compose our body, focusing on the leaves of plants.

Leaves are usually the most active and conspicuous organs of plants. The most important of their functions is absorbing sunlight and producing sugar through photosynthesis. On a global basis, leaves produce more than 200 billion tons of sugars per year. These sugars sustain virtually all life on our planet. In this chapter, you learn how the form of leaves aids their function and about the diversity of leaves adapted to different environments in the world. In the next chapter, you learn about photosynthesis.

This chapter also describes the role leaves play in the movement of water and minerals in plants. We focus on one of the two tissues specialized for commerce between organs: xylem, which carries water and dissolved minerals from roots to leaves. (Recall we learned about the other transport tissue, phloem, in the previous chapter.) The evolution of these tissues was largely responsible for the widespread invasion of land by plants and thus for the evolution of land-dwelling animals.

The Form of Leaves Is Important to Their Function.

Leaves appear green because of the hundreds of chloroplasts each leaf cell contains. Each chloroplast is a tiny factory that converts carbon dioxide and the energy of light into sugar during the process of photosynthesis (see chapter 10). To do this, leaves expose large surface areas to the environment. For example, a maple tree (*Acer*) with a trunk 1 meter wide has approximately 100,000 leaves with a combined surface area exceeding 2,000 square meters; that's roughly the area of six basketball courts. Large oak trees have approximately 700,000 leaves (fig. 9.1), and you'd better think twice before agreeing to rake a lawn shaded by American elm trees (*Ulmus americana*); when mature, these trees can each produce more than 5 million leaves per season.

The arrangement of leaves maximizes the absorption of sunlight for photosynthesis.

If you look at a plant from above, you will notice that rarely is a leaf positioned, or arranged, directly above the next lower leaf on the stem. What is the selective advantage of leaf arrangement to a plant? This positioning reduces the amount of self-shading (leaves shading other leaves of the same plant). Thus, leaf arrangement allows leaves to absorb a maximum amount of light for photosynthesis.

Most plants, including birch, oak, and *Agave,* have **alternate leaf arrangement** (fig. 9.2), meaning that they have one leaf per node. In this type of arrangement, leaves often are found alternately along two parallel ranks of the stem, as in ginger and pea (fig. 9.3). Plants with **opposite leaf arrangement** have two leaves per node, as in maple, ash, dogwood, and *Coleus* (fig. 9.4). Plants with a **whorled leaf arrangement** have three or more (and as many as 25) leaves per node. Oleander, *Peperomia,* and horsetail have whorled leaves (fig. 9.5).

Two main types of leaves are simple and compound.

Each leaf is attached to a stem at a node. The upper angle where a leaf joins a stem is the **axil.** In the axil of each leaf is an **axillary bud** that can grow and develop into a branch with leaves or flowers on it (fig. 9.6).

Two main kinds of leaves are simple and compound. **Simple leaves** have a flat, undivided blade that is supported by a stalk called a **petiole** (see fig. 9.2). The petiole is typically supported by rigid, thick-walled fiber cells. Redbud, elm, and maple have simple leaves with a petiole. In plants such as silk tree (*Albizia*), the petiole includes a jointlike swelling called a **pulvinus** that enables the leaf to respond and move to environmental stimuli such as light and gravity. Leaves of plants such as *Zinnia* and stonecrops (*Sedum*) lack petioles and are called **sessile leaves.**

FIGURE 9.1
Large trees, such as this oak (*Quercus*), produce thousands of leaves per year.

FIGURE 9.2
Alternate leaf arrangement in poplar. Notice the long petiole of each simple leaf.

FIGURE 9.3
The compound leaves of this pea (*Pisum*) plant are arranged alternately along two sides of the stem.

FIGURE 9.4
Coleus has opposite leaf arrangement. It has two leaves per node; these pairs of leaves at successive nodes form at right angles to each other. This view is from above the plant. The meristem that produced these leaves is shown in figure 9.6.

Compound leaves have a blade divided into **leaflets** that form in one plane (see fig. 9.3 and fig. 9.7). Leaflets lack axillary buds, but each compound leaf has a single axillary bud at the base of its petiole. There are two kinds of compound leaves: **pinnately compound leaves** and **palmately compound leaves.** Leaflets of pinnately compound leaves form in pairs along a central, stalklike **rachis** (fig. 9.7*a*). Ash, walnut, and some roses have pinnately compound leaves. Leaflets of palmately compound leaves attach at the same point, much as fingers that extend from your palm (fig. 9.7*b*). Examples of palmately compound leaves include horse chestnut, marijuana, and lupine. The most famous palmately compound leaves are those of the shamrock (*Trifolium*), the national plant of Ireland. Similarly, the "leaves" of a four-leaf-clover are actually four leaflets of a palmately compound leaf.

Whether simple or compound, the form of leaves (broad and thin) is important to their primary function (photosynthesis). The blade of a simple leaf or the combined surface of the leaflets in a compound leaf provide a

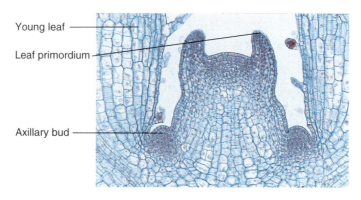

Young leaf

Leaf primordium

Axillary bud

FIGURE 9.6

Shoot tip of *Coleus* (×490). Axillary buds form in the axil where a leaf attaches to the stem. The products of the apical meristem of a *Coleus* plant are shown in figure 9.4.

FIGURE 9.5

This horsetail (*Equisetum*) has many branches arising from each node along its stem. Each branch develops in the axil of a leaf, which means that there are also many leaves at each node. Thus, the horsetail has a whorled leaf arrangement.

broad area for absorbing sunlight. While some species of plants possess thick leaves, the vast majority have thin leaves. So what's the advantage of thinness? A thick leaf may be difficult to support on a stem. Moreover, in thick leaves, sunlight may not penetrate the leaf to reach all cells, so producing thick leaves could be an inefficient use of energy for most plants. What function might a thick leaf have?

Peltate leaves are simple leaves with a petiole attached to the middle of the blade (fig. 9.8*a*). Mayapple (*Podophyllum*) is an example of a plant with peltate leaves. The most extreme examples of peltate leaves are the tubular leaves of some carnivorous plants (see fig. 9.30). **Perfo-**

liate leaves are simple, sessile leaves that surround stems (fig. 9.8*b*) such as those of yellow-wort and thoroughwort.

Many plants have juvenile leaves that form on young parts of the plant; these leaves are different from adult leaves that form on mature parts. For example, juvenile leaves of some species of *Acacia* are compound, whereas adult leaves are simple. Similarly, elms produce progressively more teeth along the edges of their leaves as the trees age.

INQUIRY SUMMARY

Leaves are the primary photosynthetic organs of most plants. The arrangement of leaves on a stem is called phyllotaxis and helps maximize absorption of light.

(a)

(b)

FIGURE 9.7

Compound leaves. (*a*) Pinnately compound leaf of black walnut (*Juglans nigra*) possesses pairs of leaflets along a central rachis. (*b*) Palmately compound leaf of buckeye (*Aesculus*).

(a)

(b)

FIGURE 9.8

(a) Peltate leaf. Petiole attaches at or near the center of the leaf (mayapple; *Podophyllum*). (b) Perfoliate leaf of *Montia perfoliata*.

Leaves are arranged in an alternate, opposite, or whorled pattern at the nodes of stems. The two types of leaves are simple and compound; each leaf has an axillary bud in its axil. The form of leaves, broad and thin, aids their function—collecting sunlight and producing sugar through photosynthesis.

Why might it not be selectively advantageous to produce a very large leaf?

FIGURE 9.9.

Welwitschia plants in the Namib Desert of southwest Africa. Each plant produces two leaves that split, thus resembling several leaves.

A Leaf's Tissues Support Its Primary Function—Photosynthesis.

Like all organs of a plant, leaves consist of epidermal, ground, vascular, and meristematic tissues. Because of their position in leaves, however, these tissues have some different functions than in other parts of a plant. It is important to remember that the primary function of leaves is photosynthesis. Thus, the structure of leaves must promote the acquisition of ingredients to make sugar—water from the roots, CO_2 from the air, and sunlight to power the chemical reactions of the process.

Leaves are the most diverse of all plant organs; they can be tubular, needlelike, feathery, cupped, sticky, fragrant, smooth, or waxy. Leaves range in size from the pinhead-sized leaves of watermeal (*Wolffia*)—the smallest known flowering plant—to the 20-meter fronds of tropical palm trees, and they range in number from the millions of leaves on American elms to those of *Welwitschia mirabilis,* a desert plant of southwest Africa that grows only two leaves during its lifetime of several hundred years (fig. 9.9). However, re-

gardless of their number or size, leaves are formed by the coordinated activities of several meristems.

Although mature leaves often consist of millions of cells (e.g., a leaf of broad bean contains about 40 million cells), the earliest stage of leaf development is a small bulge near the shoot apex called a **leaf primordium** that consists of a few hundred cells (see fig. 9.6). Continued cell divisions and cell expansion increase the length of the leaf and produce a meristem that thickens the leaf, as well as meristems along the leaf margins that form the flattened blade and the stalklike petiole, which attaches the blade to the stem.

Continued growth of a leaf involves more cell division and cell expansion. For example, leaf formation in

FIGURE 9.10

Leaves of ferns are called fronds (see fig. 7.19). Fronds unroll from base to tip. When they are young and coiled as shown in this photo, they are referred to as fiddleheads.

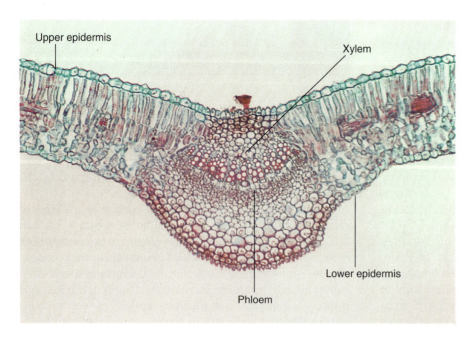

FIGURE 9.11

Cross section through the midvein of a leaf of lilac (*Syringa vulgaris*) (×100).

cocklebur involves almost 30 generations of cells; cell divisions continue until the leaf is one-half to three-fourths grown. Cell expansion then forms most of the intercellular spaces and produces a broad leaf that captures sunlight, an ingredient in photosynthesis.

Fern leaves are called **fronds.** Unlike the leaves of other plants, fronds usually form in a curled structure called a fiddlehead (fig. 9.10). During development, the upper side of the fiddlehead elongates faster than the bottom side, causing the fiddleheads to unroll and form a frond. Fronds of some ferns can live more than 50 years and be more than 15 meters long. We even eat some of them: fiddleheads of ostrich fern (*Matteuccia struthiopteris*) are an asparaguslike delicacy enjoyed by many New Englanders. What leaves are commonly in your diet?

If you hold a leaf up to the light, the pattern of vascular tissues in the leaf may be revealed. Xylem and phloem, the vascular tissues of a plant, form in strands called **veins.** Xylem forms on the upper side of a vein (on a vertical leaf, this is the side next to the stem), and phloem forms on the lower side (on a vertical leaf, this is the side away from the stem; fig. 9.11). Veins are often supported by fibers and are usually surrounded by a layer of parenchyma cells called the **bundle sheath.**

Veins are a leaf's "fingerprint" and can be used to help identify plants. Most dicots and some nonflowering plants have **netted venation,** meaning they have one or a few prominent midveins from which smaller minor veins branch into a meshed network (fig. 9.12*a*). The leaves of most monocots such as corn and other grasses have **parallel venation,** meaning that several prominent and parallel veins interconnect with smaller, inconspicuous veins (fig. 9.12*b*). Minor veins are extensive; for example, there are about 70 centimeters of minor veins per square centimeter in leaves of sugar beet (*Beta*). Each vein ends in a small neighborhood of cells where most water, minerals, and sugars are exchanged with cells of the leaf. Thus, water moves through xylem to leaf cells, where it is used to produce sugar, which is then transported through phloem to other parts of the plant.

The epidermis of most leaves is transparent, usually not photosynthetic, and contains numerous stomata, or small pores (e.g., one cabbage leaf typically has more than 11 million stomata). The frequency and distribution of stomata vary in different species and in different parts of individual leaves (fig. 9.13). For example, stomata develop in parallel rows in monocot leaves, whereas in dicots they are more scattered. In horizontally oriented leaves, there are usually more stomata on the protected lower side of the leaf than on the exposed upper side. Conversely, as shown in

(a)

(b)

FIGURE 9.12

Venation in leaves. Most water and solutes are exchanged between the veins and other cells of the leaf at vein endings. (*a*) Leaves of dicots such as persimmon (*Diospyros kaki*) have a netted venation. (*b*) Leaves of monocots such as lily (*Lilium*) have parallel venation (×0.5).

table 9.1, vertically oriented leaves (leaves with their blades perpendicular to the ground) usually have similar numbers of stomata on the two sides of their leaves.

Each stoma is surrounded by two **guard cells** (fig. 9.13), which are kidney shaped in dicots and shaped like dumbbells in monocots. Guard cells in all plants have the same function: they regulate the exchange of gases (e.g., carbon dioxide, oxygen, and water vapor) by opening and closing the stomatal pore. Stomata open and close in response to environmentally induced changes in the turgor pressure of guard cells (see fig. 9.13*b*). For instance, light may cause potassium ions (K^+) to enter guard cells, thereby drawing water into them via osmosis. This influx of water nearly doubles the volume of guard cells and causes them to elongate, bow apart, and open the stoma. Similarly, stomata close in the dark when K^+ and water leave the guard cells.

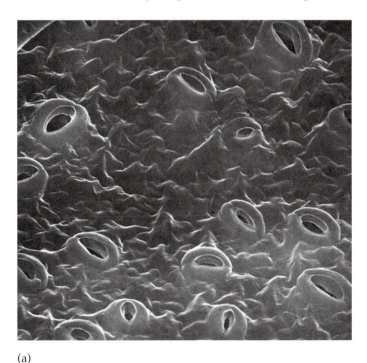

(a)

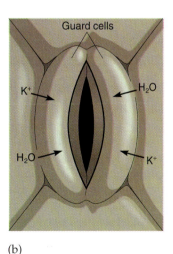
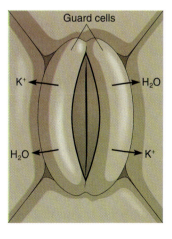

(b)

FIGURE 9.13

(*a*) Stomata on lower surface of a cucumber (*Cucumis*) leaf (×1,400). The pore formed by the two guard cells allows for gas exchange between the atmosphere and the underlying cells. (*b*) Light triggers an influx of K^+ into guard cells, which causes an influx of water (via osmosis) into the guard cells. As the cells elongate, they bow apart and form the stomatal pore. The pore closes via the reverse mechanism: K^+ exits the cells, causing water to move out of them via osmosis. This, in turn, shrinks the cells and closes the pore.

TABLE 9.1

Leaf Stomatal Frequency (stomata per square centimeter)

	UPPER EPIDERMIS	LOWER EPIDERMIS
Horizontally oriented leaves		
apple (*Malus sylvestris*)	0	38,760
bean (*Phaseolus vulgaris*)	4,031	24,806
oak (*Quercus velutina*)	0	58,140
pumpkin (*Cucurbita pepo*)	2,791	27,132
Vertically oriented leaves		
corn (*Zea mays*)	9,800	10,800
pine (*Pinus sylvestris*)	12,000	12,000
onion (*Allium cepa*)	17,500	17,500

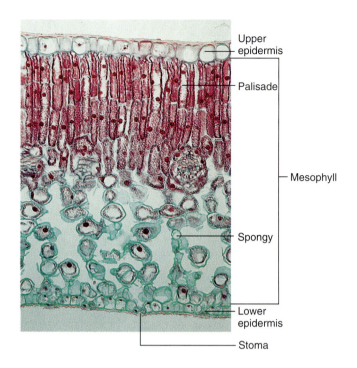

FIGURE 9.14

Cross section of a leaf of the common hedge privet (*Ligustrum*) (×50). The photosynthetic tissue consists of a densely packed palisade mesophyll and a loosely packed spongy mesophyll.

Although stomata usually occupy less than 1% of the leaf surface, huge amounts of water are lost through them to the atmosphere. The evaporation of water from plants into the atmosphere is called **transpiration** and can influence patterns of rainfall. In fact, about half of the moisture in rainfall in Amazon rain forests originates from transpiration. On the more local scene, your neighbor's 1-acre lawn can transpire more than 100,000 liters of water per week during the summer. Based on the data in table 9.1, what do you think is the selective advantage of having so many more stomata on the bottom of the leaf compared to the upper side of a horizontal leaf?

The organization of the mesophyll promotes gas exchange in a leaf.

The ground tissue of leaves is called **mesophyll.** Mesophyll contains parenchyma cells that are photosynthetic and that store the products of photosynthesis. The form of the mesophyll tissue plays an important role in its function. Examine the cross section of a privet (*Ligustrum*) leaf shown in figure 9.14. Leaves such as these have two kinds of mesophyll tissue. Along the upper side of the leaf are one or more layers of long cells called **palisade mesophyll** cells, which are densely packed (i.e., lack intercellular spaces), are arranged in columns, and perform as much as 90% of the leaf's photosynthesis. Palisade cells contain many chloroplasts and are thus specialized for light absorption and photosynthesis. Along the lower side of the leaf are **spongy mesophyll** cells, which are irregularly shaped and

green due to chloroplasts they contain. Spongy mesophyll cells are separated by large intercellular spaces connected to stomata (imagine a sponge with many holes and open spaces). Depending on the plant species, these air spaces account for 10% to 70% of the leaf volume.

Spongy mesophyll is specialized for gas exchange. The arrangement of mesophyll cells promotes the movement of water vapor and CO_2 and oxygen gases into and out of a leaf. What is the advantage to a plant of having spongy mesophyll cells in the bottom of a leaf instead of many tightly packed palisade parenchyma cells? The position of spongy mesophyll allows CO_2 in the air to diffuse into the leaf when the stomata are open. The CO_2 then diffuses into and around the spongy mesophyll cells to the palisade parenchyma cells above them. This ensures that all cells inside the leaf can obtain the CO_2 they require for photosynthesis. If cells were too tightly packed just above the stomata, flow of CO_2 into the leaf would be greatly impeded.

Plants such as corn, lawn grasses, and other monocots form leaves vertically. These leaves intercept light from all directions and lack palisade and spongy layers. Rather, most of their mesophyll cells appear similar and surround a photosynthetic bundle sheath containing large, active chloroplasts. This special combination of mesophyll and photosynthetic bundle-sheath cells is referred to as **Kranz** (German for "wreath") **anatomy** and underlies a metabolic division of labor called C_4 photosynthesis (see chapter 10). Pigweed, corn, Bermuda grass, crabgrass, and nutsedge are examples of C_4 plants that have vertically oriented leaves with Kranz anatomy.

INQUIRY SUMMARY

Leaves have many shapes and sizes and are made of ground tissue, veins, meristems, and epidermis. Mesophyll cells of the ground tissue produce sugar and receive nutrients and water from vascular tissues in veins. Leaves are covered by epidermis containing stomata that regulate gas exchange. Opening and closing of each stomatal pore is controlled by a change in turgor pressure in the two surrounding guard cells. Leaves are formed by several meristems in specific positions. All these tissues support the main function of a leaf, photosynthesis. The arrangement of mesophyll promotes gas exchange between the air and leaf cells.

Would it be an evolutionary advantage or disadvantage for a leaf to have secondary growth like stems? Explain your answer.

Most Water Absorbed by a Plant's Roots Is Lost from the Shoot.

Water is the most abundant compound in plants: it accounts for 85% to 95% of the weight of most plants and 5% to 10% of the weight of seeds. Water is used to make organic compounds (e.g., sugar), to support the plant (via turgor pressure), to serve as a solvent in which chemical reactions occur, and to function as the medium in which minerals move. Given the critical roles that water plays in plants, it seems peculiar that more than 95% of the water gathered by a plant evaporates back into the atmosphere through transpiration, often within only a few hours after being absorbed. Most transpiration from leaves is through stomata.

The arrangement of cells inside a leaf increases transpiration.

Although leaves are adapted for photosynthesis, these adaptations also enhance a plant's greatest threat—dehydration. This is the dilemma all plants face each day; stomata must remain open for CO_2 to enter the leaf and be used in photosynthesis; however, as long as stomata are open, water is lost through transpiration.

The rate of transpiration depends, among other things, on the amount of surface area available for evaporation of water. The loose internal arrangement of cells in leaves produces a large internal surface area for evaporation (fig. 9.14). Indeed, the internal surface area of a leaf may be more than 200 times greater than its external sur-

face area. This promotes transpiration by increasing the surface area for evaporation of water.

The internal surface area of a leaf is linked to the dry air by intercellular spaces that occupy as much as 70% of the volume of some leaves. The stomata in the leaf epidermis link the inside of the leaf with the atmosphere. Leaves also have an efficient plumbing system of veins that distribute water to their internal evaporative surface; 1 square centimeter of leaf can have as many as 6,000 veins. As a result of this leaf architecture, a well-watered plant can lose tremendous amounts of water via transpiration; for example, a corn plant transpires almost 500 liters of water during a 4-month growing season. If humans required the same amount of water, we would each have to drink about 40 liters of water per day (and visit the bathroom often).

Water lost via transpiration is usually replaced by water absorbed from the soil by roots. This water moves to the leaves as fast as 75 centimeters per minute, which is about the speed of the tip of a secondhand sweeping around the face of a clock. Despite their huge requirements for water, however, plants have no active mechanism for acquiring water, and they have no way to pump water up the stem to the leaves. Thus, losses of water through transpiration do not indicate an excess of water; on the contrary, water is the primary factor that limits plant growth in most habitats.

Several environmental factors affect transpiration.

In most vascular plants, transpiration is necessary to lift water and solutes up a plant. Too much water loss, however, can cause a plant to wilt and perhaps die. The rate of transpiration is greatly affected by factors in the environment:

- *Humidity.* Transpiration typically increases in dry air and slows in humid air.

- *Internal concentration of CO_2.* The CO_2 concentration *in leaves* changes considerably throughout the day, especially when stomata close and photosynthesis removes CO_2 from the intercellular spaces of the leaf. Low concentrations of CO_2 in leaves can cause stomata to open, whereas high concentrations can cause them to close. Thus, a reduced internal supply of CO_2 for photosynthesis may stimulate swelling in guard cells and open stomata and, as a result, increase transpiration.

- *Wind.* The thin, moist layer of air adjacent to a transpiring leaf is called its **boundary layer.** A thick boundary layer decreases transpiration. Wind usually replaces the boundary layer with drier air, thereby increasing transpiration; this is why laundry hanging on a clothesline dries more quickly on a windy day than a still one. Specialized cells in the leaves of many grasses are sensitive to water loss. These cells shrink when they desiccate, thereby rolling the leaf into a cylinder. This shape increases the leaf's boundary layer and reduces the amount of light and wind that reaches the leaf, thereby decreasing transpiration. In addition, many plants have hairs on the bottom of their leaves that greatly reduce the effect wind has on transpiration.

§ *Air temperature.* In direct sunlight, the temperature inside a leaf may exceed that of the air by as much as 10°C. Increasing the leaf temperature leads to faster rates of transpiration.

§ *Soil.* Any factor that affects the availability of water can also affect transpiration; therefore, transpiration is affected by the water content of soil. Plants function as wicks that evaporate water from soil, which explains why soils covered by plants can lose water faster than does bare soil. Indeed, almost all water lost from below 15 centimeters in the soil is lost via transpiration. Weeds, therefore, not only compete with crop plants for light and nutrients but also decrease the availability of water in soil.

§ *Light intensity.* Light usually causes stomata to open and therefore increases transpiration. Although stomata typically open at sunrise and close at sunset, these are not all-or-none effects; instead, stomata open gradually in the morning over a period of about 1 hour and gradually close throughout the afternoon. The regulation of transpiration by light is important because it prevents plants from needlessly losing water when there is not enough light for photosynthesis. Transpiration is greatest in plants growing in moist soil on a sunny, dry, warm, and windy day. In these conditions, transpiration often exceeds the ability of plants to absorb water. As a result, many plants wilt at midday, even though the soil in which they are growing may contain much water.

Adaptations have evolved that reduce transpiration.

Life on the land can be very drying. Those plants with adaptations to reduce transpiration have an advantage over those that do not. The following are adaptations that reduce transpiration in different species of plants:

§ *Leaf position.* Leaves of plants such as turkey oak (*Quercus laevis*) and *Eucalyptus* trees reorient, thus avoiding the midday sun (fig. 9.15). This repositioning decreases the leaf temperature and therefore reduces transpiration.

§ *Circadian rhythms and stomatal cycling.* Stomata open at dawn and close at night. This cycle of opening and closing is influenced by light and by an internal biological clock of the plant. For example, plants grown in 16 hours of light and 8 hours of dark continue to open and close their stomata in a regular, daily pattern, a **circadian rhythm,** even when they are placed in the dark.

§ *Abscisic acid.* Plants with genetic mutations have played an important role in our understanding of

FIGURE 9.15
Leaves of *Eucalyptus* hang vertically; their flat surfaces are not presented directly to the midday sun, thereby minimizing heating and water loss.

water movement in a plant. For example, tomato plants with the so-called "wilty" mutation helped identify the role of the hormone abscisic acid (see chapter 6) in transpiration. Stomata of these mutants are always open, which causes the plants to lose water constantly and appear wilty. These mutants contain only about 10% as much abscisic acid as do normal plants. Moreover, adding abscisic acid to them causes their stomata to close and the plants to regain turgor, suggesting that abscisic acid prevents desiccation. Subsequent research has supported this conclusion. Applying a very dilute solution (10^{-6} molar solution) of abscisic acid causes the stomata of most plants to close, and desiccation stimulates the synthesis of abscisic acid in the mesophyll cells of leaves. In normal plants, abscisic acid causes stomata to close and helps plants conserve water during drought.

§ *Cuticle.* The retention of water, and thus survival, would be almost impossible for plants without a waxy cuticle covering the epidermis of leaves and stems. The cuticle is an effective means of conserving water: less than 5% of the water lost by a plant evaporates through the cuticle. In general, thicker cuticles provide more protection from desiccation than thin ones. Desert plants typically have thick cuticles. Would you expect aquatic plants to have a cuticle? Why or why not?

§ *Trichomes.* Although trichomes increase the thickness of the boundary layer overlying a leaf, the primary means by which trichomes decrease transpiration is by reflecting light and thus decreasing the temperature of a leaf.

§ *Sunken stomata.* Some leaves have stomata that are sunken below the surface of the leaf. Sunken stomata increase the boundary layer surrounding guard cells, which explains why plants with this adaptation typically transpire less than plants with stomata that are flush with the epidermal surface.

§ *Reduced leaf area.* Many desert plants have greatly reduced leaves (e.g., the spines of cacti), thereby decreasing the surface area from which transpiration can occur. In these plants, succulent stems that store large amounts of water often replace leaves as the primary photosynthetic organs.

§ *Leaf abscision.* An effective means of decreasing transpiration is to reduce the surface area from which water can evaporate. One way that many plants accomplish this is leaf abscision; that is, by dropping their leaves when water or temperature becomes limiting, such as in the cold of winter or during prolonged drought (fig. 9.16).

(a)   (b)

FIGURE 9.16

Opportune production of leaves. Plants in hot, dry environments lose great amounts of water through their leaves. One adaptation to reduce transpiration is to have leaves only when water is available. The ocotillo (*Fouquieria splendens*) plant shown here does this. (*a*) During dry periods, the spiny, leafless stems appear almost dead. (*b*) When water is available, leaves form rapidly and become photosynthetic.

Deciduous plants drop all their leaves each year.

Leaves typically have a limited life span, after which they are shed from a plant. Leaf abscision also can result from injury, changes in climate such as drought, or the shortening days of autumn. Plants that drop all their leaves in the fall, such as oak and pecan, are called **deciduous.** Leaves of **evergreens,** such as pine and fir, live 3 to 5 years and can be shed anytime during the year, but not all at once.

Leaves abscise at the base of the petiole, a region called the **abscision zone** (fig. 9.17). Injury, drought, or short days alter the production of auxin and ethylene, which are plant hormones that influence abscision. As a result, a **separation layer** forms in the abscision zone. Cells along the separation layer suberize (i.e., form suberin) and, in doing so, block water from entering the leaf. Once the separation layer forms, wind or other disturbances break the dead leaf from the stem. What's left on the stem is a suberized **leaf scar** that helps the plant resist infection and desiccation.

Leaf abscision is important because it prunes plants and rids them of injured or dying leaves. Most importantly, however, it reduces the amount of surface area from which water is lost by a plant during a dry season, such as winter in much of North America. During winter in temperate regions, most of the precipitation is frozen and lies on the ground as snow and ice; liquid water is limited in the soil. If a tree such as a maple kept its leaves all winter, it could lose so much water from transpiration that it would die.

As you know, however, not all plants are deciduous. Evergreens keep most of their leaves during the winter. How is this possible, given what we know about water loss through transpiration? The answer is that leaves of evergreens such as pine and fir are relatively small, with few stomata per leaf. Thus, water lost through transpiration is low compared to that of most deciduous leaves (fig. 9.18). Deciduous and evergreen trees such as pine illustrate the idea of an **ecological trade-off** in nature. If evergreens can produce sugars for many more weeks each year compared to deciduous trees, why aren't all trees evergreens? While evergreens keep most of their leaves all year, the trade-off is that more sugar can be produced in a maple leaf than in a pine leaf during the summer, but the larger maple leaves lose more water through transpiration.

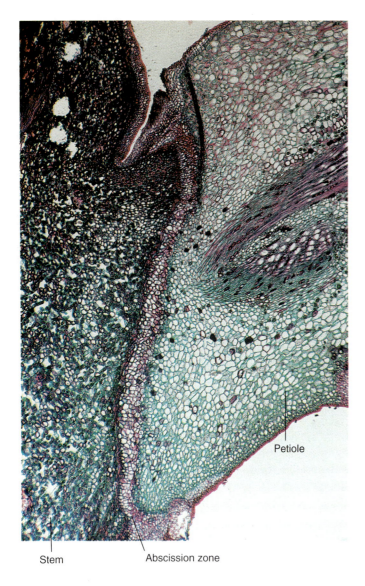

Petiole

Stem   Abscission zone

FIGURE 9.17

Leaf abscission occurs when an abscission zone forms near the base of the petiole (×10). This zone minimizes infection and loss of nutrients. How would the leaf scar that forms here affect transpiration?

FIGURE 9.18

Leaves of deciduous trees are often much larger than leaves of evergreen trees such as pine. What are the advantages and disadvantages of large leaves?

In addition, there is a selective advantage for a leaf to remain attached to a tree at least long enough for the leaf to produce more sugar than was used to construct itself. Because of their small size, evergreen leaves produce less sugar through photosynthesis than do leaves of deciduous trees in the same amount of time. Also, evergreen trees often grow in soils that are low in minerals, which contributes to a lower rate of photosynthesis. Thus, plants such as pines, firs, and spruces keep their leaves a relatively long time before they are shed—hence, the name *evergreen*. An increased life span of a leaf, however, increases the probability that the leaf will be exposed to severe weather conditions, herbivores, or disease-causing organisms.

INQUIRY SUMMARY

Transpiration is affected by humidity, temperature, light, soil, wind, and the internal concentration of CO_2 in leaves. Leaf abscission, water-use efficiency, leaf position, circadian rhythms, and the production of abscisic acid are physiological adaptations of plants that affect transpiration. Cuticles, trichomes, reduced leaves, and sunken stomata are structural adaptations that reduce transpiration. Deciduous plants drop all of their leaves during the dry season.

Would you expect to find plants with large, evergreen leaves living in a desert? Why or why not?

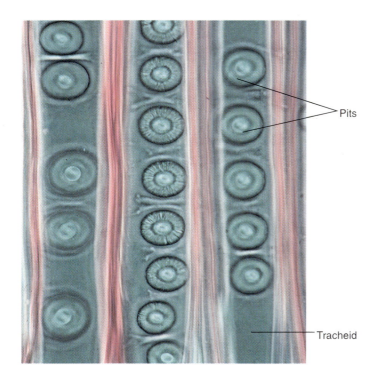

FIGURE 9.19
Light micrograph of tracheids in wood of pine (*Pinus*) (×163). Tracheids are connected by pits, which are thin areas in adjacent cell walls. The wood of most conifers resembles that of pine; the primary conducting cells are tracheids.

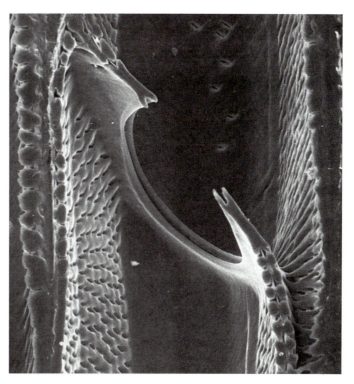

FIGURE 9.20
Scanning electron micrograph of part of a vessel in a stem. Cell walls of vessel elements are stacked end-to-end to form a vessel: the end-walls separating adjacent vessel elements are often either wholly or partially dissolved.

Water Movement in a Plant Is Aided by the Form of Xylem Cells.

Water must move rapidly through plants to replace water lost by transpiration. Through what tissue does this water move? To answer this question, consider the results of two simple experiments:

1. Removing the bark from a tree does not significantly alter transpiration.

2. When roots are exposed to a soluble dye, only xylem cells in the stem contain the dye.

These results suggest that water and dissolved minerals move from roots to leaves through the xylem.

The two kinds of xylem cells, tracheids and vessel elements (see fig. 8.23*a*), are both well adapted for conducting water: at maturity only their cell walls remain, which are hollow and therefore have no cellular organelles to retard water flow. Furthermore, the thick cell walls can withstand changes in turgor pressure associated with water flow caused by transpiration. However, tracheids are usually long (up to 1 millimeter) and thin (10 to 15 micrometers in diameter), and they overlap each other. Their walls contain many thin areas called **pits.** Pits are valvelike structures that link tracheids into long, water-conducting chains, allowing a slow flow of water through xylem (fig. 9.19).

Vessel elements are shorter and wider than tracheids: their diameters are usually 40 to 80 micrometers and may be as large as 500 micrometers. Vessel elements are stacked end to end, and the end-walls are often missing (fig. 9.20). As a result, these elements form cellulose pipes called **vessels,** ranging in length from a centimeter to more than a meter, in which water can move faster than in tracheids. This increased flow rate in vessels is one reason why angiosperms dominate today's landscapes. Most angiosperms contain both tracheids and vessels, while gymnosperms contain only tracheids (see Perspective 9.1, "The Risks of Having Vessels").

Multicellularity and colonization of land promoted the evolution of vascular tissue.

The survival of all plants depends on their ability to transport a variety of substances into, through, and out of their bodies. Primitive single-celled organisms such as algae typically rely on diffusion as a primary means of internal transport. However, the evolution of **multicellularity**—that is,

9.1 Perspective Perspective

The Risks of Having Vessels

Gymnosperms have only tracheids, while angiosperms have tracheids and vessels. This difference in xylary structure provides angiosperms with tremendous potential benefits as well as risks. For example, consider a tracheid with a diameter of 10 micrometers and a vessel with a diameter of 80 micrometers. Flow through cylindrical pipes such as these is proportional to the fourth power of their radius (i.e., radius^4), so that the flow rate through these cells would be as given in the table that accompanies this reading. As is shown in this table, water flows 4,096 times faster (i.e., 2,560,000/625) through the vessel than through the tracheid. Of what value, then, are tracheids?

The increased diameter that allows such rapid water transport in vessels of angiosperms also puts these

CELL TYPE	DIAMETER	RADIUS	FLOW RATE (RADIUS^4)
Tracheid	10 micrometers	5 micrometers	$(5)^4 = 625$
Vessel element	80 micrometers	40 micrometers	$(40)^4 = 2{,}560{,}000$

plants at risk: the larger water columns of vessels have a lower tensile strength than do the thinner columns of tracheids, thereby making them more likely to break when stressed by freezing or by wind-induced bending.

An air bubble forms where a water column breaks, which stops the flow. Capillarity then shapes the air bubble into a sphere, which prevents it from moving through bordered pits into adjacent cells. Since the walls separating adjacent vessel elements are dissolved, an air bubble stops water flow in an entire vessel rather than in a single cell, as in tracheids. When

air bubbles form in vessels, water flow is delegated to smaller vessels and tracheids of the wood. Thus, tracheids are the "backup system" of woody angiosperms: they ensure that water flow doesn't stop when air bubbles form in vessels.

Bubbles in vessel elements are common in many angiosperms. For example, the flow of water in virtually all of the large vessel elements of oak (*Quercus*) and ash (*Fraxinus*) is stopped by air bubbles by the end of a growing season. When this occurs, water moves in tracheids until the next spring, when the vascular cambium produces a new set of vessels.

---

having a body consisting of many cells—rendered diffusion woefully inadequate for moving substances throughout a plant because the time required for diffusion is extremely long. For instance, a molecule that diffuses across an algal cell in less than a second requires more than 8 *years* to diffuse only 1 meter in the watery environment of a multicellular plant.

Transport between adjacent cells of multicellular organisms improved with the evolution of cytoplasmic streaming—the continuous movement of cytoplasm within a cell. Cytoplasmic streaming speeded transport to about 5 centimeters per hour. Although cytoplasmic streaming is much faster than diffusion, it too is inadequate to meet the needs of long-distance transport in ever-larger multicellular plants.

Meanwhile, multicellularity was not the only evolutionary trend placing increased demands on the transport systems of plants. Plants colonizing the land were in a "space race" to intercept light for photosynthesis. In large part, this meant growing taller and exposing large leaves to the dry, hostile environment. However, the photosynthetic cells in leaves could function only in a highly humid environment but were far from the soil's precious water. Absorbing this water was accomplished by roots and root hairs in ferns and other plants. However, merely absorbing water was not enough, because leaves were located at ever-increasing dis-

tances from the roots. Thus, plants required a system for transporting water from the soil to their leaves. Multicellularity also required a system for transporting the sugars made in leaves to distant sites for storage and use (recall the discussion of phloem in chapter 8). Thus, multicellularity and colonization of the land were largely responsible for the evolution of xylem and phloem, the long-distance transport systems in plants. These systems gave vascular plants an evolutionary advantage, and in few instances do deficiencies in internal transport limit plant growth.

Water can be pushed up a plant a short distance by root pressure.

On many mornings you may have seen leaves with water droplets at their edges (fig. 9.21). This loss of water from leaves of intact plants is called **guttation** and is common in herbaceous plants growing in moist soil on cool, damp mornings. Guttation is caused by **root pressure** inside xylem cells that is generated when minerals are actively absorbed by roots at night. This influx of minerals causes water to move into the xylem, which, in turn, increases the pressure of water inside the xylem. Eventually, this pressure forces water out of leaves through openings at the margin of the leaf.

FIGURE 9.21

Strawberry (*Fragaria*) leaves with drops of guttation water at vein endings along the edges. Guttation is caused by root pressure.

Guttation continues as long as the plant is kept in conditions favoring rapid absorption of minerals and minimum transpiration, such as in wet soils at night. The amount of water lost by guttation from most plants is relatively small. However, a notable exception is taro (*Colocasia esculenta*), a tropical plant from which Polynesians make poi—a favorite at luaus. Individual leaves of taro plants can lose more than 300 milliliters (over a cup) of water per night by guttation. Although root pressure can push water up a short distance in a plant, it cannot push water to treetops. Thus, another mechanism must explain how water gets to the tops of tall plants.

Water is pulled up plants by evaporation from leaves.

A hypothesis was proposed more than a century ago to account for how water is transported up a tall plant. Today, this concept is referred to as the **transpiration-cohesion hypothesis** for water movement. Basically, a plant functions like a wick: water evaporating from the leaves draws water up from the roots and the soil.

The hypothesis is summarized in figure 9.22 and is described as follows:

1. As the sun heats a leaf, sunlight causes guard cells to open stomata, thereby increasing transpiration. This solar-powered loss of water dries cell walls inside the leaf as water exits through the stomata.

2. The loss of water from leaf cells causes water to diffuse from the neighboring xylem; thus, leaf cells bordering xylem cells replace their lost water with water from the xylem.

3. The water in the xylem may be thought of as an unbroken column of water. The loss of water from the xylem lifts the column of water in the xylem up the plant; water is pulled up the xylem to leaf cells to replace water lost through transpiration.

4. Water molecules are relatively "sticky"; they tend to cohere, or stick to each other. This cohesion of water molecules prevents the column of water from breaking so that it can be pulled up from the roots. Adhesion of water molecules to cell walls prevents gravity from draining the water column.

5. The cell walls of root hairs are made of cellulose fibers (as are all cell walls). Cellulose fibers imbibe, or quickly soak up, water. This is why paper towels, which are made of cellulose, are "quicker-picker uppers" of liquid spills and why cell walls of root hairs quickly soak up water from the soil. Water then diffuses from root hairs into the cortex cells of the root and eventually into the xylem. As a result, water flows passively from the soil into root cells and across cell walls of the root cortex cells.

Just before water reaches the stele, or center, of the root containing xylem and phloem, the flow of water is impeded by the endodermis (see chapter 7). The endodermis forms a single layer of cells around the stele. Each endodermis cell has a thick layer of suberin impregnating four of its six sides (see fig. 7.8)—on the top, the bottom, and the two sides that are perpendicular to the epidermis of the root—called the Casparian strip. The Casparian strip prevents water from passing through, so water and solutes are forced through the cell membrane of endodermis cells. Thus, all substances entering and leaving the stele must pass through cell membranes and be "screened" by the protoplast of endodermis cells.

Recent research places part of the transpiration-cohesion hypothesis in question. This research suggests that living cells next to the xylem push water into the column of water in the xylem and help repair breaks in the column of water. In this new hypothesis, however, transpiration is still the major pulling force of water up the stem.

The transpiration-cohesion hypothesis helps explain why flowers you buy for your loved ones often wilt soon after they are purchased, even if they are quickly placed in water. Rose petals, which are modified leaves, continue to lose water through transpiration after the rose is cut from the bush. To replace this lost water, water is pulled up as a column of water in the xylem of the flower stem. In an intact rose bush, this water would be replaced by water absorbed by the rose roots. However, the water column in the cut flowers continues to move up but, without roots, is replaced by air. When you place the flower stems in water, air bubbles in the xylem do not allow the water in the xylem

The Lore of Plants

The interior of a forest can be 10° to 15°C cooler than the surrounding countryside because of transpiration. Indeed, each large tree of a forest has the cooling capacity of about five air conditioners. Investigate how a swamp cooler operates and how its operation is similar to this cooling by plants.

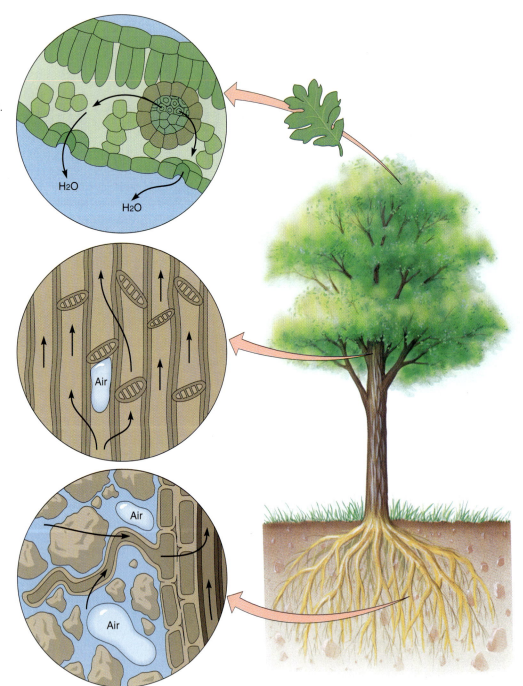

Evaporation (the driving force)

• The lower water potential of air causes evaporation from cell walls.

• This lowers the water potential in cell walls and in the protoplast.

• Water transpires through stomata pulling water up xylem.

H₂O

H₂O

Cohesion (in xylem)

• Cohesion holds water columns together in capillary-sized xylem elements.

• Air bubbles block movement of water to next element.

Air

Water uptake (from soil)

• Water is imbibed by the cell walls of root hairs.

• Water moves into root cells.

• The absorptive surface increases with the production of more root hairs.

• Water moves through endodermis.

Air

Air

Air

FIGURE 9.22

A summary of the transpiration-cohesion hypothesis for the ascent of water in plants.

to touch the water in your vase—the column of water in the xylem has been broken. Water that transpires from your flowers is never replaced, and so your flowers wilt. To prevent this problem, cut the bottom of each flower stem under water. If you cut high enough, you'll reach the level of water in the xylem and then, when you put your flowers in a vase, they can draw up water for many days.

INQUIRY SUMMARY

As plants evolved, they became multicellular—that is, their bodies consisted of many cells. Xylem and phloem helped satisfy requirements for long-distance transport of water, minerals, and sugar throughout the plant body. Water can leak from plants in a process called guttation, and it can be pushed up short distances in a plant by root pressure. Water is pulled up as a column of water in the xylem by transpiration, which is explained by the transpiration-cohesion hypothesis.

Would you expect plants that live submerged in water to possess xylem? Why or why not?

Leaves Are Affected by Environmental Factors.

Recall from chapter 2 that plants are often indicators of the natural environment in which they grow. The morphological differences that distinguish plants growing in different habitats are most striking in leaves. Indeed, leaf development is affected by several environmental factors, the most influential of which are light and moisture. In the next paragraphs, we discuss the effect of light on leaves, and in chapter 19, you will learn more about the influences of water on leaves.

Leaves of most plants must absorb light to expand and produce chlorophyll. Light that controls leaf expansion is absorbed by phytochrome, a pigment that also influences other aspects of plant growth and development (see chapter 6). This light-controlled expansion of leaves is important because it ensures that a plant does not form leaves unless light is present for photosynthesis.

The leaves of many plants respond to differing intensities of light. To best appreciate this, consider the plants growing in a dense rain forest. Leaves in the plant canopy are bathed in intense light and are called **sun leaves.** These leaves have significantly different structures than do **shade leaves,** which grow in the dim light near the forest floor. Shade leaves are often larger and thinner than sun leaves; for example, leaves of Swedish ivy (*Plectranthus*) grown in intense light are three times thicker than leaves grown in dim light. In intense light, sun leaves produce sugars faster than do shade leaves. Sun and shade leaves can form on the same plant; leaves near the shoot apex are often sun leaves, whereas leaves in the dimmer light of the lower canopy are

shade leaves. This flexibility in leaf development allows plants to exploit different and changing environments. How might knowledge of sun and shade leaves be important to the business of growing ornamental plants?

Leaves can move.

The leaves of many plants move in response to their environment. For example, the leaves of a Venus's-flytrap (*Dionaea muscipula*) rapidly close around an unsuspecting insect that has touched the trigger hairs on the leaf surface (see fig. 6.7). The insect is then slowly digested and its nutrients absorbed by the plant. Other plants have leaves that move perpendicularly to sunlight, thereby increasing the amount of light they absorb for photosynthesis. Similarly, the leaves of some desert plants move parallel to sunlight, thereby *decreasing* the heat they absorb. A most unusual kind of leaf movement occurs in the telegraph plant (*Desmodium gyrans*), a member of the legume (bean) family. Its leaves sometimes move in circles; those on one side of the plant circle in one direction, while the rest circle in the opposite direction. At other times, the leaves jerk and twitch wildly, move up and down, or move in circles—with rest periods between the spasms. We do not know the basis for these bizarre movements.

INQUIRY SUMMARY

Light strongly modifies leaf structure, which allows plants to adapt to various environments. Sun leaves develop in areas of high light intensity, whereas shade leaves develop in areas with lower light intensities. Factors in the environment also can cause plants to move, such as insects that trigger the leaf-closing actions of carnivorous plants and light that causes plants to reorient their leaves.

Some plants do not form leaves unless they are struck by light. How might this be an advantage for a plant?

The Evolution of Plants Has Produced a Wide Variety of Modified Leaves.

As with other organs, evolution has resulted in leaves that are modified for functions other than photosynthesis. Refer to figures 9.23 through 9.30 as you read about these modifications.

Stipules are small, leaflike structures found in pairs at the base of petioles. Stipules have a variety of functions. For example, stipules of sweet pea are photosynthetic, whereas those of black locust and spurge form protective spines (fig. 9.23). In plants such as greenbrier, stipules become tendrils that coil around objects they touch.

Tendrils of plants such as sweet pea and trumpet flower are leaves modified for support (fig. 9.24), coiling

FIGURE 9.23

The spines of the *Euphorbia* occur in pairs along its stem; these spines are modified stipules.

FIGURE 9.24

Tendrils on the garden pea (*Pisum*) enable the plant to cling to other objects for support.

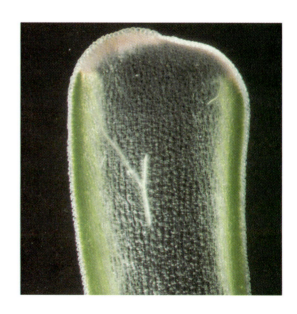

FIGURE 9.25

In window leaves of plants such as fairy-elephant's-feet (*Frithia pulchra*), the tip of the leaf is transparent (×53).

around upright objects and holding plants up off the ground. In plants such as yellow vetchling, the entire leaf is a tendril; photosynthesis in these plants is delegated to leaflike stipules at the base of each leaf. Conversely, only the petiole is a tendril in potato vine and garden nasturtium. Tendrils of some plants may be up to 30 centimeters long.

Spines of plants such as ocotillo and cacti are modified leaves that help protect plants from herbivores as well as reduce water loss through transpiration.

Bud scales are tough, overlapping, waterproof leaves that cover and protect buds from low temperatures, desiccation, and pathogens. Bud scales form over immature stems and leaves of trees and bushes before the onset of unfavorable growing conditions in winter and drop off in spring as buds open.

Some desert plants such as fairy-elephant's-feet (*Frithia pulchra*) produce **window leaves** (fig. 9.25). These leaves are shaped like tiny ice-cream cones and grow mostly underground, with only a small, transparent "window" tip protruding just above soil level. The covering soil shields window plants from the desert's drying winds and increases their chances of being overlooked by grazing herbivores. Their windows allow light to penetrate and illuminate the mesophyll tissue, which is below soil level, allowing underground photosynthesis!

Bracts are leaves that form at the base of a flower or flower stalk. They are usually small and scalelike and pro-

FIGURE 9.27
(a) A flowerpot plant (*Dischidia* species). (b) A leaf of *Dischidia* has been sliced lengthwise to reveal roots that have started to grow inside.

(a)

FIGURE 9.26

The bracts of some plants, such as this *Euphorbia punicea,* are brightly colored whereas the flowers lack attractive petals. What selective advantage might these bracts give the plant?

tect developing flowers. Some plants have colorful bracts; for example, the colorful portions of the flowers and inflorescences of poinsettia, Indian paintbrush, bird-of-paradise, *Bougainvillea,* and chaconia (*Warszewiczia coccinea,* the national flower of Trinidad) are bracts (fig. 9.26). In these plants, bracts attract pollinators, while the petals are small and inconspicuous. The tiny flowers of dogwood (*Cornus*) are inconspicuous yellow-green and are arranged in a circle (i.e., the "eye" of the "flower"); large bracts are pink or white petal-like structures that attract pollinating insects.

Storage leaves of plants such as onion (*Allium cepa*) and lily are fleshy, concentric leaves modified to store food. Onions have above ground tubular leaves with green, photosynthetic tops and below ground white bases that form the bulb (see fig. 8.13).

Flowerpot leaves are packed tightly into a flowerpot-like structure that catches falling water and debris. Many epiphytes (i.e., plants that grow completely above the ground on other plants) grow roots among the bases of their own flowerpot leaves to absorb nutrients collected by the flowerpot. More ingenious plants such as *Dischidia* have hollow leaves that function as flowerpots (fig. 9.27). These leaves do not catch falling debris, rather, they are homes for ants that bring soil and debris into the leaf. Almost one-third of the plant's nitrogen comes from feces and decaying debris deposited by ants; nitrogen from these wastes is mined by *Dischidia* roots that grow into the plant's own hollow leaves. Some species of *Dischidia* get almost 40% of their carbon from CO_2 released by ants living in the flowerpot leaves.

Succulent leaves of plants such as *Peperomia, Begonia,* rock-lettuce, and maternity plant commonly have a reproductive function (fig. 9.28). Leaves of these plants produce

(b)

tiny plantlets that become new individuals after dropping from the parent leaves.

Cotyledons are embryonic leaves—that is, leaves that are found associated with seeds and very young seedlings. Monocots, such as corn, have one cotyledon, and dicots

FIGURE 9.28

Leaves of *Kalanchoë* form tiny plants at their margins. When separated from the parent, these tiny plants can become new individuals. Is this sexual or asexual reproduction? How do you know?

have two. Recall from chapter 1 that cotyledons have several functions. In beans, they absorb the endosperm and therefore store energy and nutrient reserves used for germination. Storage products in cotyledons are usually carbohydrates and proteins, but they also may be oils, as in peanut seeds, the source of peanut oil. Cotyledons of the cacahuanache tree (*Licania arborea*) contain large amounts of flammable oils—enough that the seeds can be strung on sticks and used as torches. Oils extracted from *Licania* cotyledons are used to make candles, soaps, and grease.

Carnivorous plants (see fig. 2.18), such as the sundew, Venus's-flytrap, and pitcher plant, live in nutrient-poor soil and use other organisms as a source of the nitrogen and other minerals they require. These plants have evolved leaves that help the plant extract nutrients from the nitrogen-rich flesh of animals.

Genlisea is a plant native to South America and Africa that lacks roots and possesses little chlorophyll. It does have leaves, but these leaves grow underground. The leaves are hollow with small holes along their length about

FIGURE 9.29

A bladderwort (*Utricularia minor*) trapping a mosquito larva (×6).

the same size as the plant's main source of nutrients, protozoans. Protozoans, which are microscopic organisms, are attracted to the leaves of *Genlisea*. The structure of the leaves traps protozoans, which are then digested by the leaf; the entrances of these traps contain numerous clawlike hairs that allow an animal to enter but not to leave. As a result, prey eventually die in the trap and are digested.

Active traps move to trap prey. Sundew and Venus's-flytrap have active traps, with leaves that enclose prey and speed digestion in what Charles Darwin called a "temporary stomach." One of the most interesting traps belongs to bladderwort (*Utricularia*), an aquatic, carnivorous plant that supplements its diet with protozoa, mosquito larvae, and small arthropods (fig. 9.29). The traps of the bladderwort range in diameter from 0.25 to 5 millimeters and are elastic (hence the name bladderwort). These traps involve a trapdoor mechanism set off by trigger hairs. The trap loses water, creating a vacuum inside the trap. When an unsuspecting prey brushes against the trap's trigger hairs, the trapdoor swings open and a suction pulls the victim into the trap and closes the trapdoor. The trap then secretes enzymes that kill and digest the prey.

Some carnivorous plants possess modified leaves that contain a community of organisms.

In carnivorous plants, insect-trapping leaves are modified for attracting, trapping, and digesting small animals. These adaptations range from sticky "flypaper" surfaces of leaves, such as those of butterwort (*Pinguicula*), to the vat-like leaves of pitcher plants, such as *Sarracenia* and *Nepenthes* (fig. 9.30). Pitcher plants typically use nectar to lure insects into the leaf chamber. About halfway down the pitcher of many pitcher plants, the epidermal surface produces flaky wax. When insects step onto this wax, their legs

FIGURE 9.30
Leaves of pitcher plants (*Sarracenia*) are modified into "pitchers" that trap, kill, and digest insects.

INQUIRY SUMMARY

Some plants have evolved modified leaves involved in obtaining and storing energy and nutrients, climbing, protection, and reproduction. Carnivorous plants obtain nutrients by digesting insects they trap in their leaves. In some cases, pitcher plants contain a group of organisms inside their leaves that feed on drowned insects, eventually contributing to the nutrient supply of the plant.

Carnivorous plants get no energy from the insects they digest. How do you think they get the energy they require for growth and reproduction?

Many Defense Mechanisms Have Evolved Against Herbivores.

Because they cannot run away from herbivores, the delicate, nutrient-rich leaves of most plants would apparently be easy prey for the 300,000 known species of herbivorous insects as well as thousands of other animals. However, leaves are not defenseless against the prospect of being eaten. The defenses of some plants are obvious: lignin toughens cell walls and makes leaves hard to chew, and irritating trichomes deter many herbivores. In other plants, an attacker's first nibbles unleash a barrage of chemicals that deter other animals.

Some plants poison their attackers.

Plants such as white clover and bird's-foot trefoil are chemical minefields, containing compounds that release poisonous cyanide when eaten. Merely touching the leaves of a plant such as stinging nettle (*Urtica*) quickly gets the attention of most animals (see fig. 2.20). Even more powerful are the toxins in leaves of plants such as poison hemlock. The ancient Greeks knew about the power of poison hemlock and used it to kill prisoners, including the philosopher Socrates. Poisons of other plants are still used today. For example, the Maku Indians of Colombia and Brazil use extracts from *Euphorbia cotinifolia* and *Phyllanthus brasiliensis* to kill fish. Shoots of these plants are placed on bridges over streams and beaten so that their juices trickle into the stream and suffocate the fish. The fish are then gathered a few hundred meters downstream.

Interestingly, several insects use plant poisons for defense. For example, the grasshopper *Poekilocerus bufonius* eats only milkweeds that contain poisonous compounds. The grasshopper itself is immune to the effects of the poisons and stores these poisons in special glands. When attacked, the grasshopper sprays the poisons on its predator, thwarting the predator.

become covered with it and are transformed into unwieldy clodhoppers. Unable to cling to the flaky wax on the side of the pitcher, insects slip into the pitcher's liquid and die. They are then digested by plant enzymes in the liquid, and their nutrients are absorbed by the leaf.

Inside a species of pitcher plant in Newfoundland, a community of organisms can be found living in the liquid trap of the pitcher. Adult flesh flies, midges, and mosquitoes seek young pitcher plants in which to lay their eggs. Young insects develop from the eggs, but neither these eggs nor insects are digested by the pitcher plant. Instead, these insects actually help the plant obtain required nutrients, especially nitrogen. Insects of other species may fall into the pitcher plant and drown. The young flesh flies (maggots) chew on the floating insect carcasses, leaving small bits of body parts that can be used as food by the young midges and mosquitoes. In addition, bacteria live on the bits of dead insects that become food for developing insects inside the pitcher plant. The activities of the midge and mosquito larvae in turn produce dissolved nutrients the plant can use. Thus, a community of organisms inside the pitcher plant contributes to the nutrient requirements of the plant.

Plants can change the lives of their attackers.

Juvenile hormone is a hormone that regulates insect development. Many plants produce this hormone, and when an insect eats such plants, its normal development is altered, usually to the benefit of the plants. The discovery of substances similar to juvenile hormone in plants is an excellent example of a scientific method of investigation in action. Researchers in the United States and Europe were collaborating on a research project that involved growing an insect called *Pyrrhocoris apterus*. These insects developed normally when grown in Europe, but those reared in the United States failed to become adults. Puzzled, the researchers on different continents tried to discover why they could not get the same results. They used carefully controlled experiments to eliminate several possible causes one by one, including temperature, light, and humidity. Finally, they discovered the culprit; the filter paper on which the insects were being grown contained a juvenile hormone. The active ingredient of the paper was traced to the pulp of balsam fir, a primary source of North American (but not European) paper products, including the filter paper used in laboratories. This hormone is also abundant in leaves; it arrests insect development and enables fir trees to eliminate many herbivores.

Plants also produce chemicals that alter the lives of vertebrates. For example, chemicals in pine leaves (i.e., needles) eaten by grazing cattle double the concentration of progesterone, a hormone needed to maintain normal pregnancy. As a result, pregnant cattle that eat as little as 0.7 kilogram (about 1.5 pounds) of needles per day usually abort their calves.

Some plants are difficult for herbivores to digest.

Many plants are unappealing to their attackers because they are less digestible than other plants. For example, leaves of tomato and potato attacked by chewing insects release substances that move through the plant and induce the formation of enzymes that interfere with insects' digestion. As a result, nutrients in tomato and potato leaves cannot be extracted easily by the insects, and the insects either die or dine elsewhere.

Plants also signal each other of impending attacks by insects. Sitka willows (*Salix sitchensis*) being attacked by herbivores produce compounds that decrease the ability of these animals to digest the willow leaves. Uninfested plants up to 60 meters away also produce these chemicals, suggesting that willows use airborne compounds to signal others that the herbivores are coming.

INQUIRY SUMMARY

Certain types of leaves are protected by toxins that poison animals, by chemicals that change the life cycle of their attackers, and by compounds that interfere with the digestive process of herbivores.

Many toxins taste bitter. How might this be important to a plant in terms of its protection against herbivores, especially those which are mammals?

Many Leaves Are Economically Important.

We use leaves and their products every day. Most importantly, leaf cells produce the sugar and oxygen that we and most other organisms need to live. Leaves also have other important, though less critical, uses for us.

The leaves of plants such as cabbage, lettuce, spinach, chard, grape, celery, parsley, and bay laurel have long been parts of our diets and favorite items in our salads and stews. Many of these leaves are especially nutritious; spinach contains beta-carotene, which we use to make vitamin A in our bodies. Conversely, the wild relatives of economically important plants are inedible to humans. For example, wild lettuce contains large amounts of a bitter and narcotic milky latex (fig. 9.31). What role could artificial selection play in producing edible plants, such as cultivated lettuce, from inedible wild plants?

The spices thyme, oregano, peppermint, spearmint, wintergreen, basil, and sage are all derived from leaves. Similarly, tea is extracted from leaves of a relative of the garden *Camellia* (fig. 9.32), and *Agave* leaves are used to make tequila and mescal. It takes over 4.5 kilograms (about 10 pounds) of *Agave* leaves to produce 1 quart of mescal.

Although most dyes are now made from coal tar, plants were the original sources of most colorings. The leaves of bearberry (*Arctostaphylos uva-ursi*) contain a yellow dye, while those of henna (*Lawsonia inermis*) contain a red dye that was used to stain fingernails and cloth used to swath Egyptian mummies.

Fibers (see chapter 8) from the leaves of palms and their relatives are used to make Panama hats, clothing, brooms, and thatched huts in the tropics. Fibers from manila hemp are used to make cords and ropes (see fig. 4.25). Pine needles have been woven into baskets by some Native Americans.

The leaves of plants such as yareta (*Azorella yareta*) contain flammable resins that can be used as fuel.

Many leaves contain poisons that, when administered in small amounts, are useful drugs. For example, digitoxin and digitalis are popular drugs extracted from foxglove (*Digitalis purpurea*). These heart stimulants increase the strength of the heartbeat, thereby increasing blood circulation. Hyoscyamine, atropine, and scopolamine are derived from deadly nightshade (*Atropa belladonna*), a poisonous plant named after Atropos, the Greek god of fate who held the shears to cut the thread of human life. *Atropa* was a savior for Macbeth, whose soldiers used it to poison the

The Lore of Plants

The trees at about 10 suburban homes can produce a ton of dead leaves each fall. Burning these leaves produces smoke that is harmful to the environment. Indeed, burning these leaves releases 90 kilograms of ash and various amounts of carbon monoxide (a poison) and hydrocarbons that can irritate the eyes, nose, throat, and lungs. Some of these hydrocarbons are carcinogens (cancer-causing agents). What is the alternative? What is composting and how might it help with the leaf problem?

What is your town doing about yard wastes?

FIGURE 9.31
Plants such as wild lettuce and this milkweed (*Asclepias syriaca*) produce a bitter latex that exudes from the wounded plant. What selective advantage would the latex provide against herbivores?

Danish army during peace talks. The *belladonna* epithet is Italian for "beautiful woman," an old reference to Italian women's use of *Atropa belladonna* to dilate their eyes and make them brighter and more attractive. The chemical used in drugs for quitting smoking is derived from leaves of a relative of the garden *Lobelia*. Coffee and tea leaves contain as much as 5% caffeine, which, like the drugs from nightshade, is a stimulant. Trichomes on *Cannabis* leaves contain narcotic tetrahydrocannabinols (THC), which produces the "dopey" effect when a person smokes or eats pot. Cocaine, a drug extracted from coca (*Erythroxylum*) leaves, is a dangerous, illegal, yet popular drug.

Tobacco leaves contain large amounts of drugs, such as nicotine, that cause cardiovascular and other health problems. Thanks in part to billions of dollars in federal subsidies, we produce almost a billion kilograms of tobacco leaves annually. The return for this investment of tax dollars is that smoking kills nearly 20% of the people in the United States (the combined effects of alcohol abuse, motor vehicle accidents, and illicit drugs account for less than 10% of U.S. deaths). But the social costs of tobacco use do not end with funerals. At last count, smoking also costs Americans over $50 billion in absenteeism from work, lost production, and health care.

Carnauba wax, used to polish your car, is derived from leaves of the carnauba palm (*Copernicia cerifera*), and extracts from the lancelike leaves of *Aloe* are ingredients of medicated soaps and creams. *Aloe* is sacred to many Muslims who often hang its leaves above their doors to show they have made a pilgrimage to Mecca. We also use leaves as names for sports teams (e.g., the Toronto Maple Leafs hockey team), on national flags (e.g., the Canadian flag), and even as parts of corporate logos. For example, the shamrock on the Boston Celtics logo is a palmately compound leaf. Look around you. Where do you find people using leaves, or symbols of leaves?

SUMMARY

Photosynthesis is the main function of a typical leaf. The arrangement, or phyllotaxis, of leaves on a stem helps to maximize the amount of light that a plant can absorb for photosynthesis. The three different arrangements of leaves are alternate, opposite, and whorled. The main types of leaves, simple and compound, are produced by a variety of meristems and are composed of the same four types of tissues—meristematic, ground, vascular, and epidermal.

Modified leaves have functions other than photosynthesis. These functions include support, protection, nutrition,

(a)

(b)

FIGURE 9.32

Tea. (*a*) Tea comes from *Camellia sinensis,* a plant that originated in subtropical Asia. (*b*) Tea leaves being processed.

and reproduction. Several defense mechanisms have evolved in leaves that help reduce the likelihood they will be eaten. These mechanisms include the production of toxins that poison animals and other chemicals that reduce the ability of herbivores to digest the leaves. Leaves of some carnivorous plants contain a community of organisms that help the plant obtain the nutrients it needs.

Deciduous plants drop their leaves during dry seasons. The leaves abscise at the abscission zone. Water moves from roots to leaves in tracheids and vessels of xylem. These cells are dead at maturity, leaving thick cell walls that conduct water and can withstand the pressures characteristic of xylem transport. Leaves expose large, evaporative surface areas to the dry atmosphere. The evaporation of water from shoots is called transpiration.

Movement of water through the xylem can be explained by the transpiration-cohesion hypothesis. The driving force for water movement is the evaporation of water from the walls of leaf cells, which pulls replacement water from the xylem. This lifts water through the xylem and pulls water into the plant. The movement of water in

plants is also promoted by the strong cohesion of water molecules. The adhesion of water to the cell walls of tracheids and vessels helps to prevent gravity from draining the water columns.

Transpiration is affected by environmental factors such as humidity, internal concentration of CO_2, air movement, air temperature, availability of water in the soil, and light. Physiological adaptations include water-use efficiency, leaf abscission, dormancy, leaf position, circadian rhythms, and the synthesis of abscisic acid. Structural adaptations that reduce transpiration include the presence of a cuticle, trichomes, sunken stomata, and decreased leaf area.

Many leaves are economically important sources of food, spices, drinks, dyes, fibers, fuel, and drugs.

 *Writing to Learn Botany*

Discuss the various defense mechanisms that have evolved in leaves. How do we exploit these potential defenses?

THOUGHT QUESTIONS

1. Discuss how epidermal cells control photosynthesis and transpiration.

2. Is it a good idea to fertilize plants during a drought? Why or why not?

3. Trace the path of water from soil to plant to atmosphere. Describe the structure and function of each cell that the water molecule passes through.

4. What is the selective advantage of producing thousands of vessels and tracheids instead of one giant conduit for water transport?

5. Suppose you collected the water droplets formed by guttation. What would you expect to find dissolved in these droplets, if anything? Why?

6. During periods of reduced transpiration, almost all water moves through large vessel elements and large tracheids. Smaller tracheids are used to move water only when water flow increases. Why do you think this might be?

7. Describe how leaf architecture (size and shape) affects transpiration.

8. Plants have been likened to an open-ended tube stuck in soil. In what ways is this analogy correct? In what ways is it incorrect?

SUGGESTED READINGS

Articles

Barthlott, W., S. Porembski, et al. 1998. First protozoa-trapping plant found. *Nature* 392:447.

Canny, M. 1998. Transporting water in plants. *American Scientist* 86:152–59.

Grill, E., and H. Ziegler. 1998. A plant's dilemma. *Science* 282:252–53.

Zimmer, C. 1995. The processing plant. *Discover* (September) 93–97.

Books

Bowes, B. G. 1996. *A Color Atlas of Plant Structure*. Ames, IA: Iowa State University Press.

Flowers, T. J., and A. R. Yeo. 1992. *Solute Transport in Plants*. London: Blackie Academic and Professional Publishing.

Van de Graaff, K. M., S. P. Rushforth, and J. L. Crawley. 1998. *A Photographic Atlas for the Botany Laboratory*. Engelwood, CO: Morton Publishing Company.

ON THE INTERNET

Visit the textbook's accompanying web site at http://www.mhhe.com/botany to find live Internet links for each of the topics listed below:

Structure of leaves

Leaves as food

Types of leaves

Transpiration

Water movement in plants

Modified leaves

Carnivorous plants

Plant chemicals and herbivory

Photosynthesis occurs in the green leaves of plants. How is photosynthesis important in the flow of carbon through an ecosystem?

Photosynthesis

CHAPTER OUTLINE

LEARNING ONLINE

The Online Learning Center at www.mhhe.com/botany provides practice quizzes and key term flashcards to help you study for upcoming exams and assignments. Do you know how the release of oxygen in ancient atmospheres was important to the evolution of plants and animals? Visit http://photoscience.la.asu.edu/photosyn in chapter 10 of the OLC to learn more about photosynthesis and the formation of carbohydrates.

Photosynthesis is a light-driven series of chemical reactions that convert the energy-poor compound, carbon dioxide (CO_2), to energy-rich sugars. In plants, photosynthesis also splits water and releases oxygen (O_2). Over time, the oxygen released by photosynthesis has dramatically changed the earth's atmosphere and enabled the evolution of aerobic respiration in animals and other organisms. Today, as in the past, virtually all life depends on photosynthesis, which can be summarized with the following general equation:

$$CO_2 + H_2O \xrightarrow[\text{chlorophyll}]{\text{light}} \text{sugars} + O_2$$

Photosynthesis has real meaning for you other than just directly providing your fruits, vegetables, and grains and the oxygen in the air you breathe. In fact, today you have probably used many processed products of photosynthesis, such as cotton fiber for shirts or pants, wood to form a pencil or toothpick, and plastic or paper to make a soft drink cup. If you drove an automobile today, it most likely ran on gasoline, which is really just a modern product formed by processing the products of ancient photosynthesis that became gas, oil, and coal long ago. The next time you drive into a filling station, ask for a tank of ancient photosynthesis and see what kind of response you receive.

How does photosynthesis work in plants? Look at a leaf of a green plant. Is it carrying on photosynthesis now? This chapter will provide the information necessary to begin to answer these questions. As you study the process of photosynthesis, keep in mind its importance in ecology and in the evolution of life on earth today.

Photosynthetic Organisms Are Autotrophs.

Before the evolution of photosynthesis, virtually all organisms used organic compounds (those with carbon-to-carbon bonds; see chapter 3) as sources of energy in the planet's "primeval broth". Those one-celled organisms were **heterotrophs**—that is, organisms that make organic compounds from other energy-rich organic compounds that they absorb and digest. As these primitive organisms consumed the carbon compounds from the primeval broth, they released CO_2 into the oxygen-deficient environment. This lifestyle was adequate in the short term, but there was a long-term problem: the primeval broth was nonrenewable. This created a constant competition for the limited and ever-dwindling supply of carbon compounds.

The evolution of photosynthesis about 3 billion years ago gave many organisms a new source of energy. Photosynthetic organisms, rather than relying on the ever-dwindling amount of primeval broth for energy, began to use sunlight as an energy source. In doing so, they became solar powered and could exploit a reliable and abundant supply of energy. These organisms were the earth's first **photosynthetic autotrophs**—that is, organisms that use light energy to make organic compounds from inorganic compounds such as water and CO_2. The ability of these organisms to convert light energy to chemical energy soon supported almost all other forms of life on the planet. These autotrophs also changed the planet and its organisms by decreasing the atmospheric concentration of CO_2, diminishing the greenhouse effect, and filling the atmosphere with a waste product of photosynthesis that some organisms ultimately found essential for life: oxygen. All of the oxygen in air that we breathe has been cycled through plants via photosynthesis—"life's grand device." This oxygen allowed the evolution of aerobic respiration and the higher life-forms we are familiar with now.

Today, only about 500,000 of the earth's 3.5 million known species are photosynthetic. However, the importance of these 500,000 organisms cannot be overstated: without photosynthesis, humans and all other animals would become extinct; in fact, they never would have evolved, at least not in the form we see now.

Photosynthesis packages the energy of light into organic molecules using CO_2 and water as the raw materials to manufacture the organic products. Indeed, each year photosynthesis produces about 1.4×10^{14} kilograms of carbohydrates—enough sugar to fill a string of boxcars reaching to the moon and back 50 times. Here's another way of appreciating the annual contribution of photosynthesis: if all of the sugars made by photosynthesis were converted to an equivalent amount of coal loaded into standard railroad cars (each holding about 50 tons of coal), then the earth's photosynthesis would fill more than 100 cars per second with coal.

INQUIRY SUMMARY

The original organisms on earth were probably heterotrophs that formed organic compounds from other energy-rich organic compounds they absorbed and digested. Heterotrophs released CO_2. About 3 billion years ago, photosynthetic organisms, rather than relying on the ever-dwindling amount of primeval broth for energy, began to use sunlight as an energy source. These organisms were the earth's first photosynthetic autotrophs—that is, organisms that use light energy to make organic compounds from inorganic compounds such as water and CO_2. All of the oxygen in air that we breathe was produced by photosynthesis. This oxygen allowed the evolution of aerobic respiration and the higher life-forms we are familiar with now.

As cells evolved, what might have been the selective advantage to using sunlight for energy as opposed to organic compounds?

What Is Light?

To understand photosynthesis, you must know a little about the properties of light, which drives the process. The physics of light is beyond the scope of this book, so we will give you just the basics.

The sun is a giant thermonuclear reactor. Each minute, more than 120 million metric tons of solar matter are converted to radiant energy—light and heat. Eight minutes later, about two-billionths of this energy has traveled 160 million kilometers and hits the earth's upper atmosphere. About one-third of the light hitting the atmosphere is reflected back to space, and only about 1% of the light is harvested by photosynthesis.

In 1905, Einstein proposed that light consists of packets of energy called **photons,** which are the smallest divisible units of light. The intensity (i.e., brightness) of light depends on the number of photons (i.e., the amount of energy) absorbed per unit of time. Light intensity is important in photosynthesis because each photon carries a fixed amount of energy that is determined by the photon's wavelength. Wavelengths of visible light are measured in nanometers, or billionths (10^{-9}) of a meter (fig. 10.1). We perceive the different wavelengths as different colors; for example, violet light has a wavelength of about 400 nanometers, which is about one-fortieth the thickness of this page. The energy of a photon is inversely proportional to the wavelength of the light: the longer the wavelength, the less energy per photon. Thus, red light contains less energy than blue light, but has longer wavelengths. All of the colors of light come together and form white light, or visible light. Photosynthetic organisms harvest sunlight and use it to drive photosynthesis, the process that feeds most of the earth.

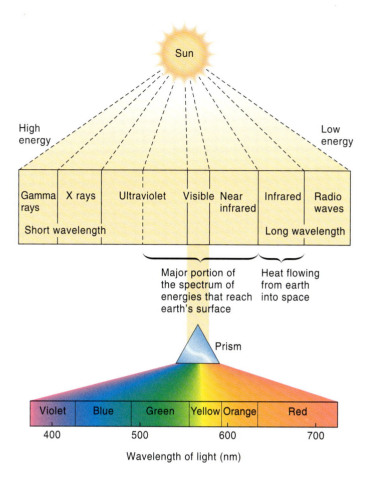

FIGURE 10.1

Properties of light. Visible light is only a small part of the electromagnetic spectrum. Visible light consists of a rainbow of colors ranging from violet to red with wavelength from 400 nm to 700 nm, respectively.

INQUIRY SUMMARY

Virtually all life depends on light, which powers photosynthesis. Light energy is contained in packets called photons. The energy of a photon is inversely proportional to the wavelength of the light: the longer the wavelength, the less energy per photon. Sunlight consists of a spectrum of colors of light having different wavelengths and energy.

Which ecosystems on earth do not depend on sunlight? Why not?

Pigments Absorb Light, Which Drives Photosynthesis.

Light (photons) striking an object can be reflected, transmitted, or absorbed. Only light that is absorbed can have an effect on the object. Light is absorbed by complex, organic

The Lore of Plants

Chlorophyll, the photosynthetic pigment that makes plants green, was wildly popular as a deodorizer in the 1950s. Pepsodent marketed a toothpaste named *Chlorodent* in 1950 and soon thereafter started selling a chlorophyll-based dog food. Other chlorophyll-based products included gum, mouthwash, deodorant, diapers, popcorn, and cologne, and there were plans for chlorophyll-based salami and beer. Although the chlorophyll industry was damaged by tests done by the U.S. Food and Drug Administration showing chlorophyll was a poor deodorizer, the folly of the craze was summarized by a magazine article that pointed out that although goats eat almost nothing but green grass, they nevertheless stink.

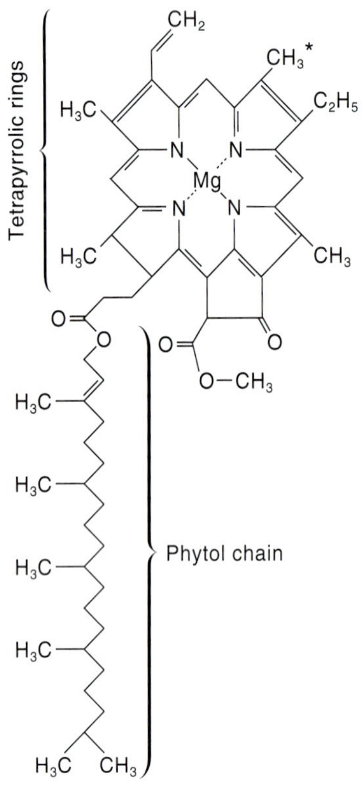

FIGURE 10.2

Chlorophyll *a*, the primary pigment of photosynthesis.

molecules called **pigments,** which appear colored because they also reflect light. For instance, black pigments absorb all wavelengths of light, while white pigments absorb no wavelengths of light. Green pigments absorb all the colors except green, which is reflected back to our eyes.

Many kinds of pigments are similar but have different functions. For example, chlorophyll pigments are made of four smaller rings (a tetrapyrrole) , each of which consists of four carbons and one nitrogen (fig. 10.2). The four rings of the pigment are linked by bridges that sequester a metal atom. If the metal is iron, the pigment is a heme, such as in hemoglobin found in your red blood cells. If the metal is

magnesium, the pigments are **chlorophyll**—the primary green pigments of photosynthesis. There are several kinds of chlorophyll, but we will occasionally refer to them collectively as chlorophyll, which is a common designation.

Chlorophylls are hydrophobic (not water soluble, with lipid characteristics; "water-hating") pigments that occur in plants, algae, and all but one primitive group of photosynthetic bacteria. Unlike the red color of hemoglobin, which is insignificant for the molecule's function as an oxygen carrier, the green color of chlorophyll is significant. Chlorophyll absorbs light maximally at wavelengths of 400 to 500 nanometers (violet-blue) and 600 to 700 nanometers (orange-red) (fig. 10.3). The synthesis of chlorophyll and several other pigments in plants is stimulated by light, which explains why plants grown in the dark (i.e., etiolated plants) contain no chlorophyll. Similarly, the light-stimulated synthesis of anthocyanin, another plant pigment, explains why apples are always redder on the sunny side of an apple tree. Recall from chapter 6 that chlorophyll and anthocyanin syntheses involve phytochrome, the plant's hourglass.

There are several types of chlorophyll, the most important of which is **chlorophyll *a*,** the primary photosynthetic pigment. Chlorophyll *a* is a grass-green pigment whose structure includes an atom of magnesium (Mg) (see fig. 10.2). It occurs in all photosynthetic organisms except photosynthetic bacteria and absorbs light maximally at 430 and 662 nanometers.

The relationship of light absorption by chlorophyll *a* versus wavelength is shown in figure 10.3. This absorption spectrum shows that chlorophyll *a* absorbs all visible light except green light; it reflects and transmits most of the green wavelengths and therefore appears green to us. This absorption spectrum for chlorophyll *a* closely matches the graph showing how photosynthesis varies with different wavelengths of light (fig. 10.4). This so-called action spectrum shows that the pattern of photosynthesis closely follows that of chlorophyll *a*, suggesting that chlorophyll *a* is the primary pigment in photosynthesis. However, the absorption spectrum of chlorophyll *a* does not perfectly match the action spectrum for photosynthesis. Therefore, we conclude that other pigments must also be involved in photosynthesis.

The Lore of Plants

What happens when trees turn to bright colors in the fall? During the spring and summer these plants simultaneously synthesize and destroy their chlorophyll at near equal rates. However, in the fall, when day lengths become shorter, the rate of chlorophyll synthesis lags behind that of its breakdown. The decreasing amounts of chlorophyll no longer mask the other pigments, which results in the red and yellow colors of autumn that are praised by poets and songwriters. In many trees, this "turning of the leaves" is spectacular. Tourists who visit New England to see the carotenoids of the trees generate a billion-dollar-per-year industry, most of which happens in just 3 weeks. States such as New Hampshire even have a Fall Foliage Hotline to keep people up-to-date on the breakdown of chlorophyll.

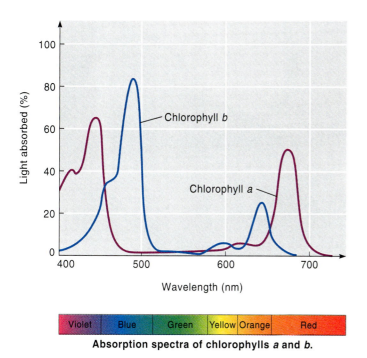

Absorption spectra of chlorophylls *a* and *b*.

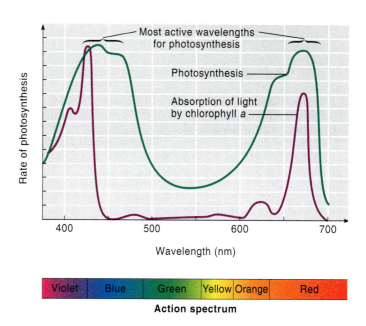

Action spectrum

FIGURE 10.3

The absorption of light by chlorophylls *a* and *b*. Chlorophylls absorb maximally at wavelengths of 400 to 500 nanometers (violet-blue) and 600 to 700 nanometers (orange-red).

FIGURE 10.4

The absorption spectrum of chlorophyll *a* shows that chlorophyll *a* absorbs maximally at 430 and 662 nanometers. The rate of photosynthesis in different wavelengths of light is similar to the ability of chlorophyll *a* to absorb those wavelengths. The similarity of the action spectrum of photosynthesis and absorption spectrum of chlorophyll *a* suggests that chlorophyll *a* is the primary photosynthetic pigment and that light absorbed by chlorophyll *a* drives photosynthesis.

Chlorophyll *b* (see fig. 10.2) is a bluish green pigment that absorbs maximally at 453 and 642 nanometers (see fig. 10.3). It occurs in all plants, green algae, and some prokaryotes. Plants usually contain about half as much chlorophyll *b* as chlorophyll *a*.

Plants also contain a rainbow of other pigments that are often called accessory pigments. They extend the range of light useful for photosynthesis by absorbing light (photons) in wavelengths not effectively absorbed by chlorophyll *a* and transmitting that energy to chlorophyll *a*. The most common of these accessory pigments in plants are the **carotenoids,** which occur in all photosynthetic organisms.

Carotenoids absorb maximally at wavelengths between 460 and 550 nanometers; therefore, these pigments are red, orange, and yellow. In plants, carotenoids extend the range of photosynthesis by absorbing light that is not absorbed by chlorophylls *a* and *b*.

The most common carotenoid is beta-carotene, a reddish yellow pigment. When split in half, beta-carotene becomes two molecules of vitamin A. This is why yellow vegetables and green leaves are good sources of vitamin A. Carotenoids occur throughout the plant kingdom and produce the colors of tomatoes, carrots, squash, bananas, avocados, and autumn's colorful leaves.

Oxidizing (adding oxygen in a chemical reaction) carotenes produces xanthophylls, which are red and yellow pigments in tomatoes, carrots, leaves, algae, and photosynthetic bacteria. Xanthophylls, which are also known as accessory pigments, are less efficient at transferring energy during photosynthesis than are carotenoids.

INQUIRY SUMMARY

Light is absorbed by pigments. Chlorophyll is the primary pigment for photosynthesis in plants. Accessory pigments such as carotenoids absorb light that chlorophylls cannot absorb, thereby extending the range of light useful for photosynthesis.

Aside from photosynthesis, how are accessory pigments important in nature?

Photosynthesis Requires Both Photochemical and Biochemical Reactions to Produce Sugars.

In 1905, F.F. Blackman provided the first evidence that photosynthesis is a two-stage process (fig. 10.5), which was described as having photochemical and biochemical reactions. The photochemical reactions became known as the **light reactions** of photosynthesis. Unfortunately, the biochemical reactions became known as the "dark reactions" of photosynthesis because initially researchers thought that light was not directly required for the reactions. Today, however, we know this is not true; the biochemical relations do depend on light to provide the chemical fuels (ATP, NADPH) necessary to form sugars from CO_2, to directly activate several of the enzymes that catalyze the biochemical reactions, and to regulate import and export of necessary molecules between the chloroplast and cytosol. Among other names, the biochemical reactions that fix CO_2 and produce sugars include the **Calvin cycle**, as explained later in this chapter. Learning about these steps of photosynthesis now will help you understand concepts discussed later in this chapter.

INQUIRY SUMMARY

The photochemical (light) reactions produce ATP and NADPH that are used in the biochemical reactions (Calvin cycle), which produce sugars from CO_2.

In ancient ecosystems, what might have been the importance of the release of O_2 on the evolution of animals?

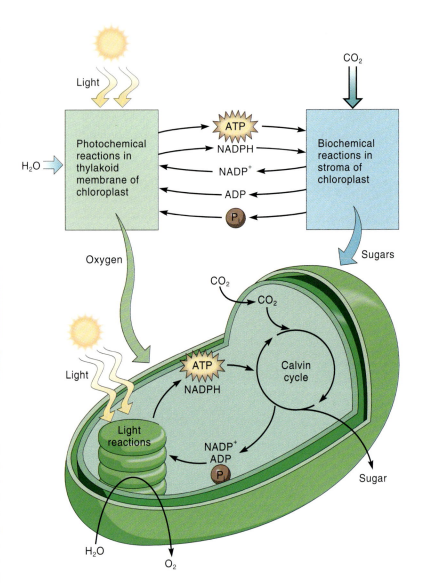

FIGURE 10.5

Photosynthesis consists of photochemical and biochemical reactions, called the light reactions and the Calvin cycle, respectively. The photochemical reactions convert light energy to chemical energy: ATP and NADPH. The biochemical reactions use the ATP and NADPH produced by the photochemical reactions to reduce CO_2 to sugars. The photochemical reactions occur in thylakoid membranes, whereas the biochemical reactions occur in the stroma. P_i is inorganic phosphate.

Chloroplasts Are the Site of Photosynthesis in Plants.

Recall from chapter 4 that chloroplasts, which are the site of photosynthesis, are membrane-bound organelles inside leaf cells (fig. 10.6). Chloroplasts are green because they contain chlorophyll, specifically chlorophylls *a* and *b*, the most important light-harvesting pigments.

Chloroplasts are usually shaped like footballs, with a diameter of 5 to 10 micrometers and a depth of 3 to

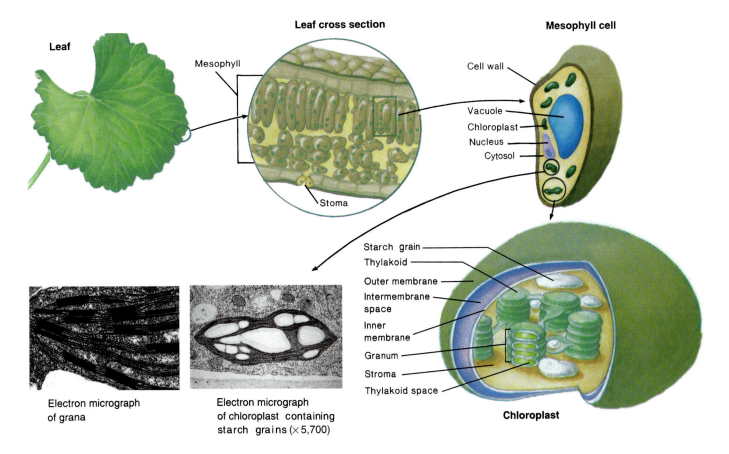

Leaf

Leaf cross section

Mesophyll

Stoma

Mesophyll cell

Cell wall

Vacuole
Chloroplast
Nucleus
Cytosol

Starch grain
Thylakoid
Outer membrane
Intermembrane space
Inner membrane
Granum
Stroma
Thylakoid space

Chloroplast

Electron micrograph
of grana

Electron micrograph
of chloroplast containing
starch grains (×5,700)

FIGURE 10.6

In plants, photosynthesis occurs in chloroplasts. Light is absorbed and converted to chemical energy in thylakoids and grana; this chemical energy is then used in the stroma to make sugars.

4 micrometers. Most photosynthetic cells have 40 to 200 chloroplasts, which amounts to about 500,000 chloroplasts per square millimeter of leaf area. To help you put the size of a chloroplast in perspective, consider that it would take about 2,000 of them to reach across your thumbnail.

Each chloroplast is surrounded by two membranes that enclose a gelatinous matrix called the **stroma** (fig. 10.6). The stroma contains ribosomes, DNA, and enzymes that make the sugars produced by photosynthesis. Suspended in the stroma are neatly folded sacs of membranes called **thylakoids** (fig. 10.6), which are membranes unique to chloroplasts. In some parts of chloroplasts, 10 to 20 thylakoids are stacked into **grana.** Thylakoids contain the chlorophylls and accessory pigments and are where light is absorbed during photosynthesis.

In all but the most primitive photosynthetic bacteria, light is captured by a network of chloroplast pigments and associated molecules arranged in thylakoids. These networks, called **antennae complexes** (fig. 10.7), consist of proteins, about 300 molecules of chlorophyll *a*, and about 50 molecules of carotenoids and other accessory pigments that gather light. Energy absorbed by antennae complexes flows energetically "downhill" to a chemically unique and special

energy-collecting molecule of chlorophyll *a* and associated proteins called a **reaction center.** Our model shows this as a "light funnel" directing photons to the reaction center, which contains a special chlorophyll *a*. Although reaction centers comprise less than 1% of the chlorophyll in plants, the special chlorophyll *a* in the reaction center is the electron acceptor that participates directly in photosynthesis; all other photosynthetic pigments function as antennae that funnel the light energy to chlorophyll *a* of the reaction center.

Today, we recognize two kinds of reaction-center chlorophyll *a* molecules. One of these chlorophylls absorbs maximally at 700 nanometers and is called P700 (for pigment 700) (fig. 10.8). The complex containing P700 is called **Photosystem I** (PS I); it contains several pigments and proteins along with the reaction center chlorophyll *a*. The second kind of chlorophyll *a* molecule occurs in a separate molecular complex called **Photosystem II** (PS II, fig. 10.8), appears identical to that of Photosystem I, but it is associated with different proteins. The associated proteins in Photosystem II shift the maximal absorption to about 680 nanometers; thus, the reaction center in Photosystem II is called P680.

When light is absorbed during photosynthesis, the energy of its photons is captured by the pigment and

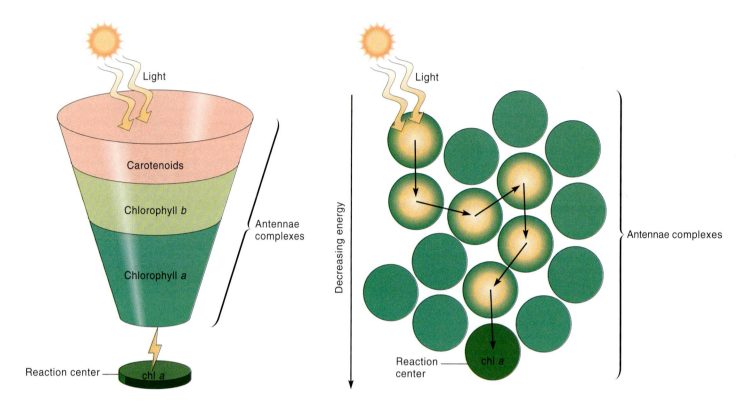

F I G U R E   1 0 . 7

The light-harvesting system in chloroplasts functions like a funnel; it collects photons of light and passes their energy to the reaction center (*Chl* = chlorophyll.). The reaction center contains a special chlorophyll *a*.

passed to a neighboring molecule in PS I or in PS II reaction centers, resulting in a flow of electrons through a thylakoid membrane between the two photosystems—like electrons flow through a battery and light a flashlight (except in photosynthesis the light causes the electrons to flow). The electrons come from water (H_2O), which, during PS II, is split into hydrogen (H_2, consisting of protons and the electrons) and oxygen (O_2). The O_2 is released as a gas from a chloroplast and then the leaf (see figs. 10.5 and 10.8). The H_2 from water can ultimately become part of the sugars produced by photosynthesis.

The light reactions of photosynthesis produce ATP.

Chloroplasts couple (i.e., connect) the light-driven flow of electrons between the two photosystems (PS I, PS II) in the thylakoid membrane to the processes that make ATP (see chapter 3) . This light-dependent production of ATP in a chloroplast is called **photophosphorylation** because light provides the energy for addition of a phosphate (P_i) to ADP, making ATP:

$$ADP + P_i \xrightarrow{\text{light}} ATP + H_2O$$

The process that directly powers ATP synthesis is the light activated accumulation of protons on one side of the thylakoid membrane. The flow of electrons, described in the previous section, drives the accumulation of protons. The resulting high concentration of protons creates a kind of chemical pressure against the membrane that when released, as protons flow back through the membrane, provides the energy to synthesize ATP (fig. 10.8). You can think of expending energy to blow up a balloon (electron flow) and then letting the air out (releasing proton pressure). As the air escapes, you can release the balloon, and the energy of the escaping air causes it to fly around the room. In a chloroplast, the energy of escaping protons drives the synthesis of ATP. The ATP produced is released into the stroma, where it is used to make sugars from CO_2. In addition, the flow of electrons results in the production of another energy-carrying molecule, nicotinamide adenine dinucleotide phosphate, **NADPH** described next.

The light reactions of photosynthesis also produce NADPH.

In light, the photon-powered flow of electrons through PS I and PS II of thylakoid membranes supplies the energy needed to produce ATP. What about the NADPH? The same flow of electrons that drives ATP synthesis in a chloroplast is used to produce this second energy-carrying compound by supplying electrons and protons (H^+) to

10.1 Perspective Perspective

The Evolution of the Light Reactions

The evolution of PS II forever changed life on earth. As PS II liberated O_2 into the atmosphere, the environment changed and allowed the evolution of aerobic respiration because the use of O_2 in respiration was a selective advantage. Why is aerobic respiration a selective advantage, and why is it important to you and other organisms? We answer these questions in chapter 11.

For nearly 35 years, botanists have been confident that they understand the major aspects of photosynthesis. Specifically, botanists have held that plants require two photosystems: PS I is essential for producing the reduced $NADP^+$ used to reduce CO_2, whereas PS II oxidizes water and releases O_2. But recently, a group of botanists at Oak Ridge National Laboratory in Tennessee questioned this long-held idea. These botanists studied mutants of *Chlamydomonas* (a single-celled alga) in which PS I does not operate. Nevertheless, the algae harvest light, fix CO_2, release O_2, and grow. These results suggest that the mutants can perform photosynthesis without PS I (i.e., with only PS II). While not all researchers are ready to accept this conclusion, the study may force botanists to consider alternatives to the model presented in figure 10.8. The study may also help us understand how photosynthesis evolved. The mutant without PS I grows better in the absence of oxygen, suggesting that PS II evolved first, under early anaerobic conditions that existed before photosynthesis filled the ancient atmosphere with O_2. That is, the earliest forms of photosynthesis used only PS II; PS I could have evolved later, as it may increase light harvest, stabilize PS II, and render electron flow more efficient.

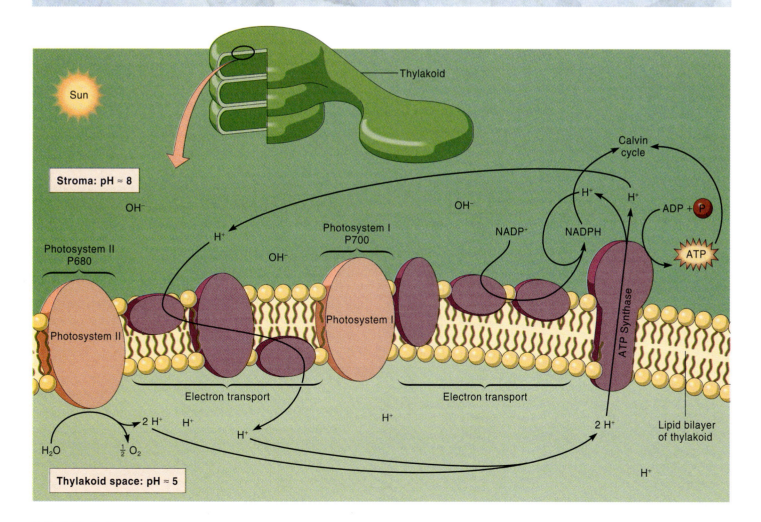

FIGURE 10.8

The light-driven transport of electrons during photosynthesis creates a proton gradient across the thylakoid membrane. This proton gradient, formed by the higher concentration of H^+ in the thylakoid space compared to the stroma, drives the formation of ATP via ATP synthase. O_2 is released as H_2O is split in light.

convert $NADP^+$ to NADPH (fig. 10.8). These events, leading to ATP and NADPH in illuminated chloroplasts, comprise the light reactions of photosynthesis. The ecological significance of the light reactions is (1) the conversion of light energy to chemical energy (ATP and NADPH), and (2) the release of O_2 into the atmosphere. Both ATP and NADPH, which can be considered the chemical energy converted from light energy, are necessary in plant chloroplasts to form the sugars produced by photosynthesis.

The biochemical reactions convert CO_2 to carbohydrates.

The biochemical reactions associated with photosynthesis, including the Calvin cycle, convert atmospheric CO_2 to carbohydrates, primarily sucrose and starch. (Sucrose, a simple sugar, and starch, a complex carbohydrate [polysaccharide], were discussed in chapter 3. We use carbohydrate and sugar interchangeably in our discussion of photosynthesis.) These carbohydrates can be considered transport and storage depots of chemical energy and carbon. Conversion of CO_2 to sugars is also called **carbon fixation**—that is, the carbon is fixed in organic form (organic molecules have carbon to carbon bonds). The carbon fixed during photosynthesis, as you learned previously, is ultimately used to make all of the organic molecules of the plant and, subsequently, of organisms that feed on the plant in an ecosystem.

Straightforward as this may sound, the actual mechanisms of CO_2 fixation and sugar production have been difficult to explain. The most important contribution to our understanding of how plants use CO_2 to produce sugars was provided by Melvin Calvin and his colleagues at the University of California at Berkeley in the early 1950s. These botanists studied photosynthesis by using a product of World War II: ^{14}C, or radioactive carbon. Calvin first exposed a dense culture of *Chlorella* (a green alga) to $^{14}CO_2$ and then traced the path of the radioactively labeled ^{14}C as it moved from $^{14}CO_2$ to ^{14}C-carbohydrate (sugars) and then other labeled compounds. When Calvin exposed the alga to radioactive $^{14}CO_2$ for 60 seconds, he found that many compounds in the alga were quickly labeled with ^{14}C. When he repeated the experiment with a 7-second exposure, most of the radioactive carbon appeared in **3-PGA** (3-phosphoglyceric acid), a three-carbon acid. Calvin began searching for a molecule that, if combined with CO_2, would produce the three-carbon acid. He found the elusive acceptor molecule in **RuBP** (ribulose-1,5-bisphosphate), a five-carbon sugar. The temporary product of RuBP (five carbons) and CO_2 is an unstable six-carbon molecule that immediately breaks down into two molecules of 3-PGA. The summary reaction for CO_2 fixation is

$$CO_2 + RuBP \longrightarrow 2 \text{ molecules of 3-PGA}$$

We shall learn more about the significance of this discovery later, but remember Calvin reported the first de-

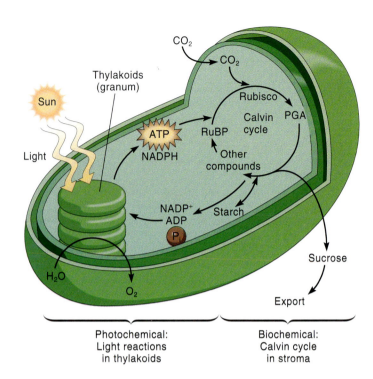

| Photochemical: Light reactions in thylakoids | Biochemical: Calvin cycle in stroma |

FIGURE 10.9

A summary of photosynthesis. The photochemical reactions, which occur in thylakoids and grana, produce O_2, ATP, and NADPH. The Calvin cycle (i.e., biochemical reactions), which occurs in the stroma, uses the ATP and NADPH to reduce CO_2 to carbohydrate. The by-products ADP, P_i, and $NADP^+$ are returned from the Calvin cycle to the light reactions.

tectable product of $^{14}CO_2$ incorporation was the three-carbon acid PGA (found as ^{14}C-PGA).

Calvin and his associates used a similar approach to figure out the rest of the biochemical reactions of photosynthesis. Later, Calvin won the Nobel Prize for this classic research in science, and the cycle of biochemical reactions that fix and assemble CO_2 is called the *Calvin cycle* to honor him and his co-workers. The carbon fixation reactions are also known as the photosynthetic carbon reduction cycle. Follow the steps of carbon fixation in figures 10.9 and 10.10 and in the following to see some details of how the Calvin cycle occurs:

- CO_2 diffuses through a stomate, into a leaf cell, and then through the chloroplast membranes into stroma, where it is fixed; that is, CO_2 is quickly incorporated into an organic compound—PGA. This occurs by the bonding of CO_2 to RuBP (five-carbon sugar) and is termed carbon fixation (or carboxylation).

- The enzyme that catalyzes the carbon fixation reaction is **rubisco** (RuBP carboxylase/oxygenase), a large, complex enzyme that has been called the most important and most abundant protein on earth. It makes up to 50% of the protein in leaves and is made

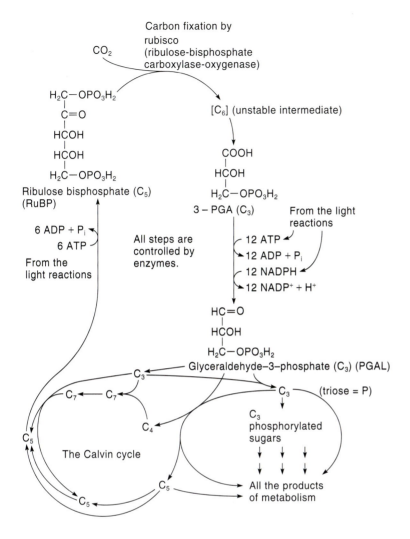

FIGURE 10.10

The Calvin cycle. PGA is the first compound formed from fixed CO_2. PGAL is the first sugar. PGAL and C_3 are triose-P. C_5 = five-carbon molecule, C_6 = six-carbon molecule, and so on.

at a rate of 4×10^{10} grams per year, which is the equivalent of about 10^6 grams per second. Every person on earth is supported by about 44 kilograms of rubisco (if the metric units are not familiar to you, convert these numbers to pounds and tons; you'll be surprised how much protein this is!).

- The assembly of rubisco inside a chloroplast is controlled by light. So is the enzymatic activity of rubisco; that is, its ability to fix CO_2 inside a chloroplast is directly dependent on activation by light and also on inhibitors and activators of rubisco that are controlled by light in various plants.

- The reaction of CO_2 with RuBP produces two molecules of 3-PGA (the 3 tells us the phosphate group is on the number 3 carbon of the acid's carbon skeleton or chain).

- In the next two steps, which are powered by ATP and NADPH from the light reactions, each molecule of

3-PGA is converted to the PGAL (glyceraldehyde-3-phosphate, also referred to as G-3-P), a three-carbon sugar. PGAL is rapidly and easily converted back and forth between another three-carbon sugar, which is not shown for simplicity in our model (dihydroxy-acetone phosphate, abbreviated DHAP). PGAL and DHAP are **triose-P** (triose-phosphates), having three-carbon atoms of the sugar backbone with a phosphate group attached to it (see chapter 3). A summary equation, with triose-P shown as $C_3 H_6 O_3$ –P, is

$$3 CO_2 + 6 H_2O \xrightarrow{\text{ATP, NADPH}} C_3 H_6 O_3P + 3 O_2 + 3 H_2O$$

Thus, the Calvin cycle produces the simple sugar triose-P, which is drained from the cycle and used to make other molecules in a plant.

- Some of the triose-P is used to re-form RuBP and keep the cycle running in the light. This requires ATP made in the light reactions, several enzyme-catalyzed steps, and a continuous supply of CO_2. Several of these enzymes, like rubisco, are activated by light: they are inactive in the dark.

- Most of the rest of the triose-P either (1) moves through a series of chemical reactions, forming starch inside a chloroplast or (2) is exported from the chloroplast to the leaf cell cytosol, where it is used to make sucrose, the disaccharide, simple sugar in plants (see chapter 3). Starch, the polysaccharide that accumulates inside a chloroplast during the day, can be broken down at night into triose-P, which is then moved to the cytosol. Here, triose-P is used to make sucrose, ensuring some consistency of sucrose supply in a plant.

- Overall, carbon fixation can be shown by the summary equation:

$$3 CO_2 + 3 RuBP \xrightarrow{\text{ATP, NADPH}} 3 RuBP + 1 \text{ triose-P}$$
$$4 \text{ triose-P} \longrightarrow 1 \text{ sucrose} + 4 P_i$$

Note the triose-P product is made from CO_2

- Sucrose provides the carbon and energy for biosynthesis of other organic compounds in all parts of a plant.

Rubisco is not perfect and fuels carbon loss via photorespiration.

Photosynthesis evolved over eons, and continues to evolve, but it has shortcomings. For example, it wastes much light energy (see Perspective 10.2) and often does a poor job of fixing carbon. The first biologist to show this imperfection of photosynthesis was Otto Warburg, who in 1920 demonstrated that increased amounts of oxygen surrounding a plant can inhibit photosynthesis. This inhibition of the rate of

10.2 *Perspective* Perspective

The Evolution of Photosynthesis: Why Aren't Plants Black?

Photosynthetic autotrophs use light energy to convert CO_2 to carbohydrate. Because the amount of energy absorbed by an organism largely determines its rate of photosynthesis, why then aren't plants black? Stated another way, what was the selective advantage for plants to reflect, and therefore waste, green light? Why didn't they evolve to utilize all of the spectrum of visible light for photosynthesis? The answers to these questions lie in the evolution of photosynthesis.

The earliest photosynthetic organisms were aquatic bacteria, several of whose descendents are around today. Chief among these is *Halobacterium halobium*, a purple bacterium that grows only in extremely salty water (three to four times saltier than seawater). No other known organisms can tolerate this environment.

Since water absorbs most light outside the visible part of the spectrum, most ultraviolet and infrared light were unavailable to the first photosynthetic organisms. Natural selection therefore favored the evolution of pigments such as bacteriorhodopsin, the photosynthetic pigment in *Halobacterium*. Bacteriorhodopsin, a purple pigment that resembles rhodopsin (the light-sensing pigment in our eyes), absorbs broadly in the middle of the visible spectrum of light. When bacteriorhodopsin absorbs light, it pumps protons out of the cell, thereby creating a proton gradient that the cell uses to make ATP. That ATP powers *Halobacterium*.

Organisms that lived on the surface of sediment faced a different problem. The bacteriorhodopsin in the *Halobacterium* swimming above the sediments absorbed most of the green light. Consequently, the light that reached the bottom-dwelling organisms was poorly suited for another bacteriorhodopsin-based type of photosynthesis. Similarly, the surrounding water absorbed all of the infrared and ultraviolet light of sunlight. Natural selection thus favored the evolution of a pigment that could absorb the remaining wavelengths: red and blue. This pigment was chlorophyll, which in addition to pumping protons out of the cell also pumped electrons into the cell. Their ability to pump electrons enabled these organisms to reduce CO_2 to sugars.

Plants and plantlike organisms have evolved other pigments that harvest a wider range of light. Accessory pigments absorb wavelengths near the center of the visible spectrum and pass this energy to chlorophyll. Cyanobacteria and red algae have phycocyanin and allophycocyanin as accessory pigments, which absorb orange light but don't cover all of the wavelengths missed by chlorophyll. These organisms also contain a red pigment called phycoerythrin, which ab-sorbs green light and thus extends the range of photosynthesis beyond that of most plants. Indeed, despite their name, many red algae are dark colored, approaching black, making them ideally suited for photosynthesis. Similarly, the dark colors of brown algae result from chlorophyll and fucoxanthin, an accessory pigment. Not surprisingly, many red and brown algae grow in deep water, where light is extremely dim. Some darkly colored red algae in the western Atlantic live more than 200 meters down, where the light is less than 1% of full sunlight.

And what about land plants? Since light is abundant on land, there has been relatively little selection pressure for the evolution of a wider range of pigments for photosynthesis. In other words, unlike red and brown algae, land plants grow in adequate light and do not have to extract as much light as possible. Consequently, land plants depend almost entirely on chlorophyll for photosynthesis. Carotenoids partly fill the gap, especially toward blue, but they do not pass this energy efficiently to chlorophyll. Since no pigment in land plants absorbs significantly in the green part of the spectrum, this light is wasted. It also creates a paradox: the most advanced plants have the least advanced system for absorbing light. Nevertheless, this is not all bad; few of us would trade a green countryside for a black one in the name of photosynthetic efficiency.

---

photosynthesis became known as the Warburg effect. You may wonder how this is relevant to plants, since the amounts of CO_2 and O_2 in air are relatively constant. The answer is that it depends on what air you're talking about. Indeed, the composition of air outside a leaf often differs significantly from that inside a leaf where photosynthesis occurs.

During photosynthesis, plants fix CO_2 from the atmosphere. Although the concentration of CO_2 in air today is only about 0.035% to 0.0365% (350 ppm to 365 ppm; 1 part per million (ppm) is analogous to 1 second in 277 hours), this represents an almost endless supply of CO_2 for photosynthesis. Indeed, there are about 7×10^{11} tons of carbon (as CO_2) in the atmosphere (for comparison, there are about 4.5×10^{11} tons of carbon in vegetation as carbohydrate). Consequently, you wouldn't think that plants would have much trouble obtaining enough CO_2 for photosynthesis. However, all of this CO_2 is available to a plant only when the plant's stomata are open. As CO_2 diffuses into a leaf for photosynthesis, water diffuses out through the stomata. As long as the plant has plenty of water, this is not a serious problem, but on hot, dry days, stomata close, at least partially, conserving water. When this occurs, the continued fixation of CO_2 (via rubisco and the Calvin cycle) in the light decreases the concentration of CO_2 inside the leaf, which greatly increases the relative amount of O_2. This has a tremendous effect on the plant: when the concentration of CO_2 inside the leaf dwindles and the concentration of O_2 remains high, rubisco tends to combine RuBP with O_2 instead of CO_2 (fig. 10.11). Rubisco is a "sloppy" enzyme

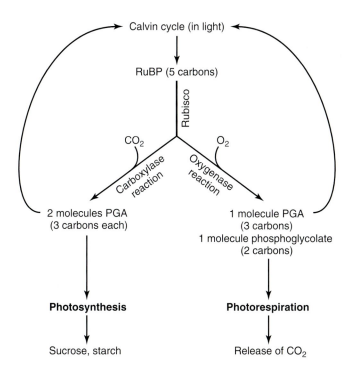

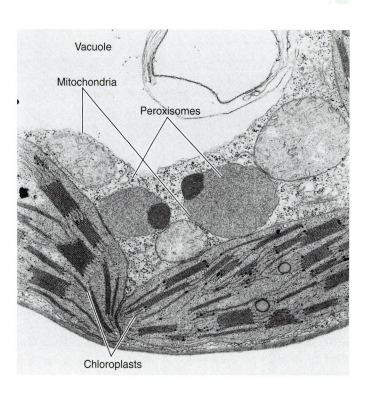

FIGURE 10.11

The relative rates of photosynthesis and photorespiration depend on the relative concentrations of CO_2 and O_2 rubisco in a chloroplast. When rubisco combines O_2 with the five-carbon acid RuBP (oxygenase reaction), the products are one molecule of PGA (three carbons) and one of phosphoglycolate (two carbons): there is no net fixation of carbon and the Calvin cycle "spins" in a futile manner, adding no carbon to the plant but using the energy from the light reactions. Subsequently, some of the carbon in phosphoglycolate is lost as CO_2 during photorespiration. Since the Calvin cycle can operate only in the light, photorespiration can operate only in the light.

and can attach either CO_2 (carboxylation) or O_2 (oxygenation) to RuBP. The relative rates of rubisco catalyzed carboxylation to oxygenation depend on the relative levels of CO_2 to O_2. Both chemical reactions occur simultaneously in a chloroplast, where millions of copies of the "sloppy" rubisco enzyme molecules are located, but the relative rates depend on the CO_2 and O_2 concentrations in a chloroplast.

Remember that rubisco is ribulose-1,5-bisphosphate carboxylase/oxygenase, and this tells us it can combine either CO_2 or O_2 with RUBP (fig. 10.11). The products of the oxygenase reaction, where O_2 is attached to RuBP (a five-carbon compound), are phosphoglycolate (a two-carbon compound) and PGA (a three-carbon compound):

$$\text{O}_2 + \text{RuBP} \xrightarrow{\text{rubisco}} \text{phosphoglycolate} + \text{PGA}$$

Note that there is no carbon (CO_2) input into the chloroplast with this oxygenase reaction. The PGA can stay in the Calvin cycle, but the phosphoglycolate leaves the cycle. In a complex series of reactions occurring in mitochondria and another organelle, the peroxisome (fig. 10.12),

FIGURE 10.12

Electron micrograph showing the close association of chloroplasts, peroxisomes, and mitochondria ($\times 58,000$). All of these organelles participate in photorespiration.

phosphoglycolate is combined with other molecules. These molecules are then split, resulting in the release of CO_2 that can be lost from the plant. The carbon lost in the released CO_2 was previously fixed during photosynthesis.

The reactions initiated by rubisco's fixation of oxygen rather than CO_2 are collectively called **photorespiration** (see fig. 10.11) because they occur only in light, consume oxygen, and release CO_2. Unlike cellular respiration in mitochondria, however, photorespiration produces no ATP. That is, photorespiration squanders large amounts of carbon originally fixed at great energy expense by the Calvin cycle. Although some of the CO_2 released during photorespiration may be recycled by the Calvin cycle, in many plants, as much as half of the carbon previously fixed during photosynthesis can escape as CO_2. Thus, photorespiration undoes photosynthesis. In most of the world's crops, photorespiration is a major problem, especially in arid lands where it is a part of the cause of much famine and starvation due to reduction in crop productivity (see Perspective 10.3). Photorespiration is a remnant process—it likely was not a problem in the ancient atmosphere when the CO_2 level was high and O_2 was low. What does this suggest about the evolution of rubisco?

INQUIRY SUMMARY

Photosynthesis in eukaryotes occurs in chloroplasts where light energy is used to power the synthesis of ATP

10.3 Perspective Perspective

How Did Such a Wasteful Process as Photorespiration Evolve?

Some botanists suggest that the oxygenase reaction of rubisco is inevitable because of rubisco's structure: its active site (that part of the enzyme that attaches to CO_2) makes it impossible for the enzyme to discriminate between the molecularly similar characteristics of CO_2 and O_2. The oxygenase activity of rubisco, leading to the destructive photorespiration, probably represents evolutionary baggage—a genetic relic—from when the ancient atmosphere contained no free oxygen and there was no competition between CO_2 and O_2; that is, there was no selective pressure against binding O_2 because there was no O_2 in the air. The subsequent buildup of oxygen, derived from millions of years of photosynthesis, led to photorespiration. The now imperfect, oxygen-binding active site of rubisco could not be corrected through natural selection because changes in the gene for the enzyme that removed the oxygenase activity also removed the carboxylating activity, and the mutations were lethal to the plant.

Photorespiration usually peaks in hot, dry conditions (especially at temperatures above 28°C), when plants begin to close their stomata to conserve water. Photorespiration causes plants to lose considerable carbon. Clearly, plants that could avoid photorespiration would have a significant evolutionary advantage, especially in those environmental conditions that worsen photorespiration. That advantage was provided by the evolution of C_4 photosynthesis.

---

and NADPH. These chemical forms of energy are used during the Calvin cycle to reduce (fix) CO_2 to sugars (carbohydrate). The initial fixation of CO_2 is catalyzed by an enzyme called rubisco. Triose-P is the primary sugar molecule moved from a chloroplast to the cytosol where it is used to make sucrose, the transportable form of energy and carbon from photosynthesis. Excess triose-P is stored as starch in an illuminated chloroplast. The primary function of photosynthesis is to provide energy and carbon sufficient to support maintenance and growth not only for photosynthetic tissues but of the plant as a whole.

Photorespiration occurs in plants when the internal concentration of CO_2 becomes low, such as when stomata close during drought. During photorespiration, rubisco fixes O_2 instead of CO_2, and CO_2 from previous carbon fixation is later released. Consequently, photorespiration undoes photosynthesis: as much as half of the carbon fixed in the Calvin cycle is released by photorespiration, especially during hot, dry days.

How might the earth's terrestrial ecosystems be different without photorespiration?

C_4 Photosynthesis Employs Two CO_2 Fixation Steps and Probably Evolved During Times of Low CO_2 Levels in the Atmosphere.

Soon after Calvin's group reported their brilliant study, other biologists began to repeat their study with a variety of plants. Most of these studies confirmed Calvin's findings, but one study of sugarcane in 1965 stood out. In this plant, a 1-second exposure to radioactive CO_2 ($^{14}CO_2$) resulted in 80% of the radioactivity being found, not in Calvin's three-carbon compound, PGA, but in a four-carbon acid called malic acid (or malate).

This new research led to the discovery and the subsequent naming of the **C_4 pathway** of photosynthesis, which occurs in C_4 plants. C_4 photosynthesis is different from **C_3 photosynthesis,** which was the type of photosynthesis described by Calvin. In C_3 photosynthesis, CO_2 is first detected in the three-carbon acid, PGA. In C_4 photosynthesis, it is the four-carbon acid malate that is detected first. Aside from the first detectable compound, just what makes C_4 different, and evolutionarily important, compared to C_3 photosynthesis?

Botanists found that in sugarcane and other C_4 plants, most of the malate appeared only in thin-walled mesophyll cells (fig. 10.13), which, strangely enough, lacked rubisco and the other enzymes of the Calvin cycle. Only after about 10 seconds exposure to $^{14}CO_2$ did the ^{14}C label appear in PGA in bundle-sheath cells, which are thick-walled cells surrounding the vascular bundles (fig. 10.13). This kind of leaf anatomy, with veins encased by large, thick-walled bundle-sheath cells that are, in turn, surrounded by thin-walled mesophyll cells, is characteristic of tropical grasses such as sugarcane and is referred to as **Kranz** (halo or "wreath") **anatomy.** More interesting, at least initially, was the fact that $^{14}CO_2$ being fixed first into malic acid was inconsistent with the idea that the Calvin cycle was the only means of fixing CO_2 in plants. What was going on here?

Careful studies of the structure and function of these leaves finally solved the puzzle: sugarcane and other C_4 plants do fix CO_2 via the Calvin cycle but only after fixing it first via another set of chemical reactions that occur

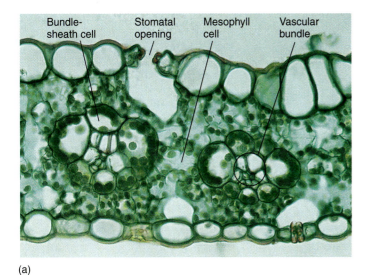

(a)

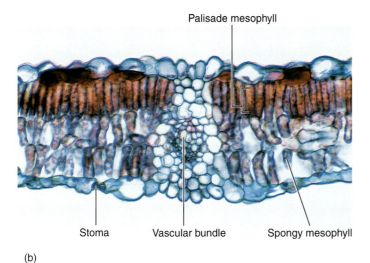

(b)

FIGURE 10.13

Structure of leaves of C_4 and C_3 plants. (*a*) Leaves of C_4 plants have a characteristic Kranz anatomy: veins are tightly surrounded by photosynthetic bundle-sheath cells, which, in turn, are surrounded by photosynthetic mesophyll cells. This is a light micrograph of a cross section of a leaf of corn (*Zea mays*) (×240). (*b*) Leaves of C_3 plants lack a prominent and photosynthetic bundle sheath (×100). Leaves of C_3 plants have a palisade mesophyll (of densely packed photosynthetic cells) and a spongy mesophyll (of loosely packed photosynthetic cells).

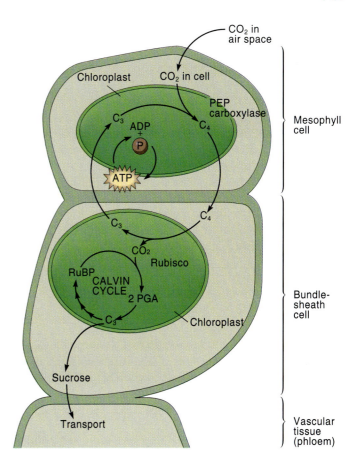

FIGURE 10.14

Summary of C_4 photosynthesis. Mesophyll cells fix CO_2 into four-carbon acids that move into bundle-sheath cells. There, the acids are decarboxylated; the CO_2 released by this decarboxylation is fixed via the Calvin cycle in the bundle-sheath cell. The three-carbon compound moves back to the mesophyll cell, where it is converted to PEP, the CO_2 acceptor.

before the Calvin cycle. The following is an explanation of C_4 photosynthesis (fig. 10.14):

- CO_2 diffuses into a leaf through stomata and into mesophyll cells, which lack rubisco and the Calvin cycle.
- CO_2 in mesophyll cells combines with a three-carbon compound, PEP (phosphoenolpyruvic acid). This reaction is catalyzed by the enzyme **PEP carboxylase**

and produces OAA (oxaloacetic acid, or oxaloacetate), a four-carbon acid as shown:

$$\text{PEP carboxylase}$$
$$CO_2 + PEP \text{ (three carbons)} \longrightarrow OAA \text{ (four carbons)}$$

OAA is quickly converted to a four-carbon compound, which then moves inward through plasmodesmata to an adjacent bundle-sheath cell.

- In bundle-sheath cells, the four-carbon acid is split into CO_2 and a three-carbon compound, which returns to the mesophyll cells. The CO_2 released in the bundle-sheath cells remains there as a raw material for subsequent fixation. Thus, the four-carbon acids function as chemical CO_2 "pumps" from mesophyll to bundle-sheath cells.

⚮ In C_4 plants, the pumping of CO_2 into bundle-sheath cells keeps the internal concentration of CO_2 in bundle-sheath cells 20 to 120-times greater than in air.

⚮ CO_2 released in the bundle-sheath cells is fixed by rubisco and continues in the Calvin cycle. Rubisco of the Calvin cycle in a C_4 plant bundle-sheath cell is sensitive to oxygen but is protected from atmospheric O_2 by the specialized anatomy of bundle-sheath cells, which can exclude external O_2. Most important, in C_4 plants, rubisco of bundle-sheath cells operates in high CO_2 due to pumping of CO_2 into bundle-sheath cells. The resulting high concentration of CO_2 allows rubisco to fix CO_2 at maximal rates (recall that in C_3 plants, rubisco operates at much less than its maximum rate because it can fix O_2—photorespiration—as well as CO_2—photosynthesis).

⚮ Thus, in C_4 plants, photorespiration is avoided, thereby enabling C_4 plants to fix CO_2 more efficiently than C_3 plants in the still relatively low concentrations of atmospheric CO_2 found today. This is the selective advantage of C_4 photosynthesis, and it is most important in hot, dry environments.

Although none of the chemical reactions of C_4 photosynthesis is unique, they can produce dramatic differences in plant growth, especially in arid climates where photorespiration tends to be most severe. The advantage of the added set of reactions in C_4 photosynthesis is that PEP carboxylase binds PEP (three-carbon) with CO_2 and does not react with oxygen (fig. 10.14). Consequently, C_4 plants, with no observable photorespiration, can fix CO_2 until the internal concentration of CO_2 almost reaches zero. This allows C_4 plants to continue to fix CO_2 even in hot, dry weather when stomata begin to close and CO_2 levels decline (fig. 10.15a). As a result, C_4 plants may need less water than C_3 plants (table 10.1). Can you explain why this is true?

In a laboratory, and probably in nature, it is almost impossible to light-saturate C_4 plants (see fig. 10.15b), and they usually "outcompete" C_3 plants in hot, dry weather. However, if given enough water, or placed in cooler growing conditions, many C_3 plants, such as buffalo gourd (*Cucurbita foetidissima*) and sunflower (*Helianthus annuus*), may fix carbon at rates comparable to those of C_4 plants. Other features of C_4 plants are summarized in tables 10.1 and 10.2.

It should not surprise you that considerable evidence indicates C_4 photosynthesis evolved in the hot conditions of the tropics; C_4 photosynthesis is common in grass species from these areas. The primary selecting mechanism for C_4 photosynthesis was probably a drop in atmospheric CO_2 concentrations about 50 to 60 million years ago. Indeed, hot and arid conditions have been common throughout the earth's 4.5 billion-year-history, but only during the

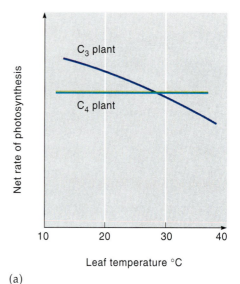

(a)

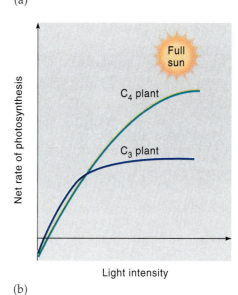

(b)

FIGURE 10.15

C_4 plants are photosynthetically more efficient than C_3 plants only in hot, dry, bright conditions. Their increased efficiency in these conditions is due to the absence of photorespiration in C_4 plants. (*a*) shows a decline in C_3, but not C_4, photosynthesis with increasing temperature (*b*) shows light saturation where increased light intensity does not increase photosynthesis, especially in a C_3 plant.

last 50 to 60 million years have CO_2 concentrations dropped to levels that gave C_4 plants a selective advantage over C_3 plants in these hot, dry regions of the globe.

Today, all known C_4 plants are angiosperms and are most common in hot, dry, and open (bright) ecosystems (see Perspective 10.4, "Why Don't C_4 Plants Dominate the Landscape?"). C_4 plants occur in at least 17 plant families, none of which includes only C_4 plants. Because these families are diverse, distantly related, and have no common

TABLE 10.1   Photosynthetic Characteristics of C$_3$, C$_4$, and CAM Plants

CHARACTERISTIC	C$_3$	C$_4$	CAM
leaf anatomy	bundle-sheath cells without dense arrangement of chloroplasts	bundle-sheath cells with dense arrangement of chloroplasts; no distinctive palisade cells	large vacuoles
primary carboxylating enzyme	ribulose bisphosphate carboxylase/oxygenase (rubisco)	PEP carboxylase, then rubisco	PEP carboxylase at night, rubisco during the day
grams of water required to produce 1 gram of dry matter	450–950	250–350	50–55
requires sodium as micronutrient	no	yes	probably
CO$_2$ compensation point (ppm)	30–70	0–10	0–5
photorespiration detectable	yes	only in isolated bundle-sheath cells	no
optimum temperature for photosynthesis	15°–25°C	30°–47°C	about 35°C
approximate tons of dry matter produced per hectare per year	20–25	35–40	usually low and variable

TABLE 10.2   Maximum Photosynthetic Rates of Major Types of Plants in Natural Conditions

TYPE OF PLANT	EXAMPLE	MAXIMUM PHOTOSYNTHESIS MG CO$_2$ DM^{-2} H^{-1}
CAM	*Agave americana* (century plant)	1–4
C$_3$		
tropical, subtropical, and Mediterranean evergreen trees and shrubs; temperate zone evergreen conifers	*Pinus sylvestris* (Scotch pine)	5–15
temperate zone deciduous trees and shrubs	*Fagus sylvatica* (European beech)	5–20
temperate zone herbs and C$_3$ pathway crop plants	*Glycine max* (soybean)	15–30
C$_4$		
tropical grasses, dicots, and sedges with C$_4$ pathway	*Zea mays* (corn)	35–70

*MG = milligram; DM = decimeter; H = hour

The Lore of Plants

Genera such as *Flaveria* (Asteraceae), *Panicum* (Poaceae), and *Alternanthera* (Amaranthaceae) contain C_3 and C_4 species as well as species intermediate between C_3 and C_4 photosynthesis. These so-called C_3/C_4 intermediates, which might be said to represent the missing link between C_3 and C_4 plants, have the following characteristics: (1) intermediate leaf anatomy—their leaves contain bundle-sheath cells, but they are indistinct and poorly developed; (2) reduced, but not eliminated, photorespiration; and (3) poor separation of CO_2 fixation pathways.

10.4 *Perspective* Perspective

Why Don't C_4 Plants Dominate the Landscape?

If C_4 plants are more efficient in photosynthesis than C_3 plants, why then don't they dominate our landscape? One answer is that C_4 plants are usually more efficient than C_3 plants only in dry, hot conditions (fig. 10.15). To best appreciate this, examine the summary equation for C_4 photosynthesis production of sucrose:

$$12\,CO_2 + 60\,ATP + 24\,NADPH + 6\,H_2O \rightarrow C_{12}H_{24}O_{12} + 60\,ADP + 60\,P_i + 24\,NADP^+ + 12\,H_2O + 12\,O_2$$

Compare this reaction with that of C_3 photosynthesis:

$$12\,CO_2 + 36\,ATP + 24\,NADPH + 24\,H_2O \rightarrow C_{12}H_{24}O_{12} + 36\,ADP + 36\,P_i + 24\,NADP^+ + 12\,H_2O + 12\,O_2$$

When it's not hot and dry, C_4 plants may be less efficient than C_3 plants because of the extra ATPs they expend to run the C_4 part of their photosynthesis. However, in a hot, dry environment in which photorespiration would otherwise remove much of the carbon fixed in photosynthesis, the additional expense of C_4 photosynthesis is the best evolutionary compromise available. Indeed, many of the plants that grow in hot climates are C_4 plants. However, this added expense is the reason they usually cannot compete effectively with C_3 plants in wet environments, when temperatures are less than 25°C, or in habitats such as dark forest floors.

Only about 5% of the 260,000 known species of plants are C_4 plants. So why all the fuss about C_4 photosynthesis? One answer is that two economically important plants—corn and sorghum—are C_4 plants. Moreover, in hot, dry conditions, C_4 plants produce more biomass than do C_3 plants. Thus, the C_4 machinery represents a potential way for botanists to increase crop yields.

C_4 ancestors, C_4 photosynthesis probably evolved independently several times in separate species. Although most C_4 plants are monocots (especially grasses and sedges), more than 300 are dicots (including some trees and shrubs). Examples of C_4 plants include members of the Poaceae family (corn, sorghum, sugarcane, millet, crabgrass, and Bermuda grass), pigweed (*Amaranthus*), and *Atriplex.*

About 85% of known plant species are C_3 plants, including cereals (e.g., barley, oats, rice, and wheat), peanuts, cotton, sugar beets, tobacco, spinach, soybeans, most trees, and lawn grasses such as rye, fescue, and Kentucky bluegrass.

INQUIRY SUMMARY

C_4 photosynthesis occurs in many plant species and fixes CO_2 into a four-carbon acid in mesophyll cells. This acid then moves to bundle-sheath cells, where it is broken down to CO_2 and a three-carbon molecule. In bundle-sheath cells, the CO_2 is fixed by rubisco and the Calvin cycle. This pumping of CO_2 into bundle-sheath cells eliminates photorespiration, which often makes C_4 plants more efficient than C_3 plants, especially during hot, dry conditions. The evolution of C_4 photosynthesis provides a classic example of natural selection.

There are no known C_4 gymnosperms, bryophytes, or algae. What does this information suggest about the environment where these organisms normally grow?

Crassulacean Acid Metabolism (CAM) Is the Third Main Type of Photosynthesis Known and Probably Evolved in Arid Ecosystems.

The renewed interest in photosynthesis stimulated by Calvin in the 1950s prompted many botanists to begin

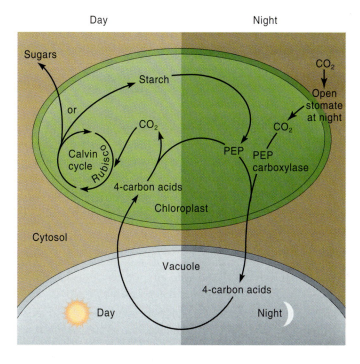

FIGURE 10.16

Photosynthesis in CAM plants. At night, CAM plants fix CO_2 into four-carbon acids, which are stored in the vacuole. During the following day, these acids are decarboxylated and the CO_2 is fixed via the Calvin cycle. CAM photosynthesis conserves water, thus enabling CAM plants to grow in dry environments (e.g., deserts) where most other kinds of plants cannot grow.

studying the biochemistry of a variety of plants. Biologists knew their predecessors in the nineteenth century had shown that some members of the Crassulaceae family, such as jade plant, become acidic at night and progressively more basic during the day. The significance of these diurnal changes was not appreciated until 1958, when botanists reported that these plants open their stomata at night and fix CO_2 into malic acid (a four-carbon compound), which is stored overnight in vacuoles of the large, succulent photosynthetic cells (fig. 10.16); the concentration of malic acid in the vacuole can reach high levels, dropping the pH to as low as 4. The next day, the stomata close, thereby conserving water. Plants exhibiting this diurnal acidity and nocturnal fixation of CO_2 are called **Crassulacean acid metabolism (CAM)** plants because their unusual photosynthesis was originally studied most extensively in the Crassulaceae family, which includes succulent plants such as *Kalanchoë*, Jade, and *Sedum*.

During the day, the acids made the previous night are split, and the liberated CO_2 is fixed via the Calvin cycle in illuminated chloroplasts (fig. 10.16). Unlike C_4 plants in which the initial fixation of CO_2 occurs in a different cell than does the Calvin cycle, CAM plants perform all of their photosynthesis in the same cell. Therefore, while the reactions of C_4 plants are separated spatially (i.e., they occur in different

cells), the reactions of CAM plants are separated both temporally (i.e., at different times of the day) and spatially (i.e., in different parts of the cell—vacuole and chloroplast).

CAM plants open their stomata at night when temperatures drop and relative humidity increases. This enables CAM plants to use water more efficiently than do C_3 plants. Indeed, although CAM plants grow slowly, they need much less water than C_3 or C_4 plants. Other features of CAM plants are summarized in tables 10.1 and 10.2.

CAM is more widespread than C_4 photosynthesis; it occurs in more than 20 families, including monocots, dicots, and primitive plants. Many CAM plants are **succulents** (i.e., fleshy plants having a low surface-to-volume ratio) associated with arid lands and desert ecosystems, including several cacti, but not all succulents are CAM plants. For example, pineapple is a CAM plant that is not succulent, and many halophytes (plants that grow in salty soil) are succulents but are not CAM. Examples of CAM plants include cacti (Cactaceae), orchids (Orchidaceae), jade plant (*Crassula ovata*), maternity plant (*Kalanchoë daigremontiana*), wax plant (*Hoya carnosa*), pineapple (*Ananas comosus*), Spanish moss (*Tillandsia usneoides*), and at least two genera of ferns.

CAM photosynthesis apparently evolved in conditions where water was at a premium. By closing stomata during the heat of the day, water is conserved. This could provide an enormous selective advantage in arid conditions. The ecological trade-off is slow growth because of limited CO_2 supply. About 10% of plant species are CAM plants, but the only ones that are agriculturally significant are pineapple and an *Agave* species used to make tequila and as a source of fibers.

INQUIRY SUMMARY

Crassulacean acid metabolism (CAM) occurs in some plants that live in extremely dry habitats. CAM plants fix CO_2 at night into a four-carbon acid that is stored in the vacuole until daylight when the acid is split and the CO_2 is used in the Calvin cycle. Fixation of CO_2 at night (when the humidity increases) helps CAM plants conserve water.

What might have been the selective pressure for the evolution of CAM in Spanish moss (Tillandsia usneoides)?

What Environmental Factors Affect the Rates of Photosynthesis?

Previously, we discussed factors that control photosynthesis: temperature and light. Other factors that control photosynthesis include the concentration of CO_2, the availability of water, and metabolic sinks:

- ✺ The concentration of CO_2 at which plants show no net fixation of CO_2 is the CO_2 compensation point

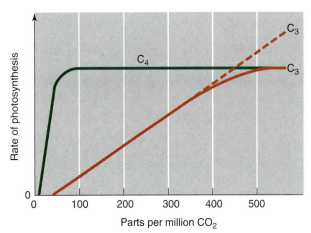

FIGURE 10.17

Effect of CO_2 concentration on the rate of photosynthesis in typical C_3 and C_4 plants. The level of CO_2 where the net rate of photosynthesis is zero (where the rate of photosynthesis equals the rate of total respiration) is the CO_2 compensation point. Maximum photosynthesis occurs at the CO_2 saturation point. The dotted line shows the escalated response of a few C_3 species to increased CO_2 concentrations in air.

(fig. 10.17). In C_3 plants such as wheat and rice, the CO_2 compensation point is usually about 50 ppm (at 25°C and 21% oxygen). For C_4 plants, the CO_2 compensation point has been reported to be as low as 0.4 ppm. Increasing the concentration of CO_2 in the air (e.g., from 0.035%, 350 ppm, to 0.07%, 700 ppm) may double the rate of photosynthesis in some C_3 plants by reducing photorespiration (such increases may not be permanent though, depending on other factors such as soil nutrients). As you'd expect, such "fertilizing" of the air with CO_2 does not significantly affect the rate of photosynthesis in C_4 plants; they often saturate at CO_2 concentrations well below atmospheric levels (fig. 10.17). The benefit of increased concentrations of CO_2 is limited, however, because stomata close and photosynthesis stops at CO_2 concentrations exceeding about 0.15% (see Perspective 10.5, "Atmospheric CO_2 and Photosynthesis").

⑨ The inability of plants to maintain enough water in their leaves causes stomata to close, thereby inhibiting photosynthesis. The lack of water, causing closed stomata, is often the most limiting factor controlling photosynthesis.

⑨ Metabolic sinks, such as roots, fruits, and other growing areas of the plant, are where sucrose from photosynthesis is used or stored (as starch). Photosynthesis decreases when sinks are removed and increases when sinks are created (e.g., by wounds or infections). In general, high rates of photosynthesis correspond with high rates of translocation of sucrose to sinks.

⑨ At night, there is no photosynthesis (note that CAM plants fix CO_2 at night but not via photosynthesis), and

cellular respiration produces CO_2 that is released from the plant (see chapter 11). At daybreak, as light intensity increases, so too does photosynthesis, eventually reaching a point at which the rate of photosynthesis equals the rate of total respiration (cellular respiration plus photorespiration). This light-compensation point varies in different plants (see fig. 10.15b). "Sun plants," such as corn and sugarcane that grow in bright sun, have a high light-compensation point and require much light to saturate photosynthesis. Conversely, "shade plants" that grow in low light (e.g., on forest floors) have a low light-compensation point and saturate at low light intensities. The rates of photosynthesis in these plants also differ significantly; shade plants may fix about one-fourth the carbon of some sun plants.

⑨ The Calvin cycle includes 12 enzymes, at least 5 of which can limit the rate of photosynthesis. These same five enzymes are all activated by light.

⑨ Often, maximum rates of photosynthesis occur near noon when light is brightest; at these times, plants usually fix about eight-times more CO_2 than is released during total respiration (photorespiration plus cellular respiration, described in the next chapter). Averaged over a day, photosynthesis typically fixes about six to ten-times more CO_2 than is released by cellular respiration, but this depends much on species and environment.

INQUIRY SUMMARY

The primary environmental factors that control photosynthesis are temperature, water availability, light, the concentration of CO_2, and metabolic sinks.

Whether C_3, C_4 or CAM, there is no photosynthesis at night, and thus, no dark reactions of photosynthesis? Explain.

What Happens to the Products of the Calvin Cycle?

The products of the Calvin cycle (photosynthate) have many fates (fig. 10.18):

⑨ Triose-P is often stockpiled as dense granules of starch. Small amounts of starch accumulate temporarily in chloroplasts, but most is stored in roots and stems. Later, when needed, starch can be broken down into simple sugars and used for making other compounds in a cell.

⑨ A small amount of the triose-P is used to make amino acids in a chloroplast.

⑨ Much of the triose-P moves from a chloroplast and is used to make sucrose in the cytosol. Sucrose may be stored in a vacuole but is more often translocated, via

10.5 Perspective Perspective

Atmospheric CO₂ and Photosynthesis

The atmospheric concentration of CO_2 has increased from about 270 ppm in 1870 to as much as 365 ppm today. Although this increase in atmospheric CO_2 could significantly enhance the greenhouse effect (see chapter 19) and promote global warming, it also provides the carbon that plants use in photosynthesis to make sugars, starch, and other compounds. How will this increase in atmospheric CO_2 affect plants?

The increased amount of CO_2 in the atmosphere could be a boon to agriculture. Although C_4 photosynthesis doesn't change significantly in response to increases in atmospheric CO_2, C_3 plants such as wheat and rice respond with a burst of photosynthesis. This increase in photosynthesis usually increases yields by as much as 60%. Horticulturists, growing commercial plants in greenhouses, have successfully used CO_2 "fertilizing."

Although increased amounts of atmospheric CO_2 increase photosynthesis of crops, we do not understand how increased amounts of CO_2 affect natural ecosystems. Indeed, in some systems, the initial burst of photosynthesis levels off after a few weeks. Despite this uncertainty, most botanists are convinced that increas-

Change in atmospheric CO_2. Since the establishment of a CO_2 monitoring station at Mauna Loa Observatory in Hawaii, a steady increase in CO_2 levels has been observed. Since 1960, the concentration of CO_2 in the atmosphere has increased by nearly 14%.

ing the amount of CO_2 in an ecosystem will affect plants and, in the process, be a potent force of global change.

Now that you know the differences and similarities of C_3 and C_4 photosynthesis, can *you explain the possible different effects of increased atmospheric CO_2 on each type of photosynthesis?*

What about the possible effects of increased temperature with the global increases of CO_2?

---

the phloem, to growing parts of a plant, such as roots and new leaves and fruits. Here sucrose may be metabolized for energy, stored in a vacuole, or converted to starch.

§ About half of all sucrose is used as fuel for cellular respiration, while the other half is used to make cellulose and other cell-wall components.

§ Plants use some of their photosynthate to make secondary metabolites (see chapters 3 and 11), such as latex. Some of these metabolites resemble crude oil; for example, the grape-size fruits of *Pittosporum undulatum* have been used for centuries as torches. Melvin Calvin and other botanists have studied how to use the hydrocarbons in the latex of some plants (e.g., members of the spurge family) as fuel. Some of the results are promising. For example, extracts from the Amazonian copa iba tree (*Copaifera langsdorfii*) are a diesel fuel that

can be burned in automobile engines with no refining. Similarly, 1 acre of gopher plants (*Euphorbia lathyris*) can produce enough latex to make 12 barrels of oil. Plantations of *Euphorbia* would be economically feasible when oil prices reach about $31 per barrel.

§ We use photosynthate of corn to make almost 90% of the ethanol used in the United States, some of which is mixed with gasoline (in a 1:9 ratio) to make gasohol. Most cars made in Brazil run on 100% ethanol, and in the United States, gasohol accounts for about 10% of gas sales. One bushel of corn will make about 3 gallons of ethanol.

§ Every year, photosynthesis produces about 1.5 billion tons of grain, the staple of the world's diet.

Each year, photosynthesis fixes about 10% of the carbon in the atmosphere. Humans, either directly or

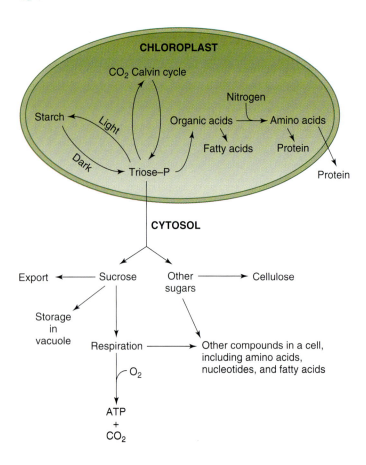

FIGURE 10.18

The products of photosynthesis provide the carbon skeletons for the assembly of all organic molecules in a plant.

indirectly, use about 40% of the net products of photosynthesis from terrestrial ecosystems. These uses affect all aspects of our lives: textiles, food (e.g., cereals such as oats and wheat provide half of the world's food), fuels such as coal and oil (formed millions of years ago), wood (the United States uses more than 3 billion cubic feet of wood per year to make paper), drugs (e.g., caffeine, quinine, cocaine, and nicotine), waxes, oils (e.g., rose and jasmine), rubber, spices (e.g., pepper, cinnamon, and peppermint), and even the perfumes that Cleopatra used to entice Marc Antony and Julius Caesar. This page, the oxygen you breathe, and all of the atoms in your body were once parts of plants.

No process can match the importance or magnitude of photosynthesis. It sustains virtually all life on earth. Without photosynthesis, all other biological reactions would be irrelevant.

a spectrum of colors. Red and blue light are most effective for photosynthesis.

Light is absorbed by pigments. Chlorophylls are the primary pigment for photosynthesis and occur in all photosynthetic organisms. Accessory pigments such as carotenoids absorb light that chlorophyll cannot absorb, thereby extending the range of light used for photosynthesis.

Photosynthesis in eukaryotes occurs only in chloroplasts, which consist of a gelatinous matrix, called the stroma, and stacks of membranes (thylakoids) called grana.

The light reactions of photosynthesis convert light energy into chemical energy. In plants, the light reactions split water, produce ATP, and NADPH, and release O_2. The ATP and NADPH made in the light reactions are used in the Calvin cycle, which begins with the fixation of CO_2 by an enzyme called ribulose bisphosphate carboxylase/oxygenase (rubisco). The Calvin cycle produces triose-P as the starting point for several metabolic pathways. Plants that use only the Calvin cycle are called C_3 plants.

Photorespiration occurs in C_3 plants and worsens when the internal concentration of CO_2 becomes low, such as when stomata close during drought. During photorespiration, rubisco fixes oxygen, and CO_2 is later released. Consequently, photorespiration undoes photosynthesis: as much as half of the carbon fixed in the Calvin cycle is released by photorespiration during hot, dry days.

C_4 photosynthesis occurs in many plant species and involves fixation of CO_2 into a four-carbon acid in mesophyll cells. The acid then moves to bundle-sheath cells, where it is broken down to CO_2 and a three-carbon molecule. The CO_2 is subsequently fixed by the Calvin cycle in bundle-sheath cells. This pumping of CO_2 concentrates CO_2 in bundle-sheath cells, thereby eliminating photorespiration. This, in turn, makes C_4 plants more efficient during hot, dry conditions.

Crassulacean acid metabolism (CAM) occurs in many plants that live in dry habitats. CAM plants fix CO_2 at night into a four-carbon acid that is stored in the vacuole until daylight. The acid is then decarboxylated, and the CO_2 is used in the Calvin cycle. Fixation of CO_2 at night helps CAM plants conserve water.

The three photosynthetic mechanisms known in nature are C_3, C_4, and CAM. Accordingly, plants with these mechanism are known as C_3, C_4, or CAM plants. The Calvin cycle operates in each type of plant.

The primary factors that control photosynthesis are temperature, water availability, light, the concentration of CO_2, and metabolic sinks. The carbon fixed during photosynthesis is used to make all the other organic molecules of a plant.

SUMMARY

Photosynthesis is the light-driven fixation of CO_2 to carbohydrates. Virtually all life depends on light. Light occurs in

 *Writing to Learn Botany*

How does photosynthesis "feed the world"? Is photosynthesis the only reason plants are considered producers in an

ecosystem? How might low CO_2 levels in the ancient atmosphere have led to the evolution of C_4 photosynthesis?

THOUGHT QUESTIONS

1. What is the most abundant protein in the world? Explain.

2. What is the significance of using ATP and NADPH in photosynthesis?

3. Geldikkop is a rare disease in sheep in which chlorophyll that gets into the blood is energized by light and causes lesions. Why would chlorophyll in blood cause such a problem?

4. Zooplankton are transparent and eat phytoplankton. Most zooplankton avoid bright light by migrating down during the day and up during the night. They also contain large amounts of carotenoids. What is the ecological and evolutionary significance of this?

5. The chloroplast genome is the same in the mesophyll and bundle-sheath cells of C_4 plants. How, then, can these chloroplasts be so different in their enzyme content?

6. Explain how the adaptations of CAM and C_4 plants enhance photosynthesis in hot, dry environments.

7. Are the products of the light reactions of photosynthesis the same in plants and bacteria? Explain.

8. How does visible light differ from ultraviolet light? From infrared light?

9. How is an absorption spectrum different from an action spectrum?

10. How does electron flow drive ATP synthesis?

11. What is photorespiration? Why is it significant?

12. Why don't C_4 plants dominate the landscape?

13. A colleague brings you a plant that has just been discovered in the tropics and asks you what type of photosynthesis takes place in the plant. You have only a microscope and slides, razorblade, and pH meter with glassware at your disposal but can still answer the question. How?

14. What are some of the important things that we do not know about photosynthesis? What experiments would you do to learn more about these mysteries?

SUGGESTED READINGS

Articles

Culotta, E. 1995. Will plants profit from high CO_2? *Science* 268:654–56.

Maroco, J., M. Ku, et al. 1998. Oxygen requirement and inhibition of C_4 photosynthesis. *Plant Physiology* 116:823–32.

Pereto, J. 1996. The Calvin cycle: A metabolic pathway still misunderstood. *Biochemical Education* 24:147–48.

Sage, R. 1995. Was low atmospheric CO_2 during the Pleistocene a limiting factor for the origin of agriculture? *Global Change Biology* 1:93–106.

Storey, R. D. 1989. Textbook errors and misconceptions in biology: Photosynthesis. *American Biology Teacher* 51:271–74.

Williams, N. 1996. Mutant alga blurs classic picture of photosynthesis. *Science* 273:310.

Books

Hopkins, W. 1999. *Introduction to Plant Physiology*, 2d ed. New York: John Wiley and Sons.

Lambers, H., F. Chapin, and T. Pons. 1998. *Plant Physiological Ecology*, 2d ed. New York: Springer-Verlag.

Lawlor, D. W. 1993. *Photosynthesis: Molecular, Physiological, and Environmental Processes*, 2d ed. New York: John Wiley and Sons.

Taiz, L., and E. Zeiger. 1998. *Plant Physiology*, 2d ed. Sunderland, MA: Sinauer.

ON THE INTERNET

Visit the textbook's accompanying web site at http://www.mhhe.com/botany to find live Internet links for each of the topics listed below:

Photosynthetic pigments

Chloroplast structure

Melvin Calvin

Photorespiration

C_3 verses C_4 efficiency

Crassulacean acid metabolism

Succulent plants

Plants and carbon dioxide uptake

Photosystems

The light spectrum and light absorption by pigments

The starch-rich endosperm of this corn kernel has been stained blue-black with iodine. Starch is fuel for the emergent seedling. How might the accumulation of starch in the corn grain (kernel) provide a selective advantage?

Respiration

LEARNING ONLINE

Check out the hyperlink www.fgi.net/~corpalt/science/glycol.html in chapter 11 of The Online Learning Center at www.mhhe.com/botany to see how plants convert food to energy. This, and other helpful information such as articles about botany from countries around the world, is available to you when you visit the OLC.

Corn is one of the most widely planted crop in the United States. To people who have driven across the midwestern United States (also called the Corn Belt) during the summer, this is no surprise, for in many areas there is little except mile after mile of corn. This crop covers a combined area about the size of Arizona and produces about 8 billion bushels of corn per year. According to one mathematically minded urbanite, that's enough corn to bury all of Manhattan Island about 5 meters deep.

Corn is the most efficient of all major grain crops; that is, corn is the champion at harvesting sunlight and using the energy to convert CO_2 to sugars and then storing the fixed carbon and energy as starch and oil in grain. We use these grains for a variety of purposes. For example, we feed about half of our corn to livestock; another 25% of the crop is exported. About 10% of the crop—optimally within 5 minutes of picking—appears on our dinner tables. The rest of the crop passes into an enormous number of pe-

ripheral products, including corn syrup, cornmeal, corn oil, cardboard, crayons, firecrackers, aspirin, wallpaper, gasohol, pancake mix, shoe polish, ketchup, chewing gum, marshmallows, soap, and corn whiskey.

In the previous chapter, you learned about the most important set of chemical reactions on earth: photosynthesis. These are the reactions that convert sunlight to chemical energy, thus providing fuel and fixed carbon for virtually all organisms. In this chapter, you learn about energy conversions and about energy use by plants. Also, you will learn about carbon flow in plants following photosynthesis and how organisms—plants included—use the energy and carbon trapped in sugars and other compounds to do work and make new compounds. In the process, you'll come to appreciate one of the themes of life: all organisms, in addition to requiring energy and carbon to stay alive, exchange energy and carbon with their environment.

The Two Primary Energy and Carbon Transformation Systems in Plants Are Photosynthesis and Respiration.

The most important energy and carbon transformation systems in plants are photosynthesis and respiration. Photosynthesis was considered in the previous chapter; respiration is discussed in this chapter. We use the term *respiration* interchangeably with the technically more correct term *cellular respiration,* which will be defined later in the chapter. Recall the summary equation for the energy-requiring, "uphill" process of photosynthesis:

$$CO_2 + H_2O + \text{light energy} \xrightarrow[chlorophyll]{enzymes} \text{sugars} + O_2$$

During photosynthesis, plants use light energy to fix CO_2 and make sugars that provide the building blocks to make proteins, fats and nucleic acids, fuel the activities of plants and all other organisms. Of course, O_2 is released during photosynthesis. This O_2 is used in aerobic respiration and is therefore important to most organisms in an ecosystem.

The energy-releasing, "downhill" stage of metabolism is **cellular respiration,** a series of chemical reactions that sustain life in all organisms. During respiration, energy-rich molecules such as sugars are broken down, or oxidized, to carbon dioxide and water. A summary equation for respiration is:

$$\text{sugars} + O_2 \xrightarrow{enzymes} CO_2 + H_2O + \text{ATP} + \text{heat}$$

Compare the summary equations for photosynthesis and respiration. Photosynthesis removes hydrogen from water and adds it to carbon; thus, carbon is reduced, or fixed, during photosynthesis, an anabolic pathway. Cellular respiration removes hydrogen from carbon and adds it to oxygen, forming water; thus, carbon is oxidized during respiration, a catabolic pathway. Although one summary equation is essentially the reverse of the other, they proceed by very different mechanisms. In addition, photosynthesis occurs in chloroplasts while respiration begins in the cytosol and is completed in mitochondria.

Now examine figure 11.1, which shows how photosynthesis and respiration are integrated in nature. This diagram illustrates an important concept about energy: energy flows through a system and is ultimately converted to heat (see chapter 3). For example, sunlight is the energy input that sustains almost all of the life on earth (an exception is the thermal vents deep in the ocean floor where geothermal energy powers the dark ecosystem). This is possible only because plants use photosynthesis to convert sunlight into chemical energy and fix CO_2 into sugars; thus, plants are producers in an ecosystem. Animals stay alive by eating or consuming plants (or by eating other animals that ate plants) and using the stored energy to power their activities. These animals are then eaten by other animals to fuel their activities. Ultimately, the organisms die and are decomposed by bacteria and fungi. In the process, the energy stored temporarily by organisms in this so-called food chain is released as heat. Organisms may temporarily store various amounts of the energy, but the net effect is that it flows through the system and is ultimately transformed to heat and lost from the system.

The Lore of Plants

On average, vegetables and fruits make up about 10% to 20% of our intake of calories, 7% to 20% of our intake of protein, most of our fiber, often over 80% of many vitamins and minerals, an unimpressive 1% of our intake of fat, and no cholesterol. That's why fruits and vegetables are high on the list when it comes to a good diet—little fat, no cholesterol, and lots of important nutrients. How do grains and cereals figure into our diet?

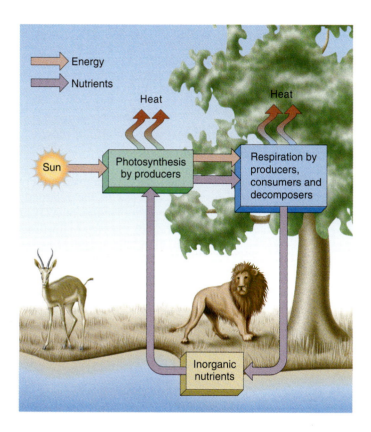

FIGURE 11.1

Photosynthesis and respiration, the major transformations of energy by organisms. Photosynthesis provides food for virtually all organisms on earth. Organisms use cellular respiration to extract energy from food. Although nutrients cycle in an ecosystem, energy flows through an ecosystem. All energy is eventually converted to heat.

Chapter 2 provides more information on plant ecology and ecosystems.

According to the second law of thermodynamics (chapter 3), all of the energy transformations at every step in the food chain are less than 100% efficient. Indeed, there is a 90% loss of usable energy at each stage. This has tremendous implications. For example, consider the corn grain shown at the beginning of this unit. Most of the energy striking the plants' leaves is reflected or converted to heat; only a small amount of the energy is trapped by photosynthesis in sugars. When these sugars are converted to starch, more of the energy is lost. When this starch is fed to a cow, only about 10% of the usable energy is stored by the cow; the rest is lost as heat. By the time we eat a steak made from the cow, another 90% of the usable energy has been lost. Thus, the amount of usable energy in the steak is only about 1% of that contained in the starch of the corn:

| 100 units of energy in a producer (corn) | → | 10 units of energy in an herbivore (cow) | → | 1 unit of energy in a carnivore (human) |

These energy transformations, as predicted by the second law of thermodynamics, are inefficient because much of the energy is converted to heat. By inserting an extra energy transformation between the corn plant and ourselves (i.e., the cow), we lose even more of the usable energy.

Look again at figure 11.1. Notice that although energy moves through the system in a one-way flow (i.e., toward heat), nutrients are cycled. That is, photosynthesis produces sugars and oxygen from carbon dioxide, light, and water. The sugars and oxygen are converted back to carbon dioxide, water and heat during cellular respiration.

INQUIRY SUMMARY

Photosynthesis and respiration are the major energy transformations in organisms. Photosynthesis uses light energy to reduce CO_2 to sugars, whereas respiration oxidizes sugars, producing ATP in the process. Together, photosynthesis and respiration form a system by which energy flows through an ecosystem and is ultimately converted to heat.

What happens to this heat?

Plants Retrieve Energy and Carbon via Respiration of Sucrose and Starch.

Most textbook discussions of respiration focus on glucose, probably because the metabolism of glucose is so similar in all organisms. However, free (unbound) glucose is usually not abundant in plants or their cells. Thus, in plants, cellular respiration begins with the conversion of energy and

The Lore of Plants

The amount of CO_2 released by running a car engine is about 100 kilograms per 40-liter tank of gasoline. If 1 mole of glucose (180 grams) yields 6 moles of CO_2 (264 grams) during aerobic respiration, then burning a tank of gasoline is equivalent to the respiration of about 68 kilograms of glucose, or the weight of an average adult human. This is roughly the amount of glucose, made from sucrose and starch, that is respired by 10,000 young sunflower plants during a warm summer night.

carbon storage compounds—such as starch—into glucose, which can then be quickly moved into cellular respiration. Such storage compounds vary from one organism to another or, in plants, even from one organ to another. For example, humans and other vertebrates store the polysaccharide glycogen. Potatoes, beans, corn, rice, yams, and bananas store the polysaccharide starch (which, like glycogen, is a polymer of glucose); a few plants, such as Jerusalem artichokes and onions, store polymers of fructose. Sucrose, from photosynthesis, is the common starter molecule for respiration in most parts of a plant, but, as suggested, respiration may begin with starch in nongreen organs such as roots and stems. In other plants, or, more specifically, in other plant parts, respiration may utilize fats, organic acids, or proteins. Nevertheless, the most well-known and probably most common sources of glucose in plants are sucrose and starch.

Sucrose, abundant in sugarcane stems and sugar beet roots, is common in plants and is commercially important because humans consume so much of it as table sugar (during an average lifetime, people in the United States consume more than 4,600 pounds of table sugar; that is over 8 million kilocalories). Besides being a common energy and carbon source in leaf cells, sucrose moves most readily through conducting cells from leaves to growing tissues, where respiration is active. The breakdown of sucrose, via hydrolysis, to its component monomers, glucose and fructose, is shown by the following equation:

$$\text{sucrose} + H_2O \xrightarrow{\text{sucrase}} \text{glucose} + \text{fructose}$$

Enzymes that catalyze this reaction are called *sucrases;* they occur in the cytosol, in the central vacuole, and sometimes in cell walls. Hydrolysis by sucrases frees glucose and fructose for respiration. In us, as in plants, fructose can be quickly converted to glucose.

Starch is a polymer of hundreds of glucose monomers (chapter 3). Because of the complexity of starch, its breakdown requires several steps and several enzymes for the complete retrieval of glucose.

$$\text{starch} \xrightarrow{\text{enzymes}} \text{glucose}$$

Recall that most of a plant's starch is made in storage organs, such as stems, roots, and seeds, from sugars originally produced in leaves.

The energy exchange and carbon economy of most organisms depend on aerobic respiration.

Cellular respiration converts the relatively large amount of energy stored in a molecule of glucose to the relatively small amounts of energy in several molecules of ATP, as discussed in chapter 3. This is one of the two main functions of respiration—ATP production. The overall equation for the cellular respiration of glucose is as follows:

$$\text{glucose} + \text{oxygen} \xrightarrow[\text{enzymes}]{\text{several steps}} \text{carbon dioxide} + \text{water} + \text{ATP}$$

Or using the chemical formula:

$$C_6H_{12}O_6 + O_2 \xrightarrow[\text{enzymes}]{\text{several steps}} CO_2 + H_2O + \text{ATP}$$

In our example, glucose (from sucrose or starch), on the left side of the equation, represents the chemical energy derived from photosynthesis. On the right side of this equation, some of the energy ends up in ATP and some in CO_2 and H_2O, but most of the energy is lost as heat (which we have not shown to keep the example simple). Molecular oxygen (O_2) is shown because the equation is for **aerobic respiration** (i.e., with oxygen)—the kind of respiration that occurs in plants and animals in the presence of ample O_2. The overall efficiency of the process, in terms of converting the energy stored in glucose into ATP for work of the cell, is not high—perhaps about 35 to 50% efficiency. Nonetheless, the total energy in the reactants on the left side is equal to the total energy of the products on the right side, including heat.

Remember, as you read further, that cellular respiration occurs in plants, fungi, protists bacteria, and animals, including you. To include respiration in all known organisms, including bacteria, we can now define *cellular respiration* as the chemically driven flow of electrons through a membrane coupled to ATP synthesis. If the acceptor of electrons at the end of flow is O_2, the process is called aerobic respiration (or often, just respiration); if some other molecule accepts the electrons, it is called **anaerobic respiration.** We return to these definitions throughout the chapter.

Cellular respiration forms ATP and carbon skeletons used in biosynthesis.

In addition to ATP production, aerobic respiration also supplies the basic building blocks for the manufacture of the other molecules in the plant. Several intermediate molecules of respiration serve as carbon skeletons for biosynthesis of amino acids, fatty acids, and nucleic acids (see chapter 3 for information on these and other molecules that originate from respiration). Thus, cellular respiration, not only produces ATP but also forms carbon skeletons for biosynthesis (which is powered by ATP): these are the two primary roles of cellular respiration.

As you read the more detailed description of the events in respiration, remember the same chemical reactions occur in you, and, in fact, they are going on right now in your cells or you would be a goner.

INQUIRY SUMMARY

In cellular respiration, chemical energy from sugars, such as glucose, is used to make ATP. Glucose can be formed from the breakdown of sucrose or starch. In plants, aerobic respiration (with oxygen) releases CO_2 and requires O_2. Cellular respiration has two roles: synthesis of ATP for energy requirements and formation of carbon skeletons for subsequent biosynthesis of complex molecules.

Plants make all their organic molecules from the carbon in CO_2. How do animals make their organic molecules?

Cellular Respiration Includes Glycolysis, the Krebs Cycle, and the Electron Transport Chain.

Cellular respiration includes three main stages: **glycolysis, Krebs cycle,** and the **electron transport chain** (fig. 11.2*a*). These are discussed in detail later but note that sugars are consumed and CO_2 is released during the process as ATP is formed. Glycolysis is a pathway; the Krebs cycle is a cycle. For convenience a series of chemical reactions in a cell is called a pathway or cycle depending on the nature of the steps involved. Pathways "end" in a product or products; cycles also form products but return molecules to the first events in the series. Each step of a pathway or cycle—that is, each chemical

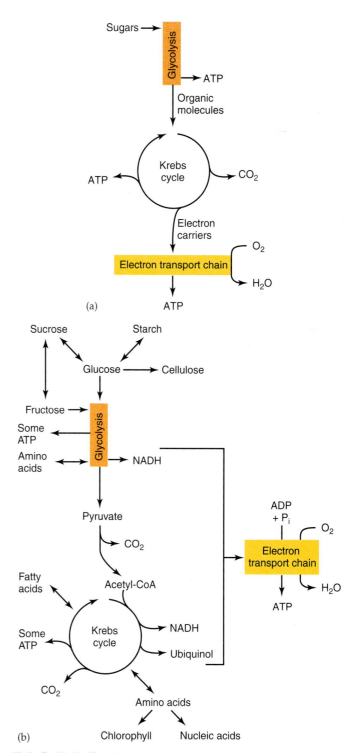

FIGURE 11.2

(a) Simplified summary of cellular respiration—glycolysis, Krebs cycle, and electron transport chain. (b) **Important steps in aerobic, cellular respiration of plants.** Sucrose or starch is converted to glucose, one starting point for aerobic respiration in plants. Upon complete aerobic respiration of a molecule of glucose, 6 molecules of CO_2 are released and up to 38 molecules of ATP can be formed. Oxygen (O_2) is reduced to water in the final step. Glycolysis occurs in the cytosol; the Krebs cycle and the electron transport chain occur in a mitochondrion. ←→ indicates molecules flow both directions. Cellular respiration in the presence of oxygen is also called aerobic respiration.

11.1 *Perspective*Perspective

The Potential Energy of Glucose

Glucose contains 686 kilocalories per mole. This means that with the addition of some energy (i.e., energy of activation) to get it going, 1 mole of glucose will yield 686,000 calories (686 kilocalories) when it is oxidized to CO_2 and H_2O. However, cells recover only a small amount of this energy during glycolysis; the rest is lost as heat. Because the energy is released in small increments during each step of respiration, the excess heat does not damage the cell. A mole refers to the amount of a substance.

reaction in the series—is catalyzed by a specific enzyme (see Perspective 11.1, "The Potential Energy of Glucose").

Respiration can be considered a series of "baby steps" that begins with a sugar and progressively releases small amounts of energy (transferred to ATP) along the way as the sugar is broken down (oxidized) and CO_2 is released. If all the energy of glucose, for example, was released at once, it would be very inefficient and even harmful. If you released all the energy in a tank of gasoline at once, your automobile would certainly suffer during the explosion. It is better to release that energy in a series of baby steps—the controlled explosions in the cylinders of your engine. The same is true for organisms. Remember, each "baby step"— that is, each chemical reaction—is catalyzed by a specific enzyme that has undergone natural selection during the evolution of the metabolic pathway or cycle.

As mentioned previously, cellular respiration involves the breakdown of glucose (or another sugar) with the release of CO_2. When this happens, electrons are released from the breakdown. These electrons flow to O_2, and ATP is produced as O_2 is used during the formation of water. Aerobic respiration is cellular respiration in the presence of oxygen (O_2). Figure 11.2*b* shows this in more detail than seen previously. Note that important organic molecules, such as amino acids and chlorophyll, are made from products of respiration.

Glycolysis is the pathway that starts cellular respiration of sugar in the cytosol (fig. 11.3). In glycolysis, glucose, a 6-carbon sugar, is rearranged, then split in half during the pathway. When glucose is split, the 3-carbon acid pyruvate is formed as the product (pyruvate is the ionized or salt form of pyruvic acid; pyruvate predominates in cells). Pyruvate then can enter a mitochondrion where it fuels the Krebs cycle (see fig. 11.2*b*). Both glycolysis and the Krebs cycle form some NADH; the Krebs cycle also forms ubiquinol, an electron carrier. Both processes can produce a small amount of ATP directly, while neither requires O_2 directly. The release of CO_2 occurs via reactions in the mitochondria, not during glycolysis in the cytosol.

Most of the ATP from cellular respiration is made by the electron transport chain in mitochondria (singular is mitochondrion) (fig. 11.2*b*). Here, energy is released as electrons flow through the membrane to O_2. The energy is used to drive the synthesis of ATP from ADP plus P_i, like electricity

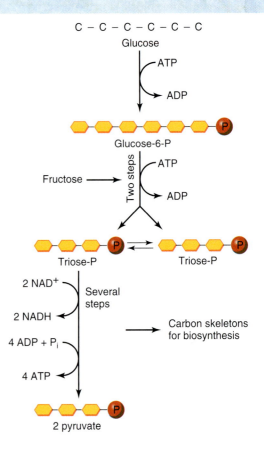

FIGURE 11.3

Overview of glycolysis. Glycolysis starts with a molecule of glucose, a 6-carbon sugar, and ends with two molecules of pyruvate, a 3-carbon acid. Two ATPs are used and four are formed in the process, along with two molecules of the electron transport molecule, NADH. Along with energy conversions, molecules formed during glycolysis can be drained from the pathway and used to make other molecules such as amino acids. is phosphate, attached to the carbon skeleton.

drives a motor. The electrons come from NADH formed during glycolysis and the Krebs cycle and also from ubiquinol, the related electron transporter from the Krebs cycle. Electrons (e^-) and protons (H^+) come from hydrogen ($H_2 \rightarrow 2 H^+ + 2 e^-$) released from NADH (NADH + $H^+ \rightarrow NAD^+ + H_2$) or ubiquinol (ubiquinol $\rightarrow$ ubiquinone + H_2).

Electrons from NADH and ubiquinol are passed through a chain of electron carrier molecules in the mitochondrial membrane to O_2, the final electron acceptor. This synthesis of ATP, via the electron transport chain, is called **oxidative phosphorylation** (compare to photophosphorylation, described in chapter 10). Thus, mitochondrial membranes play an important role in coupling energy from the oxidation of glucose to the phosphorylation of ADP to ATP:

$$ADP + P_i \rightarrow ATP + H_2O$$

Glycolysis splits glucose, producing some ATP.

During glycolysis, each 6-carbon glucose molecule is split into two, 3-carbon pyruvate molecules. The entire process requires 10 steps, all of which occur in the cytosol (fig. 11.3). The first step adds P_i (inorganic phosphate from phosphoric acid) to glucose by using one ATP. This glucose plus phosphate molecule is rearranged, and a second ATP adds another phosphate before the glucose molecule is split in half. After glucose is split, each of the two resulting 3-carbon sugars has one phosphate attached. These two triose-phosphate molecules represent the halfway point of glycolysis. By this point, energy (as two molecules of ATP) has been used, but no ATP has been made.

The ATP-producing steps of glycolysis occur in the second half of the pathway. First, the reduction of (addition of electrons to) NAD^+ provides two molecules of NADH. Then, four molecules of ATP are produced from ADP, and the final product of glycolysis, pyruvate, remains. Thus, the products of glycolysis are two molecules of NADH, two molecules of ATP (net; four are formed, but two are consumed along the pathway starting with glucose), and two molecules of pyruvate. Along the way, different steps form carbon skeletons that are drained from the pathway and used in the cell to make other, more complex, molecules. The functions of glycolysis are to produce ATP, NADH and pyruvate and also the carbon skeletons for biosynthesis (see figs. 11.2 and 11.3).

The Krebs cycle is fueled by pyruvate and forms some ATP.

The pyruvate from glycolysis can enter a mitochondrion and fuel the Krebs cycle (also known as the **citric acid cycle** because the organic acid citrate is involved in the cycle as an intermediate). However, the pyruvate that is transported into the mitochondrion is not used in the Krebs cycle directly. Instead, pyruvate first loses a molecule of carbon as CO_2, and some NADH is formed via a chemical reaction (fig. 11.4). The remaining 2-carbon group, acetyl CoA (CoA is a carrier; the acetyl group, with two carbon atoms, is the molecule of interest to us) combines with a 4-carbon molecule to maintain the cycle. Note that when CO_2 is released from pyruvate, NAD^+ is reduced to NADH as protons (H^+) and electrons (e^-) are combined with NAD^+.

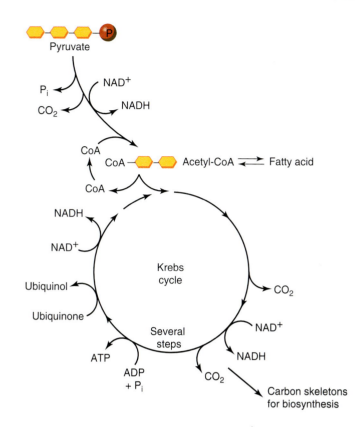

FIGURE 11.4

Overview of the Krebs cycle. In a mitochondrion, pyruvate from glycolysis undergoes a reaction that forms acetyl-CoA (CoA is just a carrier; the acetyl is a two-carbon acid group), which enters the Krebs cycle. For each pyruvate molecule that enters the mitochondria, one ATP, three NADH, and one ubiquinol can be made along with three molecules of CO_2 released (remember that twice this number can be made per glucose because each glucose splits into two pyruvate molecules). Also, carbon skeletons can be drained from the cycle and used for biosynthesis of other complex molecules.

Glycolysis and the Krebs cycle are linked by the conversion of pyruvate to acetyl CoA in mitochondria. Pyruvate is the product of glycolysis that enters a mitochondrion, but it is the acetyl group of acetyl CoA that actually enters the Krebs cycle. The 2-carbon acetyl group is the ionized form of acetic acid, or vinegar.

The Krebs cycle, named in honor of the British scientist Sir Hans Krebs (see Perspective 11.3, "Botany and Politics"), is a cycle because, unlike a pathway, the last step regenerates the starter chemicals for the first step. In all, there are eight enzyme-catalyzed steps of the Krebs cycle that occur in the mitochondria. The combined result of these steps is the production of ATP, NADH, ubiquinol (ubiquinol is the mobile product of the Krebs cycle, not $FADH_2$ as shown in many books), and carbon skeletons for biosynthesis. The production of these molecules is the function of the Krebs cycle (see figs. 11.2 and 11.4). Although CO_2 is released from the reactions in the mitochondrion, no O_2 is involved yet in

11.2 Perspective Perspective

Models and Respiration

Remember that the models shown in figures 11.3, 11.4, and 11.5 are just that, models. Just as a map is not the territory, models are not the real molecules or enzymes. We use models to represent the research that leads to our understanding of respiration and of the chemical reactions and enzymes that make up the pathways and cycles. We could present the original data, but it is easier to visualize the stages of respiration using models. Remember there are actually thousands of copies of each enzyme and millions of copies of each intermediate compound that participate in each chemical reaction of respiration within a cell. Also remember the entire oxidation of glucose to CO_2 might take less than a second; respiration is fast! One last point about models: most textbooks show the Krebs cycle as a cycle, like a Ferris wheel. However, the cycle does not actually "turn" but is a series of chemical reactions catalyzed by enzymes. The Krebs cycle does not turn clockwise any more than glycolysis runs vertical rather than horizontal. Textbooks simply show it turning to help students understand the principle.

---

respiration. Most of the energy derived from the steps of the Krebs cycle is temporarily contained in the high-energy electrons of NADH and ubiquinol. The energy in these molecules which move within a mitochondrion—they are mobile—is harvested via oxidative phosphorylation in the third phase of respiration, the electron transport chain.

ATP is formed as electrons flow to oxygen.

When compared to glycolysis and the Krebs cycle, the electron transport chain is a bonanza for ATP synthesis. Energy for ATP synthesis via the chain comes from NADH and ubiquinol, which are produced by the first two stages of respiration. The electrons of NADH and ubiquinol cannot be used directly for ATP synthesis; instead, they start a series of oxidation-reduction reactions that move electrons through several carriers, which are proteins in or on the mitochondrial membrane. This movement of electrons is called electron flow. The sequence of electron carriers is known, appropriately enough, as the **electron transport chain.** Energy from electron flow ultimately drives the synthesis of ATP—like electricity can set a lightbulb aglow. Oxidative phosphorylation refers to the combination of the oxidation-reduction reactions of electron transport that enable a cell to use the energy in NADH and ubiquinol to convert ADP plus P_i to ATP. This conversion is called phosphorylation because phosphate (P_i) is added to the molecule, in this case, ADP, producing ATP.

The electron transport chain (fig. 11.5) is like a series of tiny, successively stronger magnets; that is, each has a higher potential than its predecessor. This means that each component of the chain pulls electrons away from a weaker neighbor and gives them up to a stronger one. Thus, the strongest electron acceptor in the chain is the terminal acceptor, molecular oxygen (O_2). You can also think of the elec-tron transport chain as a bucket brigade, where electrons are the buckets and the brigade of firefighters are the carriers, passing buckets (electrons) along to the end—oxygen.

There are at least nine electron carriers in the electron transport chain of the membrane. The electron carriers are proteins and complexes of proteins and pigments. Some of these carriers also accept protons and release them into the intermembrane space. Although scientists are uncertain how these components work together during electron and proton transport, figure 10.14 shows a widely accepted model for the sequence of steps in the chain.

According to the model, as electrons continue to pass down the chain between carriers, the energy from this electron flow drives the transport of protons from one side of the inner mitochondrial membrane to the other. The electrons flow to cytochrome a/a_3 (cytochrome oxidase or the terminal oxidase), which, in turn, passes the electrons to O_2, the terminal electron acceptor during aerobic respiration. The O_2 combines with electrons from the chain and with protons from the mitochondrial matrix, producing water:

$$O_2 + 4\,e^- + 4\,H^+ \rightarrow 2\,H_2O$$

Note that the transport of protons from one side of the inner mitochondrial membrane to the other requires energy, which is provided by the movement of electrons through the electron transport chain (fig. 11.5).

ATP synthesis is driven by electron flow that generates a proton gradient.

How do the protons that are pumped by the electron transport chain provide energy for ATP synthesis? British scientist and Nobel Prize winner Peter Mitchell suspected that the membrane must have a role in ATP synthesis and

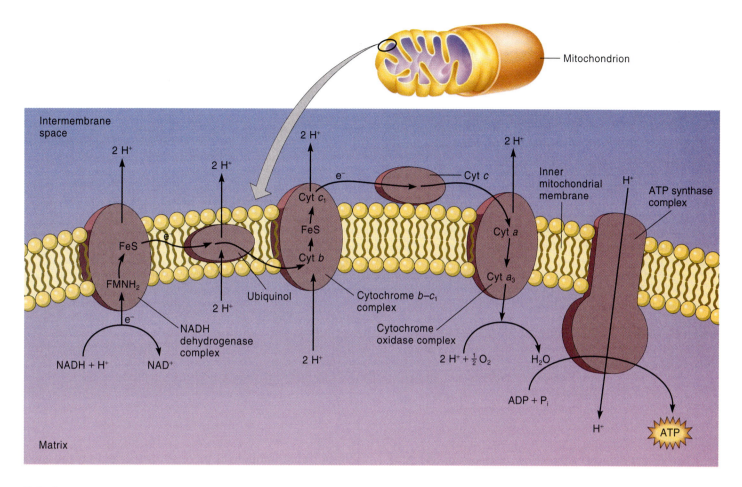

FIGURE 11.5

Overview of the electron transport chain. Moving from the Krebs cycle, NADH and ubiquinol (and from glycolysis for NADH) donate electrons to the chain. As electrons pass from carrier to carrier in or on the membrane, the energy of electron flow is used to concentrate (pump) protons into the intermembrane space. As the electrons flow to the final acceptor, O_2, the protons flow back through the ATP synthase complex and, in doing so, the synthesis of ATP—oxidative phosphorylation. In theory, each NADH and each ubiquinol can drive the synthesis of three and two ATP molecules, respectively.

developed a hypothesis to explain ATP synthesis in mitochondria (fig. 11.5). As explained, carriers of the electron transport chain are embedded in or on the inner mitochondrial membrane. The flow of electrons through this series of carriers is used to actively pump protons across the membrane into the intermembrane space. Because the membrane is not very permeable to protons, relatively few protons leak back across the membrane. Continued electron transport pumps more and more protons into the space between the membranes of the mitochondria. This creates a gradient of protons across the membrane, with a much higher concentration of protons accumulating in the intermembrane space. According to Mitchell's model, that gradient of protons drives the synthesis of ATP. The high concentration of protons functions like a battery to do work, including making ATP.

Other scientists used electron microscopy to show that ATP synthase, the enzyme that makes ATP, protrudes into the matrix. Thousands of copies of the ATP synthase occur

throughout the membrane, making the mitochondrial membrane a powerhouse for making ATP. The proton gradient is unstable, with the ever-increasing concentration of protons in the intermembrane space "pushing" out on the membrane. The pressure reaches a point where it must be relieved, like filling a water balloon to the maximum. Pressure from the gradient is relieved as protons escape through ATP synthase. The energy from the flow of escaping protons is used to make ATP by combining ADP and P_i (fig. 11.5). This is like water from the bottom of a reservoir rushing out through a drainpipe in the dam and turning a turbine that makes electricity. The water in the reservoir represents the high concentration of protons on one side of the membrane, the dam represents the mitochondrial membrane, the rush of water is the proton flow, the turbine is the ATP synthase, and the electricity is ATP. In fact, ATP synthase can be considered a tiny, molecular turbine that spins as protons rush through and uses the energy from the spin to make ATP from ADP plus P_i.

11.3 *Perspective* Perspective

Botany and Politics

Science is a search for truth. Although truth is not affected by politics, the business of science often is affected. Consider, for example, Otto Heinrich Warburg, a German biochemist. Warburg began his scientific career in the early 1900s and in 1924 announced his discovery of "iron oxygenase," the oxygen carrier that we now call cytochrome oxidase. In 1931, Warburg was awarded a Nobel Prize for this work. However, less than two years later, Hitler was appointed chancellor of Nazi Germany. Although Warburg was half Jewish, he was unharmed by the Nazis because he studied cancer. Indeed, when Nazis removed Warburg from his position in 1941, he was personally ordered back to work on cancer by Hitler, who feared dying of cancer. Warburg stayed in Germany because he thought his work would be destroyed if he left. Although Warburg's pact with the Nazis incensed his colleagues outside of Germany, his research was brilliant. For example, Warburg's 1935 discovery that nicotinamide is the active group of hydrogen-transferring molecule, NADH, would have earned him a second Nobel Prize, but Hitler forbade German citizens from accepting the award. Nevertheless, Warburg continued to study cancer, photosynthesis, and respiration until his death in 1970.

Nazi policies also affected other German scientists. One of these scientists was Hans Adolf Krebs, who assisted Warburg in Berlin from 1926 to 1930. In 1932, using much of what he learned from Warburg, Krebs announced the first metabolic cycle ever to be described. However, the next year (1933), Krebs was fired by the Nazis and fled from Nazi Germany to Britain. Only four years after arriving in Britain, Krebs announced his discovery of the "tricarboxylic acid cycle," a pathway used by almost all organisms to convert two-carbon fragments to carbon dioxide, water, and energy. This pathway, which later became known as the Krebs cycle, earned Krebs a Nobel Prize in 1953. Krebs was knighted in 1958, making him Sir Hans Krebs. Krebs worked in Britain until his death in 1981.

Just how powerful is this flow of electrons and gradient of protons? That is, how much ATP can be made by the electron transport chain? Scientists have estimated that each NADH might drive the synthesis of up to three ATPs, while each ubiquinol might drive the synthesis of up to two ATPs. When there is sufficient O_2 present, each molecule of glucose, if it goes all the way through aerobic respiration, has the net potential to form as many as 38 molecules of ATP. Up to 34 of these ATPs can come from the electron transport chain when powered by NADH and ubiquinol; however, this impressive number can be realized only when sufficient O_2 is present to act as the terminal electron acceptor. Put another way, under ideal conditions, glycolysis and the Krebs cycle can net only four ATPs per glucose; the other 34 come from the chain. The electron transfer chain is the clear champion of ATP production in cells when O_2 is present.

Note that O_2, which is essential to most life as we know it, comes into the respiration picture only at the very end of the three stages. In humans, think about all the anatomical and biochemical adaptations that have evolved to deliver that O_2 to the end of the electron transport chains in our cells; you might list noses, lungs, arteries, red blood cells, and hemoglobin among these adaptations. In plants, elaborate systems of aerenchyma (chapter 9) have evolved as a mechanism for air to reach submerged roots. Note also that CO_2 is released from glucose during respiration, but only in mitochondria, and only if O_2 is present to receive electrons from the first two stages. What happens if there is insufficient O_2 present? For us, bad things, as you know. But what happens in plants? We turn to that question next.

INQUIRY SUMMARY

During glycolysis, glucose is split into pyruvate, and some ATP and NADH are formed along with carbon skeletons for biosynthesis. In the presence of sufficient oxygen, pyruvate from glycolysis moves into mitochondria and is converted into acetyl CoA, which enters the Krebs cycle. Each "turn" of the Krebs cycle uses one molecule of acetyl CoA and produces ATP, ubiquinol, NADH, and carbon skeletons for biosynthesis. The six carbons that make up glucose are released as six molecules of CO_2.

Electrons from NADH and ubiquinol are donated to the electron transport chain in which electron carriers in the mitochondrial membrane pass electrons to O_2 and pump protons to the intermembrane space. The resulting proton gradient produced by electron transport provides the energy necessary for making ATP. The complete aerobic respiration of one molecule of glucose can produce as many as 38 molecules of ATP.

Which part of respiration probably evolved first? Explain your answer.

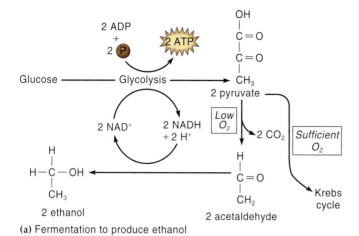

(a) Fermentation to produce ethanol

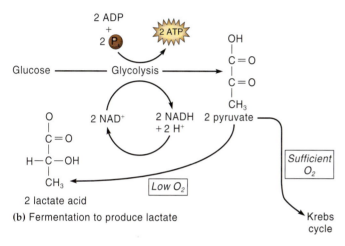

(b) Fermentation to produce lactate

FIGURE 11.6

Anaerobic respiration usually produces ethanol and carbon dioxide in plants and certain fungi (a); in other fungi, some bacteria and in animals, it produces lactate (b).

Fermentation Can Occur When O$_2$ Levels Are Low.

The foregoing discussion of respiration focuses on the main type of respiration in most organisms. This respiration is called aerobic respiration because it requires oxygen as the terminal electron acceptor. In some bacteria, other molecules have evolved as electron acceptors in a process called *anaerobic respiration* because O$_2$ is not the final electron acceptor at the end of the electron transport chain. However, in plants and animals, including us, when O$_2$ is insufficient, the pathway of cellular respiration backs up to pyruvate, and the process called **fermentation** occurs (fig. 11.6). Fermentation includes glycolysis and a step after pyruvate that forms either lactate, ethanol, or (in some bacteria) other lesser-known products of fermentation. We define fermentation as an anaerobic process (meaning no O$_2$ required) that involves the first stage of respiration (glycolysis) but

produces an organic product, such as lactate or alcohol, along with small amounts of ATP. Only lactate fermentation occurs in animals.

Although O$_2$ is required only at the end of electron transport during aerobic respiration, the electron transport chain and the Krebs cycle are both inhibited when oxygen is low or not available. For animals, certain fungi, some bacteria, and a few plants, pyruvate is converted to lactate (lactic acid, a 3-carbon organic acid). This step, which forms lactate, uses excess NADH produced during glycolysis (fig. 11.6) and returns NAD$^+$ to glycolysis. NAD$^+$ is a reactant needed in a critical step of the glycolysis pathway. The return of NAD$^+$ allows some ATP formation in the absence of O$_2$ and probably explains why fermentation has not been removed by natural selection as plants and animals have evolved.

When lactate is formed in humans, it is most often because we exercise at a rate exceeding our ability to deliver oxygen to our muscle cells. The accumulation of lactate is painful, and we stop exercising until we "catch our breath," thereby allowing the lactate to diffuse out of the muscle cells to the liver. If you are in shape, you probably have a more efficient system to deliver and use the O$_2$. If brain or heart cells are deprived of O$_2$ for more than a few minutes, stroke or heart damage may occur.

In some fungi, such as yeast, and in most plants, pyruvate from glycolysis is converted to ethanol and CO$_2$ (fig. 11.6). These anaerobic reactions are called *alcoholic fermentation*. Recall that fermentation to lactate (lactate fermentation) or ethanol (alcoholic fermentation) returns NAD$^+$ to a step of glycolysis and allows the pathway to continue (fig. 11.6). Note the release of CO$_2$ during alcoholic fermentation, making the process essentially irreversible. Alcoholic fermentation in plants and yeast leads to the buildup of ethanol, which is toxic to the organism at levels around 10%. This is why wine has a maximum natural alcohol content of about 10%.

Unless they have adaptations (such as aerenchyma— see chapters 4, 7, and 8) to facilitate diffusion of air (with O$_2$) to their roots, many plants ferment when they grow in mud or in oxygen-poor water; this is why houseplants die when they are overwatered or seeds do not germinate when submerged. Fermentation in plants can be lethal because of the potential buildup of ethanol; it is also inefficient because it produces relatively little ATP compared to aerobic respiration. In fact, fermentation of glucose to ethanol in a plant produces just two new ATP molecules per glucose molecule. Can you estimate how many ATP molecules could be made from fermentation of a molecule of sucrose? How does this compare to aerobic respiration of sucrose?

Fermentation was probably the first form of energy-converting metabolism to evolve, perhaps more than three billion years ago. Following the evolution of photosynthesis, O$_2$ became available and aerobic respiration evolved. This provided a selective advantage for the evolution of eukaryotic cells, and therefore plants. Thus, while fermentation seems less important to most organisms today, it was critical in the evolution of life. How was ATP important in this evolution?

Fermentation is economically important. For example, fermentation by bacteria and fungi is important for producing and flavoring cheese and yogurt. Similarly, fermentation by yeast is important for making alcoholic beverages and bread; wine, beer, and bread are made by different strains of brewer's yeast (*Saccharomyces cerevisiae*). The CO_2 produced during alcoholic fermentation makes bread rise and forms the holes in the loaf, but the alcohol from fermentation evaporates when the bread is baked. Wine is usually made by yeast that grows on grapes, although fermented honey (mead), apples (cider), and other carbohydrate-rich substrates also are used to make wine or winelike beverages. Likewise, although beer is usually made by fermenting barley, rice beer is common in the Orient, and corn beer is made by the Tarahumara tribe of northern Mexico.

INQUIRY SUMMARY

When O_2 levels are low, the flow of carbon is backed up to pyruvate. To cycle NAD^+, carbon from pyruvate flows to ethanol or lactate in an anaerobic process called fermentation. In plants, alcoholic fermentation can lead to production of some ATP but the buildup of ethanol can be toxic.

What might be the selective advantage of switching to fermentation when oxygen is low?

Cyanide-Resistant Respiration Is Important in Some Plants.

In early laboratory experiments, it was discovered that aerobic respiration can be strongly inhibited when the terminal electron carrier, cytochrome oxidase (see figs. 11.6 and 11.7), combines with cyanide or other poisons. However, later it was discovered that in many plant tissues, respiration continues despite the addition of cyanide. Such continued respiration was originally called **cyanide-resistant respiration** for its discovery in a botany laboratory. Certain plants, fungi, and bacteria have cyanide-resistant respiration, but it is rare in animals.

The reason respiration does not stop when cytochrome oxidase is poisoned is that plant mitochondria often have an alternative short chain of electron carriers that branches from the main chain. Because this branching chain also ends in an oxidase, it is aerobic; that is, oxygen is also the terminal electron acceptor in this alternative pathway or chain (fig. 11.7). This oxidase is called the **alternative oxidase,** and the pathway is called the *alternative respiratory pathway* or *chain*. Little or no oxidative phosphorylation is coupled to this alternative chain, however, so the energy from the oxidation of NADH and ubiquinol in this pathway produces heat instead

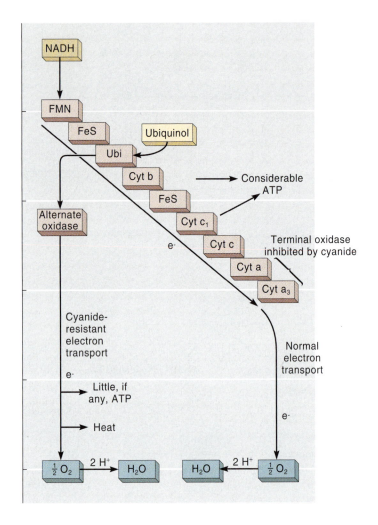

F I G U R E 1 1 . 7
Model of electron transport chain for cyanide-resistant respiration. As in cyanide-susceptible electron transport (see fig. 11.5), the final carrier of cyanide-resistant electron transport is an oxidase, because the terminal electron acceptor is oxygen.

of ATP. Although cyanide resistance was discovered over 20 years ago, the alternative oxidase has proven difficult to study, and relatively little is known about it.

In nature, alternative respiration is active in sugar-rich cells, where glycolysis and the Krebs cycle occur unusually fast. This observation provides indirect evidence that the main electron transport chain becomes saturated and the alternative pathway takes up the overflow of electrons. Both the normal and the alternative pathways may operate simultaneously and naturally. Thus, the main benefit of this alternative path of electron flow for many plants is probably that it dissipates excess energy from respiration. In some tissues, such as leaves and roots of spinach or pea, up to 40% of total respiration may occur via the alternative pathway. Remember that the application of cyanide to plant mitochondria in a laboratory led to the discovery but

11.4 *Perspective* Perspective

How Does Skunk Cabbage Get So Hot?

Skunk cabbage spends the spring and summer packing starch into its fleshy rhizome. The rhizome then acts as a furnace during late winter by providing the energy needed to heat the flowering shoot and melt the snow around it as it grows up through the snow. Skunk cabbage burns this fuel and uses oxygen at a rate comparable to that of a hummingbird. Heat is conserved by a thick, spongy structure that acts as an insulator around the flowering shoot. The overall effect is that the flowers develop in a near-tropical climate even though the plant is buried in snow.

These plants have an alternate form of respiration in their mitochondria. This alternate form of respiration generates so much heat that it is costly to the plant; moreover, it makes little or no ATP. There is no evidence that getting such an early jump on spring gives skunk cabbage an advantage over other plants. Why, then, does skunk cabbage get so hot? One possible explanation is an evolutionary one; that is, heat-generating, alternate respiration was passed down unchanged from the ancestors of skunk cabbage.

To study this hypothesis, botanists have compared skunk cabbage with its

Skunk cabbage.

tropical relatives, which include many well-known houseplants, such as caladiums, philodendrons, voodoo lilies, and dumbcanes. These plants, as well as skunk cabbage, belong to the Arum family (Araceae), most of whose members still live in the tropics. When some aroids heat up, they vaporize foul odors that diffuse into the air around the flowering shoot. These odors include such chemicals as putrescine, cadaverine, and ska-

tole. They smell like dung or decaying flesh, which attracts flies and other insects seeking suitable places for laying their eggs. Instead of finding places to lay their eggs, however, the flies and insects are often trapped temporarily by the plant and become covered with pollen. After they are released, they may get trapped again and leave pollen in another flowering shoot. In this way, they transfer pollen from one flower to another.

Unlike its tropical relatives, skunk cabbage does not seem to benefit from its vaporized fragrance by tempting pollinators. For one thing, few insects are active when skunk cabbage is hot. Also, although skunk cabbage smells a lot like its tropical relatives, it has no mechanism for trapping insects.

The heat-generating ability of skunk cabbage and some of its relatives has intrigued botanists for decades. Is the heat-generating process an adaptation for surviving cold weather, or is it merely an evolutionary remnant of a feature of some tropical plants?

What kinds of information and experiments would help you answer this question?

---

that the alternative pathway operates naturally in plants, independent of cyanide.

In some plants, such as the aroid lilies, the amount of heat generated from an alternative pathway of respiration is remarkable. For example, in the eastern skunk cabbage (*Symplocarpus foetidus*), this heat can raise temperatures in some parts of the plant more than 20°C above the air temperature. Such extreme temperatures are important in the ecology of these plants (see Perspective 11.4, "How Does Skunk Cabbage Get So Hot?").

INQUIRY SUMMARY

Some plants contain an alternative pathway of respiration that was discovered by an insensitivity to cyanide that normally is lethal to cells. The alternative pathway may be important for dissipation of excess energy in a plant. In

others, the alternative pathway may generate heat important in the growth and reproduction of the plant.

How might the alternative pathway be critical to alpine plants?

Lipids Are Respired in Some Germinating Seeds.

Lipids, which are made mostly of fatty acids (see chapter 3), are important energy and carbon storage compounds in oil-containing seeds. When such seeds germinate, their fatty acids are metabolized in organelles called *glyoxysomes* that allow the plant to utilize the energy and carbon stored in the fat, usually an oil such as in coconut, soybean, corn, or safflower. In germinating seeds, fatty acids are first removed from glycerol in stored fats (oils), then snipped into two-carbon pieces that form acetyl CoA (fig. 11.8; recall that

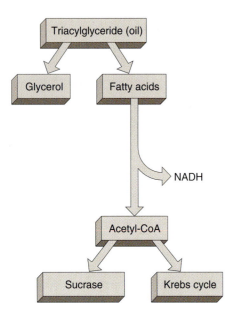

FIGURE 11.8

Respiration of lipids. Triacylglycerides are hydrolyzed into glycerol and fatty acids. Fatty acids are degraded into acetyl groups that are attached to coenzyme A, which can be used in respiration or for carbohydrate synthesis. NAD^+ is also reduced.

CoA serves as a carrier of the 2-carbon acetyl group). This reaction is repeated for every pair of carbon atoms until all of the fatty acid is converted into acetyl-CoA molecules. Thus, a molecule of stearic acid, which has 18 carbons, yields nine molecules of acetyl CoA.

Acetyl CoA from lipids, which is the same molecule described previously that is formed from pyruvate, can be used in the Krebs cycle and generate ATP. However, in germinating seeds, most of the acetyl CoA from the fatty acids is converted to sucrose and exported to developing parts of the germinated seed. In this way, the stored lipid acts as a reservoir of carbon and energy for the new seedling—the shoot and root—to use until it becomes photosynthetic. Because fatty acids are not soluble in water, they are converted to sucrose that flows readily through the vascular tissue of the young seedling. What might be the selective advantage of storing carbon and energy in oil rather than starch?

SUMMARY

Photosynthesis and respiration are the primary energy transformations in organisms. Together, they comprise a system by which energy and carbon flow through an ecosystem and energy is ultimately converted to heat. Unlike energy, nutrients cycle in an ecosystem.

Cellular respiration harvests energy from organic molecules and converts the energy to ATP. Respiration also forms carbon skeletons for biosynthesis of complex molecules in a plant. The most common kind of cellular respiration in plants is called aerobic respiration, which requires O_2 to produce ATP and release CO_2. Aerobic respiration functions to form ATP and carbon skeletons for biosynthesis.

Aerobic respiration of sugars occurs in three stages:

1. During *glycolysis,* glucose is broken down into pyruvate. This occurs exclusively in the cytosol and converts some of the energy stored in glucose to ATP and NADH.

2. The *Krebs cycle* occurs in the mitochondria. First, pyruvate is converted into acetyl CoA, which, in turn, enters the Krebs cycle—a series of reactions that produce ATP, NADH, ubiquinol, and release CO_2.

3. The *electron transport chain* receives energy-rich electrons from NADH and ubiquinol. Electrons flow through a series of carriers that release energy at each step. The terminal electron acceptor is O_2, which is reduced to H_2O. The energy of electron flow is used to pump protons that accumulate, then stream back through ATP synthase complexes, forming ATP. This synthesis of ATP from the energy of electron transport is called oxidative phosphorylation.

Fermentation occurs in low levels or the absence of oxygen and produces limited amounts of ATP. An alternative pathway that is not inhibited by cyanide may provide some selective advantage to certain plants. Lipids are stored in seeds as a source of energy and carbon. Upon germination, most of the carbon in the oil is used to make sucrose for export from the germinating seed.

Writing to Learn Botany

What are the ecological and evolutionary advantages and disadvantages of aerobic versus anaerobic respiration?

THOUGHT QUESTIONS

1. What is aerobic respiration? How does it depend on oxygen?

2. What is CO_2? How is carbon passed between organisms in an ecosystem?

3. Which contains more energy for use by a cell, a molecule of ATP or a molecule of glucose? How do you know?

4. Does the Krebs cycle turn clockwise? Explain.

5. What is anaerobic respiration and fermentation? What are the most important cellular roles of fermentation?

6. Suggest reasons why the absence of oxygen inhibits both electron transport and the Krebs cycle but not glycolysis.

7. Without distillation, wine and other fermentation products do not exceed about 10% to 12% alcohol. Suggest reasons why fermentation does not continue after this concentration of alcohol is reached.

8. How are lipids broken down in plant cells?

9. Where might an organism live if it could respire only anaerobically?

10. What is the importance of compartmentation of the reactions within organelles during respiration?

SUGGESTED READINGS

Articles

Amthor, J. S. 1991. Respiration in a future, higher-CO_2 world: Opinion. *Plant, Cell and Environment* 14:13–20.

Cammack, R. 1987. $FADH_2$ as a "product" of the citric acid cycle. *Trends in Biochemical Sciences* 12:377.

McIntosh, L. 1994. Molecular biology of the alternative oxidase. *Science* 105:781–86.

Moller, I., and A. Rasmusson. 1998. The role of NADP in the mitochondrial matrix. *Trends in Plant Science* 3:21–27.

Ryan, M. 1991. Effects of climate change on plant respiration. *Ecological Applications* 1:157–67.

Storey, R. D. 1991. Textbook errors and misconceptions in biology: Cell metabolism. *American Biology Teacher* 53:339–43.

Storey, R. D. 1992. Textbook errors and misconceptions in biology: Cell energetics. *American Biology Teacher* 54:161–66.

Wivagg, D. 1987. Research reviews: How many ATPs per glucose molecule? *American Biology Teacher* 49:113–14.

Books

Dey, P. M., and J. B. Harbourne, eds. 1998. *Plant Biochemistry.* NY: Academic Press.

Hopkins, W. G. 1999. *Introduction to Plant Physiology,* 2d ed. New York: John Wiley and Sons.

Salisbury, F. B., and C. W. Ross. 1992. *Plant Physiology,* 4th ed. Belmont, CA: Wadsworth.

Taiz, L., and E. Zeiger. 1998. *Plant Physiology,* 2d ed. Sunderland, MA: Sinauer.

ON THE INTERNET

Visit the textbook's accompanying web site at http://www.mhhe.com/botany to find live Internet links for each of the topics listed below:

Glycolysis

Krebs Cycle

Hans Krebs

Electron transport chain

Aerobic respiration

Anaerobic respiration

Fermentation

Yeast and alcohol

Respiration and the germination of seeds

Respiration and heat generation by plants

Flowers of the scarlet gilia (Ipomopsis aggregata). Flowers are the reproductive organs of flowering plants. What is the evolutionary significance of the thousands of different kinds of flowers?

CHAPTER 12

Flowers and Fruits

Learning Online
LEARNING ONLINE

Did you know that some flowers produce fruit underground? To
learn more about flowers, pollination, and the life cycle of
flowering plants, visit http://149.152.32.5/Plants_Human/
pollenadapt.html in chapter 12 of The Online Learning Center at
www.mhhe.com/botany.

What attracts us to flowers? Many people spend hundreds of hours each year tending flowering plants in their gardens and thousands of dollars to raise roses, chrysanthemums, tulips, and hundreds of landscape and indoor plants. Flowers are often brightly colored, sweet smelling, and oddly shaped. These attractants draw other animals to flowers, too, which is essential for the sexual reproduction and seed production of many flowering plants. Successful reproduction depends on adaptations to various environmental factors. Many of the adaptations among flowers involve specializations for certain animals that pollinate them by transferring pollen from other flowers. This means that features such as the colors and odors of flowers and the number of seeds in a fruit are functionally important. Plants have evolved an extraordinary diversity of features involved in sexual reproduction. Indeed, there are more flowering plant species than species of any other group of plants. For this reason, the focus of this chapter is on the reproductive morphology of flowering plants.

Flowering plants live in almost all terrestrial and aquatic habitats on earth. Except for coniferous forests and some parts of the arctic tundra, flowering plants dominate all of the major terrestrial communities. Moreover, angiosperms include some of the largest and some of the smallest plants; lilies, oak trees, corn, wheat, rice, lawn grasses, cacti, broccoli, persimmons, and duckweeds are all flowering plants. Such plants surround us and affect virtually all aspects of our daily lives. It is not surprising that this group of plants attracts the most attention from scientists and the public alike.

Most of the discussions about the genetics, physiology, ecology, structure, and economic importance of plants in this book focus on angiosperms, simply because most of our knowledge of plants is based on the flowering plants. This chapter presents a discussion of some of the most puzzling questions about plants, including how flowering plants got to be what they are today.

Flowers Are the Sexual Reproductive Organs of Flowering Plants.

Flowering plants, or **angiosperms,** are the most diverse of all plants. Like other structures, flowers consist of different tissues and parts that collectively have one main function: in this case, to produce a new generation through sexual reproduction. For flowering plants, the next generation begins with the embryo that develops in each seed.

Study the flower shown in figure 12.1, which you were introduced to in figure 2.4. As we see in this generalized drawing, flower parts are attached to the **receptacle,** which is the swollen region at the end of the **peduncle,** the stalk of the flower. The outermost whorl of floral parts is composed of **sepals,** collectively called the **calyx. Petals,** together called the **corolla,** form the next inner whorl. Sepals and petals are the sterile parts of flowers—they do not produce sperm or eggs—but they do play important roles in reproduction, especially in the attraction of pollinators.

The fertile parts of flowers are the **stamens,** which form just inside the corolla, and the **pistil** in the center of the flower (fig. 12.2). The stamens produce pollen grains, which ultimately produce sperm. Inside the ovary of the pistil, one to many ovules are produced. Inside each ovule is a single egg. After the sperm and egg fuse, an embryo forms inside the ovule, and the ovule develops into a seed. The ovary of the flower (and sometimes associated parts as well) then becomes a fruit that protects, covers, and helps disperse the seed or seeds inside.

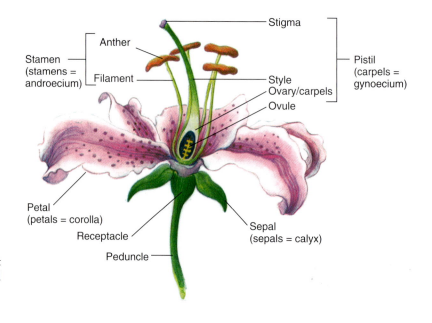

FIGURE 12.1
The parts of a flower. This is a generalized flower that has four main kinds of parts: sepals, petals, stamens, and pistils.

(a)

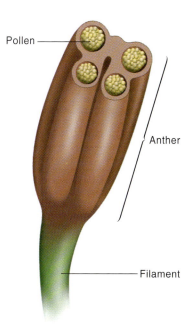

(b)

FIGURE 12.2

(a) Flower of lily (*Lilium* species) showing six stamens, each with a white filament and golden brown anther, and a pistil in the middle. (b) A single stamen, showing pollen in four pollen-containing chambers.

FIGURE 12.3

Flowers of southern magnolia (*Magnolia grandiflora*) have many carpels, stamens, and petals and usually three petal-like sepals.

Stamens produce pollen grains.

Most stamens consist of four pollen-containing chambers that are fused into an **anther,** which produces pollen grains and is generally on a stalk called a **filament** (fig. 12.2b). The stamens are often referred to as the "male" part of the flower, however, *male* actually refers to the pollen grains that produce sperm. The stamens are often collectively called the **androecium,** a term that is preferred by some botanists.

Sources of variation in flowers are the number and arrangement of stamens, the attachment of anthers to the filament, and the way the anthers open for releasing pollen. For example, most orchid flowers possess one stamen, while two stamens occur in ash, six in mustard, eight in fuchsia, and dozens in buttercups, poppies, and magnolias (fig. 12.3). In addition, the stamens of mustard, buttercup, and poppy are free from each other, but those of garden pea, cotton, and mallow are fused at their filaments (fig. 12.4). Although anthers usually release pollen by splitting along a slit in their side, rhododendrons and other members of the heath family (Ericaceae) release pollen through pores at the tip of the anther.

FIGURE 12.4
Stamens are fused in some flowers. Filaments in rose mallow (*Hibiscus* species), for example, are fused to the long style (*arrow*).

FIGURE 12.5
Flower of evening primrose (*Oenothera hookeri*), showing stigma with four lobes that indicate four fused carpels.

Pistils contain at least one ovule in their ovary.

Flowers usually bear one or more pistils, each consisting of an enlarged lower end called the **ovary,** and the **stigma,** which is the receptive area for pollen (see fig. 12.1). The stigma is often borne on a stalklike **style** that extends from the ovary, thereby elevating the stigma to a level that may enhance pollination.

A pistil has one or more ovule-bearing units called **carpels.** In the simplest pistils, there is one carpel, with one ovary containing one or more **ovules,** as in the garden pea (see chapter 13 opening photograph). Violets and many other flowers have one ovary, but several carpels that are fused at their edges. Some flowers, such as those of larkspur and magnolia (see fig. 12.11a), consist of two or more pistils that are not fused; each pistil has one ovary. Carpels and pistils are often referred to as the "female" part of the flower. However, the carpel and pistil have no gender; a specialized structure, the **embryo sac,** inside each ovule is actually the female part of the plant (see fig. 12.12 and chapter 13). Carpels are often collectively called the **gynoecium,** a term which is preferred by some botanists and used instead of the term pistil.

An ovary usually has several indicators for how many carpels it contains. The most straightforward cases have a chamber for each carpel, as in a citrus fruit. This means that each section of an orange fruit, for example, represents a carpel. In addition, stigmas or styles may also reflect the number of carpels. Accordingly, the lily has three seed chambers and three stigmas, but it has one style that probably evolved by the fusion of three styles from three carpels. Likewise, some kinds of evening primrose have four lobes on the stigma, signifying four carpels (fig. 12.5). Conversely, in saguaro and other cacti, the ovary has one chamber and several stigmas, which suggests that the ovary evolved from several carpels that fused sometime in the evolutionary past of the cacti.

Petals are often the colorful, fragrant parts of a flower.

The corolla, composed of petals, is usually the most noticeable and attractive part of a flower. Differences in number, size, color, and odor of petals distinguish many kinds of flowers. In addition, petals may be free (see figs. 12.1, 12.2a, and 12.3), fused into a short tube with large lobes, or fused into a long tube that encompasses most of the corolla (as in honeysuckle; fig. 12.6). Many of the modifications of corollas are important in pollination, which is discussed later in this chapter.

In addition to color, size, and number, petal development also influences flower symmetry. When all petals develop equally, corollas are **radially symmetrical.** However, when petals do not develop equally, corollas become **bilaterally symmetrical.** Mustard, lily, oleander, poppy, and stonecrop have radially symmetrical corollas (fig. 12.7), whereas orchids, monkeyflower, garden pea, and honeysuckle have bilaterally symmetrical corollas (see fig. 12.6).

Flowers of a plant usually have a specific number of petals. For example, wild roses have five petals. The number of petals often corresponds to the number of stamens, carpels, and sepals. However, the flowers of cacti, buttercups, and magnolias have an indefinite number of petals. Furthermore, in cacti and magnolias, petals intergrade with sepals, thereby making it difficult to distinguish petals from sepals.

Leaflike sepals protect the immature flower.

Most sepals are leaflike, but they may resemble petals. Sepals of lily are indistinguishable from the petals except for their location; both are referred to as **tepals.** The calyx in four-o'clocks looks like a corolla, but flowers of these plants

The Lore of Plants

One of the rarest of flower colors is black. The distinction of the blackest flower in the world may belong to *Lisianthius nigrescens*, which grows in southern Mexico and Guatemala, where it is called *la flor de muerte*, the "flower of death." The local name for this species comes from the tradition of planting it around graves. The plant produces flowers that are 4 centimeters long with petals the color of blackish purple eggplants. While many other species produce flowers that reflect ultravi-olet light, which can be seen by insect pollinators as a distinct color, the "flower of death" absorbs ultraviolet light, meaning it appears black to insects as well as humans. No known pollinator has yet been discovered for the flower; because of its black color, it would not appear to be too attractive to animals. One hypothesis is that the black color allows the flowers to absorb sunlight during the day, thus building up heat used to attract pollinators that visit flowers at night.

FIGURE 12.6

Flower of honeysuckle (*Lonicera* species) showing petals fused into a long tube. This flower is bilaterally symmetrical.

FIGURE 12.7

Flowers of stonecrop (*Sedum lanceolatum*) have five sepals, five petals, ten stamens, and five carpels. These flowers are radially symmetrical.

have no petals. Like petals, sepals may be fused into a tube, and the calyx may be radially or bilaterally symmetrical.

The number of sepals, like the number of petals, often corresponds to the number of other flower parts. In most flowers, the number of petals and the number of sepals are identical. Moreover, if you look directly at the center of a flower that has the same number of sepals and petals, the sepals appear to alternate with the petals.

Sepals protect the inner parts of a flower before it opens. The calyx is especially important for keeping the un-opened flower from drying out. However, sepals often fall off the flower at maturity or following pollination and fer-tilization. Open poppy flowers, for example, appear to have no sepals because they fall off just before the flower opens.

INQUIRY SUMMARY

Flowers are the reproductive structures of flowering plants. Flowers consist of two kinds of structures involved in sexual reproduction: the stamens, which produces pollen grains, and the pistil, which produces ovules contained within carpels. The calyx consists of sepals, which protect the immature flower, and the corolla consists of petals that are often colorful, fragrant, and attractive to pollinators.

What might be a disadvantage to a plant of producing many ovules in each ovary compared to producing just one ovule per ovary?

How Did Angiosperms Evolve?

Although flowering plants evolved between 120–140 million years ago, they have been the dominant group of plants only since the late Cretaceous and early Tertiary periods, some 65 million years ago. The relatively sudden and abundant appearance of angiosperms in the fossil record has been of interest to scientists since the time of Charles Darwin. To reveal this floral history, botanists have used evidence such as fossils, new ways of analyzing evolutionary relationships, comparisons of modern molecular data from nucleic acid sequences, and studies involving developmental biology and comparative anatomy. Although answers to all questions about when, where, and how flowering plants arose have not yet been found, botanists have learned much about the evolution of angiosperms in the nearly century and a half since Darwin first pointed out the "abominable mystery" of angiosperms' origins.

Early in seed-plant evolution, insects became pollen carriers as they searched for food. The earliest known insects that could have been pollinators were probably beetles. Long before the appearance of angiosperms in the fossil record, cycadeoids (see fig. 18.2), which are extinct gymnosperms with palmlike leaves, were already specialized for pollination by beetles. Cycadeoids had tiny ovules that were protected by scalelike leaves, which is a feature that is consistent with pollination by insects with chewing mouthparts, such as some beetles. Angiosperms that may have lived with the first beetle-pollinated cycadeoids were probably also adapted for pollination by beetles.

How did pollination evolve into a mutualistic relationship in which both interacting species benefit from the relationship? A recent hypothesis is that pollen consumption by ancient insects may have preceded the establishment of a mutualistic relationship. Pollen consumption could have become a mutualistic relationship if the pollen eater delivered pollen it didn't eat to the reproductive parts of a flower and did so more efficiently than alternative pollen carriers such as the wind. Thus, pollination would require a plant to produce more pollen than was eaten by the early pollinators and to sacrifice some pollen it produced for improved efficiency in its reproduction. Eventually, plants that produced floral nectar and odors attracted insects that carried pollen.

Insect pollination was not associated with rapid evolution of angiosperms until the appearance of specialized insects about 60 million years ago. The rise to dominance by angiosperms, therefore, seems to have been greatly influenced by adaptations for pollination by an increasing diversity of flying insects. Beetle-pollinated angiosperms gave way to a much greater diversity of angiosperms adapted for pollination by butterflies, moths, and bees. Insects played an important role in the evolution of angiosperms into the largest and most diverse group of plants.

Flower parts probably evolved from leaves.

Botanists consider each part of a flower to be a specialized leaf. This idea is best supported for sepals because they are mostly leaflike. Support for the origin of stamens from leaves comes from the stamen morphology of certain tropical trees, whose stamens look like leaves yet possess embedded reproductive structures (fig. 12.8). However, this issue remains unsettled.

The argument for the origin of the carpel from a leaf is the same as that for the origin of the stamen. A supportive example is the carpel of the genus *Drimys*, which is a living tropical tree related to magnolia. The carpel of *Drimys* looks like it evolved from a leaf by folding in the middle and joining at the margins, with a stigma along part of the fused area (fig. 12.9). According to this model, carpels evolved further when the margins sealed and the stigmatic surface

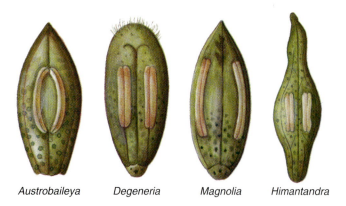

Austrobaileya   *Degeneria*   *Magnolia*   *Himantandra*

FIGURE 12.8

Stamens of *Austrobaileya* and certain other tropical trees look like leaves with pollen-producing structures embedded in them. Such leaflike stamens are indirect evidence that stamens arose from fertile leaves.

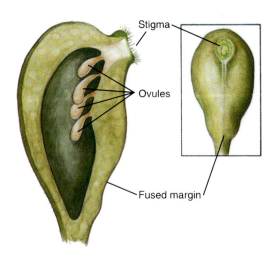

Stigma

Ovules

Fused margin

FIGURE 12.9

A carpel of *Drimys* resembles a leaf folded at the middle and fused at the edges to form a container that holds ovules. A stigma forms along part of the fused area.

shrunk to a small area at the apex of the carpel. A pea pod represents the result of such evolution.

Alternative views on the origin of the carpel come from fossil evidence. One view is that the carpel arose from the seed-bearing structure of an extinct group of plants called the **seed ferns.** Seed-bearing structures from fossil plants challenge the idea that carpels evolved from leaves. Thus, like the origin of stamens, the origin of carpels remains controversial.

The fossil record provides clues about the evolution of flowering plants.

All but a few plants from the past decomposed without leaving a trace of their existence. Indeed, botanists estimate that the fossil record of plant species may be only 1% complete and that at least 90% of the species that ever existed are extinct. The main problem, of course, is that soft tissues such as those of flowers usually decayed before they could be fossilized. Nevertheless, the fossil record of plants does provide a solid basis for some general ideas about where flowering plants came from and how they might have evolved.

The first fossils of vascular plants are more than 400 million years old, and the first seeds appeared as long as 360 million years ago. However, fossils of plant fragments that probably came from angiosperms are not known before the early Cretaceous period, about 140 million years ago. Unfortunately, most of the oldest of these fossils are fragmented and incomplete. **Paleobotanists,** scientists who study fossil plants, recently discovered a fossilized plant estimated to have lived some 120–140 million years ago in northeast China. This plant, named *Archaefructus* (*archae,* meaning "ancient," and *fructus,* meaning "fruit"), may represent the earliest known flowering plant, with podlike structures that may have been carpels and flowers without petals. Interestingly, *unfossilized* seeds dropped out when the fossil was cracked open; unfortunately, the embryo inside each seed had long before died, and therefore the seeds would not germinate. *Archaefructus* shares certain characteristics with relatives of the magnolia family (Magnoliaceae).

Another fossil stands out because it consists of all the parts of a flower attached to a reasonably intact plant. This flowering plant is from a 120-million-year-old fossil deposit near Koonwarra, Australia. Some paleobotanists think that this plant represents the ancestral type of flower. If this is true, then the features shared by the Koonwarra angiosperm and certain modern angiosperms may show which living plants are closest to the ancestral origin of the group.

Discovered in 1986, the Koonwarra angiosperm was thought at first to be a fern, but closer examination showed that the "fern" had carpel-bearing inflorescences (fig. 12.10). The fossilized portion of the plant was less than 3 centimeters long, and overall it resem-

bled a black pepper plant. The Koonwarra angiosperm had several features that are typical of many modern angiosperms. For example, it had small flowers without petals, single-carpel ovaries with short stigmas and no styles, and flowers with several bracts at their bases. These features occur in present-day members of the lizard's tail family (Saururaceae) and the pepper family (Piperaceae), which are dicots. In addition, the leaf venation in the Koonwarra angiosperm resembles that of these families as well as the leaf venation of the birthwort family (Aristolochiaceae), the greenbrier family (Smilacaceae), and the yam family (Dioscoreaceae), the latter two being monocots.

The Koonwarra angiosperm shows how the ancestor of flowering plants may have looked: a small, rhizome-bearing herb that had secondary growth, small reproductive organs, and simple flowers with bracts at their bases. Families of living plants that share several features with the Koonwarra angiosperm are believed to be primitive members of the dicots and monocots. Furthermore, the appearance of this plant near the apparent beginning of the evolution of angiosperms and its similarity to dicots and monocots suggest that the Koonwarra angiosperm may be an ancestor of all modern-day flowering plants. If this is correct, then all of the first flowers on earth were probably radially symmetrical, like poppies and buttercups, and their petals, when present, were free (i.e., unattached to each other; fig. 12.11). Flowers with distinct bilateral symmetry, such as those of modern violets and peas, and flowers with fused petals, such as those of honeysuckle, did not appear until the Paleocene period, less than 65 million years ago (see fig. 14.25).

At a recent meeting of botanists in 1999, results of 5 years of genetic analyses of over a hundred species of flowering plants were reported. These results suggest that the closest living relative of the first flowering plants is *Amborella,* a shrub with small, creamy flowers and red fruit found only on New Caledonia, an island in the South Pacific. The study of the evolution of flowering plants continues to be a major focus of research efforts by botanists around the world.

INQUIRY SUMMARY

The Koonwarra angiosperm and the fossil *Archaefructus* from China represent some of the earliest intact fossils of a flower. Both of these primitive, fossilized flowering plants have features that occur in several modern families of flowering plants. The early evolution of certain floral features is known from fossils from the Cretaceous and Tertiary periods. Flower parts may have evolved from leaves, and the first pollinators may have been pollen-eating insects.

Would you call an insect a pollinator if it eats all of the pollen from a flower? Why or why not?

(a)

(b)

FIGURE 12.10

(*a*) One of the oldest known fossil flowers, which is about 120 million years old, was found in the Koonwarra fossil beds near Melbourne, Australia. (*b*) As this drawing shows, the entire fossil resembles a small black pepper plant, less than 3 centimeters long.

Flowering Plants Produce an Embryo and Endosperm Inside a Seed.

Sexual reproduction is part of the **life cycle** of a flowering plant, which you will learn more about in the next chapter and in chapter 18. The life cycle is a series of genetically programmed developmental changes in a plant. The life cycle of a flowering plant includes growth of the embryo in a seed into a mature flowering plant, formation of sperm and egg, and fertilization. Although the life cycle of humans differs from that of plants in many regards, there are a few similarities. For instance, all of our body cells have a **diploid** number of chromosomes; that is, each cell has two sets of chromosomes—one set from the sperm of our father and one set from the egg of our mother. Likewise, each cell in the body (the sporophyte) of a flowering plant has a diploid number of chromosomes. In human cells, the diploid number of chromosomes is 46; in onions, the diploid number of chromosomes is 20; in *Ophioglossum reticulatum* (a fern), the diploid number is 1,260—the highest chromosome number known in any organism. How do you explain that the normal diploid chromosome number in most individual organisms is *even* (i.e., 2, 4, 6, 8, etc.)?

Gametes are haploid, reproductive cells.

Sexual reproduction involves the formation of an offspring through the fusion of sperm and egg—a process known as **fertilization.** Sperm cells and egg cells are **gametes,** or sex cells. Each gamete is **haploid,** containing only one set of chromosomes, and is produced through a process called **meiosis,** which you learn about in the next chapter. When a sperm fuses with an egg during fertilization, a one-celled, diploid zygote is produced that now contains two sets of chromosomes. In adult humans, males produce sperm in specialized cells in the testicles, and females produce eggs in their ovaries. Flowering plants also produce sperm and eggs in specialized reproductive structures associated with the flower. Flowering plant sperm are produced inside germinating pollen grains that were produced in the anther, and an egg is produced inside an embryo sac inside each ovule. The pollen grains and embryo sacs are the gametophytes of the flowering plant (see fig. 13.11). The sperm and egg of a

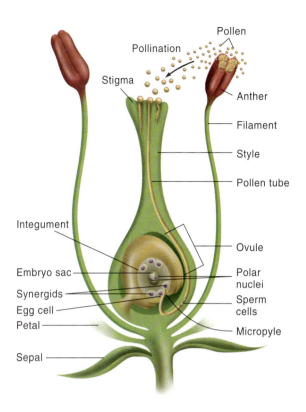

FIGURE 12.11

The earliest flowers had floral parts arranged in spirals (*a*), as in the pistils of modern magnolia (*Magnolia*), or in whorls (*b*), as in the petals of the present-day lily (*Lilium*). Many of the oldest fossil flowers also had petal-like sepals, as in the lily, and radial symmetry (*c*), like the buttercup (*Ranunculus*). More recent flower types include those with bilateral symmetry (*d*), such as the pansy (*Viola*), and those with petals fused into a tube (*e*), such as the cape primrose (*Streptocarpus*).

flowering plant each have one set of chromosomes. When these gametes fuse during fertilization, the result is a zygote that has two sets of chromosomes. This zygote grows inside the ovule and is the first cell of the next generation. In onion, each gamete contains 10 chromosomes, and the zygote (and every cell in the adult onion plant) contains 20 chromosomes. Production of gametes, fertilization, and growth of the zygote are all parts of the life cycle of a plant.

Double fertilization involves the fusion
of two sperm cells with different cells in the ovule.

Sexual reproduction in flowering plants depends on **pollination**, which is the transfer of pollen grains from an an-

FIGURE 12.12

Pollination and fertilization. Two sperm cells move down the pollen tube. One fertilizes the egg, and the other fertilizes the polar nuclei, a process called double fertilization. The pollen tube enters the ovule through the micropyle.

ther to the receptive stigma of a pistil. An embryo sac forms inside the ovule (see figs. 12.12 and 13.13), which also possesses a stalk and one or two **integuments** that later develop into the seed coat. After a pollen grain reaches the stigma of a flower, it germinates, forming a **pollen tube** that grows down the style of the flower into the ovary. The cell that forms the pollen tube is the **tube cell.** The second cell in the pollen grain is called the **generative cell** because it divides and generates two sperm cells. The generative cell divides as the pollen tube grows, and the two sperm cells move through the pollen tube to a small opening in an ovule, the **micropyle.** The two sperm cells enter the embryo sac in the ovule through one of the **synergids,** which are the cells next to the egg. The first sperm then fertilizes the egg. The second sperm fuses with **polar nuclei,** which are the two haploid nuclei near the midregion of the embryo sac. This process is called **double fertilization** because both sperm cells undergo fusion. The diploid cell produced by the fusion of sperm and egg is the zygote, which becomes the embryo. The fusion of the second sperm cell with the two polar nuclei produces a **triploid cell,** which possesses three sets of chromosomes. This cell usually forms the **endosperm,** a nutritive tissue for the developing embryo. After fertilization, the ovule matures into a seed.

What directs the pollen tube to the proper location in the ovule? Through a series of experiments involving the culture of pollen and ovules, scientists have determined that a chemical signal from the ovule attracts pollen tubes. When synergids are broken, pollen tubes do not move toward the ovule, suggesting that the source of the chemical attractant comes from these cells next to the egg.

During the lifetime of a flower, many insects may visit, bringing pollen from a number of different flowers. Much of the pollen that is deposited on the stigma of the flower may be from inappropriate sources—from plants that will not reproduce with the recipient plant. How does a plant deal with this inappropriate pollen? On plants such as *Arabidopsis* of the mustard family, pollen from individuals of the same species is held tightly to the stigma, but pollen from a different species falls off easily. The stigma of the flower apparently can detect differences in pollen from flowers of the same or different species.

Botanists are interested in flowers because of their role in reproduction. Of additional interest is the fact that flowers are important for identifying plants, classifying plants, and studying plant relationships. For instance, floral variation represents the evolutionary record of flowering plants. The next few sections of this chapter present a brief overview of floral variation and the nature of different parts of flowers.

INQUIRY SUMMARY

Specialized cells in the flower produce the haploid gametes, sperm and egg. The life cycle of a flowering plant includes pollination, double fertilization, and the formation of a seed from an ovule. After pollination, a pollen tube grows toward the ovule in the ovary, apparently attracted to the egg by chemicals from the synergid cells. Endosperm cells, which have three sets of chromosomes, usually provide nutrients and energy for the embryo.

From which parent plant (egg or sperm donor) does the endosperm get two sets of chromosomes, and from which parent does it get the third?

Reproductive Morphology Helps Explain the Diversity of Flowering Plants.

The diversity of flowering plants is greater than that of any other group of plants; perhaps as many as 250,000 species of flowering plants have been discovered and named. The diversity of angiosperms is reflected in their different sizes, shapes, and forms, from a huge perennial eucalyptus tree to a small garden phlox. The basis for this diversity comes from the incredible reproductive success of flowering plants in a wide variety of habitats. Reproductive success

FIGURE 12.13
Grass flowers are highly simplified. Each of the many flowers in these grass spikes have no petals or sepals but many stamens. Each flower of rye grass (*Lolium* species) produces about 50,000 dry, dustlike pollen grains, which readily take to the air in a light breeze.

in angiosperms is based on the evolution of the flower, which resulted in this plant group diversifying into new habitats. Such diversification was made possible by seeds, the fruit that protects them, and the pollination of flowers by insects, other animals, and wind.

The following sections describe some of the variations in flowers, fruits, and pollination mechanisms that represent angiosperm diversity. Appreciating this diversity and knowing some of the terms that botanists use to describe it will help you use field guides and other books to identify and study plants in native habitats and gardens.

Some flowers lack sepals, petals, stamens, or pistils.

A typical flower has several sepals, petals, and stamens and a pistil with several carpels. Much of the variation among flowers is based on variation of these basic parts. For example, the showy, cream-colored flowers of the southern magnolia (*Magnolia grandiflora*) have many carpels, stamens, and petals and usually three petal-like sepals (see fig. 12.3).

(a)

(b)

(c)

FIGURE 12.14

Longitudinal sections of flowers show differences in the positions of ovaries in different plants. (*a*) This photo shows dozens of stamens in St.-John's-Wort (*Hypericum* species) and its superior ovary. (*b*) Daffodil (*Narcissus* species) has an inferior ovary. (*c*) In rose (*Rosa* species), the ovary is surrounded by the receptacle.

In contrast, the bright yellow flowers of stonecrop (*Sedum* species) have five sepals, five petals, ten stamens, and five carpels (see fig. 12.7). The highly evolved flowers of many grasses have three stamens, one functional carpel (and perhaps two nonfunctional ones), and no petals or sepals (fig. 12.13). Still other grass flowers have either stamens or a carpel, but not both.

A flower that has all the major parts is a **complete flower;** that is, it possesses sepals, petals, stamens, and a pistil. An **incomplete flower** lacks one or more of these four whorls of parts. However, even when sepals or petals are missing, a flower that has both an androecium and a gynoecium is called a **perfect flower.** Thus, all complete flowers are perfect, but not all perfect flowers are complete. In contrast, flowers that have only stamens or only pistils are **imperfect flowers.**

The position of the ovary also varies among different flower types. St.-John's-wort (*Hypericum* species), for example, has a **superior ovary;** that is, the base of the ovary is attached above (i.e., is superior to) the sepals, petals, and stamens (fig. 12.14*a*). In flowers such as the daffodil (*Narcissus* species), the other three whorls grow from the top of the ovary, which is **inferior** to them (fig. 12.14*b*); the sepals, petals, and pistil rest on top of the ovary. Some members of the rose family have intermediate flowers. Flowers of rose and cherry have an ovary that is surrounded by the receptacle; the petals and stamens branch from the receptacle above the ovary (fig. 12.14*c*).

Monocots and dicots can be identified by the number of their floral parts.

Floral variation provides part of the basis for dividing the flowering plants into two major groups: the dicotyledons and the monocotyledons. Recall from chapter 1 that the informal name *dicot* is given to plants having two cotyledons (seed leaves) in each seed; *monocot* refers to plants that have

TABLE 12.1

Main Differences Between Monocots and Dicots

CHARACTERISTIC	DICOTS	MONOCOTS
Flower Parts	in fours or fives (usually)	in threes (usually)
Pollen	usually having three furrows or pores	usually having one furrow or pore
Cotyledons	two	one
Leaf Venation	usually netlike	usually parallel
Primary Vascular Bundles in Stem	in a ring	scattered arrangement
True Secondary Growth, with Vascular Cambium	commonly present	absent
Examples	rose, pea, sunflower, magnolia, ash	lily, corn, palms, pineapple, banana

one cotyledon in the seed. In monocots, flower parts occur in threes, or multiples of three; for instance, three petals, three sepals, six stamens, and a pistil that has three chambers (see fig. 12.2*a*). In dicots, flower parts usually occur in fours or fives, or multiples of four or five (see figs. 12.4 and 12.5). Although dicots and monocots may have other numbers of floral parts, many other features are unique to each group (table 12.1). Dicots include about 80% of all angiosperm species, including many herbaceous (nonwoody) plants and all woody, flower-bearing trees and shrubs. Monocots are primarily herbaceous, but they also include nonwoody trees such as palms (see fig. 8.24) and Joshua trees (see chapter 8 to review why such trees are not woody).

Groups of flowers comprise an inflorescence.

Flowers may be solitary, or they may be grouped closely together in an **inflorescence.** The spectacular inflorescences of urn plants, lupine, and snapdragon make these plants popular ornamentals (fig. 12.15*a*). Less obviously, the flowers of hazelnut, oak, and willow also occur in inflorescences (fig. 12.15*b*). Like a solitary flower, an inflorescence has one

main stalk, or peduncle. It also bears numerous smaller stalks, called **pedicels,** each with a flower at its tip. The arrangement of pedicels on a peduncle characterizes different kinds of inflorescences. Some of the common types of inflorescences are diagrammed in figure 12.16 and can be described as follows:

- **spike:** an unbranched, elongated main axis whose flowers have very short or no pedicels (examples: plantain, spearmint, tamarisk)
- **raceme:** an unbranched, elongated main axis whose flowers have pedicels that are all about the same length (examples: lily-of-the-valley, snapdragon, mustard)
- **panicle:** a branched main axis with side branches bearing loose clusters of flowers (examples: oat, rice, fescue)
- **corymb:** an unbranched, elongated main axis whose flowers have pedicels of unequal length, forming an inflorescence that appears flat topped (examples: hawthorne, apple, dogwood)
- **simple umbel:** a peduncle bearing all of the pedicels at its apex (examples: onion, geranium, milkweed)
- **compound umbel:** a cluster of simple umbels at the apex of a main axis (examples: carrot, dill, parsley)
- **head:** a peduncle bearing closely packed flowers that have no pedicels (examples: sunflower, daisy, marigold)
- **catkin:** a spikelike inflorescence that bears only unisexual flowers; catkins occur only in woody plants (examples: hazelnut, willow, birch)

Some types of inflorescences characterize different plant groups. For instance, nearly all members of the carrot family (Apiaceae) have compound umbels. All members of the sunflower family (Asteraceae) have heads, including chrysanthemum, zinnia, marigold, and dandelion. Thus, the "flower" of a sunflower is actually many individual flowers packed into a single inflorescence called a head (fig. 12.17).

Plant reproductive morphology promotes outcrossing and hybrid vigor.

No discussion of reproductive morphology and plant diversity is complete without mention of how variation in flower structure can be related to its function—the idea of form and function (see chapter 7). Reproductively, the most versatile plants have perfect flowers that are **self-compatible,** as in the garden pea and snapdragon. If a plant is self-compatible, successful reproduction can occur within a single flower or between flowers of the same plant. This type of reproduction results from **self-pollination,** in which pollen is transferred from the anthers of a flower to the stigma of the same flower or to the stigmas of flowers on the same plant. For instance, corn has two types of imperfect flowers

(a)

(b)

FIGURE 12.15

(*a*) The inflorescences of lupine (*Lupinus nootkatensis*) are spectacular because of their brightly colored flowers. (*b*) The less obvious flowers of willow (*Salix* species) are packed into catkins. Each catkin is composed of many unisexual flowers.

on the same plant: **staminate flowers** (flowers producing pollen but not ovules) compose its tassels, and **pistillate flowers** (flowers producing ovules but not pollen) develop on the corncob. Fruits can develop on a corncob when pollen from the tassels falls onto the stigmas of the pistillate flowers of the same plant (each kernel of corn is a fruit, and the "silk" of corn plants is actually the long styles of the pistillate flowers). This is still sexual reproduction because sperm in the pollen fertilize each egg inside the immature corn kernels. How much variation in the offspring would you expect with self-pollination?

Plants such as corn with staminate and pistillate flowers on the same individual are called **monoecious** ("one house"). Mulberry, cottonwood, willow, and persimmon are examples of plants that have staminate flowers and pistillate flowers on *different* plants. These plants, therefore, are **dioecious;** that is, they have "two houses" (different plants) for reproduction. This means that reproduction in such plants can occur only by **cross-pollination,** which is

the transfer of pollen from the anthers of one plant to the stigma of a flower on another plant.

Mulberry, cottonwood, willow, and persimmon are plants that depend on mating between the sperm and eggs from different plants; this is called **outcrossing.** In contrast, garden pea and snapdragon are plants characterized by **inbreeding,** which occurs when sperm and eggs from the same plant produce offspring. In general, cross-pollination leads to outcrossing, and self-pollination leads to inbreeding.

The breeding systems of plants range from complete outcrossing to complete inbreeding, but most plants use some combination of the two. For instance, flowers of the hoary plantain (*Plantago media*) have only carpels at first; stamens mature after the carpels are mature. This breeding system enhances outcrossing because most pollen is shed from a particular flower before the pistil matures; however, self-pollination is still possible as the flowers get older. In contrast, flowers of garden pea are

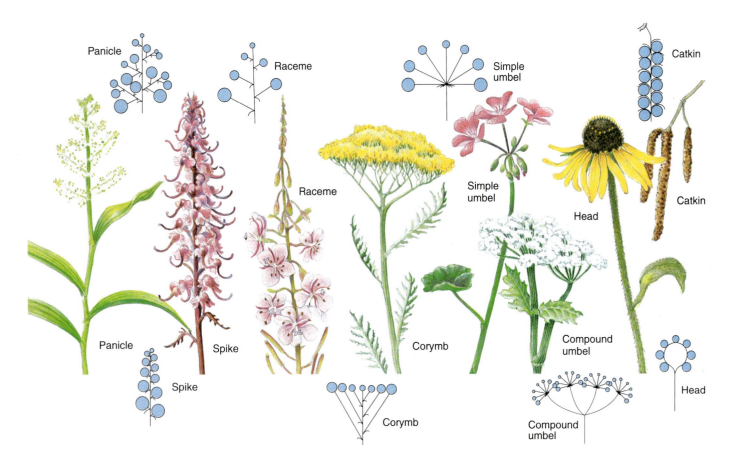

FIGURE 12.16

Common types of inflorescences. Each type is distinguished by how the flowers are arranged in the inflorescence. Each colored circle in the diagrams represents an individual flower.

Source: Insets from Kingsley R. Stern, *Introductory Plant Biology*, 7th ed. Copyright © 1997 The McGraw-Hill Companies, Inc. All Rights Reserved. Reprinted by permission.

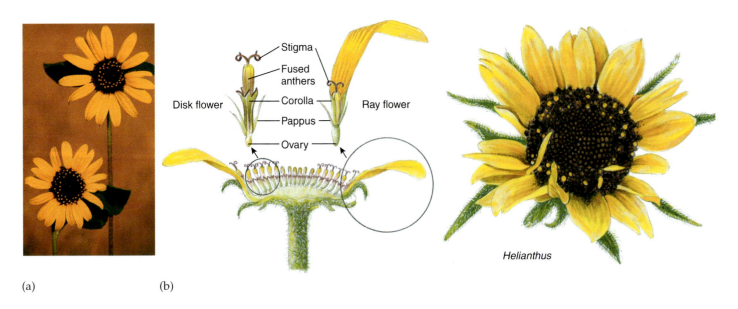

(a) (b)

FIGURE 12.17

The "flower" of sunflowers and other members of the family Asteraceae is actually a type of inflorescence called a head. (*a*) Two inflorescences of the common sunflower (*Helianthus annuus*), each composed of many small flowers. (*b*) Each sunflower inflorescence is composed of dozens of each of two kinds of flowers: ray flowers around the perimeter of the inflorescence and disk flowers toward the interior.

equally receptive to self-pollination and cross-pollination throughout their development. This feature of garden pea enabled Mendel to produce pure-breeding plants for his experiments (see chapter 13).

INQUIRY SUMMARY

Flowering plants are the most diverse group of plants, with about 250,000 species identified. Part of this diversity arises due to differences in the number, size, color, and position of different floral parts. Monocot flowers have parts in threes or multiples of three, while dicot flowers have parts in fours or fives or multiples of four or five. Groups of flowers comprise an inflorescence, which characterizes different plant families. Monoecious species of plants have two types of imperfect flowers on the same plant, while dioecious species have two types of imperfect flowers, with only one type on any plant.

Suppose a plant produced flowers with 12 petals and 12 sepals. How might you tell if it was a monocot or a dicot?

Trade-offs Occur in the Reproduction of Flowering Plants.

What are the selective advantages and disadvantages—that is, the ecological and genetic trade-offs—of inbreeding versus outcrossing? Self-pollination is very efficient; pollen from the anther of a flower can be transferred easily onto the stigma of the same flower because of the proximity of the two parts. Thus, flowers that self-pollinate may produce seed relatively easily. Cross-pollination, however, is riskier, and oftentimes flowers that rely on outcrossing mechanisms for reproduction fail to reproduce; for instance, in regions with few animal pollinators, the opportunities for cross-pollination may be greatly reduced. Yet, most flowers have mechanisms that promote outcrossing, so what is its advantage? That advantage has to do with **hybrid vigor**—the health and reproductive ability of the offspring from cross-pollination.

You may be much taller than either of your parents; if so, you could be an example of hybrid vigor, receiving a combination of beneficial genes from both your parents. In plants, offspring from an outcrossing often show hybrid vigor, being larger and producing more offspring than the parents; offspring from inbreeding are often less vigorous, less productive, and suffer from **inbreeding depression.** In general, self-pollination increases the likelihood that some seeds are produced, while cross-pollination increases the likelihood that offspring are healthy and fertile. Cross-pollination also leads to the production of offspring that are genetically different from each other, thus improving the chances that some offspring will survive if the environment changes. This is the trade-off.

Some plants, such as ground-cherry (*Physalis* species), produce perfect flowers that are **self-incompatible.** This means that even though self-pollination can occur, the pollen does not function normally on the pistils of the same plant and no seeds are formed. Thus, while pollination is successful, fertilization does not take place. Self-incompatibility in ground-cherry is controlled by genes, as in most self-incompatible plants. Depending on the plant species, these genes may work on the interaction between the pollen and stigma, thereby preventing the pollen tube from germinating, or may block the growth of the pollen tube as it grows through the style.

Plants produce clones through vegetative reproduction.

Plants are characterized by indeterminate vegetative growth; that is, they have meristematic cells that can divide indefinitely (see chapter 1). Because of this property, tissues or cells can be removed from a plant and induced to grow into a new plant that is a **clone** of the old one (see appendix A). A clone is an individual that is genetically identical to the organism from which it was produced. This kind of reproduction is **vegetative,** or asexual, because sperm and eggs cells are not involved. Most plants reproduce vegetatively to some extent, and some species reproduce almost exclusively by vegetative means. Aspen, for example, grows in large stands that consist of just one or a few clones (see discussion in chapter 7 on *Pando,* the giant quaking aspen). Strawberries and Bermuda grass also produce clones. Vegetative reproduction has advantages over sexual reproduction in those environments where pollination is difficult or the production and germination of seeds is difficult. In what types of environments would you expect that plants frequently reproduce vegetatively?

Humans exploit the ability of plants to reproduce vegetatively by using cuttings to produce many kinds of houseplants, garden plants, and agricultural crops, including roses, African violets, potatoes, bananas, and oranges. In nature, a clone may be long-lived; clones of some grasses routinely live for a few hundred years.

Some plants seem to go through the motions of sexual reproduction without actually undergoing fertilization. For example, species of cholla cacti form flowers that develop fruits without seeds. These naturally seedless fruits fall to the ground and grow into clones of the parent plant (fig. 12.18).

INQUIRY SUMMARY

Breeding systems of flowering plants include self-compatibility, which is associated with inbreeding, and self-incompatibility, which promotes outcrossing. Outcrossing leads to hybrid vigor, while self-pollination leads to inbreeding depression. With

FIGURE 12.18
Seedless fruits of the teddy-bear cholla (*Opuntia bigelovii*) fall to the ground and grow into clones of the parent plant.

self-pollination, however, a plant may have a greater chance of producing seed. Many plants produce clones, which are genetically identical to them.

Why do you think corn farmers buy seed each year instead of planting corn kernels that are produced on their own plants?

Floras Help Identify and Describe the Ecology of Flowering Plant Species of a Region.

The angiosperms are such a large group that a full appreciation of their diversity would be a monumental undertaking. **Systematists** are scientists who study the evolutionary relationships of plants, normally specializing in a group of related species, a genus, a family, or even a group of families. If even professional botanists can know only about 2,000 kinds of plants in significant detail at one time, then what hope is there of keeping track of a quarter of a million species of flowering plants? How can we even know all of the species of just one state, such as California, where more than 6,000 species occur, most of which are angiosperms? Part of the answer is to maintain lists, descriptions, classifications of plants, keys, and herbaria (a **herbarium** is a collection of dried plant specimens that are arranged systematically for easy reference) to help identify plants just as botanists have done for centuries. Books that contain detailed information about plants of a given region are called **floras.** Floras also include nonflowering plants, but these are almost always a small, but important, part of the species included, unless the floras are devoted to some special group. Floras usually include information about the ecology of each

species of the region, including details about the habitat in which it usually grows.

The most recent flora of California includes about 1,400 pages of relevant ecological, morphological, and taxonomic information of plants native to that state. If this proportion held true, then at least 40 books of that size would be needed to cover all known flowering plants of the world. The information in these floras helps botanists identify plants and study relationships of plant species to each other.

A flora has not been completed for all known species of flowering plants for at least two centuries, nor is one likely ever to be completed. Most species live in tropical areas, which occur primarily in developing countries that do not have enough money to finance the work needed to produce their floras. Even in the United States, work on an updated flora of North America began only relatively recently. It is a 10- to 20-year project being developed by a collaboration of hundreds of systematists.

The most important features for classifying plants in floras are usually those of the reproductive organs. For example, plants in the sunflower family (Asteraceae) have small flowers with no sepals, five petals fused into either a tubular or strap-shaped corolla, inferior ovaries, five anthers fused into a cylinder around the style, two stigmas, single-seeded fruits called achenes, and inflorescences in heads (see fig. 12.17). Sometimes, however, vegetative features are also important because of the wide variability in reproductive structures or the similarity of such features among families, each family containing a group of related genera (singular, genus). For instance, the pineapple family is unique because of absorptive scales on its leaves, twisted stigmas, and a haploid chromosome number of 25. These features unite into one family such seemingly disparate plants as the pineapple (*Ananas comosus*) and the Spanish moss (*Tillandsia usneoides*) (fig. 12.19).

INQUIRY SUMMARY

There are so many kinds of flowering plants that systematists usually study only a few of them at a time, depending on the number of species in a genus or the number of genera in a family. Floras contain ecological information about plant species of a region, as well as information to help identify and classify those species. Floras may cover thousands of species in a state or region, but there is no such treatment for all of the world's flowering plants or even for all of the plants in North America. Comprehensive treatments of the flowering plants mostly involve the classification of families.

Look at a flora for your part of the world. What characteristics are used to help identify plants in your flora?

FIGURE 12.19
Pineapple plants (*Ananas comosus*) showing the stiff, narrow leaves and a pineapple. The pineapple has developed from an inflorescence of 100 or more flowers whose ovaries and other structures have united into a single, multiple fruit.

Pollination of a Flowering Plant by an Animal Is an Example of Co-evolution.

All organisms on earth live in communities made up of many species (see chapters 2 and 19). These organisms interact on a daily basis, and one of the most important types of interactions is that between flowering plants and their pollinators. This interaction is an example of a mutualism, which is an interaction where both organisms benefit from their relationship. In the case of pollination by animals, both the plant and the pollinator benefit when the animal visits the flower; the pollinator receives a reward from the flower, and the plant has its pollen transferred to another plant.

When animals reproduce, the male animal brings sperm cells to the female animal, which is carrying the egg or eggs. In fertilization, a sperm and egg fuse, producing the first cell of an offspring, a zygote. However, plants are not motile; they are stuck in the place where they begin to grow. Thus, plants require a carrier to transfer the pollen from an anther to a stigma. The stigma is the entry point for

sperm via a pollen tube to the egg inside the ovule of a flower.

Flowers are often highly adapted for a specific type of pollen carrier—a single species of animal. Animals visit flowers because of a combination of attractants and rewards. **Attractants** are such characteristics as the color and odor of petals or the movement of flower parts in the wind. While attractants may get an animal to visit a flower once, the **rewards** a flower produces encourage the animal to visit other flowers of the same kind and to move from plant to plant. Flowers produce such rewards as nectar, a sugary liquid that animals drink as a source of energy and some nitrogen-containing compounds, as well as pollen, which animals eat.

When an animal visits a flower, it may pick up pollen on its body and carry it to a second flower of the same kind. Upon entering the second flower, the pollen is transferred to the stigma, thereby successfully pollinating the flower. Thus, both plant and pollinator benefit from their mutualistic relationship; the plant benefits by getting pollinated, which allows it to reproduce sexually, while the animal receives the benefit of a reward.

Flowers and the animals that are attracted to them often have evolved so closely that they are now completely interdependent. Such **co-evolution** occurs when the floral characteristics of the plant and the mouthparts, body parts, and behavior of the pollinators become mutually adapted to each other and increase the effectiveness of the plant-animal interaction. As you read about pollination mechanisms, remember that pollination biology is a large and fascinating discipline that links botany and zoology. Pollination biologists are continually discovering new and complex interactions between flowers and their pollinators. Some of the major kinds of pollination mechanisms are described next.

Pollination by wind is common; pollination by water is rare.

The most inefficient, yet common, method of pollination occurs by wind. It is inefficient because the wind blows most of the pollen grains onto the ground, into bodies of water, or into our noses—a major cause of allergic reactions in the spring—and not onto the stigma of an appropriate plant. Wind-pollinated plants produce tremendous numbers of pollen grains, thus improving their chances of sexual reproduction. The main features of wind-pollinated angiosperms are that they produce enormous amounts of lightweight, nonsticky pollen (fig. 12.20); lack showy floral parts or strong fragrances; have well-exposed stamens and large stigmas; have a single ovule in each ovary; and have many flowers packed into each inflorescence (see fig. 12.13). The energy and carbon that might have been used to produce showy floral parts or fragrances instead is used by these plants to make more pollen grains, but the lack of showy, fragrant parts presents no disadvantage because

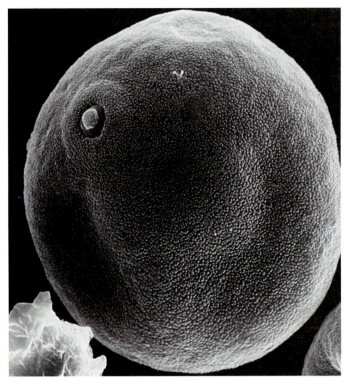

(a)

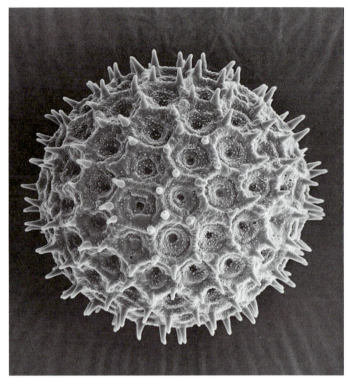

(b)

FIGURE 12.20

Scanning electron micrographs of pollen from different plants, showing variation in pollen wall morphology. (*a*) Grass pollen is smooth with a single pore (×13,000). (*b*) Pollen from a morning glory is spiny (×1,160).

these plants do not need to attract animal pollinators to their flowers. While wind pollination may not be efficient, it is still effective; grasslands and coniferous forests are dominated by plants that use this method of pollination.

Pollination by water is rare, simply because so few plants have flowers that are submerged underwater. Such plants include sea grasses (*Zostera* species), which release pollen that is carried passively by currents of water, much as wind-borne pollen is carried by wind. Other submerged aquatic plants, however, have more complicated pollination mechanisms. For example, pollen of ribbon weed (*Vallisneria spiralis*) is carried from one plant to another in "pollen boats" (fig. 12.21).

Insects pollinate brightly colored, fragrant flowers.

Insects are the most common group of animals that pollinates flowers. Figure 12.22 shows several types of flowers that are adapted for different insect pollinators. Although bees pollinate more kinds of flowers than any other type of insect, flowers can also be pollinated by wasps, flies, moths, butterflies, ants, or beetles. There is no single set of characteristics for insect-pollinated flowers because insects are such a large and diverse group of animals (table 12.2).

FIGURE 12.21

Staminate plants of the aquatic ribbon weed (*Vallisneria spiralis*) release their flowers as "pollen boats," which drift near the larger, surface-borne flowers of pistillate plants. Those pollen boats not eaten by fish may reach the edge of a dimple in the water that is created by surface tension around the flower. Once there, the pollen boats slide down into the flower and are catapulted onto a receptive stigma.

(a)

(b)

(c)

(d)

FIGURE 12.22

Insect-pollinated flowers. (*a*) Bumblebee gathering pollen from an aster. (*b*) The large nocturnal flowers of *Hydnora africana* emit foul odors that attract carrion beetles for pollination. (*c*) This elephant hawkmoth has a long tongue that reaches deep into the narrow tube of a honeysuckle flower (*Lonicera* species). (*d*) Ant pollinating a flower of *Orthocarpus pusillus*. This is one of the few examples of pollination by ants.

Rather, each plant may have a set of reproductive features that attracts mostly one species of insect.

Many kinds of brightly colored flowers look like targets to insects because in ultraviolet light (UV), their petals are darker toward the center of the flower. Such targets, or **nectar guides,** are made by UV-absorbing pigments called flavonoids, which are visible to insects in the sunlight but invisible to other kinds of animals (fig. 12.23), including humans.

Some moths are attracted to strongly scented, night-blooming flowers, which are usually white or cream colored (see fig. 12.22c). Such flowers are often tubular or trumpet shaped, which prevents all but the long-tongued

moths from reaching the nectar. The corolla of flowers such as *Penstemon* have **landing platforms** that stick out like a large bottom lip (fig. 12.24). A landing platform allows an insect to land, crawl into the flower, and drink nectar, picking up pollen grains on its body in the process. The landing platform allows an insect to visit the flower using relatively little energy compared to hovering above it.

Attractiveness and sweetness of fragrance are relative terms. Some "attractive" flowers are reddish brown and appear drab to us, and their fragrance is like that of rotting flesh (see fig. 12.22b). These flowers are referred to as **carrion flowers** and are pollinated by carrion flies or beetles that are attracted to the foul odors as they search

TABLE 12.2

Floral Features Associated with Pollination by Different Kinds of Insects

TYPE OF INSECT	FLOWER COLOR	FLOWER ODOR	NECTAR GUIDES
Beetles	dull	strong and fruity	absent
Carrion/ Dung flies	reddish to purple-brown or greenish	strong and foul	absent
Bee-flies	variable	variable	absent
Bees	variable but not solid red	usually sweet	present
Hawkmoths	white or pale	strong and sweet	absent
Small moths	variable but not solid red	usually sweet	absent
Butterflies	variable, commonly pink	moderately strong; sweet	present

FIGURE 12.24
Some flowers, such as these of the mint, *Lepechinia fragrans*, possess a large landing platform, which allows insects to crawl inside while searching for nectar.

FIGURE 12.25
Nonpollinating flower visitors. This bright yellow crab spider seems conspicuous on the fuchsia-colored petals of this orchid flower, but it is camouflaged in UV light to pollinators who are its prey.

(a)   (b)

FIGURE 12.23
Flowers of the evening primrose (*Oenothera*) are uniform in color to the human eye (*a*), but insects see a different pattern in UV light (*b*).

for a place to lay their eggs. Rotting flesh is an ideal food source for the developing young of these insects. Although the scent of carrion flowers attracts adult carrion flies and beetles, the flowers do not provide any nourishment for the immature insects.

A few animals obtain rewards from a flower but without pollinating the flower; these animals are nectar robbers. Some predatory animals use flowers as hunting

grounds, where they prey on the pollinators attracted to the blossoms (fig. 12.25). Several orchid flowers resemble the females of particular wasp species and open near the time when male wasps emerge from their nests. In addition, these orchid flowers emit a chemical **pheromone** that resembles the odor produced by the female wasps. The combination of sight and smell attracts male wasps to the orchids, triggering a response in the insect to copulate,

12.1 *Perspective* *Perspective*

Orchids That Look Like Wasps

Although many flowers reward their pollinators, other flowers use deception. Perhaps the most sensational cases of pollinator deception occur in orchids. For example, flowers of *Ophrys* species imitate female wasps. These orchids have a wasplike shape, often including a lower lip that is fringed with red hairs like the abdomen of the female wasp. Chemicals in the flower's fragrance are similar to those secreted by the female wasp as a sexual attractant. These orchids bloom before most of the female wasps have emerged in the springtime.

Wasplike flower of a species of Ophrys.

Male wasps are attracted to the wasplike orchid and try to copulate with the flower. A large pollen packet is dislodged from the flower during the vigorous movements of this pseudocopulation. The pollen packet sticks to the wasp and is carried to another flower, where it detaches and nestles snugly among three stigmas. One of the stigmas is sterile and modified into a small outgrowth with a sticky area. The sticky spot ensures that the pollen packet stays on the stigmas after the wasp has left.

---

mistakenly, with the flower (fig. 12.26). In the process, a packet of pollen is attached to the body of the wasp by the orchid flower. The male wasp then flies to another flower and "copulates" with it, transferring the pollen packet to the second flower. Although the plant gets pollinated, the wasp gains nothing from this type of pollination, called **pseudocopulation** (see Perspective 12.1, "Orchids That Look Like Wasps"). What disadvantage would it be to the wasps if all the male wasps copulated with orchids all of the time?

A few flowers attract insects by being warmer than the air surrounding the plants. Insects use the flowers as a resting spot, basking in the warmth and warming up their muscles before they fly. By being warmer than the air, the plants may increase the efficiency of their pollinators, helping insects fly more quickly from flower to flower. Some plant species such as the sacred lotus (*Nelumbo nucifera*) can regulate the temperature of their flowers between 30° and 35°C even if the air is 5°C cooler. The sacred lotus can do this by undergoing rapid cellular respiration and producing heat as a by-product of this process. Other early blooming plant species such as snow trilliums and some anemones flower in the spring when temperatures are relatively low and snow may still linger on the ground. Flowers of some species warm as the flowers follow the sun during the day. Other flowers have a shallow-dish shape with shiny upper petal surfaces that reflect sunlight into their centers. Temperatures in these flowers may be 10°C higher than the air around

them. Although the flowers present very little nectar as a reward, insects are still attracted by the flowers' warmth. As they warm themselves, insects may pick up pollen on their bodies, which they carry to other flowers.

Some bats and birds pollinate large flowers.

Like moth-pollinated flowers, flowers that attract bats and small rodents also open at night. Mammal-pollinated flowers are usually white and strongly scented, often with a fruity odor. Such flowers must be large and sturdy enough to bear the vigorous visits of these small mammals (fig. 12.27), as well as big enough to provide the pollinator enough pollen and nectar to fulfill its large energy requirements.

Hummingbirds are the most common group of flower-visiting birds in the Americas; honeycreepers are common pollinators in Africa and Asia. The long beaks of both kinds of birds can reach to the base of long, tubular corollas to obtain nectar (fig. 12.28). Hummingbirds are mostly attracted to bright red flowers, which are not usually attractive to insects. Birds also have a relatively poor sense of smell, and hummingbird-pollinated flowers are generally odorless. Columbines, penstemons, and scarlet monkeyflowers are examples of flowers that attract hummingbirds. Some flowers such as *Fuchsia* commonly hang upside down and have an inferior ovary. This position helps the hummingbird drink nectar from the flowers, and the structure of the flower protects the delicate ovary and

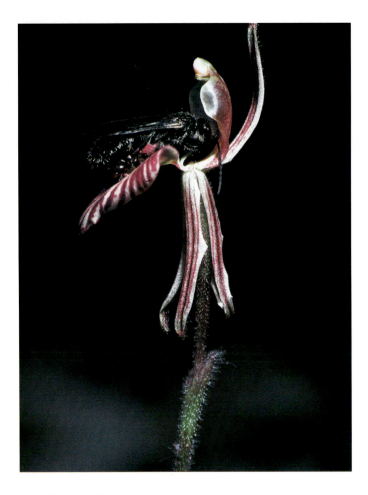

FIGURE 12.26

The flower morphology of this zebra orchid makes its pollen accessible to only a few kinds of pollinators such as this wasp. The flower releases sex pheromones that prompt the pseudocopulation by this wasp.

ovules inside when the hummingbird inserts its beak into the flower (fig. 12.28*a*).

INQUIRY SUMMARY

Flowers are adapted for pollination mainly by wind or animals. A few kinds of flowers rely on pollination by water. Wind-pollinated flowers produce abundant pollen in nonshowy flowers. Most animal-pollinated flowers are showy or strongly scented. Pollination is a mutualistic relationship in which the plant may reproduce sexually while the animal receives a reward such as food or warmth. Pollinators have co-evolved with the plants that they pollinate.

What might be the advantage to a plant of having only a single kind of animal as its pollinator compared to many different pollinators?

A Fruit Develops from the Ovary of a Flower.

A fruit is a seed container that develops from the ovary of a flower and, often, other tissues that surround it (fig. 12.29). As such, fruits are products of flowers and therefore occur only in flowering plants. A fruit is important because it protects developing seeds and may help disperse mature seeds from the parent plant. (Recall from chapter 2 the importance of fruits in the life of the persimmon tree and the importance of fruits in the dispersal of seeds.) Angiosperms have a remarkable diversity of fruit types, classified on the basis of the characteristics of the mature ovary. For example, a fruit may

(a)

(b)

FIGURE 12.27

Mammal-pollinated flowers. (*a*) A greater short-nosed bat feeds on the pollen and nectar of a banana plant (*Musa* species). (*b*) The tiny Australian honey-possum pollinates plants such as this coral gum (*Eucalyptus* species) as it forages for pollen and nectar.

(a)

(b)

FIGURE 12.28

Bird-pollinated flowers. (*a*) A rufous hummingbird gets nectar from flowers of *Fuchsia*. (*b*) The yellow-plumed honeyeater has a brush-tipped tongue for lapping up nectar and pollen from the bell-fruited mallee (*Eucalyptus pressiana*).

FIGURE 12.29

Angiosperms are plants whose seeds are enclosed in fruits, as shown here from papaya (*Carica papaya*).

be fleshy or dry at maturity, or the ovary may be fused to other kinds of tissues. These adaptations are often related to how the fruit and its seeds inside are dispersed. Examples of different types of fruit and their main features follow.

Types of fruit can be identified using a dichotomous key.

Fruits are classified by several main features, including whether they are fleshy or dry; split open at maturity; develop from one or more ovaries in a single flower or an inflorescence; or consist solely of mature ovary. Based on

these and other differences, the major types of fruit can be presented in a **dichotomous key** (table 12.3). By using this key, you can identify the major types of fruit by choosing between successive pairs of features (dichotomies) at each of several steps. Examples of each fruit type are listed in the key, and several are illustrated in figure 12.30. The key should enable you to determine the fruit types of, for example, bananas, peaches, and peanuts.

Ovaries that have matured into a fleshy fruit often consist of three regions: the skin (**exocarp**), the fleshy part (**mesocarp**), and the interior (**endocarp**) that surrounds the seeds. Collectively, these three regions constitute the **pericarp.** Stone fruits, such as apricots and peaches, fit this description, as do tomatoes. The main difference between an apricot and a tomato is that the endocarp in an apricot is hard and stony, whereas the endocarp of a tomato is soft and moist. The mesocarp and endocarp are fused and indistinct from each other in some types of fruit. In dry fruits, all three layers are fused into one pericarp, which is often a thin layer around the seeds.

Although the classification of fruits is informative, it is also inexact because many variations in fruits do not fit into the key to major fruit types. For example, the fleshy tissue of strawberries is an expanded receptacle, which makes it an **accessory fruit**—that is, a fruit whose flesh is not derived from the ovary. **Pomes,** such as apples, are also accessory fruits. Other fruits, such as the coconut, are modified so that they partially fit the description of a certain fruit type but not perfectly. A coconut is a **drupe,** and drupes normally have a fleshy mesocarp such as of a peach; however, the mesocarp of a coconut is a fibrous husk (fig. 12.31). Only the hard endocarp containing a single, large seed ends up on the grocery shelf, but it is difficult to separate the endocarp from the seed. Blackberry and magnolia form

TABLE 12.3	Dichotomous Key to the Major Types of Fruit

Complex fruits (i.e., from more than one ovary)

    fruit from many carpels on a single flower (magnolia, strawberry, blackberry):   AGGREGATE FRUIT

    fruit from carpels of many flowers fused together (pineapple, mulberry):   MULTIPLE FRUIT

Simple fruits (i.e., from a single ovary)

    fleshy fruit

        flesh derived from ovarian tissue

            endocarp hard and stony; ovary superior and single-seeded (cherry, olive, coconut):   DRUPE

            endocarp fleshy or slimy; ovary usually many-seeded (tomato, grape, green pepper):   BERRY

                outer layer of berry a leathery skin containing oils (orange, grapefruit, lemon):   HESPERIDIUM

                outer layer of berry a thick rind not containing oils (watermelon, pumpkin, cucumber):   PEPO

        flesh derived from receptacle tissue (apple, pear, quince):   POME

    dry fruits

        fruits that split open at maturity (usually more than one seed)

            seeds released through longitudinal seams

                split occurs along one seam in the ovary (magnolia, milkweed):   FOLLICLE

                split occurs along two seams in the ovary

                    seeds borne on one of the halves of the split ovary (pea and bean pods, peanuts):   LEGUME

                    seeds borne on a partition between halves of the ovary (mustard, radish):   SILIQUE

            seeds released through pores or multiple seams (poppies, irises, liles):   CAPSULE

        fruits that do not split open at maturity (usually one seed)

            pericarp hard and thick, with a cup at its base (acorn, chestnut, hickory):   NUT

            pericarp soft and thin, without a cup

                ovaries often together in pairs (parsley, carrot, dill):   SCHIZOCARP

                ovaries occur singly

                    pericarp winged (maple, ash, elm):   SAMARA

                   pericarp not winged

                      single seed attached to pericarp only at its base (sunflower, buttercup):   ACHENE

                      single seed fully fused to pericarp (cereal grains):   CARYOPSIS

FIGURE 12.30

Examples of some of the types of fruit described in table 12.3: (*a*) grapes (berry), (*b*) apple (pome), (*c*) acorns (nut).

(a)   (b)   (c)

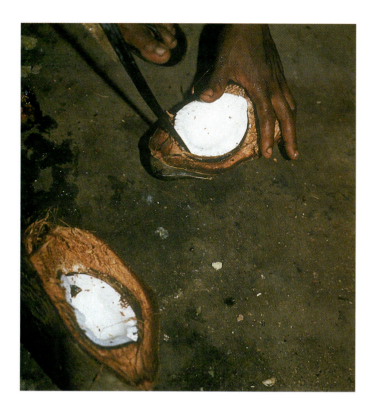

FIGURE 12.31
The fruit of a coconut (*Cocos nucifera*) is a drupe, but the mesocarp is fibrous instead of fleshy.

aggregate fruits, although the individual ovaries in blackberry develop into tiny drupes, and in magnolia, they develop into follicles. A cob of corn is a **multiple fruit,** but each kernel is a single fruit, called a caryopsis, or **grain.**

   The importance of the different types of fruits is related to how they, and the seeds they contain, are dispersed from the parent plant (revisit chapter 2). Fleshy fruits are typically dispersed by hungry animals that carry off and eat the fruit, dispersing the seeds in the process. Dry fruits are typically dispersed by the wind or contain seeds that are dispersed by the wind; these fruits and seeds possess features that aid their dispersal, such as wings or hairs.

Hormones promote fruit development.

Fertilization and the production of seeds are usually prerequisites for the development of fruits. Chemical signals, called hormones, are secreted by seeds as they develop. These hormones induce the ovary to expand and mature into a fruit, an example of cell signaling (see chapter 6). In agriculture and horticulture, naturally produced hormones or their synthetic counterparts are applied to some crops so that fruits will form and mature in synchrony, which makes harvesting more efficient and economical. Treatment of flowers with artificial hormones can also induce the formation of seedless fruits (e.g., seedless grapes) in the absence of fertilization.

INQUIRY SUMMARY

The ovary of a flower develops into a fruit after fertilization, aided by hormones. Fruits are either fleshy or dry, which determines how they are dispersed. A fleshy fruit has three distinct regions of ovary tissue, but some fleshy fruits are surrounded by accessory tissue that is not derived from the ovary. Dry fruits may remain tightly closed, or they may split open at maturity. A dichotomous key helps identify fruit types.

What do you think is the advantage to the plant that its seeds are mature at the time the fruit is taken from the plant?

The Economic Value of Flowers and Flowering Plants Is Enormous.

Most of this book is devoted to flowering plants. As we learned in the first chapter, plants are essential to human existence on earth, and most of the plants that are important to us produce flowers. We rely on flowering plants as a source of food, medicines, clothing, and building materials and for a wide variety of other uses. Flowers themselves, however, are of great economic importance. Imagine holding the Rose Parade on New Year's Day without flowers or celebrating birthdays, your anniversary, or Mother's Day without them. The business of growing and selling ornamental potted plants, flowers, fruits, and seeds is called **floriculture.** In the Netherlands, known for growing and exporting tulips, floriculture is a $4-billion-dollar industry and makes up 20% of the total agricultural productivity of the country. In the United States, in 1999, the total production of cut flowers such as roses sold at florist shops, potted flowering plants, bedding plants for gardens, turf grasses for lawns, and other greenhouse and nursery plants totaled more than $8 billion dollars.

   Production of perfumes and scented lotions and shampoos, mixtures that depend on natural and artificial fragrances, contribute to a multibillion-dollar industry in the United States. Fragrances from natural perfumes come mainly from the oils of flowers, which are extracted through a process called steam distillation. A vat is filled with flower petals, and then steam is forced through the vat, causing the fragrant oils from the flowers to evaporate. The steam is collected and allowed to condense, after which the fragrant oil is separated from the water. Six hundred bushels of rose petals are needed to produce about 500 grams of concentrated rose oil. The flower oil is then diluted in alcohol and added to other fragrances and chemicals to make the perfume or lotion. Perfume contains approximately 20% fragrant oils, eau de toilette between 11% and 18%, and cologne between 5% and 10%. Plant hunters from perfumeries travel the world in search of undiscovered flowers with potentially important, new fragrances. The prize for finding a new flower could be the next "Obsession" or "Chanel."

SUMMARY

Flowers are the reproductive structures of angiosperms. A flower is a stem tip that bears some or all of the following kinds of appendages on a receptacle: sepals, petals, stamens, and pistils. The pistil consists of an ovary, which includes one or more ovule-bearing carpels, and a stigma that is usually borne on a style. During pollination, pollen is transferred from an anther to a stigma. A pollen tube grows into an ovule and carries sperm to the egg inside the ovule. After fertilization, ovules become seeds, and ovaries become fruits.

The angiosperms, which are the most dominant plants on earth, have flowers that include seeds in a carpel. Fossils of carpels and other parts of flowers are known from Cretaceous deposits that are at least 140 million years old. A 120-million-year-old fossil flower resembles several plants in different families of monocots and dicots on the basis of simple features of the flower and its herbaceous type of body. Evidence suggests that flower parts evolved from leaves. The main force behind the rapid evolution of angiosperms may have been pollination by insects. The first flowers were probably pollinated by beetles; later angiosperms attracted butterflies and bees.

The great diversity in angiosperm species is based largely on variation in reproductive morphology. Angiosperms are divided into two large groups: the monocots and the dicots. Flowers may be complete or incomplete, perfect or imperfect, regular or irregular. They also may be solitary or arranged in an inflorescence with several other flowers. The diversity of angiosperms is so great that most of our knowledge about this group is divided into regional floras or taxonomic treatments of smaller groups of plants.

Most flowers are adapted for pollination by wind or by animals. Wind-pollinated flowers are usually incomplete and not showy, whereas insect-pollinated flowers and bird-pollinated flowers are often colorful. Night-blooming flowers attract nocturnal mammals or insects and are usually aromatic and white or cream colored.

Fruits may be either fleshy or dry. Some kinds of dry fruits remain sealed at maturity, whereas others split open at maturity. Fleshy fruits that are colorful and sweet tasting attract animals that eat them and disperse the seeds. Dry fruits, or the seeds from them, are also adapted for dispersal, often by the wind.

Writing to Learn Botany

Describe the role animals have played in the evolution of different kinds of flowers.

THOUGHT QUESTIONS

1. Dinosaurs and most other animals went extinct at the end of the Cretaceous period about 65 million years ago. Some scientists have proposed that such mass extinctions were caused by geological catastrophes. If this is true, how might angiosperms have escaped such catastrophes?

2. In addition to co-evolution with insect pollinators, what other kinds of co-evolutionary interactions may have occurred between flowering plants and animals?

3. Does the fact that angiosperms do not appear in the fossil record until the Cretaceous period mean that they were not around before then? Explain your answer.

4. Angiosperms usually have a shorter life cycle than do other plants. How might this have affected the evolution of angiosperms?

5. Why are strawberries, raspberries, and mulberries not berries? Since they are not berries, what are they?

6. Many seeds store their food reserves mostly as oil instead of carbohydrate. What are some examples of plants that produce oily seeds?

7. Why is it incorrect to refer to commercial sunflower "seeds" (those still in their shells) as seeds? Since they are not seeds, what are they?

8. Apples are fleshy, sweet-tasting fruits that attract and are eaten by fruit-eating animals that disperse their seeds. How could you explain the fact that the seeds of such an attractive fruit contain cyanide-producing chemicals?

9. What might the advantages be for self-compatibility versus self-incompatibility?

10. Recent research suggests that the production of more flowers decreases a plant's chances of mating with others by increasing the likelihood of self-pollination. How would you test this hypothesis?

11. Do plants compete to attract pollinators? Write a short paragraph to explain your answer.

SUGGESTED READINGS

Articles

Bakker, R. T. 1986. How dinosaurs invented flowers. *Natural History* 11:35–38.

Beattie, A. J. 1990. Seed dispersal by ants. *Scientific American* 263:76.

Burd, M. 1994. Bateman's principle and plant reproduction: The role of pollen limitation in fruit and seed set. *Botanical Review* 60:83–139.

Cox, P. A. 1993. Water-pollinated plants. *Scientific American* 269:68–74.

Fleming, T. H. 1993. Plant-visiting bats. *American Scientist* 81:460–67.

Milius, S. 1999. The science of big, weird flowers. *Science News* 156:172–74.

Natural History. 1999. The Flower Issue. 108 (4).

Strauss, E. 1998. How plants pick their mates. *Science* 281:503.

Books

Barth, F. G. 1991. *Insects and Flowers: The Biology of a Partnership.* Princeton, NJ: Princeton University Press.

Bell, A. D. 1990. *Plant Forms: An Illustrated Guide to Flowering Plant Morphology.* New York: Oxford University Press.

Fenner, M., ed. 1992. *Seeds, the Ecology of Regeneration in Plant Communities.* Wallingford, England: CAB International.

Lloyd, D. G. and S. C. H. Barrett, eds. 1996. *Floral Biology.* New York: Chapman & Hall.

ON THE INTERNET

Visit the textbook's accompanying web site at http://www.mhhe.com/botany to find live Internet links for each of the topics listed below:

Characteristics of the flower

Fruit types

Flower evolution

Angiosperm life cycle

Monocots and dicots

Inflorescence types

Plant systematics

Floriculture

Flowers, pollinators and coevolution

Carl Linne

Garden pea (Pisum sativum) with flowers, pods (fruit), and seeds. Gregor Mendel used the garden pea to establish the basic postulates of the theory of inheritance. Genes govern inheritance, but what are genes?

CHAPTER 13

Genetics

LEARNING ONLINE

The Online Learning Center at www.mhhe.com/botany contains a world map on which markers represent articles about current botanical news in that particular area. This is just one of the many useful resources available to you on the OLC.

"Like begets like." This saying has been around for thousands of years because people have long recognized that, for example, only roses can make more roses and only camels can make more camels. This means that critical information passes from one generation to another, thereby defining what each organism can be. The inheritance of such information is the foundation for the discipline of genetics. This chapter helps you understand how heredity, reproduction, and genes work.

People have always been interested in heredity and its importance in their families, crop production, and animal husbandry. The earliest farmers, thousands of years ago, selected bigger, more desirable plants for continued cultivation. By so doing, they selected herita-ble characteristics, even though they had little or no understanding of inheritance.

Many theories on the nature of heredity, some based on religion or mythology and some based on factual observations, have held sway at one time or another during the past few centuries. The modern theory of inheritance was established in the mid-nineteenth century by Gregor Mendel in Austria. Once science fully appreciated Mendel's work, many discoveries involving the chemical and biological nature of heredity followed. Among the more recent of these discoveries are that DNA is the hereditary material and that genes are the basic units of inheritance. The history of genetics serves as an excellent introduction to modern genetics and how we use plant genetics to feed ourselves and other animals.

Gregor Mendel's Landmark Work Led to the Theory of Inheritance.

Gregor Mendel (1822–84) was an obscure amateur botanist working alone in a monastery garden in Austria (fig. 13.1). He studied the simple genetic system of green peas (*Pisum sativum*) and recognized the importance of his results for understanding patterns of inheritance that are basic to all organisms. Remarkably, Mendel chose an appropriate plant species, designed powerful experiments, and interpreted his results without knowing about chromosomes or genes. By doing so, he derived a set of correct assumptions about heredity that form the theory of inheritance for all higher organisms, including you. Subsequent sections of this chapter describe how Mendel achieved such a significant scientific advancement and how his work fits into the framework of modern genetics.

We use a historical approach to explaining genetics because of its significance in biology and also because it is interesting and important. Remember, the basic principles of genetics described in the following pertain to plants and animals. If you wish to breed a champion show dog or horse, you must understand the basic principles of genetics. Or if someone you know has a genetic disease, such as lupus or Down syndrome, knowledge of genetics can help you understand the condition.

Before Mendel, the most extensive studies of plant hybridization were done by Karl Freidrich von Gaertner, who was from Germany. Beginning in the 1820s, Gaertner's experiments established the role of pollen in transmitting **traits** of the pollen-producing parent to the offspring. A trait is a genetically determined characteristic or condition; that is, it is a distinguishing feature. Also, Gaertner discov-ered that certain traits in garden peas are expressed preferentially over others. For example, purple flowers are **dominant** (as later proposed by Mendel) over white flowers, where flower color is a trait. Mendel suggested a dominant trait is one that masks the alternative trait for a particular feature. The masked trait is said to be **recessive.** Without suggesting the terms, Gaertner actually determined the dominant and recessive forms of several traits, including flower color (purple over white), seed shape (round over wrinkled), and pod color (green over yellow) in the garden pea. By confirming the tendency of these dominant traits to be expressed more commonly among offspring, Gaertner's work gave Mendel a strong foundation for developing the theory of inheritance and Mendel's subsequent suggestion of the terms dominant and recessive.

Armed with a strong background in biology and mathematics, Mendel tried to measure the results of Gaertner's experiments—to "ascertain their numerical interrelationships." His work is an excellent example of how expertise in one discipline (mathematics) can provide insight into another (biology). Although many people had done the same experiments as Mendel, it was his use of mathematics that led to his great discoveries.

Mendel studied several kinds of plants, mostly from the pea family, that were easy to grow and manipulate in hybridization experiments. The focus of the following discussion is on one of these, the garden pea, because it is the only one that showed patterns of inheritance that Mendel could fully explain.

Mendel grew his pea plants for 2 years, choosing nine sets of traits based on the ease of distinguishing one trait from another. Three of these traits (flower color, seedling color, and seed-coat color) were so well correlated, however, that he regarded them as expressions of the same *factor,*

The Lore of Plants

The first plant breeders were probably ancient Babylonians and Assyrians who cultivated date palms (*Phoenix dactylifera*) in the region of modern-day Iraq more than 4,000 years ago. Like many plants, date palms are dioecious, meaning that each plant produces either pollen or fruits, but not both. Date growers discovered that pollen from just a few trees could be transferred by hand to several hundred fruit-producing trees, causing them to make dates. This practice was important economically because

only a few fruitless (pollen) trees had to be maintained in each plantation. Unknown to the growers, pollen transfer was also important biologically because it promoted variation in the size, flavor, shape, and color of the dates. As a result, desirable variations were selected for replanting through many generations, producing more than 400 varieties of dates. Even though the patterns of inheritance were exploited, how plants inherited traits from their parents remained unknown for many centuries.

FIGURE 13.1

Gregor Mendel (1822–84). Mendel's experiments with the garden pea established the basic postulates for the theory of inheritance.

which today he would call a gene. Thus, Mendel followed the inheritance of seven simple, easily distinguished, and separate pairs of traits shown in figure 13.2.

In his initial experiments, Mendel crossed (hybridized) pairs of plants that differed in only one visible trait. For example, he crossed purple-flowered plants with white-flowered plants, and wrinkled-seeded plants with smooth-seeded plants. In so doing, he made reciprocal crosses between parental strains; that is, he transferred pollen from purple flowers to white flowers and pollen from white flowers to purple flowers. He then allowed the offspring of the various crosses to self-pollinate for several generations and counted the plants with each trait in every generation. The same techniques are used today by plant geneticists.

The first generation of offspring is called the **F_1 generation.** (The *F* comes from filial, which is derived from the Latin word for "daughter or son"; we use the term **offspring** instead.) In every experiment that Mendel did, only the dominant trait was expressed in the F_1 generation. However, when F_1 plants were allowed to grow and then self-pollinate, the recessive trait reappeared in the F_2 generation. By counting the offspring in the F_2 generation, Mendel found that the ratios of dominant to recessive traits were the same for all seven pairs of traits, about 3:1 (table 13.1). Mendel concluded that "factors" from each parent somehow controlled the traits he observed. Moreover, he suggested that each offspring gets one of two factors from each parent. In the F_1 generation, all of the offspring have both factors, but the factor for the dominant trait masks the factor for the recessive trait. Mendel also predicted that any plant having two different factors would donate the recessive factor to one-half of the offspring and the dominant factor to the other half. Thus, the probability for F_2 plants to receive two recessive factors would be one-half times one-half, or one-fourth. This prediction matches the observed ratio of three dominant traits (3/4) to one recessive trait (1/4) in the F_2 generation (fig. 13.3).

13.1 *Perspective* Perspective

Artificial Hybridization in Plants

The offspring of parents with different traits is called a **hybrid.** Reproductive structures of many kinds of flowers allow easy manipulation of the pollen-containing anthers, which means that plant hybrids can be made artificially—that is, by human hands. In hybridization experiments, anthers are typically removed from each flower to prevent self-pollination. This is necessary only when a plant is self-compatible—that is, when it can fertilize itself. Not all plants are self-compatible. Mendel took the precaution of removing pollen-producing anthers because the garden pea is a self-compatible species.

After the anthers are removed, pollen from another individual is brushed onto the stigma, which is the receptive surface of the seed-producing organ the gynoecium (collectively, all of the carpels of a single flower, chapter 12). The pollen can be from any other plant, but successful fertilization depends on genetic compatibility between the pollen parent and the seed parent. It is impossible to fertilize a garden pea with pollen from an orchid because peas and orchids are too dissimilar genetically. However, pollen from one variety of garden pea can fertilize another variety of garden pea (see figure). The limits of interfertility are inconsistent from one species of plant to another. For example, many species of pines cannot be hybridized artificially, but hybrids of orchids, mustards, carnations, and many other plants are common.

Plant breeders develop many new kinds of plants by selecting flowers having the most desirable traits (size, color, etc.) for making artificial hybrids. Most commercially available garden plants are the products of many generations of hybridization. These plants are referred to as cultivars to denote cultivated varieties that are not found in nature. Botanists have produced thousands of particularly remarkable or unique forms. The most valuable of these are patented.

A pea flower (longitudinal section). The petals enclose the anthers and stigma, thereby promoting self-pollination.

Artificial hybridization in pea flowers. Cross-pollination occurs when anthers from one flower are brushed against the stigma of another flower.

Mendel accounted for heritable factors and the probabilities of inheritance for each trait only because he counted the offspring of each cross. Several assumptions can be derived from his initial experiments. Taken together, these assumptions are referred to as *Mendel's theory of inheritance.* Note, however, the assumptions of this theory, which are listed next, are modernized versions of Mendel's ideas; Mendel did not actually state them as assumptions, nor did he propose a theory of inheritance. In fact, Mendel barely mentioned the ideas behind numbers 7 and 8.

1. Each trait is controlled by heritable factors (today these can be called genes).

2. Factors are passed from parent to offspring by reproductive cells (sperm and egg).

3. Each individual contains pairs of factors in every cell except reproductive cells, which contain only one factor of each trait.

4. Paired factors separate during the formation of reproductive cells so that each reproductive cell gets one of the factors of a pair.

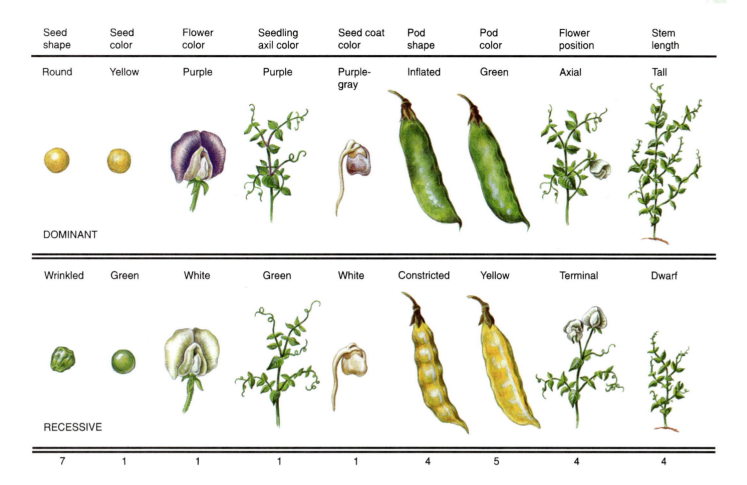

Seed shape	Seed color	Flower color	Seedling axil color	Seed coat color	Pod shape	Pod color	Flower position	Stem length
Round	Yellow	Purple	Purple	Purple-gray	Inflated	Green	Axial	Tall
DOMINANT								
Wrinkled	Green	White	Green	White	Constricted	Yellow	Terminal	Dwarf
RECESSIVE								
7	1	1	1	1	4	5	4	4

FIGURE 13.2

Characteristics of the garden pea studied by Mendel. Mendel regarded flower color, seedling axil color, and seed coat color as expressions of the same factor. The number below each characteristic is the number of the chromosome in garden peas that bears the gene for the characteristic. This information was not discovered until the twentieth century.

TABLE 13.1

Mendel's Experimental Results for Seven Characteristics of Garden Pea (these data represent the F$_2$ generation)

	SAMPLE SIZE	DOMINANT FORM	RECESSIVE FORM	RATIO
	7,324 seeds	5,474 round	1,850 wrinkled	2.96:1
	8,023 seeds	6,022 yellow	2,001 green	3.01:1
	929 plants	705 purple flowered	224 white flowered	3.15:1
	1,181 plants	882 with inflated pods	299 with constricted pods	2.95:1
	580 plants	429 with green pods	152 with yellow pods	2.82:1
	858 plants	651 with axial flowers	207 with terminal flowers	3.14:1
	1,064 plants	787 with long stems	277 with short stems	2.84:1
TOTAL:	19,959	14,949	5,010	2.98:1

Note: Two additional characters, seedling axil color and seed-coat color, were also examined by Mendel. However, purple seedling axils and purple-gray seed coats were always associated with purple flowers, so Mendal regarded these three features as expressions of the same factor. He therefore recorded data for flower color and disregarded the other two.

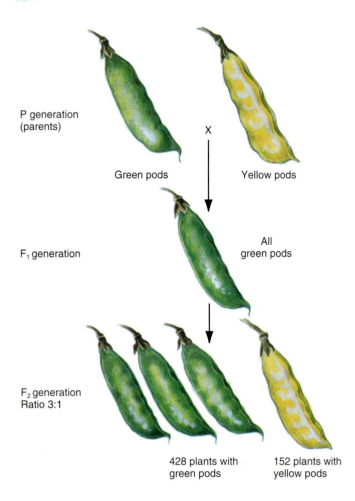

P generation
(parents)

Green pods   X   Yellow pods

F₁ generation   All
                              green pods

F₂ generation
Ratio 3:1

428 plants with   152 plants with
green pods   yellow pods

FIGURE 13.3

The inheritance pattern of one of Mendel's crosses (green versus yellow pods) through two generations. All plants in the F₁ generation had green pods. About three-fourths of the individual plants in the F₂ generation had green pods, and about one-fourth had yellow pods.

5. There is an equal chance that a reproductive cell will get one or the other factor of a pair.

6. Each factor from one parent has an equal chance of combining either with the identical factor or with the other factor from the other parent during fertilization.

7. Sometimes one factor dominates the other factor; in such cases, the dominant factor controls that feature of the plant.

8. When two or more traits are under consideration, the factors for each trait move independently to the reproductive cells. This is known as independent assortment.

The significance of Mendel's work is now clear, but it remained obscure to science until the beginning of the twentieth century.

In modern genetics, each of Mendel's paired factors represents a **gene,** which we first discussed in chapter 4. One fairly standard definition of a gene is that it is a sequence of

DNA that codes for a gene product, usually mRNA that directs the synthesis of a protein. Genes also code for tRNA and rRNA. Each pair of genes is an **allele,** which is an alternative form of the gene; in the previous examples, this could be the dominant and recessive genes for a trait. This means that in true-breeding parents, the genes controlling flower color or other features used by Mendel had two identical alleles for each gene; for example, two alleles for purple flower color. A plant that has the same alleles for a gene is said to be **homozygous** for that gene. In contrast, a plant that has different alleles for a gene is **heterozygous** for that gene. Thus, the parental generation in Mendel's experiments was homozygous for each gene, either dominant or recessive. Thus, one parent contained two alleles for purple flower color and the other parent two alleles for white flower color. The F₁ generation was therefore heterozygous for each gene: each offspring had one allele for purple and one allele for white flowers. The F₂ generation included both homozygous and heterozygous plants.

Genotypes are composed of genes, and phenotypes are traits.

For clarity, we distinguish between an organism's genes, individually or collectively, which we call its **genotype,** and the observable characters or traits they control, which we call its **phenotype** (what the organism looks like and how it functions). For example, in garden peas, phenotypes for flower color are purple and white. The genotype can be designated in several ways, and one is by the first letter of the dominant trait. Accordingly, the gene for flower color in garden pea is P for the dominant allele (purple) and p for the white recessive allele (fig. 13.4). Consequently, the genotype for flower color is PP (homozygous) or Pp (heterozygous) for purple and pp (homozygous) for white. In this case, the only way white flowers can be produced is if the plant has two recessive alleles—pp. Of course we can observe many phenotypes, but we cannot observe genotypes—the types of alleles an individual has—even if we look at the chromosomes of a cell. Thus, botanists do genetic crosses to study patterns of inheritance.

Note that Mendel counted phenotypes and calculated phenotypic ratios. The phenotypic ratios for all seven traits of garden pea in the F₂ generation were the same: about 3 dominant to 1 recessive (fig. 13.4). However, what do we know about the genotypic ratios? We know that the genotype for white flowers (pp) occurs 25% of the time because of the probability of matching two recessive alleles (pp) from heterozygous parents (Pp) is 1 in 4. Similarly, the probability of matching two dominant alleles (PP) would be the same. The remaining 50% of the genotypes are therefore heterozygous (Pp). Thus, the 3:1 phenotypic ratio in the F₂ generation is based on a genotypic ratio of 1 PP to 2 Pp to 1 pp (1:2:1).

Genotypic and phenotypic ratios can be conveniently calculated by using a grid called a Punnett square. Recall that gametes are haploid sex cells, called sperm and egg. A Punnett square is a checkerboardlike diagram that has the genotype of one parent's gametes across the top and the genotype of the other parent's gametes down one side (fig. 13.4). In this way, we show that the gametes (i.e., sperm and egg) from a cross between two heterozygotes for flower color ($Pp \times Pp$) would be P and p across the top and P and p down the side. When the gametes of such a cross fuse (fertilize), adding the genotypes yields the expected ratio of 1:2:1 in the genotypes of the offspring. This ratio is perhaps as simple to obtain by direct inspection as it is by using a Punnett square. Nevertheless, the Punnett square can be useful for calculating ratios from multiple traits, as discussed next.

Mendel also studied multiple traits.

Mendel also did experiments focusing on two or three pairs of traits at a time. For example, he studied combined inheritance of seed shape (dominant round versus recessive wrinkled) and seed color (dominant yellow versus recessive green) in the same plants. One set of parents was homozygous for both dominant traits ($RRYY$), and another set of parents was homozygous for both recessive traits ($rryy$); this means he crossed plants that were true-breeding for round and yellow seeds with plants that were true-breeding for wrinkled and green seeds. All of the seeds produced by plants from this cross were round and yellow. Moreover, since the seeds received dominant alleles from one parent and recessive alleles from the other parent, these seeds (F_1) were heterozygous for both genes ($RrYy$). This pattern of inheritance conforms to predictions from experiments using one pair of traits at a time. (Note that a Punnett square of this cross would be simple: one parent has only gametes with the genotype RY, and the other parent has only gametes with the genotype ry. Thus, the sole genotype of their offspring is—the F_1 generation—$RrYy$.)

When the hybrid seeds were grown into mature F_1 offspring and allowed to self-pollinate, the plants produced the F_2 generation of seeds, which included four phenotypes in the quantities listed in table 13.2. The approximate ratios among these phenotypes are nine round yellow to three round green to three wrinkled yellow to

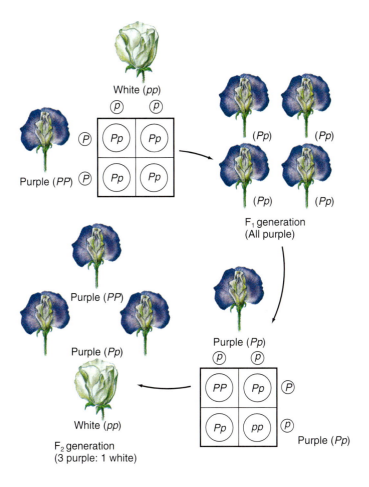

FIGURE 13.4

Use of a Punnett square to show the inheritance of phenotypes and underlying genotypes in Mendel's cross with flower color. P and p at the top and side of the Punnett squares represent gametes that have dominant (P) or recessive (p) alleles of the gene for flower color. The genotypes of the offspring are predicted by combining alleles from different gametes into the boxes of the Punnett square. Each square of the box indicates the genotype of one type of offspring.

TABLE 13.2	Mendel's Results from a Dihybrid Cross		
	SAMPLE SIZE	PHENOTYPES	RATIO
Seed shape–Seed color	556 seeds	315 round yellow	9.84
		108 round green	3.38
		101 wrinkled yellow	3.16
		32 wrinkled green	1.00

Note: Ratios are calculated by using the wrinkled green phenotype as the common denominator.

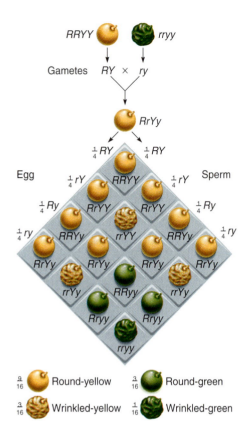

FIGURE 13.5

Inheritance in the garden pea. This Punnett square shows the expected pattern of inheritance in a dihybrid cross according to independent assortment of alleles for seed color and seed shape. It predicts four phenotypes in the F_2 generation in a ratio of 9:3:3:1.

one wrinkled green (9:3:3:1). This is the expected result of a **dihybrid cross,** which follows two genes that are both heterozygous (fig. 13.5). Using the pea seed example, a dihybrid cross is written $RrYy \times RrYy$. In a Punnett square of this cross, each parent has four different kinds of gametes: RY, Ry, rY, and rY. This means that the Punnett square of a dihybrid cross will have 16 boxes (4 × 4). These 16 boxes will contain nine different genotypes that underlie the four phenotypes of the F_2 generation. What are these nine genotypes?

The phenotypic ratios from Mendel's dihybrid cross can be explained only if the segregation of one pair of traits (green versus yellow seeds) is not influenced by the other pair of traits (round versus wrinkled seeds)—that is, if different genes are inherited independently of each other (segregation is the separation of paired genes (alleles) during meiosis, so that the members of each pair of alleles appear in different gametes). In other words, both dominant traits don't necessarily wind up in the same gamete all the time. Although Mendel did not recognize the 9:3:3:1 ratio of F_2 phenotypes, he did note that different pairs of traits were inherited independently of one another. From this observation, we derive assumption 8 (independent assortment) of Mendel's theory of inheritance.

INQUIRY SUMMARY

Twentieth-century geneticists derived several assumptions from Mendel's experiments on inheritance in garden peas. These assumptions make up Mendel's theory of inheritance. A dominant trait is one that masks the alternative trait for a particular feature. The masked trait is said to be recessive. By counting the offspring in the F_2 generation, Mendel found that the ratios of dominant to recessive traits were the same for all seven pairs of traits in peas, about 3:1. Each one factor of a pair of genes is an allele, which is an alternative form of the gene. A plant that has the same alleles for a gene is said to be homozygous for that gene. In contrast, a plant that has different alleles for a gene is heterozygous for that gene. An organism's genes, individually or collectively, is its genotype, and the observable characters or traits they control is its phenotype (what the organism looks like or how it functions).

How does genotype determine phenotype?

Meiosis Separates "Heritable Factors" During Gamete Formation.

Plants and animals produce **gametes**—that is, sex cells called the **sperm** (male) and **egg** (female). Gametes are termed *haploid* (the *n* number of chromosomes) because they have only one of each pair of chromosomes. More importantly, each gamete has one of each pair of chromosomes and therefore one of each of the pair of genes. The example of X and Y chromosomes in humans is perhaps most familiar to you. Your father has one X and one Y chromosome in all of his diploid cells, his genotype is XY, and his phenotype is male. Upon the production of sperm cells, the X and Y chromosomes separate, and each sperm cell has either an X or a Y, but not both. Each sperm contains either the alleles on the X chromosome or the alleles on the Y chromosome. How does this happen? The answer is **meiosis.**

Meiosis is a type of division in a cell's nucleus that produces offspring nuclei with half of the alleles of the parent nucleus. This feature of meiosis explains how haploid cells are produced from **diploid** (the 2*n* number of chromosomes) cells in certain parts of flowers. Haploid cells have only one set of chromosomes in the nucleus compared to that in the nucleus of the parent cells. Cells that have the full amount of nuclear genetic material—that is, two sets of chromosomes—are diploid cells. Mendel's peas had two factors for each feature of the plant because they were diploid, and because of meiosis, each diploid parent passed on only one of the factors in each haploid gamete.

By the start of the twentieth century, biologists had learned that chromosomes exist in pairs and that the pairs separate during meiosis. The parallel separation between pairs of chromosomes and between pairs of alleles during sexual reproduction was the first indication that genes were associated with chromosomes. This was the main component of the chromosomal theory of heredity, which accounts for the separation of two alleles for each gene on **homologous chromosomes**—chromosome pairs that have alleles for the same genes. The meiotic separation of homologous chromosomes and their alleles is shown in figure 13.6 for one pair of traits from the garden pea. This figure shows homologous chromosomes (which contain genes) separate during meiosis, and come together during fertilization. We will return to meiosis and sexual reproduction later in this chapter.

INQUIRY SUMMARY

Genes occur on chromosomes. Chromosomes and alleles segregate during meiosis. Haploid nuclei have one set of chromosomes and one allele for each gene. Meiosis is a type of division in a cell's nucleus that produces offspring nuclei with half of the alleles of the parent nucleus. Meiosis is necessary for sexual reproduction, which means that it has many important consequences for inheritance.

Where would meiosis fit into a cell cycle (chapter 5)? Would this same model for cell division hold for meiosis? Explain.

DNA is the Hereditary Material.

In the decades following Mendel's experiments, new methods and improvements in light microscopy permitted detailed studies of the nucleus and chromosomes, while advances in chemistry led to the discovery of DNA and nuclear proteins. Although the role of chromosomes in heredity was suspected, it was not demonstrated until after 1900. In fact, the importance of DNA in inheritance was not confirmed until halfway into the twentieth century.

By the 1880s, nuclei were known to consist mostly of proteins and DNA, together referred to as **chromatin** because this material can be stained with various dyes. The dual chemical nature of chromatin created a difficult puzzle: of which of the two components, protein or DNA, are genes made? Without evidence one way or another, the logical choice seemed to be proteins because they consist of complex chains of amino acids that form many different kinds of proteins. Complex genetic processes, it was thought, must be controlled by complex protein molecules. Conversely, because there were only four different nucleotides in DNA, chromosomal DNA was thought to be too simple a molecule to meet complex cellular demands.

By 1928, British microbiologist Frederick Griffith and his colleagues discovered that pathogenic (i.e., disease-causing) strains of the bacterium now called *Streptococcus*

pneumoniae could transform nonpathogenic strains into infectious strains. Later, in 1944, a group of researchers led by Oswald Avery discovered that the "transforming principle" of this bacterium is DNA. Avery's experiments were the first indication that DNA is the hereditary substance; from there, landmark discoveries in science were made. Nonetheless, general acceptance did not come until nearly a decade later, when direct evidence for the hereditary role of DNA was obtained in experiments using viruses.

Viruses have even simpler genetic systems than do bacteria. By the 1950s, viruses were already characterized as being made of nucleic acids surrounded by a protein coat. Certain types of viruses, called bacteriophages, parasitize bacteria: they attach to bacterial cells, inject their genes into the host, and cause the host cell to make more viruses. In other words, viruses transform bacterial cells into miniature factories that make more viruses. In 1952, Alfred Hershey and Martha Chase grew a strain of bacteriophage and found only DNA, not protein, entered the bacteria and caused their transformation to produce more viruses. These bacteriophage experiments confirmed the role of DNA in heredity and added to our rapidly growing knowledge of this molecule. Just one year later, Watson and Crick published their famous paper describing the structure of DNA, and the age of molecular genetics and DNA was launched (see chapter 5).

Genes are made of DNA and form the units of inheritance described by Mendel.

Mendel's results can be readily appreciated in light of the accepted role of DNA as the hereditary material and of the nature of genes as sequences of nucleotides. We now know that each of the traits Mendel studied in garden pea is controlled by a unique sequence of nucleotides—that is, by a single gene. The position of a gene on a chromosome is referred to as that gene's **locus** (plural, *loci*). Thus, Mendel's experiments dealt with the genes at seven loci.

At the molecular level, each gene is a code for making a specific protein via mRNA or for tRNA or rRNA. The size of a locus ranges from a few dozen to several thousand nucleotides for different genes. Details about how DNA works as a code for RNA and therefore protein were discussed in chapter 5.

How does a gene produce a phenotype? Although many genes have been described both by their base sequences and their coded products, we do not understand how the sum of this molecular information becomes a complex organism. Our understanding is generally limited to biochemical reactions catalyzed by enzymes, such as when a single enzyme in flowers of garden pea speeds up the reaction that makes a colored pigment. This reaction, controlled by the product of one gene, explains how that gene influences flower color. However, we are still learning how other genes made the flower, and other parts of a plant, in the first place.

MEIOSIS

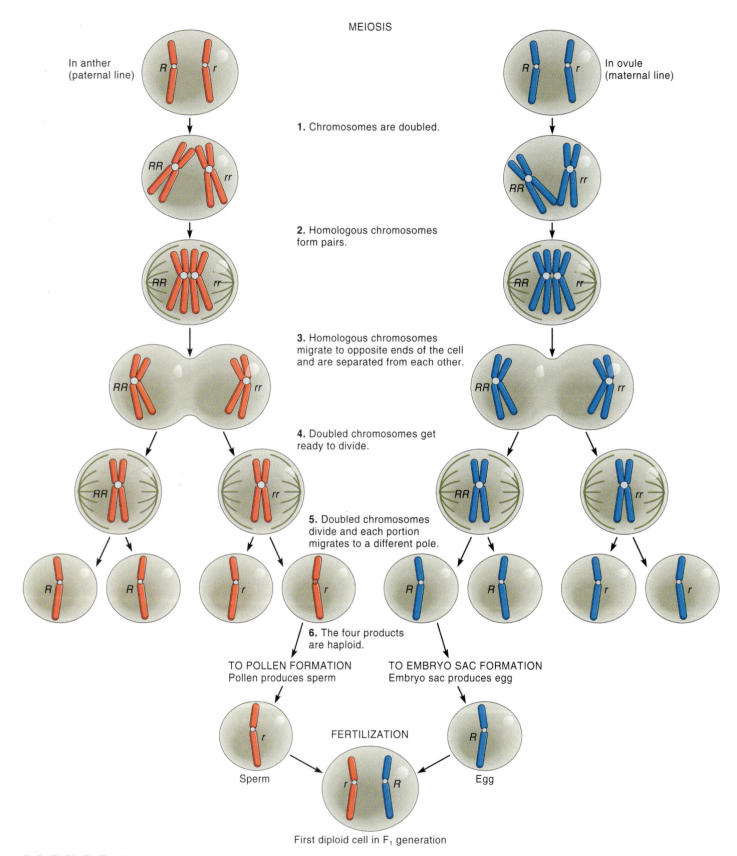

In anther
(paternal line)

In ovule
(maternal line)

1. Chromosomes are doubled.

2. Homologous chromosomes
form pairs.

3. Homologous chromosomes
migrate to opposite ends of the cell
and are separated from each other.

4. Doubled chromosomes get
ready to divide.

5. Doubled chromosomes
divide and each portion
migrates to a different pole.

6. The four products
are haploid.

TO POLLEN FORMATION
Pollen produces sperm

TO EMBRYO SAC FORMATION
Embryo sac produces egg

FERTILIZATION

Sperm

Egg

First diploid cell in F_1 generation

FIGURE 13.6

In anthers, alleles, shown as R and r, separate during meiosis and become the single gene, R or r, in the sperm. Similarly, in ovules, alleles separate during meiosis and form the egg. Alleles are reunited by fertilization. The parental genotype (*Rr*) is reproduced in the F_1 generation in this example, but different combinations of sperm and egg alleles will yield *RR* and *rr* genotypes in other offspring of the same parents.

INQUIRY SUMMARY

The importance of DNA in inheritance was not confirmed until halfway into the twentieth century when it was demonstrated that DNA is the genetic material. Genes are codes for manufacturing RNA and protein and are considered to make up the genotype of an organism.

Why was it difficult to demonstrate experimentally that DNA is the genetic material?

Several Forms of Complex Inheritance Have Been Discovered.

The traits Mendel used and the patterns of inheritance he discovered are often described as **Mendelian inheritance.** Many other traits in Mendel's peas, however, are inherited differently than the ones he studied. Moreover, the same kinds of traits that Mendel studied in peas may be inherited differently in other plants, and some hereditary patterns do not conform to Mendelian predictions at all. Such complex and non-Mendelian patterns of inheritance are discussed in the next few sections of this chapter. As you will see, Mendel was lucky to have avoided some pitfalls that would have greatly complicated his task.

There are three types of dominance: complete, incomplete, and codominance.

The seven genes studied by Mendel all exhibit **complete dominance,** which is a relatively rare type of inheritance. Complete dominance occurs when one gene of the allele completely masks the recessive gene of the allele. In other words, the dominant gene is expressed, and the recessive gene is not expressed. More frequently, one gene of the allele is only partly masked by the other, a condition called **incomplete dominance.** Incomplete dominance occurs when hybrids have a phenotype intermediate between those of the two parents. For example, the allele for red flowers in camellia (*Camellia japonica*) is incompletely dominant over the allele for white flowers. As a result, the F$_1$ progeny from a cross between red-flowered and white-flowered camellias all have pink flowers. The phenotypic ratio in the F$_2$ offspring is 1:2:1 (25% red, 50% pink, 25% white) (fig. 13.7). Accordingly, in cases of incomplete dominance, the phenotypic and genotypic ratios are the same.

    Codominance occurs when both alleles of a heterozygote are expressed equally, so there is really no dominance at all. Codominance is common for heterozygous genes that code for two equally functional enzymes. This means that there is more than one form of the same enzyme. The differing forms of enzymes made by different genes of alle-

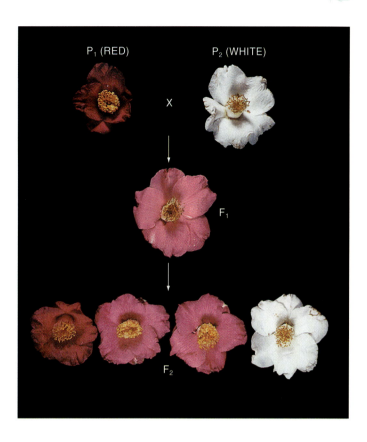

FIGURE 13.7

Camellias (*Camellia japonica*) show incomplete dominance in flower color. Pink flowers are heterozygous for flower color (*Rr*); red and white are homozygous for flower color (*RR* and *rr*, respectively). What are the genotypes of the F$_2$ offspring shown here?

les at the same locus are called **allozymes.** Although allozymes catalyze the same reaction, they differ from each other by one or a few amino acids, which makes them slightly different from each other. For example, in wild sunflower (*Helianthus debilis*), there are allozymes of phosphoglucoisomerase, which catalyzes one of the first reactions in glycolysis (see chapter 11). Heterozygotes produce both forms of the enzyme, but homozygotes produce only one or the other (see Perspective 13.2, "Examining Proteins by Gel Electrophoresis"). How is this important to a plant?

There can be multiple alleles of the same gene.

Diploid plants have only two alleles at a single locus. If the genes of two alleles produce functional enzymes (allozymes, described previously), they may have slightly different properties. Also, some populations of plants have more than two enzymes for some loci. This means that even though each individual, diploid plant can have a maximum of two allozymes, a population of plants can have more than two allozymes.

    Botanists are not sure why populations of plants have multiple forms of the same enzyme, but such variation may

13.2 *Perspective* Perspective

Examining Proteins by Gel Electrophoresis

Gel electrophoresis is used to separate large protein molecules by their different rates of movement through a gel in an electric field. The gel is often made of polyacrylamide. A protein extract is put into a small well at one end of the gel, which is placed on a Plexiglas plate and immersed in an aqueous solution. Direct current is supplied to a negative electrode at the same end as the protein well and to a positive electrode at the other end of the gel. Proteins move toward the positive end because their overall charge is negative. A protein's rate of movement is determined by its size and charge: larger proteins are slowed down by the gel, while proteins having greater negative charges move faster. Complications between size and charge are generally eliminated by treating extracts with a detergent that gives all proteins nearly the same charge. Detergent-treated proteins are therefore separated on the basis of size alone.

A protein extract may contain thousands of proteins, none of which is visible on the gel without some kind of chemical modification. To analyze allozymes, gels are treated with a substrate appropriate for the enzyme being studied. The enzyme reacts with the substrate to produce a colored band on the gel. Specific color-generating substrates are known for only about 40 enzymes, which means that the inheritance of most enzymes cannot be studied by gel electrophoresis.

Inheritance of enzymes typically involves codominant alleles. This means that, for simple enzymes, heterozygous and homozygous genotypes can usually be identified directly by inspecting the gel. Homozygotes have one band for a gene, and heterozygotes often have two bands. Enzymes that consist of more than one polypeptide chain usually show more complicated patterns.

Nucleic acids also can be analyzed by gel electrophoresis. To learn more about the analysis of DNA by gel electrophoresis, see appendix B.

---

be adaptive. There are two kinds of indirect evidence for this explanation. One is that different allozymes work at different optimum pHs, temperatures, or other conditions. We imply from this evidence that allozyme variation enables plants to thrive under a range of environmental conditions. The second kind of evidence is that certain allozymes occur more frequently in some populations, for example, at higher elevations, in wetter soils, or within shadier forests. In this case, allozymes are thought to be adaptive because their occurrence is correlated with where certain populations live. Nevertheless, we have no strong direct evidence for the adaptiveness of allozymes. Allozymes are studied by gel electrophoresis, described in Perspective 13.2, "Examining Proteins by Gel Electrophoresis," and appendix B.

Multiple genes are common in plants.

Allelic variation of a single gene is often complicated by the presence of more than one gene for the same enzyme. The same enzymes from different genes is called an **isozyme** to distinguish it from an allozyme. Multiple isozymes are common in plants. For example, sunflower (*Helianthus debilis*) has two nuclear genes and two chloroplast genes for a chloroplast enzyme (making it an isozyme). The four genes for this enzyme in this sunflower make a total of nine forms of the enzyme. Isozymes can also be studied by electrophoresis.

INQUIRY SUMMARY

The traits Mendel studied, and the patterns of inheritance he discovered, are often described as Mendelian inheritance. Complete dominance occurs when one trait completely masks its recessive allele. More frequently, the phenotype for one allele is only partly masked by the other, a condition called incomplete dominance. Codominance occurs when both alleles of a heterozygote are expressed equally, so there is really no dominance at all. Plants have multiple forms of the same enzyme (allozymes). Other plants may have more than one gene for the same enzyme, producing isozymes.

What might be the evolutionary significance of incomplete dominance?

Non-Mendelian Inheritance Does Not Produce Mendelian Ratios.

The inheritance of many traits does not yield Mendelian ratios. The most common causes of non-Mendelian inheritance are linkage, cytoplasmic inheritance, mutations, and transposable elements.

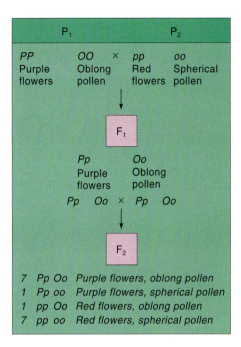

F I G U R E   1 3 . 8

A dihybrid experiment with the sweet pea does not yield the expected 9:3:3:1 ratio of phenotypes in the F₂ generation because of linkage between the loci for flower color and pollen shape. Instead, most of the offspring are like the parents (P₁ generation). However, linkage is not perfect, since a small number of nonparental combinations (red-oblong and purple-spherical) appear in the F₂ generation.

Linked genes are inherited simultaneously.

Many genes occur on each chromosome. When two or more genes occur on the same chromosome, they are normally inherited together, not independently. The simultaneous inheritance of genes on the same chromosome is called **genetic linkage.**

Linkage was first reported in the sweet pea (*Lathyrus odoratus*), a relative of the garden pea. In 1906, geneticists at Cambridge University discovered that genes for flower color and pollen shape did not fall into the expected 9:3:3:1 phenotypic ratio in the F₂ generation. Instead, the ratio was 7:1:1:7 (fig. 13.8). This means that almost 44% (7/16) of the F₂ plants had the dominant flower color (purple) and the dominant pollen shape (oblong), while an equal proportion had the recessive flower color (red) and the recessive pollen shape (spherical). This pattern indicates that the purple oblong and red spherical phenotypes are inherited together, which means that the genes for these traits are linked.

The main puzzle of linkage, however, was that some of the F₂ plants were red oblong and some were purple spherical, unlike the purebred parents or the hybrids of the F₁ generation. If linkage is perfect, it seems that these phenotypes should not occur. Imperfect linkage occurs when chromosomes exchange complementary fragments of DNA during meiosis. As a result, those fragments that have different alleles for linked genes may be rearranged to produce combinations of alleles not found in the parents. This process, which occurs during meiosis, is called *crossing-over* and is discussed more completely later.

Cytoplasmic genes are inherited independently of sexual reproduction.

You learned in chapter 4 that chloroplasts and mitochondria contain DNA. Genes in these organelles control certain aspects of photosynthesis and respiration, respectively. Inheritance of these genes is independent of sexual reproduction because they are transmitted to offspring with the cytoplasms, often that of the maternal parent (because an egg is larger and contains much more cytoplasmic material than a sperm).

One example of cytoplasmic gene control occurs in certain forms of the cultivated four-o'clock (*Mirabilis jalapa*) that have yellowish white leaves instead of green leaves. This difference in leaf color is caused by defective genes in chloroplasts. This is an example of the cytoplasmic inheritance of non-nuclear genes.

The cooperation of genes from organelles and the nucleus is often necessary for normal metabolism. For example, the photosynthetic enzyme rubisco (chapter 10) has two subunits of polypeptides, one derived from a nuclear gene and one from a chloroplast gene. Similarly, some ATPases (chapter 11) have a dual genetic origin. In each case, the final product—that is, a complete and functional enzyme—depends on genes from the nucleus and mitochondria in the same cell.

Mutations are inheritable changes in genes.

Dutch botanist Hugo de Vries was one of the discoverers of Mendel's work. De Vries studied patterns of inheritance in several kinds of evening primrose (*Oenothera* species). Most of his results conformed to Mendelian inheritance, but occasional characteristics appeared that were not present in either parent. De Vries called these spontaneous hereditary changes **mutations,** a term that remains in widespread use today. In modern genetics, the term **mutation** includes a variety of genetic changes, including chromosomal rearrangements and changes in DNA. Thus, a mutation is an inheritable change in a gene. Mutations are often lethal, but some have no effect, at least under the current environmental conditions in which the organism lives. Occasionally, mutations may provide some benefit to the organism because the change in the gene is inherited and passed on to offspring. In this case, the mutation can provide some variation within members of a population and provide a basis for natural selection to act (see chapter 14).

Transposable elements are mobile pieces of DNA.

One of the most important kinds of mutations involves sequences of DNA that seem to multiply and move spontaneously among an organism's chromosomes. These movable pieces of DNA are called **transposable elements** and have been estimated to account for as many as half of a plant's spontaneous mutations. Movable DNA may contain one or more genes that can function as inhibitors, modifiers, or mutators, thereby affecting the action and phenotypic expression of ordinary genes in many ways. Inserting transposable elements into or near other genes produces unpredictable patterns of inheritance.

Transposable elements were first identified in the 1940s by Barbara McClintock (see chapter 1 and fig. 1.21), who observed their effects on pigment patterns in intact corn kernels (grains). Specifically, McClintock noticed that a fully pigmented grain was often flanked by a totally unpigmented grain on one side and on the other side by a grain having pigment over only a part of its area. Thus, some cells contained pigment, and some did not with an intact kernel. If all of the cells in the endosperm of a given grain derive from a single cell, why weren't they all alike? McClintock proposed that portions of kernels remained white, even though genes for pigment synthesis were present, because the pigment genes were disrupted by "controlling [transposable] elements." McClintock noted that reversion to the wild, nondisrupted genotype occurred when these elements moved away from the affected genes. Normal pigmentation appeared in cells in which reversion occurred. The pigment-inhibiting mutation was so unstable that many groups of cells reverted to the original, wild-type pigmentation as each kernel developed. This produced a patchwork of colored spots and streaks, distributed in seemingly random patterns among colorless portions of the kernel. For McClintock, examining an ear's variously colored grains was like reading the history of the movement of transposable elements in cells.

Recent research has shown that coloration patterns in snapdragons (*Antirrhimum majus*) and morning glories (*Ipomoea purpurea*) are also caused by transposable elements. Variegated flower colors result from complex interactions between mobile genes and genes for pigment synthesis (fig. 13.9). Similar color patterns are also probably caused by transposable elements in many other species of plants.

Transposable elements are now accepted as a general feature of many organisms and are widely used to induce mutations. Scientists think, however, that since not all genes are active at the same time, the potential effects of transposable elements on gene regulation vary. For example, transposable elements may help regulate cancer-causing genes, which are widespread among humans but are normally not activated. The importance of transposable elements led to the awarding of a Nobel Prize to Barbara McClintock in 1983.

All of the mechanisms of inheritance discussed depend on sexual reproduction. Why is sex important in plants? What are evolutionary advantages of sex? We discuss these fundamental questions in the next section.

INQUIRY SUMMARY

Each chromosome has many genes. Genes on the same chromosome may not assort independently because they are linked. Because of crossing-over, however, linked genes can be rearranged into nonparental combinations of alleles and mimic independent assortment.

Many kinds of non-Mendelian inheritance patterns cannot be explained by linkage. Changes in the base sequences of DNA, either by random mutation or by the insertion of transposable elements, alter inheritance patterns in unpredictable ways. Also, the inheritance of organellar genes does not conform to Mendelian predictions. In many plants, chloroplast and mitochondrial chromosomes are inherited with the cytoplasm of the female gamete. A mutation is an inheritable change in a gene. Transposable elements are mobile pieces of DNA.

Develop a hypothesis to explain how linkage might be involved in the evolution of plants?

Sexual Reproduction Provides Variation in Plants.

Sexual reproduction in all types of organisms involves meiosis, which produces haploid cells from diploid cells. Sexual reproduction also involves **fertilization,** the union of two haploid cells (gametes) into a diploid cell. The processes of meiosis and fertilization are called a *life cycle* because they are repeated every generation. We can return to our example of human XY chromosomes for an example. Males have the XY chromosomes in diploid cells; females have XX. In males, meiosis produces haploid sperm cells containing either X or Y chromosomes; 50% are X, and 50% are Y. Meiosis in females produces haploid egg cells with only X chromosomes; all the egg cells have an X chromosome. Fertilization brings either the X or the Y of a sperm together with the X of an egg. If a Y sperm and X egg combine, the offspring has the XY genotype and is a male (phenotype). If an X sperm and X egg unite, the offspring has the XX genotype and is female.

In plants, the haploid cells produced by meiosis are called **spores,** which are cells that can divide by mitosis to produce a multicellular haploid structure. This haploid (*n* number of chromosomes) structure makes gametes, either eggs or sperm, so it is called the **gametophyte.** The diploid (2*n* number of chromosomes) plant that bears

FIGURE 13.9

Photograph of a morning glory flower shows the effects of transposable elements on flower color. White sections of the flower occur where transposable elements disrupt the synthesis of pigments.

© Evelyne Cudel

spore-producing cells that undergo meiosis is called the **sporophyte.** Such a life cycle is known as an **alternation of generations** because it alternates between two distinct multicellular haploid and diploid "generations," which are phases of the same life cycle.

Figure 13.10 shows the generalized life cycle of a plant with alternating sporophyte (2n) and gametophyte (n) generations. Compare the generalized life cycle of a plant (fig. 13.10a) to the life cycle of a typical flowering plant (fig. 13.11) described in the following sections. We focus on flowering plants for this discussion because they are the most recently evolved, diverse, familiar, and widespread group of plants on earth. Our emphasis is on the features of reproduction that all angiosperms (flowering plants) share. We present reproduction in other plant

groups and in other types of organisms in chapters 16 through 18. In these later chapters, compare the life cycles illustrated there to those shown here in figures 13.10 and 13.11. Such a comparison will help you understand the evolution of structures in plants.

In angiosperms, indeed in all plants, the gametophyte, which is haploid (n), produces male and female gametes—sperm and egg. The sporophyte generation is diploid, following union of the sperm and egg. Plants are distinguished from algae by the retention of the embryo (2n) within the nourishing cells of its parents' tissue. In all flowering plants, this structure has evolved into a seed within a fruit that develops from a flower (fig. 13.11). Next, we will examine the reproductive process of angiosperms in some detail.

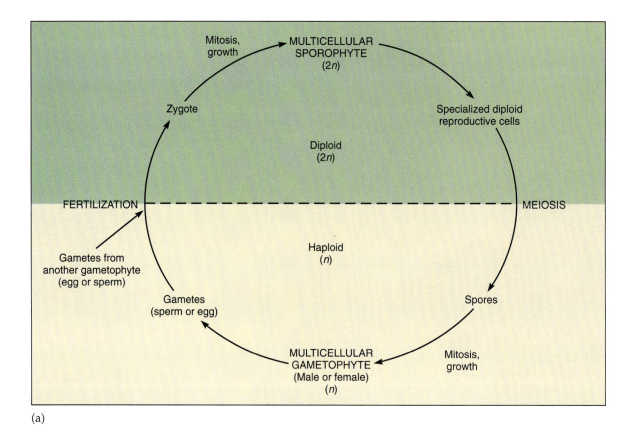

(a)

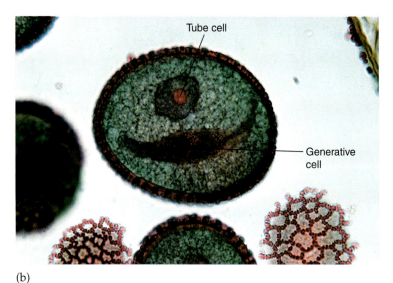

(b)

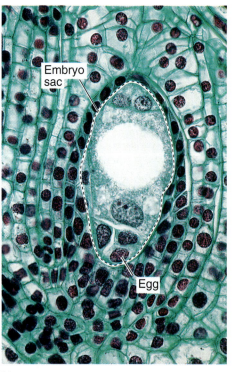

(c)

FIGURE 13.10

Alternation of generations in plants. (*a*) The plant life cycle is an alternation of generations between diploid sporophytes and haploid gametophytes. Each sporophyte produces haploid spores by meiosis. Spores divide and grow into gametophytes, which produce gametes. The union of the two gametes (i.e., fertilization between sperm and egg) forms a diploid cell, the zygote, which restores the sporophyte generation. (*b*) Immature pollen of lily. These pollen consist of a two-celled male gametophyte; the generative cell will divide to become two sperm cells; the tube cell will produce the pollen tube that grows to the female gametophyte, where the sperm are released for fertilization (×450). (*c*) Section through an immature lily seed (ovule) with an embryo sac containing an egg cell (×100) (not all cells are visible in this section). The development of male and female gametophytes is shown in figures 13.11 and 13.12.

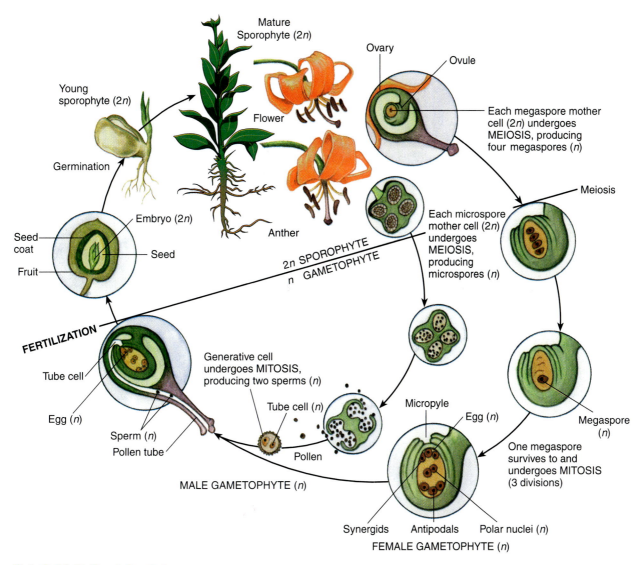

Mature Sporophyte (2n)

Young sporophyte (2n)

Germination

Seed coat

Fruit

Embryo (2n)

Seed

FERTILIZATION

Tube cell

Egg (n)

Sperm (n)

Pollen tube

Generative cell undergoes MITOSIS, producing two sperms (n)

Tube cell (n)

Pollen

MALE GAMETOPHYTE (n)

2n SPOROPHYTE
n GAMETOPHYTE

Flower

Anther

Each microspore mother cell (2n) undergoes MEIOSIS, producing microspores (n)

Ovary

Ovule

Each megaspore mother cell (2n) undergoes MEIOSIS, producing four megaspores (n)

Meiosis

Megaspore (n)

One megaspore survives to and undergoes MITOSIS (3 divisions)

Micropyle

Egg (n)

Synergids   Antipodals   Polar nuclei (n)

FEMALE GAMETOPHYTE (n)

FIGURE 13.11
Life cycle of a typical flowering plant. Flowering plants, of course, produce flowers, which distinguishes them from other plants. They also produce seeds, a trait shared with gymnosperms such as pine.

Meiosis produces haploid spores that become gametes.

Meiosis results in the production of two kinds of spores in flowering plants. Spores produced in the anthers of flowers are called **microspores** (fig. 13.12) while spores produced in ovules are called **megaspores.** Microspores are produced from specialized diploid cells called **microspore mother cells.** Each microspore mother cell is diploid and divides via meiosis to form four haploid microspores. Following meiosis, each microspore divides by mitosis and cytokinesis (chapter 5) and produces an immature male gametophyte. At this stage, the structure is called a **pollen grain.** The mature male gametophyte—the germinated pollen grain—contains two **sperm cells,** which are the male gametes and a tube cell.

Spores produced in **ovules** are called **megaspores.** In each ovule, only one of the diploid cells undergoes meiosis to form four haploid megaspores. The fate of the megaspores varies among different flowering plants. The following description is based on lily (*Lilium* species), whose development of the female gametophyte is more commonly known than it is for other plants (fig. 13.13). Three of the four megaspores in lily disintegrate, leaving the ovule with only one functional megaspore. The nucleus of this megaspore usually divides by mitosis three times, which produces eight free nuclei. These eight nuclei are then partitioned into seven cells: six have one nucleus, whereas the other cell contains two nuclei. The seven-celled, eight-nucleate structure is the **female gametophyte,** also called the **embryo sac.** Hence, the sexually mature, full-fledged female gametophyte stage consists of seven cells. The two nuclei that remain free are called

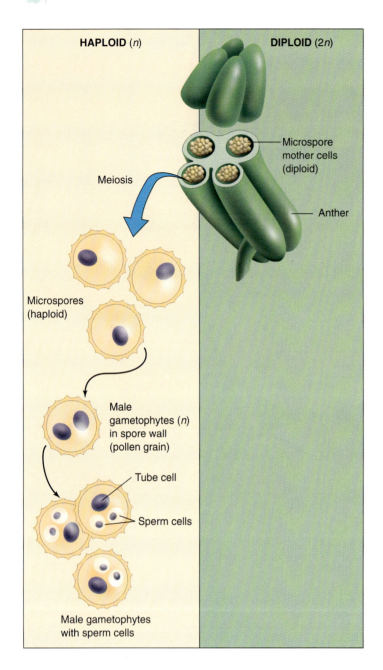

FIGURE 13.12

Development of the male gametophyte. Microspore mother cells in the anther undergo meiosis. Microspores grow into two-celled male gametophytes, each producing two sperm cells.

the polar nuclei because they come from opposite ends of the embryo sac.

Pollination and fertilization bring gametes together in sexual reproduction.

As described in chapter 12, sexual reproduction in flowering plants depends on *pollination,* which is the transfer of pollen grains from an anther of a flower to the receptive

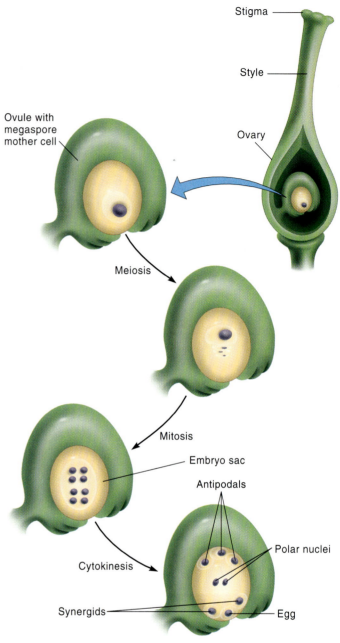

Embryo sac (7 cells, 8 nuclei)

FIGURE 13.13

Development of the *Lilium*-type female gametophyte. A single megaspore mother cell in each ovule undergoes meiosis. Three of the megaspores disintegrate, leaving one functional megaspore. Three mitotic divisions in the functional megaspore produce the embryo sac, which consists of eight nuclei that are walled off into seven cells. One of the cells is the egg, which is flanked by two synergids. Three cells are called antipodals, and the remaining large cell contains two polar nuclei.

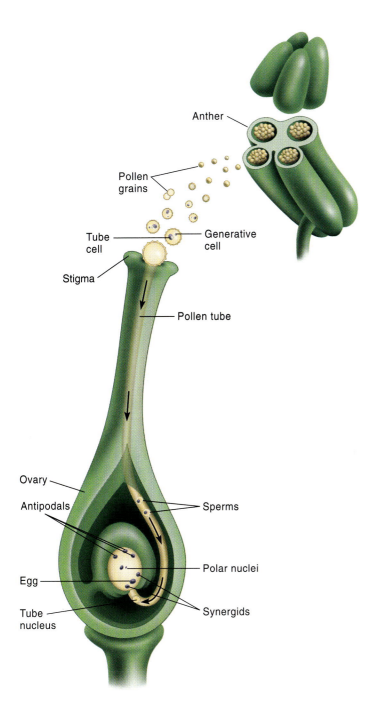

FIGURE 13.14
Pollination and fertilization. During pollination, pollen is transferred from an anther to the stigma. Two sperms move from the pollen grain through the pollen tube to an opening in the ovule. One sperm fertilizes the egg, and the other fertilizes the polar nuclei.

stigma region of a flower (fig. 13.14). In nature, pollen is usually transferred from anther to stigma by animals or wind (see chapter 12). After fertilization (the union of sperm and egg), each ovule matures into a diploid seed. Remember, a key evolutionary feature of sexual reproduction is the resulting genetic variation within a population of plants that develop from the seeds.

Some of the genetic variation among offspring of the new sporophyte generation comes from mixing the genes inherited from different parents. For example, ovaries from several flowers of the same plant may receive nonidentical pollen from many other plants. Variation occurs because pollen from different plants contains different sets of genes. Thus, 10 sources of pollen will produce embryos with at least 10 different genotypes. However, 10 pollen-producing plants will produce many more than 10 different pollen genotypes. Such additional genetic variation develops as a result of meiosis. The remainder of this chapter describes how meiosis works and how it produces genetic variation. Before you move on, we suggest you review figure 13.11 to see how the steps come together in the life cycle of a typical angiosperm.

INQUIRY SUMMARY

Sexual reproduction in plants involves regular alternation between diploid and haploid generations. The diploid generation produces the haploid generation by meiosis. The haploid generation restores the diploid generation by fertilization. The diploid seed can develop into the adult plant.

What is the biological significance of gametes coming together during fertilization?

Meiosis Involves Two Nuclear Divisions.

The basic events of meiosis were introduced earlier (see fig. 13.6) in this chapter to help you understand the significance of Mendel's work. Meiosis is important because it produces haploid gametes that can come together in new combinations from the parents during fertilization. New combinations make individuals within a population different. This diversity may allow the population to survive, indeed to evolve, as environmental conditions change over time.

In flowering plants, meiosis occurs in the micro- or megaspore mother cells. Meiosis involves the events of the cell cycle described in chapter 5 except here the M phase is for meiosis. As with mitosis, DNA and chromosomes are replicated during the S phase portion of the meiotic cell cycle. Figure 13.15 shows the steps of meiosis in greater detail than figure 13.6. After duplication, chromosomes enter the first of two successive nuclear divisions: **meiosis I** and **meiosis II.** The phases of meiosis are distinguished by a I for the first division and a II for the second. Accordingly, meiosis I consists of prophase I, metaphase I, anaphase I, and telophase I; in meiosis II, nuclei go through prophase II, metaphase II, anaphase II, and telophase II. The names

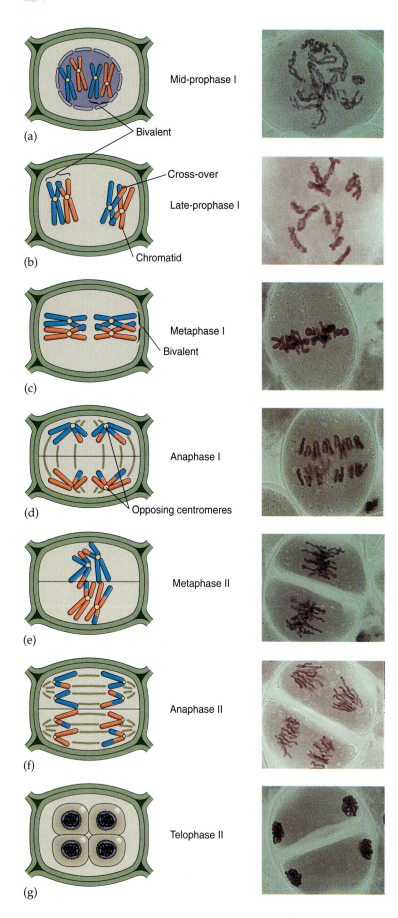

Mid-prophase I

Bivalent

(a)

Cross-over

Late-prophase I

Chromatid

(b)

Metaphase I

Bivalent

(c)

Anaphase I

Opposing centromeres

(d)

Metaphase II

(e)

Anaphase II

(f)

Telophase II

(g)

FIGURE 13.15

Stages in meiosis. (All photos ×250) (*a*) Mid-prophase I. Homologous chromosomes, composed of two chromatids, form pairs called bivalents. Pairing is called synapsis. (*b*) Late-prophase I. Chromatids cross over between homologous chromosomes, making it possible to exchange genetic material. (*c*) Metaphase I. Chromosomes align along the metaphase plate. (*d*) Anaphase I. Homologs separate and move to opposite ends. Chromosome fragments that were exchanged during prophase move with recombined chromosomes. (*e*) Metaphase II. Chromosomes align along metaphase plates in a plane that is perpendicular to the direction of the plate in metaphase I. (*f*) Anaphase II. Centromeres divide, and former sister chromatids move to opposite poles of the spindle. (*g*) Late telophase II. Chromosomes decondense, and nuclear envelopes form around each haploid daughter nucleus. At this stage the products of meiosis are called spores, which form spore walls and are later released from the parent cell. Light micrographs are of meiosis in the anthers of the Easter lily (*Lilium longiflorum*). Meiosis is a continuous process.

of the phases in meiosis are the same as those of mitosis because chromosomes behave similarly in both types of nuclear divisions. Nevertheless, starting with prophase I, meiosis exhibits several major differences from mitosis. The most obvious distinction is that homologous chromosomes form pairs in meiosis but not in mitosis. Chromosomes are homologous to one another when they have the same genes (but not necessarily the same alleles). Starting with chromosome pairing in prophase I, meiosis is a much more complicated and time-consuming process than mitosis.

In summary, here are the detailed steps in the process of meiosis (fig. 13.15):

1. As prophase I begins, the chromosomes condense into long "threads." Each chromosome in prophase I is actually a pair of sister chromatids—the replicated strands of DNA still attached to each other by a protein complex—even though they may look like one chromatid (fig. 13.16). Paired chromosomes in prophase I are called bivalents (figs. 13.15, 13.16, and 13.17). Because each member of a bivalent has doubled during interphase, a bivalent consists of four chromatids, two centromeres, and two chromosomes.

2. Soon after chromosomes pair, nonsister chromatids may twist around each other and exchange genetic material by **crossing-over.** Crossing-over occurs when two intertwined, nonsister chromatids break, and each broken end re-fuses with its nonsister chromatid instead of with its original chromatid (fig. 13.17). Thus, fragments of chromosomes switch places between homologs. Crossing-over changes the genetic makeup of two chromatids when the alleles on one fragment differ

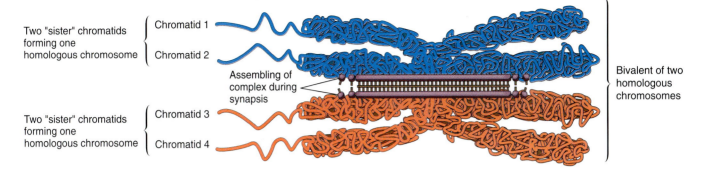

Two "sister" chromatids forming one homologous chromosome { Chromatid 1 / Chromatid 2

Assembling of complex during synapsis

Two "sister" chromatids forming one homologous chromosome { Chromatid 3 / Chromatid 4

Bivalent of two homologous chromosomes

FIGURE 13.16

Bivalents form by synapsis during Prophase I of meiosis. Following replication in the S phase, each homologous chromosome is made of two identical, attached "sister" chromatids (in this diagram, 1 and 2 are sister chromatids and 3 and 4 are sister chromatids of two homologous chromosomes). Homologous chromosomes are joined at a complex and form pairs of homologous chromosomes called a bivalent. Synapsis is the pairing of homologous chromosomes into bivalents. During synapsis, crossing over between "non-sister" chromatids results in different combinations of genes along homologous chromosomes (fig. 13.17).

from those on the other. There are four possible combinations of two bivalents in metaphase I (fig. 13.18). Such chromosomal rearrangements during crossing-over are a type of **genetic recombination,** which is the general term for producing offspring that are genetically different from the parents; this is genetic diversity. Several kinds of genetic recombination are derived from meiosis in addition to chromosomal rearrangements (next section).

3. As prophase I continues, the complex dissolves and the nuclear envelope disintegrates. At this time, the arms of homologous chromosomes seem to repel each other, but they are still attached at their centromeres.

4. Metaphase I is characterized by the formation of a meiotic spindle apparatus, which moves the bivalents to a metaphase plate at the center of the cell (see fig. 13.15c). Because of this pattern of spindle attachment, opposing centromeres are pulled apart during anaphase I (see fig. 13.15d). This means that homologous chromosomes separate in meiosis I, and the haploid offspring nuclei begin to form. This is why meiosis I is called the reduction division of meiosis; it reduces the number of chromosomes in half.

5. When chromosomes reach their destinations in the haploid nuclei, each offspring nucleus has a haploid set of chromosomes, but each chromosome still consists of two chromatids.

6. The second meiotic division is the same as in mitosis. The spindle apparatus forms and aligns chromosomes in a plane (see fig. 13.15e). Once they are aligned, chromosomes enter anaphase II, when centromeres divide and sister chromatids become individual chromo-

somes. The new centromeres are pulled apart during anaphase II, thereby separating duplicate chromosomes from each other (see fig. 13.15f).

7. In telophase II, chromosomes decondense and a nuclear envelope forms around each of the four offspring nuclei (see fig. 13.15g).

8. By the end of meiosis II, two divisions have been completed. The diploid parent nucleus that entered meiosis has divided twice to form four haploid nuclei. After the second meiotic division, cytokinesis divides the cytoplasm into four spores.

Meiosis is usually an orderly process that produces four haploid nuclei from one diploid nucleus. Cytokinesis follows and divides the cytoplasm into distinct cells. However, "mistakes" in meiosis have apparently occurred many times during the evolution of plants. The evidence for such meiotic errors is that many plants are **polyploid,** which means that they have more than two sets of chromosomes. As explained in Perspective 13.3, "Polyploidy in Plants," meiotic "mistakes" have played an important role in plant evolution and, more recently, in the origin of cultivated plants.

Genetic recombination can occur without crossing-over.

Meiosis always produces genetic recombination, much of which comes from the exchange of chromosomal fragments during prophase I. However, genetic recombination would occur even without crossing-over. New combinations of traits would still appear in offspring because of chromosomal separation in meiosis.

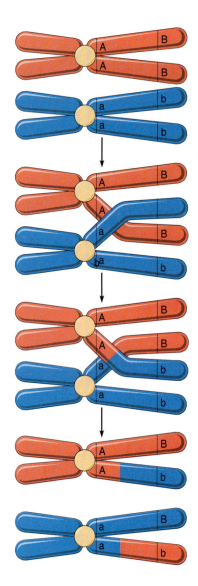

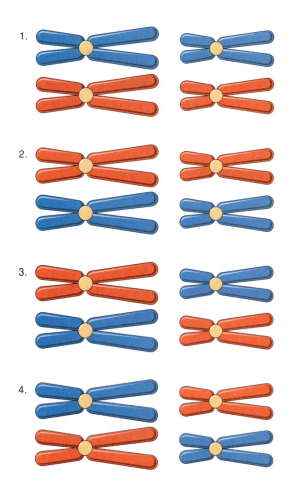

FIGURE 13.18

There are four possible orientations of two bivalents in metaphase I. Two of them (*1* and *2*) will segregate into parental combinations in the gametophyte generation. The other two (*3* and *4*) will result in new combinations of chromosomes in the gametophyte generation.

FIGURE 13.17

Homologous chromosomes align during prophase I, enabling nonsister chromatids (indicated by different colors) to cross over. After breakage and refusion, fragments of chromosomes switch places. In this way, nonsister chromatids exchange genetic material, resulting in genetic recombination. How might this be important in sexual reproduction?

Homologs are not oriented along the metaphase plate until spindle fibers attach to each centromere. This means that, before spindle attachment, each homolog has an equal chance of moving to either pole. Furthermore, the direction of movement of each homolog is independent of all other bivalents. This is analogous to independent assortment in Mendelian genes, but for entire chromosomes.

To illustrate the potential for new genetic combinations based on chromosome separation, consider the gametes of the desert sunflower, *Machaeranthera gracilis*. Cells of the sporophyte have only four chromosomes, which is one of the smallest chromosome numbers known in plants. This means that there are only two bivalents in prophase I. If we color-code the

homologs of these bivalents, we see that they can be oriented in four (i.e., 2×2) combinations in metaphase I (fig. 13.18). In anaphase I, two of the combinations will keep same-colored chromosomes together, whereas the other two combinations will produce nuclei with mixed colors. Thus, by chance, half of the spores in the desert sunflower will have different combinations of chromosomes; that is, they will differ genetically from their parents.

INQUIRY SUMMARY

Prophase I begins when already replicated, homologous chromosomes come together along their entire lengths. While fused, chromosomes exchange genetic material by crossing-over. By the end of prophase I, homologs are paired only at their centromeres. After prophase I, homologous chromosomes align on a metaphase plate and then segregate into two haploid nuclei for the remainder of meiosis. After meiosis I,

13.3 Perspective Perspective

Polyploidy in Plants

About 35% of the species of flowering plants are polyploid. The most common level of polyploidy is tetraploidy, which means that plants have four sets of chromosomes instead of two. This also means that the gametes are diploid instead of haploid. Most of the easily observed effects of polyploidy are associated with increased size, and they are usually observed in plants that have been induced by humans to become polyploid by treatment with chemicals such as colchicine. In nature, however, polyploid plants are often indistinguishable from diploid plants, at least upon initial inspection. Polyploids may be better adapted to temperature or water stress than their diploid relatives. Thus, polyploids can often survive in habitats that are not as suitable for their diploid counterparts.

Many cultivated plants are polyploids. One of these is bread wheat (*Triticum aestivum*), the most commonly cultivated crop in the world. Bread wheat probably arose in the Middle East about 12,000 years ago. It evolved by hybridization between a durum wheat, which is a tetraploid that has 28 chromosomes, and one of the goat grasses, which are wild diploid relatives of wheat that have 14 chromosomes. Thus, bread wheat has 42 chromosomes, which means that it is hexaploid. Durum wheat arose as the polyploid descendant of a diploid hybrid between two kinds of einkorn wheat and is still cultivated as the principal grain used in macaroni products.

---

chromosomes condense once again as they enter meiosis II. They then align on a plate, their centromeres split, and the former sister chromatids move to opposite poles of the cell. The result of the two meiotic divisions is four haploid nuclei from one diploid nucleus.

What is the importance of crossing-over during sexual reproduction?

Why Sex in Plants?

Sexual reproduction requires considerable energy; some would say wasted energy. Almost all higher organisms reproduce sexually. What, then, is the selective advantage of sexual reproduction in plants? Why don't all plants, like members of the poplar family, reproduce almost exclusively by asexual reproduction? One compelling answer: sexual reproduction leads to genetic diversity within a population or species. Asexual reproduction does not provide diversity but can be a fast and efficient way to produce another individual, and therefore population. Still, sex, and the resulting diversity of offspring, is critical to natural selection (chapter 14).

Sexual reproduction results in variation among members of a population.

The cycle of meiosis and fertilization depends on so many variables that it seems to flirt with failure every time sexual reproduction is attempted. For instance, most male gametes are wasted since most pollen grains do not arrive (at all or on time) at the stigma of a flower. This is especially true of wind-pollinated species, such as oaks, walnuts, and birches. In addition, sexual reproductive structures require substantial amounts of energy. In some species, such as the small monkey flower (*Mimulus kelloggii*), more than 20% of the energy from photosynthesis is used to make flowers. Moreover, genetic recombination during sexual reproduction disrupts adaptive gene combinations. This means that although an individual plant may be well adapted to its environment, it is unlikely that its gene combination will remain intact in the offspring. In a constant, unchanging environment, it is also unlikely that genetic recombination will produce offspring that are more fit than the parents, since the parents are already well adapted. But environments do change. Sexual reproduction generates genetic variability, which is important for the evolutionary adaptation of a population to new or changing environments. A basic characteristic of organisms is reproduction, but it can be sexual or asexual.

Asexual reproduction produces clones with uniform genetic information.

Sexual reproduction is so common in plants that there must be enormous evolutionary advantage to it. However, for rapid production of individuals, especially in an unchanging (or very slowly changing) environment, **asexual reproduction** (also known as vegetative reproduction) has advantages of being relatively rapid and efficient. Also, in harsh environments, the uniformity of the gene pool of a population can be important. Asexual reproduction requires only one parent, so one individual can produce many offspring and even an entire population, as with aspen, described in the next paragraph. These clones can

FIGURE 13.19

This stand of aspen is a clone of asexually reproducing angiosperms. The stand is adapted to a high-mountain, extreme-cold, and summer-arid environment. Aspen groves often are one individual plant that reproduce via rhizomes.

be just as successful in the current environment as the parent because of unchanged genomes. Asexual reproduction can be an advantage in habitats with few pollinators (see chapter 12). Finally, asexual reproduction can rapidly produce many offspring, thereby enabling the species to invade and dominating a suitable habitat more quickly than plants that reproduce only sexually.

In asexual reproduction, each new plant is identical to the parent. Asexual reproduction involves only mitosis, not meiosis or fertilization. Thus, the offspring are clones: genetic replicates and asexual reproduction may be an advantage in such cases. Quaking aspen (*Populus tremuloides*) is a well-known flowering plant that reproduces asexually by special underground roots that periodically send up shoots called (rather disrespectfully) suckers (fig. 13.19). Strawberries, violets, and many grasses also reproduce asexually. The roots of cherry (*Prunus*), pear (*Pyrus*), apple (*Malus*), and teak (*Tectona grandis*) produce adventitious buds, which form aerial shoots or suckers. When separated from the parent plant, suckers become new individuals. Adventitious buds are a common means of propa-

gating many other plants. For example, most groups of creosote bushes (*Larrea tridentata*) are clones derived from a single plant. Some of these clones are more than 12,000 years old.

Remember that asexual reproduction does not produce much, if any, diversity among individuals of a population, and variation is what natural selection and evolution work on. In fact, the only obvious disadvantage of asexual reproduction is the lack of genetic diversity. If the environment changes, as it inevitably will given enough time, all of the identical individuals of a cloned population will be equally susceptible to any new environmental stress, such as drought, cold, heat, insects, and pathogens the change may bring. Sexual reproduction, even with its disadvantages, initiates genetic diversity and thus provides the best opportunity for long-term survival of the species when environments change. This is the important material of evolution that we introduced in chapter 2 and will expand in chapter 14. For comparison, figure 13.20 summarizes sexual and asexual reproduction in a grass.

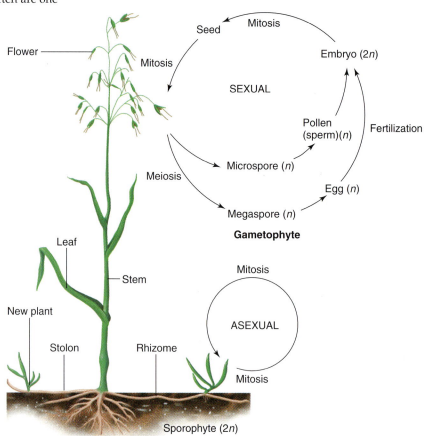

FIGURE 13.20

A model of sexual and asexual reproduction in a grass. Asexual reproduction (vegetative reproduction) occurs via stolons or rhizomes. Sexual reproduction occurs in flowers.

SUMMARY

The basic postulates for the theory of inheritance were derived from experiments by Gregor Mendel. Exceptions to the basic postulates result from linkage, cytoplasmic inheritance, mutations, and transposable elements.

Chromosomes consist of DNA and proteins. Genes were originally thought to be made of proteins, but experiments with bacteria and viruses, which do not have DNA associated with protein, showed that the hereditary substance is DNA.

Genes that occur on mitochondrial or chloroplast chromosomes control cytoplasmic inheritance. The expression of nuclear genes occasionally depends on these genes in organelles.

Mutations are inheritable changes in genes. One type of mutation that occurs frequently involves spontaneous movement of DNA segments called transposable elements. The expression of genes is often influenced by their proximity to these mobile elements.

The life cycle of plants is an alternation between diploid and haploid generations. The diploid generation produces the haploid generation by meiosis; the haploid generation produces the diploid generation by fertilization. The stages of meiosis have the same names as the stages of mitosis: prophase, metaphase, anaphase, and telophase. However, meiosis consists of two divisions.

New genetic variation occurs during meiosis because of chromosomal separation and genetic exchange between homologs. Chromosomal separation produces new combinations of parental chromosomes in the offspring. Genetic exchange also occurs during meiosis.

Sexual reproduction, via meiosis and fertilization, produces genetic variation. Asexual reproduction, via mitosis, is fast, efficient, and often adapted to local environments.

Writing to Learn Botany

We assume that sexual reproduction and genetic variation are advantageous since most plants and animals have developed it. However, sex is not necessary for reproduction; many plants do without it, either periodically or permanently. What are the ecological advantages and disadvantages of sexual and asexual reproduction?

THOUGHT QUESTIONS

1. Suppose you have a purple-flowered garden pea. What would its genotype be if it were crossed with a white-flowered individual and the phenotypic ratio of F_1 plants was 1:1?

2. Assume that a red and a white allele exist for flower color and that a blue and a yellow allele exist for pollen color. What phenotypic ratios would be predicted in a dihybrid cross in which both traits showed incomplete dominance?

3. What would the genotypic ratios be in the cross described in question 2?

4. Explain why a 9:7 phenotypic ratio in the F_2 generation indicates control by two genes.

5. Why is the inheritance of chloroplast and mitochondrial genes non-Mendelian?

6. Describe how the movement of transposable elements could produce patchwork pigmentation in corn kernels shown in figure 1.21. Account for fully pigmented kernels, unpigmented kernels, and kernels having varying degrees of pigmentation.

7. What are the major differences and similarities between meiosis and mitosis? What is the significance of each process?

8. Gene is a common term. What does it mean?

9. Based on your reading to date, can you develop a simple explanation of how natural selection works on genes and traits?

10. What might be the selective advantage of alternation of generations in plants?

SUGGESTED READINGS

Articles

Anderson, A. 1992. The evolution of sexes. *Science* 257:324–26.

Dahl, Hans-Henrik M. 1993. Things Mendel never dreamed of. *Medical Journal of Australia* 158:247–54.

Hartl, D. L. 1992. What did Mendel think he discovered? *Genetics* 131:245–54.

Huckabee, C. J. 1989. Influences on Mendel. *American Biology Teacher* 51:84–88.

Leitch, I., and M. Bennett. 1997. Polyploidy in Angiosperms. *Trends in Plant Science* 2:470–76.

Maes, T., P. De Keuleleire, and T. Gerats. 1999. Plant tagnology. *Trends in Plant Science* 4: 90–96.

Maguire, M. P. 1992. The evolution of meiosis. *Journal of Theoretical Biology* 154:43–55.

Mol, J., E. Grotewold, and R. Koes. 1998. How genes paint flowers and seeds. *Trends in Plant Science* 3:212–17.

Books

Corcos, A. F., and F. V. Monaghan. 1993. *Gregor Mendel's Experiments on Plant Hybrids: A Guided Study.* New Brunswick, NJ: Rutgers University Press.

Fowler, C., and P. Mooney. 1990. *Shattering: Food, Politics, and the Loss of Genetic Diversity.* Tucson: University of Arizona Press.

Klug, W. S., and M. R. Cummings. 1995. *Concepts of Genetics.* New York: Macmillan.

Silvertown, J., M. Franco, and J. Harper. 1997. *Plant Life Histories: Ecology, Phylogeny and Evolution.* Cambridge: Cambridge University Press.

Slatkin, M., ed. 1995. *Exploring Evolutionary Biology: Readings from Scientific American.* Sunderland, MA: Sinauer.

ON THE INTERNET

Visit the textbook's accompanying web site at http://www.mhhe.com/botany to find live Internet links for each of the topics listed below:

Gregor Mendel

Plant genetics

Crossing-over in genetics

Meiosis and genetics

Gel electrophoresis

Genetic variation in plants

Polyploidy in plants

Sexual reproduction

Genetics and the pea experiment

Chromosome theory

Evolution is the grand unifying theme of biology, linking all of the organisms that have ever lived on earth and all that scientists know about them. This fossilized Ginkgo leaf dates from the Cretaceous Period over 65 million years ago. A comparison with its modern counterpart raises tantalizing questions about change. How much? How little? What happened during those 65 million years? Do you think organisms change at the same rate through time?

Evolution

LEARNING ONLINE

To explore the concepts of evolution and connect to other sites, visit www.ucmp.berkeley.edu/history/evolution.html in chapter 14 of the Online Learning Center at www.mhhe.com/botany. The OLC is full of helpful resources for you as practice quizzes, key term flashcards, assistance with writing assignments, and hyperlinks.

For the general public, the word *theory* often suggests an educated guess, as in the sentence, "I have a theory about why she got sick." In this case, it would be more appropriate to say, "I have a hypothesis about why she got sick." In science, the word *theory* refers to an explanation that organizes all of the known evidence about a particular subject, has predictive powers, and is well supported by scientific investigations. Thus, the scientific **theory of evolution** explains how the first forms of life diversified into the organisms of today; it is the most comprehensive and powerful theme in biology. As the famous biologist Theodosius Dobzhansky said, "Nothing in biology makes sense except in the light of evolution." The theory of evolution provides a scientific explanation about how the many different kinds of plants, animals, and microorganisms arose on earth. This explanation incorporates evidence gathered from thousands of biologists, geologists, chemists, and other scientists around the world.

Evolution happens. It's a fact—like gravity—but it is also an organizing theory. Are there alternative explanations about the origin of diversity of life on earth? Certainly, but none stands the test of science; that is, none accounts for all the information we have about living things, has predictive power, or has been tested again and again over many years by thousands of scientists using a process of scientific investigation.

One of the goals of evolutionary biology is to discover the history of life on earth and the relationships among all species that have ever lived. A second goal is to understand the causal processes of evolution—that is, how the processes of evolution operate in the world. In this chapter, we learn about both goals.

Studies of the Natural History of Organisms Contributed to the Theory of Evolution.

In 1809, Jean Baptiste Lamarck, a famous French naturalist and plant taxonomist, formally proposed his theory of the **inheritance of acquired characteristics.** According to this theory, traits acquired by an individual during its life were passed to its offspring. This theory was caricatured by many detractors in drawings showing giraffes obtaining their long necks from previous giraffes who had stretched to eat the leaves of high tree branches. According to Lamarck, stretching increased the length of their necks, and this *acquired* characteristic was passed to the next generation. Giraffe necks were stretched further in each successive generation. Using similar reasoning, Lamarck also proposed that the use and disuse of a feature governed the fate of that feature in successive generations—sort of a "use it or lose it" idea. Organs of the body that were used extensively to cope with the environment became larger and stronger, while organs that were not used deteriorated.

Evidence, however, refuted Lamarck's theory that individuals inherited acquired characteristics. Inheriting acquired characteristics would mean that a feature obtained during an organism's lifetime would somehow be transmitted to the genes of that organism's gametes. This doesn't happen. For instance, if you dye your hair blond or if a plant has all of its flowers torn off by the wind, neither of these characteristics is passed on to offspring. Traits are passed to offspring through the parents' genetic material.

Charles Darwin's work laid the foundation for the modern theory of evolution.

Evolution implies change in living things over time, and its definition in dictionaries is usually something like the following: "The development of complex forms of life from simple ancestors via genetic change." Unfortunately, this definition is oversimplified to the point of being only partially correct. Another definition of evolution is "a change in the **gene pool** of a population over time." The gene pool represents all of the genetic information in a population; when this genetic makeup of a population changes, evolution has happened. Also, it is important to note that while the genetic information changes, so do the molecules produced by these genes, such as proteins. A third definition is that "evolution is the change in the inheritable characteristics of a group of organisms over the course of generations." All of these definitions are similar in that they describe changes in organisms through time. There is no simple definition of evolution because evolutionary studies may involve the study of an organism's genetics, ecology, morphology, anatomy, biochemistry, physiology, taxonomy, and biogeography, as well as several other aspects of its life history.

Throughout recorded history, people have wondered how the great variety of earth's organisms came to be. Among the earliest thinkers was Aristotle, who thought that each species—that is, a group of similar, interbreeding individuals—arose independently from inorganic matter and did not change. The foundation for understanding how biological change occurs and how such change forms new species rests on the *theory of evolution.* This theory is most closely identified with Charles Darwin (1809–1882), the most famous and influential naturalist-philosopher in history (fig. 14.1). His ideas

FIGURE 14.1

Charles Darwin. Darwin was a keen observer of nature who developed the theory of descent with modification and the theory of natural selection.

about evolution may have impacted a wider array of human endeavors than any other scientific advancement in history.

How Darwin arrived at his ideas and how they have affected science, philosophy, religion, and attitudes is one of the most fascinating stories of human achievement. This subject occupies a significant portion of hundreds of books spanning the past century. For this book, the most important aspect of the story is how Darwin's work provided the first scientific explanation for the diversity of life and a mechanism for evolutionary change—that is, natural selection (see chapter 2).

Evolution implies change, but it specifically refers to the modification of forms of life and their reproductive success. The foundation for modern evolutionary thought was described by Charles Darwin more than a century ago. Darwin also described natural selection as the main force that drives evolution. Although Darwin provided evidence for evolution by the mechanism of natural selection, he and his contemporaries did not know how parental traits could change and be passed to offspring. Beginning with the work of Gregor Mendel (see chapter 13), however, genetics provided some clues as to how genetic variation arose and how it was inherited. In the 1900s, evolutionary biologists began to learn about genetics of individuals within populations. More recently, molecular biology has contributed greatly to the study and understanding of evolution.

Although the theory of evolution is popularly attributed to Charles Darwin, the word *evolution* did not appear in his famous book, *On the Origin of Species,* published in 1859. It is more accurate to think of Darwin's work as a set of three theories. Two of his theories are still accepted: the **theory of descent with modification,** which explains the pattern of biological diversity via evolution, and the **theory of natural selection,** which explains the primary mechanism by which evolution occurs. The third theory proposed

by Darwin, the **theory of pangenesis,** was an incorrect explanation of inheritance. This theory was soon discarded and later replaced by the **chromosome theory of heredity** and the **theory of inheritance** (see chapter 13). Darwin's theory of pangenesis and Lamark's theory of the inheritance of acquired characteristics are good examples of how science is self-correcting. As we learn more, some theories must be altered or discarded altogether because of new evidence. Evolutionary biologists have spent a great amount of time trying to test, refine, reinterpret, and update Darwin's work. As a result, we now have the **genetic theory of evolution.** The components of this theory include many of Darwin's views on evolution, as well as modern ideas about inheritance and genetic change within populations. Today, while scientists debate the pace and mechanisms of evolution, they do not debate the existence of evolution. Evolution occurs and, as you'll see, is a fascinating process.

Charles Darwin was a natural historian.

Charles Darwin was born in 1809 and studied to become a physician and a clergyman, but he became neither. He was fascinated with nature, had keen observational skills, and because most naturalists at that time were also clergymen, Darwin's pursuit of natural history was an accepted practice. He joined the crew of the HMS *Beagle* whose mission included surveying lands around the world. The *Beagle* sailed from England in 1831 to chart, among other areas, some of the remote islands and coastline of South America (fig. 14.2). The 22-year-old Darwin spent much of his time on shore collecting plants and animals and making geological and biological observations. The plants and animals of South America seemed diverse and exotic to Darwin, and his collections provided examples of the varied adaptations of organisms to life in mountains, rain forests, grasslands, and inhospitable islands such as the Galápagos Islands and Tierra del Fuego of South America. **Adaptations** are characteristics of an organism that increase its chances for survival and for reproduction.

The geographic distribution of South American organisms both fascinated and puzzled Darwin, and he wondered why they were so diverse and different from organisms in England. During the voyage, Darwin also read Charles Lyell's *Principles of Geology* and realized that the earth was very old and constantly changing (fig. 14.3). This idea, along with his collections and observations, convinced Darwin to embrace the idea that the organisms that fascinated him may have changed (that is, evolved) along with a slowly changing environment over a long period of time.

Darwin made some of his most profound observations on the Galápagos Islands. These islands are near the equator and about 950 kilometers west of South America (see fig. 14.2). Many of the plants and animals found there are **endemic** (i.e., they live nowhere else) to the islands, but they strongly resemble species from the closest mainland. Among

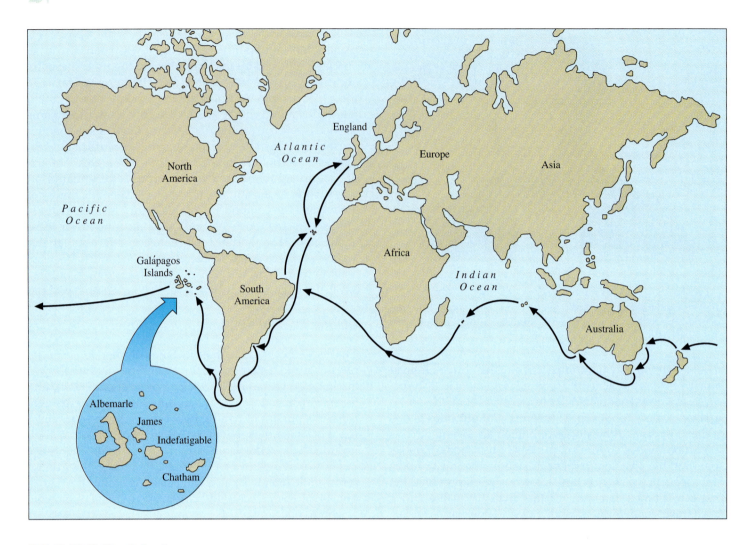

FIGURE 14.2
The *Beagle* voyaged around the world and covered 60,000 kilometers between 1831 and 1836. At each of many stops, Darwin observed and chronicled variation among and within species. Later he used this evidence to support his theories of speciation and natural selection.

the more famous examples of unusual organisms on these islands are birds called finches (fig. 14.4). Darwin collected 14 types of finches that were similar but appeared to be different species; that is, there were 14 different populations of finches that looked and behaved differently from each other and that did not produce offspring with birds of other populations. Furthermore, these finches were curiously similar to the mainland species, yet slightly different. If species were unique and unchanging creations, then why, Darwin wondered, did so many of the Galápagos finches resemble nearby South American species rather than European or African organisms? If finches on the islands had been specially created, why didn't they all look the same? Darwin later concluded that the island finches had descended from a population of ancestral finches from the mainland. The population of mainland finches changed and diversified after becoming isolated on the islands, giving rise to 14 species, 13 of which are endemic to the islands today.

Since the time of Darwin, many other groups of organisms around the world have been discovered to be descended from a single ancestral type. One of the best-known examples of such diversification in plants is the genus *Phlox,* which evolved from an ancestral bee-pollinated species to a variety of species that undergoes pollination by beetles, flies, bats, hummingbirds, moths, or butterflies (fig. 14.5).

Darwin and Wallace independently formulated a theory of natural selection.

During his 5-year voyage on the *Beagle,* Darwin shipped many crates of carefully packed specimens back to England. After returning to England in 1836, Darwin did not immediately begin writing his book on the origin of species. Instead, he organized his observations and analyzed a tremendous number of specimens and data.

(a)

(b)

FIGURE 14.3

(*a*) Geological processes have exposed layers of ancient soils and fossils in this canyon. These layers of rock hold "snapshots" of plants and plant communities of the past and record the sequence of ancient organisms. The deeper the layer, the older the fossils it contains. Preserved impressions of ancient plants hold the most tangible clues to evolutionary history. (*b*) This fossil of a Triassic seed plant, *Dicroidium* species, is about 200 million years old. The gradual accumulation of sediments that surrounded and fossilized such plants indicated to Charles Lyell in the early 1800s that the earth was millions rather than thousands of years old.

In 1838, two years after returning to England, Darwin encountered an inspiration for his theory of natural selection from a book entitled *An Essay on the Principle of Population,* by Thomas Malthus. Malthus was an economist and clergyman who wrote that populations had an inherent tendency to increase in geometric proportions. Malthus also claimed that resources to support this growth may increase slowly or not at all. He therefore reasoned that because continued growth of a species would outstrip needed resources (especially the food supply), there would inevitably be a struggle for existence (fig. 14.6). Specifically, Malthus warned of the explosive growth of the human population. He pessimistically

FIGURE 14.4

Darwin theorized that the variety of finches on the Galápagos Islands arose from mainland species. He believed that many generations after their ancestors' arrival, they had adapted to the islands' various environments and food sources and had become unique, reproductively isolated species. Thirteen of the fourteen species of finch are endemic to the Galápagos Islands. Similar examples of evolution after separation from ancestral populations are common among plants.

(a)   (b)

FIGURE 14.5

Evolution has produced a variety of phloxes with different pollinators. (a) *Phlox longifolia* is adapted to moth pollinators. (b) *Phlox sibirica* is adapted to butterfly pollinators.

warned that humans were reproducing so fast that war, famine, disease, and other human sufferings must eventually limit growth. He believed that humans were doomed to suffering, and he foresaw the need for birth control and social change, such as delayed marriage.

Darwin adopted Malthus's ideas that populations tend to increase geometrically and outstrip their resources. This meant that some newborn organisms will die in competitive environments. Darwin observed that wild organisms have variable traits, and he reasoned that in a resource-limited environment, the hardiest or best-suited individuals have a competitive and reproductive advantage over "weaker" individuals. As a result, the best-suited organisms would live longer and be more likely to reproduce than "weaker" organisms. Thus, traits of well-adapted individuals would increase in successive generations, and the traits of poorly adapted individuals would decrease. Darwin called this process *natural selection*, where the best-adapted organisms in a population produce more viable offspring than those poorly adapted. He saw a similar mechanism in the modern world among plant and animal

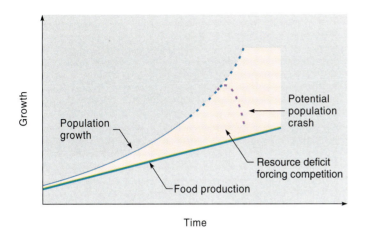

FIGURE 14.6

The predictions of Malthus guided Darwin's formulation of natural selection theory. Malthus reasoned that the deficit between population growth and food production forced competition and would later crash the human population. Darwin examined these predictions closely and concluded that expanding competition creates strong selection pressures that shape future generations. Are there any examples of the predictions of population crashes in the world today?

breeders who practiced *artificial selection* (see chapter 2). These breeders changed the population by breeding organisms with desirable traits and destroying or consuming those with undesirable traits.

In Darwin's view, the history of changes of organisms could be represented as a tree with multiple branches. The trunk and bases of the branches represent common ancestors that gave rise to all the subsequent species represented by branches farther from the trunk. The diversity we see today is represented by all the tips of the branches, which may become extinct or may produce new branches through the formation of new species (fig. 14.7). Most branches of the evolutionary tree (possibly as many as 99%) have been dead ends; that is, they did not produce new species but became extinct. Darwin often referred to "descent with modification" when describing differences between generations of organisms through time. This pattern of descent includes many branches, and modern biologists have redrawn the "tree of life" to look more like a bush, with many major branches to reflect our current understanding of evolution.

Darwin was convinced that a group of similar individuals—a population—could give rise to a new species by accumulating adaptations to a new environment. New and different adaptations were especially likely when a species became separated into populations isolated by geographical barriers such as mountain ranges or bodies of water. Each of the separated populations might then reproduce and change in response to local environmental conditions in the different locations. As an isolated population became adapted to its new environment, its members would become increasingly different from the ancestral population.

After many generations, an isolated population would be different enough to be a new species. Darwin hypothesized that this had happened to the Galápagos finches. The populations of these birds on the Galápagos Islands gradually adapted to each island environment and its food supply and, in the process, became different from their mainland ancestors. In addition, the populations on different islands became different from one another.

Within 5 years after returning to England, Darwin had established the major components of his theory of natural selection. But he hesitated to publish his idea that species arose by adaptations of organisms to their environment. His idea strongly contradicted public opinion in Victorian England, even though the concept of a changing earth and changes in its inhabitants was gaining credibility among naturalists. Darwin overcame his hesitation in 1858 after receiving a letter from Alfred Russel Wallace containing a manuscript about a theory of natural selection that was remarkably similar to his own. Darwin was shocked and discouraged. He wrote to Lyell, "I never saw a more striking coincidence; all my originality, whatever it may amount to, will be smashed." However, Darwin and Wallace were interested in the truth, and their theory was presented jointly at a scientific conference, after which Darwin quickly finished and published his *On the Origin of Species* in November, 1859. The first edition sold out on the day it appeared. It is important to note that these two naturalists, living in different parts of the world (Wallace was living in the East Indies) independently came to the same conclusions based on their own observations of plants and animals they had intensively studied.

Darwin explained and documented his ideas so much more extensively than Wallace that Darwin is known as the leading author of the theory of natural selection. The theory was revolutionary not only because it was unique but also because Darwin presented a strong argument based on overwhelming evidence. To Darwin, evolution was the gradual and selective accumulation of adaptations that were beneficial to a species in its environment—no more, no less.

INQUIRY SUMMARY

Lamarck recognized change in organisms but incorrectly explained it as the inheritance of acquired characteristics. Acquired characteristics are not heritable. According to Darwin, evolution included gradual and selective accumulation of the beneficial adaptations of a population to its environment and a reduction of detrimental characteristics. Natural selection is a mechanism that accounts for these heritable changes in populations.

How does the fact that Wallace and Darwin worked independently, yet came to the same conclusions, strengthen the case for natural selection and evolution?

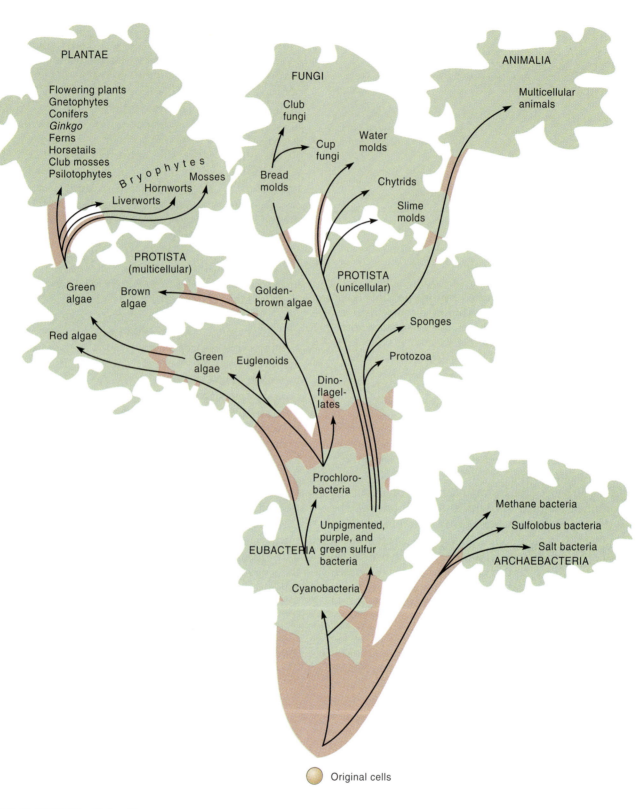

FIGURE 14.7

The evolutionary history of organisms has been portrayed as a tree with ancient forms low on the tree. At the tips of branches are modern forms, and the base of diverging branches represents the common ancestry of the newly evolved kinds of organisms. Prokaryotes (eubacteria and archaebacteria) are the most ancient organisms and probably gave rise to eukaryotic organisms that were ancestors to plants, animals, and fungi. Compare this tree to a classification scheme of plants in figure 15.14 and more modern phylogenies such as that shown in figure 16.5.

Investigations Since Darwin's *On the Origin of Species* Provide Extensive Evidence for Descent with Modification.

Darwin meant to publish a full explanation of evolution based on a thorough analysis of the evidence he had gathered, but he never did. *On the Origin of Species* was an abstract of his work (although over 400 pages long!), which he rushed into publication when Wallace's work appeared. Although not complete, Darwin's work provided abundant evidence for the theory of descent with modification. Work since Darwin's has added much new and supporting information.

Evolution explains the unity of living organisms—that is, why many characteristics are shared by all living things, including the composition of proteins and DNA. Evolution also explains the diversity of life on earth—how millions of species of organisms have descended, with modification, from different lineages of common ancestors. This suggests that all forms of life are related by unbroken chains of descent and that all living species can be traced back through time to fewer and fewer common ancestors. The following sections provide evidence for the evolution of living things.

Ample evidence indicates the earth is over 4.5 billion years old.

If evolution occurs gradually, then the earth must be old enough to allow it to happen. Many people in Darwin's day believed that the earth was only a few thousand years old, a period that seemed too brief for all living things to have evolved from a single common ancestor. Today, we estimate the age of the earth by comparing ratios of radioactive to nonradioactive elements in rocks. For example, physicists have evidence that the ratio of uranium-238 (^{238}U) to lead when the solar system was formed was about double the current ratio. We know that the **half-life** of ^{238}U—that is, the time in which half of the ^{238}U in a sample decays to something other than ^{238}U—is about 4.5 billion years. This indicates that the earth is between 4.5 and 5.0 billion years old, since the ratio of ^{238}U to lead has decreased to about half of what it was when the earth was formed. Incidentally, meteorites found on earth are also estimated to be about 4.5 billion years old, which supports the idea that the earth was formed simultaneously with other parts of the solar system.

Vascular plants have left fossil evidence about their evolution.

The oldest organisms probably either were not preserved as fossils because they were too small or too soft, or we can't recognize their preservation in rocks. The oldest recogniza-

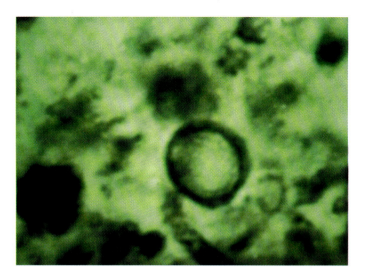

FIGURE 14.8

Fossilized bacterial cells such as these have been found in ancient rocks of Australia and South Africa. These rocks are about 3.5 billion years old and predate the earliest known fossils of eukaryotes by at least 1 billion years. Chemical evidence within these fossils suggests that photosynthesis was occurring 3.5 billion years ago.

ble fossils are of bacteria that lived at least 3.8 billion years ago. This age establishes a minimum time span for biological evolution. Biologists have examined fossil and molecular evidence and calculated that this is long enough for the evolution of all of the organisms of today.

Darwin predicted that if evolution happened, the fossil record would contain links among related organisms, from ancient to progressively more recent species (fig. 14.8). The information about fossils that subsequently has been gathered supports Darwin's prediction. Some of the best fossils are from bones and teeth, so zoologists can construct nearly complete series of fossils for some groups such as horses, humans, other mammals, and reptiles. The most complete record of plants is for those with vascular tissue; however, there is not a clear lineage of plants—one giving rise to the other—as there seems to be for other organisms. Nevertheless, 410-million-year-old fossils of the first vascular plants, if not directly ancestral to present-day groups, look like they share a common ancestry with certain groups of living vascular plants (fig. 14.9). More details about the evolutionary relationships of living plants are presented in chapters 15 through 18 on the diversity of organisms. Why do you think the best fossil hunting is often in locations that are now very dry?

Homologous structures are derived from a common ancestor.

Other evidence supporting evolution comes from comparing the body parts of organisms. If different species have evolved from a common ancestor, they may possess

FIGURE 14.9
This modern plant (*Psilotum*) resembles the fossils of ancient ancestors. Although the fossil lineage of vascular plants is incomplete, the earliest fossils show common ancestry with some simple plants such as *Psilotum*. This modern *Psilotum* has forked branching, sporangia, simple vascular tissue, and no roots or leaves.

features that are **homologous,** that is, the features have the same evolutionary origin and come from the same ancestor. In flowering plants, for example, the ovary surrounding the seeds of magnolia resembles that of lily, poppy, oak, and orange, suggesting that all of these plants inherited the ovary from a common ancestor (see fig. 12.3). Likewise, the red pigments of cactus flowers, beet roots, *Portulaca* stems and flowers, and spinach petioles are homologous. These pigments are of the same chemical type, called betalains, which is unlike other types of red pigments. Most of the major features of plants that we've discussed, such as their vascular structure and their hormonal responses to stimuli, probably descended from a common ancestor of the plant kingdom. Plants also have features that are homologous with features of other kinds of organisms. Cells with nuclei, chromosomes with histones, and genes with introns were inherited from the common ancestor of plants, animals, and fungi.

Convergent characteristics occur in unrelated organisms living in similar habitats.

Whether similar features are homologous is often unclear because not all similarities are due to homology. The issue of homology is clouded because similar features may arise in distantly related organisms that live in similar environments. One of the most common examples, which often fools all but the most astute observers, is the stem succulence of cacti and euphorbias (fig. 14.10). Instead of homol-

ogy, this feature shows **convergence,** which is one explanation for a similarity that does not come from a common ancestor. In this case, succulent stems provided descendants of a cactus ancestor an adaptation to arid environments in the Americas. Likewise, descendants of a euphorbia ancestor with succulent stems became adapted to the arid environments of Africa. Thus, the succulence seen today in cacti and euphorbias is not a homologous characteristic, but a convergent one because the two distantly related groups of plants evolved similar characteristics in similar environments.

Similarities among distantly related organisms are not likely to result from coincidence. Instead, such features of plants indicate that different organisms have evolved in similar ways in similar environments found in different areas of the world. Natural selection has helped shape these organisms in similar ways.

Biogeography also provides evidence for evolution.

Biogeography is the study of the geographic distribution of organisms. The distribution of today's organisms provides strong evidence for evolution because biogeography reflects the history of living species; that is, biogeography reveals the movement of organisms and their change through time. From biogeographical information, we can determine the ecological factors that control a species' distribution and the events that led to this distribution. In other words, the biogeography of a species partially records the evolution of that species. For instance, the study of the biogeography of *Magnolia* trees reveals that this primitive flowering plant currently grows naturally in the eastern United States and Central America as well as in distant China, yet nowhere in between—at least without the help of humans. This distribution can be explained if magnolias were once widespread throughout the world (several million years ago). Then, as other flowering plants evolved that were better competitors, the magnolias died, surviving in only two parts of the world today (fig. 14.11). Fossil evidence supports this hypothesis; fossil remains of ancient magnolias have been discovered throughout Europe and in Greenland.

What explains the oddly disjunct distributions of other organisms such as creosote bushes (*Larrea*), which live in the deserts of western North America and southern South America; or skunk cabbage (*Symplocarpus foetidus*) and the tulip tree (*Liriodendron tulipifera*), which occur in eastern Asia and eastern North America; or cacti, which are exclusively American except for one species in Madagascar? How have these patterns of distribution come about? The answers lie in the study of biogeography.

Island biogeography has been particularly informative about the movement and evolution of organisms. *Island* refers not only to land surrounded by water but also to any area surrounded by a different habitat, such as a mountain surrounded by desert or a clump of trees surrounded

(a)

(b)

FIGURE 14.10

Stem succulents: (*a*) cactus and (*b*) *Euphorbia*. These plants are strikingly similar not because they share a common ancestry, as you might suspect, but because they have similar adaptations to arid environments. Their similarity illustrates convergent evolution, unrelated plants may accrue similar adaptations due to similar environmental pressures.

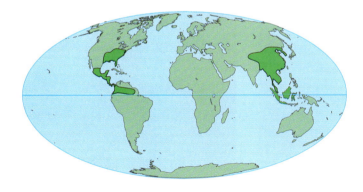

FIGURE 14.11

World distribution of the magnolias (darker areas of map), illustrating a disjunct distribution. What are alternative explanations for such a distribution? What evidence would support each alternative?

by prairie. Islands often have species found nowhere else in the world, and the closest relatives of these species may inhabit a different environment on the nearby mainland. This distribution is best explained by dispersal of a plant to an island, adaptation by the plant's offspring to the island's

new environment, and the subsequent formation of new species through time. Individuals living in a particular place (e.g., the mainland) often spread, or radiate, to other habitats (e.g., islands). These new habitats include different environmental conditions that promote the selection and evolution of new adaptations. Migration to new environments followed by adaptation and formation of new species is called **adaptive radiation** (fig. 14.12).

The finches that Darwin found on the Galápagos Islands are an example of adaptive radiation. Botanists also have found many examples of adaptive radiation in plants, most notably on the islands of Hawaii. A well-studied example involves the flowering plants called sticktights (*Bidens* species), so named because their fruits stick to socks or other clothing when you walk among the plants (fig. 14.13). This stickiness explains how these plants may have attached to and been transported by seabirds from the mainland to the Hawaiian Islands. The sticktights in Hawaii apparently evolved from a single weedy plant that colonized Hawaii thousands of years ago and then radiated into the more than 40 species that now occur only on the islands. The most recently appearing species in Hawaii may be *Bidens ctenophylla*,

The Lore of Plants

Islands can be hot spots of evolution. For example, environmental conditions on New Caledonia, an island distant from the east coast of Australia, have produced one of the world's most unusual floras. The forests of New Caledonia contain more than 1,575 species of plants, of which an astonishing 89% live nowhere else in the world. Unfortunately, disruption of the environment by logging, mining, brush fires, and other human activities has reduced the amount of undisturbed forest to less than 1,500 square kilometers (less than 600 square miles) covering less than 9% of the island. What do you expect will be the long-term consequences for the plant species on this island?

which is a long-lived, woody tree whose fruits are larger and fewer than those of its weedy ancestor and do not stick to anything. This species no longer has the weedy lifestyle that characterized its earliest island ancestor.

Biologists now also use molecular evidence to study the evolution of organisms.

Since the late 1960s, evolutionary biologists have been comparing molecules such as proteins, DNA, and RNA of various species, much as they have done with anatomy and morphology for more than a century (fig. 14.14). This molecular evidence, like evidence from tissues and organs, helps reveal the evolutionary relationships of plants. One of the most widely used comparisons is based on cytochrome *c* oxidase, an enzyme of the electron transport chain in cellular respiration (see chapter 11). All living organisms contain this enzyme. The universal occurrence of this enzyme indicates that it evolved early in the origin of life and has been inherited by all of the descendants from a single ancestor; that is, it is evidence of a single origin of life. In addition, the amino acid sequence of this enzyme has slowly changed as organisms evolved through time. Related organisms have similar amino acids in this enzyme, revealing the pattern of descent among organisms. For example, if the cytochrome *c* oxidase in two organisms differs by only a few amino acids, this indicates a relatively recent common ancestry and a close evolutionary relationship between the organisms. The examination of amino acid sequences of cytochrome *c* oxidase shows that wheat, and thus all plants, actually is more closely related to animals than to fungi. This conclusion is based on the fact that there are fewer dif-

(a)

(b)

(c)

F I G U R E   1 4 . 1 2

Isolated and diverse environments such as the Hawaiian Islands fuel adaptive radiation and divergence. Radiation has been striking and rapid for the three species shown here, (*a*) *Argyroxiphium sandwicense,* (*b*) *Wilkesia gymnoxiphium,* and (*c*) *Dubautia ciliolata.* These and twenty-five other closely related species probably descended from a single species arriving from the Pacific Coast of North America. Adaptive radiation on the islands in diverse environments has produced diverse morphologies even though the species differ in only a few critical genes.

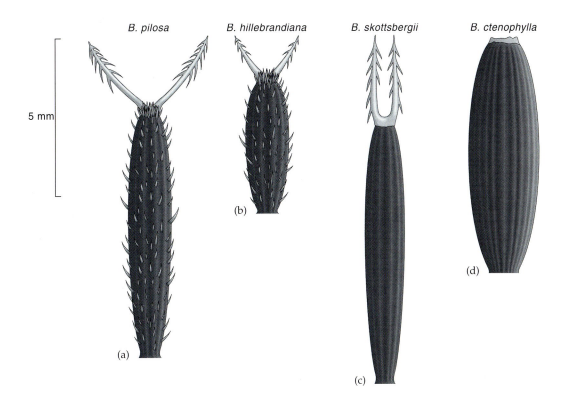

B. pilosa B. hillebrandiana B. skottsbergii B. ctenophylla

5 mm

(a) (b) (c) (d)

FIGURE 14.13

Variations in fruit morphology show adaptations to new environments and the pattern of colonization and inland invasion by Hawaiian species of *Bidens*. (*a*) *B. pilosa* was apparently the ancestral colonizer. It was easily dispersed from the American mainland by birds carrying its sticky fruits with upward-pointing bristles and spreading barbs. (*b*) Today, a resident species on the coast of Hawaii, *B. hillebrandiana*, has ornamented fruit similar to *B. pilosa*. (*c*) More inland, the low-elevation species *B. skottsbergii* has fruit with reduced dispersal ability due to loss of bristles and closing of barbs. (*d*) Still farther inland, the central mountain habitats have *B. ctenophylla* with no bristles or barbs. Apparently, habitats progressively inland do not favor fruit with strong features for animal dispersal; these fruits may be dispersed only short distances.

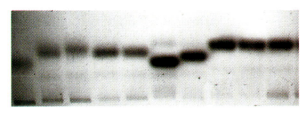

FIGURE 14.14

Results of a molecular analysis in which the same enzyme from 10 different plants was compared to determine the relatedness of the plants. The results indicate that the plants are different genetically, otherwise all the spots would have been at the same level.

ferences in the cytochorome *c* oxidase of plants compared to animals than in plants compared to fungi.

More recently, evolutionary biologists also have begun to compare sequences of DNA and RNA, as well as many other kinds of molecular information, to deduce patterns of evolutionary descent. Research by Svante Pääbo and others at the University of Munich has shown that genetic information is retrievable from DNA in fossils many thousands of years old. The results of these fossil DNA comparisons are generally consistent with the patterns of other homologous features such as anatomical and morphological characteristics.

Evolution can be observed.

Some evolutionary changes happen fast enough for people to observe over a relatively short period of time. For instance, changes in the food supply due to drought on the Galápagos Islands caused a change in the beak size of a species of finch within just a few years. Also, hundreds of species of crop-eating insects have evolved resistance to DDT and other insecticides since World War II. Certain disease-causing organisms in humans have rapidly evolved resistance to drugs, including the bacterium that causes gonorrhea and the virus that causes AIDS (see Perspective 14.2, "Unstoppable Killers").

Evolution also can be observed as a result of experimentation, especially with rapidly growing organisms such as bacteria that are exposed to high temperatures, chemicals, and antibiotics, which are used as selective agents to identify tolerant individuals.

INQUIRY SUMMARY

The earth is over 4.5 billion years old. Living organisms have been on earth for about 3.8 billion years. Modern evidence for evolution includes the age of the earth, the fossil record, homology, the convergence of adaptations of different species in similar environments, the similarity of species in different parts of the world, similarities of nucleotide and amino acid sequences among species descended from a common ancestor, and direct observations of experimental results.

Some people have called the state of California an island because it is surrounded by mountains and an ocean. How do you think these mountains and ocean might have affected the evolution of plants and animals in the state?

Natural Selection and Evolution Are Not the Same Thing.

Although the theory of descent with modification was revolutionary, even during Darwin's time it was supported by convincing evidence. Darwin's greatest contribution to evolutionary biology, however, was not the concept of change but his explanation of a *mechanism* for change. To Darwin, observable evidence verified the results of evolution, but the specific *process* was apparently so subtle that it had eluded other naturalists. That process was natural selection.

Even if the reproductive advantages of some characteristic are only slightly greater than those of other characteristics, the favorable characteristic inevitably accumulates in a population *after many generations* of being "selected" in a particular environment; that is, those individuals with favorable characteristics produce more offspring than those with less favorable characteristics. This accumulation of characteristics favorable in a particular environment is the heart of evolution. Of course, the same mechanism *decreases* the occurrence of unfavorable characteristics. However, evolution is not about progress; evolution has no "goal" to produce highly complex organisms like humans. In fact, because the environment changes over the years—warmer summers, more carbon dioxide in the air, greater snowfall some years—those individuals that are highly adapted to the environment of a particular habitat may be unsuited to the same habitat when the environment changes. The characteristics that made the individuals, and their offspring, reproductively successful in one environment might not help them in the future if the environment changes (fig. 14.15).

Remember that natural selection and evolution are not identical: natural selection is a mechanism that results in evolution. Natural selection adapts a population to a particular set of environmental factors. When the environment changes, so do the characteristics that are favored in the environment.

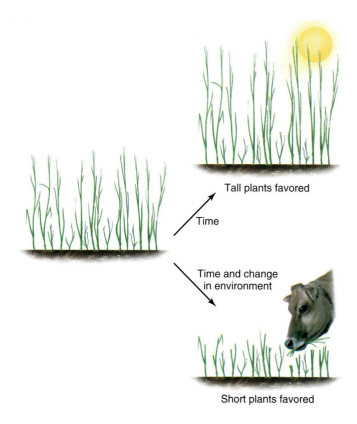

Tall plants favored

Time

Time and change in environment

Short plants favored

FIGURE 14.15

Environments change through time. Depending on the environment, different organisms will be favored. In this example, tall plants may have a selective advantage over short plants, which are shaded until a change in the environment occurs—in this case, a herbivore that eats tall plants wanders into the area, eating tall plants and leaving short ones.

Recall from chapter 2 the main principles of natural selection are:

1. Organisms tend to produce more offspring than survive.
2. Organisms compete for limited resources essential for life.
3. Individuals in a population are not all the same—some have heritable variations.
4. Individuals having favorable traits that allow them to compete successfully for resources produce, on average, more offspring than those with unfavorable traits, which produce fewer offspring.

Fitness is a measure of an organism's evolutionary success but does not strictly refer to its health, survival, or adaptation to the environment. Rather, an individual's fitness is measured by its number of surviving offspring relative to the number of offspring from other individuals in the same population. Thus, fitness represents the extent to which an individual's genes are passed to succeeding generations. Individuals with high fitness produce more surviving offspring than do individuals with low fitness. Thus, the phrase "survival of the fittest" does not necessarily refer to the strongest, fastest, or smartest individuals; it

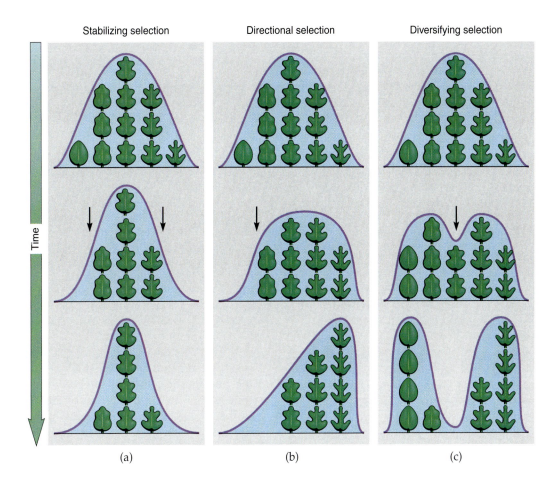

FIGURE 14.16

Stabilizing, directional, and diversifying selection for leaf margination will change phenotypic frequencies over time. The height along the curve represents the number of individuals with that phenotype, and the arrows indicate individuals that are selected against. (All three types of selection begin with the same population of plants in the top row.) (*a*) Stabilizing selection reduces variation and preserves a narrow phenotype best adapted to the environment. (*b*) Directional selection promotes phenotypes at one end of the range of variation, and the distribution of phenotypes moves toward that end. (*c*) Diversifying selection promotes extremes of a characteristic and may produce two distinct populations.

refers to those individuals that can produce the most surviving offspring.

The three commonly considered kinds of natural selection are stabilizing, directional, and diversifying selection.

Genetic traits usually do not occur in clear categories such as the tall or short traits studied by Mendel (see chapter 13). Instead, populations often contain a continuum of values for many traits; for example, some individuals may be very tall and some very short, but most are various heights in between these extremes. Natural selection can change the number of individuals in a population with different characteristics along this continuum. There are three kinds of selection: stabilizing, directional, and diversifying.

Stabilizing selection promotes the most common phenotype and reduces both extremes (fig. 14.16*a*). This

kind of selection operates in most populations, and organisms with characteristics near the extremes of variation may die or not reproduce and therefore not pass their extreme traits to the next generation. As a result, the frequency of individuals with the common phenotype increases. Stabilizing selection in most populations eliminates extreme individuals (including mutant forms) and is most common in stable, unchanging environments. For instance, some plants in a population may have long stems that topple over in the wind. Other plants may have short stems and die because they are shaded by taller plants. Stabilizing selection then results in a population of plants with medium-length stems.

Directional selection is a primary mechanism of change in changing environments. In directional selection, individuals at one extreme of population variation don't produce as many offspring as those at the other extreme. Over many generations, the average condition in the population moves toward the extreme characteristic that is favored (fig. 14.16*b*). For example, plants studied in a grazed

14.1 Perspective *Perspective*

Darwin's Misreckoning

In seeking a genetic mechanism for traits passing from one generation to another, Darwin addressed the previously unquestioned assumption that the parent's body somehow influenced the form of the offspring. In this regard, Darwin subscribed to Lamarck's theory of the inheritance of acquired characteristics. But Darwin took this idea of heritable traits a step further. He knew that there must be some physical connection between parent and offspring. He explained this connection in his theory of pangenesis. According to this theory, every part of a mature individual produces tiny packets of heritable information, which Darwin called gemmules. To be transmitted to offspring, gemmules from all over the body were somehow transported to the reproductive organs and packed into the gametes before fertilization. The gemmules of the two parents would blend together in the offspring, where they would sprout wherever appropriate to determine the features of each part of the new organism. Darwin's ideas about pangenesis probably arose from his belief that the use and disuse of organs influenced heredity, which was a widely held view at the time. In that context, pangenesis was a perfectly logical hypothesis, although it was inaccurate. Darwin's idea of gemmules has been refuted through scientific evidence, however, investigations have continued to support his other theories.

---

pasture in Maryland were shorter than the same species in ungrazed pastures. Were they short due to grazing or due to genetic differences? To separate the effects of grazing from genetics, some of the short plants were planted away from grazing herbivores. Surprisingly, some of the transplants remained short even without grazing. Apparently, directional selection promoted by grazers in the past had altered the population, favoring short plants less susceptible to chewing herbivores (see fig. 14.15).

Diversifying selection (also called *disruptive* or *catastrophic* selection) occurs in environments that are suddenly or drastically changing. These environments become unfavorable to individuals that were previously successful living there. Sometimes organisms with extreme characteristics survive and reproduce because they live at the periphery of the population. As you might expect, drastic environmental changes are usually short-lived, but they significantly increase genetic diversity and temporarily decrease population size.

As an example, genetic diversity of sugar maples in Ohio is maintained by diversifying selection. The southern variety of sugar maples can survive occasional droughts and cool winters, while the northern variety is better adapted to the coldest winters. In Ohio, a cool, dry winter may favor the southern plants, but the following year, a severe winter may select for the northern ones. Diversity is maintained as both types intermingle in the Ohio population. Sugar maples with intermediate tolerances are not favored in either cold or warm climates.

INQUIRY SUMMARY

Natural selection is the process by which the best-adapted organisms in a particular environment successfully reproduce more than other organisms in the population. Natural selection is the mechanism that fosters the reproduction of advantageous traits in the next generation. Changes in the environment can lead to changes in the best-adapted organisms in the habitat. Fitness refers to an organism's ability to produce surviving offspring. Stabilizing selection reduces variation and promotes the most prevalent phenotype; directional selection promotes an extreme phenotype; and diversifying selection promotes overall variation.

If an organism is strong and healthy but produces no offspring, do you think this individual is "fit" in evolutionary terms? Why or why not?

Genetic Variation Is Necessary for Natural Selection.

Natural selection requires genetic variation in a population. That is, if all organisms of a population are genetically identical, we might expect all of them to die or reproduce equally in the same environment; therefore, no variation, no natural selection. Where does genetic variation originate? Where do new alleles come from? You've already learned about a major source of genetic variation—mutation (see chapter 13). In this section, we describe other sources of variation upon which natural selection can operate. Understanding these sources is essential to understanding evolutionary theory, because natural selection does not affect all genes in the same way. Genetic variation within populations results, in part, from gene flow from one gene pool to another.

FIGURE 14.17

Large populations such as these white prickly poppies (*Argemone albiflora*) share a common gene pool and frequently cross-pollinate. Population geneticists study the genetics of populations rather than the genetics of an individual.

The genes of an organism are part of the gene pool.

Among the most important developments in evolutionary biology since Darwin is the application of genetics to the theory of natural selection. A major problem with the initial understanding and acceptance of evolution was a lack of knowledge of the genetic basis of variation (see Perspective 14.1, "Darwin's Misreckoning"); Mendel's work was unknown.

Individuals do not evolve; populations or species do. Modern evolutionary biologists think in terms of populations and frequencies of genes or alleles in a population. The integration of Mendelian genetics and natural selection is called **population genetics.** A unifying property of a population is its gene pool, which is all of the genes of all the individuals in the population. Members of the next generation get their genes from this gene pool, and individual organisms temporarily contain a small portion of the gene pool. To population geneticists and to evolutionary theory, the gene pool, its variations, and the mechanisms causing change are more important than the individual. Indeed, because populations often include thousands or millions of organisms, the fate of one individual usually has little influence on the fate of the whole population (fig. 14.17).

Recall from chapter 13 that each gene may have several different alleles; for instance, the gene for flower color may have alleles for red, white, and blue flowers. Evolutionary theory presumes that some alleles in natural populations increase in frequency from generation to generation, while others decrease. Thus, favorable alleles in a genotype are more likely to occur in greater frequencies in the next generation. In contrast, unfavorable alleles are less likely to be passed to the next generation, and their frequency will be reduced. The change in the frequency of alleles within a gene pool over many generations is a population geneticist's definition of evolution.

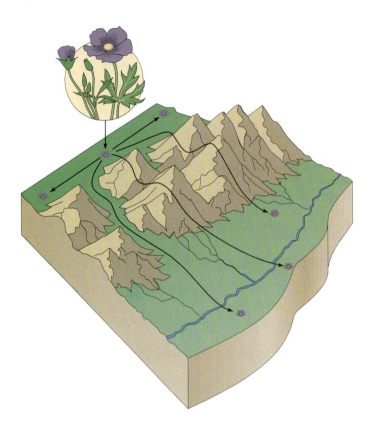

FIGURE 14.18

Gene flow is inhibited by geographic barriers between populations. Dots represent distant populations. Arrows show directions
alleles may flow to these populations. Although distance and barriers slow the flow of alleles to a distant population, even occasional foreign alleles may maintain significant genetic variation
within that population and increase its similarity to neighboring
populations.

*Genetic variation is promoted by gene
flow from one population to another.*

Gene flow is the transfer of genetic material from one population to another (fig. 14.18), that is, from one gene pool to
another gene pool. This typically occurs when animals migrate from one population to another or when seeds or
pollen of plants are dispersed to neighboring populations
(fig. 14.19). For example, the grass *Bothriochloa intermedia*
seems to have incorporated genes from many other
grasses, including plants from Pakistan, eastern Africa, and
northern Australia. This could have happened when seeds
of the grasses were carried to a distant population and the
seeds grew into mature plants that reproduced with their
new neighbors.

  Gene flow boosts the variation within a population
and decreases genetic differences between populations.
Genes may flow frequently between neighboring populations; gene flow between distant or isolated populations is

FIGURE 14.19

Gene flow occurs when pollen or seeds are dispersed to neighboring populations. The seeds of this cattail (*Typha* species) are
dispersed by the wind, carrying genetic information to new
populations.

less frequent, which allows their gene pools to diverge over
time. Geographic barriers that prevent gene flow are important in the formation of new species. Significant reduction in gene flow partially explains why islands isolated by
water are more likely to produce new species than are vast
expanses of grassland. The lack of gene flow also helps explain why widely separated lakes and streams contain
more variation among populations than do oceans in
which organisms can freely migrate.

  Separated populations of a species are seldom genetically identical; often, the greater the distance between
populations, the greater are the differences between them
genetically. This is because distant populations have a
good chance of living in different environments; as a result,

individuals with different characteristics are favored in each environment.

Gene flow in plant populations is difficult to measure, but it can be experimentally estimated by planting plants that are recessive homozygotes at various distances from plants with a dominant allele and then examining the distribution of heterozygous progeny. Using this technique, A.J. Bateman measured pollen dispersal in wind-pollinated (e.g., corn) and insect-pollinated (e.g., radish) crops. The proportion of corn plants receiving the dominant allele by gene flow decreased exponentially with distance and was reduced to 1% at only 13 to 16 meters from the pollen source. Similarly, pollen of insect-pollinated plants is often carried only a few meters from the source. However, the small proportion of pollen that is carried farther may contribute importantly to gene flow.

The genetic variation of small populations may be affected by genetic drift.

In addition to natural selection and gene flow, another process that can affect the genetic makeup of a population is genetic drift. **Genetic drift** is the process in which changes in the genetic makeup of a population are due to chance alone. Although genetic drift can happen in any population, its effect is greatest on the gene pool of a small population. In small populations, random events such as mutations or weather may significantly affect the gene pool and change gene frequencies independently of natural selection. Suppose, for example, one individual in a small population possesses the allele to produce red flowers and all other plants in the population have the allele for yellow flowers. If a hailstorm kills the red-flowered plant, the allele for red flowers won't be passed to the next generation. In a large population, however, we expect to find more individuals with alleles for red flowers, and thus there is a greater chance that some of these plants would survive a storm and reproduce. With genetic drift, alleles in a small population can be eliminated, or become a greater proportion of the gene pool, by chance alone.

Sexual reproduction maintains genetic variation in a population.

Sexual reproduction is probably the most important factor that promotes genetic variation in plants and animals (fig. 14.20). Sexual reproduction produces new combinations of genes, so offspring are always different genetically from, but similar to, their parents. New combinations arise through crossing-over and genetic recombination, independent assortment of chromosomes during meiosis, and the combination of genetic material from two different parents during fertilization (see chapter 13). In contrast, the offspring of asexual organisms are clones of the parents and genetically identical to each other. In these organisms,

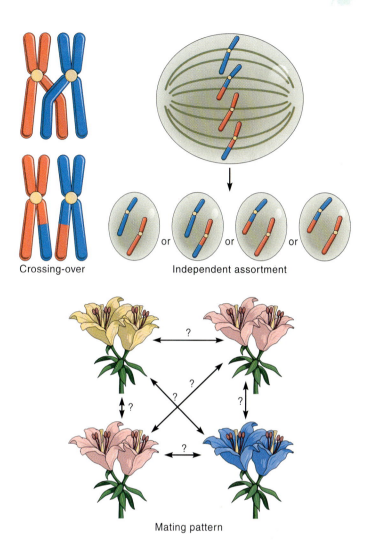

Crossing-over   Independent assortment

Mating pattern

F I G U R E   1 4 . 2 0

Three mechanisms of sexual reproduction enhance genetic variation. (1) Crossing-over recombines alleles. (2) Independent assortment varies combinations of chromosomes in each gamete. (3) Different mating patterns (possible crosses) vary combinations of parents. Mating pattern is subject to selection pressure, chance, and spatial distribution, and increases exponentially with the number of individuals in the population (see chapter 13).

genetic variation occurs only through random mutation. By *random,* we mean that the mutations occur irrespective of their possible consequences for survival or reproduction.

Outcrossing—reproduction with other individuals—promotes new genetic combinations. Although many plants are **self-compatible** (see chapter 12) and can thus produce seeds by themselves, several mechanisms have evolved that promote sexual reproduction with other individuals (fig. 14.21). For outcrossing to occur, pollen must land on the stigma of a flower of another plant. Holly trees and date palms have male flowers on one tree and female flowers on another (i.e., they are **dioecious;** see chapter 12); as a result, no individual tree can produce seed by itself. In other plants such as avocado (*Persea americana*), production of gametes is separated by time: when a flower

FIGURE 14.21

The flower of the bird-of-paradise (*Strelitzia reginae*) has colors and shapes that attract bird pollinators. Successful attraction of flying pollinators promotes outcrossing, which allows for new genetic combinations.

FIGURE 14.22

The flowers of bogbean (*Menyanthes trifoliata*) may be either long-styled ("pin" flower on right) or short-styled ("thrum" flower on left). This difference increases the chances of cross-pollination because a pollinator that is dusted with pollen from a short-styled flower is more likely to successfully pollinate a long-styled flower.

is producing pollen, the stigma of the same flower is not receptive. Thus, self-pollination cannot occur; in these plants, stigmas are usually receptive to pollen only after the flower, or all flowers on the same plant, have shed their pollen. Other plants such as the primrose and the bogbean have flowers with two arrangements of styles and anthers (fig. 14.22). The "pin" flowers have a long style and low-set anthers, whereas the "thrum" flowers have a short style and elevated anthers. The contrasting positions of style and anthers ensure that pollen is deposited on pollinators so that the pollen from pin flowers is more likely to pollinate thrum flowers, and vice versa. This promotes outcrossing.

Genetic systems for self-incompatibility also promote outbreeding. Pollen in **self-incompatible** plants will not germinate properly on the stigma of any flower that has the same genetic makeup, such as the same flower or flowers of the same plant. For example, plants such as evening primrose (*Oenothera*) cannot fertilize themselves because the pistil and pollen have the same allele for self-incompatibility. They can fertilize another individual only if it has a different allele for self-incompatibility.

Heterosis, sometimes called *hybrid vigor* (fig. 14.23), also helps maintain genetic variation in a population. Hybrid corn, for example, bred from two different strains, is extremely vigorous and hardy because it has alleles from two genetically different parents. Commercial developers of corn seed use a controlled process of corn improvement initially involving 5 to 7 years of inbreeding and selection of desired features. This inbreeding produces purebred and strongly homozygous strains of corn. After repeated in-

breeding, purebred strains of corn are crossed to produce a hybrid strain of seeds with strong hybrid vigor.

INQUIRY SUMMARY

The gene pool of a population contains the genes of all the individuals in that population. Gene flow occurs when individuals from different populations reproduce. Genetic variation within a population may be increased by gene flow. Genetic drift reduces variability by the loss of alleles. Genetic variation is maintained by sexual reproduction, outbreeding, self-incompatibility, and heterosis.

Why do you think plants that only reproduce asexually (vegetatively) are called evolutionary deadends?

The Study of Evolution Keeps Growing.

The foregoing discussion is a brief synopsis of some of the best-supported ideas about evolution. It is not the final word, however, because our knowledge of evolutionary biology itself is changing. While the evidence that evolution has happened (and continues to happen) is overwhelming, a heated debate continues among biologists about *how* the process of evolution occurs and how quickly it occurs.

The fossil record shows there have been relatively short periods of explosive diversification in large groups of

FIGURE 14.23

Hybrid corn may be superior to inbred strains because outbreeding promotes heterosis (hybrid vigor). Hybrids shown in the background are taller and more robust than the inbred strains shown in the foreground. Corn ears (inset) produced by the hybrids are much larger than those produced by inbred strains.

FIGURE 14.24

Modern corn probably evolved from teosinte. Note the double row of seeds in a single husk.

organisms, followed by longer periods of little or no evolution. Some organisms remain relatively unchanged for millions of years and then almost disappear, only to be replaced over tens of thousands of years by other groups. Giant ferns and dinosaurs, for example, existed for tens of millions of years and dominated the earth, but these plants dwindled rapidly, and the dinosaurs disappeared about 65 million years ago. A large diversity of flowering plants and mammals quickly took their place. This pattern of evolutionary change is one of stasis (no large-scale change) punctuated with short periods of rapid change. Such a pattern of change is referred to as **punctuated equilibrium,** which is a focus of many current investigations in evolution.

Some of the most interesting evidence for the claim that minor genetic changes can produce large evolutionary leaps was reported in 1996 by Jane Dorweiler, a graduate student at the University of Minnesota. Dorweiler studied teosinte, the presumed ancestor of domesticated corn. Teosinte is a bushy weed whose kernels are encased in a lignified armor that is hard enough to crack the teeth of even the most ardent corn lover (fig. 14.24). Dorweiler found that giving teosinte a single portion of a chromosome from modern corn alters flower development and results in kernels that are exposed like those of domesticated corn. That hybrid, according to Dorweiler, is what teosinte may have looked like 7,000 to 10,000 years ago, when archaeologists believe teosinte was domesticated in what is now Mexico. Scientists suggest that this is an example of how small genetic changes can have major changes in the growth and development of an organism, which can lead to significant evolutionary changes.

Macroevolution is the origin or extinction of large groups of organisms. These events are on a grand scale and include evolution of novel, adaptive characteristics (e.g., tracheids and vessels; see chapter 4), broad diversification of organisms (e.g., the explosive radiation of flowering plants), and sometimes massive extinctions (fig. 14.25). Punctuated equilibrium is considered a pattern of macroevolution.

Microevolution is the progressive change in gene frequencies within a population over many generations. Natural selection, mutation, and genetic drift are primary forces of microevolution and can culminate in **speciation,** the formation of new species. Although much indirect evidence exists for macroevolution, direct evidence is limited because macroevolution deals with large-scale events that occur over long periods of time. Conversely, microevolution is a robust experimental discipline that draws on an abundance of evidence, including direct evidence from experiments on such diverse species as fish and flowering plants. Studies continue today on both the macroevolution and microevolution of organisms.

The study of the evolution of plants and their characteristics is a focus of many botanists. Plants are photosynthetic, eukaryotic, multicellular organisms that also possess cellulose cell walls and embryos, produced through sexual reproduction. The time line (fig. 14.25) shows a few of the major events in the evolution of life,

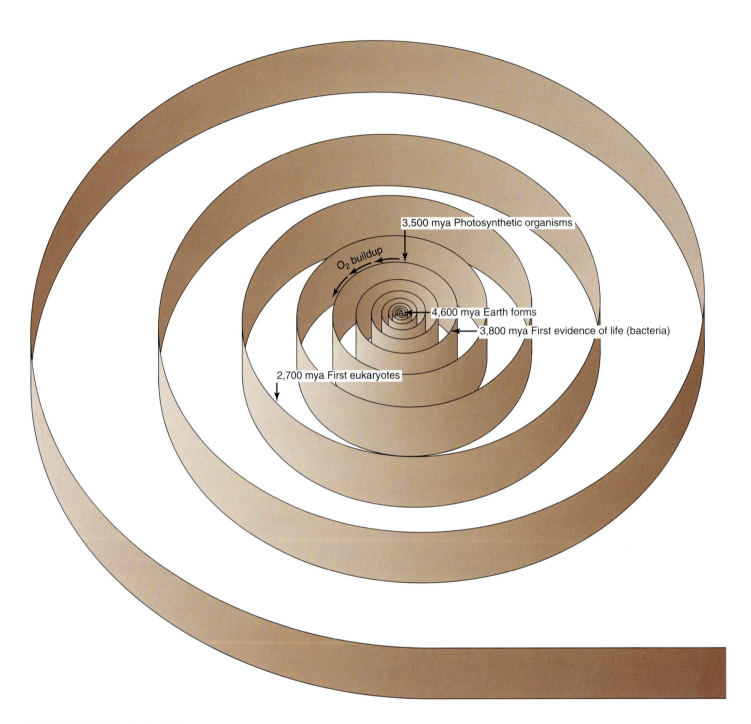

FIGURE 14.25

A time-line illustrating some of the major events in the evolution of life on earth beginning with the formation of the earth about 4,600 mya (million years ago) to the present. Every centimeter in the timeline represents several million years of time. (The timeline is coiled over two pages to accommodate the earth's long history.) Discoveries constantly move back dates of earliest evidence of organisms. How might you explain the difference in the length of time between the first photosynthetic organisms and the first land plants compared to the time between the first land plants and the first flowering plants?

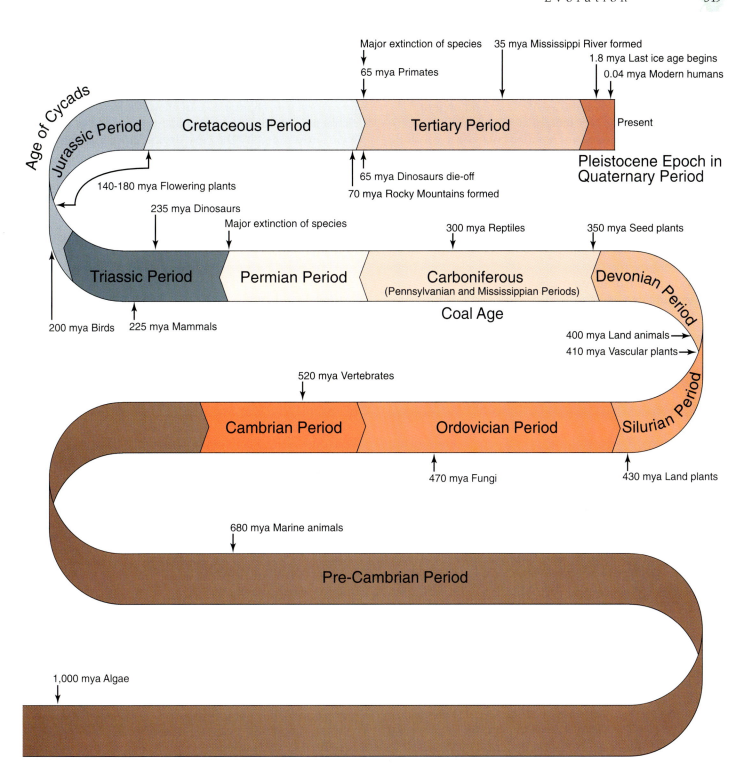

Major extinction of species
65 mya Primates
35 mya Mississippi River formed
1.8 mya Last ice age begins
0.04 mya Modern humans

Age of Cycads

Jurassic Period

Cretaceous Period

Tertiary Period

Present

Pleistocene Epoch in
Quaternary Period

140-180 mya Flowering plants

65 mya Dinosaurs die-off
70 mya Rocky Mountains formed

235 mya Dinosaurs

Major extinction of species

300 mya Reptiles

350 mya Seed plants

Triassic Period

Permian Period

Carboniferous
(Pennsylvanian and Mississippian Periods)
Coal Age

Devonian Period

200 mya Birds   225 mya Mammals

400 mya Land animals →
410 mya Vascular plants →

520 mya Vertebrates

Cambrian Period

Ordovician Period

Silurian Period

470 mya Fungi

430 mya Land plants

680 mya Marine animals

Pre-Cambrian Period

1,000 mya Algae

F I G U R E   1 4 . 2 5
continued

including plants, on earth. Throughout this book, we have discussed events in the evolution of plants, which have contributed to a group of accumulated characteristics that are shared by plants today.

Evidence based on the fossil record and on morphology, anatomy, physiology, molecular studies, and the reproductive biology of organisms have led botanists to conclude that plants evolved from green algae ancestors. What is this evidence? Some green algae living today are similar to plants; certain species of green algae are photosynthetic and possess the photosynthetic pigments chlorophyll *a* and *b* (see chapter 10), are multicellular, and have cell walls made out of cellulose, in addition to possessing other similarities to plants. Plants differ from algae in that plants develop from an embryo (or young sporophyte of plants), which is a multicellular offspring produced from the sexual fusion of a sperm and egg and that remains attached to and is protected by the tissue of the parent plant—at least for a while (see chapter 16). Thus, in their evolutionary history, plants retained many characteristics of their green algal ancestors while developing adaptations that provided a selective advantage for life on the land. Note that it is possible for plants and their ancestors to co-exist on earth today. Such is the nature of evolution; it is not necessary for one organism to become extinct when it evolves into another organism. As seen in this chapter, it is possible for a population of one species, in response to a local environment, to evolve into a different species while the original species continues to exist.

Which of the characteristics of plants evolved first? The first evidence of life on earth is approximately 3.8 billion years old (see fig. 14.25). Life began in the oceans, and the first living things were primitive bacterialike organisms. The first photosynthetic organisms evolved approximately 3.5 billion years ago and most closely resembled modern-day cyanobacteria. These early photosynthetic organisms changed the earth forever, releasing large quantities of oxygen gas into the atmosphere for the first time and thereby creating an environment where larger organisms could live. Evidence indicates that it was millions of years after this before the first eukaryotic organisms evolved, probably as a result of cells taking in, but not destroying, other organisms that became the first organelles. (This endosymbiotic hypothesis of the origin of chloroplasts and mitochondria was discussed in chapter 4.) The next major event in the evolution of life on earth was the evolution of multicellular organisms, including algae that evolved approximately 1 billion years ago. Larger, multicellular organisms possess a greater surface area that allows them to absorb necessary resources for growing and reproducing and to eliminate wastes, which might have given multicellular plants an advantage over single-celled algae. It was, however, necessary for cells in multicellular organisms to communicate with each other and to transport resources from one cell to another—in the case of plants, the means of communication are plasmodesmata (see chapter 4).

The colonization of land by multicellular, photosynthesizing organisms may have occurred many times, but the first such organisms to wash up on the land probably resembled algae more than plants. The first organisms to colonize land and reproduce there successfully were land plants that probably resembled small, primitive mosses or liverworts (see chapter 17). In chapter 18, we discuss the characteristics that evolved as plants invaded the land and were surrounded by the very drying air. These characteristics include a waxy cuticle covering the surface of the plant that prevents it from losing too much water through transpiration. The evolution of vascular tissue allowed plants to grow tall while distributing water to all parts of the body, as noted in chapter 8. Natural selection favored those characteristics that provided an advantage in terrestrial environments.

Following the evolution of vascular plants, came the evolution of seed plants, and then flowering plants. In chapter 12, we outlined the possible evolution of flowers and flowering plants. The study of how and when the major characteristics of land plants evolved is based in large part on the fossil history of plant ancestors, and this history is filled with ancient specimens; that is, with transitional fossils of vascular tissue and seeds that were sufficiently hard to become fossilized occasionally. The first land plants, the first embryos, and the first flowers, however, were probably too soft to fossilize except on rare occasion. Thus, just as the details of the evolution of flowering plants are incomplete, the evolution of the plant embryo and the first land plants is still not completely resolved. The complete story about the evolution of plants remains hidden in rocks around the world containing even more transitional fossils of ancient plant parts.

Many evolutionary studies are economically important.

Agricultural scientists, geneticists, molecular biologists, and evolutionary biologists have worked together to breed improved varieties of plant crops and domesticated animals. Recent work has focused on finding genes that control such characteristics as fruit size and sugar content to improve the quality of crops and on transferring desirable genes into plants that lack them.

The evolution of disease-causing organisms (see Perspective 14.2, "Unstoppable Killers") and crop pests continues to concern scientists around the world. More than 500 species of insects, including crop pests, are known to have evolved resistance to insecticides in the last 40 years. The evolution of pest resistance costs American farmers and foresters over $1 billion a year. Some of these insects destroy crops and stored grains or are carriers of plant diseases. Investigators are studying ways to delay or prevent the evolution of insecticide resistance in insects. In addition, scientists hope to find natural, or biotic, predators or parasites of the insect pest species that might be useful in controlling the pests without insecticides.

14.2 Perspective Perspective

Unstoppable Killers

In 1998, events occurred that had been long feared by scientists and physicians; the health of patients suffering from an infection caused by the deadly bacterium *Staphylococcus aureus* (commonly known as staph) did not improve when the patients were given the once-reliable antibiotic vancomycin. *Staphylococcus aureus* is a major cause of hospital-acquired infections, including those affecting blood, bone, or skin. Around the world, many strains of *S. aureus* are now resistant to all antibiotics *except* vancomycin, a powerful antibiotic considered to be the drug of last resort. Thus, the discovery of a vancomycin-resistant bacterium suggests that disease-causing organisms have evolved that are resistant to *all* known antibiotics. Scientists worry that more pathogenic bacteria will become unstoppable killers with no known treatment. Already, death rates for some diseases such as tuberculosis have started to rise that had been declining for many years.

What has propelled this evolution of pathogens? Part of the answer has to do with the overuse of antibiotics and their release into the environment. The Centers for Disease Control and Prevention estimate that one-third of the 150 million prescriptions for antibiotics every year are unnecessary; patients with a cold, flu, or other viral infections (which are untreatable with antibiotics) request and are given antibiotics. In addition, patients suffering from bacterial infections often do not complete the full course of their antibiotic treatment. In either case, bacteria that are susceptible to antibiotics are killed; however, those that have some resistance survive and pass their genes to their descendants. This evolution of antibiotic-resistant organisms is also promoted by ranchers and farmers.

Over 40% of antibiotics manufactured in the United States are given to animals to treat infections and promote growth. Antibiotics also are sprayed on fruit trees to control bacterial infections. While high concentrations of antibiotics can kill disease-causing bacteria on crop plants and in animals raised for meat, the antibiotic residues in the environment promote the evolution of resistant bacteria. These resistant bacteria then find their way into the intestinal tracts of humans through the food chain. One solution to this growing problem is to use antibiotics only when absolutely necessary.

What do you think is the long-term effect of using antibacterial soaps and other household products that claim to be antibacterial?

SUMMARY

Evolution refers to all the changes of life on earth from its earliest beginnings to the diversity of today. Today's organisms arose through the modification of ancient forms of life, which was fueled by accumulations of genetic changes through many generations—descent with modification. Evolution involves a change in the frequencies of alleles in a population's gene pool from one generation to the next.

Lamarck tried to explain changes in species by inheritance of acquired characteristics; however, such changes in characteristics are not heritable. Charles Darwin observed remarkably diverse adaptations and distributions of species. From these observations and ideas adopted from Lyell and Malthus, Darwin conceived of a mechanism called natural selection to explain the development of diverse species through time. At the same time, Alfred Russel Wallace also developed a theory of natural selection, and both naturalists presented their work in 1858. The next year, Darwin published his explanation for natural selection in *On the Origin of Species.*

Support for evolutionary theory includes both indirect and direct evidence. Direct evidence comes from experiments with living organisms, and indirect evidence comes from the age of the earth, the fossil record, homology, convergence, biogeography, and molecular biology.

The essence of natural selection is the differential reproduction of genetic variants. Organisms often produce more offspring than are needed to replace themselves and more than environmental resources can support. Some offspring die before maturity, and those with genes that confer the best adaptation to their environment reproduce the most. Stabilizing selection promotes the most prevalent phenotype and reduces variation, directional selection fosters an extreme phenotype, and diversifying selection reduces the most frequent phenotype and promotes the extremes thereby maintaining overall variation.

Evolution and natural selection are genetic processes. Population genetics is concerned with the study of entire gene pools rather than individuals. Genetic variation as the raw material for natural selection is introduced into populations by mutation and gene flow. Gene flow is the transfer of genetic material from one population to another by migration of individuals or reproductive cells. Genetic drift may produce significant changes in small populations that are subject to chance events such as natural catastrophes and mutation.

Genetic variation is maintained by sexual reproduction through the combining of genes from different parents. Outbreeding promotes new genetic combinations by ensuring sexual reproduction with other individuals rather than within the same individual. Heterosis also maintains diversity by promoting the heterozygous condition.

More recent ideas about evolution include macroevolution versus microevolution. Microevolution explains the origin of species, while macroevolution explains the origin of higher taxonomic categories. Many evolutionary investigations result in economically important plants, animals, or microbes. Scientists can create a timeline based on fossil evidence that outlines the history of life on earth.

Writing to Learn Botany

Like all important ideas, evolution attracts controversy. It has attracted more controversy than other scientific ideas because it has affected not only science but also philosophy, religion, and attitudes. Unfortunately, evolution is often seen as competing with religious ideas as a way of explaining the origin and diversity of life. This competition is inappropriate because evolutionary theory is scientific, and religions are not. Evolution is evaluated according to a scientific method of investigation; the existence of a divine creator is a product of faith and cannot be evaluated scientifically. Write about the time in history when people believed that the earth was the center of the universe and the transition to the understanding that the earth revolved around the sun. Why did people have such beliefs?

THOUGHT QUESTIONS

1. In earlier chapters of this book you've seen examples of evolution; for example, how plants form symbioses with other organisms (e.g., fungi, ants, and bacteria). You've also seen more subtle examples of biochemical adaptations, such as how plants make compounds that deter herbivores and even how other organisms such as insects have evolved ways of using plants' defenses for their own survival. What would be the selection pressures for these examples of biochemical evolution?

2. Are gametes subject to the environment? Explain.

3. What would be the consequences if a population reproduced without the influence of natural selection?

4. Under what circumstances might a single mutation significantly affect a population?

5. How would you explain positive selection for survival beyond the oldest reproductive age?

6. How is evolution different from natural selection?

7. Why can't individuals evolve?

8. What was the importance of artificial selection to the formulation of Darwin's theory of natural selection?

9. How would you determine if most mutations were harmful?

10. Herbert Spencer was an eighteenth-century English sociologist who believed in applying Darwin's concepts to human society. He popularized the phrase *survival of the fittest* and suggested that the unemployed, the sick, and other "burdens on society" be allowed to die rather than be objects of public assistance and charity. What do you think of this idea and of the validity of his interpretation of the concept of natural selection?

11. Hypotheses to account for evolutionary events such as the pace of evolution and the cause of mass extinctions are often conflicting. Those who do not believe that evolution has taken place sometimes perceive these conflicts among scientists as signs of doubt that evolution occurs. Scientists, however, maintain that conflicting hypotheses do not argue against evolution at all but demonstrate the process of scientific thinking. Why are hypotheses particularly important in understanding evolutionary processes compared to other fields of biology?

12. James Watson, who along with Francis Crick, originally described the structure of DNA (see chapter 5), said, "Today, the theory of evolution is an accepted fact for everyone but a fundamentalist minority, whose objections are based not on reasoning but on doctrinaire adherance [sic] to religious principles." Do you agree with Watson? Why or why not?

SUGGESTED READINGS

Articles

Hartman, H. 1990. The evolution of natural selection: Darwin versus Wallace. *Perspectives in Biology and Medicine* 34:78–88.

Max, E. E. 1998. "New" persuasive evidence for evolution. *American Biology Teacher* 60:662–70.

Mlot, C. 1996. Microbes hint at a mechanism behind punctuated equilibrium. *Science* 272: 1741.

Moore, R. 1997. The persuasive Mr. Darwin. *BioScience* 47:107–14.

Niklas, K. J. 1994. One giant step for life: Simple law-abiding plants led the invasion of hostile lands. *Natural History* 103:22–25.

Pääbo, S. 1993. Ancient DNA. *Scientific American* 269:86–92.

Science. 1999. June 25 issue, several articles devoted to evolution.

Books

Darwin, C. 1967. *On the Origin of Species,* facsimile 1st ed. of 1859. New York: Atheneum.

Hancock, J. F. 1992. *Plant Evolution and the Origin of Crop Species.* Englewood Cliffs, NJ: Prentice-Hall.

Liebes, S., E. Sahtouris, and B. Swimme. 1998. *A Walk Through Time.* New York: John Wiley & Sons, Inc.

Patterson, C. 1999. *Evolution.* Ithaca, NY: Cornell University Press.

ON THE INTERNET

Visit the textbook's accompanying web site at http://www.mhhe.com/botany to find live Internet links for each of the topics listed below:

Charles Darwin

History of evolutionary thought

Pangenesis

Biogeography

Galapagos Islands

Adaptive radiation

Population genetics

Macroevolution and microevolution

The Hardy-Weinberg Law

This community includes a sample of the diversity of flowering plants in the world. How did so many species of plants evolve?

The Diversity and Classification of Plants

CHAPTER OUTLINE

LEARNING ONLINE

The Online Learning Center at www.mhhe.com/botany contains
practice quizzing, key term flashcards, and hyperlinks to help
you study the material in this chapter. For example, did you
know that the formation of a river or mountain may also result in
the formation of a new species? To learn more about plant
diversity and classification, visit http://phylogeny.arizona.edu/
tree/phylogeny.html in chapter 15 of the OLC.

Perhaps the first requirement for classifying plants—that is, arranging them into groups—is to name them. Once a plant has a name, anyone using that name can expect other people to associate it with a specific plant. Although there are few records of plant names from longer than about 4,000 years ago, you can nevertheless imagine that plant names were important to early humans for talking about plants used for medicines and foods. Plant names are still important for these reasons, but they are also important for tracking the diversity of the plant kingdom, monitoring the effects of environmental change, reporting botanical research, understanding evolutionary relationships, cultivating crops and ornamental plants, and all other aspects of the uses of plants.

Evolutionary research focuses on **speciation,** the origin of new species, to explain the diversity of life. New species are the foundation of biological diversity. Today, 2 to 3 million species of organisms have been named and perhaps 15 times that many remain to be dis-

covered. As vast as this diversity may appear, extrapolations from the fossil record indicate that today's species probably represent less than 1% of all those that have ever existed; more than 99% of all species have become extinct. Obviously, speciation has occurred countless times in all sorts of environments; it is as much a property of life as growth, reproduction, and death.

Ideas about how species arise have produced as many heated arguments as there are ideas about how to define a species. Many of these arguments are about the relative importance of the different mechanisms in nature that produce new species. As noted by Charles Darwin in *On the Origin of Species* (see chapter 14), the creation of new species by human activities through artificial selection is indirect evidence that the same mechanisms occur in nature. The first objective of this chapter is to explain what a species is and how species are formed and organized into groups. The second objective is to explain how plants are named and how scientists study the evolutionary history of related organisms.

We Organize Plants into Groups Using Many Different Criteria.

Taxonomy is the science of discovering, describing, naming, and classifying organisms. If naming plants is a natural human endeavor, then everyone is to some degree or another a plant taxonomist. Many people share a tendency to organize plants into groups; edible plants, medicinal plants, and roses, for example, are recognized as groups of plants. Rationales for grouping plants have varied over time, with some systems of classification influenced by religious or ethnic considerations. In northern Mexico, for example, the Tarahumara tribe classified some plants on the basis of ceremonial uses involving hallucinogens. Tarahumarans grouped botanically unrelated plants such as *hikuli*—known to us as peyote cactus (*Lophophora williamsii*)—and *dowaka* (*Tillandsia benthamiana,* a member of the pineapple family) because these plants are used together in ceremonies in the belief that their combined psychoactivity is greater than that of either plant by itself (fig. 15.1). Although informal classifications are widespread, they are generally not recorded, nor are they used outside of the tribes or small groups of people who invented them.

What is a species?

The **species** is the fundamental category of biological classification. A species can be defined as a group of individu-

FIGURE 15.1

Peyote (*Lophophora williamsii*) is the source of various hallucinogens.

als that are morphologically similar to each other and capable of breeding successfully with each other but that are different from members of other groups and unable to breed with them. Beginning with Darwin, the recognition of species has commanded a great deal of attention from biologists because of its importance in classifying and understanding speciation and, therefore, evolution. *Species concepts*—that is, perceptions and definitions of species—remain a hot topic today. At least a dozen species concepts have been proposed during the past two centuries. Among

(a)

(b)

(c)

FIGURE 15.2

These three lilies are easily recognized by color and shape; that is, the morphological species concept clearly distinguishes them as separate species based on their general morphology. However, this concept gives no insight into their degree of reproductive isolation or their genetic similarity. Are they truly different species? (*a*) *Lilium regale.* (*b*) *Lilium auratum.* (*c*) *Lilium martagon.*

the most widely used are morphological species, biological species, and evolutionary species.

Taxonomists often identify species based on the morphology of the organisms.

The **morphological species concept** states that a species is the smallest group of organisms that can be consistently distinguished by the presence or absence of body parts and the shapes, sizes, and numbers of these parts. This is the most practical and widely used species concept among taxonomists, who deal with the identification and classification of organisms. Taxonomists use this concept mostly by default because they know most of the species on earth only from their morphology. This concept is often called the **classical species concept** because it generally has been used to describe species since people first began to classify organisms.

New species are usually named on the basis of the morphological species concept; morphological descriptions are all that we need to distinguish new species from previously known ones. Virtually all of the names of plant species used in this book are of morphological species as are all the names used in a flora describing plants of a particular region.

Most of the morphological species you are familiar with are intuitively recognizable as distinct from one another. For example, you can distinguish a species of lily from one of pine or oak. With a little study, you could also learn to distinguish some of the 100 species of *Lilium* (lilies), the 93 species of *Pinus* (pines), and the 600 species of *Quer-*

cus (oaks) from one another (fig. 15.2). In so doing, you would be learning these plants as morphological species.

At this point, it might seem that the morphological species concept allows you to recognize all the species that you need or want to know. So why go any further? The answer is that speciation has not always formed well-defined species; the lines between species are often blurred. In such cases, taxonomists often disagree on which plants are the same or different species. As an example, the morphological species concept of the cactus genus *Opuntia* includes anywhere from 400 to 1,000 species, depending on whose judgment is followed, because of morphological variation in *Opuntia* plants. In addition, the morphology of plants is usually affected by the environment, which can reduce the reliability of morphological characters in identifying plants. For instance, when the buttercup *Ranunculus peltatus* grows completely in water, its leaves are thin and finely divided, but if stems develop above the water, the leaves are broad and flat (fig. 15.3). A person who was unfamiliar with this buttercup species might think, incorrectly, that one plant in the water was a member of a different species than a plant growing partly in water. Thus, the morphological species concept is not a perfect system.

Biological species are defined by their inability to reproduce with other species.

In contrast to a morphology-based concept of species, the **biological species concept** uses reproductive biology to define species. According to this concept, a species is a group (all

FIGURE 15.3

The leaves of this buttercup (*Ranunculus hyperboreus*) are broad and flat above water but finely divided underwater.

the populations) of similar individuals that can reproduce with each other but cannot reproduce with individuals of other groups. The biological species concept was first proposed by ornithologists (bird specialists) because birds commonly occur in populations that are hard to distinguish morphologically. Most animal species probably fit this concept. However, many plants do not fit the biological species concept. For example, morphologically different plant species that are geographically isolated may hybridize (interbreed) readily when they are grown together (fig. 15.4). Are they really separate species? This is an important question for biologists who study the evolution of plants, for conservation biologists who want to preserve species, and for people who need to identify proper sources of medicines and food, but the answer may differ depending on which botanist is responding.

Plants often reproduce asexually or by self-pollination, which means that reproduction with *any* other organisms may be absent. Dandelions and blackberries, for example, have played particular havoc with the biological species concept because they produce seed without fertilization. The biological species concept is useless in defining certain crop species, such as the commercial banana or the navel orange, because many of these species are exclusively asexual. Also, this concept is not very useful with fossils because information about the reproductive biology of extinct organisms is often missing. So, neither the morphological species nor biological species concept is perfect. In fact, many botanists have abandoned the biological species concept because of problems trying to apply the concept to plants.

Others use an **evolutionary species concept,** which defines a species as a group of individuals with a common evolutionary lineage; that is, a group of individuals with a common evolutionary fate through time. Most botanists today use a wide variety of evidence including morphological, molecular, reproductive, ecological, genetic, and geographic characteristics to help them distinguish one species from another and to study the evolutionary relationships of organisms. However, because the concept of a species is important to many individuals other than taxonomists, botanists often rely on morphological evidence as the primary evidence to describe species so that species can be determined readily by nonspecialists.

INQUIRY SUMMARY

Ideas about speciation are based on what kinds of species exist and on the different ways that species arise. Biologists argue about how to define a species. The morphological species concept defines species by their observable morphological features. It is the most widely used species concept because most species are

Liriodendron tulipifera

(a)

Liriodendron chinensis

(b)

Hybrid, *L. tulipifera* × *L. chinensis*

(c)

FIGURE 15.4

These two species of *Liriodendron* are geographically isolated and morphologically different from one another, but they readily hybridize when they are grown together. (*a*) *L. tulipifera* is a common tree of the eastern United States. (*b*) *L. chinensis* occurs in small populations in eastern China and Vietnam. (*c*) The hybrid occurs in botanical gardens where trees of both species are cultivated near one another. What characteristics of each parent does the hybrid possess?

known only by their morphology. A biological species is a group of interfertile populations that are reproductively isolated from other populations. Plants often do not fit this species concept because plant species may be predominantly asexual. The evolutionary species concept considers evolutionary lineages of organisms.

How is the morphological species concept more practical in its use than is the biological species concept?

Biologists Debate How New Species Are Formed.

New species evolve from preexisting species as populations adapt to local conditions. This idea, first proposed by Charles Darwin, is now uniformly accepted in biology. However, the widespread agreement that speciation has occurred and continues to occur is spiced with plenty of arguments about the pace and mechanisms of speciation. Natural selection (see chapters 2 and 14) provides some of the answers, at least in a general way. Many possible mechanisms of speciation have been described, some of which are based on much scientific evidence and some of which are rational explanations but are based on little evidence.

Many biologists are alternately excited or frustrated by the *species problem*—that is, the difficulty of defining a species. No single definition fits all species. But this does not keep biologists from using the term *species* or from using species names, as we have done throughout this book.

If a species is defined as a group of organisms capable of breeding successfully with each other but not with members of other groups, the formation of a new species is a genetic event first involving isolation of the gene pool of a population. In other words, a population of one species may evolve into another species if individuals in the population reproduce only with each other and not with individuals in other populations. Such **reproductive isolation,** the restriction of gene flow into a population, is a critical characteristic of a species.

Physical and biological barriers can reproductively isolate populations from each other.

Reproductive isolation is the foundation of the species concept. Reproductive isolation protects the integrity of a species by preventing the gene pool from being changed by genes from other species. Mechanisms that prevent the production of fertile offspring from different populations are **reproductive barriers** that promote reproductive isolation. The most obvious barrier is geographic separation: if two populations are separated by a great distance, they will not interbreed (fig. 15.5).

Development of reproductive barriers is critical for speciation, and natural selection promotes these barriers.

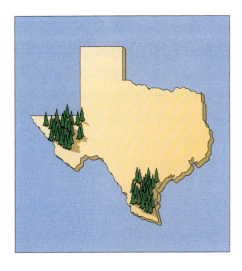

FIGURE 15.5
New species may form when populations are geographically separated and the individuals in the populations adapt to local environmental conditions.

Parents from different species that circumvent barriers often produce unfit offspring. Indeed, offspring resulting from parents that are very different and that somehow circumvent reproductive barriers have genes from both parents that provide the growing embryo with mixed signals. This combination of genes often disrupts normal development and kills the embryo. Even if the embryo survives, the mature organism is often weak and sterile. In this way, natural selection favors organisms that neither mate nor reproduce outside their species. The barriers to reproduction are summarized in table 15.1.

Speciation requires a population to diverge from its parent population.

To form a new species, the gene pool—that is, the collective genetic makeup—of a population must become different from that of the parent (i.e., ancestral) population. Individuals in such a diverging population become different: they usually look different from the ancestral population, exploit dissimilar habitats, and respond to the environment differently. The mechanisms controlling this divergence are not predictable or easily observed because some forces homogenize, mix, and minimize variation, while other forces enhance and promote divergence (table 15.2). The first step in speciation, however, is partial reproductive isolation, which initiates genetic divergence. Once the gene pool of a population is isolated, it evolves further when its gene frequencies change because of natural selection, genetic drift, and mutations in subsequent generations (see chapter 14). For example, mutations occur randomly, and isolated populations of the same species will randomly accumulate different mutations, even if the environments are identical. Enough mutations may accumulate that the two populations diverge and no longer interbreed. Similarly, genetic

The Lore of Plants

The flora of Colombia, Ecuador, and Peru has the greatest known plant diversity in the world; over 40,000 plant species occur on just 2% of the world's land surface. In the rain forest near Iquitos, Peru, Alwyn Gentry found about 300 tree species in each of two 1-hectare plots.

Peter Ashton discovered over 1,000 species in a combined census of 10 selected 1-hectare plots in Borneo. In contrast, only about 850 native species of trees occur in all of the United States and Canada. What factors contribute to the abundance of species in the tropics?

TABLE 15.1

Mechanisms of Reproductive Isolation

1. Mechanisms prevent mating or successful fertilization.
   a. Geographic isolation: Populations live in distant geographic areas.
   b. Microhabitat isolation: Populations live in different microhabitats (local environments) and do not meet.
   c. Temporal isolation: Individuals are receptive to mating or flowering at different times.
   d. Mechanical isolation: Flowers are structurally different, and pollination is prevented.
   e. Gametic isolation: Female and male gametes from two species are incompatible.
2. Mechanisms prevent production of fertile adults.
   a. Hybrid zygotes die before developing to sexual maturity.
   b. Hybrids do not produce viable gametes, so cannot reproduce.
   c. The progeny of fertile hybrids are weak or infertile and die before reproducing.

TABLE 15.2

Forces That Promote or Limit Genetic Variation Within a Population

FORCES THAT PROMOTE GENETIC VARIATION	FORCES THAT LIMIT GENETIC VARIATION
mutation	random sexual mating
abnormal meiosis (polyploidy)	stabilizing selection
sexual recombination	asexual reproduction
genetic drift	
disruptive selection	
environmental gradients	
geographic separation	
geologic events	

Geographic isolation can lead to new species.

One obstacle to interbreeding among organisms is **geographic isolation.** Populations that are separated geographically may become genetically different from each other because of different mutations in the two populations. If there is no interbreeding between individuals in the two populations, no mutations can be transferred between populations. This genetic divergence initiates speciation and is promoted by different environments. Because the two populations occupy different environments, advantageous adaptations may be different for the two populations. Eventually, the phenotypes of the two populations diverge and the populations may become different species. For instance, recall from the last chapter that the finches Darwin observed on the Galápagos Islands were geographically isolated from their ancestral populations on the mainland of South America.

drift and environmental variation can enhance the divergence of recently isolated populations. Sexual reproduction does not usually promote divergence directly because it mixes genetic information within a population and can slow isolation; however, sexual reproduction does supply new combinations of genes. New genetic combinations arising in a splintered population may allow some individuals to exploit a new or different patch of environment. Once established in a new environment, a population's morphology, competitiveness, or reproductive life history may change and thereby enhance isolation from the parent population. Separation of a gene pool of a population from other populations of the parent species is often a crucial event in the origin of a species.

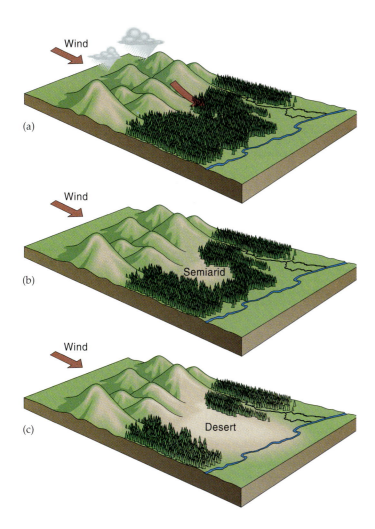

FIGURE 15.6

The formation of a desert can separate populations and promote their genetic divergence. (*a*) Heavy rainfall on the left of the mountains leaves a dry wind on the leeward side. (*b*) A desert can form and expand to divide a population into two reproductively isolated populations. (*c*) Over time, accumulations of adaptations and incompatibilities by the two populations can lead to speciation.

Geological processes can split a population. For example, the formation of mountain ranges, canyons, deserts, and bodies of water can separate organisms. Such features of the land are geographic barriers that may prevent an organism or its spores, pollen, or seeds from reaching other populations, thus allowing genetic divergence to occur between populations (fig. 15.6).

Small, isolated populations may become extinct; a bad storm, cold weather, fire, or pests may wipe out the few individuals in such a population. Also, recall that genetic drift (see chapter 14) and mutations are more significant in small populations. In a large population, genetic drift is insignificant, and natural selection may require many generations to replace alleles and change the phenotype of the population. For example, large populations of North American and European sycamore (*Platanus occidentalis*) trees have been sep-

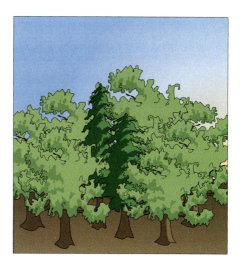

FIGURE 15.7

Reproductive isolation may occur in the midst of intermingled populations.

arated for 30 million years, but they can still interbreed; that is, they have not evolved into separate species (although some botanists call them different species). In contrast, the gene pool of a small population can change in relatively few generations. A few reproductively successful individuals with favorable gene combinations can greatly affect the evolution of a small population and may lead to speciation in a few hundred or few thousand generations, over a period of several hundred or several thousand years.

Species may arise even when populations are not geographically isolated.

Populations that overlap geographically may also become distinct species (fig. 15.7). Speciation of overlapping populations may seem unlikely because individuals in the two populations might easily exchange genes. There are, however, mechanisms that reproductively isolate groups within the same environment. For example, a slight change in color, shape, or chemical attractants in a mutated flower may make it unattractive to a pollinator. Such a mutation can prevent the individuals from being pollinated by the same pollinator that visits the surrounding plants and may effectively prevent the plants with mutated flowers from reproducing with the parent population. The isolated plants may reproduce through self-fertilization, reproduce with limited members of their own kind, or hybridize with a nearby population of a different species. Once the small subpopulation is isolated, it may diverge genetically and form a new species.

An example of speciation without geographic barriers is found in the two species of monkeyflower that live in the western United States, *Mimulus cardinalis* and *Mimulus lewisii*. *Mimulus cardinalis* flowers are bright red and produce copious amounts of nectar in a narrow corolla tube. These flowers are pollinated by hummingbirds. *Mimulus*

15.1 Perspective Perspective

Polyploids in the Kitchen

In 1924, a Russian geneticist had a tasty idea: crossing a cabbage and a radish to produce a new kind of vegetable called a cabbish, having the crispy leaves of cabbage and the red, sharp-tasting root of radish. The geneticist knew that he could probably produce a hybrid, because cabbages and radishes each have 18 pairs of chromosomes. Unfortunately, the hybrid that he produced was sterile because the chromosomes of radishes differed so much from those of cabbage that they did not pair during meiosis. Consequently, the hybrid could not produce viable gametes.

However, the hybrid spontaneously produced a tetraploid having a total of 72 chromosomes: two sets from radish and two from cabbage. In this tetraploid, each chromosome of cabbage and radish had a homologous partner, which allowed pairing to occur during meiosis. The gametes produced by the tetraploid each had a complete haploid set of chromosomes from radish and cabbage. Therefore, the hybrid was a polyploid containing the chromosomes of different species. Although the polyploid was fertile, it never reached our dinner tables because it was a

"radbage," not a "cabbish": it had distasteful, radishlike leaves and uninteresting, cabbagelike roots.

Although the cabbish experiment was a culinary failure, other polyploids have been successful. For example, the development of Western civilization depended largely on wheat, a polyploid that is a combination of three species that each contributed two sets of chromosomes (see Perspective 13.3, "Polyploidy in Plants"). Thus, while you may never sink your teeth into a cabbish, you probably eat a polyploid every day.

---

lewisii flowers are pollinated by bumblebees attracted by the pink flowers with yellow nectar guides, landing platforms, and a small amount of concentrated nectar. The floral differences between these species are a result of their genetic differences. Natural hybrids have never been reported even though the plants sometimes grow near each other and flower at the same time, and hybrids can be formed in greenhouse experiments. Thus, a natural barrier to gene flow between these two *Mimulus* species has developed because of their different types of flowers and pollinators. Eventually, these two species may not be able to produce hybrids with each other.

Plants frequently possess multiple sets of chromosomes, which can lead to speciation.

A widespread cellular process affecting plant evolution is **polyploidy,** a condition in which an organism possesses more than two sets of chromosomes. Some polyploids (*poly,* meaning "many," *ploid,* meaning "sets") result from nondisjunction, an aberrant chromosomal separation during cell division (fig. 15.8). Domesticated grains such as durum wheat (a tetraploid; having four sets of chromosomes), oats (hexaploid; having six sets), and rye (hexaploid) are polyploids, as are cotton, tobacco, and potato. Polyploids also include garden flowers such as chrysanthemums, pansies, and daylilies. Polyploidy is rare among animals, fungi, and most gymnosperms but occurs frequently among angiosperms and ferns. The rarity of polyploidy among animals may be related to their having distinct sex chromosomes; extra sex chromosomes in animals may fatally disrupt hormonal and sexual development.

Estimates of polyploidy in angiosperms have ranged from 30% to 80%. In 1994, Jane Masterson of the University of Chicago tested these estimates by studying a feature affected by polyploidy: cell size. Because cell size correlates positively with DNA content and thus with chromosome number, Masterson could estimate ploidy levels by studying cell size in fossils of extinct plants. Masterson's clever research enabled her to infer that approximately 70% of angiosperms have polyploidy in their history.

The most common polyploids are tetraploids, with four sets of chromosomes. Different species in the same genus often have different numbers of chromosome sets. For example, various species of wheat (*Triticum*) are diploid, tetraploid, and hexaploid (see Perspective 13.1, "Artificial Hybridization in Plants" and Perspective 15.1, "Polyploids in the Kitchen").

Some domesticated plants have been induced to become polyploids for the desired characteristics that result. Polyploid plants can be produced in the laboratory by using colchicine, a drug extracted from the meadow saffron (*Colchicum autumnale*) that disrupts normal mitosis. Some desired characteristics of polyploids include larger cells, thicker leaves, increased water retention, slower growth, delayed flowering, and flowering over a longer period (fig. 15.9). Cold tolerance is also enhanced, and the number of polyploid species increases with increasing latitude. For these reasons, natural polyploids are most common in harsh environments, where selection pressures are intense. Enhanced tolerance to harsh environmental factors is often the competitive edge that allows a polyploid species to establish a permanent population.

Some polyploids have two or more distinct sets of chromosomes typically derived from different, and not

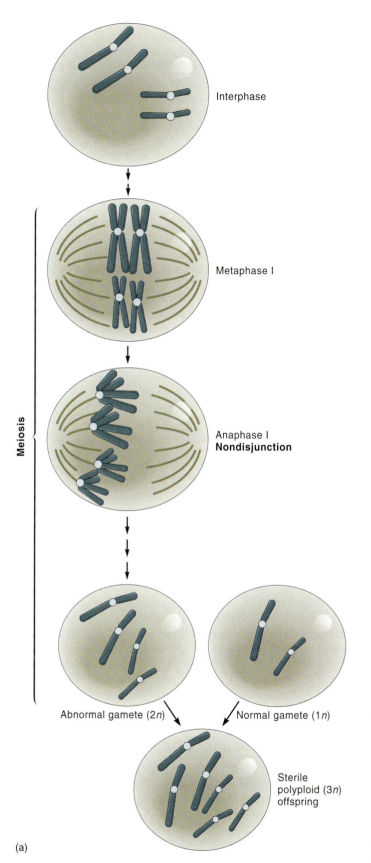

Interphase

Metaphase I

Anaphase I
Nondisjunction

Meiosis

Abnormal gamete (2n)

Normal gamete (1n)

Sterile
polyploid (3n)
offspring

(a)

R. insignis
2n = 48

×

R. insignis
2n = 48

R. insignis
2n = 96

(b)

FIGURE 15.8

Polyploidy by nondisjunction of chromosomes. (*a*) Nondisjunction during meiosis leads to the production of an abnormal gamete that may lead to sterile polyploid offspring. (*b*) Three species of *Ranunculus* live on the North Island of New Zealand. Two of these species have 24 pairs of chromosomes ($2n = 48$). The hybrid from the two parent species probably produced the third species with twice the chromosome number ($2n = 96$) as the hybridizing parents.

closely related, individuals. Many plant hybrids are polyploids, including cultured hybrids that have been artificially selected for their large cells, plump plant parts, high water content, and drought resistance.

As opposed to most other processes that form species, polyploidy can be an immediate mechanism of species formation in higher plants. A hybrid between two species is often sterile because chromosomes are not paired at meiosis and thus cannot form viable gametes. Polyploidy, however, restores paired chromosomes and thus fertility (fig. 15.10). Because the polyploid cannot readily reproduce with either parent, some biologists consider the polyploid a new species. Bread wheat (*Triticum aestivum*) is a polyploid species with 42 chromosomes that probably arose about 8,000 years ago by hybridization between a wheat species with 28 chromosomes and a grass with 14 chromosomes. The polyploid could produce offspring on its own but did not with either of its parents; thus, it is considered to be a different species.

INQUIRY SUMMARY

Biologists recognize that different mechanisms of speciation help explain biological diversity. Reproductive isolation is critical to speciation and is maintained by a variety of reproductive barriers, including those that prevent successful fertilization. The formation of new species involves reproductive isolation of a gene pool

FIGURE 15.9

The daylily on the left is diploid, whereas the one on the right is tetraploid. Such polyploids often have exceptionally robust petals, leaves, and stems. In terms of their effect on plants, how is polyploidy similar to hybrid vigor?

and results from genetic divergence of populations. Reproductively isolated populations continue to evolve through mutation, genetic drift, and natural selection. Populations separated geographically (e.g., by mountains or rivers) may form new species as genetic variations accumulate and reproductive barriers form. When successful reproduction between individuals of different populations can no longer occur, the populations represent different species. Polyploids have more than two sets of chromosomes. Polyploidy is a common evolutionary mechanism that can produce immediate speciation and restore fertility to hybrids.

How could the time when a plant flowers be a reproductive barrier?

A Species May Show Ecological Variation over Its Range.

Some species extend over a large geographic area, its range, having different climatic features that cause individuals to diverge genetically and morphologically. Plants in different areas are subject to local environmental factors, and combinations of adaptations may evolve that are best suited to the local conditions. A species with an extensive range may grow in cooler temperatures in the northern parts and increasingly warmer temperatures in the southern parts. This gradient of

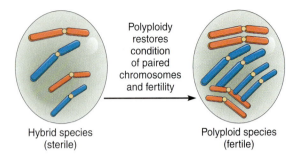

FIGURE 15.10

Polyploidy can restore fertility. Hybrids are often sterile because their chromosomes are unpaired (note there is only one large red and blue chromosome and one small red and blue chromosome), and so only abnormal gametes form. Polyploidy can restore the condition of paired chromosomes necessary for successful meiosis and sexual fertility.

temperatures may also produce a gradient of plants: taller ones at the southern end may grade into shorter plants toward the north. This continuum of a characteristic is a **cline** of morphology. Members of a cline are reproductively compatible, and plants may reproduce throughout the cline (fig. 15.11). Such species show **ecological variation.** Individuals of the species with ecological variation may differ physiologically (e.g., in terms of frost tolerance, time of flowering, or photosynthetic rates) but not look different. Individuals with ecological variation also may look markedly different, even though they are all reproductively compatible and produce viable offspring. Thus, defining a species with ecological variation may be difficult.

Different morphologies along a cline may be due to genetics; that is, short northern plants transplanted to the south may not grow as tall as the southern plants, indicating a true genetic difference. Cline formation is a simple form of genetic divergence, and the degree of divergence is proportional to the length of the cline; the longer the cline, the greater the differences. These differences may lead to the development of new species.

Ecological variation may be the first step toward the formation of new species.

A well-known description of ecological variation was presented by Jens Clausen, David Keck, and William Hiesey, who studied the genetic adaptation of isolated populations along a variety of climatic zones in California. These zones included a sea-level station near the Pacific coast, a mid-altitude station in the Sierra Nevada Mountains, and a timberline station at 3,000 meters elevation. One of their experimental plants, *Potentilla glandulosa*, a member of the rose family, grew at all three locations and showed genetic adaptations to each habitat (fig. 15.12). Because this plant reproduces asexually (as well as sexually), clones could be cut into equal parts and used in the experiments. By using clones, the scientists

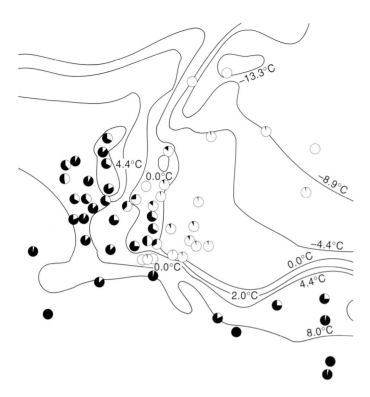

FIGURE 15.11

Clines. Darker circles indicate increased cyanide production by a European species of clover; this cyanide discourages herbivores. A cline of the cyanogenic phenotype occurs along a temperature gradient: the production of cyanide occurs only when temperatures are above freezing (0°C). This cline is apparently determined by a balance between the advantage of being unpalatable to herbivores and the disadvantage of frost rupturing cellular membranes, releasing cyanide in the plant's tissues.

FIGURE 15.12

Cinquefoil (*Potentilla glandulosa*), one of the California plant species used to study ecological variation.

were able to control for genetic variation and determine if differences were due to the habitat or not. Individual plants were gathered from a variety of habitats along the gradient and grown side by side at a field station in each climatic zone. Plants from the three stations had distinct morphological traits that were adapted to their original environment; that is, these subpopulations had diverged genetically. Thus, *Potentilla glandulosa* is a species showing ecological variation; such variation could lead to the formation of new species through time.

Most aspens (*Populus tremuloides*) can produce viable seeds and reproduce sexually as well as asexually. In the higher elevations of the Utah mountains, however, aspen seeds rarely germinate because of the climate, and aspen trees apparently reproduce only asexually by producing suckers from roots (see fig. 7.14). These aspen show ecological variation, producing leaves earlier in the growing season than other aspen as well as reproducing asexually, and successfully colonizing higher elevations of the Utah mountains. It is possible that some of these clones of aspen are 8,000 years old.

INQUIRY SUMMARY

A cline consists of plants that diverge morphologically and physiologically along an environmental gradient, but gene flow occurs throughout the population. Plants at the extremes of a cline usually appear different and may have different physiologies. Genetic divergence may occur along an environmental gradient across the range of a population. In a patchy or graded environment, fragmented subpopulations with significant genetic divergence may form, causing a species to develop ecological variation.

If you were raising a crop of plants, how would knowledge about the ecological variation of the plant species be important to you?

Classification Is a Continuously Changing Process.

The history of classification of life is one of continual change. Change, which still occurs today, has come from the discovery of new species, the acquisition of new knowledge about plants and other organisms, as well as the development of new philosophies and methods of investigation. To a large extent, new methods are due to the development of technologies that are applicable to classification, especially technologies involving computers and molecular genetics. After reading this chapter, you will probably realize that classification has been, and continues to be, one of the most dynamic aspects of biology. You will also see that, partly because of its dynamic nature, the subject of classification encourages a diversity of ideas and opinions about how organisms should be classified.

Why is classification important? To scientists, being able to communicate with other scientists around the world is essential. It is natural for humans to organize information, and identifying and naming organisms helps bring order out of the chaos for the diverse organisms on earth. Scientists are also keenly interested in understanding the relationships of organisms to each other to piece together the evolutionary history of life. The pieces of this evolutionary puzzle include which characteristics closely related species share and which characteristics distantly related organisms do not share. Also, if medicines have been found in some members of a plant family, then identifying other members of the same family may reveal new drugs. In addition, people around the world have the same common name for different plants. For instance, the word "corn" is used for wheat by people in Great Britain and for oats by people in Scotland. If, however, anyone in the world writes about *Zea mays*, this communicates to others exactly which plant is being investigated—what people in the United States call corn.

Classification is often viewed by nonspecialists as the discipline of biology that places organisms in their proper categories, where they remain permanently. Nothing could be further from the truth. Taxonomists, whose specialty is the discovery and classification of organisms, throughout history have revised and modernized their classification systems as new species were discovered and as new knowledge about plants and other organisms was obtained in different disciplines of biology. The first classifications of plants, for example, made more than 2,000 years ago, dealt with hundreds of species and were based only on features that could be seen with the naked eye. The number of known plant species is now in the hundreds of thousands, and taxonomists must integrate knowledge from molecular biology, genetics, ecology, anatomy, chemistry, physiology, morphology, and reproductive biology when devising their system of classification.

For many centuries, the first major dividing line among all living things was between plants and animals. We now know much more about life and, as a result, also recognize archaebacteria, eubacteria, fungi, and other kinds of organisms in our classification schemes. The history of classification, therefore, is one of continual change. The subject of classification is more dynamic now than at any time in the past.

Linnaeus developed a system of binomial nomenclature that we still use today.

Swedish physician and naturalist Carolus (Carl) Linnaeus (1707–78) merits special attention in any discussion of classification because he is credited with giving two-part scientific names to organisms, which taxonomists still do today. His system of naming organisms is called the **binomial system of nomenclature** because all of the scientific names of organisms have two parts; that is, they are binomials ("two names"). Linnaeus's classification system was published in 1753 in a two-volume set called *Species Plantarum* ("Species of Plants"), which included about 7,300 kinds of plants. Linnaeus's classification was based on only a few reproductive features. He related the reproductive structures of plants to those of humans and referred to flowers as a "bridal bed" in which plants reproduced. Because his classification system did not reflect natural, evolutionary relationships between organisms, it is referred to as an **artificial classification.** Nevertheless, the system was more convenient and more comprehensive than any other system available at that time. Although some botanists want to replace Linnaeus's classification system with one based on evolutionary relationships, Linnaeus's binomial system remains popular and widely used for naming plants and other organisms (see Perspective 15.2, "Is It Time to Change the System?").

When Linnaeus began his work, it was customary to use lengthy, descriptive Latin phrase names for plants and animals. The first word of the phrase constituted the genus to which the organism belonged. For example, all known poplars were given phrase names beginning with the word *Populus*. Similarly, the phrases for willows began with *Salix*, those for roses with *Rosa*, and those for mints with *Mentha*. In Linnaeus's time, the complete phrase name for peppermint was *Mentha floribus capitatus, foliis lanceolatis serratis subpetiolatis*, or "Mint with flowers in a head, leaves lance shaped, saw-toothed, with very short petioles." Although such more-than-a-mouthful names were specific, they were far too cumbersome to be useful.

Linnaeus used similarities to place groups of plants in the same genus; each member of a genus was called a species. Linnaeus began to add a word that, when combined with the genus name, formed a convenient abbreviation for a species. For example, peppermint was designated *Mentha piperita* (fig. 15.13). Linnaeus eventually replaced all

15.2 PerspectivePerspective

Is It Time to Change the System?

Carolus Linnaeus classified plants according to their reproductive parts, thereby endowing them with sex lives. For example, Linnaeus's system was based on *nuptiae plantarum*—the marriages of plants—which involved plants functioning as "husbands" and "wives" and enjoying sexual relations in "bridal beds." Linnaeus's system of classification also reflected eighteenth-century values; for example, male parts of flowers were given more taxonomic significance than were female parts. This judgment was based not on empirical evidence but rather on traditional tenets of gender-related bias. Indeed, Linnaeus's system was artificial as well as explicit (one opponent called it "loathsome harlotry"); that's why most of the details of the system have largely been abandoned. What survives of Linnaeus's system is its method of hierarchical classification and its custom of binomial nomenclature.

Although Linnaeus's classification system has been a bedrock of biology for more than two centuries, some biologists now want to scrap the system. These biologists argue that Linnaeus's system is not only outdated but also misleading. Whereas Linnaeus classified organisms according to appearances, biologists today view organisms as related by evolutionary history. Because Linnaeus knew nothing of evolution, his rules and taxonomic categories have no biological meaning. As a result, his system of classification may confuse the evolutionary relationships of organisms.

Will biologists abandon Linnaeus's system? Although most biologists acknowledge the limitations of the system, they aren't ready to try something new. Indeed, abandoning Linnaeus's system of classification would involve a confusing transition period; moreover, many taxonomists wonder if any new system—regardless of its strengths—could displace the entrenched system of Linnaeus. The debate continues.

FIGURE 15.13

Peppermint plants (*Mentha piperita*). *Mentha* is the genus name that means this plant is a mint; the full name, *Mentha piperita*, indicates what kind of mint.

the phrase names with abbreviated ones; that is, with binomials. Today, scientific names of all organisms, including plants, are binomials. Plants are named after people, places, or special characteristics. For example, there are flowering plant genera named *Lewisia* and *Clarkia* after the famous explorers of the western United States, Lewis and Clark. The names *Gutierrezia texana* (from Texas) and *Penstemon oklahomensis* (from Oklahoma) let you know where you might expect to find these plants growing. *Sequoiadendron giganteum* aptly describes the size of the giant redwood trees in California. Linnaeus often named his favorite plants after friends and unappealing plants after his critics. For more information about plant names, see Perspective 15.3, "How to Name a New Plant Species."

Modern classification systems are based on evolutionary relationships.

Until the latter half of the nineteenth century, virtually all classification systems were based on the belief that organisms had not changed and would not change in the future. However, as the ideas of natural selection from Charles Darwin and Alfred Wallace (see chapters 2 and 14) spread and became accepted because of enormous supporting evidence for it, these systems fell into disfavor, and classifications based on evolutionary relationships began to be developed.

Systematics is the scientific study of organisms and their **phylogenetic,** or evolutionary, relationships. Scientists who specialize in systematics are called **systematists.** Recently, systematists worldwide have begun to organize a

15.3 *Perspective* *Perspective*

How to Name a New Plant Species

New species are discovered every day, usually in little-explored regions of the world such as tropical rain forests. New plant species, however, may be found almost anywhere that wild plants grow. When a botanist finds a plant that does not match the description of any known plant species, a new species may have been discovered. Because a new species usually fits into an existing genus, the botanist needs to show how the new plant differs from other species in the genus and give it a name. Botanists are guided in such a task by the *International Code of Botanical Nomenclature*, which is the rulebook for naming new plants. The new name must be one that has not been used for another plant and must be accompanied by a description, written in Latin, of the species or an explanation (diagnosis) of how the new species is different from other, similar species. The name and description or diagnosis must be published to establish the validity of the new name. Editors of journals or books for this purpose send the species proposal to specialists to verify that the species is a new one and that the name is unique.

The name of each new species must also be represented by a single specimen, called the type specimen, that was used to describe the species. Specimens (including type specimens) are generally made by pressing the freshly collected plants as flat as possible between sheets of newspaper and cardboard, drying them, and then mounting them on large sheets of paper for safekeeping in a herbarium. A herbarium is like a library of preserved plant specimens. Large herbaria, such as that of the New York Botanical Garden, or the Gray Herbarium of Harvard University, maintain millions of specimens, thousands of which are type specimens.

Herbarium collections are vital for scientific research. They are especially useful for comparing new species with previously known ones, keeping records of the discovery of new species, keeping track of species migration and extinction, and knowing which species are becoming rare or endangered due to human activities.

---

campaign to classify all organisms on earth in a comprehensive phylogenetic system. This monumental effort has several goals, including to study the diversity of life on earth and to guide the search for new crop species, novel and useful genes, more biological controls of pests, and new products from organisms. An additional goal is to determine the role of human activities in the extinction of life on earth. The goals of this multinational project are especially ambitious because there may be several million species yet to be discovered, hundreds of thousands of which may be plants.

Systematists estimate that the research needed to obtain a phylogenetic classification of life within the next 25 years will cost at least $3 billion per year. Currently only about $500 million is spent annually on global systematics. At the present rate of research funding and species extinction, systematists estimate that more than half of the species of organisms now living will be extinct before the project can be completed. Moreover, they estimate most extinctions will be of undiscovered species that could hold untold benefits for humans.

Classification systems emphasize ancestral versus derived characteristics.

Classification systems that try to reflect evolutionary history are phylogenetic, and sytematists consider these systems to be superior because the systems more accurately represent the natural history of organisms. However, botanists do not agree on which particular phylogenetic system is the best. Accordingly, various phylogenetic classifications have been proposed, each differing in its basic assumptions about the features that best reflect the evolution of plants.

All of these systems stress what their authors believe are the ancestral features of plants versus those that have undergone evolutionary change; that is, they emphasize *ancestral* features versus *derived* features. For instance, if a characteristic is found in all species of a genus, it is assumed to be an ancestral feature. However, if a characteristic occurs in only a few species of a genus, it may be derived, developing in these species as they evolved; therefore, derived characteristics would not be expected in the older species from which the new species evolved. More recent classifications also show lines of descent between related groups of plants (fig. 15.14).

INQUIRY SUMMARY

Taxonomy is the study of describing, naming, and classifying organisms. Classification continues to change because of the discovery of new species and the new knowledge about species that have already been described. Classification is important for many reasons, including the ability of scientists to communicate with

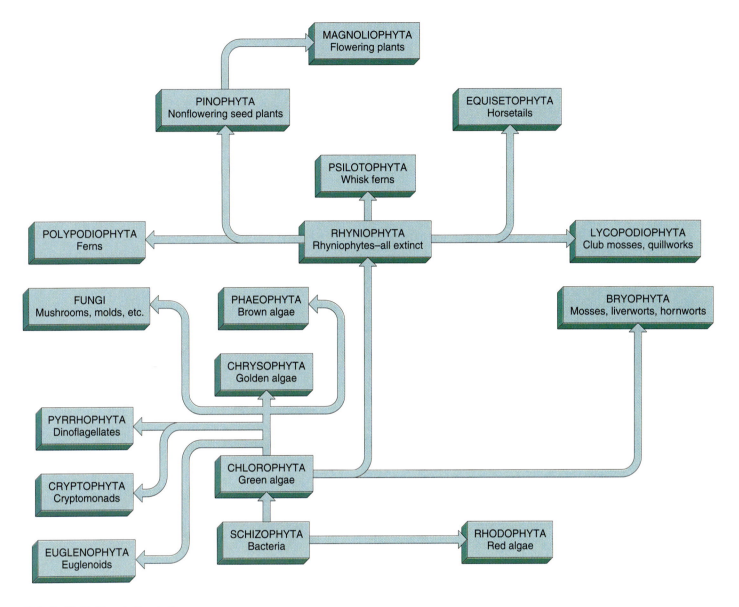

FIGURE 15.14

Example of a traditional classification and phylogeny of plants (by the late Arthur Cronquist, formerly of the New York Botanical Garden). Classification schemes are updated as new plants and knowledge are discovered. Thus, not all of the names in this classification are still generally accepted. Compare this to a newer tree of life based on molecular data (see fig. 16.5).

each other and to understand evolution. Linnaeus is credited with the binomial system of nomenclature for all species. Because Linnaeus's system was based on just a few reproductive features, it is considered to be an artificial system of classification. Today, all organisms are named using the binomial system. Modern classification systems are based on the evolutionary relationships of organisms. These systems emphasize ancestral characteristics versus derived characteristics.

Many scientific names, such as Petunia *have become common words in the English language. What other scientific names are commonly used?*

Scientists Disagree About Which System of Classification Is the Best.

At this point, you may be wondering why there are different phylogenetic classifications of plants. In other words, if there is general agreement about the desirability of phylogenetic systems, why is there no general agreement about which system is best? There are at least two good answers to this question. One is that classification involves a subjective selection of characteristics (characters) and a subjective evaluation of their importance for designating species, genera, families, or any other taxonomic grouping. A good example of this subjectivity is the classification of flowering plants whose fruits are legumes, such as pea and bean. Legume-producing plants can be divided into three groups on the basis of different flower types. One type is exemplified by the common pea (*Pisum*) flower, which is bilaterally symmetrical and has five petals, the middle petal being larger and exterior in the bud to the other four. A second type, represented by honey locust (*Gleditsia*), also has a bilaterally symmetrical flower with five petals, but the middle petal is not larger and is interior in the bud to the other four. Flowers of the third type, represented by *Acacia* and *Mimosa*, are radially symmetrical, which means you could slice the flower through its center along any radius and always get two equal halves. Taxonomists who consider the fruit (i.e., legume) to be the unifying feature of this group recognize a single family, the Fabaceae, with three subfamilies based on flower type. Other taxonomists who consider differences in flower types more significant divide the group into three families: the Fabaceae (*Pisum*-type), the Caesalpiniaceae (*Gleditsia*-type), and the Mimosaceae (*Mimosa*-type). Taxonomists who habitually take the more conservative view, such as considering all leguminous plants in one family, are often called *lumpers.* Those who divide families or other groups more liberally are often called *splitters* (fig. 15.15). Both views are widely accepted, although heated debates over lumping or splitting specific groups are common.

A second explanation for the variety of phylogenetic classifications of plants comes from differing views of taxonomists on how to derive evolutionary relationships from the features, or characters, of plants. Some taxonomists make intuitive judgments whose underlying assumptions about the evolution of characters often are not obvious and cannot easily be scientifically evaluated. Many taxonomists, therefore, consider traditional classifications to be artistic as well as scientific, despite the fact that most of the recent classifications of flowering plants have been traditional.

New investigative approaches help systematists study phylogenetic relationships.

Systematists have developed new approaches to phylogenetic classifications. Keep in mind, though, that ultimately all classifications are human constructions. Nevertheless, these newer approaches are more explicit in their assumptions and are more testable scientifically. Perhaps the most widely adopted of the new approaches of the past two decades is one called **cladistics.** Cladistics is generally defined as a set of concepts and methods for determining **cladograms,** which graphically depict branching patterns of diversity and, therefore, evolution (fig. 15.16). These branching patterns represent hypotheses about evolutionary relationships among organisms.

FIGURE 15.15

This diagram shows how Dr. Lumper and Dr. Splitter would differ in their classifications of the legume-bearing plants *Pisum*-type, *Gleditsia*-type, and *Mimosa*-type.

A goal of cladistics is to improve classification schemes

For cladistics, a **character state** is the form or value of a character. For example, the character could be flower color, and the state could be red or blue. Or the character could be the pigment chemical type for red pigments, whose states are either a *betalain-type* red chemical or an *anthocyanin-type* red chemical (fig. 15.17). The states, or values, of each character must be explicit. Furthermore, once characters and their states are declared, much emphasis in cladistics is placed on assumptions about which states are ancestral

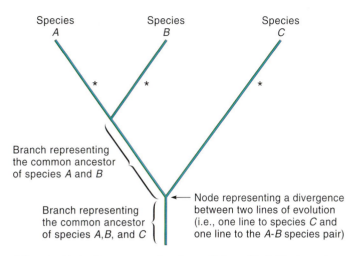

Branch representing
the common ancestor
of species *A* and *B*

Branch representing
the common ancestor
of species *A*,*B*, and *C*

Node representing a divergence
between two lines of evolution
(i.e., one line to species *C* and
one line to the *A*-*B* species pair)

* Terminal branches representing the evolution of individual species
after divergence from ancestors shared with other species

FIGURE 15.16

This cladogram shows the phylogenetic relationships of three
species, all of which arose from a single ancestor. As shown here,
species *A* and species *B* are more closely related to each other than
either is to species *C*.

and which are derived (i.e., more recently evolved). If the
anthocyanin type of pigment is assumed to be ancestral,
then the betalain type is derived, and vice versa. By assuming the anthocyanin type to be ancestral, we can see
that plants with betalain-type pigments have a recent common ancestor in which betalain pigments first appeared

(fig. 15.18a). In this case, the occurrence of betalain pigments indicates a shared evolutionary change (shared der-
ivation) that was passed from an ancestral betalain producer to its descendant betalain producers.

What if, instead, we assume the betalain type to be ancestral? The same cladogram would then show two evolutionary origins of the anthocyanin type, one in the flowering plants and one in the nonflowering seed plants (fig.
15.18b). Alternatively, another cladogram based on the evolution of pigment chemical types could be made (fig.
15.18c). This alternative cladogram is important because it
represents an alternative hypothesis about phylogenetic relationships: it minimizes the number of times the anthocyanin type arose and shows instead that flowering plants
had two evolutionary origins. As an alternative hypothesis,
it can be tested by evaluating additional characters. In
cladistics, testing hypotheses entails finding additional
characters whose states may corroborate some cladograms
but not others. Those that are not corroborated by additional data can be rejected, and those that are corroborated
can be maintained for further testing.

How do we choose the most accurate cladogram?
The alternative cladograms based on pigment chemical
type show that betalain biosynthesis arose once in plant
evolution (fig. 15.18a), that anthocyanin biosynthesis
arose twice in evolution (fig. 15.18b), or that anthocyanin
biosynthesis arose once, but flowering arose twice (fig.
15.18c). Anthocyanin biosynthesis involves a complex
metabolic pathway, and it is intuitively hard to think
that it could have two independent evolutionary origins
or that flowers and all their associated features arose

(a)

(b)

FIGURE 15.17

Two possible states of the character, chemical type of red pigment. (a) The flowers of this rose (*Rosa* species) contain an anthocyanin-type
red pigment. (b) The red flowers of this hedgehog cactus (*Echinocereus triglochidiatus*) contain a betalain-type red pigment.

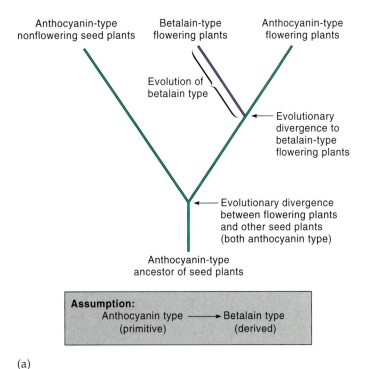

(a)

(b)

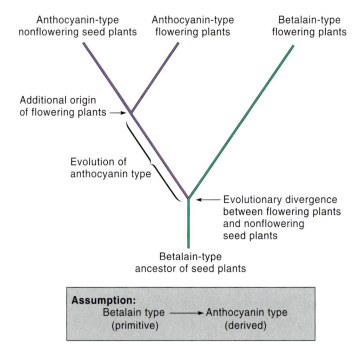

(c)

FIGURE 15.18

Alternative cladograms for plants with anthocyanins or betalains. (a) Cladogram built by assuming that anthocyanins were primitive and betalains were derived. (b) Cladogram showing how character evolution would look if betalains were primitive and anthocyanins were derived twice. (c) Alternative cladogram showing that betalains were primitive and anthocyanins were derived once. Which of these cladograms is the simplest scenario?

because there are often exceptions to the principles that we discover (fig. 15.19).

In cladistic terms, the principle of parsimony can be restated to say that the best cladogram requires us to make the fewest assumptions about evolutionary changes. Parsimony is especially powerful when there can be many possible cladograms for a group of organisms. The number of possible cladograms would be easy to evaluate for three kinds of organisms (3 possibilities), for four kinds (26 possibilities), or maybe even for five kinds (125 possibilities). However, the number of imaginable cladograms increases more than exponentially as the number of different organisms increases. Indeed, there are more than 7 *trillion* possible cladograms that can be made for 15 kinds of organisms. What role do you think computers have played in modern systematics?

The goal of cladistics or any other approach to making phylogenies is to evaluate and improve classifications and to help us understand the evolutionary relationships of organisms (fig. 15.20). Opinions vary, however, as to how the branching points on a cladogram should be used to designate taxonomic levels and usually will not convince splitters to lump or lumpers to split. In spite of its competing ideas about how it should be applied to classification, how-

twice. The simpler scenario, that each of these arose only once, is somehow more satisfactory. Such intuition is consistent with a principle of logic called *Occam's razor,* which states a person should not make more assumptions than the minimum needed to explain anything. The principle is often called the principle of **parsimony,** which is what systematists call it when it is applied to cladistics. Nevertheless, understanding nature is tricky

FIGURE 15.19

The red-capped *Amanita* (*A. muscaria*) mushroom. Although the red pigment in this fungus is a betalain that resembles the pigment that occurs in the plants shown in figure 15.17, this feature alone does not mean that fungi (see chapter 16) are closely related to plants.

ever, cladistics has been used to infer the phylogenetic relationships of many different groups of organisms.

INQUIRY SUMMARY

Modern systems of classification have attempted to be phylogenetic. Several different phylogenetic classifications of flowering plants currently are used widely by different taxonomists. Cladistics produce cladograms, which depict patterns of diversity and evolution through the comparison of primitive, or ancestral, versus derived characteristics.

In studying relationships of organisms, explain how a characteristic could be considered to be ancestral in some situations and derived in other situations.

Similar Organisms Are Grouped in Successively Higher Categories.

We can make several basic assumptions, or postulates, about how we classify organisms. Some of these assumptions are more obvious than others, but the main ones are as follows:

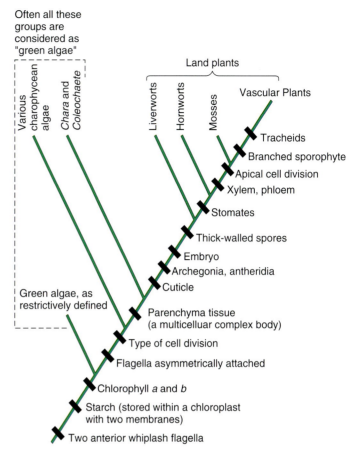

FIGURE 15.20

A phylogenetic tree of green plants and green algae, the likely ancestors of green plants. Major features of green plants are located along the tree where they are thought to have evolved. (Mosses are not traditionally considered vascular plants even though some species possess tissues that may be homologous [see chapter 14] to xylem and phloem.)

Source: Based on W. S. Judd et al., *Plant Systematics*, 1999, Sinauer Associates, Sunderland, MA.

1. Organisms exhibit various degrees of similarities and differences among individuals and groups.

2. Those organisms that are similar in nearly all respects are a species.

3. Species that share some of their features comprise a genus.

4. On the basis of their shared features, similar genera can be organized into a family; likewise, families and larger groups can be organized into successively higher levels of a taxonomic hierarchy.

5. The greater the similarity among organisms and among groups, the closer their evolutionary relationship or, the greater the number of shared derived characters among organisms, the closer is their relationship.

Taken together, these assumptions function as the theoretical foundation of biological classification. These assumptions are constantly being tested. Systematists, for example,

TABLE 15.3   The Taxonomic Hierarchy for Four Species of Plants

Kingdom:	Plantae	Plantae	Plantae	Plantae
Division:	Magnoliophyta	Magnoliophyta	Magnoliophyta	Pinophyta
Class:	Liliopsida	Liliopsida	Magnoliopsida	Pinopsida
Order:	Zingiberales	Commelinales	Fagales	Pinales
Family:	Musaceae	Poaceae	Fagaceae	Pinaceae
Genus:	*Musa*	*Hordeum*	*Quercus*	*Pinus*
Species:	*Musa acuminata*	*Hordeum vulgare*	*Quercus alba*	*Pinus ponderosa*

Note: The common names of these species are, left to right, banana, barley, white oak, and ponderosa pine. The Magnoliophyta are the flowering plants; the Pinophyta are the conifers. Liliopsida is the current class name for the monocots; Magnoliopsida is the class name for the dicots, although modern classification systems often divide the dicots into several major groups.

are especially interested in what a species is; there is no general agreement about how many similarities are enough to call a group of organisms a species. However, many definitions of a species include a reproductive component: species are an interbreeding or potentially interbreeding population.

According to such a taxonomic hierarchy, species are organized into genera (singular, *genus*), genera into families, families into orders, orders into **classes,** classes into **divisions,** divisions into **kingdoms,** and kingdoms into **domains.** Divisions are called *phyla* by zoologists, a term which is sometimes used by botanists. A debate is occurring among some botanists about which term—division or phylum—should be used. Each category is a **taxon** (plural, taxa), which is a general term for any level of classification, such as species, genus, or family. A good way to envision a taxonomic hierarchy is to see how a few examples are organized, as shown in table 15.3.

Some systems of classification identify three domains.

Botany does not exist in a vacuum; plant biology is best understood as it relates to the biology of other kinds of organisms. In finding out what plants are and where they came from—that is, in defining the plant kingdom—it is helpful, therefore, to consider relationships of members of the plant kingdom with members of other kingdoms.

A popular classification system generally includes plants, animals, fungi, and bacteria. These kingdoms are referred to as the kingdom **Plantae,** the kingdom **Animalia,** the kingdom **Fungi,** and the kingdom **Monera,** respectively. A hodgepodge of organisms are placed in a fifth group, the kingdom **Protista.** In alternate classification systems, the kingdom Monera is divided into two king-

doms, the Eubacteria (true bacteria) and Archaea (see chapter 16), to produce a six-kingdom system. Characteristics of the organisms in a six-kingdom system are summarized in table 15.4.

Molecular data, such as nucleotide sequences of genes or amino acid sequences of proteins, can be compared in the same way that morphological data are compared, either by overall similarity or by cladistic methods. Nucleotide sequences for the same gene or RNA molecule or amino acid sequences for the same protein can be compared after they are obtained from organisms in a group. A phylogeny based on molecular data is called a **molecular phylogeny.** A system based on molecular phylogeny groups kingdoms into one of three domains: Eubacteria, Archaea, and Eukaryota (eukaryotes: protists, algae, fungi, plants, and animals)(see chapter 16). Thus, based on molecular data, bacteria and archaebacteria are as different from each other as they are from plants, animals, fungi, and algae. Recent molecular-based studies and the discoveries of several new species of archaebacteria have complicated the picture of classification and brought into question the organization of life on earth into three domains.

Results of other recent molecular studies, in combination with morphological and fossil data, have forced botanists to re-examine longstanding ideas about the relationships among plants and other organisms. At a botanical conference in 1999, scientists suggested that plants should not be grouped in just one kingdom but divided into three kingdoms—red plants, brown plants, and green plants—that evolved from different one-celled ancestors. In this organization, slime molds (see chapter 16) are grouped in the brown plant kingdom. This organization has not received widespread acceptance within the scientific community. Thus, the phylogeny of organisms remains in a dynamic state; as noted

FIGURE 15.21

Bark of Pacific yew (*Taxus brevifolia*) is a source of taxol, an anti-cancer drug.

before, classification is a continuously changing process and will undoubtedly change as more data accumulate.

Many systematic studies are economically important.

According to the World Health Organization, over 20,000 different kinds of plants have been used for medicinal purposes by humans. It is estimated that over 75% of people living in developing countries still use traditional medicines, many of which are extracted from wild plants, as an important means for their health care. Drugs derived from plants generate billions of dollars in sales each year, and discovering new medicines often relies on a careful systematic search of plants with medicinal properties and their close relatives. As one example, taxol is a powerful anticancer agent that is extracted from the bark of the Pacific yew (*Taxus brevifolia*). However, three trees must be killed to produce enough taxol for each cancer patient (fig. 15.21). Scientists predicted that close relatives of the Pacific yew might also contain taxol, and they discovered that taxol also can be obtained from the European yew (*Taxus baccata*). This process of discovery could have taken many years if the search was conducted randomly. Serendipitously, a new fungus species (*Taxomyces andreanae*) was discovered that also produces taxol and that lives in the bark of the Pacific yew.

Other examples of the potential economic importance of systematic studies focuses on biological control of plant pests. Knowing the correct species of a pest is often important in locating effective biological control agents. One example is the sugar-beet leafhopper (*Circulifer tenellus*), which is a major pest of crop plants and was originally thought to be a member of a genus native to South America. When biologists went to South America to look for a potential predator of this pest, no control agent could be found. Later, systematists discovered that this leafhopper belonged to the genus *Circulifer,* which comes from the Mediterranean. Several natural enemies of the sugar-beet leafhopper were found there and introduced to control it in crop fields of California.

Systematists also have discovered close relatives of important crop species, such as corn and tomatoes. Wild relatives of corn have been discovered in Mexico, and relatives of tomato have been found in the Andes mountains of South America. The relative of corn, teosinte, possesses genes that make corn resistant to seven types of viral diseases. Because corn is nearly a $60-billion business around the world, this discovery could have huge economic benefits (see fig. 1.8). The recently discovered wild relatives of tomato possess genes that increase flavorful content of the red fruits, making them more delicious, a characteristic that may be worth several million dollars a year.

SUMMARY

Speciation—the origin of new species—is a central concept of evolutionary theory and is the mechanism that explains the vast diversity of organisms. Speciation is difficult to describe because it includes a variety of mechanisms; moreover, a species is difficult to define clearly. The most general definition of a species is a group of similar individuals that can successfully breed with one another but not with members of other groups. This definition emphasizes reproductive isolation. Distinguishing between species and between the mechanisms that produce new species is further confounded because all members of a species are not identical.

During speciation, segments of the gene pool of a plant population diverge, sometimes rapidly through processes such as mutations, polyploidy, and hybridization, and sometimes slowly through responses to natural selection pressures. As parts of the population become different, they lose some features and accumulate others that prevent interbreeding. This genetic accumulation and divergence is promoted by environmental variation, mutations, and different rates of gene flow and genetic drift. Ultimately, members of two different populations acquire different traits and are reproductively isolated. These populations no longer produce fertile offspring (hybrids), and, therefore, the two populations represent two species.

TABLE 15.4	**Major Characteristics of the Six Kingdoms**		
	PLANTAE	ANIMALIA	FUNGI
Cell type	eukaryotic	eukaryotic	eukaryotic
Genetic material	DNA associated with protein in chromosomes	DNA associated with protein in chromosomes	DNA associated with protein in chromosomes
Gene structure	introns present	introns present	introns present
Nuclear envelope	double	double	double
Membrane-bound organelles	present	present	present
Chloroplasts	present	absent	absent
Cell wall	cellulosic	absent	chitinous or cellulosic
Means of genetic recombination	fertilization and meiosis	fertilization and meiosis	fertilization and meiosis
Mode of nutrition	autotrophic	heterotrophic by ingestion	heterotrophic by absorption
Flagella/cilia	absent or 9 + 2 microtubules in gametes of some groups	9 + 2 microtubules	absent
Multicellularity/ cell specialization	present	present	present
Nervous system	absent	present, often complex	absent
Respiration	aerobic	aerobic	aerobic or anaerobic
Life cycle	alternation of generations	diploid except for gametes	mostly haploid, often two nuclei per cell; some alternation of generations
Unicellular spore formation	present in all groups	absent	present in all groups

Biologists need a working definition of a species as a framework for questions and research. Morphologically, a species is a group of organisms that look the same and appear significantly different from other organisms. This definition is easily applied and communicated. Biologically, a species is a group of organisms whose members can reproduce with each other in nature to produce fertile progeny but cannot successfully reproduce with members of other species. Genetically, a species is a group of organisms that exchange genes.

Reproductive isolation is maintained by barriers, including geographic isolation of populations in distant areas. For a subpopulation to form a new species, the gene pool must diverge from that of the parent population. Divergence of geographically separated populations is usually gradual and results from genetic differences accumulated through mutation, genetic drift, and natural selection. When divergence is great enough to eliminate interbreeding, then the two populations are different species.

Speciation also may begin with the formation of a cline, which is the pattern of ecological variation among individuals of the same species distributed along an environmental gradient. If the environment is distinctly patchy, the cline may develop subpopulations subject to local selection and that diverge both genetically and morphologically. Significant divergence can lead to speciation.

EUBACTERIA	ARCHAEA	PROTISTA
prokaryotic	prokaryotic	eukaryotic
DNA not associated with histone protein in chromosomes	DNA associated with some proteins; one chromosome	DNA associated with protein in chromosomes
introns absent in most groups, present in some	introns present	introns present
absent	absent	double or single
absent	absent	present
absent	absent	present or absent
noncellulosic; peptidoglycan	noncellulosic; various types different from eubacteria	absent or present; cellulosic or various types
conjugation, transduction, or transformation	conjugation (?), transformation, transduction (?)	fertilization and meiosis in most groups
autotrophic or heterotrophic by absorption	heterotrophic or chemoautotrophic (many are extremophiles)	autotrophic or heterotrophic by absorption or phagocytosis
solid, rotating	solid, rotating	9 + 2 microtubules
absent	absent	absent in most, present in algae and water molds
absent	absent	absent or simple
aerobic or anaerobic	anaerobic or aerobic	aerobic
haploid; fission	haploid; fission	mostly haploid; some forms diploid
present in some groups	absent	present in some groups

Polyploids have more than two sets of chromosomes; they result from the nondisjunction of chromosomes during cell division. Polyploidy, which is common in higher plants, is an instantaneous mechanism of speciation because polyploids are reproductively incompatible with the parent population. Polyploidy may restore fertility to a hybrid by providing paired chromosomes.

Carolus Linnaeus's system of classification is important because it designated each species by a Latin binomial. The binomial system of nomenclature is still used today for all species of organisms. Once the ideas of Darwin and Wallace became widely accepted, classifications were revised to show the evolutionary relationships among plant groups.

These classifications are called phylogenetic classifications. These systems are modified and revised repeatedly.

The abundance of different phylogenetic classifications of plants has prompted taxonomists to develop more explicit methods of building phylogenies. The most widely adopted of these newer methods comes from cladistics, whose goal is to determine evolutionary branching patterns among hierarchies of organisms.

At present, biologists classify organisms into six kingdoms or three domains by phylogenetic methods. The methods of cladistics, especially as they are applied to molecular data (e.g., DNA, RNA, proteins), provide evidence for these phylogenies.

Writing to Learn Botany

Paleontologist Stephen Jay Gould said the following about taxonomy:

"Taxonomy is often regarded as the dullest of subjects, fit only for mindless ordering and sometimes denigrated within science as mere 'stamp collecting.' If systems of classification were mere hat racks for handling the facts of the world, this disdain might be justified. But classifications both reflect and direct our thinking. The way we order represents the way we think. Historical changes in classification are fossilized indicators of conceptual revolutions." What was Gould saying? Do you agree or disagree? Why?

THOUGHT QUESTIONS

1. Few, if any, intelligent persons would mistake a giraffe for an elephant, or a tomato for an oak, nor would most know or care about the scientific names of these organisms. Why, then, should we be concerned about all organisms having scientific names?

2. Most early taxonomists used the form and structure of major plant parts in their systems of classification. What additional features of plants can modern taxonomists use? How might they be used for classification?

3. Linnaeus recognized 3 kingdoms of organisms; others have recognized up to 13 kingdoms. Is there a "correct" system of classification? Why or why not? How might you explain the abundance of classifications of plants that exists today?

4. The late Isaac Asimov wrote, "The card-player begins by arranging his hand for maximum sense. Scientists do the same with the facts they gather." How does Asimov's analogy relate to plant classification? Explain your thinking.

5. What role could a mountain range play in speciation?

6. Why are many hybrids sterile?

7. In what ways does the environment affect speciation?

8. Why are hybrids often more successful in disturbed environments?

SUGGESTED READINGS

Articles

Brown, K. S. 1999. Deep Green rewrites evolutionary history of plants. *Science* 285:990–91.

Doyle, J. J. 1993. DNA, phylogeny, and the flowering of plant systematics. *BioScience* 43:380–89.

Gaffney, E. S., L. Dingus, and M. K. Smith. 1995. Why cladistics? *Natural History* 104:33–35.

Gibbons, A. 1996. On the many origins of species. *Science* 273:1496–99.

May, R. M. 1992. How many species inhabit the earth? *Scientific American* 267:42–49.

Pennisi, E. 1999. Is it time to uproot the tree of life? *Science* 284:1305–07.

Schiebinger, L. 1996. The loves of the plants. *Scientific American* 274:110–15.

Systematics Agenda 2000. 1994. Systematics agenda 2000: Charting the biosphere. Technical report. 21 pp.

Books

Crawford, D. J. 1990. *Plant Molecular Systematics: Macromolecular Approaches.* New York: John Wiley and Sons.

Judd, W. S., C. S. Campbell, E. A. Kellogg, and P. F. Stevens. 1999. *Plant Systematics.* Sunderland, MA: Sinauer Associates.

Wilson, E. O. 1992. *The Diversity of Life.* New York: W. W. Norton.

ON THE INTERNET

Visit the textbook's accompanying web site at http://www.mhhe.com/botany to find live Internet links for each of the topics listed below:

Plant taxonomy and classification

The species concept

Reproductive barriers

Plant speciation

Polyploidy

Cladistics

Phylogenetics

Molecular systematics

Systematics and economic botany

Plant variation

A kelp from along the coast of California. Kelps are brown algae; giant kelps can grow to more than 60 meters long. Where do algae fit into the evolution of plants?

Bacteria, Fungi, and Algae

CHAPTER OUTLINE

Bacteria are the most abundant organisms on earth.

> *Bacteria are prokaryotes.*
> *Bacteria are more diverse metabolically than eukaryotes.*
> *The generalized life cycle of bacteria is simpler than in plants.*
> *The classification of bacteria has recently changed.*
> *Bacteria-like cells were probably the first living organisms.*
> *Chloroxybacteria provide evidence for the evolution of chloroplasts.*
> *Cyanobacteria were once called blue-green algae because they resemble algae and are photosynthetic.*
> *Bacteria affect every aspect of our lives.*
> *Perspective 16.1 Deadly Food! Beware!*

Fungi are nonphotosynthetic eukaryotes with diverse roles in ecosystems.

> *Like bacteria, fungi grow just about everywhere.*
> *Perspective 16.2 Disappearing Fungi*
> *Fungi are traditionally classified into five main groups: Zygomycota, Ascomycota, Basidiomycota, Chytridiomycota and Deuteromycetes.*
> *Lichens are fungi and photosynthetic organisms living together.*
> *Some fungi are important to plant growth.*
> *Fungi have considerable economic and ecological importance.*

Algae include some of the smallest and largest photosynthetic eukaryotes.

> *Green algae are the ancestors of plants.*
> *Most brown algae are large marine organisms.*
> *Red algae are also mostly marine organisms.*
> *Diatoms are important in ocean ecosystems.*
> *There are other divisions of algae with ecological importance.*
> *The fossil history of algae provides insights into their origins.*
> *Algae are important in many ecosystems.*
> *Algae have widespread economic importance.*

Summary

LEARNING ONLINE

To learn more about bacteria, fungi, and algae, visit web1. manhattan.edu/fcardill/plants/intro/index.html in chapter 16 of the Online Learning Center at www.mhhe.com/botany. Key hyperlinks like this one in the OLC provide research sources for writing assignments and additional studying.

Although offering insight into evolutionary events, many kinds of organisms have posed problems in the classification of plants. For example, bacteria, fungi, and algae were once classified as members of the plant kingdom. This was a time when biologists classified all organisms as either plants or animals, and it was thought that bacteria, fungi, and algae fit better with plants—probably because they have cell walls, many are photosynthetic, and none move about like animals. Biologists now agree, however, that none of these organisms share a sufficiently close evolutionary relationship with plants to be classified as plants. Nevertheless, tradition among botanists dictates that any discussion of the diversity of plants is not complete without some accounting of these organisms and their importance. Accordingly, this chapter presents a brief overview of bacteria, fungi, and algae with a focus on their structure, evolution, and ecological significance. Perhaps what best links these organisms is their history of being considered plants and their relatively early evolution.

Bacteria Are the Most Abundant Organisms on Earth.

Bacteria live almost everywhere (fig. 16.1): at the bottom of oceans, in our mouths, in acid and alkaline hot springs, and in puddles on glaciers. Most people know something about bacteria because some bacteria make people sick. This limited awareness gives bacteria an undeservedly bad reputation, because most of them do not cause diseases. For example, Robert J. Price, a seafood technology specialist at the University of California at Davis, told the press in 1990 that about once per month he receives a report of seafood that glows in the dark. The glow usually comes from *Photobacterium phosphoreum,* a bacterium that produces light as it respires. Such bacteria occur on the skin and in the intestines of many fish and shellfish, and they glow when salt is added to cooked seafood. These luminescent bacteria are harmless; no one has ever reported getting sick from eating seafood on which the bacteria were growing. Bacteria also live in our intestines, and we depend on them for normal digestion. They are usually so abundant that the dry weight of normal feces can consist of up to 80% bacterial cells.

Bacteria can grow in such extreme environmental conditions that they must be considered the most hardy of all organisms. Many can tolerate hot acids, others survive temperatures below freezing for years, and still others thrive in boiling hot springs. There are deep-sea bacteria that live near volcanic vents, where the temperature approaches 360°C and the pressure can be more than 50–times greater than on the surface. Their tolerance of high temperature requires methods of sterilization that include a combination of high heat (120°C) and high pressure.

(a)

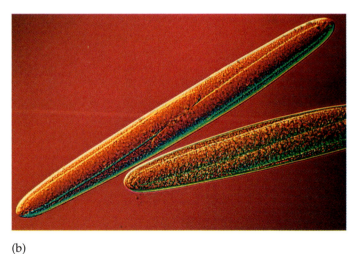

(b)

FIGURE 16.1

Mycoplasmas and bacteria have various shapes, but their biochemical and metabolic diversity far surpasses their morphological variation. (*a*) *Pneumonia mycoplasma* bacteria (×62,000). (*b*) *Epulopiscium fishelsonii,* a very large bacterium discovered in the guts of surgeonfish. Each cell is about 0.3 millimeters long.

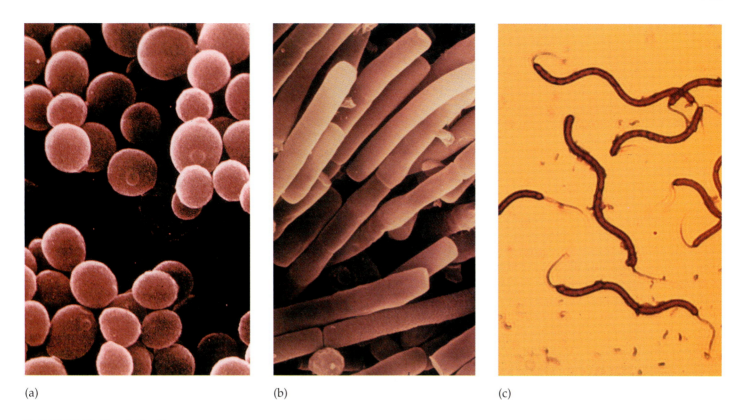

(a)   (b)   (c)

FIGURE 16.2

The three basic shapes of bacteria. (*a*) Spherical (*Micrococcus*) (×40,000). (*b*) Rod (*Bacillus*) (×60,000). (*c*) Corkscrew (*Spirilla*) (×450).

Bacteria play important ecological roles in every habitat on earth. For example, photosynthetic bacteria help maintain the global carbon balance, and nitrogen-fixing bacteria (chapter 7) account for about one-fourth of the total nitrogen fixed in oceans. Some of the most important bacteria in agriculture live and fix atmospheric nitrogen in the root nodules of crops such as alfalfa, soybean, and pea (see fig. 4.20).

Decomposer bacteria—those that obtain nourishment from dead organic matter—are partially responsible for the breakdown and recycling of organic material in the soil. Decomposition by some bacteria and fungi produces as much as 90% of the CO_2 in the atmosphere (via respiration, see chapter 11). Without decomposers, the earth's surface would be covered by dead organisms in a matter of weeks. So much carbon would be tied up in organic matter that less atmospheric CO_2 would be available, which would limit photosynthesis. Why would this be an ecological problem?

Bacteria are prokaryotes.

An important distinguishing feature of bacteria is that they are prokaryotic, which means that they have no membrane-bounded organelles; that is, they have no nucleus, no mitochondria, and no chloroplasts (chapter 4). Bacterial DNA, which occurs as a single, circular ring of double-stranded DNA, is rarely associated with histone proteins as is always found in nuclear chromosomes of eukaryotes. Cell walls, which occur in bacteria, differ markedly in their composition from those of plants. Indeed, the cell walls of bacteria are usually made of peptidoglycans, in which carbohydrate polymers are interconnected with short chains of amino acids. In addition, most bacterial cell walls also contain muramic acid, which is not found in walls around plant cells. Many bacteria are strictly anaerobic and are killed by oxygen. Others can live anaerobically or aerobically, and still others require oxygen for respiration. Different bacteria may also have flagella for mobility or shorter, thinner filaments called fimbriae for attachment to solid surfaces. Some bacteria also have pili, through which DNA can move between two bacteria. While genetic material can move from the donor bacteria to the recipient bacteria through pili, reproduction leading to new bacterial cells is asexual: one cell simply splits in half and forms two cells.

The distinguishing features of different prokaryotes—the bacteria—are their metabolism, chemical composition, and shapes. In fact, bacterial cells typically have one of three shapes (fig. 16.2): **spherical** (also called cocci), cylindrical **rod shaped** (bacilli), or long, curved or spirical rods often **corkscrew shaped** (spirilla). Bacteria can be further distinguished by a diagnostic test called the Gram stain, named after Danish physician H.C. Gram. This stain distinguishes two bacterial groups: gram-positive bacteria and gram-negative bacteria.

Although most bacteria are 1 to 5 micrometers in diameter (much smaller than plant cells), bacteria known as mycoplasmas are often only 0.2 micrometer across (see fig. 16.1*a*). At the other extreme, some photosynthetic cyanobacteria are up to 3 millimeters long, and the recently discovered *Epulopiscium fishelsonii*—the largest known nonphotosynthetic bacterium—is as large as a small hyphen in this book (0.3 millimeter long) (see fig. 16.1*b*). This should help you appreciate the size of a typical bacterium: it would take about 9 trillion average-sized bacteria to fill a box the size of a five-stick package of chewing gum.

Bacteria are mostly simple, single cells, but there are some exceptions. For example, some species have cells that cohere in chains; similarly, filamentous forms occur among the photosynthetic group of bacteria called **cyanobacteria.** Others have flagella and live as motile individuals or colonies, and a group called the myxobacteria often form upright, multicellular reproductive bodies. These two groups are thought to be the most complex forms among bacteria.

Bacteria are more diverse metabolically than eukaryotes.

The metabolic diversity of bacteria is reflected in the range of energy sources that bacteria can use. Different bacteria can use sulfide ions, iron, or methane gas for energy in metabolism. In addition, some soil bacteria, such as *Nitrosomonas*, are nitrifying; that is, they convert ammonia (NH_3) into nitrite (NO_2^-) for energy. Still other soil genera, such as *Nitrobacter,* obtain energy by converting nitrite ions to nitrate (NO_3^-), an important plant nutrient.

Cyanobacteria are photosynthetic, using light to convert CO_2 to sugars, and are important producers in many ecosystems. Bacteria also can ferment a variety of sugars and organic acids, thereby producing important commercial products such as ethanol, methanol, acetone, lactic acid, acetic acid, and propionic acid. Bacteria have also been used to clean up oil spills because they can digest many of the chemicals in petroleum oil. Still other bacteria can detoxify polychlorinated biphenyls (PCBs), a class of human-made pollutants that have become a worldwide environmental problem.

The generalized life cycle of bacteria is simpler than in plants.

As mentioned, bacteria reproduce mainly by a simple form of cell division: when a cell reaches a certain size, its DNA is replicated and the cell pinches in half by ingrowth of the cell membrane and cell wall, thereby forming two identical cells. The bacterial genome (the circular strand of DNA that is called a bacterial chromosome) attaches to the cell membrane, where it is replicated. Cell division usually follows quickly, and some bacteria, such as *Escherichia coli* in the human intestine, can divide every 20 minutes after replication is completed (fig. 16.3). Following division, each new bacte-

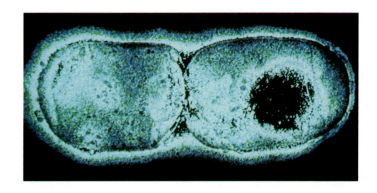

FIGURE 16.3

Electron micrograph showing division of *Escherichia coli,* a common bacterium (×75,000). In bacteria, cell division is a form of asexual reproduction because neither mitosis nor meiosis are involved.

rial cell contains an identical molecule of DNA; that is, each new bacterium has one main chromosome. Because bacteria have no nuclei and are haploid, their cell division is not called mitosis, and they cannot undergo meiosis and fertilization.

Populations of bacteria can evolve quickly. About 5,000 genes are present in the chromosomal DNA of a typical bacterium. Mutations occur randomly (approximately once in every 200,000 genes of a strain of bacteria). This means that about 1 of every 40 bacteria in a given population may spontaneously develop a mutant characteristic. Even at this relatively low rate, as many as 50 million mutant bacteria could be present among the more than 2 billion bacteria inhabiting a single gram of garden soil. Mutation in bacteria is important to the medical industry because random mutations often lead to populations of bacteria that are resistant to antibiotics and disinfectants. Such drug-resistant bacteria often spread when susceptible strains of a population are eliminated by the overuse of antibacterial agents, such as antibiotic drugs in humans and other animals. Recently, scientists have directly observed evolution using a microscope to monitor the gene pools of populations of bacteria that change repeatedly following altered test-tube environments.

Some bacteria are known to transfer DNA and genes via direct, cell-to-cell contact or bacterial conjugation. This form of "bacterial sex" can be responsible for exchange of genetic information between bacteria and can generate a higher than expected change in bacterial populations than mutation alone would explain. Plasmids, discussed next, are often involved in rapid spreading of specific genes throughout a bacterial population.

Bacteria contain DNA outside of their main genome (chromosome). These smaller pieces of circular, double-stranded DNA are called **plasmids** (fig. 16.4). Plasmids can exist and replicate separate from the main chromosome. Plasmids are usually not required for growth and development of the bacterium but often carry genes for resistance to antibiotics. For example, some disease-causing bacteria are resistant to penicillin because of a gene in a plasmid.

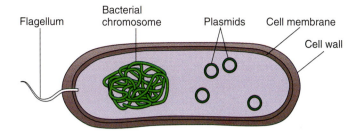

Flagellum   Bacterial
             chromosome   Plasmids   Cell membrane
                                                    Cell wall

FIGURE 16.4

A typical bacterial cell with plasmids. The tiny plasmids often contain genes that make the bacterium resistant to antibiotics. Some biologists do not consider the haploid genome of bacteria to be a chromosome because of the simple structure compared to eukaryotic chromosomes.

The classification of bacteria has recently changed.

Taxonomic disputes abound regarding the kingdom Monera (see chapter 15), which are the prokaryotes. Different biologists estimate the number of bacterial species to be anywhere from 2,500 to 100,000 or more. The most widely used manual for identifying bacteria, *Bergey's Manual of Systematic Bacteriology,* traditionally divides the bacteria into four divisions and seven classes. However, recent proposals of taxonomic arrangements may be more consistent with evolution.

According to Carl Woese at the University of Illinois, who is one of the leading proponents of using molecular characteristics for classifying prokaryotes, traditional classifications of bacteria like that in *Bergey's Manual* rely heavily on characters that are almost useless for determining phylogenetic relatedness. Nevertheless, traditional and molecular views do agree on the classification of prokaryotes into two main groups, the **Eubacteria** (*eu,* meaning "true") and the **Archaebacteria** (*archae,* meaning "ancient"; some microbiologists prefer the spelling *Archaeobacteria*). Molecular systematists often prefer to call them **Bacteria** and **Archaea,** respectively. The two taxonomic groups have traditionally been classified as subkingdoms of the Monera, although there is growing and sizable sentiment to separate the Eubacteria and Archaebacteria into their own kingdoms. Moreover, Woese, and other microbiologists, consider them to be domains—Bacteria (the Eubacteria) and Archaea (the Archaebacteria)—and classify all eukaryotes as the **Eukarya** (also spelled **Eucarya**), making three domains in nature (fig. 16.5).

Archaea differ at the molecular level from bacteria in such structures as their cell walls, membrane lipids, transfer RNA, ribosomes and RNA synthesis machinery (see table 15.4). The Archaea, or ancient bacteria, may have been the first cells to evolve on earth. Many bacteriologists suggest the ability of members of the Archaea to survive in extremely harsh environments (such as highly acidic, hot springs and thermal vents), which likely resemble conditions on a young Earth, is consistent with the origin of the earliest cells.

While Archaea live in the harshest environments—including hot springs, salty lakes, and oxygen-starved swamps—they are also abundant in normal habitats. Interestingly, the waters of Antarctica are a surprisingly rich source of Archaea. Edward F. DeLong and his colleagues at the University of California reported that Archaea constitute 34% of the prokaryotes in coastal Antarctic surface water. Archaea are not just biological oddities; rather, they represent a major component of the ocean's biota.

Bacterial-like cells were probably the first living organisms.

Fossil evidence for bacteria goes back about 3.8 billion years, which is much older than the oldest fossil evidence for eukaryotes (about 2.7 billion years). Moreover, cyanobacteria-alike probably lived at least 3.0 billion years ago, which is evidence that oxygen-producing photosynthesis might be that old. Scientists believe that before the appearance of cyanobacteria, prokaryotes lived in an atmosphere that was probably no more than 1% oxygen. Once the cyanobacteria appeared, they began to transform the air into an oxygen-rich atmosphere that may have reached as much as 30% oxygen (the present level is about 21% oxygen). This transformation set the evolutionary stage for the diversification of aerobic bacteria and the origin of oxygen-respiring eukaryotes.

Hypotheses about the origin of Archaea are essentially ideas about the origin of life, since these prokaryotes seem to be the oldest form of life that is still living. Scientists have long speculated that prebacterial cells arose by the spontaneous aggregation of simple molecules into polymers. Recent discoveries about the catalytic properties of RNA have led to the idea that "prebacteria" were probably regulated by small RNA molecules (see chapter 4). According to this hypothesis, the first Archaebacteria evolved from RNA-based ancestors—the prebacteria. RNA, along with its catalytic abilities, also can store information, so RNA is a good candidate for the key molecule in evolving early life. RNA could have catalyzed chemical reactions and carried genetic information, including the direction of synthesis of polymers. All extant cellular organisms, however, are DNA based, so molecular phylogenies comparing the RNA-based relatives of modern-day bacteria cannot be made. DNA is a better information-storage molecule than RNA, and proteins are better catalytic enzymes than RNA. Thus, as early cells evolved, RNA became most important in protein synthesis.

Recent discoveries have indicated that modern eukaryotic cells arose from prokaryotes as long as 2.7 billion years ago (see chapter 4). The most widely accepted explanation for the origin of eukaryotic organelles, and therefore eukaryotic cells, is the **endosymbiotic hypothesis** discussed in chapter 4. According to this hypothesis, chloroplasts and mitochondria and perhaps even nuclei and flagella, became organelles after smaller bacteria moved into larger bacteria.

Ancestral eukaryotic cells may have evolved from a fusion of ancient Eubacteria and Archaebacteria. This could

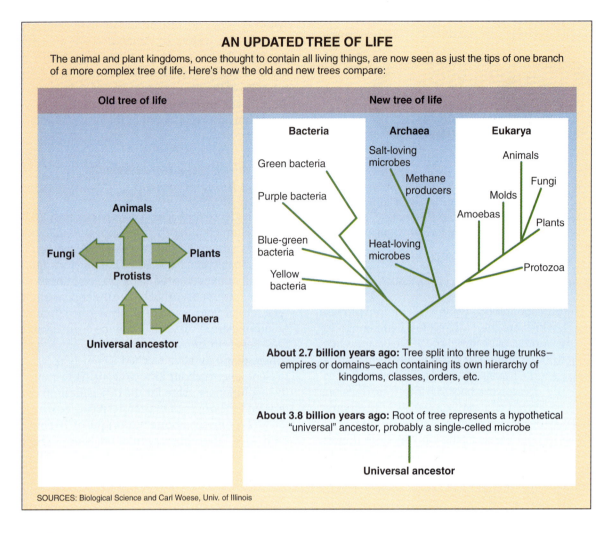

FIGURE 16.5

A universal phylogenetic tree, or the "tree of life," favored by a growing number of biologists. This classification is based primarily on analyses of the organisms' RNA.

have formed a eukaryotic cell with a symbiosis developing between the larger bacterial cell and the smaller ones. The smaller prokaryote lost its cell wall and became an organelle.

Chloroxybacteria provide evidence for the evolution of chloroplasts.

In 1976, Ralph A. Lewin of the Scripps Institute of Oceanography announced the discovery of a unicellular, prokaryotic organism living on or in small invertebrate animals called sea squirts found in the intertidal region of Baja California, Mexico. This prokaryote, although similar to cyanobacteria in structure and chemistry, possesses chlorophylls *a* and *b*, like higher plants. Lewin named the new organism *Prochloron*. In 1984, a group of Dutch scientists discovered a similar organism in shallow lakes of the Netherlands. This organism, which they named *Prochlorothrix*, also has chlorophylls *a* and *b*. *Prochloron* and *Prochlorothrix* are now classified together,

sometimes as cyanobacteria and sometimes as the only two members of their own class or own division, the *Chloroxybacteria*. The structure and functions of these organisms are similar to chloroplasts and have been suggested as additional evidence for the endosymbiont theory of organelle evolution, particularly of cyanobacteria as the most likely ancestors of chloroplasts. Recently, an endosymbiotic cyanobacterium that lives inside a small protist organism has been found and shown to perform photosynthesis while inside the host, providing more evidence for endosymbiosis.

Cyanobacteria were once called blue-green algae because they resemble algae and are photosynthetic.

Even a brief summary of examples from all groups of bacteria is beyond the scope of this book, but one group is of special interest because of its mechanisms of photosynthesis. These are the cyanobacteria, which are named for the

FIGURE 16.6

A hot spring at Yellowstone National Park is brilliantly colored by cyanobacteria. The smooth lining of the spring is a travertine layer (carbonate) precipitated by cyanobacteria that tolerate high temperatures in the spring.

blue-green pigment **phycocyanin.** Phycocyanin, along with chlorophyll *a*, gives some cyanobacteria a blue-green appearance. This, along with their photosynthetic capabilities and similar appearance to algae, led to these bacteria being called blue-green algae before the kingdom Monera was recognized (see previous discussion).

Cyanobacteria live in a wide variety of habitats, including snowfields; frozen lakes of Antarctica; extremely acidic, basic, salty, or pure (i.e., nutrient-poor) water; deserts; inside rocks; and hot springs, where water temperatures approach 75°C. In Yellowstone National Park, a different species of cyanobacterium lives in each temperature of the hot-springs range. In such habitats, the cyanobacteria precipitate chalky, carbonate deposits, which become a rocklike substance called travertine. The travertine is often marked with brilliantly colored streaks of cyanobacteria (fig. 16.6). Cyanobacteria are usually the first photosynthetic organisms to appear on bare lava after a volcanic eruption. They also live on the shells of turtles and snails, as symbionts in lichens (which are dis-

cussed later in this chapter), and in protozoans, amoebae, aquatic ferns, the roots of tropical plants, sea anemones, and a variety of other hosts. Cyanobacteria that live symbiotically in other organisms often lack a cell wall and function essentially as chloroplasts inside their hosts.

The more than 1,500 known species of cyanobacteria are diverse in form (fig. 16.7); they range from single-celled spheres to filaments and colonies. Many of the cyanobacteria have a gelatinous sheath around their cells, which protects them and binds elements required for their metabolism. In some parts of the open ocean, cyanobacteria sometimes reach densities of 10 million cells per liter. By themselves, cyanobacteria account for about 20% of the photosynthetic production of organic matter in the seas and provide food for the rest of the marine food chain.

If the substrate they grow in is low in nitrogen, most cyanobacteria produce **heterocysts,** which are larger, thicker-walled cells that can fix nitrogen (fig. 16.7*a*). Cyanobacteria may also produce thick-walled spores that

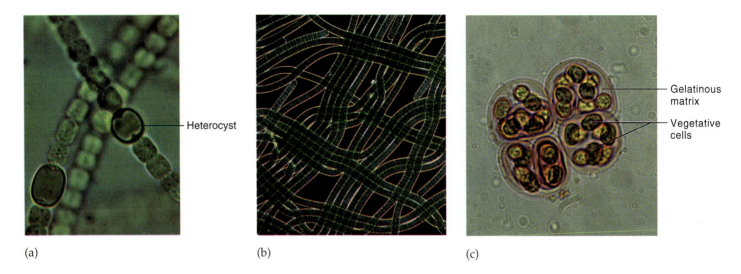

(a)   (b)   (c)

FIGURE 16.7

Light micrographs of cyanobacteria. (*a*) *Anabaena* (×850). (*b*) *Oscillatoria* (×400). (*c*) *Gleocapsa* (×300). Heterocyst cells can fix nitrogen.

resist desiccation and freezing. These spores can lie dormant during long, dry periods and then germinate to form a new chain of cells when water becomes available. This is one reason these algae come back when a dry lake is flooded.

Bacteria affect every aspect of our lives.

Bacteria cause many diseases of plants and animals, including humans, and they are the microscopic laborers in several multimillion-dollar industries. Thus, we can think of the economic importance of bacteria as both positive and negative. The following paragraphs discuss a few examples of each.

The cyanobacterium *Spirulina* is cultivated for human consumption. This organism, which is common in saline lakes, has a dry-weight protein content of about 70%. In the United States, *Spirulina* is sold as a nutritional food supplement in most health food stores, but there is essentially no scientific evidence this "pond scum" provides any unusual nutritional benefit for us.

The genus *Bacillus* is important to humans. For example, three species have been approved by the U.S. Department of Agriculture for biological control of pests. *Bacillus thuringiensis* (BT) reproduces only in the intestinal tracts of caterpillars, which are killed by a toxin from the bacterium. BT is being used to control more than 100 species of plant pests, including tomato hornworms and corn borers.

The discovery in 1994 of a mutant of *Bacillus stearothermophilus* may eventually enable us to produce ethanol from agricultural wastes such as corn shucks and potato peels as cheaply as we refine gasoline from crude oil. This could dramatically increase the proportion of plant waste converted to fuel. Several species of *Lactobacillus* are mainstays of the dairy, beverage, and baking industries.

Lactobacilli are used to make acidophilus milk, kefir, cheeses, yogurt, and related foods. These bacteria also are used in the production of wine, beer, sourdough bread, sauerkraut, and many other commercial products.

Antibiotics from actinomycete bacteria (i.e., nonphotosynthetic bacteria with branched filaments) account for about two-thirds of the more than 4,000 antibiotics that have been discovered. *Actinomycetes* was the original source of such well-known antibiotics as tetracycline, neomycin, aureomycin, erythromycin, and streptomycin. Streptomycin also is produced by *Streptomyces,* which is common in soil. The various species of *Streptomyces* play a significant role in breaking down and recycling plant and animal products such as lignin, latex, and chitin. These bacteria also produce the characteristic and often pleasant odor of damp soil.

Bacteria can also cause damage and be harmful. In water supplies, cyanobacteria frequently clog filters, corrode steel and concrete, soften water, and produce undesirable odors or coloration. Cyanobacteria also produce toxins. Fish that eat the bacteria but are immune to the toxins can in turn become toxic to their predators. In summer months, various cyanobacteria and algae often become abundant in bodies of freshwater, especially if the water is polluted. When this happens, a large floating mat called a **bloom** may form and cause numerous environmental problems.

Some bacilli cause diseases such as, diphtheria, periodontal disease in humans, and anthrax in cattle. The most dangerous of the bacilluslike bacteria may be the *clostridia:* a species of *Clostridium* that causes tetanus, gas gangrene, and botulism (see Perspective 16.1, "Deadly Food! Beware!"). Recently, other bacteria have been implicated in ulcers and in heart disease.

16.1 *Perspective* Perspective

Deadly Food! Beware!

The bacterium *Clostridium botulinum* can grow in foods and produces a toxin called botulinum, the most toxic substance known. Microbiologists estimate that 1 gram of this toxin can kill 14 million adults! The good news is that *C. botulinum* requires anaerobic conditions for growth, which limits its prevalence. The bad news is that *C. botulinum* is extremely tolerant to stress; it can withstand boiling water (100°C) for short periods but is killed at 120°C in 5 minutes. This tolerance makes *C. botulinum* a serious concern for people who can vegetables from their garden or the market. If home canning is not done properly, *C. botulinum* will grow in the anaerobic conditions of the sealed container and be extremely poisonous. Several adults and infants die every year from botulism in the United States.

Tolerance to stress is enhanced in *C. botulinum* and many other bacteria by the formation of thick-walled spores. These highly resistant spores may later germinate and grow after decades or even centuries of inactivity. The spores of *C. botulinum* can germinate in poorly prepared canned goods, so never eat food from a swollen (gas-filled) can of food; you risk contracting botulism, which can lead to nerve paralysis, severe vomiting, or death.

INQUIRY SUMMARY

Bacteria reproduce only asexually. New genotypes can arise by high rates of mutation. The prokaryotes have recently been classified into the Eubacteria and Archaebacteria. Some microbiologists have suggested there are three domains of life: Bacteria, Archaea, and Eukarya. The Archaebacteria may have been the first cells to evolve on earth and are much older than the eukaryotes. After prokaryotic cells were established, eukaryotic cells may have evolved by endosymbiosis, which formed organelles.

Bacteria are prokaryotes with diverse metabolism and some unique structures. Most are one celled, and either spherical, rod shaped, or corkscrew shaped. Bacteria can be useful to humans and important in various ecosystems. Bacteria cause diseases in various organisms, including us.

Can you suggest an ecological role for plasmids in the evolution of bacteria?

Fungi Are Nonphotosynthetic Eukaryotes with Diverse Roles in Ecosystems.

Fungi are eukaryotes that occupy an odd place in our perspective on living organisms. We smile at the aroma of yeast in fresh-baked bread, yet we are irritated by the itching of athlete's foot and disgusted by the welts of a ringworm infection. Rusts and smuts ruin some of our valuable crops, but we rarely consider the vital role of fungi as decomposers, allowing new plants to replace the old. Probably most confusing of all is that some fungi look like plants, while others look more like used motor oil.

All fungi are eukaryotic, and most are nonmotile, spore-producing filaments. Cell walls surround fungal cells during some or most stages in their life cycle, a characteristic that was partly responsible for their original classification in the plant kingdom. Fungal cell walls are characteristically made of chitin combined with other complex carbohydrates (chitin is also the main component of the exoskeletons of insects, spiders, and crustaceans). While plants are autotrophic, fungi are heterotrophic. Fungi lack chlorophyll and therefore photosynthesis, a basic characteristic of plants. Unlike plants, fungi have only one diploid cell in their life cycle. They also store glycogen as an energy and carbon reserve. What compound serves this role in plants?

The main vegetative feature of most fungi is that they grow as a tubular, threadlike, whitish or colorless filament called a **hypha** (plural, hyphae). Hyphae, which usually vary in thickness between 0.5 and 100 micrometers, grow only at their tips and can grow indefinitely in favorable conditions. The hyphae of most fungi are divided by crosswalls. Hyphae lacking crosswalls are coenocytic and multinucleate—meaning many nuclei within a common cytoplasm comprising the filament. The hyphae of most fungi branch repeatedly, intertwine, or fuse with other hyphae, forming a mass known as a **mycelium** (plural, mycelia). Mycelia can be compact and may take on the form of parenchymalike tissue (chapter 3), in which individual hyphae are indistinguishable from one another. Such compact mycelia form mushrooms and similar spore-bearing structures that have traditionally been referred to as fruiting bodies (fig. 16.8). When growth is undisturbed, a mycelium tends to grow more or less equally in all directions from its point of origin, sometimes for hundreds of years. Recent estimates of single-mycelium mats in the states of Washington and Michigan, each of whose mycologists claim world records for the largest fungus, put them at several metric tons for a single mycelium.

Fungi absorb nutrients through the cell wall and cell membrane. To do this, they secrete digestive enzymes into their immediate environment and absorb the liquified nutrients

The Lore of Plants

Biologists agree that fungi are not plants and now separate these two groups into their own kingdoms. Indeed, recent molecular analyses suggest that fungi are more closely related to animals than to plants. Nevertheless, confusion about the evolution and taxonomy of fungi and plants has a long tradition. In fact, we have included a section on fungi in this book primarily because of this long tradition and because many botany courses continue to study fungi.

In what ways are fungi different from, and similar to, plants?

(a)   (b)   (c)

(d)   (e)   (f)

FIGURE 16.8

The diversity of fruiting bodies in fungi. (*a*) Oyster mushrooms (*Pleurotus ostreatus*). (*b*) A jelly fungus (*Tremella mesenterica*). (*c*) A bird's-nest fungus (*Cyathus striatus*). (*d*) A shelf, or bracket, fungus (*Grifola sulphurea*). (*e*) Puffballs (*Lycoperdon* species). Note the spores being released. (*f*) A common stinkhorn fungus (*Phallus impudicus*). Fruiting bodies produce spores and are quite different from the fruits of flowering plants (chapter 18).

produced by the secreted enzymes. Thus, fungi are particularly important decomposers, aiding the breakdown of dead matter and the recycling of inorganic as well as organic molecules in an ecosystem. Many species of fungi are either decomposers or symbionts (i.e., live with other organisms). As symbionts, they may be parasitic and harm the host, they may benefit their host, or they may cause no mutual harm or benefit.

Like bacteria, fungi grow just about everywhere.

The distribution and ecology of fungi resemble those of bacteria. Fungi and bacteria both decompose and recycle plant and animal remains, and they have probably done so for millions of years. Fungi attack virtually all organic materials. They also can etch the lenses of cameras, tele-

16.2 *Perspective* Perspective

Disappearing Fungi

European gourmets with a taste for the subtle flavors of fresh wild mushrooms are discovering that these delicacies are increasingly harder to find. A few years ago, it was easy to pick a basket of the most prized fungus of all, the apricot-scented chanterelle. However, not only are these mushrooms becoming scarce, they are also getting smaller: it takes 50-times as many chanterelles to make up a kilogram than it did in 1958. Other fungi also are becoming rare. For example, the average number of fungal species in Holland has dropped from 37 to 12 per 1,000 meter^2. These and many similar observations by other mycologists suggest that a mass extinction of mushrooms is happening all over Europe. But why?

Ecologists have recently noticed a negative correlation between the abundance and diversity of fungi and the amount of pollution. This negative correlation is stronger for fungi than for plants or other kinds of organisms. Apparently, fungi are more sensitive to air pollution than are plants because fungi have no protective covering, whereas the aerial parts of plants are protected by cuticles and bark. This distinction is a functional one: some fungi absorb water directly from air, along with whatever else is in the air, but

A prized chanterelle mushroom.

plants get water from soil through their roots. It seems, therefore, that fungi are being driven to extinction by bad air.*

*Colonies of luminescent fungi are sometimes maintained aboard spaceflights. Sensitive to escaped fuel or other noxious fumes, the fungi stop luminescing when exposed to as little as 0.02 parts per million of fuel. Thus, the fungi serve as an early warning system for noxious fumes, much as the death of canaries warned miners of the lack of oxygen or the presence of dangerous gases such as methane.

---

scopes, and binoculars. In lichens (discussed later), fungi can even degrade rocks.

Most diseases of living plants are caused by fungi. For example, diseases called rusts, smuts, and other delightful names frequently attack crops and stored foods, causing millions of dollars of losses annually. Fungi also cause several diseases in humans, especially diseases of the skin and lungs. Although most fungi are not poisonous, the few exceptions include mushrooms that are relatively common and widespread. Consumption of such fungi has often proven fatal; no effective antidote for human poisoning by mushrooms has been found.

Nearly 70,000 species of fungi are known, and descriptions of more than 1,000 new species are published each year. Moreover, biologists suspect that more than half of all existing fungi have yet to be described. Mycologists also suspect that unknown numbers of undiscovered species have already become extinct, due primarily to human activities such as the clearing of tropical rain

forests and other natural habitats. In addition, many fungi are disappearing in industrialized countries because of pollution (see Perspective 16.2, "Disappearing Fungi"). Their disappearance is important because many fungi have provided us with sources of medicines and food.

Fungi are traditionally classified into five main groups: Zygomycota, Ascomycota, Basidiomycota, Chytridiomycota and Deuteromycetes.

Fungi may be classified into four phyla. Three are known to reproduce sexually: the **Zygomycota** (zygomycetes), the **Ascomycota** (ascomycetes), and the **Basidiomycota** (basidiomycetes). The fourth phyla, the **Chytridiomycota** (chytrids), apparently cannot reproduce sexually (that is, no sexual spore has been found). These are the four phyla of fungi recognized by most mycologists today, but the chytrids have only recently been included in the fungi. **Deuteromycetes,** the conidial fungi, (sometimes named fungi imperfecti because sexual

The Lore of Fungi

Ants have grown crops of fungi for about 50 million years. When the ants moved into new areas, the fungus was brought by the queen ant. The ants then used this "starter packet" to clone a new crop, thereby explaining why many species of ants are nourished by the same fungus as were their ancestors millions of years ago. The dependence of the ants and fungus is mutual: the ants can't survive without the fungus, and the fungus can't propagate without the ants.

reproduction remains unknown) are often put in their own artificial conglomeration with over 15,000 types of fungi that may only reproduce asexually. Most deuteromycetes, however, are believed to be ascomycetes but remain in the fungi imperfecti for various reasons. The taxonomy of fungi remains in flux, but the distinguishing features of each of these groups primarily involve differences in reproduction (table 16.1).

More than 750 species of zygomycetes are known, examples are the hat-throwing fungi (*Pilobolus*), which grow on horse dung; the fly fungi (*Entomophthora*), which are most commonly seen growing from dead flies on window panes; and most of the fungi that are symbiotic with plant roots. One of the best known and most widespread of the zygomycetes is *Rhizopus stolonifera,* a common bread mold that is also the chief cause of "leak," a disease of strawberries that appears while they are being transported to market.

There are about 30,000 known species of ascomycetes, the most famous of which are brewer's or baker's yeast, bread mold, truffles, and morels (fig. 16.9). Truffles and morels are gourmet delicacies, primarily in North America and Europe. Yeast is probably the most commercially important ascomycete; bread mold is famous primarily as a research organism. People have used yeast to make beer since the fourth millennium B.C. Ascomycetes are also of interest because they cause plant diseases such as chestnut blight, Dutch elm disease, apple scab rot, apple bitter rot, stem rot of strawberries, powdery mildew, and brown rot of peaches, plums, and apricots. Ascomycetes are the most common type of fungi that occur in lichens (see next section).

Basidiomycetes, like the ascomycetes, are filamentous, with single-nuclei or multinuclei cells forming hyphae. Basidiomycetes form spores on structures called basidia. The fruiting body of basidiomycetes is best known in the form of mushrooms, bracket fungi, jelly fungi, puffballs, earth stars, coral fungi, bird's-nest fungi, and stinkhorns (see fig. 16.8). Basidiomycetes also include rusts and smuts, which cause plant diseases.

Deuteromycetes are mostly free-living and terrestrial, but some are pathogenic. The best known of the pathogenic deuteromycetes include the causal agents of a respiratory disease called aspergillosus, athlete's foot, ringworm, and candida "yeast" infections. The most famous deuteromycetes are species in the genus *Penicillium:* *P. notatum* for its role in the discovery of penicillin, *P. chrysogenum* for the commercial production of penicillin,

P. griseofulvum for the production of griseofulvin (the only effective antibiotic against ringworm and athlete's foot), and *P. roquefortii* and *P. camembertii,* which are used to make Roquefort and Camembert cheeses, respectively.

Because their asexual reproduction usually resembles that of the ascomycetes, most of the deuteromycetes probably descended from an ascomycete ancestor that lost the ability to reproduce sexually. This suggestion is supported by the observation that whenever sexual reproduction is discovered in a deuteromycete, it is usually of the ascomycete type.

Lichens are fungi and photosynthetic organisms living together.

Lichens (fig. 16.10) are symbiotic relationships consisting of a fungus and a green alga, a fungus and a cyanobacterium, or a fungus with both, in a body called a **thallus** (plural, thalli). Within this symbiosis, where perhaps both organisms may benefit, the fungus gets carbohydrates from both the algae and cyanobacteria and fixed nitrogen from the cyanobacteria; the photosynthetic organisms apparently receive nutrients and a place for protected growth within the body of the surrounding fungus. Recently, lichenologists have suggested the relationship is more a parasitism than a mutualism.

Lichen thalli grow slowly, sometimes up to 1 centimeter or as little as 0.1 millimeter per year. On the basis of such growth rates, lichenologists estimate that some lichens have lived at least 4,500 years. Furthermore, lichens tolerate environmental conditions that are too extreme for most other forms of life. Indeed, lichens live on bare rocks in the blazing sun or bitter cold in deserts, in both arctic and antarctic regions, on trees, and just below the permanent snow line of high mountains where nothing else will grow. One species grows completely submerged on ocean rocks. Some fungi even attach to artificial substances such as glass, concrete, and asbestos.

Although lichens can withstand many environmental extremes, they are, like other fungi, sensitive to air pollution. Indeed, lichens have been considered to be effective monitors of air pollution: lichens in or near large cities (and around polluting industries such as power plants) throughout the world disappeared during the twentieth century. Similarly, they have recolonized areas in which air quality has improved.

TABLE 16.1

Selected Characteristics of the Groups of Fungi

GROUP	CHARACTERISTICS
Zygomycota	The name zygomycetes refers to the thick-walled, sometimes elaborately ornamented spore that zygomycetes form during sexual reproduction.
Ascomycota	Asco means "sac," which refers to the saclike filamentous structures where spores form in ascomycetes. The spore sacs are called asci (singular, ascus).
Basidiomycota	Like ascomycetes, these fungi are filamentous with uninucleate or multinucleate hyphae. Basidiomycetes are so named because they form spores on structures called basidia (singular, basidium), which are hyphal swellings that bear spores on tiny pegs.
Chytridiomycota	The chytrids, which are varied in form, are predominately aquatic distinguished by motile cells (spores and gametes), usually propelled by a single, whip-like flagellum. Some are plant pathogens and cause major diseases of crops.
Deuteromycetes	The "fungi imperfecti" or "imperfect fungi" are defined by the absence of sexual reproduction (the sexual stage remains unknown, if it exists). They reproduce almost exclusively by conidia.

Slime molds and water molds were classified as fungi by early botanists. These organisms have funguslike, animal-like, and plant-like features, but they are now classified in kingdom Protista. Multinucleate hyphae lack cross walls and are coenocytic.

(a)

(b)

FIGURE 16.9

Fruiting bodies of ascomycetes. (*a*) Truffles. (*b*) Morel.

The Lore of Fungi

The stinkhorn fungus *Dictyophora* is one of the world's fastest-growing organisms: it pushes out of the ground at a rate of about 0.5 centimeters per minute. The growth is so fast that a crackling can be heard as the fungus swells and stretches. During growth, a delicate, netlike veil forms around the fungus (this is the basis for the other common name of the fungus, "the lady of the veil"). The fungus then decomposes and, in the process, produces a strong odor of decaying flesh. This odor attracts flies, which crawl over the fungus and collect spores on their feet, thereby ensuring that the spores are carried to new areas.

FIGURE 16.10
Lichens are able to colonize on rocks.

Some fungi are important to plant growth.

The ecology of fungi is especially important for plants. For example, tiny orchid seeds cannot germinate until they are invaded by hyphae of the soil fungus *Rhizoctonia,* and plants of all kinds are healthier when their roots associate with soil fungi. Members of all 400 or so families of flowering plants, with the exception of possibly fewer than a dozen, form **mycorrhizal associations** (see chapter 7). A **mycorrhiza,** meaning "fungus root," is an association between a fungus and the underground parts of a plant. The association is mutually beneficial: the plants provide a source of carbon used by the fungus, and the fungus absorbs phosphorus or other minerals the plant might not otherwise obtain from the soil.

Mycorrhizae are divided into two groups, depending on whether the fungal component penetrates the plant (endomycorrhizae) or forms only an external mantle around the plant's roots (ectomycorrhizae) (fig. 16.11). Endomycorrhizae occur in about 80% of all vascular plants, usually forming balloonlike structures (vesicles) or treelike structures (arbuscles). Endomycorrhizae are especially important to tropical plants because the fungi can help plants obtain phosphorus from the phosphate-poor soils typical of tropical habitats. Ectomycorrhizal fungi, which are primarily associated with the roots of trees and shrubs in temperate regions, apparently replace root hairs, which are often absent in ectomycorrhizal roots.

Fungi have considerable economic and ecological importance.

Fungi have greatly benefited human societies as sources of industrial chemicals such as antibiotics, medicines, and vitamins. They are the mainstay of the brewing and baking industries and are also important for making certain dairy foods, including gourmet cheeses. Fungi also cause many plant and animal diseases.

Fungi produce gallic acid, which is used in photographic developers, dyes, and indelible black ink and in the production of artificial flavoring and perfumes, chlorine, alcohols, and several acids. Fungi also are used to make plastics, toothpaste, and soap and in the silvering of mirrors. In Japan, almost 500,000 metric tons of fungus-fermented soybean curd (tofu, miso) are consumed annually.

Many fungi are important ecologically: together with some bacteria, several types of fungi are the main decomposers in all known ecosystems. Decomposers are just as critical to the existence of our world as producers because decomposers break down dead organisms and release nutrients and minerals to the soil and CO_2 to the air.

Different strains of the rust fungus (*Puccinia graminis*) cause billions of dollars of damage annually to food and timber crops throughout the world. Plant breeders are constantly faced with the challenge of developing rust-resistant varieties of crops. As a result of this effort, for example, Donald Winkelmann of the International Maize and Wheat Improvement Center in Mexico City recently announced that the defeat of wheat rust is imminent. Scientists at the center have found a Brazilian-grown wheat plant that somehow controls the disease by slowing its growth. The "slow-rust" genes of this plant have been incorporated into cultivated wheat by hybridization experiments, and the new rust-resistant wheat is now being grown in more than 100 countries.

Another plant disease that has had a significant impact on human society is caused by the ergot fungus,

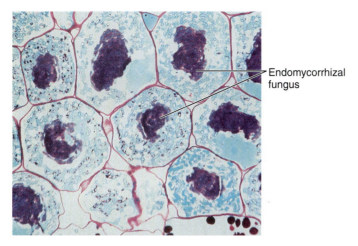

(a)

Endomycorrhizal fungus

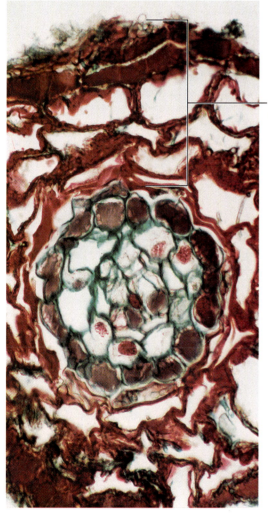

(b)

Ectomycorrhiza fungus

FIGURE 16.11

Mycorrhizae. (*a*) Light micrograph of V A endomycorrhizal fungus in plant root cells (×400). (*b*) Transverse section of a plant root surrounded by ectomycorrhizal fungus. What is the ecological role of mycorrhizae?

Claviceps purpurea. This fungus infects rye and other grain crops. Ergot seldom damages a crop significantly, but it produces several powerful drugs in the maturing grain. If infected rye is harvested and milled, a disease known as ergotism may occur in those who eat the bread made from the contaminated flour. The disease can affect the central nervous system, often causing hysteria, convulsions, and even death. Another form of ergotism causes gangrene of the limbs, and cattle that eat infected grass often abort their calves. Regardless, ergot-derived drugs have been used since the sixteenth century to hasten childbirth by stimulating uterine contraction. Because other ergot drugs constrict blood vessels, they are used to stop the bleeding often associated with childbirth. Ergot drugs also have been used to treat migraine headaches, heart palpitations, nervous stomach, menopausal disorders, and several other medicinal problems.

Fungal infections cause many serious human diseases, such as thrush and pneumonia, while other infections are less serious, such as ring worm and athletes foot. Thus many fungi are pathogens, and attack or decompose living, rather than dead, organisms.

INQUIRY SUMMARY

Fungi are heterotrophs, they lack chlorophyll, and their cell walls usually contain chitin. The main body of a fungus is a hypha or an aggregate of hyphae called a mycelium. Fungi are decomposers or symbionts. Fungi are classified into these four phyla: zygomycota, ascomycota, chytridiomycota and basidiomycota. A fifth group, deuteromycetes, does not reproduce sexually. Lichens are symbiotic associations of fungi and photosynthetic organisms, including algae and bacteria. Ecologically, the most significant association of fungi with plants is the formation of mycorrhizae. The fungal symbiont in a mycorrhiza gets carbohydrates from the plant host, and the plant gets minerals from the fungus. Most plants depend on this association. Fungi provide us with useful commodities but also cause considerable damage to crops.

What future benefits to us might be lost because of the disappearance of fungi?

Algae Include Some of the Smallest and Largest Photosynthetic Eukaryotes.

Although algae live in a diversity of habitats (fig. 16.12), most are aquatic and live in water. The great diversity of algae has led to many conflicting ideas about how they should be classified, but three main groups (with up

The Lore of Fungi

Flyfishers must keep their artificial dry fly sufficiently free of water for the imitation fly to float. Long ago flyfishers reported that a fungus called *Amadou* was excellent for drying a fly because of its superior ability to absorb water. The angler simply wraps the waterlogged fly in a wad of *Amadou* and then squeezes the wad a little, and the fungus soaks up the water. This dries the imitation fly, leaving it ready to float on the surface of a lake or stream. Whether the *Amadou* is better than a paper towel for this purpose is unknown, but flyfishers have a unique folklore about this fungus.

(a)

(b)

FIGURE 16.12
Diverse habitats of algae. (*a*) *Pleurococcus* growing on a yew tree and stone. (*b*) Algae growing on the fur of a three-toed sloth in Costa Rica.

to seven divisions within these groups) are consistently recognized: the **green algae,** the **brown algae,** and the **red algae.** Many, but not all, biologists include algae with other simple organisms in the kingdom Protista, a "catch-all" group of organisms that usually just do not fit well into the other large kingdoms (evolution is not a tidy process!).

Thus, protists are eukaryotic organisms not considered to be fungi, plants, or animals. Algae, unlike plants, do not form embryos that are protected in a sterile jacket of cells called an ovary (see chapters 12, 17, 18).

The unifying feature of algae that traditionally led botanists to classify them with plants is their ability to

TABLE 16.2	Comparison of the Main Features of Algal Divisions*				
DIVISION	HABITAT	PHOTOSYNTHETIC PIGMENTS	CELL-WALL COMPONENTS	CARBOHYDRATE STORAGE	FLAGELLA
Chlorophyta (green algae)	mostly freshwater, some marine, terrestrial, or airborne	chlorophylls *a* and *b*, carotenoids	polysaccharides, including cellulose	starch	none, 1–8, or dozens; whiplash
Phaeophyta (brown algae)	almost all marine, rarely freshwater	chlorophylls *a* and *c*, fucoxanthin and other carotenoids	cellulose, alginic acid, and sulfated polysaccharides	laminarin, mannitol	2, lateral; forward tinsel, behind whiplash
Rhodophyta (red algae)	mostly marine, some freshwater	chlorophyll *a*, carotenoids, phycobilins	cellulose, pectin, calcium salts	floridean starch	none
Chrysophyta (diatoms, yellow-green and golden-brown algae)	marine and freshwater, some terrestrial or airborne	chlorophylls *a* and *c*, fucoxanthin and other carotenoids	cellulose wall or silica shell; sometimes absent	chrysolaminarin	none, 1, or 2; whiplash or tinsel
Euglenophyta (euglenoids)	marine or freshwater, some airborne	chlorophylls *a* and *b*, carotenoids	absent	paramylon	1–3; tinsel
Pyrrhophyta (dinoflagellates)	marine and freshwater, some airborne	chlorophylls *a* and *c*, peridinin and other carotenoids	armorlike plates that may be cellulosic	starch	none or 2; tinsel
Cryptophyta (cryptomonads)	mostly freshwater, some marine	chlorophylls *a* and *c*, carotenoids, phycobilins	absent	starch	2; tinsel

*The 7-division system represents the simplest traditional classification of algae still in use today; some treatments divide the algae into as many as 15 divisions. *Carbohydrate storage*—laminarin is a unique polymer of glucose stored in the vacuoles of brown algae; mannitol is a sweet, sugar-alcohol compound often used as a dietary supplement; floridean starch resembles glycogen and is stored in the cytoplasm; chrysolaminarin is a water-soluble polysaccharide stored in vacuoles; paramylon is a unique polysaccharide resembling starch. *Flagella* are thin (about 0.2 micrometers across), long (2 to 150 micrometers), threadlike organelles used for movement and feeding—tinsel flagella bear minute lateral hair-like threads and resemble a thin, ostrich feather; whiplash flagella appear smooth in an electron micrograph as they lack these hair-like threads (chapter 4).

photosynthesize. Certain groups of algae also store starch for energy storage and contain cellulose in their cell walls. Nonetheless, algae are an informally defined group of eukaryotes. The names and main features of the divisions of algae are given in table 16.2. At a recent meeting of the International Botanical Congress it was decided, after much debate, that division, not phylum, should remain the correct designation despite the preference for phylum by some biologists who study algae or plants. Either term is considered acceptable (see chapter 15).

Although the simplest algae are unicellular (fig. 16.13*a*), the most complex algae just might rival the giant redwoods as the largest of all photosynthetic organisms. Most algae are somewhere between these two extremes in size. Colonial algae are those with groups of cells that are loosely attached to each other and sometimes surrounded by a slimy sheath. Filamentous algae are either branched or unbranched and have either one or several nuclei per cell. Some filamentous algae are termed coenocytic because they have no cross walls. Thus they form a multi-nucleate cytoplasmic mass enclosed by a single cell wall.

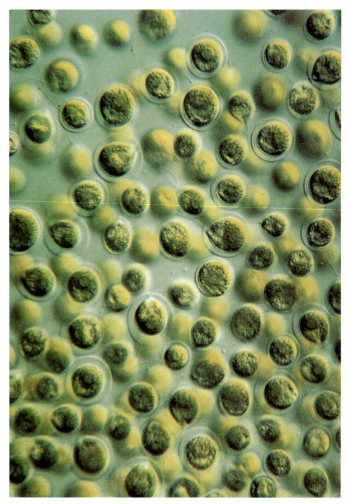

(a)

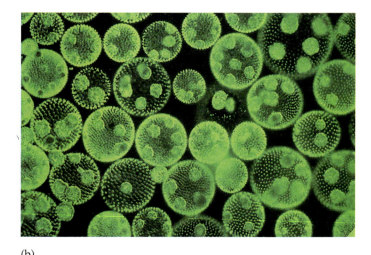

(b)

(c)

FIGURE 16.13

Green algae vary greatly in size and shape. (a) *Chlamydomonas,* is a single-celled green alga that is less than 100 micrometers long (×150). Compare the microscopic size of *Chlamydomonas* with the kelp shown in the opening photograph of this chapter. (b) *Volvox* is a colonial green alga (×50). (c) *Cladophora,* a green algae, forms branched filaments that consist of multinucleate cells (×100).

Green algae are the ancestors of plants.

The division Chlorophyta, or green algae, includes about 7,500 species and is more diverse in vegetative organization, life cycles, and habitats than any other group of algae. Most green algae live in freshwater, but different species also occur in marine habitats, clouds, snowbanks, or soil or on the shady moist sides of trees, buildings, and fences. Green algae also live symbiotically with several different kinds of animals and with the fungi of lichens. Green algae and plants share many important characteristics that support the hypothesis that plants arose from a green algalike ancestor. For example, the zygotes of both are retained in the gametophyte, and both make pigments called flavonoids. It now seems apparent the green algae, which are the most plantlike of the algae groups, evolved into the lower plants discussed in the next chapter.

The most well known example of a motile green alga is *Chlamydomonas* (fig. 16.13a). *Chlamydomonas* swims by means of two flagella. The dominant feature of each cell is a single, large chloroplast. Most cells in a population of *Chlamydomonas* are products of asexual reproduction by mitosis and cell division. Sexual reproduction involves meiosis and produces haploid zoospores (fig. 16.14). Environmental conditions such as nitrogen availability or day length influence reproduction of this green alga.

In dry conditions, *Chlamydomonas* retracts its flagella and becomes dormant. When water becomes available again, *Chlamydomonas* regrows its flagella, enlarges, and reproduces. After a rainfall, *Chlamydomonas* cells that have been dormant grow quickly in puddles and drainage ditches. After they grow, the cells live independently most of their life. An example of colonial green algae, *Volvox,* is shown in figure 16.13b.

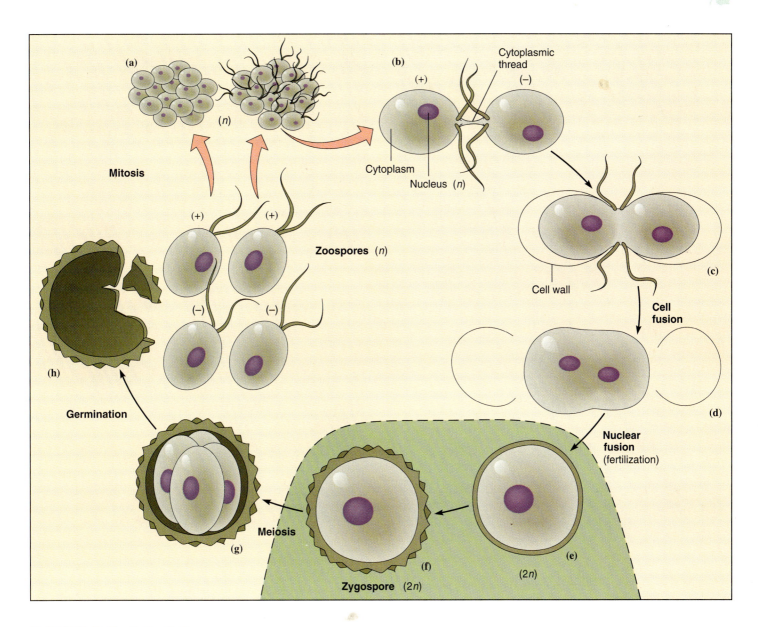

FIGURE 16.14

Stages in sexual reproduction in *Chlamydomonas*. (*a*) Cells from compatible mating strains aggregate into clumps. (*b*) A cytoplasmic thread forms between cells of opposite mating strains, indicated by "+" and "−". (*c–d*) Cytoplasm of both cells fuses; the old cell walls are discarded. (*e*) Nuclei fuse (fertilization occurs), and a new cell wall forms around the zygote. (*f*) The cell wall surrounding the zygote thickens and becomes spiny, becoming a zygospore. (*g*) Four haploid cells are produced by meiosis and cytokinesis of the zygote. (*h*) Two of the new cells after meiosis are of one mating strain, and two are of the other mating strain.

Filamentous green algae are microscopic and are either branched or unbranched (see figs. 16.13c and 16.15). Filamentous green algae often grow on aquatic flowering plants; they also attach to rocks or other objects underwater. Some filaments are free-floating. One of the most common representatives of filamentous green algae is *Spirogyra*, which is named for the spiral chloroplasts in each cell (fig. 16.16). Species of this genus grow as frothy or slimy green masses of unbranched filaments that float in the water of small ponds and lakes. Flagellated cells are absent in all species of *Spirogyra*, and asexual reproduction is restricted to fragments of filaments that form new filaments.

Sexual reproduction in *Spirogyra* is called **conjugation.** During conjugation, filaments growing side-by-side in a dense mass form small protuberances that grow toward each other. Upon contact, the walls create a tube. The contents of the cell of one filament then move through the tube, and the cytoplasm and nuclei of both cells fuse. The diploid nucleus in each cell then undergoes meiosis, producing the haploid condition of a gametophyte.

Figure 16.7a shows the filamentous green algae, *Chara*, a stonewort. Note the branched, leaf-like structures of the filaments are whorled, as in some plants. In appearance, stoneworts are the most complex and plantlike of the

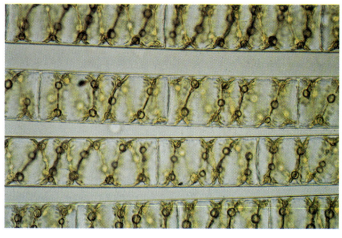

(a)

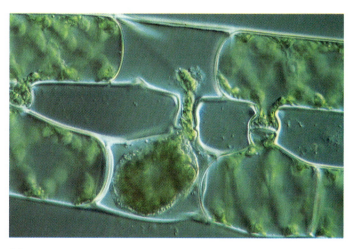

(b)

F I G U R E   1 6 . 1 5

Light micrograph of an unbranched filament of the green algae *Oedogonium* (×780). The enlarged cell is an oogonium where the egg forms.

F I G U R E   1 6 . 1 6

Spirogyra, a green algae. (*a*) Light micrograph of *Spirogyra* filaments showing spiral chloroplasts (×150). (*b*) Phase-contrast light micrograph of *Spirogyra* filaments in a late, sexual-reproduction stage of conjugation (×250).

green algae. Also, stoneworts have unique sexual organs among the algae, as the antheridia and oogonia are multicellular and surrounded by a layer of sterile (infertile) cells. If *Chara* is ancestral to plants, the sterile jacket of cells surrounding the reproductive organs is a shared characteristic. Stoneworts, like plants, produce flavonoid pigments. No other algae is known to do this.

Most brown algae are large marine organisms.

Almost all of the approximately 1,500 species of brown algae are marine organisms, but a few species live in freshwater. The division Phaeophyta—the brown algae—includes a wide range of morphological types, from the most complex of all the algae, the kelps (see the opening photograph of this chapter), to microscopic, branched filaments. There are no unicellular, colonial, or unbranched filamentous organisms among the Phaeophyta. The dominant

form of large brown algae is a thallus—a simple, largely undifferentiated vegetative body.

Ectocarpus, perhaps the best known of the filamentous brown algae, grows on rocks or on larger marine algae along ocean shores worldwide. Kelps and rockweeds dominate shorelines and nearby offshore habitats in cool climates worldwide. Some kelps are free-floating; they can grow rapidly by vegetative propagation into massive populations in the open ocean. Kelps and rockweeds are of interest to biologists because of their vegetative organization and the distinctiveness of their life cycle in comparison with those of green algae, which are predominantly haploid.

The life cycle of kelps is dominated by a large sporophyte (chapter 13). The huge sporophytes of giant kelp (*Macrocystis*), for example, consist of stipes, blades, and branching holdfasts that anchor them to the ocean floor (fig. 16.17*b*). The leafy blades

(a)

(b)

FIGURE 16.17

(a) *Chara* is a green alga, or stonewort, that has branched filaments with whorls of leaf-like structures. The small, yellow-green or brownish globes are the unique antheridia and oogonia. (b) Giant kelp (*Macrocystis*) is a structurally complex brown alga with a stalk-like, support trunk (stipe), leaf-like, photosynthetic blades, and large, round, air bladders. Giant kelps are anchored to the ocean floor by root-like structures called holdfasts. What might be the selective advantage of air bladders?

often have air bladders that keep the kelps afloat. In the rockweed genus, *Fucus,* the fertile swollen tips of branches bear specialized oogonia and antheridia that produce haploid egg and sperm respectively (fig. 16.18). These structures, common to algae, are not found in bacteria or fungi. *Fucus* is a common brown alga found near the beaches of southern California.

Red algae are also mostly marine organisms.

Like the brown algae, the red algae—division Rhodophyta—are mostly marine organisms that are either microscopic filaments or macroscopic leafy branches. One group, the coralline algae, have cell walls impregnated with calcium carbonate and magnesium carbonate, which makes the walls hard and crusty. Coralline algae are an important part of coral reefs.

The most significant features of the approximately 3,900 species of red algae are their phycobilins (pigments). Red algae are red because they have an abundance of **phycoerythrin,** a red phycobilin. Phycoerythrin absorbs blue light, which penetrates more deeply into water than do other colors of light. This means that red algae can photosynthesize at greater depths than other algae and explains why some red algae can grow at depths of more than 200 meters. Can you explain how this might be an evolutionary advantage?

Diatoms are important in ocean ecosystems.

The Chrysophyta form the largest division of algae, with more than 11,000 species. About 10,000 species are **diatoms,** almost all of which are unicellular and either bilaterally symmetrical or radially symmetrical (fig. 16.19). Diatoms lack flagella, except for some male gametes, but some diatoms can be motile (move) by secreting a sticky substance that sticks to other objects, enabling the diatom to pull itself toward them. Diatoms are unique because of their exquisitely ornamented glass shells. Chrysophytes occur in both freshwater and salt water and are important primary producers in the food chain of many aquatic ecosystems, especially the oceans. Filter feeders, such as some whales, depend on diatoms, and on tiny organisms that feed on diatoms, for food.

Reproduction has been more thoroughly studied in diatoms, probably because of their large number and their great economic importance (see the last section in this chapter). During asexual reproduction, diatoms divide mitotically while they are still encased in their rigid glass walls. The walls consist of two parts, called valves, one of which fits tightly over the other like the lid of a petri dish. The new cells inside the rigid wall expand and force the valves to separate. New inner valves form inside the old valves so that one of the new cells is smaller than the parent cell and the other new cell is the same size. Division of the smaller cell again produces two new cells, one of which will be even smaller. Thus, in a population of diatoms, some cells are about half the size of others. When the lowest size limit is reached, the smallest cells become sexual and produce new large cells.

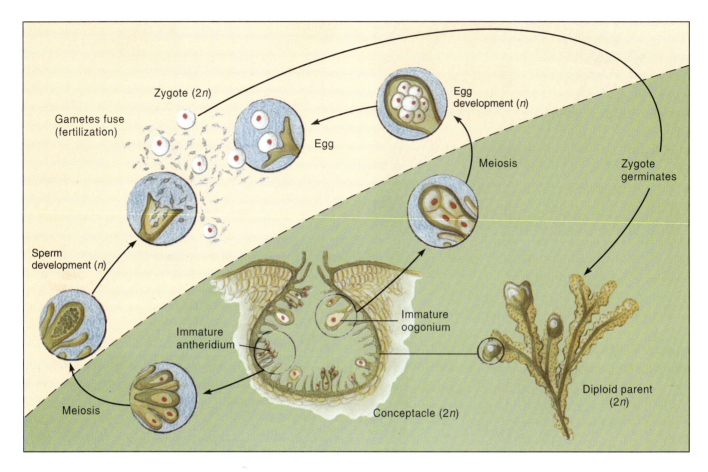

FIGURE 16.18

Conceptacles of *Fucus*, a brown alga, are embedded containers that bear oogonia and antheridia. Meiosis in oogonia and antheridia produces eggs and sperm, respectively, which are the only haploid cells in the life cycle of *Fucus*.

(a)

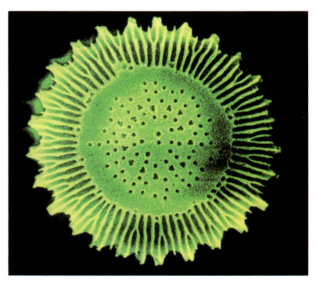

(b)

FIGURE 16.19

Diatoms. (*a*) Scanning electron micrograph of bilaterally symmetrical diatoms (×1,000). (*b*) Scanning electron micrograph of a radially symmetrical diatom (×1,000).

(a)

(b)

FIGURE 16.20

(*a*) Colored scanning electron micrograph of the euglenoid, *Euglena* (×2,800). *Euglena* cells are bounded by a special, flexible cell membrane or periplast instead of a cell wall. (*b*) Many dinoflagellates (pyrrophyta) have cellulose-containing armor plates that give these organisms a highly sculptured appearance (×3,000).

There are other divisions of algae with ecological importance.

In addition to the four divisions already discussed, other algae are included in three divisions that have only unicellular species. Each of these three divisions has one or more features that distinguish it from other algae (see table 16.2). Each division also shares important biochemical and anatomical structures with the green, brown, or red algae. Some of the unicellular species lack chlorophyll and engulf their food like single-celled animals.

Three divisions of algae consist exclusively of flagellated, unicellular organisms: the **euglenoids** (Euglenophyta, 800 species), the **dinoflagellates** (Pyrrhophyta, 3,000 species), and the **cryptomonads** (Cryptophyta, 100 species). These three divisions are traditionally studied by botanists and zoologists alike because of the plant-like and animal-like features of different species.

Euglenoids and cryptomonads have no cell walls. Instead, their cells are bounded by a flexible cell membrane or periplast that has extra inner layers of proteins and a grainy outer surface (fig. 16.20*a*). Euglena have been called algae and also animals because they share characteristics of both. Dinoflagellates also lack cell walls, but most species have cellulose plates interior to the plasma membrane. These plates are grouped into armorlike arrangements that are useful for identifying different genera (fig. 16.20*b*). The dinoflagellates have the most distinctive arrangement of flagella among the unicellular algae: one flagellum coils around the cell like a belt, while the other trails the cell as a rudder. The rhythmic beating of these flagella propels the dinoflagellate through the water like a spinning top. Most dinoflagellates live in salt-water habitats.

The fossil history of algae provides insights into their origins.

The earliest photosynthetic organisms were undoubtedly prokaryotic; their fossils are more than 3 billion years old.

FIGURE 16.21
Deposits of diatomaceous earth such as this one at Lompoc, California, are mined for commercial applications of diatoms.

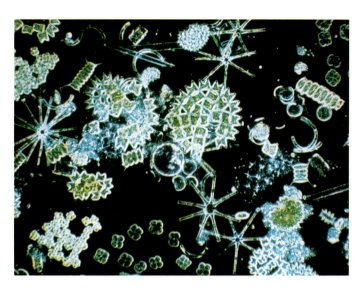

FIGURE 16.22
Light micrograph of a drop of pond water (×40). Microscopic organisms, including algae, are the plankton that are eaten by many aquatic and marine animals.

The first photosynthetic eukaryotes (see chapter 3 on the evolution of eukaryotic cells) may be represented by fossils of the Precambrian period (1.2 to 1.4 billion years old) from Bitter Springs, Australia. These fossils have subcellular structures that resemble those of algae cells, but not all paleobotanists agree that these are truly eukaryotic organelles. Nevertheless, green algae certainly existed in the Paleozoic era, approximately 500 million years ago.

Algae that are commonly found in the fossil record include dinoflagellates and diatoms. Dinoflagellates were preserved because they formed thick, protective plates around themselves. The oldest fossils that are confirmed as dinoflagellates are about 430 million years old. The best-known and perhaps the most widespread fossil algae are diatoms. The shells of these algae become "instant fossils" when the protoplast dies because silica of their cell walls does not decompose. Diatoms have been identified in sediments deposited about 120 million years ago, but organisms that seem to be diatoms lived almost 200 million years ago. Approximately 40,000 species of fossil diatoms, all of which are probably extinct, have been named from different cell-wall patterns. Diatom cell walls sometimes settle in large deposits, the largest of which is about 900 meters deep and 2 kilometers long, in Lompoc, California (fig. 16.21). Diatoms in this deposit are harvested commercially and used in many ways. Unicellular organisms that resemble dinoflagellates also occur in the Bitter Springs deposit.

Algae are important in many ecosystems.

Algae are nearly ubiquitous and occupy a wide variety of habitats. They live primarily in marine and freshwater habitats, either free-floating or attached to rocks, wood pilings, shells of shellfish, or other algae. Many species are terrestrial on moist soil, rocks, stone roofs and walls, or tree bark. Some of their more unusual habitats include clouds and airborne dust, snow, and the fur of certain animals. Some unicellular species are the food-producing endosymbionts inside the cells of protozoans, sponges, sea slugs, sea anemones, and salt-water fish. Green algae live symbiotically with fungi and cyanobacteria in lichens.

The most significant ecological role of algae is as **plankton** (fig. 16.22). Algae and other unicellular organisms are eaten by small animals, which, in turn, are eaten by larger animals. Thus, algae are the primary producers that support food webbs in marine and freshwater habitats. Commercial fish farmers often exploit this relationship by fertilizing tanks and ponds to enhance the growth of plankton. The oxygen that plankton produces is equally important to life; perhaps 50% to 70% of the earth's atmospheric oxygen comes from unicellular marine algae.

Brown and red algae form "forests" in intertidal zones and shallow coastal waters because the nutrients required for growth are more abundant there than in open seas. These marine forests are habitats for teeming populations of many kinds of animals. Coral reefs are special habitats whose primary production comes from coralline red algae. Some red algae precipitate calcium carbonate around them or bond the calcium carbonate from marine sponges. These interactions of red algae, sponges, and other organisms form coral reefs when the organisms die. Corals and many other kinds of nonphotosynthetic organisms also harbor green algae, yellow-green algae, or dinoflagellates as food-producing endosymbionts. Coral reefs can form extensive, highly complex ecosystems that support many

The Lore of Algae

Scientists at Ohio State University have grown batches of a single-celled algae (*Chlamydomonas reinhardtii*) that makes a metal-binding protein useful in cleaning up toxic wastes. The algae have been used to remove zinc, mercury, lead, copper, and other metals from contami-nated water samples and show promise in removing the metals from the Great Lakes. Practical application is in the future but perhaps this alga, which has been geneti-cally engineered, may be the natural water-pollution sponge long sought in environmental science.

species of marine animals. The most spectacular example of a coral reef is the Great Barrier Reef off the eastern coast of Queensland, Australia.

Blooms of some dinoflagellates are called red tides be-cause of the abundance of these red-pigmented organisms. Unlike other algal blooms, the cause of red tides is unknown. Red tides usually occur at least 16 kilometers offshore and are especially common along the Gulf coast of Florida and the coast of central California. One of the most interesting fea-tures of many red tides is that they are often bioluminescent; they glow in the dark just like fireflies. Red tides kill millions of fish each year because of a nerve toxin called brevetoxin se-creted by dinoflagellates. Brevetoxin is responsible for the red or brown discoloration of water during a red tide. Shellfish that eat dinoflagellates appear to be immune to these toxins, but humans or other vertebrates that eat these shellfish may suffer poisoning and death. Each year, red tides kill hundreds of endangered manatees off southwest Florida. Red tides have also been associated with pollution.

FIGURE 16.23

Irish moss (*Chondrus crispus*) is a red alga that is commercially im-portant as a source of carrageenan.

Algae have widespread economic importance.

As a group, the diatoms are perhaps the most economically important algae. Indeed, they have a substantial role for the fishing industry because they are food for freshwater and marine animals. Diatom cell walls are also harvested from large deposits as diatomaceous earth (fig. 16.21), which is used in many industries. For example, diatomaceous earth is an abrasive in metal polishes and a few brands of toothpaste; it is also used in filters for cleaning swimming pools and clar-ifying beer and wine. Reflective paint on highways, road signs, and license plates also contains cell walls of diatoms. Diatomaceous earth is an absorbent in some kitty litter, and one of the recipes for dynamite calls for diatomaceous earth as an inert absorbent for mixing with nitroglycerin. Because much of the earth's fossil oil originates from or is absorbed by diatoms, the presence of diatoms in sample cores of the earth is often a useful indicator of oil deposits.

Red and brown algae produce cell-wall polysaccha-rides that have many industrial uses. The main red algal polysaccharides are carrageenan from seaweeds such as Irish moss (*Chondrus crispus*) (fig. 16.23) and agar from sev-eral other seaweeds, including species of *Gracilaria*. Car-rageenan is used to stabilize or emulsify paints, cosmetics,

cream-containing foods, and chocolate. The main use of agar is in preparing culture media to grow bacteria and other microorganisms for laboratory research. Agar is also used as a gel for canning fish and meat and for making desserts. Some medicines contain agar as an inert carrier.

A promising use of algal polymers is in medicine. Polysaccharides extracted from certain unicellular and colonial green algae stimulate the immune systems of ani-mals, which may enhance resistance to disease. Although several patents for these potential medicines have been awarded in Japan, algal polysaccharides are not yet ap-proved for human medicinal use.

Many algae have been used traditionally as food, al-though most are not particularly nutritious. Some, perhaps most, seaweeds, however, contain iodine, which is an im-portant mineral in the thyroid gland. Although much of the cell-wall material of algae cannot be digested by humans, some seaweeds contain useful amounts of protein. These include nori (*Porphyra tenera*), a red alga that is used mostly as a flavoring in soups and salads and as a wrapping around small rolls of rice (sushi). Species of *Ulva*, *Laminaria*, and other seaweeds are also eaten fresh or pickled.

In recent years, some algae and cyanobacteria have been advertised as sources of vitamins, industrial chemicals, food additives, and fertilizer. Algae have increasingly important roles in treating sewage and industrial wastes. These uses of algae as sewage treatment "plants" are especially important in areas that cannot afford basic sanitation services.

Outbreaks of some unicellular algae are on the increase, and some can bring severe consequences to humans and other animals. Ingestion, and in some types, just skin contact, can be dangerous. Among the most recently identified villains are a dinoflagellate, *Pfiesteria piscicida*, which has been called "the cell from hell." *P. piscicida*, and related algae, can release a toxin that kills fish and then dissolves the tissue. A diatom in the genus *Pseudonitzschia* produces a compound called domoic acid that is a deadly neurotoxin.

SUMMARY

Bacteria are prokaryotic, do not reproduce sexually, and are the most metabolically diverse group of organisms. Some biologists divide bacteria into two separate kingdoms: Eubacteria and Archaebacteria. More recently, microbiologists have suggested life can be separated into three primary groups called domains: Eubacteria, Archaea (Archaebacteria), and Eukarya (Eukaryotes). The Eubacteria, or "true bacteria," comprise the vast majority of prokaryotes and include photosynthetic organisms. These photosynthetic prokaryotes are the cyanobacteria (formerly called the blue-green algae) and have chlorophyll *a* and phycobilins as their main photosynthetic pigments. Archaebacteria, or "ancient bacteria," differ from eubacteria in such things as their cell wall, membranes, and RNA structure. Bacteria are probably more closely related to the ancestor of life than any other group of organisms.

Fungi have nongreen multicellular filaments that look plantlike. Unlike plants, though, fungi have only one diploid cell in their life cycle, they can have chitin in their cell walls, they store glycogen, and they can reproduce asexually by spores. All fungi are heterotrophic; none are photosynthetic. Most plant diseases are caused by fungi. The two most complicated kinds of fungi, the ascomycetes and the basidiomycetes, are useful for making many commercial products. Mushrooms and truffles, which are basidiomycetes and ascomycetes, respectively, include species that are eaten as delicacies.

The vegetative organization of algae includes microscopic unicellular, colonial, and filamentous organisms and complex macroscopic seaweeds. Most algae reproduce both sexually and asexually. Green algae are the ancestors of plants. The fossil record of algae contains representatives that are at least 500 million years old. Some fossils that may be algae are up to 1.4 billion years old, but there is no general agreement that these fossils are actually eukaryotic. The most common fossil algae are diatoms because their glass cell walls do not decompose in most sediments. The first likely brown algae are represented by giant species that are about 400 million years old.

The most important ecological role of algae is as plankton. In this role, algae are the primary producers in the food webs of marine and freshwater habitats; plankton feeders are then eaten by other animals. Certain red algae cause calcium carbonate to precipitate around them, or they cement the calcium carbonate from marine sponges. The interaction of red algae with sponges, corals, and other sea life forms coral reefs.

Diatoms are probably the most economically important group of algae because they produce large deposits of discarded cell walls. These fine glass particles are used in many commercial, nonfood products. Polysaccharides from kelps and red algae are used to stabilize and emulsify many foods and nonfood products. Some algae are used directly as human food.

Writing to Learn Botany

Botanists have long known that some bacteria called thermophiles (such as those at Yellowstone National Park) thrive in water as hot as 93°C. However, in 1982, a German microbiologist discovered the first organisms that thrive above 112°C in shallow hot springs off the coast of Sicily. What kinds of molecules can withstand such brutal temperatures? Write a paper that summarizes the evolutionary significance of the research on thermophiles.

Some biologists refer to bacteria as the most primitive organisms on earth, while others believe that the persistence of prokaryotes in the environment for millions of years makes them the most advanced forms of life. How would you characterize bacteria and their complexity?

Go to the library and read an article about the classification of algae. Summarize the main points of the article. Considering what you know about algae and evolution, were you convinced by the arguments presented by the author(s)? Explain your answer.

THOUGHT QUESTIONS

1. If bacteria can reproduce only asexually, how can new forms arise?

2. Since many of the diseases of plants and animals are caused by bacteria, explain how it would or would not benefit humans if a virus that would eliminate all bacteria could be developed through genetic engineering.

3. If you could develop the most useful bacterium known to humans from various existing bacteria, what features would you combine in your new bacterium?

4. The morphology of cyanobacteria has changed relatively little during the past billion or so years; indeed, cyanobacteria have persisted two to ten times longer than other "living fossils" (e.g., crocodiles). How would you explain this observation?

5. Paleontologist Stephen Gould claims that bacteria, not humans, are evolution's most important product. Do you agree? Write a short essay to explain your answer.

6. If a single mushroom can produce a trillion spores, each of which can germinate and develop, why are we not overrun with mushrooms and other fungi that also produce prodigious numbers of spores?

7. Seeds planted in fertilized, sterilized soil may germinate when watered but do not grow nearly as well at first as their counterparts in the forest. What might explain this phenomenon?

8. Why are only a few algae well preserved in the fossil record?

9. What features of algae are plantlike? What features of algae are not plantlike?

10. It has been reported that some algae are a good source of human nutrients. How would you test this hypothesis?

11. What features of fungi prompted early botanists to classify them as plants? Why are fungi no longer considered to be plants?

SUGGESTED READINGS

Articles

Anderson, D. M. 1994. Red tides. *Scientific American* 271 (August):62–68.

Carmichael, W. W. 1994. The toxins of cyanobacteria. *Scientific American* 270:64–72.

Fryxell, G., M. Villac, and L. Shapirio. 1997. The occurrence of the toxic diatom genus *Pseudonitzschia* on the West Coast of the USA. *Phycologia* 36:419–37.

Lewis, R. 1994. A new place for fungi? Molecular evolution studies suggest fungi should be taxonomically transposed. *Bioscience* 44:389–91.

Steidinger, K., J. Burkholder, et al. 1996. *Pfiesteria piscicida* gen sp. nov., a new toxic dinoflagellate with a complex life cycle and behavior. *J. of Phycology* 32:157–64.

Zimmer, Carl. 1995. Triumph of the Archaea. *Discover* (February):30–31.

Books

Bhattacharya, D., ed. 1997. *Origins of Algae and Their Plastids.* New York: Springer-Verlag.

Carlile, M., and S. Watkinson. 1994. *The Fungi.* San Diego: Academic Press.

Prescott, L., J. Harley, and D. Klein. 1999. *Microbiology*, 4th ed. Dubuque, IA:WCB/McGraw-Hill.

Sze, P. 1997. *A Biology of the Algae,* 3d ed. Dubuque, IA: WCB/McGraw-Hill.

Varma, A. 1998. *Mycorrhiza: Structure, Function, Molecular Biology and Biotechnology,* 2d ed. New York: Springer-Verlag.

ON THE INTERNET

Visit the textbook's accompanying web site at http://www.mhhe.com/botany to find live Internet links for each of the topics listed below:

Economics of algae

Bacterial classification and morphology

Endosymbiont hypothesis

Bacteria and medicine

Decomposition

Fungal classification and morphology

Economics of fungi

Fungi as food and medicine

Green algae and plant evolution

Mycorrhizae

The discharge of spores from the capsule of the moss Polytrichum. What are the ecological advantages and disadvantages of producing so many spores?

Bryophytes and Ferns:
The Seedless Plants

LEARNING ONLINE

Did you know that bryophytes will grow on the south side of trees, on some animals, and at least one species will even grow on rolling stones? To learn more about the bryophytes and ferns, visit web1.manhattan.edu/fcardill/plants/intro/index.html in chapter 17 of the Online Learning Center at www.mhhe.com/botany.

Plants dominate the ecosystem of our planet and are essential to our lives, thanks largely to their diversity and remarkable adaptations. In previous chapters, we discussed how evolution produces this diversity. In this and the next chapter, we discuss the nature and extent of the diversity of lower and higher plants. We include as plants those eukaryotic organisms that are multicellular, form embryos produced via sexual reproduction, and have cellulose-rich cell walls, chloroplasts containing chlorophylls *a* and *b*, flavonoids and carotenoids, and starch as their primary food reserve.

Plants probably evolved from green algae and, for convenience, can be divided into two main groups: seedless and seed plants. The seedless plants, which, as the name states, do not produce seeds, include bryophytes and ferns. Bryophytes are considered to lack vascular tissue (they have no true xylem or phloem), while ferns and their "allies", along with gymnosperms and angiosperms, are vascular plants; that is, they do contain xylem and phloem.

The seedless plants are the bryophytes and the seedless vascular plants, (primarily ferns). Seed plants, which are the gymnosperms and angiosperms, do, of course, produce seeds and contain vascular tissue. You will learn about seed plants in chapter 18.

In the previous chapter, we discussed algae (because of their presumed relationship to plants) and certain other organisms—bacteria and fungi—that are traditionally included in botany textbooks. In this chapter, we discuss the two major groups of seedless plants: (1) bryophytes and (2) ferns and their related "allies." These organisms are primarily important today because of the ecosystems they occupy and their evolutionary history.

Plants Evolved from Aquatic to Terrestrial Ecosystems.

Although most biologists agree that early photosynthetic organisms evolved in water, true plants with embryos are virtually absent in the oceans, and those that occupy freshwater habitats are highly specialized. Aquatic habitats offer tremendous stability when compared to land; there is always plenty of water, and the temperature range is narrow. Yet this watery medium has its drawbacks. Light does not penetrate far into water, so most stationary plants cannot grow at depths greater than a few meters. Furthermore, somewhat limited amounts of CO_2 (for photosynthesis), nutrients, and O_2 for their submerged organs can sometimes make freshwater a very inhospitable environment. Clearly, movement to land opened vast new evolutionary opportunities to obtain nutrients, CO_2, and light.

The first organisms to survive on the land were probably simple organisms with no organized vascular systems, perhaps like the **bryophytes** (mosses, hornworts, and liverworts). After all, there was no selection pressure for any wasteful vascular tissue while these organisms were living in the water. Can you explain why?

When we consider these organisms, we are tempted to think about wet habitats where the bryophytes are close to water, basking in the sun of a bog or cooling off in the spray of a waterfall. Certainly, these are habitats where bryophytes are common. But keep thinking. What about the rocks on a cliff or the sand of dunes? Are bryophytes found there? In fact, some of these organisms can survive large changes in moisture and temperature, even within a single day, and occupy habitats where no vascular plants survive.

Even with so many diverse habitats occupied by plants today, we still consider the move from water to land to have been a major one. The greatest challenge to plants colonizing land was to keep their cells wet. Land plants responded to this challenge in two ways. In some—the ones we call vascular plants—lignin, a complex water-transport system, and a waxy, waterproof cuticle evolved. These adaptations enabled land plants to maintain a watery environment in their leaves, despite the fact that they were suspended in desertlike air. Other plants, the bryophytes, developed other strategies that we are only beginning to understand. They lack lignin, and the highly-evolved transport system of seed plants. They have a thin cuticle, if any, and their photosynthetic tissues are only one cell thick, making it easy to lose water. Yet, there are about 16,000 species of bryophytes, more than any other group of plants besides flowering plants.

Add to these simple survival challenges the problem of transferring gametes from a male organ to a female organ when the male gamete, the sperm, requires free water to swim! It seems that one of the best evolutionary "solutions" was to produce gametes only when water was available, but that solution required developing the gamete-producing gametangia well in advance to be ready on time. Something had to trigger the plants to stop using all their energy for growth and put some of it into making gametangia; that is, a method of receiving and responding to environmental signals was necessary. The life cycles of bryophytes have evolved in a way suited to their individual environments.

INQUIRY SUMMARY

Plants dominate our planet because of their diversity and adaptations. Plants are multicellular eukaryotic organisms that form embryos and have cellulose-rich cell walls, chloroplasts containing chlorophylls *a* and *b*, flavonoids and carotenoids, and starch as their primary food reserve. The seedless plants include bryophytes and ferns.

Are bryophytes dependent on water for reproduction? Why or why not?

Bryophytes Are Small, Seedless, Nonvascular Plants.

The classification of bryophytes usually include mosses, liverworts, and hornworts (fig. 17.1) although some biologists divide the bryophytes into different groups. Bryophytes are most noticeable when they grow in dense mats, but they can grow just about everywhere, including on bark and exposed rocks, where other plants cannot grow. Bryophytes are especially common in moist areas and lack the specialized vascular tissues that characterize other groups of plants. Gametophytes dominate the life cycle of bryophytes. Sporophytes are short-lived and obtain their food from the gametophytes to which they are attached. Bryophytes have the following general features:

- Most bryophytes are small, compact, green plants. Like green algae, they produce chlorophylls *a* and *b*, starch, cellulose cell walls, and swimming sperm cells. Many bryophytes grow very slowly.

- Bryophytes lack highly-evolved vascular and lignified tissues. As a result, they grow low to the ground and primarily absorb water by osmosis.

- Organs such as leaves and roots of vascular plants are defined by the arrangement of their vascular tissues. Since bryophytes lack true vascular tissues, they lack true leaves, stems, and roots. However, many bryophytes have structures without vascular tissues that are similar to leaves, so they are often referred to as being "leafy."

- Bryophytes get their nutrients from dust, rainwater, and substances dissolved in water at the soil's surface. Tiny **rhizoids** (hairlike extensions of epidermal cells) along their lower surface anchor the plants but absorb only small amounts of water and minerals. Water and dissolved minerals move by diffusion over the surface of bryophytes.

- The haploid gametophyte, which dominates the bryophyte life cycle, is often associated with mycorrhizal fungi (see chapter 16) and is usually perennial. Gametes form by mitosis in multicellular gametangia (gamete-forming tissue) called **antheridia** ("male") and **archegonia** ("female") (fig. 17.2; to understand where these structures occur in bryophytes, see fig. 17.3). Each flask-shaped archegonium produces one egg, and each saclike antheridium produces many sperms. Gametangia are protected by a sterile sheath or "neck" of cells that surrounds the archegonium or antheridium.

- Sperms produced by antheridia (singular, antheridium) use flagella to swim through water, down an archegonium (singular, archegonium) to eggs. Thus, bryophytes need free water for fertilization and sexual reproduction (in this regard they are like amphibians). Unlike those of nonplants (e.g., algae), fertilized eggs in bryophytes develop in, and are nourished by, protective organs. This is an important characteristic that helps us classify bryophytes as plants. Despite this, most reproduction in bryophytes is asexual.

- The diploid sporophyte is often short-lived, unbranched, and produces a terminal sporangium, where spores are formed. Although photosynthetic for some or most of its life, the sporophyte is permanently attached to and partially dependent on the gametophyte. This contrasts with the situation in green algae such as *Ulva* and *Ulothrix* in which the sporophyte and gametophyte live independently of each other. The sporophytes of bryophytes have no direct connection to the ground. Haploid spores, which are produced via meiosis in the sporangium of the sporophyte, are usually dispersed by wind and then germinate into haploid gametophytes.

The life cycle of bryophytes involves dominant gametophytes and small sporophytes.

Figure 17.3 shows the life cycle of a moss as an example of the most common bryophyte. Unlike many green algae, bryophytes have distinctly different generations of the gametophyte and sporophyte. The gametophyte generation begins with germination of haploid spores that were previously formed by meiosis. In most bryophytes, the germinated spore initially forms a protonema, which later develops into a "leafy" gametophyte plant. These haploid gametophytes are more diverse in bryophytes than in any other group of plants. Moreover, they are the dominant (and most visible) structure of bryophytes that you are likely to encounter. Rhizoids (hairlike extensions of epidermal cells that serve as anchors and aid water and mineral absorption) attach the gametophyte to its substrate, which is usually moist soil. The gametophyte grows and eventually produces antheridia and archegonia. Antheridia produce sperm, and archegonia produce eggs. Sexual reproduction in bryophytes requires free water, because the sperm must swim to the egg. Since sperm cannot swim far, antheridia and archegonia must be close to each other for sexual reproduction to occur. Sperm released from an antheridium do not swim randomly; rather, they are attracted

(a)

(b)

(c)

(d)

FIGURE 17.1

The diversity of bryophytes. (*a*) A hornwort (*Anthoceros*) growing in a highland rain forest. (*b*) A liverwort (*Conocephalum*) growing near the entrance of a cave. (*c*) A moss (*Dawsonia*) growing on Mount Kinabalu, Borneo. (*d*) Another moss (*Hylocomium splendens*) growing in a rain forest in Olympic National Park, Washington.

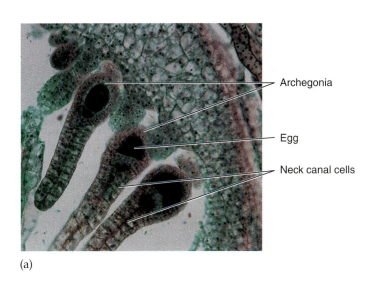

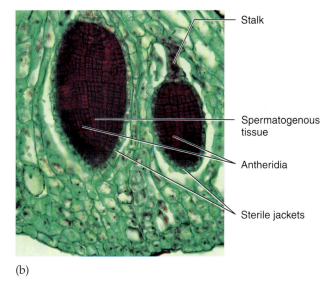

FIGURE 17.2

Gametangia of the liverwort *Marchantia*, a type of bryophyte. (*a*) Archegonia, each containing an egg (×100). (*b*) Antheridia, each containing spermatogenous tissue, which produces haploid sperm (×100).

by still-unidentified substance(s) produced by an archegonium. This is a form of cell-to-cell signaling (see chapter 5) in seedless plants. A sperm fertilizes an egg in the archegonium, thereby forming a diploid zygote and beginning the diploid sporophyte generation of the life cycle.

The developing sporophyte eventually depends on the gametophyte for nutrition and water. Although sporophytes develop differently in mosses, liverworts, and hornworts, they all ultimately produce **sporangia** containing **sporogenous** (spore-producing) **tissue.** This tissue undergoes meiosis to produce haploid spores, which are released to the environment. If a spore lands in a dry area, it can remain dormant, often for several decades. When water becomes available, the spore germinates (sometimes within a few hours) and forms the gametophyte, thus completing the sexual life cycle.

There are three major groups of bryophytes: mosses, liverworts, and hornworts.

Bryophytes are variously treated as classes of a single division, **Bryophyta,** or as three separate divisions. In the latter case, they are division Bryophyta (mosses), division Hepatophyta (liverworts), and division Anthocerotophyta (hornworts). The classification of all bryophytes in a single division implies they have descended from one evolutionary ancestor. As you will see again later in this chapter, this idea is in dispute.

Mosses are the largest and most familiar group of bryophytes.

Mosses are remarkably successful plants that thrive alongside more conspicuous vascular plants. The approximately

12,000 species of mosses make up the largest and most familiar group of bryophytes. Moss morphology is very diverse (fig. 17.4). One of the smaller mosses is the pygmy moss, which is only 1 to 2 millimeters tall and completes its life cycle within a few weeks. Luminous mosses such as *Schistostega* (cave moss) and *Mittenia* often grow near entrances to caves and glow an eerie golden green. The upper surface of these mosses is made of curved, lenslike cells that concentrate the cave's dim light onto chloroplasts for photosynthesis.

Review the life cycle of a typical moss as shown in figure 17.3. A haploid moss spore germinates and develops into a tiny protonema, which is often strikingly similar to a filamentous green alga. As the protonema grows, it forms buds that become "leafy," haploid male and female gametophytes. "Leafy" means in the general shape of a leaf.

Moss gametophytes, which are green and photosynthetic, are occasionally mistaken for a low-growing grass or other flowering plants (fig. 17.5). Mature gametophytes produce the sperms and egg in the antheridia and archegonia, respectively. Mature sperm cells swim down the archegonia and fuse with the egg. Swimming sperm cells require moisture; this is why mosses usually inhabit wet environments. Following fertilization in the female archegonium, the diploid zygote is formed and develops into a young sporophyte, which can resemble a thin, brown stalk with an enlarged head, or **capsule** (see fig. 17.3 and the chapter-opening photograph). The mature sporophyte is brown, yellowish, or reddish and has three parts: a **foot,** a **seta** (stalk), and a capsule covered by a **calyptra.** The foot, which penetrates the base of the venter (the swollen base of an archegonium containing an egg) and grows into the gametophyte, absorbs water, minerals, and nutrients from the

FIGURE 17.3

The general life cycle of a moss, a type of bryophyte. Spores germinate and produce a protonema, which becomes a leafy gameto-phyte. This gametophyte is the dominant structure that you are most likely to see.

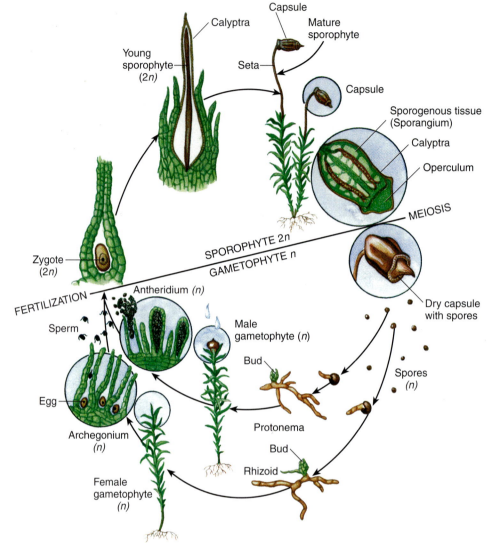

gametophyte. The wiry seta elongates and raises the capsule as much as 15 centimeters above the gametophyte. As this occurs, the protective cover-ing, or calyptra, is around the capsule. The capsule is actually a sporangium that begins developing while within the calyptra. Specialized sporangium cells (sporogenous tissue) of the cap-sule undergo meiosis, forming as many as 50 million haploid (n) spores per capsule. When these spores are mature, the calyptra falls off, thereby exposing the operculum (the lid of the cap-sule when present). Contraction forces associated with dry-ing of the capsule then release the operculum (see fig. 17.3), and its spores are usually dispersed by wind (see chapter-opening photograph). Spores that land in suitable environ-ments germinate, forming protonemata and thus complet-ing the life cycle.

Mosses actually contain internal conducting tissues, which have been called hydrome and leptome. In the mid-1980's, it was suggested these tissues are homologous to xylem and phloem in vascular plants (ferns, gymnosperms and angiosperms). If this suggestion is correct, then mosses could be considered vascular plants. However, in many species the conducting tissue has been reduced (and there-fore is non-functional), and most water conduction occurs along the surface of these small plants. Until further botan-ical evidence is available, we continue to classify mosses in the traditional manner as non-vascular plants. Nonethe-less, from morphology and DNA studies, it is clear mosses are more closely related to vascular plants than to liver-worts and the long-standing recognition of the bryophytes as a group may soon change.

Liverworts are low-growing plants usually found in wetlands.

Liverworts were named during medieval times, when herbalists followed the now discounted Doctrine of Signa-tures. A few liverworts are lobed and thus have a fanciful re-semblance to the human liver, so the word liver was com-bined with wort (herb) to form the name liverwort. Although we now know that the Doctrine of Signatures is invalid and that eating liverworts will not help an ailing liver, the common name liverwort has endured. Some 8,500 species of liverworts have been named. What is the now dis-proved Doctrine of Signatures?

Liverworts range in size from tiny, leafy filaments less than 0.5 millimeters in diameter to a thallus more than 20

(a)

(b)

(d)

(c)

(e)

FIGURE 17.4

The diversity of mosses. (*a*) Mosses often dominate wet places, such as this waterfall. (*b*) Moss (*Racomitrium*) colonizing a lava bed in Iceland. (*c*) Cushion moss growing in Antarctica, where daily summer temperatures range from −10° to −30°C. (*d*) *Splachnum luteum,* (mammal dung moss) a moss that resembles a small flowering plant. The colors of the moss and the chemicals it produces attract insects that disseminate its spores. (*e*) *Grimmia,* a rock moss that survives on bare rocks, often in scorching sun. In this photo, *Grimmia* is growing among larger, more conspicuous lichens.

FIGURE 17.5

Gametophytes of the hairy-cap moss (*Polytrichum*), showing antheridial heads. Note the radial symmetry and differentiation into "leaves" and "stems."

centimeters wide (fig. 17.6). All liverworts have a prominent gametophyte, which sometimes has a waxy cuticle. Also, the spores of liverworts have thick walls. These waterproofing features are believed to be important adaptations that enable liverworts to live on land in moist environments, but a few liverworts are adapted to dryer environments. Liverworts also have the following distinguishing features:

- Their rhizoids are unicellular.

- Their gametophytes are leafy or **thallose** and are often lobed and bilaterally symmetrical. They lack a midrib. The entire thallus is often photosynthetic; the lower side may be adapted for storage in a few species.

- Their sporangia are often unstalked.

- They shed spores from sporangia for a relatively short time.

Liverworts frequently reproduce asexually. One way they do this involves the death of older parts of the plant. When this occurs, the growing areas are left isolated from the parent plant. A second means of asexual reproduction involves the production of ovoid, star-shaped, or lens-shaped pieces of haploid tissue called **gemmae** (singular, gemma). In some species (e.g., *Lunularia, Marchantia*), gemmae form in small "splash cups" on the upper surface of the gametophyte (figs. 17.7 and 17.8). Gemmae become detached from their cups by falling raindrops and are often splashed as far as a meter away from the parent plant. Gemmae grow into mature gametophytes. Pieces of gametophytes that are broken or torn from the parent plant can also regenerate entire plants.

As mentioned, the haploid gametophytes of liverworts have two shapes: "leafy" and thallose (a plant body undifferentiated into stem, root, or leaf; see fig. 17.6*a*, and *e*). Approximately 80% of liverworts are "leafy"; they usually grow in wetter areas than mosses and are abundant in tropical rain forests and fog belts. Thallose liverworts have a flat, ribbonlike gametophyte. *Marchantia* is a thallose liverwort (fig. 17.8) and is the most intensively studied liverwort. The gametophytes of *Marchantia* are perennial and branched (fig. 17.8). Each branch grows from the notch at the tip of a lengthwise groove. Many liverworts, including *Marchantia,* are unisexual: sperm- (antheridia) and egg- (archegonia) producing organs are on separate gametophytes (figs. 17.1a, 17.8).

The diploid sporophyte of liverworts lacks stomata. The capsule and its stalk is attached to the gametophyte by a knoblike foot (fig. 17.8). The sporophytes of some thallose liverworts are spherical, unstalked, and held within the gametophyte until they shed their haploid spores.

Hornworts are the smallest group of bryophytes.

There are only about 100 species in six genera of hornworts. The most familiar hornwort is *Anthoceros*, a temperate genus (fig. 17.9). Hornworts have several features that distinguish them from most other bryophytes:

- The sporophyte is shaped like a tapered horn (fig. 17.9), hence the common name.

- Each photosynthetic cell contains one to only a few chloroplasts. Each chloroplast is associated with a starch-storing body called a pyrenoid, as are the chloroplasts of green algae and *Isoetes*, a vascular plant.

- Archegonia and antheridia are enclosed snugly in the sporophyte thallus and are in contact with the surrounding vegetative (nonreproductive) cells of the thallus.

- The flat, dark green gametophytes of hornworts are structurally simpler than those of other bryophytes. They are flattened and may superficially resemble those of thallose liverworts. Hornwort gametophytes are either annual or perennial and are anchored to the substrate by rhizoids.

In most hornworts, sex organs form on the upper surface of the thallus. One or more antheridia resembling those of liverworts form in roofed chambers in the upper portion of the thallus, and archegonia form in rows beneath the surface. Hornworts reproduce asexually by fragmentation.

The pores and cavities of some hornwort gametophytes are filled with mucilage instead of air. Nitrogen-fixing cyanobacteria such as *Nostoc* live in this mucilage and release nitrogenous compounds to the hornwort. What other associations with nitrogen-fixing bacteria exist in nature?

The diploid sporophytes of hornworts differ remarkably from those of other bryophytes. Hornwort sporophytes are long, green spindles (1 to 4 centimeters long) having

(a)

(b)

(c)

(d)

(e)

FIGURE 17.6

The diversity of liverworts. (*a*) Leafy liverworts growing on a leaf of an evergreen tree in the Amazon Basin in Brazil. (*b*) *Calypogeia muelleriana,* a "leafy" liverwort. (*c*) *Bazzania trilobata,* a "leafy" liverwort having dichotomous forking. (*d*) An aquatic thallose liverwort (*Ricciocarpus natans*) floating on the shallows of a pond. (*e*) *Conocephalum conicum,* a thallose liverwort.

tapered tips (see fig. 17.1). They have a distinct epidermis and stomalike structures. The sporophyte remains photosynthetic and can live for several months while spores are released over time. This semi-independence is viewed by many botanists as an evolutionary step toward the independent sporophytes that characterize vascular plants. Indeed, some botanists no longer consider hornworts to be bryophytes; they consider them to be more closely related to ferns.

Bryophytes grow in many different habitats.

Bryophytes live in almost all places that plants can grow and in many places where vascular plants cannot grow (see

(a)

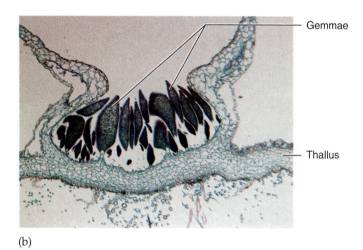

(b)

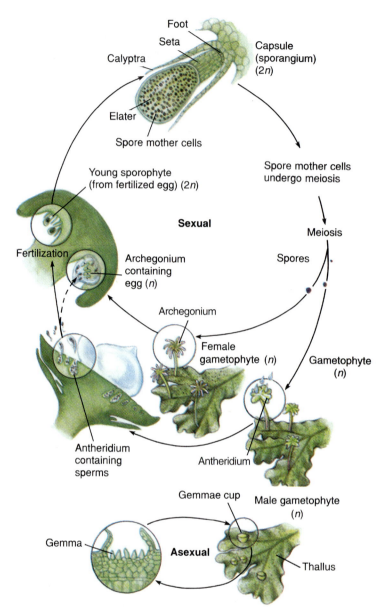

FIGURE 17.7

Gemmae cups of liverworts. (*a*) Gemmae cups ("splash cups") containing gemmae on the gametophytes of *Lunularia* (×1). Gemmae are splashed out of the cups by raindrops, after which the gemmae can grow into new gametophytes, each identical to the parent plant that produced it by mitosis. (*b*) Cross section of a thallus showing a gemmae cup (×10). How is reproduction via gemmae similar to that via a stolon?

FIGURE 17.8

Life cycle of *Marchantia,* a thallose liverwort. During sexual reproduction, spores produced in the capsule germinate to form independent male and female gametophytes. The archegonium contains an egg; the antheridium produces many sperm. After fertilization, the sporophyte develops within the archegonium and produces a capsule containing spores. *Marchantia* also reproduces asexually by fragmentation and gemmae.

fig. 17.4). This is one of the ecological advantages of being small and nonvascular: bryophytes can live on impermeable substrates, thereby avoiding competition with vascular plants. Indeed, bryophytes grow in cracks of sidewalks, on moist soil, rooftops, the faces of cliffs, tombstones, and birds' nests; they carpet forest floors, dangle like drapery from branches, and sheathe the trunks of trees in rain forests. Examples of the extremes of bryophyte habitats include exposed rocks, volcanically heated soil (up to 55°C), and Antarctica, where summer temperatures seldom exceed minus 10°C. However, the most unusual habitat for bryophytes is reserved for *Splachnum,* the mammal dung moss (see fig. 17.4*d*). This moss produces a colored capsule and releases a putrid odor that attracts flies. Unlike the spores of other mosses, those of *Splachnum* are sticky and

adhere to the visiting flies. The spores are disseminated when the flies move from the moss to piles of dung.

Bryophytes, which are often the first plants to invade an area after a fire, grow at elevations ranging from sea level to 5,500 meters. There are no marine bryophytes, but some, such as dune mosses, grow near the seashore. Bryophytes dominate the vegetation in peatlands. Mosses are especially abundant in the arctic and antarctic, where they far out-

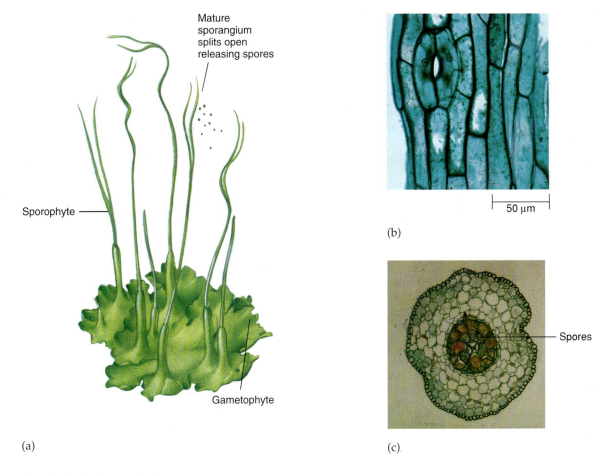

Mature sporangium splits open releasing spores

Sporophyte

Gametophyte

(a)

50 μm

(b)

Spores

(c)

FIGURE 17.9

Anthoceros, a hornwort. (*a*) *Anthoceros* with mature sporangia. (*b*) Stoma-like structure on the gametophyte of *Anthoceros;* such structures are absent from the gametophytes of all other plants. (*c*) Spores form along the central axis of the sporophyte. (See fig. 17.1*a.*)

number vascular plants. Some bryophytes can withstand many years of dehydration and often grow in deserts. Other bryophytes can withstand prolonged periods of dark and freezing, which explains why they are the most abundant plants in Antarctica. Although generally widespread, some bryophytes grow only in specific habitats, such as on the bones and antlers of dead reindeer or on animal dung. Bryophytes, along with lichens, are among the first organisms to colonize bare rocks and volcanic upheavals and are, therefore important as pioneer species in ecological succession (chapter 19). For example, *Andreaea* is a black to reddish brown moss that grows on exposed rocks. These organisms slowly convert rock to soil, thus paving the way for colonization by other organisms.

Many bryophytes, such as the genus *Hypnum,* are notoriously sensitive to pollution, especially sulfur dioxide. As a result, most bryophytes are rare in cities and industrialized areas; for example, 23 species of bryophytes that grew in Amsterdam in 1900 no longer grow in that area. Bryophytes increase the humus in soil and often indicate the presence or absence of particular salts, acids, and minerals. The liverwort *Carrpos* grows only on gypsum-rich

"salt pans," and *Mielichhoferia* and *Scopelophila* are mosses that grow only on copper-rich substrates.

Bryophytes usually grow very slowly. However, *Sphagnum* peat moss accumulates at rates exceeding 12 metric tons per hectare annually (fig. 17.10); for comparison, yields of corn and rice average about 6 and 5.5 metric tons per hectare, respectively. Peat moss includes various mosses of the genus *Sphagnum,* usually found growing in very wet places. The partly carbonized remains of these plants are used as a mulch and for improving soil texture and structure. Peat moss can absorb 20 to 30-times its weight in water; this is why peat moss is an excellent soil conditioner that retains water and helps to prevent flooding and minimize erosion. Peat moss produces large amounts of acids and antiseptics that kill decomposers. As a result, dead peat moss and other organic material accumulates and forms large deposits called peat bogs (fig. 17.11). The pH of these bogs often approaches 3.0 (i.e., about the same pH as vinegar) and thus eliminates all but the most acid-tolerant plants, such as cranberry, a few carnivorous plants, blueberry, and black spruce, which is often a climax species (see chapter 19) in acidic communities. The acidity of peat bogs also makes

The Lore of Plants

Contrary to what you may think, not all organisms referred to as mosses are true mosses. For example, Irish moss and other sea mosses are red algae, Iceland moss and reindeer moss are lichens, club mosses (i.e., a type of lycopod) are seedless vascular plants, and Spanish moss is a flowering plant in the pineapple family. Fishermen are probably the worst "offenders;" they seem to constantly complain about the "moss" in a lake or stream, yet the vegetation there is not moss. In fact, there are very few extant, aquatic species of moss. Most of what the fisherman complain about in lake and stream vegetation is either algae or aquatic flowering plants.

FIGURE 17.10

Sphagnum. (*a*) Carpet of peat moss. (*b*) Close-up of photosynthetic (i.e., chloroplast-containing) and water-retaining cells of *Sphagnum* (×100).

these communities very stable; some peat bogs are estimated to be more than 50,000 years old. Peat bogs cover approximately 1% of the earth's surface, an area that is equivalent to half of the United States. In Ireland, peat sod is an important fuel.

The acidity and anaerobic environment of peat bogs also preserve dead plants and animals. Paleobotanists have used pollen preserved in peat to determine which plants grew in a particular area and how they have changed over time. The preservative effects of peat bogs have also yielded horrifying secrets about past civilizations (see Perspective 17.1, "Secrets of the Bog").

The fossil record of bryophytes is poor and shows relatively little diversity.

Some fossils dating from the Carboniferous period (286 to 360 million years ago) have been interpreted as being mosses. The most ancient liverworts (*Pallaviciniites devoni-cus*) date to the Devonian period (360 to 408 million years ago), and other fossils that resemble liverworts appear in the Carboniferous and succeeding periods.

In spite of the meager fossil record, botanists suspect that bryophytes diverged from an ancestor common to vascular plants more than 430 million years ago (i.e., sometime during the Silurian period). There is no general agreement as to what the common ancestor was or even whether bryophytes diverged from one group. Among the green algae, the most likely ancestral group may be the charophytes (e.g., *Chara,* a stonewort, see chapter 16) because they are the only group of algae with a metabolic pathway that produces flavonoid pigments, a feature of plants. Another possibility is that the ancestor of bryophytes is represented by the green alga *Coleochaete,* which retains the zygote on the gametophyte, as do plants, but does not produce flavonoids.

Bryophytes today have limited economic importance.

Bryophytes are generally not edible, although liverworts were often soaked in wine and eaten in the 1500s. Today, the earthy aroma of Scotch whiskey is partly due to the fact that the malted barley is dried over peat fires.

Mosses are used as stuffing in furniture, as a soil conditioner, for fuel, as an absorbent in oil spills, and for cushioning. Florists use peat moss as a damp cushion when shipping plants. *Sphagnum* also has been used by aboriginal people for diapers and as a disinfectant. Because of its acidity and absorbing powers, peat moss is an ideal dressing for wounds. Indeed, the British used more than 1 million such dressings per month during World War I, and the Red Cross refers to *Sphagnum* as a wound dressing in its publications. North American Indians used *Mnium* and *Bryum* to treat burns. *Dicranoweisia* has been used to waterproof roofs in Europe.

The rapid growth of peat moss suggests that peat bogs may be an important source of renewable energy. Peat and lignite (a soft, brown type of coal) have several properties that make them good sources of fuel. For example, peat has a caloric value of 3,300 calories per gram, a value that is greater than that of wood (but only half that of coal). Furthermore, peat is abundant. The United States (excluding Alaska) has more than 60 billion tons of peat, an

(a)

(c)

(b)

FIGURE 17.11

Peat and peat bogs. (*a*) A peat bog showing the diversity of plants that grow on peat moss. (*b*) A family in Ireland stacking peat sods to dry. This peat will later be used as fuel to heat their home. (*c*) The bucolic image of peat fires smoldering on small hearths misrepresents peat's use. Indeed, more than 95% of the world's peat harvest is burned to generate electricity. Peat-fired power plants are preferred over coal-fired power plants because of peat's low sulfur and ash content, which help reduce air pollution from burning.

amount of fuel equivalent to approximately 240 billion barrels of oil. Countries of the former Soviet Union annually harvest more than 200 million tons of peat, which is used as fuel for nearly 80 power plants. Ireland obtains more than 20% of its energy from peat, and the United States annually harvests more than a million tons of peat in 22 states. Considerable peat is used in the horticulture industry and by home gardeners as a potting soil additive and to improve soil structure and texture mostly by adding moisture-holding capabilities and organic material.

INQUIRY SUMMARY

Bryophytes are small plants lacking vascular tissues and supporting tissues. Their life cycle is dominated by free-living, photosynthetic gametophytes that produce archegonia and antheridia. Bryophytes require free water for sexual reproduction.

Mosses are the largest and most familiar group of bryophytes. Their haploid gametophytes, which produce sperms and eggs, are leafy and photosynthetic. Sperm from antheridia swim through the neck of an archegonium to reach an egg. Following fertilization, a diploid sporophyte grows and produces haploid spores via meiosis in a capsule. Spores germinate and form a filamentous protonema, which grows into a haploid gametophyte. Mosses may contain early, primitive vascular tissue.

Liverworts are low-growing plants often found in wetlands. Most liverworts have a prominent gametophyte, which occasionally has a waxy cuticle. Also, the spores of liverworts have thick walls. These waterproofing features are believed to be important adaptations that originally enabled liverworts to live on land. Hornworts comprise the smallest group of bryophytes. Their sporophytes are shaped like tapered horns, and their photosynthetic cells each contain one chloroplast associated with a pyrenoid.

How might the harvesting of peat effect a moist, terrestrial ecosystem?

17.1 *Perspective* Perspective

Secrets of the Bog

In the 1950s, several countries in northern Europe began mining peat as fuel. This peat from bogs not only provided valuable fuel but also yielded some 2,000-year-old secrets that were as grotesque as they were valuable. Horrified workers uncovered several hundred human bodies. These bodies had been tanned by the bog's acids (i.e., much as we use acids to tan leather), and many were in such good condition that the workers suspected they had been recently dumped into the bogs. Furthermore, many of these bodies were mutilated; they had slit throats, severed vertebrae, and nooses around their necks.

Further study showed that the bodies were actually 2,000 to 3,000 years old. The acidity and anaerobic environment of the bogs had preserved the bodies by inhibiting the growth of bacteria and fungi, which normally decompose dead organic material. Archaeologists gave the bodies that were pulled from the peat bogs names, such as Lindow man and Tollund man, referring to the geographical locations of the bogs that became graves. The most intensively studied of these bog people is Lindow man, a

Lindow man.

fellow who lived about the time of Aristotle. The preserved stomach contents of Lindow man showed that his last meal was barley and linseed gruel. Thanks to the preservative effects of the bog, archaeologists have also learned about musical instruments and household items of the 2,000-year-old civilizations and determined that bogs were the sites of human sacrifice in religious ceremonies.

Ferns and Related Plants Are Seedless Vascular Plants.

Seedless vascular plants—those that have xylem and phloem but do not produce seeds—include ferns and their related plants, often called fern "allies." These plants share features with bryophytes, including the same types of pigments, the basic life cycle, and the storage of starch as their primary food reserve. However, the evolution of functional vascular tissue enabled vascular plants to invade and dominate the drier habitats on land more effectively than could nonvascular plants. The sporophyte of ferns dominates the life cycle; the gametophyte is always smaller, sometimes microscopic, and nutritionally independent of the parent sporophyte. The ancestors of modern seedless vascular plants dominated the earth's vegetation for more than 250 million years, eventually giving way to seed plants. Most species of seedless vascular plants are true ferns, but they also include horsetails, whisk ferns, and club mosses—the three groups of fern "allies."

Vascular plants, such as ferns (fig. 17.12), are not watertight, nor do they use or conserve water efficiently. Water escapes when stomata are open and gases are exchanged during photosynthesis. However, if sufficient moisture is present, vascular plants can continuously replace water that is lost and keep the entire plant moist; water lost by transpiration is replaced by absorption into the roots (chapter 9). When water loss increases, water is absorbed and transported faster to compensate for the loss. When the demand for water uptake exceeds the ability of the roots to absorb it, then stomata close and prevent further loss. Bryophytes, which lack vascular tissue, lack such control over their use of water.

Although many features of seedless vascular plants are not found in the bryophytes, these two groups do have many characteristics in common. Some of these features are shared with a few algae as well. For comparison with the

FIGURE 17.12
The ferns, represented here by bracken fern (*Pteridium aguilinum*), are the largest group of seedless vascular plants.

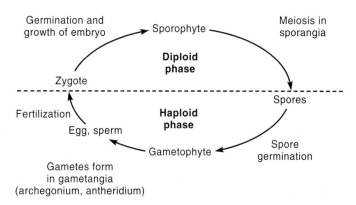

FIGURE 17.13
The generalized life cycle of seedless vascular plants. Alternation of generations (see chapter 13) is shown.

algae and bryophytes, the general features of seedless vascular plants are summarized as follows:

- The life cycle of seedless vascular plants is similar to those of bryophytes and algae, which produce haploid spores following meiosis in the sporophyte (fig. 17.13). Each spore germinates and grows into a haploid gametophyte that produces gametes. The gametes (eggs and sperms) fuse to form a diploid zygote that develops into a sporophyte.

- Eggs are produced in archegonia, and sperm cells are produced in antheridia.

- The zygote germinates into a multicellular embryo that depends on the gametophyte for its nutrients. In turn, the embryo grows into a mature, diploid sporophyte. A multicellular embryo also characterizes the bryophyte life cycle, but it is absent in algae; this is one character that distinguishes algae from plants.

- Seedless vascular plants produce chlorophylls *a* and *b*, carotenoids, starch, cellulose cell walls, and swimming sperm cells. These features are also shared with bryophytes and many algae.

- Each sporangium is protected by a multicellular jacket of nonreproductive (sterile) cells. Spores are dispersed from sporangia by wind and are covered by cutin that resists desiccation.

- Seedless vascular plants may have a well-developed cuticle that minimizes water loss. Stomata have evolved that allow gas exchange during photosynthesis. Although hornworts and mosses have stomata, they are not well developed and do not function efficiently to prevent water loss.

- Flagellated sperm cells swim through water to eggs. Like bryophytes, seedless vascular plants require free water for sexual reproduction.

- Unlike bryophytes, seedless vascular plants have well-developed vascular tissues. Xylem transports water and dissolved minerals great distances from the soil. Carbohydrates are transported up and down throughout the plant's phloem.

- Many seedless vascular plants have lignin and cellulose in their secondary cell walls. Lignin strengthens cellulosic microfibrils (chapter 3), thereby enabling these plants to grow taller than bryophytes.

- Sporophytes and gametophytes of seedless vascular plants are largely nutritionally independent of each other. Sporophytes are photosynthetic, and gametophytes are either photosynthetic or they obtain nutrition from dead and decaying organic matter.

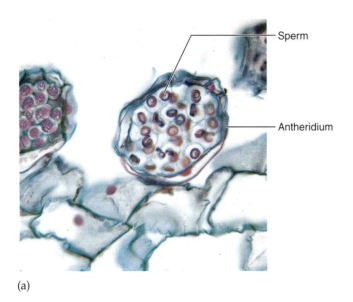

Sperm

Antheridium

(a)

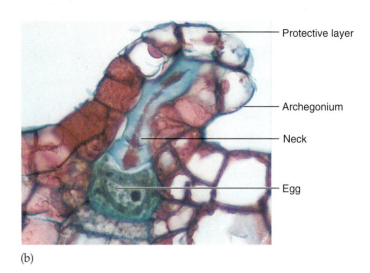

Protective layer

Archegonium

Neck

Egg

(b)

FIGURE 17.14

The sexual organs of ferns are similar to those of bryophytes: antheridia and archegonia are multicellular and include a protective layer of sterile cells. (*a*) Antheridia are nearly spherical, with each antheridium producing many flagellated sperm cells (×125). (*b*) Archegonia are flask-shaped, with a narrow neck on top of an expanded region containing a single nonmotile egg (×150).

§ The sporophyte of seedless vascular plants, which dominates the life cycle, is long-lived and often highly branched.

The life cycle of seedless vascular plants involves dominant sporophytes and small gametophytes.

As in bryophytes, sexual reproduction in seedless vascular plants entails alternation of generations (chapter 13), which consist of diploid and haploid phases. Meiosis produces haploid spores in the sporangium of diploid sporophytes. Spores then germinate and grow into gametophytes that produce eggs in archegonia and sperms in antheridia. Sperm cells must have water to swim to archegonium, where fertilization occurs to produce a diploid zygote. The zygote starts another diploid phase.

The life cycle of seedless vascular plants is dominated by the sporophyte, which is the "plant" that everyone thinks of when they think of plants. Sporophytes become nutritionally independent of the gametophyte soon after the zygote grows out of the archegonium. In many genera, the sporophytes are perennial; new growth sprouts from underground rhizomes year after year. For example, the branching rhizomes of horsetails (see following section) form extensive underground networks, and in ferns, new leaves usually grow from the same rhizome every year. Sporophytes commonly reproduce asexually, producing populations consisting entirely of clones.

Often the gametophyte of seedless vascular plants is short-lived and small, sometimes microscopic. With few exceptions, it is nutritionally independent of the sporophyte as soon as the spore leaves the sporangium. The gameto-

phytes of most whisk ferns and club mosses, along with some ferns, are nutritionally dependent on the sporophyte. In horsetails and most ferns, the gametophytes are photosynthetic. The gametophytes of many ferns look somewhat like liverworts; these gametes sometimes reproduce asexually and, like many sporophytes, can produce extensive clonal populations where the sporophytes are absent or rare. One example that has fascinated botanists for many years is the "Appalachian gametophyte" of the fern genus *Vittaria,* whose sporophyte is thought to be extinct. The gametophyte is distributed over several thousand square kilometers in the eastern half of the United States, but it is not known to produce a sporophyte. Can you develop a hypothesis to explain this observation?

Organs that produce gametes and spores in seedless vascular plants are similar in several ways to those of bryophytes (fig. 17.14). Antheridia and archegonia are multicellular and protected by a layer of sterile (infertile), nonreproductive cells. Each antheridium produces many sperm cells, but each archegonium makes only one egg.

The sporangia of many vascular plants are aggregated into cones, called **strobili** (singular, **strobilus**). The spores produced in strobili are usually dispersed by wind. A strobilus is essentially a stem tip with several closely packed sporangia. The strobilus occurs primarily in club mosses (fig. 17.15), spike mosses, and horsetails but not in ferns. However, the strobilus is one of the most significant developments in reproductive organization among plants because it foreshadows cones in gymnosperms and flowers in angiosperms. The flowering plants, which are the most highly developed group of plants, bear sporangia exclusively in flowers that are highly modified strobili (see chapter 18).

There are four groups of seedless vascular plants: whisk ferns, club mosses, horsetails, and ferns.

We distinguish the four divisions of living seedless vascular plants by significant variation among their branching patterns, leaf morphology, vascular organization, and underground absorptive organs. The types and arrangements of sporangia on the sporophyte can also be used in defining divisions. Gametophytes of the four groups vary in their origin by different types of spore and in their mechanism of obtaining nutrients. No single characteristic defines each division; rather, classification depends on sets of features summarized next.

Division Psilotophyta: Whisk ferns. The whisk ferns are the simplest, most primitive (the oldest or earliest to evolve) vascular plants, primarily because they have no roots and no leaves. Their traditional classification as a separate division is controversial because some botanists have suggested whisk ferns are simplified ferns.

Instead of roots with root hairs, the psilotophytes have rhizomes with absorptive rhizoids. The larger of the two genera in the division, *Psilotum,* has enations (an outgrowth on the surface of an organ) instead of leaves (fig. 17.16). Botanists hypothesize that the enations of *Psilotum* may be the reduced remnants of leaves, which were larger and more leaflike in the ancestors of the psilotophytes. The stems of *Psilotum* are green and photosynthetic.

Probably the name *whisk ferns* comes from the highly branched stems of *Psilotum,* which give the plant the appearance of a whisk broom (fig. 17.16). *Psilotum* is widespread in subtropical regions of the southern United States and Asia, and it is a popular and easily cultivated plant that is grown in greenhouses worldwide.

Division Lycopodiophyta: Club mosses. This division, which despite the common name is not made of true mosses, consists of 10 to 15 genera and more than 1,100 species that live in various habitats worldwide. Lycopods are primarily tropical but also form a conspicuous part of the plant life in temperate regions. Most of the species are included in two genera, the club mosses (*Lycopodium,* about 400 species) and the spike mosses (*Selaginella,* about 700 species), both of which get their common names from their club-shaped or spike-shaped strobili (see fig. 17.15). Most species are terrestrial, but many are epiphytic (growing on other plants). One species of *Selaginella,* called the resurrection plant (*S. lepidophylla*) because of its ability to defy drought conditions, occurs in the deserts of southwestern United States and Mexico. During periods of drought, this plant forms a tight, dried-up ball; when rain comes, its branches absorb water, expand and resume photosynthetic activity.

The sporophytes of club mosses are differentiated into leaves (called microphylls), stems, and roots. The roots branch from perennial rhizomes that sometimes grow outward from a central point to form "fairy rings" (as many mushrooms do). One such fairy ring of a *Lycopodium,* when

FIGURE 17.15

Strobili, or cones, are aggregations of closely packed sporangium-bearing branches or leaves. Shown here are strobili of *Lycopodium obscurum,* a club moss.

Xylem and phloem provide vascular plants with a selective advantage on land.

Most likely the evolution of functional vascular tissue in ferns—xylem and phloem—which makes up much of the tissue in stems, roots, and leaves, provided two selective advantages to plants: conduction and support. The arrangement of vascular tissues is referred to as a **stele,** which is the central cylinder of tissue inside the cortex of stems and roots. The stele includes xylem, phloem, and pith, if any. The basic terminology of stem and root structure was presented in chapters 7 and 8. Those chapters focused on seed plants, but the basic cell and tissue types are the same in the seedless vascular plants.

(a)

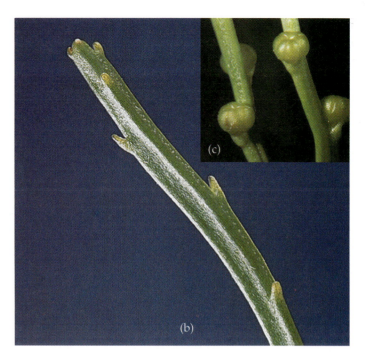

(c)

(b)

FIGURE 17.16

(*a*) Whisk fern (*Psilotum nudum*). (*b*) The stems of *Psilotum* are leaf-less but bear lobed sporangia, (*c*) which are called synangia.

measured for its size and annual growth rate, was calculated to have started growing in 1839.

Lycopodiophyta also includes the quillworts (*Isoetes*), so named because of their narrow, quill-like leaves (fig. 17.17). Quillworts may be aquatic or they may grow in small lakes or pools that dry out during some seasons. These plants live in freshwater habitats on almost every continent. Most of the leaves of quillworts are fertile and do

FIGURE 17.17

Quillworts (*Isoetes* species) are so named because of their narrow, quill-like leaves.

not aggregate into strobili; some leaves produce sporangia that abort before they mature. Quillworts are also distinctive in the division because they have CAM photosynthesis (see chapter 10). Some lake-dwelling species in the Peruvian Andes have no stomata in their leaves and obtain carbon dioxide for photosynthesis from the muddy substrate where they grow.

Division Equisetophyta: Horsetails. *Equisetum* is the only living genus of the *Equisetophyta*. Some of the 15 species of *Equisetum* have branched stems, and some have unbranched stems. *Equisetum* species are called horsetails (fig. 17.18) or scouring rushes, the latter because their epidermal tissue contains abrasive particles of silica. Scouring rushes were used by Native Americans to polish bows and arrows and by early colonists and pioneers to scrub their pots and pans. *Equisetum* occurs worldwide in moist habitats along the edge of streams or forests. Its rhizomes are highly branched and perennial. Because its rhizomes can grow rapidly and its aerial stems are poisonous to livestock, *Equisetum* can be a serious problem for farmers and ranchers. Gardeners have often chopped up the rhizome while trying to remove it from the soil, only to have new plants arise from each of the fragments left behind.

Although *Equisetum* has true leaves, the stem is the dominant photosynthetic organ of the plant body. The most conspicuous feature of the stem is the series of "joints" formed by whorls of small leaves (fig. 17.18). The leaves are fused along most of their length, but their brown tips give the appearance of a collar around the stem just above each node. When the stems are pulled apart, they break easily at

(a)

(c)

(b)

FIGURE 17.18

Horsetails. (*a*) A species of horsetail with photosynthetic stems and whorled branches (in background) and nonphotosynthetic reproductive stems. (*b*) Diagram of a horsetail. (*c*) The conspicuous "joints" of *Equisetum* stems consist of whorls of small leaves. These leaves, whose brown tips give the appearance of a collar just above the node, are fused into a sheath around the stem.

the nodes to yield pipelike internodal pieces. In addition, *Equisetum* stems are notable for a branching pattern that is unique among vascular plants. Instead of growing in the axils of leaves, the lateral branches of horsetails sprout from between the leaf bases.

Gametophytes of *Equisetum* are photosynthetic, pincushion-shaped plants that can grow up to 1 centimeter in diameter. The sexuality of *Equisetum* gametophytes is not well understood because it is variable and appears to be related to environmental conditions.

Division Polypodiophyla: Ferns.   Ferns include approximately 12,000 living species, making this division by far the largest among the seedless vascular plants. Ferns are primarily tropical plants, but some species inhabit temperate regions, and some even live in deserts. North or south of the tropics, the abundance of ferns decreases because of decreasing moisture. In tropical Guam, for example, about one-eighth of the species of vascular plants are ferns, but ferns constitute only about one-fiftieth of the total species in mostly dryer California.

Some genera of ferns have leaves that are the largest and most complex in the plant kingdom. For example, one species of tree fern in the genus *Marattia* has leaves that are up to 9 meters long and 4.5 meters wide, which is nearly the size of a two-car garage (fig. 17.19*a*). At the other extreme, the aquatic ferns *Salvinia* and *Azolla* have relatively tiny leaves

(fig. 17.19*b*). For most people, however, a typical example of a fern is the bracken fern, *Pteridium aquilinum* (see fig. 17.12). The bracken fern, like most ferns, is classified in the order Pteridales, which includes about 10,000 species. Because the bracken fern and its relatives are typical examples of the largest order of ferns, their characteristics are used in the following discussion to illustrate the general features of ferns.

Often the most conspicuous parts of a fern are its compound leaves, called **fronds.** A pinna is the leaflet of a frond. New leaves grow from a fleshy rhizome. Early leaves are curled because they grow faster on their lower surface than on their upper surface. This growth pattern, which is called circinate vernation, produces young leaves that are coiled into **fiddleheads** (fig. 17.20). New fiddleheads arise close to the growing tip of the rhizome at the beginning of each growing season. The leaves of most ferns die back each year, but the leaves of the walking fern (*Asplenium rhizophyllum*) can form new plants. Near the tip of each leaf, certain cells revert to meristems and grow into new roots, leaves, and a rhizome (see fig. 17.19*c*). Interestingly, the fiddleheads of certain ferns are considered a delicacy by some and can be found in health food stores and restaurants in this country.

Fern leaves are usually fertile but do not form strobili. The fronds characteristically have dark, spotlike structures, often on their lower surfaces, each of which is a collection of sporangia. Together sporangia are called a **sorus** (plural,

(c)

FIGURE 17.19

The diversity of ferns. Ferns range from (*a*) *Marattia* (×1/10) and other tree ferns, some of whose leaves are the largest known among plants, to (*b*) tiny-leaved aquatic ferns such as *Azolla* (×3/4). (*c*) The tips of leaves of the walking fern (*Asplenium* species) can become meristematic and form new plants. In this photograph, the tip of the long leaf on the plant to the left has formed the new plant (actually a clone) to the right.

sori). The sori of some species are covered by an outgrowth from the leaf surface called an **indusium,** while the sori of other species either are not covered or are enfolded by the edge of the leaf (fig. 17.21). Some ferns can produce millions of spores because of the number of sporangia per sorus, large number of spores per sproangia and the enormous number of sori per leaf. For example, one mature plant of *Thelypteris dentata* can produce more than 50 million spores each season.

A life cycle of a typical fern is shown in figure 17.22. Spores of some ferns are catapulted from their sporangia by a fascinating method that has attracted much attention from botanists. The flinging action comes from the behavior of an incomplete ring of cells, called the **annulus,** that encircles the sporangium. The thin outer cell walls of the annulus slowly contract as they dry, creating a pulling force that ruptures the thin-walled lip cells and the outer walls of the sporangium. As drying continues, the water tension increases in the annulus, sometimes exceeding 300 atmospheres of pressure. Ultimately, the water evaporates completely, and the tension is broken. The sporangium then snaps back into its original position, ejecting the spores forcefully to a distance of up to 1 meter. If you view a sorus under a microscope, the heat from the light often accelerates this process, and you can see the fern's spore "sling shot" in action.

In most ferns, haploid (*n*) spores germinate and eventually differentiate into green, often heart-shaped prothallus gametophytes that are anchored to the soil by rhizoids (fig. 17.22). The haploid gametophytes are usually bisexual with the sex organs on the lower surface. Archegonia are sunken in the gametophyte tissue near the notch of the "heart," their

FIGURE 17.20

The coiled "fiddleheads" of fern leaves are rolled-up leaf buds. Fiddleheads are formed by a pattern of growth called circinate vernation, as shown in this *Blechnum* fern.

(a)   (c)

FIGURE 17.21

Fern sporangia. Most ferns have sporangia aggregated into clusters, called sori, on the undersides of the leaves. (*a*) In some ferns, such as the marginal woodfern (*Dryopteris marginalis*), each sorus is covered by a flap of leaf tissue called an indusium. (*b*) Other ferns bear uncovered sori, as shown here in *Alsophila sinuata*. (*c*) In still other ferns, as in the giant maidenhair fern (*Adiantum trapeziforme*), sori occur along leaf edges and are enfolded by the edge of the leaf itself.

necks sticking out slightly; antheridia protrude from the surface near the tip and are intermingled with rhizoids. The multiflagellate sperms released by the antheridia swim into the neck of the archegonium to reach the egg. Once fertilization occurs, the diploid (2*n*) zygote germinates into a young sporophyte that quickly becomes independent and lives separately from the host gametophyte. Shortly thereafter, the short-lived gametophyte dies.

Some ferns are decidedly unfernlike in their overall appearance. *Azolla*, for example, is a floating aquatic fern having tiny leaves that are crowded onto slender stems (see fig. 17.19*b*). Other genera, such as *Marsilea*, grow from roots buried in the muddy bottoms of ponds or ditches; only their leaves reach the water surface and float upon it. These aquatic ferns also have a very different reproductive biology.

Seedless vascular plants coevolved with other organisms.

One of the most interesting ecological aspects of any group of plants involves their interactions with other organisms. For example, the occurrence of mycorrhizal fungi in the gametophytes of *Psilotum* and *Lycopodium* and in the rhizomes and roots of most plants suggests a general symbiotic relationship between fungi and plants. Fungal filaments extract certain minerals from the soil better than do rhizoids or root hairs. Plants obtain such minerals from the fungi associated with them, while the fungi, in return, receive sugars and other nutrients from their hosts. Similar fungi found in fossil plants probably had the same relationship with their hosts more than 400 million years ago. Indeed, this association with fungi is thought to have hastened the invasion of land by plants because of increased efficiency of water and mineral absorption.

Several tropical species of ferns harbor ant nests in their rhizomes (fig. 17.23). Roots grow into the ant-carved tunnels and absorb nutrients from the decaying organic matter brought in by the ants, and ants use the sporangia as food. Spores are often shaken loose and dispersed from the parent plant as the ants handle the sporangia. Botanists speculate that the loss of spores to ants is sufficiently offset by the advantage of having at least some of the spores dispersed.

Many ferns can be noxious weeds. For example, when people burn native forests to establish pastures, bracken ferns can quickly invade the newly available habitats (see fig. 17.12). Populations of this plant spread rapidly from an extensive network of fast-growing rhizomes. The problem of bracken infestation is worsened by the toxicity of this plant to the cattle raised in such pastures. Herbicides can alleviate the problem somewhat, but this damages other plants and leaves a toxic residue in the soil. Weed scientists are now studying how to control bracken fern by introducing pathogenic fungi into invasive populations.

Kariba weed (*Salvinia molesta*), an aquatic fern, can be a more serious problem than bracken fern, even though kariba weed is not poisonous (fig. 17.24). Trouble with this plant began after it became established in Papua, New Guinea, north of Australia. The invasion of kariba weed overtook waterways and eliminated native species of plants, eventually threatening the availability of water to 80,000 people. Kariba weed is sterile but reproduces asexually very rapidly, doubling its mass approximately every 2 days. From a genetic viewpoint, millions of tons of this weed may represent a single individual, making it a candidate for the largest organism on earth. In the 1980s, botanists made a significant breakthrough in understanding the ecology of this organism. Its native range was discovered in southeastern Brazil, where a new beetle species was found that feeds exclusively on kariba weed. This beetle is now being used successfully to control invasions of kariba weed. In

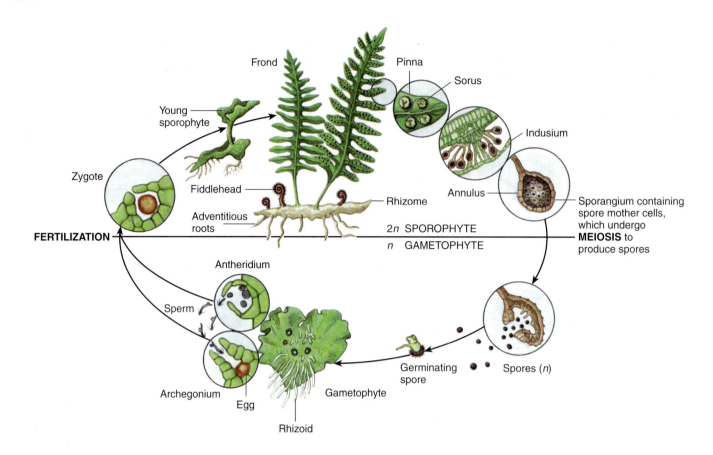

FIGURE 17.22

Life cycle of the fern *Polypodium.* Many ferns are like *Polypodium* and have similar life cycles. In most ferns the diploid sporophyte is dominant.

Papua, New Guinea, the beetles have eaten about 2 million metric tons (2 billion kilograms) of the plant, reducing the water surface it covered by more than 90%.

Seedless vascular plants of today have limited economic importance.

Seedless vascular plants have their greatest economic impact in fossil fuel deposits. Their spores are easy to identify and are associated with oil deposits. Many seedless vascular plants, especially ferns, are often found in greenhouses or are grown as houseplants and ground covers. Western pioneers used *Equisetum* to scrub their dishes, and before the invention of flashbulbs, photographers used flash powder that consisted almost entirely of dried *Lycopodium* spores. A pound of spores can still be purchased for only a few dollars from some scientific supply companies. In China, where petroleum-based fertilizers are not affordable, *Azolla* is substituted as a rotated crop in rice paddies. This aquatic fern hosts a cyanobacterium, *Anabaena azollae,* that fixes nitrogen from the air, thereby acting as a fertilizer to replenish the nitrates removed from the soil by other crop plants (see fig. 7.25).

The evolution of biochemical complexity in land plants coincided with their morphological advancements. Besides structural compounds such as biochemically complex lignin in secondary cell walls (see chapter 3), chemicals also evolved that protected the plant from ultraviolet light, parasitic fungi, protozoa, and other predators. The continued biochemical diversification of land plants has resulted in a wealth of chemicals that have been useful to humans. For example, Native Americans treated wounds and nosebleeds with spores from one species of club moss, *Lycopodium clavatum,* which has antibiotic and blood-coagulant properties. Resin from the rhizomes of the marginal fern *Dryopteris marginalis* was once taken internally to get rid of intestinal tapeworms. As is true for most medicinal plants, the exact identity of the active ingredient from these plants has not been determined. We do know, however, that many *Lycopodium* species synthesize complex, nitrogen-containing chemicals called alkaloids that are potent animal poisons. The dried and powdered leaves containing these chemicals are used as pesticides in parts of eastern Europe.

Fossils of vascular plants are more abundant than those of bryophytes.

Vascular tissues, cuticles, and spores are particularly well preserved because of their resistance to decomposition by bacteria and fungi. Although the fossil record is still far from

(a)

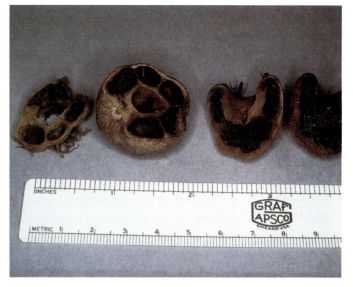

(b)

FIGURE 17.23

Ants live symbiotically with some ferns. (*a*) *Solanopteris brunei*, a fern native to Costa Rica, harbors nests of ants. (*b*) Ants live in podlike chambers of rhizomes that are formed by the plant.

complete, several thousand extinct species of ferns have been discovered. Most of these are known only from spores or fragments of the plant body and cannot be matched with an entire plant (see Perspective 17.2, "How Does a Plant Become a Fossil?"). The oldest fossil genera, consisting of cuticles and spore tetrads, were discovered in late Ordovician rocks in Libya. Many botanists are not convinced that these plants had vascular tissue, however, since tracheids or other evidence of vascular tissue has not been found with them.

The oldest fossils that are unquestionably seedless vascular plants consist of well-preserved vegetative and reproductive structures. These plants, named *Cooksonia,* were rootless and leafless, with slender, dichotomously branched stems (fig. 17.25). They produced spores in sporangia that developed from the expanded tips of their branches. *Cooksonia* first appeared in the late Silurian period, about 420 million years ago. These first known invaders of land were abundant in the tidal mudflats of New York State, South Wales, the Czech Republic, and Podolia (Ukraine). Their distribution probably included many other areas that experienced periodic flooding, but fossils in these areas were either not preserved or have not yet been discovered.

Cooksonia represents the ancient opening of the floodgates of diversification for vascular plants. When there was little competition for colonizing immense areas of land. Although *Cooksonia* was the most successful of the land invaders, it gave way to larger and more complex kinds of plants during the Devonian period. Plants grew taller and thicker, and they began to develop leaves that had a larger surface area for photosynthesis. The development of roots and stem cuticles allowed plants to move even farther from their swampy origins into increasingly drier habitats. The development of mycorrhizae (chapter 16) probably played an important role in enabling plants to obtain nutrients from soil. Spores became more resistant to desiccation, which enabled them to tolerate dispersal by wind without drying out.

Depending on their quality and the extent of preservation, fragments of fossil plants can occasionally be matched so that a significant portion of the original plant can be reconstructed. Perhaps 200 or so fossil species fit this category. Of these, not more than two dozen are known in detail from a large number of specimens. Our picture of the earliest vascular plants relies heavily on this small number of well-known plants.

Devonian plants reveal two distinct evolutionary lines. Plants in one line, referred to as the zosterophyllophytes (division Zosterophyllophyta), superficially resemble the living genus *Zostera,* the marine eelgrasses. Zosterophyllophytes grew in clusters of upright, leafless branches that arose from horizontal stems, much as *Cooksonia* did (fig. 17.25).

The second evolutionary line includes *Cooksonia* and its relatives in division Rhyniophyta. This division is named

(b)

(a)

FIGURE 17.24

The kariba weed (*Salvinia molesta*) reproduces so fast that it can (*a*) crowd out other kinds of plants and (*b*) overtake waterways.

for the remarkably preserved assemblage of Devonian fossils near the village of Rhynie, Scotland. Although *Cooksonia*, a rhyniophyte, is the oldest known vascular plant, it is unlikely that the rhyniophytes gave rise to the zosterophyllophytes. These two divisions must have shared a common ancestor at some point in their history, but their exact relationship to one another is unknown.

INQUIRY SUMMARY

Seedless vascular plants share many of their reproductive and vegetative features with bryophytes. However, the evolution of a resistant cuticle, complex stomata, vascular tissue, and lignin enabled these plants to dominate habitats on land.

Seedless vascular plants reproduce by spores; the sporophytes, unlike those of the bryophytes, is the dominant phase of the life cycle. Most seedless vascular plants are ferns, division Polypodiophyla. The remainder are classified in three divisions: the Psiloto-

phyta (whisk ferns), Lycopodiophyta (club mosses), and Equisetophyta (horsetails). All seedless vascular plants produce flagellated sperm cells and therefore require free water to reproduce sexually.

The first vascular land plants are known from fossil sporophytes of the Silurian period, approximately 420 million years ago. Like present-day *Psilotum*, they had no leaves or roots but instead had photosynthetic stems and underground rhizomes with absorptive rhizoids. Descendants of these plants evolved leaves, roots, and strong structural support tissues.

How was the development of functional xylem and phloem important to the evolution of ferns?

Coal-age trees were giant club mosses, giant horsetails, and tree ferns.

Evidence shows the first land plants were relatively small and herbaceous. With the advent of leaves, thicker cuticles,

17.2 *Perspective* *Perspective*

How Does a Plant Become a Fossil?

Archaeologists trying to piece together evidence of past civilizations search for telltale shards of pottery and other artifacts. Paleobotanists use similar direct evidence in the form of fossils.

The public perception that fossils are just old plants or animals in rocks doesn't do justice to the wide variety of things that are called fossils. Different types of fossils form, depending on where the organisms grew and how fast they were buried in sediment. The preservation of most organisms has occurred in water as sedimentary particles were deposited on plant parts that fell into the water. Heavy sediment flattened leaves and other plant parts, squeezing out water and leaving only a thin film of tissue, called a compression fossil (see photo). Cellular structure rarely survived this process, but well-prepared cuticles have occasionally been found in deposits of compression fossils. Intact cells (and even DNA) have also been found in compressions, as in the Miocene deposit (ca. 20 million years old) at a fossil site named Clarkia Lake in Idaho.

Other types of fossils usually lack plant tissue. An impression, for example, is an imprint of an organism that is left behind when the organic remains have been destroyed. Only the contour of the plant remains.

In a third type of fossilization, tissues became surrounded by hardened sediment and then decayed. The hollow negative of the original tissue is called a mold. In conditions that allowed the mold to become filled with other sediment that conformed to the contours of the mold, the resultant fossil is called a cast. Fossil molds and casts, although formed on a geological timescale, are

Types of fossils include (a) compressions, which are made when heavy sediment squeezes out water and leaves only a thin film of tissue; (b) impressions, which are the imprints left in rocks after the organic remains of an organism have been destroyed; and (c) petrifactions, which form when cell contents are replaced with minerals, thereby transforming the organic material into rock. Petrifactions are so common in some areas that they form "petrified" forests, such as the Petrified Forest of Arizona.

analogous to those that modern artists make of their subjects.

An interesting but poorly understood process of fossilization involves the replacement of cell contents with minerals. The compaction of such mineralized tissues essentially transforms the organic material into rock. Such fossils, called petrifactions, make areas like the Petrified Forest of Arizona famous among botanists and tourists alike.

Plant fossils are rarely found as whole plants. Different plant parts in the same fossil bed are given separate species and genus names, referred to as form species or genera. Sometimes two or more parts are connected to each other, and the plant they came from can be drawn as a reconstruction of the original. One such example, the genus *Lepidodendron*, is actually the name given to stem fragments; the rest of the plant is derived from several unconnected organs and other fragments, each of which was given a separate name when it was discovered.

roots, and perhaps most important, lignin, plants could grow larger and could more effectively exploit the year-round growth conditions during the Carboniferous period. By the middle of the Carboniferous period (about 300 million years ago), some plant groups had evolved into large trees, and some of their herbaceous progenitors were well

on their way to becoming extinct. These first trees included the earliest seed plants (discussed in the next chapter) as well as several groups of seedless vascular plants. These huge plants formed the extensive coal deposits that characterize the Coal Age. When you heat your home with coal or even fill your car with gasoline, you are filling it with the

(a)

(b)

(c)

FIGURE 17.25

(*a*) Reconstruction of a Devonian habitat. The low-growing seedless vascular plants of this period included *Cooksonia, Zosterophyllum,* and *Psilophyton.* (*b*) Swamp forest of the Carboniferous period. Most of the trees in this reconstruction are giant club mosses. Plants with frondlike leaves (e.g., left foreground) are seed ferns. The tree on the right that has bottlebrush-appearing branches is *Calamites,* a giant horsetail. (*c*) A fossil specimen of the zosterophyll (*Bathurstia denticulata*) from the early Devonian Period, found on Bathurst Island in the Northwest Territories of the Canadian Arctic. This specimen has numerous curled side-branches developing from its rhizome. The zosterophylls are most likely ancestors of the lycophytes, including modern-day club mosses and *Selaginella.*

SUMMARY

Plants are eukaryotic, multicellular, photosynthetic organisms with cellulose based cell walls and embryos produced via sexual reproduction. Their cells contain chloroplasts, chlorophylls *a* and *b,* carotenoids and flavonoids. Starch is the major storage carbohydrate. In plants with mobile sperm, only the gametes, not the vegetative structures, have flagella—this includes mosses, liverworts, ferns, cycads and *Ginko.*

Bryophytes, which include mosses, liverworts, and hornworts, are small plants that produce chlorophylls *a* and *b,* starch, cellulosic cell walls, and swimming sperm cells. They lack complex vascular tissues and supporting tissue, and they absorb water by capillarity. Bryophytes have an alternation of generations dominated by a photosynthetic and independent gametophyte. Eggs and sperms form in gametangia called archegonia and antheridia, respectively, each of which are surrounded by a protective jacket of sterile cells.

Bryophytes require free water for sexual reproduction, so that flagellated sperm cells can swim to eggs. The fertilized egg begins the diploid sporophyte generation, which is attached to and nutritionally dependent on the gameto-

products of ancient photosynthesis performed by these early plants.

The swamp-dwelling giant club mosses grew nearly 40 meters tall and dominated the Carboniferous period. The massive stems were supported by extensive periderm tissue and a small amount of secondary xylem. Underground they had a modified branch system with rootlike appendages. These giant club mosses also had large strobili. A single strobilus, up to 0.5 meters long from one of these plants, may have produced as many as 8 billion spores. Although the waste must have been enormous, spore dispersal from tall trees enabled the trees to colonize large areas of land successfully. Giant horsetails such as *Calamites* lived in swampy forests alongside the giant club mosses. Secondary xylem enabled these organisms to grow to almost 20 meters tall.

The Carboniferous period also saw the origin and diversification of tree ferns, though today the particular genera of tree ferns from the Carboniferous period are extinct. Still, as a larger group, these primitive plants are well represented in tropical regions of the world today. At the end of the Carboniferous period, the giant club mosses began to disappear rapidly and were gone completely after the mass extinction that characterized the end of the Permian period, approximately 250 million years ago. In contrast, the giant horsetails continued to dominate the landscape for another 100 million years or so, until the mid-Cretaceous period, when they also died out. Nevertheless, many of the primitive features of these extinct trees are maintained in the smaller, modern members of these groups.

phyte. Haploid spores formed by meiosis are released from sporangia and are usually dispersed by wind. Spores germinate and form gametophytes, thus completing the life cycle. Bryophytes have modest growth requirements and grow almost everywhere that other plants grow. Often reproduction in bryophytes is asexual.

Mosses are the largest and most familiar group of bryophytes. They produce rhizoids and have "leafy" gametophytes. Haploid spores form in sporangia that open and release the spores. A spore produces a filamentous and branched protonema, which develops into the gametophyte. Sperm cells released from antheridia swim to archegonia and to the egg. The zygote begins the diploid sporophyte generation that consists of a foot, seta, and capsule (sporangium). The foot penetrates the gametophyte and absorbs food and water for the sporophyte. The stalk-like seta raises the spore-producing capsule above the gametophyte.

Liverworts have leafy or thallose gametophytes with unicellular rhizoids and frequently reproduce asexually. Hornworts have the fewest number of species among the bryophytes. Their sporophytes are shaped like tapered horns. The thallus has stomatalike structures. Some botanists do not consider hornworts to be bryophytes.

Bryophytes reduce erosion, condition soil, and are often among the first organisms to invade disturbed areas. Many bryophytes grow in specific habitats and are sensitive to pollution.

The evolutionary history of bryophytes is vague. They probably diverged after plants invaded land. Botanists do not agree on the relationships of bryophytes to other groups of plants.

Seedless vascular plants are primarily ferns, but they also include whisk ferns (which are not true ferns), club mosses, and horsetails. Features of seedless vascular plants that enable them to thrive on land include a resistant cuticle, complex stomata, vascular tissue, absorptive root hairs, and desiccation-resistant spores. Vascular tissue in ferns is more complex than in plants of the other groups of seedless vascular plants.

Sexual reproduction in seedless vascular plants is similar to that of bryophytes; that is, haploid spores germinate into gametophytes. The life cycle of seedless vascular plants is dominated by the diploid sporophyte phase. Spores are generally dispersed by wind, although the spores of some ferns are dispersed by ants. All horsetails and some club mosses bear sporangia in densely packed cones, or strobili. Many seedless vascular plants can reproduce vegetatively, usually by the sporophyte reproducing asexually.

The seedless vascular plants are of little direct economic value. Some species were used as folk medicines, pesticides, or photographic flash powder. Fossil plants have some indirect economic value because geologists use the location of spores in the earth's strata to judge where oil deposits might occur. Some seedless vascular plants cause problems because they diminish the usefulness of pastureland, clog waterways, and eliminate native plant species.

The current fossil record of vascular plants begins with *Cooksonia,* a leafless and rootless progenitor of most modern plant groups. Giant horsetails, giant club mosses, and tree ferns dominated the vegetation of the Carboniferous period. Although these seedless trees ultimately gave way to seed plants, their descendants live on as herbaceous lycopodophytes and as modern tree ferns.

Writing to Learn Botany

- Consider the results of the following experiment: if the zygote is removed from an archegonium of a moss, it forms a protonema and a leafy gametophyte. If a piece of the gametophyte is transplanted into an archegonium, it will form a sporophyte. What do you conclude from these results?
- Go to the library and read a recent article about the use of *Azolla* in rice production in China. Is *Azolla* used in rice production in the United States? Why or why not?
- What is the potential evolutionary significance of the discovery of *Cooksonia?*

THOUGHT QUESTIONS

1. Mosses and liverworts have been used extensively to monitor radioactive fallout from the Chernobyl reactor accident that occurred in April 1986. What features of these organisms make them ideal for this?

2. What is the evidence that bryophytes and vascular plants had a common ancestry? What does this evidence suggest about the common ancestor of bryophytes and other plants?

3. Although bryophytes require free water for sexual reproduction, several bryophytes grow in deserts. How do you think these bryophytes reproduce in spite of their mostly dry environment?

4. Why are there so few fossils of bryophytes?

5. What might be the adaptive significance of unisexual versus bisexual gametophytes?

6. What does the fossil record show regarding the relationships of bryophytes to other plants?

7. How are the gametophytes of seedless vascular plants similar to the gametophytes of bryophytes? How are they different?

8. Why is *Cooksonia,* which is the oldest confirmed plant fossil, not considered to be the oldest ancestor of all modern plant groups?

9. What are the major distinguishing features of the divisions of extant seedless vascular plants?

10. If you had only a microscope, how could you tell a bryophyte from a fern stem?

SUGGESTED READINGS

Articles

Levanthes, L. E. 1987. Mysteries of the bog. *National Geographic* 171(3):396–420.

Raghavan, V. 1992. Germination of fern spores. *American Scientist* 80:176–85.

Wang, T. L., and D. J. Cove. 1989. Mosses—lower plants with high potential. *Plants Today* (March-April): 44–50.

Books

Chopra, R. N., and P. K. Kumra. 1988. *Biology of Bryophytes.* New York: John Wiley and Sons.

Gifford, E. M., and A. S. Foster. 1989. *Morphology and Evolution of Vascular Plants.* New York: W. H. Freeman.

Kenrick, P., and P. Crane. 1997. The Origin and Early Diversification of Land Plants—a Cladistic Study. Washington D.C.: Smithsonian Institution Press.

Iwatsuki, K., and P. Raven. 1997. *Evolution and Diversification of Plants.* New York: Springer-Verlag.

ON THE INTERNET

Visit the textbook's accompanying web site at http://www.mhhe.com/botany to find live Internet links for each of the topics listed below:

Major groups of bryophytes

Doctrine of signatures

Classification and evolution of bryophytes

Economics of bryophytes

Classification and evolution of ferns and allies

Alternation of generations

Fern allies

Economics of ferns

Movement from water to land

Bryophytes as pollution indicators

A community in northern Manitoba, Canada. The flowering plants in the foreground and coniferous trees in the background represent the two main kinds of seed plants. What characteristics allow these plants to dominate the land?

CHAPTER 18

Gymnosperms and Angiosperms: The Seed Plants

CHAPTER OUTLINE

In gymnosperms and angiosperms, characteristics have evolved that allow them to dominate life on land.

Gymnosperms produce naked seeds.
Gymnosperms produce both male and female gametophytes.
Many gymnosperms produce seeds and pollen in cones.
The pollen grain of a gymnosperm is a reduced male gametophyte.

Compared to angiosperms, few gymnosperms live in the world today.

Maidenhair trees are native to China but are cultivated around the world.
Cycads are tropical gymnosperms.
Many conifers produce seeds in woody cones.
The most unusual gymnosperms are gnetophytes.
Many gymnosperms are adapted to temperate climates.

Gymnosperms have many economically important uses.

Perspective 18.1 A Tree That Saves Lives
Perspective 18.2 How Much Paper Do We Use?

Flowering plants are the most diverse plant group.

Could dinosaurs be responsible for the diversity of flowering plants today?

The life cycle of flowering plants is different from that of all other plants.

Flowering plants are economically, culturally, and historically important to humans.

Cereals feed the human population.

The struggle to produce enough food in the world continues.

There are promising prospects for the future beyond the Green Revolution.

Summary

LEARNING ONLINE

The Online Learning Center at www.mhhe.com/botany contains a world map on which markers represent articles about current botanical news in that particular country. This is just one of the many useful resources available to you on the OLC. Also helpful are the hyperlinks that correlate to the listings at the end of every chapter. Visit www.ucmp.berkeley.edu/seedplants/seedplants.html in chapter 18 of the OLC to learn more about gymnosperms and angiosperms.

When astronauts circle the earth, they see great oceans covering our planet and great expanses of forests and grasslands covering the land. These plant communities consist mostly of gymnosperms and angiosperms. What makes these plants so successful in comparison to others such as mosses and ferns? What adaptations have evolved that enable them to live on land? Part of the answer to both of these questions comes from one of the features that characterizes these plants—the seed.

Gymnosperms are mostly cone-bearing trees such as pines, spruces, firs, and their relatives; angiosperms are the flowering plants. Both groups produce seeds at some point during their life cycle; no other plants have seeds. A seed protects and nourishes the immature, vulnerable offspring—the embryo—from the hostile environment until the embryo can produce roots, stems, and leaves and grow on its own. Seeds, in combination with other characteristics you will learn about in this chapter, allowed cone-bearing and flowering plants to colonize and dominate the land. However, while there are many individual gymnosperms in the world, they are represented by only a few hundred species of cone-bearing plants. How can we explain this paucity of gymnosperm species compared with over 250,000 species of flowering plants? What has led to the evolution of so many species of angiosperms?

By far, gymnosperms and angiosperms are also the most important groups of plants to humans as sources of lumber, paper, food, medicine, and many other useful products. We have learned about some of these products from both groups throughout the book, and in this chapter, we learn many more. Gymnosperms and angiosperms are essential to our lives.

In Gymnosperms and Angiosperms, Characteristics Have Evolved That Allow Them to Dominate Life on Land.

Gymnosperms and angiosperms were not the first plants to evolve; they were not even the first plants to populate the bare terrain of a primitive earth. These seed-producing plants, however, quickly covered much of the earth with their offspring. Life began in the water, and the land that rose above the ancient seas was initially devoid of life. The earliest life-forms on earth had no problem getting the water they needed because they were surrounded by it. Also, because no competition existed on the land at that time, any plant that could live in this wide-open habitat could quickly dominate the landscape. However, the main problem for early colonizers of the land was a lack of water. Any characteristic that would allow a plant to obtain and maintain water would give it a great selective advantage over other plants. In addition, as the land became populated with plants, taller individuals with larger leaves could absorb more light for photosynthesis than shorter plants with small leaves, and therefore they could grow faster than their smaller competitors. So, natural selection favored taller plants that could get enough water to all of their parts. In chapters 8 and 9, you learned about the importance of vascular tissue for transporting water and nutrients to all parts of the plant body. Ferns, gymnosperms, and angiosperms all have vascular tissue, which has allowed them to grow taller and live in drier places than most mosses, liverworts, and other nonvascular plants.

In the dry environment on land, the evolution of other vegetative features such as sunken stomata and thicker cuticles enabled land plants to minimize water loss. These characteristics, in addition to xylem and phloem, helped vascular plants retain the water they absorbed through roots. However, while cone-bearing and flowering plants share the characteristic of vascular tissue with ferns, they have one important characteristic that ferns and their relatives do not—seeds.

Among the most successful of the early seed plants were *seed ferns* (fig. 18.1), extinct gymnosperms that modern scientists first mistakenly identified as ferns because of their fernlike leaves—until it was discovered that these plants also produced seeds. Seed ferns dominated the landscape for about 70 million years until the climate changed during the early Permian period. After the seed ferns disappeared, three other groups of gymnosperms thrived: (1) cycads, ancient gymnosperms, some of which exist today; (2) conifers, the cone-bearing plants; and (3) cycadeoids, seed-bearing gymnosperms with palmlike leaves (fig. 18.2). Of these, only a few descendants of the cycads and conifers remain today. The cycadeoids and cycads were so abundant during the Jurassic that this period is known as the Age of Cycads, at least among botanists (other biologists usually refer to it as the Age of Dinosaurs). The cycadeoids went extinct about the same time as the dinosaurs, about 65 million years ago (see fig. 14.25).

Gymnosperms produce naked seeds.

The term *gymnosperm* derives from the Greek word roots *gymnos,* meaning "naked," and *sperma,* meaning "seed." **Gymnosperms** are plants whose pollen is carried by wind directly to ovules (unfertilized seeds) instead of to a stigma (as in the flowering plants) and whose seeds are naked (i.e., not enclosed in fruits). Thus, by definition, gymnosperms

FIGURE 18.2

Vegetatively, the now-extinct cycadeoids such as *Cycadeoidea* were similar to cycads. Internally, like cycads, they had a broad pith, a small amount of xylem, a broad cortex, and a tough outer protective layer formed by persistent leaf bases.

FIGURE 18.1

Seed ferns, represented here by this reconstructed *Medullosa,* were abundant in the late Carboniferous period.

are all seed plants without fruits. Examples of gymnosperms include the maidenhair tree (*Ginkgo*), cycads, junipers, pines, redwoods, and members of the Gnetophyta (e.g., *Ephedra, Gnetum*) (fig. 18.3).

A major change occurred in plant evolution when seeds evolved about 350 million years ago, a time when most continents were flooded by tropical swamps. Seeds develop from ovules, and the ovule protects the **female gametophyte,** which houses the egg. This protection of the female gametophyte increases the likelihood that an egg will be fertilized by a sperm cell and that the resulting embryo will survive. This gives seed plants a tremendous selective advantage over seedless plants that simply release their offspring into the environment with minimal protection. What advantages does an embryo gain by developing inside a seed?

In addition to seeds, gymnosperms are characterized by secondary growth that usually forms woody trees or shrubs, but some species are vinelike. Most gymnosperms lack vessels in their xylem; in this regard, the gymnosperms are like most ferns and their relatives.

Considering the relatively small number of living gymnosperm species (about 720 species in 65 genera), they are remarkably diverse in their reproductive structures and leaf types. Pollen-bearing structures may be either loosely arranged, packed into multiple clusters, or flowerlike (fig.

18.3). Leaf types range from simple, flat blades to needles, compound frondlike leaves, and highly reduced scalelike leaves (fig. 18.4). To what kind of environments do you think plants with small leaves are adapted?

Gymnosperms produce both male and female gametophytes.

Recall the generalized life cycle of flowering plants that was introduced in chapter 13. In the life cycle of gymnosperms, the alternation between sporophytic and gametophytic phases is similar to that in other plants (fig. 18.5). Like the angiosperms, gymnosperms produce two types of spores, and their gametophytes are unisexual. This means that gymnosperms have male gametophytes that produce only sperm cells and female gametophytes that produce only eggs.

Ovule development begins when a single, diploid cell—the megasporocyte—in the megasporangium (or nucellus) undergoes meiosis (see chapter 13), forming four haploid megaspores (fig. 18.5). Three of these megaspores usually disintegrate, and the remaining **functional megaspore** undergoes repeated cell divisions that are not followed immediately by cytokinesis. The result is a female gametophyte composed of one cell with many nuclei. The number of nuclei (per cell) in the female gametophyte can be as low as 256 in some species of *Ephedra* to as high as about 8,000 in the maidenhair tree (*Ginkgo biloba*). Cell walls later form around each nucleus, after which two or more

(a)

(b)

(c)

(d)

F I G U R E   1 8 . 3
Pollen-bearing cones of gymnosperms. (*a*) *Ginkgo.* (*b*) *Dioon,* a cycad.
(*c*) Pine (*Pinus*). (*d*) *Ephedra.*

F I G U R E   1 8 . 4
Scalelike mature leaves of juniper (*Juniperus*). Each branch is surrounded by many tiny green leaves.

(up to 200) archegonia develop near one end of the female gametophyte, inside an ovule. Each archegonium—a female sex organ—contains a single egg, and all eggs may be fertilized. Most of the mass of a mature gymnosperm seed consists of an integument layer that later develops into the seed coat, as well as a multicellular female gametophyte, and usually one, but possibly more, embryos.

Many gymnosperms produce seeds and pollen in cones.

The simplest seed-bearing structures among gymnosperms are those of the maidenhair tree (*Ginkgo biloba*) (fig. 18.6*a*) and the yew family (e.g., *Taxus* and *Torreya*). Seeds of these groups are borne singly at the ends of stalks; *Ginkgo* ovules start out in pairs, but usually one aborts early in development. The mature ovules of yews and California nutmeg (*Torreya californica*) are surrounded by a fleshy, cuplike structure (an aril) that is often bright red (fig. 18.6*b*). In contrast, most other gymnosperms produce seeds in a complex reproductive structure consisting of several scales grouped together at the end of a stem and commonly called **cones** (the botanical term is **strobilus**).

Most gymnosperms have two types of cones: "male" cones (microstrobili), in which pollen grains (male gametophytes) are produced, and "female" cones (megastrobili), in which ovules are produced. Technically, there is no such thing as a male or female cone because gender should be applied only to those structures, the gametophytes, that produce gametes. Each ovule contains a female gametophyte, which produces at least one egg. When the sperm cell from the pollen grain fertilizes an egg inside a female gametophyte, the ovule develops into a seed. This means that sexual reproduction in most gymnosperms requires the transfer of gametes from a male cone to a female cone.

The smallest female cones include those of the junipers (*Juniperus*), which have fleshy scales that are fused into a berrylike structure (fig. 18.6*c*). In many gymnosperms, the scales of the female cones become woody. The largest seed cones, which may be up to 1 meter long and weigh more than 15 kilograms (over 33 pounds), are produced by cycads (fig. 18.6*d*). Ovules are borne on reduced branches that occur in the axils of spirally arranged bracts of the cone (fig. 18.7). Pollen cones usually do not become woody and are much smaller than seed cones, except in cycads, where the pollen and seed cones may be similar in size.

The pollen grain of a gymnosperm is a reduced male gametophyte.

Most gymnosperms produce huge amounts of pollen. For example, each male cone of a pine tree releases 1 to 2 million pollen grains. Similarly, the immense conifer forests of Sweden have been estimated to release 75,000 *tons* of pollen

every spring. It is no wonder that gymnosperms contribute to some people's allergy problems each year.

The gymnosperm pollen grain is a reduced, immature male gametophyte combined with and contained within a microspore wall. Pollen grains evolved by a continued reduction in size of the male gametophyte and a more secure retention of the gametophyte inside the protective spore wall. Compare this to the seedless plants such as mosses (see fig. 17.3), which release their spores into the wind and their gametophytes must grow on their own.

Each immature pollen grain contains two cells, the generative cell and the tube cell. The generative cell of a pollen grain divides to form a stalk (or sterile) cell and a body (or spermatogenous) cell, which is the only sperm-producing cell in the male gametophyte (see fig. 18.5). Immediately before fertilization, the body cell divides to form two sperm cells.

Seedless plants produce sperm cells having **flagella;** a flagellum is a whiplike structure essential for movements such as swimming. Thus, for sexual reproduction to occur in seedless plants such as mosses, liquid water is required in order for sperm to swim to an egg. Gymnosperms and most angiosperms do not require water to get their sperm cells from one plant to another because the pollen grain carries the sperm to the egg. The only gymnosperms with swimming sperm cells are the cycads and the maidenhair tree, but even these gymnosperms must be pollinated by wind, animals, or water. This means that the movement of sperm to egg in seed plants relies not on water transport but instead on transport in pollen; the sperm cells are produced inside the pollen grains, which are carried from plant to plant by wind or pollinating insects. Thus, seed plants don't require water in the environment for sexual reproduction to occur. How would pollen give a seed plant a great advantage over a seedless plant in colonizing dry land?

Pollination in gymnosperms involves a **pollination droplet** that protrudes from the micropyle when pollen grains are being shed (fig. 18.8). This droplet provides a large, sticky surface that catches wind-borne pollen grains of gymnosperms so that the ovule is more likely to be fertilized. During pollination, dozens of pollen grains may stick to each droplet. After pollination, the droplet dries and contracts, carrying the pollen grains, with sperm cells, into contact with the ovule. The tube cell of the pollen grain grows into the ovule, serving as a passageway for the sperm cells, one of which fertilizes the egg.

INQUIRY SUMMARY

Gymnosperms and angiosperms are seed plants that dominate communities on land. They possess characteristics that allowed them to colonize land, including vascular tissues, a cuticle, pollen grains, and seeds, which also allowed them to become highly successful,

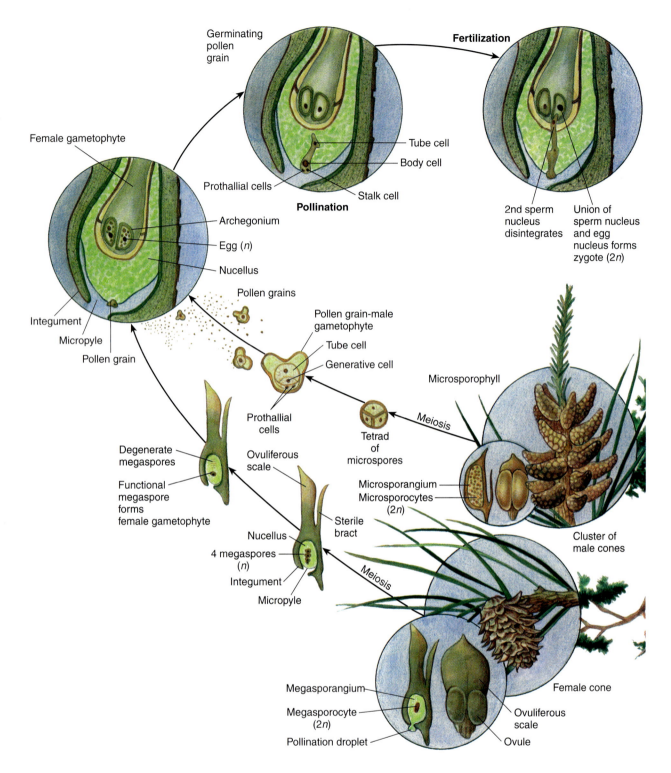

FIGURE 18.5

Life cycle of pine (*Pinus*), a representative gymnosperm. In pine, pollination occurs more than a year before the ovule produces a mature female gametophyte. Seeds may mature more than 6 months after fertilization, which would be more than 18 months after pollination.

Female gametophyte
(*n*)

Proembryo
(2*n*)

Female gametophyte

Developing embryo

Suspensor

Rosette cells

Female cone
matures,
seeds
disperse

Female gametophyte

Seed
coat

Plumule

Cotyledon

Cotyledon

Shoot apex
(apical meristem)

Hypocotyl

Root apex
(apical meristem)

Root cap

Remnants of
nucellus

Germinating
seed

Seedling

Mature tree
(*Pinus sylvestris*)

Young
tree

FIGURE 18.5

Continued

(a)

(b)

(c)

(d)

FIGURE 18.6

Gymnosperms. (*a*) *Ginkgo* seeds attached to tree. (*b*) Yew (*Taxus cuspidata*) seeds. The fleshy, bright red coverings are arils. (*c*) Juniper (*Juniperus osteosperma*) "berries." The "berries" of junipers are not fruits because they do not develop from an ovary. (*d*) Seed-bearing cone of a cycad (*Zamia* species).

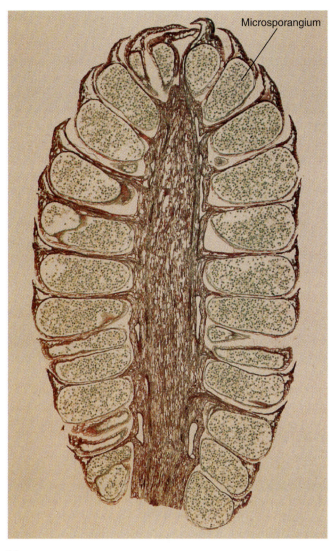

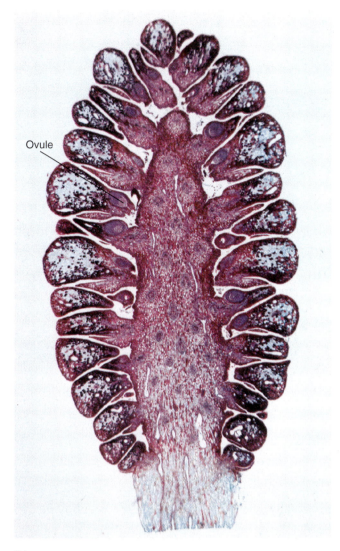

(a)

(b)

FIGURE 18.7

Pinecones. (*a*) Longitudinal section through a pollen-bearing cone (microstrobilus) (×3.2). (*b*) Longitudinal section through an ovule-bearing cone (megastrobilus) (×4). An intact pollen-bearing cone of *Pinus* is shown in figure 18.3*c*.

competitive plants. The most significant development leading to the evolution of gymnosperms was the origin of the seed. The life cycle of gymnosperms is characterized by two types of gametophytes, the male and female gametophytes. Most gymnosperms produce cones in which seeds or pollen are formed. Pollen grains evolved by reducing the male gametophyte to fewer cells and retaining it within the microspore wall.

Would you expect sperm cells in the pollen grains of pines to have flagella? Explain.

Compared to Angiosperms, Few Gymnosperms Live in the World Today.

There are considerably fewer species of gymnosperms than there are of angiosperms. Most classifications of gymnosperms include about 65 genera and 720 species. For centuries, the gymnosperms were lumped into a single class of seed plants, but many botanists consider them sufficiently diverse to be separated into four divisions: maidenhair tree (Ginkgophyta), cycads (Cycadophyta), conifers (Pinophyta), and gnetophytes (Gnetophyta).

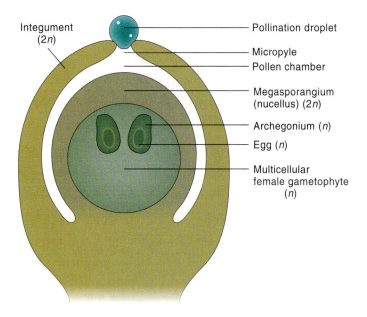

Integument (2n)
Pollination droplet
Micropyle
Pollen chamber
Megasporangium (nucellus) (2n)
Archegonium (n)
Egg (n)
Multicellular female gametophyte (n)

FIGURE 18.8

Diagrammatic representation of a longitudinal section through an ovule of pine (*Pinus*). The illustration shows which structures are haploid (*n*) and which are diploid (*2n*).

Maidenhair trees are native to China but are cultivated around the world.

Ginkgo biloba, the maidenhair tree (see fig. 18.6*a*), derives its common name from the resemblance of its leaves to leaflets of the maidenhair fern. *Ginkgo biloba,* which has remained virtually unchanged for 80 million years, is the only living representative in its division. Its distinctive, fan-shaped leaves with branched venation are produced on two types of shoots, or stems: relatively fast-growing long shoots produce leaves with a distinct apical notch (hence, *biloba*), while slow-growing short shoots produce leaves without a notch.

Ginkgo is exclusively dioecious; that is, having individual "male" trees that produce pollen but not ovules and other individual "female" trees that produce ovules and seeds but not pollen. After wind carries pollen grains to ovule-bearing trees, the pollination droplet of an ovule withdraws with its load of pollen. Pollen tubes begin to grow by digesting the tissue of the megasporangium. Once the female gametophyte is mature, the egg cells protrude toward the pollen. The pollen tubes convey sperm part of the way to the egg. The sperm cells of *Ginkgo* are flagellated, and after their release from the pollen tube, the sperm swim the rest of the way to the egg.

Seeds of *Ginkgo* include a massive integument that consists of a fleshy outer layer, a hard and stony middle layer, and an inner layer that is dry and papery. Mature seeds have the size and appearance of small plums, but these are not fruits because *Ginkgo* has no ovary surrounding its ovules. The fleshy integument has a nauseating odor and irritates the skin of some people. Nevertheless, pickled *Ginkgo* seeds are a delicacy in some parts of Asia.

Ginkgo trees are deciduous—that is, their leaves turn golden yellow during the fall and then drop off, a feature that otherwise occurs only in a few gymnosperms. *Ginkgo* is a popular cultivated tree, but it is apparently extinct in nature. Many cultivated *Ginkgo* trees are descendants of plants that were grown in temple gardens of China and Japan, although in the 1950s, there were reports of some individuals still living in the wild in eastern China. Much of the genetic diversity of *Ginkgo* has probably been lost in cultivation because most nurseries propagate only cuttings of male trees to avoid the stinky and messy seeds of female trees. *Ginkgo biloba* recently has been praised for its ability to improve memory; however, this has not been confirmed by experimental testing.

Cycads are tropical gymnosperms.

There are about 10 genera and 100 species of cycads, distributed primarily in the tropical and subtropical regions of the world. Like *Ginkgo,* all species of cycads are dioecious, meaning that they have male and female plants. Although cycads are planted as ornamentals throughout the milder climatic areas of the world, only two species are native to the United States. Both are in the genus *Zamia* and native to Florida. Many species of cycads are threatened with extinction, and some may soon remain only in cultivation.

Cycads have palmlike leaves that bear no resemblance to the leaves of other living gymnosperms. Under favorable conditions, cycads usually produce one crown of leaves each year. In some cycads, the roots grow at the surface of the soil and develop nodules containing nitrogen-fixing cyanobacteria. Cycads usually have large cones with simple, shield-shaped modified leaves (see fig. 18.6*d*) that may be covered with thick hairs. The seeds of cycads are more like those of *Ginkgo* than those of any other gymnosperm; they have a three-layered integument, but the inner layer is soft instead of papery.

Also like *Ginkgo,* cycads have flagellated sperm cells (fig. 18.9). The sperm cells of cycads are the largest among plants (up to 400 micrometers in diameter), and each sperm cell can have 10,000 to 70,000 spirally arranged flagella.

Many conifers produce seeds in woody cones.

The common name of this group, *conifers,* signifies plants that bear cones, even though other divisions of gymnosperms also include cone-bearing species. Members of the genus *Pinus,* considered a typical conifer, are the most abundant trees in the Northern Hemisphere; many of the species in conifer forests are pines. They have also been widely planted around the world, but only the Merkus pine (*Pinus merkusii*), whose distribution barely extends south of the equator in Sumatra, is native in the Southern

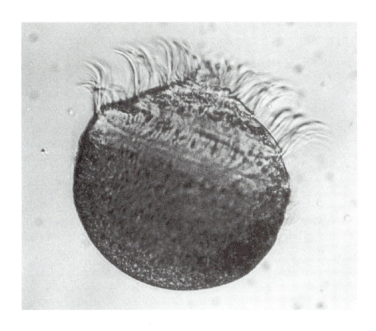

FIGURE 18.9
Light micrograph of a cycad (*Zamia*) sperm cell with flagella (×400).

Hemisphere. Also included in this genus is the bristlecone pine (*P. longaeva*) of the western United States, some of which are the oldest known plants that are not clones (see fig. 8.18). Some of these trees are over 4,700 years old; this means that the seeds of these ancient plants germinated before the famous pyramids of Egypt were constructed!

Like *Ginkgo*, pines have short shoots, long shoots, and two kinds of leaves. The more obvious type of leaf is the pine needle, which occurs in groups called **fascicles** of generally two to five needles (fig. 18.10*b*). A fascicle always forms a cylinder and is actually a short shoot that is surrounded at its base by small, nonphotosynthetic, scalelike leaves, which usually fall off after 1 year of growth. The needle-bearing fascicles are also shed a few at a time, usually every 2 to 10 years, so that any pine tree, while appearing evergreen, has a complete change of leaves every 3 to 5 years or so. Bristlecone pines are an exception; their needles last an average of 25 to 30 years.

Other conifers in the Northern Hemisphere also have narrow leaves that often have a small point at the tip, but they do not occur in fascicles. These include yews, firs, larches, spruces, and the coastal redwood (fig. 18.11). In contrast, the leaves of cypresses and juniper are scalelike at maturity (see fig. 18.4). Like pines, most other conifers are evergreen; however, larches, the bald cypress, and the dawn redwood are deciduous (fig. 18.12).

Diversity within the conifers is reflected in a wide variety of reproductive structures and variations of the reproductive cycle. As already noted, seed cones may be absent, as in yews, or have a berrylike appearance, as in junipers. One of the most distinctive variations, however, involves the timing of different reproductive stages in pines. Unlike other gymnosperms in which pollination, fertilization, and seed maturation occur within the same year, the pines have an extended reproductive cycle that lasts about 1½ years. The exact timing varies among species and in different localities.

At the time of pollination, the seed cones of conifers are small (see fig. 18.10*a*), and the ovule-bearing scales are slightly separated from one another, enabling pollen grains to reach the pollination droplet (see fig. 18.8). After the pollen grains are drawn into the ovule, the micropyle closes, and the seed cones seal up. Meiosis begins in the megasporangium about a month later. The development of the female gametophyte from the functional megaspore is extremely slow, going through an extensive multinuclear stage and finally becoming cellular about 13 months after pollination.

(a)

(b)

FIGURE 18.10
Pinecones. (*a*) First-year seed cones open for pollination. (*b*) Second-year pinecones at time of fertilization. Note the pine needles in fascicles.

FIGURE 18.11
Leaves of coastal redwood (*Sequoia sempervirens*). These leaves do not occur in fascicles.

In pines, the pollen tube transports sperm cells to the female gametophyte. The seed cone becomes larger and woodier than it was the previous year, but it remains closed while the embryos develop. Following fertilization, the embryo undergoes division and then matures before the following winter. By this time, the seed cones have grown into the mature, woody structures that we recognize as pine cones. The time between pollination and seed maturation is usually about 18 months.

When mature, the cones of most pines are dry, the scales are woody, and they open, releasing the winged seeds. The cones of some pines do not open so gracefully; these cones often require a fire to cause them to open and release their seeds. The cones of some pines explode like popcorn when heated. How might pine seeds released after a major forest fire have an advantage over pine seeds that were released in the middle of a dense forest?

The most unusual gymnosperms are gnetophytes.

The gnetophytes include some of the most distinctive (if not bizarre) of all seed plants. There are three clearly defined genera, each in its own order, and 71 woody species. These genera are *Ephedra* (40 species), *Gnetum* (30 species), and *Welwitschia* (1 species).

Ephedra, whose species are either monoecious (male and female cones on one plant) or dioecious (male and female cones on separate plants), is known as *Mormon tea, ma huang,* or *joint fir* (see fig. 18.3*d*). The first two names come from its use as a stimulating or medicinal tea. Ephedrine, used in nasal and bronchial remedies for relieving congestion, was originally extracted from *Ephedra* stems until the synthetic form of the drug was produced. Ma huang is an Asian species, *E. sinica,* which contains chemicals that are

FIGURE 18.12
These bald cypress (*Taxodium distichum*) trees lose all of their leaves in the fall; they are deciduous gymnosperms.

similar to human neurotransmitters. The joint fir appears leafless but possesses small leaves that lose their photosynthetic capability as they mature. Therefore, most photosynthesis in *Ephedra* occurs in green stems.

Members of the genus *Gnetum* inhabit tropical forests. These plants, which are dioecious, are either climbing vines or trees, all with broad, simple leaves similar to those of woody dicots (fig. 18.13*a*).

Welwitschia mirabilis is the sole living representative of its genus and looks more like something out of science fiction than a real plant (fig. 18.13*b*). This slow-growing species is confined to the Namib and Mossamedes deserts of southwestern Africa, where most of its moisture is derived from fog that rolls in from the ocean at night. The woody stem, which is concave and bark encrusted, may be as much as 1.5 meters in diameter and is connected to a large taproot. Mature plants have a pair of large,

(a) (b)

FIGURE 18.13

Gnetophytes. (*a*) Leaves and immature seeds of *Gnetum*. (*b*) *Welwitschia* plants in the Namib Desert of southwestern Africa. Each plant produces only two leaves, which split as they grow. The cones produce seeds.

strap-shaped leaves, which persist throughout the life of the plant. These leaves often split lengthwise, thus giving the appearance of more than two leaves per plant. Each leaf has a meristem at its base, which constantly replaces tissue that is lost at its drier, aging tip.

Welwitschia is dioecious, meaning it produces male and female cones on different plants. Fertilization in *Welwitschia* is unique in that tubular growths from the eggs grow toward and unite with the pollen tubes emerging from pollen that has been carried to a female cone. Fertilization occurs within these united structures. Reproduction is otherwise similar to that of other gnetophytes.

Gnetophytes are unique among gymnosperms because one of the sperm cells from a male gametophyte fertilizes an egg, while the second sperm cell fuses with another cell in the same female gametophyte. Thus, gnetophytes undergo *double fertilization,* a process otherwise known only in the angiosperms (see chapter 12). Unlike double fertilization in angiosperms, however, double fertilization in gnetophytes is not followed by the formation of a triploid endosperm, the food storage tissue in seeds of flowering plants. Instead, the diploid cell from fertilization by the second sperm disintegrates.

Many gymnosperms are adapted to temperate climates.

Ecology is the study of an organism and its environment. The ecology of gymnosperms is, in some ways, like that of flowering plants, but the gymnosperms live in a narrower range of habitats. Pines and their relatives are mostly trees and shrubs that are well adapted for temperate or cold cli-

mates, especially where water is scarce for part of the year. Accordingly, northern forests in some areas consist mostly of these gymnosperms, and southern forests consist mostly of pinelike relatives, including Norfolk Island pine (a common household plant) and the monkey puzzle tree. Pines are adapted to such regions by their narrow leaves, sunken stomata, thick, waxy cuticle and several other characteristics. Some gymnosperms, such as *Ephedra, Welwitschia,* and piñon pines, live in arid lands. Only a few gymnosperms are tropical; the cycads and *Gnetum* are the main examples. The ecology of gymnosperms is also addressed in the next chapter.

INQUIRY SUMMARY

Cycads and *Ginkgo* are dioecious and have flagellated sperm and fleshy integuments. Cycads, however, usually have large cones and persistent, palmlike leaves, whereas *Ginkgo* has loose small cones, single-seeded stalks, and simple deciduous leaves. Pines are the most abundant species of northern conifer forests. Many conifers have needlelike leaves that occur in fascicles, whereas others have broad, flat leaves. Reproduction in pines can require 18 or more months between pollination and the formation of mature seeds in cones. Gnetophytes are the most distinctive group of gymnosperms because of their many similarities with the angiosperms and their bizarre shapes. Various gnetophytes undergo double fertilization.

How would you define a gymnosperm?

18.1 Perspective Perspective

A Tree That Saves Lives

In the late 1980s, researchers at the Johns Hopkins University Oncology Center reported that tumors caused by ovarian cancer, which had not responded to traditional therapies (including, in some cases, surgery), had shrunk or disappeared in several patients after treatment with taxol. Taxol is a drug obtained from the bark of the Pacific yew (*Taxus brevifolia*), a gymnosperm that grows in the Pacific Northwest and is included among the trees that give shelter to the rare northern spotted owl. Unlike most other cancer drugs, which keep cancer cells from reproducing by damaging their DNA, taxol "freezes" the cancer cells early in the process of cell division. Unable to divide, the cells eventually die.

Unfortunately, there was far less taxol available than required to meet the need. Pacific yew trees are small and do not occur in extensive stands, and they grow so slowly that they take more than 70 years to attain their full size. Three whole trees produce only enough taxol for a single treatment; a pound of the drug originally cost $250,000. Furthermore, the extraction of taxol requires removal of the bark, which kills the tree.

Scientists discovered the potential of taxol to fight ovarian cancer. Because of its limited availability, medical science launched several studies to find alternative sources of taxol. These included searching for relatives of the yew that also produced taxol, investigating methods of synthesizing taxol in the laboratory, developing ways to grow taxol-producing cell cultures from the plant, and looking for the genes of taxol biosynthesis, which could be transferred into bacteria for large-scale production of taxol in bacterial culture (in the controlled environment of a laboratory). Scientists also discovered a fungus that grows in the bark of Pacific yew that produces taxol. More impor-

tantly, the fungus produces taxol in culture, which means that it has its own genetic machinery for making the drug. The fungus, which turned out to be a new genus, was named *Taxomyces andreanae*.

Today, taxol is chemically synthesized, beginning with extracts mostly from needles and twigs of plants related to the Pacific yew, as well as extracts from the Pacific yew's bark. Annual sales of the drug were greater than $1 billion in 1999.

The potential for large-scale production of taxol is expected to satisfy additional demand for the drug in treating cancers of the breast, neck, and head. Worldwide, hundreds of thousands of cancer cases may be treatable by taxol. If the Pacific yew were the only source for this drug, the species would be destroyed long before the demand for taxol could be satisfied. And if all the trees had been destroyed by lumbering, we might never have learned about taxol in the first place.

Gymnosperms Have Many Economically Important Uses.

The gymnosperms are second only to the angiosperms in their daily impact on human activities and welfare. A detailed account of all that gymnosperms contribute would occupy many volumes, but a few of their main uses are described here. The greatest economic impact of gymnosperms in the Northern Hemisphere comes from our use of their wood for making paper and lumber. Some products are obtained from the bark of gymnosperms, including a potential medicine to treat ovarian cancer (see Perspective 18.1, "A Fungus That May Save the Yews"). Conifers produce about 75% of the world's timber and much of the pulp used to make paper. The chief source of pulpwood for newsprint and other paper in North America is the white spruce (*Picea glauca*). A single midweek issue of a large metropolitan newspaper may use an entire year's growth of 50 hectares of these trees (about the size of 50 football fields), and that amount may double for weekend editions (see Perspective 18.2, "How Much Paper Do We Use?").

Douglas fir (*Pseudotsuga menziesii*) was once widespread in the foothills and mountains of the West. In the Pa-

cific Northwest, it grew into giant trees, which were second in size only to the redwoods. Douglas fir is probably the most desired timber tree in the world today. The wood is strong and relatively free of knots as a result of rapid growth with less branching than in most other conifers; it is heavily used in plywoods and is a major source of large beams for construction. The lumber industry has nearly eliminated native old-growth stands of Douglas fir, but large numbers of new trees are being grown in managed forests, where they are usually logged before they reach 50 years of age.

Redwoods (*Sequoia sempervirens*) are also prized for their wood, which contains substances that inhibit the growth of fungi and bacteria. The wood is light, strong, and soft, but it splits easily. It is used for some types of construction, furniture, posts, greenhouse benches, decks, and hot tubs.

Wood from the red spruce (*Picea rubens*) is especially important to the music industry. The tracheids of spruces have spiral thickenings on their inner walls, and those of red spruce apparently give its wood a resonance that makes it ideal for use as soundboards in violins and related musical instruments.

Important wood products from conifers besides lumber and pulp include **resin,** a sticky, aromatic sub-

18.2 Perspective Perspective

How Much Paper Do We Use?

Before the 1860s, most newspaper was made from linen or cotton rags; states such as Massachusetts appointed officials to help get rags to mills. Although many people suggested unusual sources for paper (including cabbage, dandelions, and asbestos), the most unusual suggestion was to make paper from the cloth used to wrap Egyptian mummies. Indeed, because each mummy was wrapped in about 14 kilograms of cloth, investors calculated that they'd need only 13.5 million mummies per year to provide for American paper mills! In the 1860s, Augustus Stanwood of Gardiner, Maine—to the dismay of archaeologists—started importing mummies. People in Maine were soon taking their lamb chops home in paper made from mummy cloth.

When mummies became hard to obtain, investors began looking elsewhere for sources of paper. They quickly settled on wood, especially spruce. Wood pulp was first used to make newspapers in the 1860s, and the first all-wood issue of the *New York Times* was published on August 23, 1873. Today, Americans use almost 200,000 tons of paper each day, enough to cover an area the size of Long Island. Our uses of paper are many: food packaging, newspapers, cardboard, toilet paper, and paperback novels are just some of the throwaway products that require a constant supply of wood pulp. Millions of trees are harvested every year to meet this demand. It is only a matter of time before the increasing demand for these products outstrips the supply of wood pulp. How can we help? The management of a large American publishing company, in an attempt to find ways of reducing paper consumption in the United States, tried trimming 2.5 centimeters (1 inch) from the width of all rolls of toilet paper in its building. It found that the employees still used the same number of rolls per month as they had previously. From this, the company calculated that if all rolls of toilet paper in the United States were similarly trimmed, 1 million trees would be saved each year. This is just one example of a simple and painless way to reduce our huge demand for paper goods made from woody trees.

stance produced by specialized cells. Resin is a combination of a liquid solvent called **turpentine** and a waxy substance called **rosin.** Both turpentine and rosin are useful products, and a large industry centered in the southern United States and the south of France is devoted to their extraction and refinement. Turpentine and rosin are often referred to as *naval stores*, a term that originated when the British Royal Navy used large amounts of resin for caulking and sealing its sailing ships and for waterproofing wood, rope, and canvas (most naval stores and one-third or more of the lumber used in the United States today come from a group of southern yellow pines, particularly slash pine, [*Pinus elliottii*]). Pine resin was used by sailors in ancient Greece, Egypt, and Rome. Egyptians sealed their mummy wrappings with pine resin, and the Greeks lined their clay wine vessels with pine resin to prevent leakage. Pine flavoring is still added to Greek wines, giving retsina its distinctive flavor. The unappreciative liken the taste to turpentine.

Turpentine is the premier paint and varnish solvent and is used to make deodorants, shaving lotions, drugs, and limonene—the lemon flavoring in commercial lemonade, lemon pudding, and lemon meringue pie. Ballerinas dip their shoes in rosin to improve their grip on the stage; violinists drag their bows across blocks of rosin to increase friction with the strings. Baseball pitchers use rosin to improve their grip on the ball, and batters apply pine tar to the handles of bats to improve their grip.

The kauri pines (*Agathis australis* and *A. robusta*) of New Zealand, which are genetically different from true pines, are the source of a mixture of resins called *dammar* (fig. 18.14). Dammar is used in high-quality, colorless varnishes and was the resin originally used to make linoleum. Dammar, also called *amber,* is the only jewel of plant origin. It comes primarily in fossil form from former or present kauri pine forests and occurs as lumps of translucent material with a deep orange-yellow tint. These lumps, which weigh up to 45 kilograms, were believed in ancient times to protect the wearer from asthma, rheumatism, and witchcraft. The best amber came from Russia; however, that supply, which was at its peak at the turn of the century, is now nearing exhaustion. Remarkably lifelike preservations of prehistoric insects in amber still have intact DNA.

Gymnosperms also have an economic importance related to their value as part of national forests, parks, and other recreational areas as well as their importance as symbols. Although the origin of the Christmas tree is uncertain (suggestions range from Germans to ancient Druids), today many Americans consider decorated coniferous Christmas trees to be an important part of their holiday season. Although the oldest mention of an American Christmas tree dates to 1747, the first Christmas tree didn't appear in the White House until 1856, decorated by the family of Franklin Pierce. People in different parts of the United States have differing preferences for their Christmas trees. Scotch pine is the favorite in the North-Central United States; the West prefers Douglas fir;

The Lore of Plants

"What the apple is among fruits . . .," wrote horticulturist Liberty Hyde Bailey, "the pines represent among conifers." This may explain why 10 states have chosen the pine as their state tree. Some states (Arkansas and Minnesota, for example) boast a generic pine. The state tree of Alabama is the longleaf pine; Michigan chose the eastern white pine; Nevada and New Mexico claim the piñon pine, Idaho the western white pine, and Montana the ponderosa pine. The state tree of Maine, the Pine Tree State, is also the eastern white pine, and its state "flower" is the pinecone, which, of course, is not a flower at all.

(a)

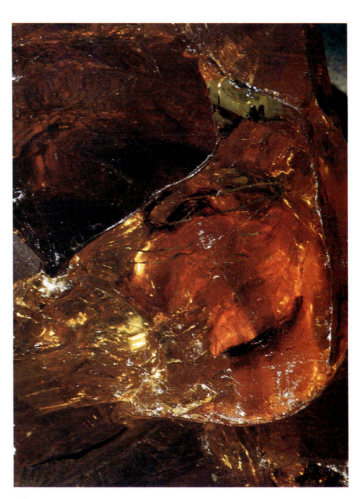

(b)

FIGURE 18.14

(*a*) Kauri pine (*Agathis australis*) of New Zealand. (*b*) Kauri pine is the source of dammar, also called amber, which is made into jewelry.

most Californians prefer Monterey pine; Southerners like white pine; and New Englanders prefer balsam fir.

INQUIRY SUMMARY

In the Northern Hemisphere, gymnosperms account for most of the wood that is needed for lumber and paper. Worldwide, conifers are the main source of resin and amber. Gymnosperms are also the source of some important medicines.

Look around you—in your classroom and your home. In what ways are gymnosperms important in your life? On what basis do you make this claim?

Flowering Plants Are the Most Diverse Plant Group.

Angiosperms live in almost all terrestrial and freshwater habitats on earth. Except for coniferous forests and moss-lichen tundras, angiosperms dominate all of the major terrestrial communities of vegetation. Moreover, angiosperms include some of the largest and smallest plants (fig. 18.15); eucalyptus, duckweeds, lilies, oak trees, lawn grasses, cacti, broccoli, and magnolias are all angiosperms. Such plants surround us and affect virtually all aspects of our daily lives.

How can we explain the incredible diversity of flowering plants—over 250,000 species as compared with 720 species of gymnosperms? One hypothesis involves dinosaurs. Dinosaurs included some of the most spectacular animals on earth, a few of which reached massive size. They lived throughout the Mesozoic era (Triassic, Jurassic, and Cretaceous periods) in the Age of Dinosaurs, or the Age of Cycads (as noted earlier, the term preferred by botanists for the same period), and during the rise of angiosperms. Of these three groups of organisms (dinosaurs, cycads, and angiosperms), only the angiosperms remain as a highly diversified and dominant life-form today. The dinosaurs were all extinct by the early Tertiary period (about 65 million years ago), which coincided with the disappearance of the cycadeoids. By this time, the number of different cycads and other gymnosperms had also begun to diminish. Recent studies by paleobotanists and paleontologists have focused on how the gymnosperms, angiosperms, and dinosaurs may have interacted and even influenced their respective evolutionary histories. The main question asked by botanists is, How did the angiosperms become dominant at the expense of the gymnosperms? One possible explanation involves adaptation to insect pollination by angiosperms, as already discussed in chapter 12. The shorter life cycle of angiosperms allows relatively rapid evolution compared to gymnosperms with their relatively long life cycle. Another hypothesis involves dinosaurs.

Could dinosaurs be responsible for the diversity of flowering plants today?

The largest herbivores ever to walk the earth were the enormous sauropods, such as *Diplodocus, Brachiosaurus,*

(a)

(b)

(c)

FIGURE 18.15

Angiosperms. (*a*) *Eucalyptus,* one of the tallest plants. (*b*) Duckweeds are among the smallest flowering plants; in this photo they are the light green plants (arrows) floating among the leaves of an aquatic fern. (*c*) Bunchberry (*Cornus canadensis*).

The Lore of Plants

The informal competition among scientists to find the oldest fossil DNA seemed to be won in 1990, when 17-million-year-old DNA was found in fossils of *Magnolia* leaves. In 1992, however, the record was broken by 30-million-year-old DNA from insects that had been preserved in amber. Even this record was surpassed by a 1993 report of DNA from a 120 to 135-million-year-old fossil weevil. Some scientists suggest that, even under optimal conditions, DNA is unlikely to survive more than 100,000 years and that most DNA breaks down rapidly after the death of an organism. They suspect that reports of DNA in fossils older than 100,000 years may be contaminants from the laboratories of the researchers.

FIGURE 18.16

What role could plant-eating dinosaurs have on the evolution of flowering plants?

heads with flat, grinding teeth for chewing plant tissues. There were more kinds of dinosaurs toward the end of the Cretaceous period than before, although by this time, large dinosaurs of the Jurassic period were extinct. The greater diversity of smaller, late-Cretaceous dinosaurs may indicate a higher degree of specialization in plant-dinosaur relationships than in earlier periods.

The long-term decrease in body size and increase in diversity of the dinosaurs coincided with the evolutionary radiation of angiosperms and the decline of gymnosperms. Gymnosperms were probably devastated by the low browsers, which would have eaten small seedlings before they were able to reach maturity and produce seeds. Because the first angiosperms were most likely smaller and herbaceous, they probably grew and reproduced more rapidly than woody gymnosperms and

and *Apatosaurus* (once known as brontosaurus). These animals probably roamed in herds, browsing on the tops of coniferous trees and other tall gymnosperms. Such browsing would not have killed large trees because the herbivores would take only a small proportion of all the leaves. Large herbivores may not have destroyed seedlings by browsing either, because they could not reach down to them. Thus, gymnosperms could flourish without being destroyed by large herbivores.

The coexistence of large, herbivorous dinosaurs and tall gymnosperms began to disintegrate at the beginning of the Cretaceous period, when the average body size of dinosaurs mysteriously began to shrink. By the late Cretaceous period, large herbivores had been replaced as smaller, lower browsers of the ornithischian type (e.g., *Ankylosaurus, Parasaurolophus,* and *Montanoceratops*) evolved. These dinosaurs had relatively large, muscular

therefore stood a better chance of producing seeds before being eaten. Furthermore, the destruction of gymnosperms by small dinosaurs opened up new habitats for the invasion and evolution of angiosperms (fig. 18.16).

While the dinosaur-gymnosperm-angiosperm connection may seem plausible, the suggestion that angiosperms diversified by adapting to insect pollination is more so. The history of insects begins approximately 200 million years before that of flowering plants. The great diversity of insects provided a rich opportunity for co-evolution between them and the flowering plants they pollinated (see chapter 12). It may be that each of these explanations is correct and that dinosaurs and insects both influenced the rapid evolution of angiosperms in the Cretaceous period. Although the greatest diversification of angiosperms occurred in the Tertiary period, after the dinosaurs were already extinct, these large animals may have promoted the

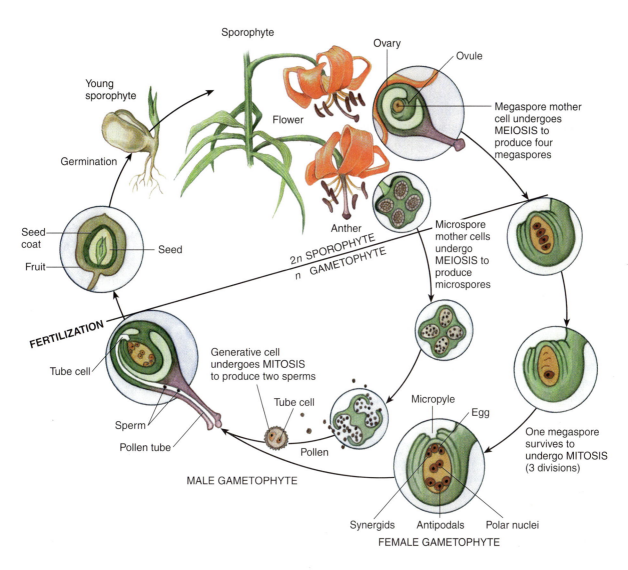

FIGURE 18.17

Generalized life cycle of flowering plants. The seeds of flowering plants are enclosed in an ovary, which becomes a fruit. In addition, the reproductive parts are borne in flowers. Both the male and female gametophytes are reduced to a few cells.

early evolution and diversification of flowering plants. It is likely that the success of flowering plants is due to a combination of characteristics, including the evolution of the flower, their comparatively short life cycles, and their adaptations to life on land, as discussed at the beginning of the chapter.

The life cycle of flowering plants is different from that of all other plants.

Today, flowering plants are the most successful of all plant groups in terms of their diversity. The group includes more than 250,000 identified species and at least 12,000 known genera. This group is called the angiosperms because these plants possess seeds enclosed in an ovary, the reproductive structure of the flower that contains ovules (fig. 18.17). The life cycle of flowering plants is different from that of other

plants because of the flower and features described in the next paragraphs.

The carpel is a primary feature that distinguishes angiosperms from gymnosperms and all other plants because it becomes part of the fruit. In addition, angiosperms have double fertilization in which one of the two sperm cells in a pollen grain fertilizes the egg to produce the embryo and the other sperm cell fertilizes the two polar nuclei to produce the endosperm (see chapter 12). In some ways, the generalized life cycle of angiosperms is similar to the life cycle of other seed plants (see fig. 18.17). Double fertilization is a feature that angiosperms share with gnetophytes, as previously indicated. A difference in life cycles, however, is that in flowering plants, the triploid nucleus (sperm plus polar nuclei) divides repeatedly and produces endosperm; the embryo in seeds of angiosperms usually derives its nutrients from the endosperm and not the female gametophyte as in gymnosperms. Another difference is that flowering plants have no antheridia or archegonia.

In the evolution of plants, there has been an overall increase in the size and dominance of the sporophyte phase of the life cycle and a corresponding decrease in the size and complexity of gametophytes. In fact, angiosperms, the most evolutionary advanced plants, produce sporophytes that are large, long-lived, structurally complex, and independent from the gametophyte. In contrast, the gametophytes are small, short-lived, consist of a few cells, and, in the case of the female gametophyte, are contained within the sporophyte. The mature female gametophyte, which produces an egg, consists of only seven cells in most species and is retained inside the ovule, which is located inside the ovary. The male gametophyte, which is the pollen grain that produces sperm, consists of only two or three cells, depending on the stage of maturation. The decrease in the size of the gametophyte in the evolution of plants can be traced by comparing different groups of plants. In mosses, the haploid gametophyte dominates the life cycle. In ferns, the gametophyte is often short-lived and much smaller than the sporophyte. In gymnosperms, the female gametophyte is reduced to a few hundred to a few thousand cells. In angiosperms, the gametophytes, both male and female, are just a few cells. In addition, in both gymnosperms and angiosperms, the female gametophyte remains within the larger sporophyte.

What's the evolutionary advantage of smaller gametophytes with fewer cells? One hypothesis is that because the cells of gametophytes are haploid, only one allele for each gene is present. Thus, any mutations or deleterious alleles will be expressed in the gametophyte, which may be lethal. In sporophytes, however, each cell has two sets of chromosomes; each cell is diploid. Thus, each gene has two alleles, and if one allele is a mutation, it may be masked by the other allele. This may protect the sporophyte from any lethal alleles it possesses.

INQUIRY SUMMARY

Flowering plants are the most diverse group of plants in the world, with more than 250,000 species. Although dinosaurs may have played an important role in the early evolution of flowering plants, their great diversity may be attributed to a combination of factors, including the co-evolution with insect pollinators and characteristics that allowed them to colonize land. The evolution of plants has involved the decrease in the size of the gametophytic phase and an increase in the sporophytic phase.

What advantage might a flowering plant gain if it had only one specific kind of pollinator instead of many different species of pollinators? What might be the disadvantage of having only a single, highly specialized species of pollinator?

Flowering Plants Are Economically, Culturally, and Historically Important to Humans.

Today, nearly half of all prescription drugs contain chemicals manufactured by plants, fungi, or bacteria, and many other medicines contain compounds that were synthesized in a laboratory but modeled after plant-derived substances. The many medicines that we derive from plants are a compelling reason why we must stop destroying the world's tropical rain forests, where plant life is so abundant and diverse that all of the species have not even been seen or identified, much less studied.

In addition, all of our food comes either directly or indirectly from plants. In terms of both food and medicine, flowering plants are the most important sources. The following are some of the flowering plant families that we use for economic or cultural purposes or have historic uses (fig. 18.18):

- **Laurel family** (Lauraceae: camphor, sassafras, sweet bay, avocados). We use laurel leaves to crown winners of athletic contests and to bestow academic honors. Sassafras is used to make toothpaste, gum, mouthwash, tea, beer, and gumbo, whereas sweet bay is used as a spice. Avocados, which contain up to 60% oil, contain more energy per unit weight than red meat.

- **Mint family** (Lamiaceae: spearmint, peppermint, coleus). These plants are important sources of oils, such as peppermint oil. Menthol from peppermint is used to make candy, gum, and cigarettes.

- **Carrot family** (Apiaceae: dill, celery, carrot, parsnip, parsley). Poison hemlock contains coniine, the Athenian state poison used to kill Socrates in 399 B.C. Ever

(a)   (b)

FIGURE 18.18

Economically important plants. (*a*) Sassafras (*Sassafras albidum*) is used to flavor gum and mouthwash and in gumbo. (*b*) Spearmint (*Mentha spicata*) is the source of spearmint oil.

the scientist, Socrates insisted on having the stages of his poisoning accurately observed and recorded. Elixir of Celery, advertised in the 1897 Sears and Roebuck catalog, was used to cure nervous ailments, and crushed celery seeds mixed with soda are still marketed as Cel-Ray by Canada Dry.

- **Cactus family** (Cactaceae: cacti). The fruits of many cacti taste like pears, which explains their common name, prickly pear. Peyote (*Lophophora williamsii*), a small, spineless cactus with carrotlike roots, is the source of mescaline. The Aztecs used dried slices (i.e., buttons) of this hallucinogenic plant for religious purposes, as do some Indians in Mexico and some Native Americans in the southwestern United States.

- **Pumpkin family** (Cucurbitaceae: pumpkin, squash, cantaloupe, watermelon). These plants have been used to make everything from jack-o'-lanterns and dishes to laxatives.

- **Lily family** (Liliaceae: asparagus, sarsaparilla, aloe, chives, garlic, meadow saffron). Sarsaparilla is used to make soft drinks, and aloe, which was once used as a source of phonograph needles, is used to treat burns; aloe is also a common ingredient of many cosmetics. Meadow saffron (fall crocus; *Colchicum autumnale*) is the source of colchicine, an important drug used in biological research. Meadow saffron is different from the true saffron, a member of the iris family and the source of the world's most expensive spice. It takes 4,000 stigmas to produce only an ounce of the spice.

- **Buttercup family** (Ranunculaceae: columbine, monkshood, larkspur). Columbine, the state flower of Colorado, has been used as an aphrodisiac. Monkshood and wolfbane contain aconitine, a powerful poison that humans have used to kill a variety of animals, including wolves. Monkshood, an attractive plant grown in many gardens, was used to kill Roman Emperor Claudius and Pope Adrian VI and was thought to help witches fly (fig. 18.19).

- **Spurge family** (Euphorbiaceae: cassava, Pará rubber tree, candelilla, poinsettias). The Pará rubber tree is a source of latex for rubber, and candelilla is the source of several waxes, including candle wax. Cassava, whose roots are used to make tapioca, alcohol, and acetone, is a staple in the tropics. Despite its widespread use in December, the poinsettia is one of the newest Christmas plants. Joel Poinsett became interested in the plant while serving as American ambassador to Mexico from 1825–29. He brought the plant home, where it was named in his honor.

- **Poppy family** (Papaveraceae: bloodroot, opium poppy). Many members of this family produce drugs, the best known of which are morphine and codeine.

- **Nightshade family** (Solanaceae: tomato, potato, tobacco, petunia, chili pepper). These plants produce

FIGURE 18.19

Monkshood (*Aconitum columbianum*) is the source of aconite, a poisonous drug once used as a sedative.

drugs such as atropine (once used to dilate pupils and as an antidote to nerve-gas poisoning), hyoscyamine (a sedative), and scopolamine (a tranquilizer). (During World War I, scopolamine was also used in "truth serum" in futile attempts to make spies talk.) Hot peppers that are dried and crushed yield cayenne pepper, and sweet peppers are used to make paprika. Fiery chili peppers get their heat from capsaicin, a chemical that is perceived not by our taste buds but rather by pain receptors in our mouth. We can taste it at concentrations as low as one molecule per million. That's why only a trace of capsaicin can make an unwary diner grab frantically for water.

Many of the pilgrims considered tomatoes, which evolved in the Andes, to be evil—on a par with dancing and card playing. Interestingly, tomato juice is said to neutralize butyl mercaptan, the nose-shriveling ingredient of skunk spray. Joseph Campbell began selling his famous tomato soup in 1897, soon after chemist John Dorrance, at a weekly salary of $7.59, figured out how to condense it. In 1962, Andy Warhol's painting of a can of that soup sold for $50,000.

Cereals feed the human population.

Cereals such as wheat, rice, and corn include nine of the ten most economically important groups of plants; these cereals provide most of the carbohydrates and about half of all the protein in our diets. These plants are all members of the grass family (Poaceae). Rice is the primary source of food in Asia, whereas corn is a major part of the human diet in many parts of Central and South America.

In the United States, many of our foods are based on wheat. Wheat (*Triticum aestivum*) grows in a range of climates, from the Arctic to the equator, from below sea level to 3,050 meters (10,000 feet) above it, and from areas with less than 5 centimeters of annual rainfall to places with more than 178 centimeters (70 inches) of rain per year. Most wheat is used for food. The grain of "hard" bread wheat contains 11% to 15% protein, mostly of a type called *gluten.* When ground and mixed with water, gluten forms an elastic dough that is excellent for making bread. The grain of "soft" bread wheat contains 8% to 10% protein and is best for making cakes, cookies, crackers, and pastries. Today, the United States produces about 12% of the world's wheat—about 70 million metric tons per year, valued at about $8 billion.

Wheat has been cultivated for at least 12,000 years. Today, there are three kinds of wheat, based on their number of chromosomes: 14 (diploid), 28 (tetraploid), or 42 (hexaploid) (fig. 18.20). Wheat plants with 28 and 42 chromosomes are polyploids. Bread wheats are hexaploid, and the durum wheat that we use to make macaroni is tetraploid. Geneticists study the number of chromosomes in modern wheats to reconstruct the evolution of these important crop plants. It is a fascinating and complex story. Wheat as we know it apparently evolved from an accidental cross between a distant wheat ancestor and a weedy goat-grass with inedible seeds. It may have happened as follows: thousands of years ago, wandering peoples in what is now Jericho, in Israel, came upon a rich oasis, a spring in the desert ringed by hills covered with wild grass. While looking for food, the people discovered that the grain held in the grass, when ground, made a fine flour. For many years, they did not know how to encourage the grass's growth intentionally, so each season they would simply forage for whatever nature provided.

The grass that the people had grown fond of, really an ancient wheat called *einkorn,* crossbred with another type of grass that was not good to eat. The hybrid grass then underwent a genetic accident (nondisjunction; see chapter 13) that prevented the separation of chromosomes in some of

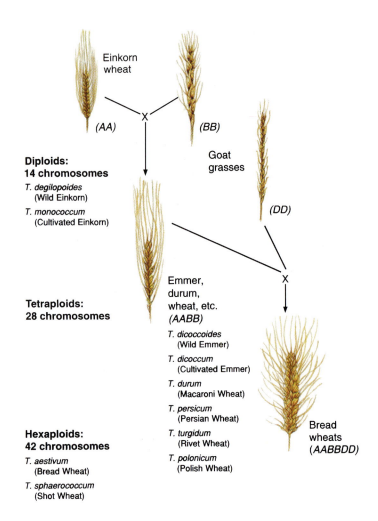

FIGURE 18.20

Evolution of wheat. What role has polyploidy played in the evolution of wheat?

the developing gametes. The result was a new type of plant that had twice the number of chromosomes as either parent plant (i.e., 28 chromosomes total). This plant was called *emmer wheat,* and it was a better source of food than the einkorn. The doubled number of chromosomes in each cell produced a plant with larger grains. Also, the grains were attached to the plant in such a way that they could be easily loosened and spread by the wind, so that the new wheat was soon plentiful. It was probably about this time that early farmers learned to select seed from the most robust plants to start crops of the next season.

Then, about 6,000 years ago, another "mistake" of nature further improved the quality of wheat. Emmer wheat crossed with another weedy goat-grass, and another fortuitous "accident" of chromosome doubling led to bread wheat, which has 42 chromosomes (see Perspective 13.3, "Polyploidy in Plants"). Bread wheat has even larger grains than the emmer wheat that produced it, but at a cost. Its reproductive structures are so compact that, on its own, the grain cannot be released. However, farm-

FIGURE 18.21
Triticale, a popular grain that is a hybrid of wheat (*Triticum*) and rye (*Secale*). Cells of triticale are polyploid: they each contain a complete set of chromosomes from wheat as well as a set from rye. Triticale can tolerate the harsh environment of rye while producing the high yields of wheat.

ers collected the rich seed each season for food and kept a certain percentage to be planted the next season. The interdependence of humans and crops that is the basis of modern agriculture was thus born. Today, interesting hybrids, such as triticale, a combination of wheat (*Triticum*) and rye (*Secale*), add variety to our diets (fig. 18.21). The cells of triticale are polyploid, containing a complete set of chromosomes from wheat and one from rye. Triticale has the high yield of wheat and can cope with the harsh environments of rye.

The tasty, sweet, or starchy kernels of corn (*Zea mays*) sustained the Incas, Aztecs, and Mayan Indians of South America and the pilgrims of colonial Massachusetts. Today, corn continues to be a dietary staple from Chile to Canada. The corn crop in the United States presently exceeds 9 billion bushels a year and is used to make food products, drugs, and industrial chemicals, as well as to feed animals. Corn is also being used to help solve a troubling environmental problem: plastics. After years of being dumped into our oceans, plastics that were made to last forever are coming back to haunt us. For example, merchant ships dump millions of plastic containers into international waters every year. These and other plastics maim and kill marine life (each year, thousands of seals die after becoming entangled in plastics). Other plastics are

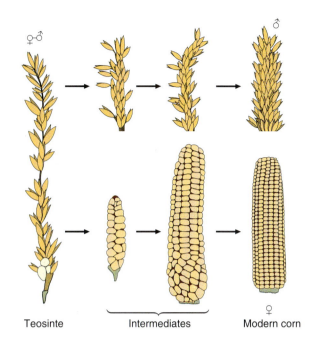

Teosinte   Intermediates   Modern corn

FIGURE 18.22
Corn and teosinte. Teosinte (*Zea mexicana*), a wild grass, has structures that may have given rise to modern corn (*Zea mays*). The seeds of teosinte evolved into corn kernels, and the tassels eventually became the familiar pollen-producing tassels of corn.

filling our landfills and polluting our water. In one of the many examples of how plants are used to combat such problems, one company has patented a technique for making biodegradable plastics containing corn starch. This technique involves inserting tiny pellets of starch into the plastic, which helps the plastic to decompose in only a few months. If this technology becomes popular, biodegradable containers and cups will soon be common.

Like wheat, modern corn probably arose from a naturally occurring wild grass that initially produced a grasslike type of corn having small ears. Fossil evidence exists of such a primitive corn. A clue to the origin of modern corn comes from *teosinte*, a group of wild grasses in the genus *Zea* that grow in Mexico today and probably have for thousands of years. The Indians call teosinte *madre de maize*, which means "mother of corn." Unlike other cereals, corn does not look like its wild ancestors (fig. 18.22). However, there are a few similarities. Both species have 20 chromosomes and produce fertile hybrids when they are crossed, suggesting that at one time they were closely related. Wild teosinte sometimes grows on the outskirts of cultivated cornfields.

Charles Darwin was interested in the origin of corn. He used breeding experiments in his greenhouses to show that when corn plants are continually self-fertilized, they produce increasingly weaker offspring; that is, inbred plants are more prone to disease and produce ears with fewer rows of kernels. Darwin noted, however, that the offspring of cross-pollinated plants are vigorous and healthy, with many rows of plump kernels. Although Darwin knew little about the genetic discoveries of his contemporary, Gregor Mendel, what he had demonstrated was the genetic phenomenon of hybrid vigor (see chapter 12). Plants with unrelated parents are often more vigorous than self-fertilized plants because they inherit new combinations of genes.

George Shull, at the Station for Experimental Evolution in Cold Spring Harbor, New York, extended Darwin's greenhouse experiments, thanks to the additional insight provided by the "rediscovery" of Mendel's laws of inheritance in the first decade of the twentieth century. Shull continually bred plants from the same ears of corn, artificially selecting highly inbred lines. The inbred corn plants were sickly looking, with small ears. But when Shull crossed two different inbred lines to one another, the offspring showed hybrid vigor. Shull had developed hybrid corn, which revolutionized corn output worldwide. In the United States, corn production has jumped from 21.9 bushels per acre in 1930 to 95.1 bushels per acre in 1979 and well beyond that today, thanks largely to hybrid corn. Indeed, hybrid corn is the largest seed business in the world today. Many of us start our days with flakes made from this tasty grain (see Perspective 1.2, "Breakfast at the Sanitarium: The Story of Breakfast Cereals").

About half of today's 9-billion bushel harvest of corn is fed to animals. Two centuries ago, however, much was used to make corn whiskey, which was consumed as a cure for colds, coughs, toothaches, and arthritis. These ailments must have been rampant: between 1790 and 1840, people drank an average of 5 gallons of whiskey per person per year.

Based on the number of people who rely on it, rice (*Oryza sativa*) is the most important food in the world (see fig. 7.25). It's been cultivated in Thailand for more than 12,000 years, and in Asia, it comprises 80% of the human diet. The 400 million metric tons of rice produced each year are grown in many types of environments. Today's rices include 20 wild species plus many cultivars, which are domesticated species produced by artificial selection rather than by the evolutionary forces of natural selection (i.e., cultivars do not grow in the wild). The oldest species of rice for which we have fossil evidence, *Oryza sativa*, originated about 130 million years ago in parts of South America, Africa, India, and Australia, which were then joined into one landmass. Unlike wheat and corn, rice evolved in a tropical, semiaquatic environment.

Human cultivation of rice may have begun as early as 15,000 years ago in the area bordering China, Myanmar (Burma), and India. By 7,000 years ago, efforts to grow rice had spread to China and India; by 2,300 years ago, the crop was growing in the high altitudes of Japan. By 300 B.C., rice was a staple throughout Asia. It has been over only the past 700 years that rice has been eaten in West Africa, Australia, and North America. As migrating peoples took their native rices with them to new lands, the plants adapted to a wide range of environments. From the 1930s to the 1950s, many

nations collected hundreds of native rice varieties, growing small amounts of each type every season, just to keep the collections going.

Other cereals used extensively by humans include barley, millet, oats, and rye. Barley (*Hordeum vulgare*) was first grown in the eastern Mediterranean and today is grown mostly in Europe, North America, and Australia. Most barley is used as cattle fodder and to make malt used in distilling and brewing. Millet (several different species of grasses) was first cultivated in China in about 2700 B.C. Although it is now grown mostly for cattle fodder in the United States, its resistance to drought is also being exploited to make it a food crop in tropical Africa. Oats (*Avena sativa*) probably originated as a weed growing with other cereals, such as barley or wheat. It was domesticated about 2,500 years ago. Today, it is used primarily as cattle fodder and in breakfast cereals. Eating oats has been touted as a way to reduce one's blood cholesterol. Rye (*Secale cereale*) originated as a weed growing among other cereals. It was first cultivated in southwest Asia in about 1000 B.C. and now is used as flour in rye bread and as food for cattle.

INQUIRY SUMMARY

Flowering plants are our major source of food and medicines. In terms of feeding people, cereals are the most important group of plants in the world. Wheat has been important in the development of ancient civilizations in what is today the Middle East, corn has been important in Central and South America, and rice in the Far East. The evolution of wheat includes crosses between different grasses and subsequent formation of polyploids.

Why do you think that plant domestication occurred in so few places around the world?

The Struggle to Produce Enough Food in the World Continues.

Throughout history, famines have killed millions of people. For example, famine in India killed 10 million in 1769–70, 1 million in 1866, 1.5 million in 1869, 5 million in 1876–78 (as another famine in China killed 10 million), and another million people in 1900. Similarly, the Irish potato famine of 1846–47 killed 1.5 million people and caused a mass exodus from Ireland. If you're American of Irish descent, chances are this famine is responsible for your ancestors emigrating to North America.

There are many factors involved in feeding the world's population, including the size of the population. Our population is growing extremely fast. For example, at the beginning of agriculture 12,000 to 15,000 years ago, only about 5 million people lived on earth. By 1 A.D., the population had grown to 250 million. Thereafter, it doubled to 500 million in 1650, doubled again to 1 billion in 1850, doubled again to 2 billion in 1920, and doubled again to 4 billion in 1976. Today's ever-increasing population of over 6 billion requires huge amounts of food.

Botanists have struggled to produce enough food for the human population. One of the pioneers in this struggle was Norman Borlaug, a plant geneticist. Borlaug began creating new varieties of plants in 1944 and soon achieved remarkable results. For example, in 1944, Mexico imported wheat to feed its citizens. By 1964, however, Mexico was *exporting* wheat by growing new varieties. Since 1950, production has quadrupled, but Mexico's population has increased tremendously. Similar advances in crop production have been made in other countries. This dramatic increase in food production was called the **Green Revolution.** For his work, Borlaug received a Nobel Prize in 1970—not so much for the technology that produced the high-yielding crops as for his humanitarianism when he tried to help feed the world. However, even his Green Revolution hasn't been able to keep up with the ever-increasing demand for food. Moreover, some critics claim that the Green Revolution actually worsened the problem because it created social and environmental havoc. For instance, in some countries, only wealthy landowners could afford to grow the new varieties of plants because growing the new crops required expensive irrigation systems and fertilizers. In addition, because new varieties are more productive than the old ones, farmers rejected the genetically diverse crops in favor of planting fields of genetically uniform crops. This loss of genetic diversity could present future problems because a single parasite could wipe out entire fields of crops. What problems might occur if enough food were produced and distributed to feed all the hungry people in the world?

The Green Revolution and similar programs have not eliminated world hunger because they have not addressed the problems that drive world hunger: overpopulation and poor distribution of the food that is available. Every year, 100 million people—more than three people per second—are added to the earth's population. That's the equivalent of adding about 25 cities the size of Los Angeles. Earth's population has doubled in less than 35 years and now grows at an annual rate of 1.8%. This means that each day there are 260,000 more people to feed.

The consequences of these facts are devastating. For example, in 1998, more than 1.2 billion people—that's one in every five—lived in abject poverty, living on less than $1 per day. More than 500 million people were getting less than 80% of the recommended intake of calories; during the hour or so that it takes for us to eat our Thanksgiving Day meal, more than 1,600 people (mostly children) die of hunger. Moreover, our efforts to feed those people are damaging the environment: we're quickly losing topsoil (see fig. 8.26) and are polluting the soil and water with herbicides and insecticides. As we bring more land into cultiva-

FIGURE 18.23
Destruction of a rain forest by fire, which destroys habitats and many native species.

tion, we destroy habitats and threaten many native species of plants and animals (fig. 18.23). Earth cannot continue to support such an increasing population.

There are promising prospects for the future beyond the Green Revolution.

Although many of the wonders you have learned about in this book have greatly increased crop yields and our ability to produce food, more people are starving than ever before. Our rapidly growing population, combined with ineffective food-distribution methods, has overwhelmed our agricultural system. Solving this problem is a tremendous—perhaps impossible—challenge. However, all hope is not lost.

Throughout this book, you've read about one of the most promising tools for helping to feed the world: biotechnology. Just as producing hybrid corn in the 1930s doubled the corn harvest, biotechnology is being hailed as the "second green revolution," which will produce plants that protect and nourish themselves and, in the process, help feed the world. For example, a group of researchers used genetic engineering to identify and manipulate a self-incompatibility gene (also called the S gene). A plant having an active S gene recognizes and rejects its own pollen and therefore cannot fertilize itself. The ability for growers to prevent plants from fertilizing themselves could double the yield (and cut labor costs by half) of many vegetables and flowers. The explanation is that self-compatible plants may produce offspring that suffer from inbreeding depression, while self-incompatible plants, by mating with another individual, produce offspring with hybrid vigor.

Other botanists are now using genetic recombination to create high-yielding crops that resist disease, drought, and pests. Still others are improving the caloric and nutritional value of crops. In years ahead, genetically engineered plants will become a leading source for increasing food production. However, biotechnology is not the only tool in the war on hunger.

FIGURE 18.24
Amaranth is grown for its cereal-like grain and edible leaves. The plants shown here are growing in Pennsylvania.

linity), yields many seeds, and comes in many varieties. Problems with cultivating amaranth include controlling weeds and pests and harvesting the tiny seeds, but nutritionally, amaranth is superb. Its seeds contain 18% protein, compared with 14% or less for wheat, corn, and rice. Amaranth is relatively rich in amino acids that are poorly represented in the major cereals. The seeds can be used as a cereal, a popcornlike snack, and a flour to make graham crackers, pasta, cookies, and bread. The germ (embryo) and bran (seed coat and ovary wall) together are about 50% protein, making them ideal to add to prepared foods and animal feeds. The broad leaves of amaranth are rich in vitamins A and C as well as the B vitamins riboflavin and folic acid. They can be cooked like spinach or eaten raw in salad.

Botanists have also established gene banks and pollen banks to help conserve rare plants and to increase the world food supply (fig. 18.25). For example, to offset a potential disaster due to reliance on only a few types of rice, the International Rice Research Institute was founded in 1961 in the Philippines. It soon became a clearinghouse for the world's rice varieties, storing the seeds of 12,000 natural variants by 1970 and of more than 81,000 by 2000. Representatives of other important crops are being banked as well. The National Germplasm System in the United States is the world's largest distributor of germplasm and seeds: in 2000, it distributed more than 150,000 samples to scientists in the United States and over 100 other nations. Similarly, potato cells are stored at the Potato Center in Sturgeon Bay, Wisconsin, and wheat cells are banked at the Kansas Agricultural Experimental Station. Pollen and seeds from 250 endangered species of flowering plants have been frozen at a plant gene bank at the University of California at Irvine.

Plant banks offer three priceless services to humanity: a source of variants in case a major crop is felled by disease or an environmental disaster; the return of endangered or extinct varieties to their native lands; and, perhaps most important, a supply of genetic material from which researchers can fashion useful plants in the years to come, even after the species represented in the bank have become extinct.

Establishing a seed or pollen bank is one solution to a growing problem in traditional plant agriculture—the

Another strategy to increase our supply of plant foods is to look for new crops among the many naturally occurring plants. Fewer than 30 of the more than 250,000 species of flowering plants provide about 90% of plant-based foods eaten by people. (Why do you think that people don't eat more kinds of plants?) One of the most promising plants that botanists are studying as a new source of food is the majestic-looking amaranth, a member of the pigweed family (fig. 18.24). These plants stand over 2 meters tall and have broad purplish green leaves and massive seed heads. Each plant produces about a half-million mild, nutty-tasting, protein-rich seeds, each the size of a grain of sand. The flowers are a vivid purple, orange, red, or gold. Interestingly, amaranth was cultivated extensively in Mexico and Central America until the arrival of the Spanish conquistadors in the early 1500s, who banned the plant from use as a crop because of its importance in the Aztec religion.

Amaranth grows rapidly, can tolerate a wide range of environmental conditions (high salt, high acid, high alka-

FIGURE 18.25
Banking plant genes at the cryogenic gene bank at the University of California at Irvine. Pollen is frozen in small, plastic ampules. How does this compare to a human sperm bank?

decrease in genetic diversity that makes a crop vulnerable to disease or a natural disaster because it does not have resistant variants. For example, the genetic uniformity of the Irish potato crop allowed a fungus, *Phytophthora infestans,* to wipe out most of the crop between 1846 and 1848. (Today in Ireland, more than 60 varieties of potatoes are actively farmed to prevent a similar kind of disaster.) However, such lessons are not always heeded. In 1970, more than 70% of the North American corn crop was restricted to just six varieties. A new strain of southern corn leaf blight fungus destroyed 15% of that crop, at a cost of $1 billion. Gene banks and pollen banks may provide the diversity to avoid such disasters in the future.

Our ability to feed ourselves will depend largely on our ability to preserve the genetic diversity of our plants, increase the quality of our crops, produce new hybrids, and develop new crops. Although biotechnology will play a major role in this, we will never be able to eliminate world hunger unless governments reform their food-distribution policies and we greatly slow the growth of our population. Plant biotechnology is a powerful tool, but it alone is not the answer to world hunger.

SUMMARY

Seed plants possess several characteristics that have allowed them to dominate the land. These include seeds, which protect the embryo until environmental conditions are favorable for germination and growth, and pollen grains, which transport sperm cells from plant to plant without the need for water. In addition, seed plants have roots, which absorb water, and cuticles, which help maintain the water inside their bodies.

Pollen arose more than 150 million years after the origin of seeds. Seed plants diverged and flourished throughout the Carboniferous period as seed ferns; diversification of ancestral pines and cycads followed. The Jurassic period was dominated by cycads, cycadeoids, and primitive pines. This period is also called the Age of Cycads.

Most of the features used for classifying living gymnosperms involve reproductive structures. Most gymnosperms bear seeds produced in cones. Pollination requires that pollen be transferred from microstrobili to ovules. Ovules are generally exposed (i.e., naked) or temporarily enclosed by sporophylls or other branches from the main axis of the strobilus.

The Pinophyta, Cycadophyta, Ginkgophyta, and Gnetophyta are four divisions of gymnosperms with living representatives. Pines are abundant in northern coniferous forests and are planted throughout the Southern Hemisphere. Bristlecone pines are long-lived. Other conifers dominate similar habitats in the Southern Hemisphere. Pine leaves form on short shoots called fascicles, which usually consist of two to five needles with deciduous, non-photosynthetic, scalelike leaves at the base. The fascicles themselves are intermittently deciduous, usually lasting for less than 5 years.

Cycads are slow-growing gymnosperms of warmer climates. Their reproduction is similar to that of the pines, except that their sperm are motile by means of many flagella and their megastrobili can be massive. All cycads are dioecious.

There is only one living species of *Ginkgo,* a tree with fan-shaped, dichotomously veined leaves. *Ginkgo* is dioecious and produces seeds that stink.

There are three distinct genera of gnetophytes. *Ephedra* species are monoecious or dioecious and native to northern drier areas. These plants are mostly shrubby and otherwise resemble horsetails in having whorled branches and small, essentially functionless leaves. Their cones consist of paired bracts, some of which contain microsporangia. Their life cycle is similar to that of conifers. *Gnetum* species are tropical vines or trees with broad leaves; their reproduction resembles that of *Ephedra. Welwitschia* is a bizarre, dioecious plant of southwest African deserts. It produces two large, strap-shaped leaves with basal meristems. The leaves arise from a concave, bark-encrusted, trunkless stem from which a taproot extends into the ground.

Gymnosperms, especially pines and their relatives, are an important source of lumber, wood pulp, and resin. Resin is used to make turpentine and rosin. Wood pulp is used as raw material for making paper.

Angiosperms, which are the dominant plants on earth, have a flower that includes seeds in a carpel. The main force behind the rapid evolutionary radiation of angiosperms may have been pollination by insects and the availability of habitats left open by the disappearance of many gymnosperms. The first flowers were probably pollinated by beetles; later angiosperms attracted butterflies and bees.

Another hypothesis that explains the rapid evolution of angiosperms involves the influence of dinosaurs. Large, high-browsing dinosaurs mysteriously gave way to smaller, low-browsing dinosaurs by the end of the Mesozoic era. The low-browsing dinosaurs probably devastated gymnosperms by eating young seedlings, which allowed low-growing, herbaceous angiosperms to survive and diversify.

Flowering plants provide us with food, clothing, shelter, and medicine. Cereals such as corn, wheat, and rice are members of the grass family, whose edible grains can be stored for long periods. Modern wheats contain two, four, or six sets of chromosomes, and they probably evolved from a natural cross between ancient einkorn wheat (a diploid) and a wild grass (also a diploid), followed by chromosome doubling in some germ cells thousands of years ago. This eventually yielded tetraploid emmer wheat. Emmer wheat also crossed with a wild grass to produce the first hexaploid wheats, which today are used to make bread and other baked goods.

Corn probably evolved from an ancient grass relative, teosinte. Both corn and teosinte have 20 chromosomes, they sometimes grow in the same fields, and they can crossbreed. About 7,500 years ago, an environmental stress may have selected teosinte plants with large kernels, and early farmers may then have cultivated these individual plants. Breeding experiments led to the development of hybrid corn, which results from crosses between separate inbred lines to yield bountiful crops.

Rice is perhaps the most widely consumed modern cereal crop, and fossils of the plants date back some 130 million years. Today, rice grows in a wide range of environments. In 1961, a rice seed bank was started in the Philippines to preserve different varieties and prevent reliance on a few types. Seed and pollen banks have since been founded for other valuable plant species.

Although the Green Revolution increased food production in many countries, overpopulation and poor distribution of food still remain problems around the world.

Writing to Learn Botany

Would you eat foods that came from plants that had been genetically altered? You may already have eaten prepared foods with ingredients that had been genetically modified. Investigate and write about the process, future, and concerns involved with genetically engineered plants that are raised for human consumption.

THOUGHT QUESTIONS

1. International agricultural officials have debated for more than a decade about whether Western countries should pay royalties on genes that germplasm banks have given them for free. The genes can be extremely valuable; for example, a barley gene imported for free from Ethiopia protects the $160-million U.S. barley crop from the yellow dwarf virus. Should countries pay royalties for such genetic resources? Why or why not?

2. Describe the ways in which scientists hypothesize that modern varieties of wheat, corn, and rice arose.

3. Why is reliance on only a few varieties of a crop plant dangerous? Cite an example, other than corn or potatoes, of when such reliance led to devastation.

4. You are planning your vegetable garden and intend to purchase five packets of carrot seeds. The seed catalog describes a new variety of carrot that is resistant to nearly every known garden pest and produces long, highly nutritious carrots in a variety of climates. It sounds too good to be true, but the company that publishes the catalog has had a lot of experience in plant breeding. If you want to obtain as many carrots as possible, would you be better off buying five packets of the new variety or five different types of carrot seeds? Give a reason for your answer.

5. We use different plants for different purposes. For example, wood of poplar does not splinter and therefore is used to make toys, tongue depressors, and Popsicle sticks. How are other plant products suited for their function?

6. Many tropical cultures depend heavily on plants such as sugarcane, bananas, and coconuts. Why haven't these plants had the same worldwide impact as cereals?

7. Dinosaurs and most other animals went extinct at the end of the Cretaceous period. Some scientists have proposed that such mass extinctions were caused by geological catastrophes. If this is true, how might angiosperms have escaped such catastrophes?

8. In addition to co-evolution with insect pollinators, what other kinds of co-evolutionary interactions may have occurred between flowering plants and animals?

9. Angiosperms usually have a shorter life cycle than do other plants. How might this have affected the evolution of angiosperms?

10. The forms of gymnosperm leaves include tiny scales, huge strap-shaped leaves with basal meristems, needles, palmlike leaves, fan-shaped leaves, and simple, broad leaves. Can you think of any functional or ecological significance of the various forms?

11. Cycads, yews, and several other gymnosperms are dioecious; pines are monoecious. Can you think of any survival or other value to the plants in having their reproductive structures on separate plants?

SUGGESTED READINGS

Articles

Crane, P. R., E. M. Friis, and K. R. Pederson. 1995. The origin and early diversification of angiosperms. *Nature* 374:27–33.

Donoghue, M. J. 1994. Progress and prospects in reconstructing plant phylogeny. *Annals of the Missouri Botanical Garden* 81:405–18.

Galen, C. 1999. Why do flowers vary? *BioScience* 49:631–40.

Gura, T. 1999. New ways to glean medicines from plants. *Science:*1347–49.

Klee, K. 1999. Frankenstein Foods? *Newsweek* (September 13):33–35

Poinar, G. O., Jr. 1993. Still life in amber. *The Sciences* 34 (2):34–39.

Stone, R. 1993. Surprise! A fungus factory for taxol? *Science* 260:154–55.

Books

Beck, C. B. 1988. *Origin and Evolution of Gymnosperms.* New York: Columbia University Press.

Bernhardt, P. 1999. *The Rose's Kiss: A Natural History of Flowers.* Washington, DC: Island Press.

Jones, D. 1993. *Cycads of the World.* Washington, DC: Smithsonian.

Simpson, B. B., and M. Conner-Ogorzaly. 1995. *Economic Botany: Plants in Our World,* 2d ed. New York: McGraw-Hill.

van Geldenen, D. M., and J. R. P. van Hoey Smith. 1996. *Conifers.* Portland: Timber Press.

ON THE INTERNET

Visit the textbook's accompanying web site at http://www.mhhe.com/botany to find live Internet links for each of the topics listed below:

Evolution of seeds

Life cycle of pines

Classification of the gymnosperms

Pines: lumber and paper

The gnetophytes

Yews and cancer

Plant families

Green revolution

Ginkgo biloba and memory

Pinnacles and plants on a dry lake bed in south central California comprise part of a desert ecosystem. In what ways might these plants be interacting with each other and with their environment?

CHAPTER 19

Ecology

LEARNING ONLINE

Did you know that in the time it would take you to read a single page in this book, about 40 hectares (100 acres) of rain forest will be destroyed? To learn more about ecology, visit http://esa.sdsc.edu/links.htm in chapter 19 of the Online Learning Center at www.mhhe.com/botany.

You live in a community of organisms—many of them microscopic and inside your intestine, on your skin, and throughout your house, apartment, or dorm room. In your community are also other animals and many plants; even in the city, you find plants wherever you find an open lot or often in cracks in the sidewalk. If you drive a short distance outside of your town or city, you will soon begin to see the plants that dominate your community and animals that live among them. You are a part of a community, dependent on plants, with community members connected to each other by their need for energy and mineral nutrition. Because of these connections, what happens to one organism in a community affects the lives of other organisms, and because humans, as a species, use more energy, minerals, and other resources than any other species on earth, we have the greatest impact on the communities of the earth.

In this chapter, we examine the interrelationships among plants, animals, other organisms, and their environment. This chapter builds on the basic ecological concepts you've read about earlier in this text and focuses on communities and ecosystems. The discussions include an examination of ecosystems and the flow of energy within them, the cycling of water and nutrients, and the impact that humans have on the natural world around us. You'll read about the interactions of producers, consumers, and decomposers as well as the factors involved in ecosystems that change through time during succession.

The chapter also includes a discussion of large ecosystems such as forests, grasslands, and deserts, which are dominated by particular kinds of plants (fig. 19.1), and ends with a discussion of the consequences of human disruption of ecosystems, including acid rain, ozone depletion, and the invasion of pests.

F I G U R E   1 9 . 1

In the foreground, a grassland ecosystem in spring. In the background, a coniferous forest ecosystem. Ecosystems are composed of more or less interdependent species living in the same location, interacting with their environment.

Energy Flows Through an Ecosystem.

Ecology (from the Greek *oikos*, meaning "home"—this is the same root as that for the word *economy*, which also involves resources and interactions) is the study of organisms in relation to their environment. The environment includes both biotic, or living, factors such as the dust mites in your pillow and the dandelions in your front yard, as well as abiotic, or nonliving, factors such as rain and sunshine. The living and nonliving factors combine to form an ecosystem, an organism's home (fig. 19.2). **Ecologists** are scientists who investigate processes such as the change in numbers of individuals of a particular species, how species are adapted to their environment and affected by climate changes, how organisms interact with other species, and how nutrients and energy move through an ecosystem. Virtually all subdisciplines of biology are involved in the study of ecology, from investigations of molecules to studies of ecosystems. Ecologists also contribute to the study of the evolution of species.

Energy is lost from an ecosystem as heat.

Energy in the form of light enters a food chain at the producer trophic level (see chapter 2) and flows to subsequent levels of consumers and decomposers in an ecosystem. Only about 1% of the total light energy striking a temperate-zone community is used in photosynthesis to convert CO_2 into sugar. This organic material —sugar—and its energy pass to subsequent trophic levels, and as the organisms at each level

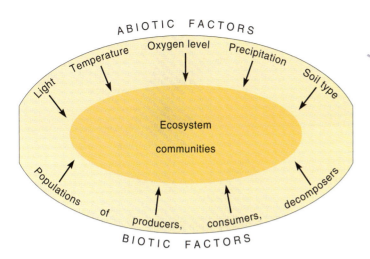

FIGURE 19.2

The basic components of an ecosystem interact to form an integrated unit.

respire, energy in the organic molecules is gradually lost as heat into the atmosphere (see chapters 3 and 11). Some energy is stored in organisms that are not consumed and is released when they decompose. For example, the energy in leaves that fall from a plant is released when decomposers such as fungi and bacteria degrade the leaves.

Only a very small portion of energy stored in one trophic level will flow to the next level. Most energy is lost as heat during respiration. This heat cannot be used by organisms for growth, maintenance, or reproduction because it cannot be captured and used to produce high-energy molecules. As a result, life on earth relies on the constant supply of energy from the sun.

Generally, less than 10% of the energy stored in plants that are eaten by cattle is converted to animal tissue; most of the remaining energy dissipates as heat during the normal activities of the animal during the day. Some energy remains tied up in the waste products of animals; settlers of the American west burned as fuel "buffalo chips" that the bison deposited over the plains, and people around the world use dried "cow pies" as a fuel.

When we eat beef, our bodies use less than 10% of the beef's stored energy for our growth, maintenance, and reproduction. Thus, over 90% of all the energy in the beef is converted to heat. If more than 90% of the energy is lost as heat at each level of a food chain, then the greater the number of carnivores in a community, the greater the number of producers is necessary to provide energy for the herbivores that are eaten by the carnivores. Consider the following example: Assume that for every 100 units of light energy striking corn plants each day, 10 units are converted to plant tissue (an optimal figure). Then suppose that the corn is fed to cattle. Less than 10%, or 1 unit, of the original energy is converted to beef. When we eat beef, our bodies, in turn, use less than 0.1% of the energy originally converted

TABLE 19.1

How to Make a Cow

Ingredients		Fertilizer Recipe	
1	80-pound calf	Combine:	
8	acres grazing land	170	pounds nitrogen
12,000	pounds forage		
125	gallons gasoline and various petroleum by-products	45	pounds phosphorus
		90	pounds potassium
305	pounds fertilizer		
1.5	acres farmland		
2,500	pounds corn		
350	pounds soybeans		
	insecticides		
	herbicides		
	antibiotics		
	hormones		
1.2	million gallons water, to be added regularly throughout		

Take one 80-pound calf—allow to nurse and eat grass for 6 months, then wean. Over the next 10 months, feed 12,000 pounds of forage. Use about 25 gallons of petroleum to make fertilizer to add to the 1.5 acres of land. Set aside rest of gasoline to power machinery, produce electricity, and pump water. Plant corn and soybeans—apply insecticides and herbicides. At 24 months, feed cow small amounts of crop and transfer to feedlot. Add antibiotics to prevent disease and hormones to speed up fattening. During next 4 months, feed remaining crop mixed with roughage. Yields about 440 usable pounds of meat—1,000 7-ounce servings. Option: Bake the 2,500 pounds of grain and 350 pounds of soybeans into bread and casseroles—18,000 8-ounce servings.

Source: (Excerpt from *The Cousteau Almanac* by Jacques-Yves Cousteau. Copyright © 1980 1981 by the Cousteau Society, Inc. Reprinted by permission of Doubleday & Company, Inc.)

to plant tissue. If we eat the corn directly instead of the beef, we receive the same amount of energy from much less corn than it took to produce the beef (table 19.1).

Each ecosystem can support only a limited number of organisms.

The concept just described—that is, the flow of energy through an ecosystem—has important implications. For example, a vegetarian diet uses energy more efficiently than does a diet based on meats because no energy is lost through respiration in the meat-producing animal. Consequently, where food is scarce or humans are abundant (as in India or Ethiopia), many humans are forced to become vegetarians. In a given part of the ocean, billions of microscopic algal producers may support millions of tiny crustacean consumers, which, in turn, support thousands of small fish, which meet the food needs of scores of medium-sized fish, which are finally eaten by one or two large fish. In other words, one large fish may depend on a billion tiny algae to meet its energy needs every day (fig. 19.3). This is the same reason you find so few hawks (carnivores) along the highway compared to the many hundreds of mice (herbivores) in a field. These hundreds of mice must be supported by the growth of thousands of plants.

The number of any kind of organism that an ecosystem can support is limited. The maximum number of individuals that can survive and reproduce in an ecosystem is the ecosystem's **carrying capacity.** As a population increases, the competition for nutrients, water, and light increases. When the germination and survival of offspring equal the death rate of mature plants in a population, population growth stops. Thus, every ecosystem has a limit to the number of plants, animals, and trophic levels that it can support.

Ecosystems exhibit considerable variation in **net productivity,** which is the energy produced by photosynthesis minus that lost in respiration. Productivity may be measured in terms of biomass produced in grams per square meter. Forest ecosystems, for example, produce 3 to 10 grams of dry matter per square meter of land per day, whereas grasslands produce 0.5 to 3 grams, and deserts less than 0.5 gram. Crops grown throughout the year (e.g., sugarcane) produce up to 25 grams. The greater the plant productivity in an ecosystem, the greater the number of animals that can live there.

Highly productive ecosystems often have a high diversity of species.

Species diversity is defined by the number of species and the number of individuals per species in an ecosystem. Harsh climates with poor soils, such as those in arctic regions and deserts, limit species diversity, whereas mild or tropical climates with fertile soils are characterized by higher diversity. Indeed, about half of all living species occur in tropical rain forests, which occur on less than 7% of the earth's landmasses. In rain forests, there is intense competition for light, minerals, and other resources, but each species is well adapted to its habitat, which allows for great diversity. For instance, in a 0.5-square-kilometer area (about 125 acres) of rain forest in Panama, 238,000 trees in nearly 300 different species were counted.

Humans and other organisms depend on ecosystems for a variety of important services such as preserving soil fertility, controlling pest outbreaks, and influencing

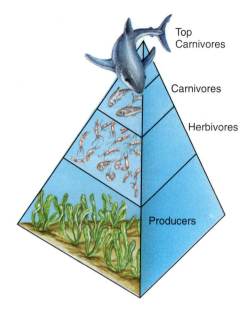

FIGURE 19.3

This pyramid represents how much energy from lower levels of the food web is required to support a limited number of trophic levels. In this example, each kilogram of the shark (top carnivore) indirectly requires as much as 1,000 kilograms of algae (producers). In terms of energy, why do you think there are a limited number of trophic levels?

weather. But how can we preserve a working ecosystem? How does the loss of biodiversity—currently a hot topic among environmentalists—change an ecosystem's functioning? The answers to these and many related questions are being sought by researchers in the separate, but related, fields of environmental engineering, environmental science, land management, and ecology.

In 1981, Stanford University ecologists Paul and Anne Ehrlich proposed the "rivet popper" hypothesis about the importance of species diversity in an ecosystem. According to this hypothesis, the diversity of life is something like the rivets on an airplane; each species plays a small but important role in the working of the whole ecosystem. The loss of each rivet weakens the plane by a small but noticeable amount; the loss of enough rivets causes the plane to break apart and crash. Several studies now support this hypothesis. One experiment in controlled-environment chambers tested the effect of diversity on productivity by setting up 14 artificial ecosystems. The result: as species numbers went up, the mean production of plant biomass also increased. A possible explanation is that more species usually generate a more diverse plant architecture, which allows the system to capture more light and produce more plant biomass. This conclusion is supported by studies of multi-crop agriculture: the best way to increase the productivity in a cornfield is not to pack in more corn plants but to add other plants such as melons and nitrogen-fixing legumes. Can you suggest reasons why this might be so?

Species diversity also increases an ecosystem's resilience: diverse ecosystems recover from stress (e.g., drought) much faster than do ecosystems with less diversity. This is supported by the work of David Tilman and his colleagues at the University of Minnesota, who studied how 250 species of plants could thrive in midwestern grasslands. The researchers created 207 plots distributed among one native prairie and three abandoned fields of different ages. Each season, they clipped a different section in each plot and analyzed its species diversity and biomass. They left some plots alone and added nitrogen fertilizer to others; nitrogen fertilizer tends to reduce species numbers even as it boosts productivity. During a drought, the productivity in all of the plots fell drastically. However, the drop in species-rich plots was only 25% of that of species-poor plots. Moreover, the species-rich plots recovered in only one season rather than in the four seasons required for species-poor plots to recover. Tilman concluded that "biodiversity is a way to hedge bets against uncertainty, even in managed systems." Thus, we should be concerned about the loss of biodiversity and extinctions that humans are now causing in biomes around the world.

INQUIRY SUMMARY

In any ecosystem, producers, consumers, and decomposers are part of food chains that interconnect in a food web, with producers at the base. The interrelationships result in food webs, with a balance between producers and consumers. Energy cannot be recycled and is lost at each level. Huge numbers of producers may be required to meet the needs of a single consumer at the end of a long food chain. Ecosystems have a carrying capacity for any species, meaning they can support only a limited number of individuals. Highly productive ecosystems often have a high species diversity.

Fields of crops are usually highly productive. How would you describe and explain their diversity?

Nutrients Cycle Through an Ecosystem.

Plants absorb minerals from the soil and air and incorporate them into their bodies. Then, animals eat plants and pass these nutrients through the food chain and food webs of the community. Eventually, decomposers break down the bodies of plants and animals and the waste products of animals. In this process, the decomposers return, or recycle, the nutrients into the air and the soil from where the nutrients came. The cycling of water and of minerals such as nitrogen and carbon is essential to all living things in an ecosystem.

Microorganisms play critical roles in the nitrogen cycle.

Much of the mass of cells—other than water—is protein, and the nitrogen in that protein comes from the most abun-

dant element in our atmosphere, nitrogen gas. Therefore, the availability of nitrogen is critical to growing plants. Ironically, there are nearly 69,000 metric tons of nitrogen in the air over each hectare of land, but the total amount of nitrogen in the soil seldom exceeds 4 metric tons per hectare and is usually much less. This discrepancy exists because nitrogen in the atmosphere is chemically inert; that is, it does not combine readily with other molecules. Thus, atmospheric nitrogen is unavailable to plants and animals.

Most of the nitrogen in plants and, indirectly, in animals comes from soil in the form of inorganic ions absorbed by roots. These ions are released by bacteria and fungi, which break down the complex molecules in dead plant and animal tissues. Some nitrogen from the air is also *fixed*—that is, converted to ammonia or other nitrogenous substances by *nitrogen-fixing bacteria* (see chapter 7) including cyanobacteria (see chapter 16). Some of these nitrogen-fixing organisms live symbiotically in plants, particularly in members of the legume family (Fabaceae; e.g., peas, beans, alfalfa), whereas many others live free in the soil.

As shown in figure 19.4, nitrogen flows from dead plant and animal tissues into the soil and from the soil into plants. Bacteria and fungi can break down huge amounts of dead leaves and other tissues to tiny fractions of their original volumes within a few days to a few months. If the activities of bacteria and fungi were to stop abruptly, the available nitrogen compounds would be exhausted within a few decades, and the supply of CO_2 needed for photosynthesis would be seriously depleted as the respiration of decomposers ceased. Forests and prairies would die as accumulations of shed leaves, bodies, and debris buried the living plants and prevented light from reaching their leaves.

In many ecosystems, much nitrogen is leached from soil by water. In agricultural systems, more nitrogen is removed with each harvest (the average crop removes about 25 kilograms per hectare per year). This nitrogen can be recycled and used as fertilizer if vegetable and animal wastes are returned to the soil. However, weeds and stubble are frequently controlled by burning, which depletes soil nitrogen. The annual combined loss of nitrogen from the soil in the United States from fire, harvesting, and other causes exceeds 21 million metric tons, and only 15.5 million metric tons are replaced by natural means. To offset the net loss, some 32 million metric tons of inorganic fertilizers are applied each year.

Photosynthesis and respiration are key processes in the carbon cycle.

Bacteria and fungi also recycle carbon and other substances. As noted in chapter 10, one of the two raw materials of photosynthesis is CO_2, which constitutes about 0.036% of our atmosphere. All the algae in the oceans and plants on the land use about 14.5 billion metric tons of carbon obtained from CO_2 every year. This is replaced by the cellular respiration of

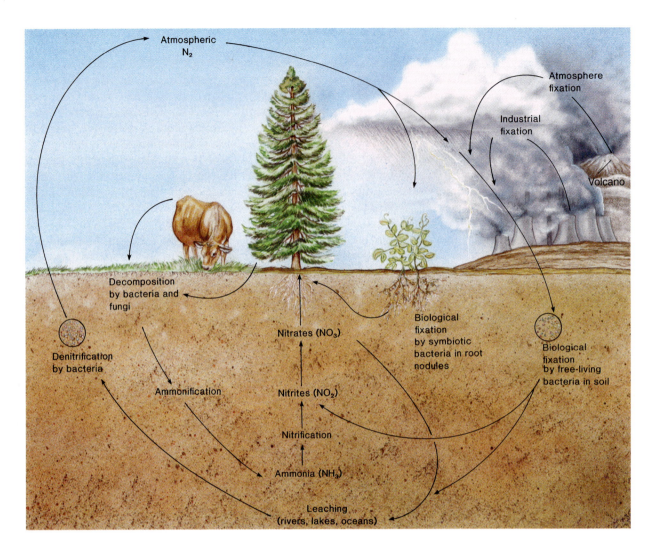

FIGURE 19.4

The nitrogen cycle. Most nitrogen occurs as atmospheric nitrogen (N_2), which is not usable by plants and animals. Nitrogen fixation and decomposition make nitrogen available for plant metabolism as nitrates. Nitrogen from these sources is incorporated into organic compounds in plants and other organisms in the food web. Denitrification by bacteria returns N_2 to the environment, whereas ammonification and nitrification by other bacteria help plants obtain nitrates.

all organisms, with perhaps as much as 90% or more being released by bacteria and fungi as they decompose tissues.

All of the CO_2 in our atmosphere would be used in photosynthesis in about 22 years if it were not constantly replenished through respiration (fig. 19.5). Oceans act as a buffer by absorbing and storing much of the excess CO_2 as carbonates, but the capacity of oceans is limited. Some biologists believe that if this storage capacity is exceeded, the CO_2 in the atmosphere will rise even more dramatically than it is rising now (see Perspective 19.1, "Curbing Methane Emissions to Curtail the Greenhouse Effect").

The atoms that make a plant's body have cycled through other organisms.

More than two-thirds of the earth's surface is covered by water, which is accumulated in oceans, freshwater lakes,

ponds, glaciers, and polar ice caps. Evaporation occurs as surface waters are warmed by the sun. The water vapor rises into the atmosphere, where it cools, condenses, and returns as precipitation. Air currents move the moisture-laden air around the globe.

When precipitation occurs over land, some of the water returns to the oceans, lakes, and streams in the form of runoff, and some percolates through soil to underground water tables. When plants absorb water from the soil, all but about 1% of it is transpired through the leaves and stems. Any water retained by the plants is stored in cells, where it is used in metabolism and the maintenance of turgor. The transpired water vapor combines with other evaporated water to form precipitation, and the cycle is repeated (fig. 19.6).

For millions of years, carbon, nitrogen, water, phosphorus, and other molecules have been passing through cycles. Some molecules that were a part of a primeval forest that became compressed and turned to coal may have

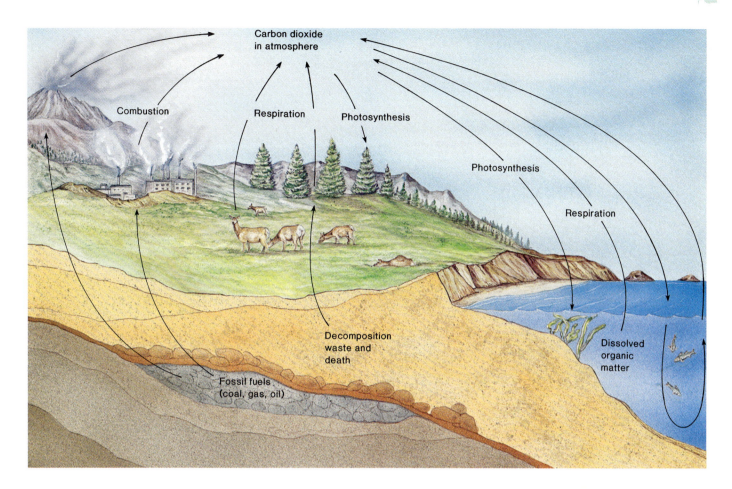

FIGURE 19.5

The carbon cycle. Carbon dioxide is formed from respiration by plants and animals, decomposition of organic matter by bacteria and fungi, burning fossil fuels and wood by humans and natural fires, and volcanic activities. Carbon dioxide is used by plants and other photosynthetic organisms for photosynthesis.

become part of another plant after the coal was burned, releasing CO_2. Then the new plant may have been eaten by an animal, which, in turn, contributed molecules to yet another organism. Molecules in our own bodies may have been a part of some prehistoric tree, a dinosaur, a woolly mammoth, or even all three.

INQUIRY SUMMARY

Microorganisms and fungi cycle nitrogen from dead tissues into the soil and back into plants. Precipitation returns a small portion of nitrogen to the soil from the atmosphere and is an important part of the cycling of water. Decomposers, through respiration, constantly produce most of the CO_2 needed by green organisms for photosynthesis. All other living organisms, as well as abiotic processes, are the sources of the remaining CO_2 produced. Excess CO_2 is stored in oceans.

How does the burning of fossil fuels today contribute to the cycling of nutrients?

Ecosystems Change Through Time.

Although nutrient cycling helps maintain an ecosystem, there are forces in nature that change ecosystems over time. After a volcano spews lava over a landscape or an earthquake or a landslide exposes rocks for the first time, there is initially no life on the lava or rock surfaces. Within a few months, or sometimes within a few years, organisms appear, and a sequence of events known as primary succession occurs. **Succession** is a cumulative change in the biotic and abiotic components of an ecosystem over time. **Primary succession** refers to those changes occurring on rock, lava, sand, and other areas that have never been covered by plants. During primary succession, the first plants, called pioneers (see chapter 7), gradually alter their environment as they grow and reproduce. Over time, accumulated wastes and decaying organic material promote the formation of soil. Other changes (e.g., water content in the soil) favor different species, which may replace the original ones. These new

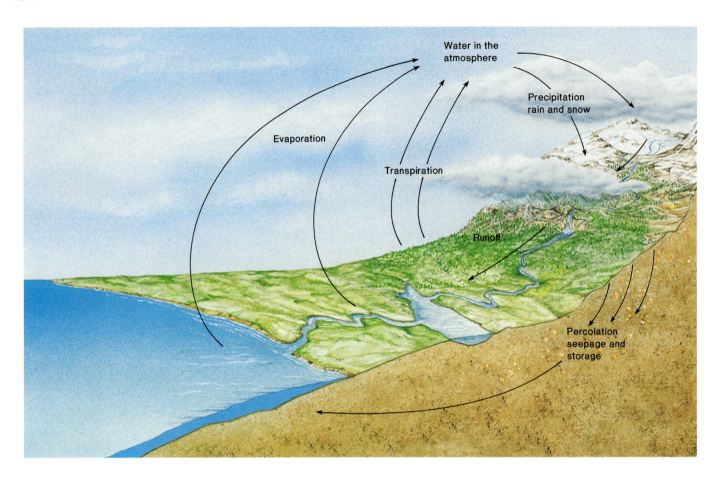

FIGURE 19.6

The water cycle. Atmospheric water falls mostly as rain or snow and either evaporates, runs off, or percolates into the soil. Some of the percolated water reaches underground storage, and some of it is absorbed by plant roots. About 99% of all water taken in by roots is re-released into the atmosphere by transpiration.

species, in turn, modify the environment further so that still other species become established (fig. 19.7).

When primary succession begins, there is little or no soil. Bare rocks are sometimes subjected to alternate thawing and freezing, at least in temperate and colder areas, which crack or flake the rocks, beginning the long, slow process of soil formation. This process may take hundreds or thousands of years. Lichens often become established on such surfaces and produce acids, which slowly dissolve the rocks. As the lichens die and contribute organic matter, they may be replaced by rock mosses adapted to long periods of desiccation. A small amount of soil begins to accumulate, augmented by dust and debris blown in by the wind.

Eventually, enough of a mat of lichens, mosses, and soil is present to permit some ferns or even seed plants to become established, and the pace of soil buildup and rock breakdown accelerates (fig. 19.8). If deep cracks appear in the rocks, the larger seeds may widen them further as they germinate and their roots expand. Indeed, seedlings have been known to split rocks that weigh several tons (fig. 19.9).

As roots respire, they produce CO_2, which combines with water to form carbonic acid. This acid helps break the rock down to small particles. The combination of tiny rock particles and organic material is a major part of soil.

As the buildup of soil continues, larger plants may take over, and eventually the vegetation may reach an equilibrium in which the kinds of plants and other organisms remain essentially the same until another disturbance occurs. Such relatively stable plant associations are often referred to as **climax communities.** These communities remain relatively stable until they are disrupted by some sort of disturbance, such as a fire, which keeps the association of species in a constant state of flux. The climax vegetation of deciduous forests in eastern North America is dominated by maples, beeches, oaks, and hickories. In desert regions, various cacti form a conspicuous part of the climax vegetation, while in the Pacific Northwest, large conifers predominate, and in parts of the Midwest, prairie grasses and other herbaceous plants form the climax vegetation. As the climate changes, however, a climax community may change as well, thus many ecologists do not think the term climax community is appropriate.

19.1 *Perspective* *Perspective*

Curbing Methane Emissions to Curtail the Greenhouse Effect

In an effort to limit global warming, the U.S. government proposed that a comprehensive framework for limiting greenhouse gases would be preferable to concentrating on a single gas. Hogan, Hoffman, and Thompson (*Nature* 354 [1991]:181–82) argued that reducing methane emissions would be relatively easy and could significantly reduce the greenhouse effect and global warming. They pointed out that cutting methane emissions would be 20 to 60 times more effective in reducing the potential warming of the earth's atmosphere this century than decreasing CO_2 emissions. They also noted that methane released by human activities is generally a wasted resource, opening the possibility that reductions might even be profitable.

Much methane is generated in landfills by the anaerobic decomposition of wastes. By collecting this gas, existing recovery systems can reduce emissions by 30% to 60%. Coal mining emits similar amounts of methane, although relatively few mines emit most of the gas that is released. If vertical wells are used before and during the mining operations, methane emissions from underground mine areas could be cut by more than 50%, and the cost of mine ventilation would also be lessened.

The production of oil and natural gas also produces much methane. Improved technologies such as leak-proof pipelines can reduce methane emissions cost-effectively. The second largest source of methane related to human activities involves the cattle industry; the cattle themselves produce large amounts of the gas as they digest their food. Some feeds that may reduce methane emissions while increasing productivity have been identified. One hormone (bovine somatotropine) can re-

duce methane emissions while increasing milk production, however, its use is being challenged for human health reasons. Recovery systems can profitably recover 50% to 90% of the methane generated from animal wastes. Such recovery systems are already in use in India, where they generate enough electricity to power lights for a few hours each evening. However, animal wastes might be more appropriately used as soil fertilizer.

Rice cultivation is considered to be the single largest source of methane emitted by human activities. An integrated approach to irrigation, fertilizer application, and cultivar selection could reduce methane emissions by as much as 30%. Alternative approaches to agricultural practices involving the clearing of land and crop stubble by fire could also cut methane emissions, but trade-offs clearly would be involved.

Primary succession often starts in water.

Hydrophytes are plants such as water lilies (*Nymphaea*) that grow in water and have modifications that adapt them to their aquatic environment. For hydrophytes, the problem is not getting enough water but getting enough oxygen to roots growing in mud. In the water, oxygen is often a limiting factor. Leaves of hydrophytes often have thin cuticles and more stomata on their upper surface than on their lower surface (fig. 19.10). (Recall from chapter 9 that many kinds of plants possess leaves with stomata mainly on their lower surface.) Hydrophytes also possess *spongy tissue,* which is a tissue containing large air spaces between its cells that allow easy passage of gases. These differences in the form of hydrophyte leaves and tissues suggest an important role in their function. In water lilies, oxygen diffuses through stomata on the top of a water lily leaf and through the spongy tissue to its roots.

In the northern parts of midwestern states such as Michigan, Wisconsin, and Minnesota, ponds and lakes of various sizes abound. These ponds and lakes shrink each year as a result of primary succession. This succession often begins with algae either carried in by the wind or transported on the muddy feet of waterfowl and wading birds. The algae multiply in shallow water near the margin, and

with each reproductive cycle, the dead organisms sink to the bottom. Floating plants such as duckweeds may then appear, often encircling the water. Next, water lilies and other hydrophytes with floating leaves become established. Cattails, sedges, and other flowering plants that produce their inflorescences—their flowering stalks—above the water often take root around the edges. Thus, plants grow toward the center of the pond or lake, and the surface area of exposed water gradually shrinks.

Each group of plants contributes to the organic material on the bottom, which slowly turns to muck—dark, highly organic mud. Dead organic material accumulates and fills in the area under the sedge mats, and herbaceous and shrubby plants then invade. As the edge becomes less marshy, coniferous trees whose roots can tolerate considerable moisture (e.g., tamaracks or eastern white cedars) may gain a foothold, eventually growing across the entire area as the pond or lake disappears. Trees continue to help form soil, and finally the climax vegetation begins to dominate. No visible trace of the pond or lake remains, and the only evidence of its having been there lies beneath the surface, where pollen grains, bits of wood, and other materials reveal the area's history. Such succession may take hundreds of years (fig. 19.11). How might a person study such a prolonged event?

FIGURE 19.7
Plant succession near Mount St. Helens, Washington. Douglas fir seedlings and other vegetation reappeared after complete destruction by volcanic ash. This photo was taken just 12 years after the volcano exploded in 1980.

FIGURE 19.8
Moss and lichens living on rocks. Lichens and mosses are colonizers on rock surfaces in early successional stages, helping to break down the rock and build up the soil.

FIGURE 19.9
Expansion in girth of the roots of this yellow birch has split a large rock.

FIGURE 19.10
Water lily (*Nymphaea*), a hydrophyte. The floating leaf blades have thin cuticles; nearly all their stomata are on the upper surface. How do you explain that these plants often grow near the edges of a pond or lake and not in the middle?

FIGURE 19.11

Succession in a pond. The pond basin slowly fills in as algae and rooted plants contribute organic matter.

FIGURE 19.12

Quaking aspen (*Populus tremuloides*) trees often become established in mountainous areas after fires. As succession proceeds, climax vegetation later may replace the aspens.

Secondary succession happens after human and other disturbances.

Succession often begins when a natural area is disturbed. Succession proceeds at varying rates, depending on the climate, soils, and organisms in the vicinity. When an area is disturbed by fire, floods, or landslides, some of the original soil, plant, and animal material may remain. The pattern of changes that follow such disturbances is **secondary succession.** Secondary succession, which has fewer phases and may arrive at climax vegetation faster than primary succession, occurs if soil is present and some species survive in the vicinity.

Secondary succession often follows human disturbances, such as what happens in abandoned farmlands. Other secondary successions follow fires. Grasses and other herbaceous plants become established on burned (or logged) land. These are usually followed by weeds and trees and shrubs that have widely dispersed seeds (e.g., aspen and sumac in the Midwest and East and chaparral plants such as gooseberries in the West). After going through fewer stages than are typical of primary succession, the climax vegetation may take over, often in less than 100 years (fig. 19.12), however, climax vegetation may never recur after secondary succession. In 1988, over 1.6 million acres burned in Yellowstone National Park. Many

people believed this spectacular park would be destroyed forever. Based on what you now know about fire, what do you predict the park looks like today?

INQUIRY SUMMARY

Succession is the change in community composition that occurs as vegetation develops in an area. Primary succession is initiated on freshly exposed rocks or lava or in bodies of water. Secondary succession occurs after disturbances occur in areas where soil already exists. A successional series may end in a climax community that is in a state of flux.

Succession is change in a community through time. How is this similar to and different from evolution, which also is a change through time?

Biomes Are Large Ecosystems Dominated by Particular Groups of Plants.

Terrestrial ecosystems in different parts of the world are so similar that ecologists classify them into larger ecosystems called **biomes.** The same biome may therefore span several continents. Although the distribution of biomes is determined mostly by climate, each biome is characterized

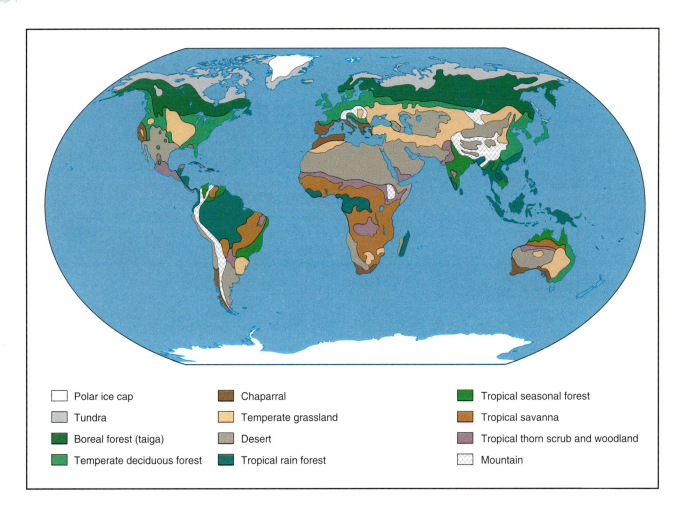

F I G U R E   1 9 . 1 3
Major biomes of the world. These biomes are widely distributed, and their distribution is primarily a product of climate and soil type. Try to name a major city in each biome of North America.

primarily by its vegetation. In North America, the biomes include the tundra, taiga, deciduous forest, grasslands, deserts, chaparral, and tropical rain forests (fig. 19.13).

The tundra is the treeless plain.

The word **tundra** is derived from a Russian word meaning "treeless, marshy plain." Arctic tundra is a vast, mostly flat biome whose terrain is marshy in the short summer period and frozen for much of the rest of the year (figs. 19.13 and 19.14). It occupies about 25% of the earth's land surface, primarily north of the Arctic Circle, with some extending farther south. Alpine tundra, which occurs in patches above timberline on mountains south of the Arctic Circle, is seldom flat and also differs from arctic tundra in having less annual variation in day length, less humidity, and more direct solar radiation (fig. 19.15). The climate and soil of tundra are usually not suitable for agricultural activities.

Fierce, drying winds and freezing temperatures can occur in the tundra on any day of the year, but tempera-

tures can also reach 27°C (80°F) or higher during a midsummer day. In addition to low temperatures, the tundra is characterized by shallow (5.0 to 7.5 centimeters deep), nutrient-poor clay soils that are waterlogged, and therefore oxygen-deficient, during the growing season, which lasts only 2 to 3 months. The cold, low-oxygen conditions of the soil prevent any significant recycling of the extensive organic components, which are produced primarily by peat mosses. Although annual precipitation averages less than 25 centimeters (10 inches) per year, waterlogging occurs primarily because **permafrost** (permanently frozen soil) beneath the surface prevents water from draining deep into the soil.

The vegetation of arctic tundra consists primarily of grasses, sedges, mosses, liverworts, and lichens and is mostly evergreen. Tundra is usually treeless, but low shrubs survive and are even abundant in many parts of the landscape. Such plants include willow, birch, and blueberry, which are rarely more than 1 meter tall at maturity. Leaves, which tend to be small, are protected by a thick cuticle and dense hairs. The stems are often covered with

The Lore of Plants

The southernmost recorded flowering plant is the Antarctic hair grass (*Deschampsia antarctica*), which was found at latitude 68° south on Refuge Island, Antarctica. The northernmost flowering plants are the yellow poppy (*Papaver radicatum*) and the Arctic willow (*Salix arctica*), which grow on the northernmost land at latitude 83° north. The highest altitude at which any flowering plants have been found is 6,400 meters (21,000 feet) on Kamet in the Himalayas; the plants, *Ermania himalayensis* and *Ranunculus lobatus*, live in the alpine tundra biome.

FIGURE 19.14
Shallow pools of water dot the arctic tundra during the summer as seen here in the Northwest Territories of Canada.

FIGURE 19.15
Alpine tundra in Glacier National Park, Montana. Tundra plants have a short but spectacular growing season near and above the timberline, where trees are sparse or absent.

lichens, which may help protect the plants from drying winds and cold temperatures. When temperatures drop below freezing, ice crystals form in intercellular spaces of the plants. The cells are exceptionally tolerant of the dehydration that occurs as the freezing draws water from the cells.

The flora of tundra also includes low perennials that produce brightly colored mats of flowers during the brief growing season (fig. 19.15). Some plants growing on the tundra reproduce vegetatively much of the time, producing seeds only during exceptionally mild and long growing seasons (which occur once or twice every century). Many of these perennials are adapted to the harsh growing conditions by having as little as 10% of their mass aboveground, the remainder being in the form of rhizomes, tubers, bulbs, or fibrous roots below the surface and above the permafrost. During midsummer, photosynthesis can occur throughout the 24 hours of light each day, and virtually all of the sugars produced by photosynthesis are produced within a few weeks each year.

Some of the bowl-shaped flowers that bloom in the tundra "follow" the sun throughout the day so that the sun's energy reflects off the shiny petals and is concentrated on the stamens and pistils. Because of the sunlight, the temperature inside such flowers may be as much as 25°C higher than the air temperatures outside. The warmth attracts insect pollinators and helps fruits and seeds of these arctic plants mature in as little as 3 weeks. Annual plants are rare on the tundra because the cold environment and short growing season usually inhibit the germination of seeds and the subsequent completion of the life cycle in such plants.

The tundra is exceptionally fragile. A vehicle driven across it compresses the soil enough to kill lichens, mosses, and roots of other plants, and the tracks remain for decades. Occasionally, sheep grazing on the tundra pull up patches of the matted vegetation, leaving exposed edges. High winds catch the exposed edges and rip away larger segments of mat, leaving barren patches called blowouts (fig. 19.16).

Taiga, found south of the tundra, is dominated by conifers.

Taiga, also referred to as *northern conifer* or *boreal forest*, occurs mostly adjacent to the arctic tundra across large areas of North America and Eurasia (see figs. 19.13 and

FIGURE 19.16
A blowout on the Alaskan tundra. High winds have torn away the thin layer of vegetation.

FIGURE 19.17
A taiga scene. Plants that grow in the taiga are adapted for long, cold winters and heavy snowfall.

19.17). Similar vegetation occurs in high mountains of many parts of the world, just below the alpine tundra. The soils of the taiga are usually acidic and nutrient poor, making them unsuitable for most agricultural activities other than timber and cranberry farming. Snow accumulates in the taiga during the winters, which are long and cold. In midwinter there may be only 6 hours of sunlight per day, and temperatures drop to −50°C (−58°F) or lower during the dark, colder months. In summer, the temperatures may reach 27°C (80°F), and daylight can last up to 18 hours.

The vegetation of the taiga is relatively uniform and dominated by a few genera of coniferous trees, including spruce, fir, and, in some places, pine. Deciduous trees such as birch, poplar, aspen, willow, alder, and tamarack often occur in some of the wetter areas, such as the margins of the many lakes, ponds, and marshes in the taiga. Nitrogen-fixing bacteria associated with the roots of the alders enable these trees to survive in otherwise comparatively barren soils. A thick-walled epidermis, sunken stomata, and a thick cuticle have evolved in the leaves of taiga conifers, which adapt these trees to the rigors of the harsh winter climate. Many (mostly bulbous) perennials and a few cold-hardy shrubs occur, but annuals are generally prevented from becoming established by the severe climatic conditions.

Although not considered part of the taiga, coniferous forests also occupy vast areas of the Pacific Northwest and extend south along the Rocky Mountains, the Sierra Nevada, and the California coast range. Trees in these forests tend to be large, particularly in and west of the Cascade Mountains of Oregon and Washington and on the western slopes of the Sierra Nevada. Part of the reason for the huge size of trees such as Douglas fir (*Pseudotsuga menziesii*) is the high annual rainfall, which exceeds 250 centimeters (100 inches) in some areas of the Pacific Northwest. The world's tallest trees, the coastal redwoods of California (*Sequoia sempervirens*; fig. 19.18), however, also depend on moisture from fog for their size and longevity. The fog, which condenses on the foliage and drips into the soil, also reduces transpiration rates.

Most of the North American mountain forests have comparatively dry summers, and lightning often starts fires during this season. Several tree species are well adapted to survival after being partially burned. Douglas fir, for example, has a thick, protective bark that can be charred without transmitting enough heat to the interior to kill the living cells (fig. 19.19). Moreover, Douglas fir seedlings thrive in open areas after a fire. When the

FIGURE 19.18
Coastal redwoods (*Sequoia sempervirens*) in northern California rely on consistent moisture from fog and rain.

FIGURE 19.19
A large Douglas fir (*Pseudotsuga menziesii*). Note the fire scars on the trunk.

bark of the giant redwoods (*Sequoiadendron giganteum*) of the Sierra Nevada is burned, the trees are rarely killed. This has contributed to the great age of many of the trees. The cones of some pine trees, such as knobcone pine (*Pinus attenuata*; fig. 19.20), remain closed and do not release their seeds until a fire causes them to open. Similarly, the seeds of several other species germinate best after exposed to fire.

Temperate deciduous forests grow in many parts of the eastern half of the United States.

Deciduous trees are broad-leaved species that shed all their leaves annually during the fall and remain inactive during the winter. In North America, **temperate deciduous forests** occur from the Great Lakes region south to the Gulf of Mexico and from near the Mississippi River to the eastern seaboard (see figs. 19.13 and 19.21). Temperatures within the area vary greatly but normally fall below 4°C (40°F) in midwinter and rise to above 30°C (86°F) in the summer.

The trees, which usually have thick bark and become dormant before the onset of cold weather, are well adapted to subfreezing temperatures, particularly if the cold is accompanied by snow cover, which prevents the ground from freezing down to the root zone. Precipitation averages between 90 and 225 centimeters (36 inches to 90 inches) per year and occurs mostly during the summer.

Some of the most beautiful of all the broad-leaved trees occur in temperate deciduous forests. Sugar maple (fig. 19.22), American basswood, and the stately American beech predominate in moist, temperate climates. In drier parts of the deciduous forests, oak and hickory are dominant.

During the summer, trees of deciduous forests form a relatively closed canopy that prevents most direct sunlight from reaching the floor. Many of the showiest spring flowers of the region, such as bloodroot, hepatica, Dutchman's breeches, buttercups, trilliums, and violets, open before the trees have leafed out fully and block out light, completing most of their growth within a few weeks (fig. 19.23).

FIGURE 19.20
The closed seed cones of knobcone pine (*Pinus attenuata*) do not open until seared by fire. How would this be advantageous to the seeds?

FIGURE 19.21
An opening within the eastern deciduous forest.

Grasslands once supported huge populations of herbivores.

Natural **grasslands** cover the interior of most continents (see figs. 19.13 and 19.24). Grasslands tend to intergrade with forests, woodlands, or deserts at their margins, depending on the amounts and patterns of precipitation. Grasslands usually receive less precipitation than do deciduous forests—as little as 25 centimeters of rainfall or as much as 100 centimeters annually in some areas. Air temperatures can range from 50°C (120°F) in midsummer to −45°C (−49°F) in midwinter. Fires often clear the standing, dead grass leaves and young tree seedlings in the dry season. These fires do not kill the grass plants because of their underground parts, but they usually destroy saplings of trees. Fires and inadequate rainfall prevent the establishment of trees in this biome.

In North America, the natural grasslands, or prairies, were once grazed by huge herds of bison. The bison disappeared as settlers cultivated more land and hunters slaughtered more of the large animals. By 1889, fewer than 600 bison remained from a population of some 50 million. Today, large areas of prairie are now used for growing cereal crops and for grazing cattle. Thus, the native grasses of the prairie have been replaced by economically important grasses such as wheat and corn, and native herbivores have been replaced by domesticated animals.

Before it was destroyed, the American prairie was a remarkable sight. In Illinois and Iowa, the grasses grew over 2 meters tall during an average season and another meter taller during a wet one. A dazzling display of wildflowers began before the young perennial grasses emerged in spring and continued throughout the growing season. Even today, more than 50 species of flowering plants can be observed flowering simultaneously in the middle of spring on as little as 1 hectare of undisturbed natural grassland. Mammoths also once roamed the prairies of North America. What may have caused such organisms to become extinct when the biome did not disappear?

FIGURE 19.22

Sugar maples (*Acer saccharum*) in the fall. Sugar maples are tapped to make maple syrup (see Perspective 19.2, "Making Maple Syrup").

Plants adapted to the dry climate of the desert biome conserve and store water.

Sand, heat, mirages, oases, and camels are features commonly associated with the most well known **deserts** (see figs. 19.13 and 19.25) such as the Sahara and other large deserts that occur both north and south of the equator in the interior of Africa and Eurasia. Neither camels nor oases, however, are typical today of the deserts, which cover roughly 5% of North America.

Deserts form whenever precipitation is consistently low or where water passes quickly through the soil. Many deserts get less than 10 centimeters (4 inches) of precipitation per year, but some desert areas with porous soil, such as those of the Sonoran and other deserts in the southwestern United States and northwestern Mexico, may receive 25 centimeters or more annually (see Perspective 19.3, "Desertification: What Is It and What Causes It?"). Some grasslands also receive only 25 centimeters of precipitation per year, but these areas are not deserts. What do you think the difference is between the areas with deserts and grasslands?

The low humidity of deserts causes large fluctuations in daily temperatures. During the summer, for example, daytime temperatures often can exceed 38°C (over 100°F) and often fall below 15°C (60°F) the same night. Many desert plants have adapted to these conditions through the evolution of crassulacean acid metabolism (CAM) photosynthesis and C_4 photosynthesis (see chapter 10), which help the plants conserve water while producing sugar. Other adaptations that have evolved in desert plants include thick cuticles, fewer stomata, water-storage tissues in stems and leaves, leaves with a leathery texture and/or reduced size, and even the absence of leaves. Roots of the bizarre *Welwitschia* plants of southwest African deserts (see fig. 18.13) get water from fog that collects on leaves and drops to the ground. Cacti often have small leaves or nonphotosynthetic leaves in the form of spines, which reduce transpiration (fig. 19.26); their stems are photosynthetic. These plants have widespread, shallow root systems that can absorb water rapidly after the infrequent rains and then store water in stems for use between rainstorms. Some desert trees, such as mesquite (*Prosopis*), have long taproots that can grow several meters down to the underground water. Such plant species adapted to low precipitation and high temperatures are called **xerophytes.** In arid climates, water is one limiting factor (see chapter 2) that controls the growth and development of plants.

Annuals in deserts provide a spectacular display of color and variety, particularly during an occasional season with above-average precipitation (fig. 19.27). The seeds of annuals often germinate after a rain and then grow and reproduce within a month before the water in the soil dries up. Hundreds of different species of desert annuals can occur within a few square kilometers of desert in the southwestern United States.

Some desert areas experience very low temperatures during the winter and are often referred to as cold deserts. In the Great Basin of western North America, found between the Sierra Nevada and Wasatch mountain ranges, a cold desert is dominated by widely scattered shrubs, such as sagebrush (*Artemisia tridentata*), and in some places, a mixture of sagebrush and perennial grasses.

Desert ecosystems, like those of the tundra, are fragile, and recovery from disturbances often takes many years. Large numbers of off-road vehicles and feral donkeys and horses have devastated several desert areas in the southwestern United States; current efforts to curb the destruction thus far have often failed. What does this suggest about the pace of succession in these biomes?

The chaparral frequently burns.

Areas with hot, dry summers and cool, wet winters occur along parts of the west coasts of North and South America. In this Mediterranean-type climate, unique scrubby vegetation that is either evergreen or deciduous in summer has

19.2 PerspectivePerspective

Making Maple Syrup

From early March through early April of each year, many farms in western Massachusetts display the same sign: "Maple Syrup for Sale." Although the product is delicious, most people are no longer accustomed to pure maple syrup and prefer the commercial syrup sold in most grocery stores, which is cheaper. Commercial syrup contains only about 2% maple syrup (the rest is corn syrup and sugar) and tastes different from pure maple syrup.

Maple syrup is harvested from the trunks of sugar maple trees, *Acer saccharum*. Unlike the production of other crops, the production of maple sugar requires widely fluctuating temperatures. From late autumn to early spring, a brief period of daily freezes and thaws triggers the flow of sap in sugar maple trees. This freezing and thawing is important because the flow stops if temperatures are constantly below or just above freezing. During cold nights, starch made during the previous summer and stored in wood is converted to sugar. The next day's warm temperatures create a positive pressure in the xylem's sapwood. When tubes called spiles are driven into the sapwood, this pressure pushes the sugary sap out of the trunk at a rate of 100 to 400 drops per minute (on a good day). Some trees produce as much as 150 liters of sap per season; attaching vacuum pumps to the tubes can increase the seasonal yield by nearly threefold. The sugar content of maple sap ranges from 1% to 10% but is usually about 3%. Once collected, the sap is boiled to remove the excess water; about 40 liters

*This maple "rancher" is collecting sap from a sugar maple (*Acer saccharum*).*

Sap drips from a metal tube driven into the sapwood of a sugar maple.

of sap are required to make 1 liter of syrup. The characteristic taste of maple syrup is due to this heating and the presence of several amino acids in the sap.

The production of maple syrup has several other unusual characteristics. For example,

- Flow usually stops in the afternoon and does not start again until the temperature rises above freezing the next morning. Maple "ranchers" harvest most sap between about 9:00 A.M. and noon.

- Because it depends on the weather, the flow of sap is often sporadic. There may be 5 to 10 "runs" of sap flow during a single spring.

- The concentration of sugar in sap is low early in the season, rises to a maximum, and then gradually decreases at season's end.

- Farmers use power drills with half-inch bits to drill about three holes into each tree. Each hole is 5 to 7 centimeters deep. Because this tapping removes less than 10% of a maple tree's sugar, trees can be tapped repeatedly without producing significant damage. Some of the trees tapped today were probably also tapped by pilgrims in the 1600s.

Production of maple syrup has been changed little by science and modern technology. We still do not understand precisely how the freeze-thaw cycle stimulates flow of sap, and experiments aimed at increasing sap flow have, for the most part, failed. Despite remaining a primarily regional product, maple syrup is enjoying a renewed popularity and is now included in foods ranging from pumpkin soup to barbecued pork ribs.

---

evolved (see figs. 19.13 and 19.28). Most growth occurs in the relatively short, wet winters and early spring; plants are dormant for the remainder of the year.

The lower elevations of the Pacific Coast hills in southern California are covered by **chaparral,** a word derived from *chabarro,* the Basque name for scrub oak (*Quercus dumosa*). (Some ecologists refer to this biome as Mediterranean scrub and woodland.) The chaparral biome is similar to one that occurs in southern Europe and char-

acterizes the Basque culture of northern Spain. Dominant chaparral species include buckbrush, scrub oak, silk-tassel bush, chamise, manzanita, California bay, and poison oak.

Many plants in the chaparral are well adapted to fires, which often occur. Such plants have thick roots or a crown at the top of the roots that resprout after the aboveground parts have been burned (fig. 19.29); others have rhizomes that survive underground when a fire races through the erect vegetation above them.

FIGURE 19.23

These trilliums (*Trillium* species) have carpeted the floor of part of the eastern deciduous forest before the leaves of the trees have expanded to form a dense canopy that blocks sunlight from reaching the ground.

Fires can be beneficial.

Natural fires started primarily by lightning and the activities of indigenous peoples have occurred for thousands of years in different biomes in North America and other continents. Trees such as giant redwood and ponderosa pine, although scarred by fire, often survive, and the dates of fires can be determined by the proximity of the scars to specific annual rings. Growthring studies from western North America indicate that forests of ponderosa pine burned about every 6 to 7 years.

Fires benefit grasslands, chaparral, and some forests by converting accumulated dead organic material to mineral-rich ash, whose nutrients are recycled within the ecosystem (fig. 19.30). If the soil is burned, the composition of microorganisms present is likely to change. Nitrogen-fixing soil bacteria, including cyanobacteria, may increase after a fire, whereas fungi that cause plant diseases decrease. In addition, competition for light, space, and water is reduced because most plants are killed, and animals that eat seeds are often killed by the fire.

At least some of the North American grasslands originated and are maintained by fire. Because grassland fires have largely been controlled, many of these areas have now been invaded by shrubs. In some areas, such as the prairies of the Midwest, grasses are better adapted to fire than are woody plants. These grasses produce seeds within a year or two after germination. Perennial grass buds at the tips of rhizomes often survive the most intense heat of fires, producing new growth the first season after a fire. Thus, a grassland fire may destroy only one season's growth of grass, often after reproduction has been completed.

Shrubs, however, have much of their living tissue aboveground, and a fire may destroy several years' growth. Also, woody plants, unlike grasses, often do not produce seeds until several years after germination. Many shrubs sprout from burned root crowns, particularly in chaparral areas. This process of regeneration is called **stump-sprouting** and allows shrubs to grow quickly after a fire and to produce stems and leaves before other plants have a chance to develop from seeds

19.3 *Perspective* Perspective

Desertification: What Is It and What Causes It?

Desertification is the conversion of grasslands or other biomes into deserts. Deserts have been expanding for the past 15,000 years due to worldwide drying since the most recent period of glaciation; thus, desertification is not new. Nevertheless, it has been accelerated in certain parts of the world because of human activities. One of the causes of desertification is the irrigation of arid habitats for farming. Irrigation water always contains salt, in some cases more than a ton (ca. 900 kilograms) per acre-foot.* As crop plants absorb the water, the salt is left behind. Salt has accumulated in some areas over the years to the point that these areas are about to undergo desertification. For instance, many farmlands in the southwestern United States that are irrigated with water from the Colorado River are becoming polluted with salt and are expected to be useless for farming within a few years.

Desertification in the Sahel region of the southern Sahara Desert correlated with the apparent drying out of adjacent savanna ecosystems. Desertification has caused tremendous human suffering, most notably the widespread famines of 1972 and 1984. The most widely accepted hypothesis to explain desertification in the Sahel is the drought hypothesis: the Sahel is drying out because of the lack of rain. *Drought* and *famine* are used in the same breath so often that cause and effect are almost fixed in our minds. But let's look a little closer. Widespread famine is a relatively recent phenomenon, yet the low rainfall of the 1970s and 1980s has occurred several times earlier in this century without causing desertification. Furthermore, a ranch of more than 1,000 kilometers² in the heart of the Sahel has remained

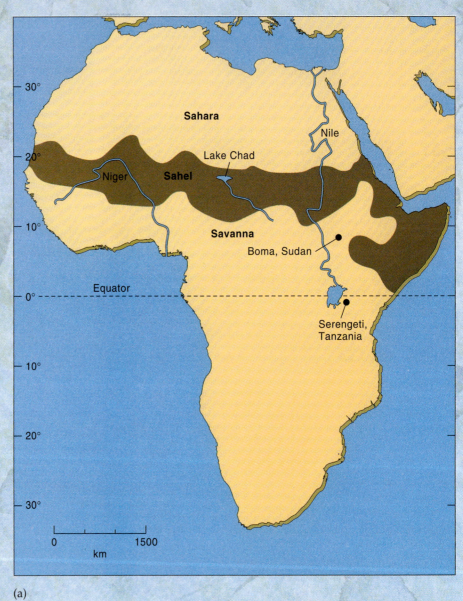

(a)

(a) The Sahel is a semiarid belt that spans Africa between the Sahara Desert and the savanna. (b) Rainfall records from the Sahel show that, since 1900, periods of below-average rainfall have been relatively frequent, although widespread drought and famine in the area have happened only since 1973.

(see fig. 19.29). Most chaparral species, both woody and herbaceous, are so adapted to fire that their seeds will not germinate until fires remove accumulated litter on the ground. Seeds of some species germinate after being stimulated by gases emitted by fires.

Trying to eliminate fires caused by lightning, at least in certain areas, disrupts natural habitats more in the long run than allowing them to occur. Agencies such as the U.S. National Park Service and the U.S. Forest Service now allow some fires to burn under prescribed conditions.

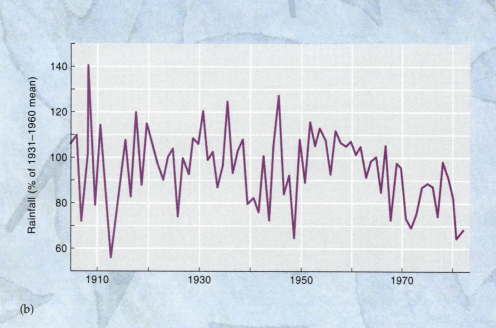

(b)

green since it was started in 1968, despite receiving the same low rainfall as adjacent areas undergoing desertification. These observations contradict the drought hypothesis.

What is the difference between the green ranch area and the desertifying areas of the Sahel that might explain why desertification is occurring? Several interrelated factors, all pointing to human activities, cause desertification. This explanation is referred to as the settlement-overgrazing hypothesis. Native people of the Sahel used to migrate seasonally, driving small herds of cattle to the north when rains came and the annual grasses grew, then back to the wetter southern areas as the rainy season passed. The cattle gained weight and reproduced based

on a diet of the high-protein annual grasses that grew in the fertile soils of the north. When water became scarce, the people and their cattle returned to the low-protein perennial grasses that grew in the nutrient-poor soil of the south; this diet is not nutritious enough for growth and reproduction.

The pattern of seasonal migration of people and their cattle has been replaced by settlements and agriculture over the years since the Second World War, when Western countries increased their aid for agricultural development in African countries. Unfortunately, crops in the southern areas of the Sahel were unsuccessful because of the poor quality of the soils there. With the exception of the green ranch, cattle were allowed to overgraze the northern

annual grasses to compensate for the destruction of these areas by failed farms until the grasses could no longer survive. Both the north and the south are now so severely denuded that they look like they have suffered an extensive drought. However, the effects of the change in life-style of the native people from a migratory one to a sedentary one are evidence that overgrazing is the real culprit.

*An acre-foot is the amount of water that covers an acre to a depth of 1 ft. It is the preferred unit of measure for water usage in agriculture. There is no metric equivalent to acre-foot, unless you convert acre to hectare and foot to centimeter, but there is no use for such a unit.

(text) Reproduced by permission of the National Research Council of Canada from the *Canadian Journal of Zoology*, Volume 63, pages 987–994, 1985.

Tropical rain forests are storehouses of species diversity.

The greatest biological diversity on earth occurs in **tropical rain forests.** In North and South America, the principal tropical rain forests occur in the Amazon Basin of northern South America, much of Central America, and the Caribbean. About 7% of the earth's surface, representing nearly half of the forested areas of the earth and 25% of the earth's species, are included in this biome (see fig. 19.13). Tropical rain forests have existed for about 200 million years and, unlike other biomes, were not decimated by glaciation during the ice age.

Rain forests occur throughout areas of the tropics where annual rainfall normally ranges between 200 and 400 centimeters (80 inches to 160 inches) and daytime

FIGURE 19.24
A mixed-grass prairie in South Dakota may also include a variety of plants with showy flowers.

FIGURE 19.25
Sand dunes and an oasis in the Sahara Desert, Algeria.

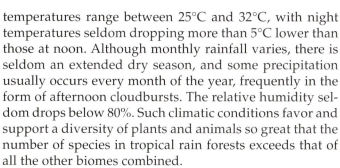

FIGURE 19.26
Barrel cactus (*Ferrocactus covillei*), a xerophyte. Evolution has resulted in modified leaves called spines that protect the plant and reduce the surface area available for transpiration.

FIGURE 19.27
Spring wildflowers in the Mojave Desert grow and flower rapidly during brief seasons with rain.

temperatures range between 25°C and 32°C, with night temperatures seldom dropping more than 5°C lower than those at noon. Although monthly rainfall varies, there is seldom an extended dry season, and some precipitation usually occurs every month of the year, frequently in the form of afternoon cloudbursts. The relative humidity seldom drops below 80%. Such climatic conditions favor and support a diversity of plants and animals so great that the number of species in tropical rain forests exceeds that of all the other biomes combined.

Rain forests are dominated by broad-leaved evergreen trees, whose trunks are often unbranched for as much as 40 meters or more, with luxuriant tops that form a beautiful dark green and multilayered canopy (fig. 19.31). In small parts of the forest, particularly along rivers, the canopy is so dense that little light penetrates to the floor. The few herbaceous plants that survive are generally confined to openings in the forest. There are hundreds of species of trees, each usually represented by widely scattered individuals. Root systems are shallow, and the trees are often buttressed; broader bases compensate for lack of root depth (see fig. 7.18). Organic matter is

sparse in tropical soils, which tend to be acidic. The soils are often deficient in important nutrients such as potassium, magnesium, calcium, and phosphorus, or the minerals (particularly phosphorus) are in forms that the plants cannot use. Despite the lush growth, there is little accumulation of litter or humus, because decomposers rapidly degrade leaves and other organic material on the forest floor; the nutrients released by decomposition are quickly recycled or leached by the heavy rains.

Most of the plants in rain forests are woody flowering plants; no conifers have evolved in these ecosystems. Many **lianas** (woody vines that are rooted in the ground) hang from tree branches, and even more numerous **epiphytes** (especially orchids and bromeliads) are attached to limbs and trunks (fig. 19.32). Epiphytes are plants that are attached to other larger plants; however, the plants are sustained entirely by their own photosynthesis and by rainwater that accumulates in their leaf bases. Traces of minerals necessary for growth accumulate in the rainwater as it trickles over decaying bark, bird droppings, and dust; these minerals are absorbed by the roots of epiphytes, which grow in crevices of the bark.

FIGURE 19.28

A chaparral scene. California lilac (*Ceanothus integerrimus*) is flowering.

INQUIRY SUMMARY

Terrestrial ecosystems are classified into biomes on the basis of the similarities of their vegetation and physical environments. The tundra is characterized by freezing temperatures, permafrost, the virtual absence of trees and annuals, and the dominance by low-growing plants. The taiga consists of conifer forest ecosystems with hardy plants adapted for survival in low-nutrient soils and extraordinary ranges of temperature and moisture. The temperate deciduous forest biome, which occurs where summers are warm and winters are relatively cold, is characterized by broad-leaved trees that lose their leaves each fall. The most common plants in the grassland biome are grasses, which are often replaced by economically important species. Deserts occur in climates wherever precipitation is consistently low. Plants of the desert biome include succulents and plants with small, tough leaves and deep taproots or extensive, shallow root systems. Ecosystems with hot, dry summers and cool, wet winters are dominated by scrubby vegetation types, which comprise the chaparral biome. Fires in certain biomes are common and can benefit the plants that survive the burning. Tropical rain forests include broad-leaved, mostly

FIGURE 19.29

Chaparral shrubs often resprout from the base a year after a fire has swept through the area and destroyed the exposed stems and leaves.

F I G U R E   1 9 . 3 0
Forest fires convert dead organic matter to mineral-rich ash, whose nutrients are recycled within the ecosystem.

F I G U R E   1 9 . 3 1
Levels of plant life in the rain forest. A few tall trees comprise the emergent layer, which overgrows the shorter trees of the dense canopy. Shrubs and low trees occupy the shady understory, and fallen leaves and seedlings are scattered on the forest floor.

evergreen plants in a hot, humid environment. More species occur in tropical rain forests than in any other kind of ecosystem.

Which three biomes do you think are the most economically important to us in the United States? On what basis do you make your claim?

F I G U R E   1 9 . 3 2
Orchids and other epiphytes on the branches of a tree in a South American rain forest.

Humans Drastically Affect the Earth's Ecosystems.

Since the 1960s, **environmental biology** has become increasingly popular as a field of study because the human population continues to grow. During the past four decades, topics such as the effects of pollution on land, water quality, and people's standard of living have filled thousands of books. During this same period, attempts to stop environmental damage have forced major construction projects to file environmental impact reports before proceeding. These reports provide information that helps various agencies evaluate the wide-ranging effects of development on the flora, fauna, and physical environment and to outline modifications that must be made before a project can be approved. Environmental impact assessment has often produced much controversy and emotional debate between industrial and land developers and those who feel that preservation of the environment should be our highest priority. Effectively resolving these environmental controversies requires an understanding of ecological principles.

The human population now exceeds 6 billion. The earth remains constant in size, but humans have occupied more of its land over the past few centuries, and their population density has also greatly increased. When feeding, clothing, and housing themselves, humans have greatly affected the environment. Humans have cleared natural vegetation from vast areas of land and drained wetlands. We have polluted rivers, oceans, lakes, and the atmosphere, and we have killed pests and plant-disease organisms with poisons that also have killed natural predators and other useful organisms. Since 1955, nearly one-third of the world's cropland (an area larger than India and China combined) has been abandoned because its overuse led to the depletion, degradation, or loss of soil. Finding substitutes for this lost acreage accounts for 60% to 80% of the world's deforestation. In general, we disrupt every ecosystem we inhabit. How do human activities affect the productivity and succession of ecosystems?

Acid rain contributes to the degradation of forests.

Acid rain occurs after the burning of fossil fuels releases sulphur and nitrogenous compounds into the atmosphere. There, sunlight converts these compounds to sulphur and nitrogen oxides, and they combine with water to become **acid rain** (mostly sulphuric acid and nitric acid). Acid rain can lower the pH of lakes and streams and kill many organisms in them. It also affects the soil and injures plants upon which it falls (fig. 19.33). About half of the Black Forest in Germany has succumbed to its effects. Acid rain has also stunted or killed trees growing downwind from industrial sites. Acid rain also affects nonliving materials. For example, the natural weathering of ancient Mayan ruins in southern Mexico, the Parthenon in Greece, and monuments in Washington, D.C., has been accelerated by acid rain during the past decades.

Acid rain is not responsible for all dead or dying trees in the world's forests. Some trees have perished as a result of insufficient rainfall during successive dry years. Others have succumbed to insect infestations or salt scattered to melt ice and snow on roads, and still others have been weakened by disease.

The thinning ozone layer allows more ultraviolet light to strike the earth.

Ozone (O_3) forms a thin layer in the stratosphere that is more effective than ordinary oxygen (O_2) in shielding organisms from intense ultraviolet radiation. Ultraviolet light triggers a chemical reaction with oxygen to produce

FIGURE 19.33
Damage caused by acid rain to spruce and fir trees.

ozone. Gases such as chlorofluorocarbons (CFCs) are broken down by sunlight at high altitudes into active compounds that destroy ozone. CFCs are also involved in producing the greenhouse effect (see chapter 10) and are the relatively recent by-products of the manufacture of refrigerants, plastics, and aerosol propellants. Destruction of ozone in the stratosphere results in increased exposure to ultraviolet radiation, which increases the incidence of skin cancers, genetic mutations, and damage to vegetation—especially crops.

The accelerating destruction of the ozone layer has been recognized as a serious global problem by the United States, the European Economic Community (EEC) and other industrial nations. While the production and use of CFCs has declined worldwide, many developing nations have contended that a ban on CFC use would place them at an economic disadvantage. Because global cooperation on this matter is urgently needed, the major industrial nations

FIGURE 19.34
Destruction of tropical rain forest in Belize, Central America. This biome is rapidly disappearing.

seek viable alternatives to chlorofluorocarbons. One such alternative is the non-CFC refrigerants that became standard in the air-conditioning systems of new automobiles sold in the United States and elsewhere.

Humans help pests invade new territories.

Despite the efforts of quarantine officers, exotic (nonnative) species continue to invade new territories. Although the Africanized honeybee and zebra mussel grab the headlines, a variety of exotic plants has disrupted ecosystems. Kudzu (*Pueraria lobata*) is an Asian vine that was intentionally planted throughout the South in the 1930s to control erosion. Today, it covers between 2 and 4 million acres and grows as fast as 30 centimeters per day. It grows over almost everything that it encounters. Kudzu is responsible for more than $50 million in lost farm and timber production each year. A recent idea for using kudzu is to harvest the starch from its huge roots. Purple loosestrife (*Lythrum salicaria*) is a European weed with pretty flowers

that was brought to the United States in the late 1800s. Although the plant is used in some areas to make bouquets, people elsewhere are splashing purple loosestrife with herbicides in hopes of saving native wetlands plants that the loosestrife is replacing. The U.S. Agriculture Department hopes to check the spread of the invader with insects now being tested as biological controls. What plants are your local pests, and from where did they come?

The future of the tropical rain forest biome is threatened.

Many parts of the Amazon rain forest have been converted into large farms, hydroelectric plants, and mines. In the 1990s, tropical rain forests were being destroyed or damaged for commercial purposes at the rate comparable to an area of about 45 football fields *per second* (fig. 19.34). This damage and destruction has had a devastating effect on wildlife. For example, a study found that populations of low-flying bird species in a 10-hectare section of the Amazon rain forest fell by 75% just 6 weeks after adjacent land had been cleared; 10 of the 48 bird species studied disappeared completely. This situation has been repeated on many occasions. Indeed, dividing large areas of rain forest into smaller pieces separated by as little as 10 meters of cleared land can have disastrous effects on the ecology of the entire forest. Most of the bird and other animal species originally present disappeared permanently.

Only a small part of the rain forest biome is now protected from commercial development. The rapid loss of biodiversity results in the permanent loss of gene pools with great potential for medicinal and agricultural plants. The cleared land, with its poor and eroded soils, often becomes unfit for farming after only 2 or 3 years. Even if the land were now to be left fallow, all the plant and animal life will never return. The loss of massive amounts of vegetation directly affects photosynthesis, respiration, and transpiration. If losses continue on a large enough scale, global climate will also be affected. Many more organisms are doomed to extinction before they have even been seen or described for the first time. The rain forest biome will vanish within 20 years if governments and individuals do not stop or slow the large-scale destruction. What can you do to help save the tropical rain forest? What can you do to keep the biome in which you live from disappearing?

SUMMARY

Ecology is the branch of biology that examines the relationships of organisms to one another and to their environment. Ecologists study populations, which are groups of individuals of the same species that form a community, and communities and their physical environment that form an ecosystem.

Ecosystems typically sustain themselves entirely through photosynthesis and interactions among producers, consumers, and decomposers. In any ecosystem, energy flows through the different levels of producers, consumers, and decomposers, but energy is not recycled in an ecosystem; rather, it flows through an ecosystem. The producer level of a food chain converts only about 1% of the available light energy to organic material, and the energy gradually escapes in the form of heat as it passes from one level to another.

Atmospheric nitrogen is fixed by bacteria living in roots of legumes and other plants. Nitrogen flows from dead plant and animal tissues into the soil and from the soil back to the plants. Water leaches nitrogen from the soil and also carries it away when erosion occurs. More nitrogen is lost when crops are harvested, but the loss can be reduced if wastes are decomposed and returned to the soil. Carbon, water, and other minerals also undergo cycling.

Succession is a pattern of change in the species composition of a given area over a period of time and occurs whenever natural areas have been disturbed. Primary succession involves the formation of soil. Secondary succession occurs in areas previously covered with vegetation. A relatively stable vegetation (climax vegetation) may become established and remain until a disturbance disrupts the balance. Fires alter the nutrient and organic composition of the soil and may benefit surviving plants.

Biomes are land-based groupings of ecosystems considered on a global or a continental scale. Tundra, which occurs primarily above the Arctic Circle, includes shrubs, many lichens and grasses, and tufted flowering perennials. It is also characterized by the presence of permafrost below the surface. Taiga is dominated by coniferous trees such as spruce, fir, and pine with birch, aspen, tamarack, alder, and willow in the wetter areas.

Temperate deciduous forests are dominated by deciduous trees. In North America, such trees include sugar maple, American basswood, beech, oak, and hickory. In early spring, a profusion of wildflowers carpets the forest floor before tree leaves expand.

Grasslands occur primarily in temperate areas toward the interiors of continents. Many grasslands have been converted to agricultural use. Deserts have low annual precipitation and widely fluctuating daily temperatures. Chaparral has unique, mostly evergreen vegetation adapted to cool, wet winters and hot, dry summers. The vegetation is also adapted to frequent fires.

The tropical rain forests contain more species of plants and animals than all the other biomes combined. Many woody plants and vines form multilayered canopies, which keep most light from reaching the forest floor. Soils are poor, and nutrients released during decomposition are rapidly recycled.

Human populations have grown rapidly. The disruption of ecosystems by activities directly or indirectly associated with feeding, clothing, and housing billions of people threatens the survival not only of humans but other organisms as well. Acid rain, which damages or kills organisms, occurs when sulphur and nitrogen compounds released by the burning of fossil fuels are converted to nitric and sulphur oxides by sunlight and then mix with water to become acids. In the stratosphere, sunlight converts chlorofluorocarbons into compounds that destroy the ozone shield, which protects us from intense ultraviolet radiation. Tropical rain forests are being destroyed so rapidly that they will disappear within 20 years if the destruction is not slowed or stopped.

Writing to Learn Botany

As you have read in this chapter, the tropical rain forest biome is being destroyed at a rate that could soon render it extinct. How could you educate the public about this serious problem, and what could you do to slow or halt the destruction?

THOUGHT QUESTIONS

1. What is the greenhouse effect? Why should we be concerned about it?

2. What concept or idea discussed in this chapter do you think is most important? Why?

3. Energy and nutrients move differently in ecosystems. What are the consequences of this difference?

4. Today, it is fashionable for some people to refer to themselves as environmentalists. What does this term mean to you?

5. In industrialized countries, humans have disrupted ecosystems everywhere they have settled. Could humans also improve an ecosystem? If so, how?

6. If a plant species is unnecessary for the functioning of an ecosystem, should we be concerned if it is eradicated? What might be some economic, moral, or aesthetic reasons for protecting an "unnecessary" plant?

7. How have human activities affected the distribution of plants and animals where you live?

8. How would you best protect biodiversity: by conserving species or by conserving ecosystems? Explain your answer.

9. It has been estimated that at least 25% of all plant species are likely to become extinct in the next 30 years. So far, we seem to be getting along without a number of plants that have become extinct in the past 10 years. Considering what you learned about biodiversity, should we try to slow the rate of extinction? Why, or why not?

10. Much of the tropical rain forest biome is in developing countries, which have weak economies. Bearing this in mind, what could be done to halt the destruction of this biome?

11. Consider the areas in which cradles of civilization once flourished, such as the Fertile Crescent, Egypt, Greece, and Central America where the Aztecs and Mayans ruled. What role did the tendency of humans to disrupt their ecosystem contribute to the degradation of the land and the downfall of these civilizations?

SUGGESTED READINGS

Articles

Baskin, Y. 1994. Ecologists dare to ask: How much does diversity matter? *Science* 264:202–3.

Baskin, Y. 1999. Yellowstone fires: A decade later. *BioScience* 49:93–97.

Holloway, M. 1993. Sustaining the Amazon. *Scientific American* 269:90–99.

Kerr, R. A. 1998. Acid rain control: Success on the cheap. *Science* 282:1024–27.

Mann, C. C. 1999. Crop scientists seek a new revolution. *Science* 283:310–14.

Monastersky, R. 1990. The fall of the forest. *Science News* 138:40–41.

Pennisi, E. 1994. Tallying the tropics: Seeing the forest through the trees. *Science News* 145:362–66.

Repetto, R. 1990. Deforestation in the tropics. *Scientific American* 262:36–42.

Ricketts, T. H., E. Dinerstein, et al. 1999. Who's where in North America. *BioScience* 49:369–81.

Science. 1999. 285:1834–43. Several articles on biological invaders.

Wildt, D. E., W. F. Rall, et al. 1997. Genome resource banks. *BioScience* 47:689–98.

Zimmer, C. 1995. How to make a desert. *Discover* (February):50–56.

Books

Barbour, M. G., and W. D. Billings. 1988. *North American Terrestrial Vegetation.* New York: Cambridge University Press.

Molles, Jr., M. C. 1999. *Ecology.* New York, NY: McGraw-Hill.

Moore, R., and D. S. Vodopich. 1991. *The Living Desert.* Piscataway, NJ: Enslow.

Myers, N. 1992. *The Primary Source: Tropical Forests and Our Future,* rev. ed. New York: W. W. Norton.

Richards, P. W. 1994. *The Tropical Rain Forest.* New York: Cambridge University Press.

Silvertown, J., and J. L. Doust. 1993. *Introduction to Plant Population Ecology,* 3d ed. Cambridge, MA: Blackwell Scientific.

ON THE INTERNET

Visit the textbook's accompanying web site at http://www.mhhe.com/botany to find live Internet links for each of the topics listed below:

Ecology and the laws of thermodynamics

Carrying capacity

Nutrient cycling

Succession

Biomes

Aquatic ecosystems

Species diversity and tropical rainforests

Destruction of ecosystems

Human population growth and ecosystems

Trophic levels in ecosystems

WHAT IS GENETIC ENGINEERING?

Genetic engineering, as discussed in chapter 5, is based on recombinant DNA technology. This technology relies on vectors (DNA carriers), such as viruses or bacterial plasmids (independent molecules of DNA apart from the primary bacterial genome; see fig. A.1). Viruses and plasmids are vectors because they carry inserted or foreign DNA as their own DNA, which is then replicated. In this way, DNA from plants or animals is cloned by inserting it into bacteria or viruses. DNA cloning is an important tool in gene technology. It is used to find and study important genes and to make genetically engineered plants.

DNA CLONING AND RESTRICTION ENZYMES ARE FUNDAMENTAL TO GENETIC ENGINEERING.

Before recombination, both the DNA molecules must be made receptive to each other. This job is done by bacterial restriction enzymes, which cut (i.e., restrict) DNA. The use of these elements in DNA cloning is discussed in the next few paragraphs. Different bacteria make hundreds of restriction enzymes, each of which recognizes a specific DNA sequence of four to eight nucleotides. In nature, these enzymes protect bacteria by destroying foreign DNA. To prevent self-destruction, bacterial DNA is chemically modified to be inert to its own restriction enzymes.

An example of a restriction enzyme is *Eco*RI, which is harvested for commercial sale from *Escherichia coli. Eco*RI recognizes the nucleotide sequence GAATTC and then cuts the DNA between the guanine (G) and the adenine (A) of the sequence. The six-nucleotide *Eco*RI site has identical sequences on both DNA strands because its complement is CTTAAG. This means that by cutting between guanine and adenine on each strand, the enzyme makes a zigzag cut that leaves short, single-stranded ends on each DNA double strand (fig. A.2). Such single-stranded ends are called *sticky ends* because they can be "glued" by hydrogen bonding to complementary sticky ends of other DNA molecules.

After cutting up a whole genome with a restriction enzyme, the next step in cloning DNA is to insert the DNA into a virus or plasmid vector. One type of vector is a bacterial plasmid, which is a small, circular molecule of DNA (see fig. A.1 and chapter 16). After the foreign DNA is inserted into isolated plasmids, the plasmids are absorbed back into bacteria and reproduced (i.e., cloned) during normal DNA replication of the bacterial genes.

Foreign DNA and plasmid DNA are receptive to each other when they are cut by the same restriction enzyme. For example, an enzyme such as *Eco*RI makes the same sticky ends on both the foreign DNA and the plasmid DNA. When the two kinds of DNA are mixed, they anneal (come together and match) because their sticky ends complement each other.

At first, plasmid DNA is weakly held to foreign DNA by hydrogen bonds between complementary nucleotides. The linkage between them is strengthened when sugar-phosphate bonds form between the two strands or molecules of DNA. Sugar-phosphate bonds are made by the enzyme DNA ligase. When this enzyme is added to the DNA mixture, it joins (i.e., ligates) phosphate groups to deoxyribose between adjacent strands of DNA.

Methods for cloning DNA in plasmids also work with viruses. Like plasmid DNA, viral DNA is cut by restriction enzymes, and foreign DNA is inserted into it. Since viruses are parasites, however, they must be reintroduced into their hosts before they can replicate the recombined DNA. For cloning DNA in viruses, the most convenient hosts are bacteria such as *E. coli* that can be easily cultured in the laboratory. The most commonly used viruses for cloning DNA are bacteriophages.

Biologists Can Make Libraries of Genes.

After a genome is cut by restriction enzymes, many of the resulting DNA fragments can be inserted into vectors. This has been done for many plants, animals, and fungi. When inserted into a vector, a fragment or gene of interest from any of these organisms can be retrieved from storage by culturing the appropriate vector. This storage and retrieval of genetic information is like having a library of genes, so scientists call the set of cloned fragments of a genome a *genomic library* (fig. A.3); such libraries may have several thousand to several million entries, each a recombined plasmid or virus that contains a different restriction fragment.

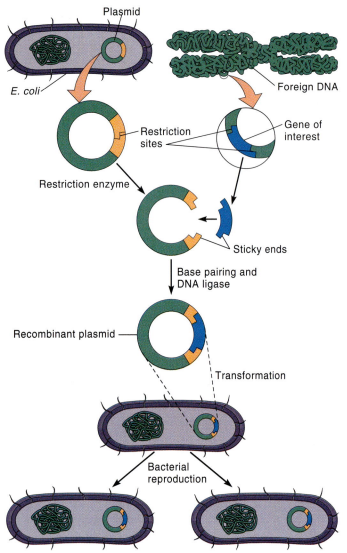

FIGURE A.1

Gene cloning by bacteria. The gene of interest from another organism is cut out of its genome by a restriction enzyme and inserted into a bacterial plasmid that has been cut with the same restriction enzyme. Sticky ends of the foreign fragment bind to the complementary sticky ends of the plasmid, followed by the formation of new sugar-phosphate bonds through the action of DNA ligase. Finally, the altered plasmid is taken up by the bacteria, which make many copies of the foreign gene by duplicating the plasmid during reproduction.

Libraries are also made using complementary DNA (cDNA). To make a cDNA library, mRNA is reverse-transcribed (using an RNA template to make DNA) to make cDNA. This reaction from RNA back to DNA is catalyzed by the enzyme reverse transcriptase, a viral enzyme that is available commercially (fig. A.4). The cDNA is later modified to have the appropriate sticky ends, after which it is annealed and ligated to a vector. Libraries of cDNA are always incomplete because they contain only genes whose mRNA

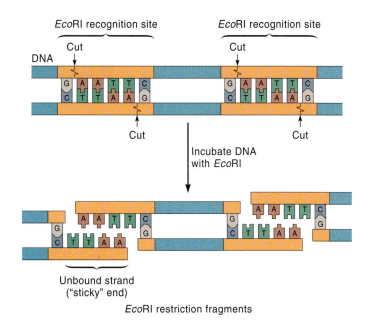

FIGURE A.2

Like most restriction enzymes, *Eco*RI makes a zigzag cut wherever it finds a recognition site. Single-stranded portions of restriction fragments are called *sticky ends* because they can form hydrogen bonds with complementary sticky ends on other molecules of DNA.

has been reverse-transcribed. However, this can also be an advantage of cDNA libraries, because it produces a library of genes that are expressed in a single type of tissue or organ. Leaf mRNA or root mRNA, for example, would yield a leaf cDNA library or a root cDNA library, respectively. In addition, cDNA genes are easier to manipulate because they are generally smaller than in their native state, since their intervening sequences (introns, see chapter 4) are missing.

In addition to being clonable, some genes can make polypeptides in their host vectors. This means that intron-free genes in a genomic library and genes from a cDNA library may be translated in the vector. The products of these genes can therefore be made in large amounts for research or industrial uses by culturing the vector. The first such product to be available commercially was human insulin, but many proteins are now being made by cloning. Unfortunately, however, intron-containing genes in genomic libraries cannot be translated in a bacterial vector, because bacteria (prokaryotes) lack the enzymes to process (modify after transcription) RNA (see chapter 5).

Making and screening gene libraries is a tedious and time-consuming procedure. A plasmid or a virus can replicate only a few thousand nucleotides of foreign DNA, so many clones are needed to replicate whole genomes. For example, by chance alone, *Eco*RI will find the sequence GAATTC once out of every 4,096 nucleotides. This is 4^6, which is the number of ways four different nucleotides can be arranged in a six-base sequence. This means that the genome of *Arabidopsis thaliana,* the smallest known in plants

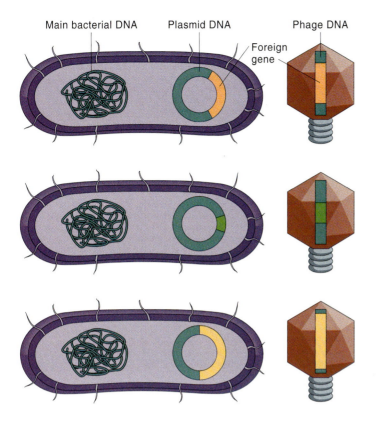

Main bacterial DNA   Plasmid DNA   Phage DNA

Foreign gene

FIGURE A.3

Genomic libraries are clones of DNA fragments, usually consisting of a single gene, that are stored in vectors. Cloning may be by plasmids or by phages. The bacterial cells on the left and the viruses on the right represent the cloning of three foreign genes by each type of vector.

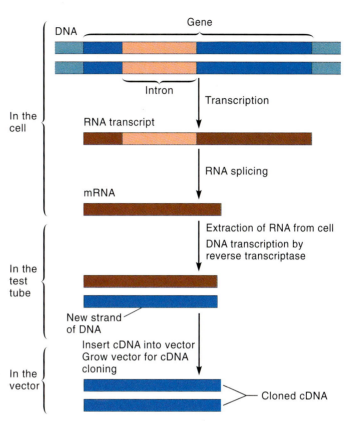

FIGURE A.4

Steps for making complementary DNA. After transcription and RNA editing in the cell, mRNA is extracted and used as a template for making DNA by reverse transcription in a test tube. The cDNA from reverse transcription is then inserted into a vector for cloning. Unlike native genes, complementary DNA lacks introns because it is made from processed RNA.

(70 million nucleotide pairs), will be cut into more than 17,000 fragments with an average size of 4,096 nucleotides each by *Eco*RI (70,000,000/4,096 = 17,089). Thus, *Arabidopsis* would require more than 17,000 clones in its complete genomic library. By the same calculation, genomes of the garden pea (10.2 billion nucleotide pairs) would require about 2.5 million clones, and that of Easter lily (90 billion nucleotide pairs) would require about 22 million clones! However, gene libraries are incomplete because only part of an organism's genome can be efficiently taken up by vectors.

Artificial Chromosomes Can Be Made in a Test Tube.

One of the most important recent advances in genetic engineering is the construction of artificial chromosomes that can be cloned in yeast (*Saccharomyces cerevisiae*). These yeast artificial chromosomes (YACs) are made by putting fragments of foreign DNA into native yeast chromosomes. The YAC contains non-yeast DNA, a centromere, at least one replication origin, and two telomeres, which are DNA sequences that counteract chromosome condensation before mitosis (fig. A.5). When such artificial chromosomes are introduced into yeast cells, they are replicated like native chromosomes.

Unlike bacterial or viral vectors, YACs can unite with and replicate millions of nucleotides of foreign DNA. Large fragments of foreign DNA are obtained by lightly digesting genomic DNA from the source organism so that only some of the restriction sites are cut. The average size of restriction fragments made by *Eco*RI may therefore be in the millions of nucleotides instead of 4,096. A complete genomic library of 3 billion nucleotide pairs may require about a thousand YACs. Besides requiring fewer clones by this method, the construction of YACs has the advantage that intron-containing genes from the source organism can be expressed because yeast (a eukaryote) has the appropriate enzymes for processing RNA. However, the use of YACs is so complicated and expensive that the only extensive work currently being done with them involves the human genome.

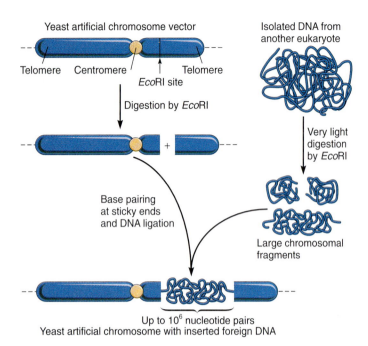

FIGURE A.5

Construction of a yeast artificial chromosome (YAC) with inserted foreign DNA. Once a YAC is made, it can be taken up by a yeast cell and replicated like a normal chromosome when the cell reproduces. YACs can replicate fragments of foreign DNA that are millions of nucleotides long.

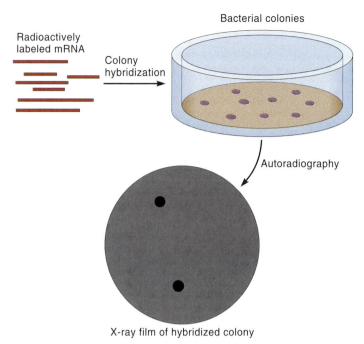

FIGURE A.6

Colony hybridization. Only those colonies containing DNA of the target gene, which is complementary to the radioactive mRNA, will show up on X-ray film. This diagram shows that two colonies have the target gene.

Finding the Gene of Interest Can Be Difficult.

Perhaps the most difficult and time-consuming step in genetic engineering is finding the gene of interest. One method is to isolate specific mRNA and then make a cDNA copy of it for cloning. For example, fungal infection stimulates soybeans to make large amounts of mRNA from a gene that is part of a biosynthetic pathway that makes compounds to fight the infection. After it is induced to make mRNA, the gene of the soybean can be cloned as cDNA.

Genes are found in gene libraries by using probes to search for them. A probe is a sequence of radioactive DNA that matches part of the gene of interest. In such cases, the amino acids are used to predict the DNA sequence that might code for the polypeptide. A short fragment containing this DNA sequence is made chemically and radio labeled, and then used as a molecular probe to explore the genome.

By using a probe, we can detect the gene of interest in bacterial cultures by colony hybridization. In this procedure, an agar plate containing bacterial cultures is blotted with filter paper to make a replica of the pattern of colonies. The replica is then covered in a solution containing the probe, which hybridizes (i.e., anneals) to its DNA complement in those colonies that have the target gene (fig. A.6). Only those colonies that have the target gene will bind to the probe and become radioactive. Their location is marked by exposing

the filter paper to X-ray film, which reacts with radiation from the probe-bound colonies. The radioactive colonies are then picked out of the culture and grown individually, thereby cloning the gene of interest for further study.

Gene libraries can be screened more directly by searching for specific polypeptides. This method is similar to colony hybridization, except that instead of probing for a DNA sequence, antibodies are used to probe for a particular protein. For example, in the molecular genetic study of round versus wrinkled seeds in the garden pea (see chapter 13), the first step was to make an antibody against the purified starch-branching enzyme. When a solution of the antibody was infused into the vector culture of the gene library from RR plants, it reacted only in colonies containing the starch-branching enzyme, SBEI, thereby revealing the presence of the gene.

The Polymerase Chain Reaction Has Contributed to Genetic Technology.

One of the newest techniques for both finding and cloning a gene of interest is the polymerase chain reaction (PCR), in which DNA polymerase assembles free nucleotides, thereby making a complementary sequence from a strand of template DNA. In cells, the DNA polymerase replicates the entire genome during a cell cycle (see chapter 5), but the goal of PCR is to replicate one gene, not the entire genome.

Therefore, a specific gene is targeted by including primers for that gene in the reaction mixture. A primer is a small sequence of DNA, usually 10 to 30 nucleotides long, that is complementary to one end of the target gene. Once it is annealed to the gene, the DNA polymerase adds nucleotides to it along the template. Both strands of the target gene are copied by using two primers, one for the end of the coding strand and one for the end of the complementary strand.

One cycle of the PCR takes about 10 minutes and requires three steps:

1. Denature the DNA at high temperature.
2. Cool the mixture to allow the primers to anneal to the target gene.
3. Incubate the mixture to allow the polymerase to make new DNA (fig. A.7).

The amount of a target gene doubles during each cycle. Normally, the PCR runs for 30 cycles, which means that several million copies of a gene are made in a few hours. By comparison, a plasmid or a virus takes several weeks to make the same amount of a specific gene.

WHAT ARE SOME INDUSTRIAL USES OF GENETICALLY ENGINEERED BACTERIA?

As previously described, bacterial vectors are used to store and retrieve genes in modern genetic research. Genetically engineered bacteria are also important for making foreign gene products on a commercial scale. The polypeptide of interest is either harvested from bacteria that are grown in large vats, or the live bacteria are used directly (fig. A.8). The synthesis of a eukaryotic polypeptide in a vector is often enhanced by linking the gene to a highly active promoter (a section of a DNA molecule to which RNA polymerase binds, initiating the transcription of messenger RNA), which helps to make large amounts of the polypeptide at relatively low cost.

In medicine, genes that code for the synthesis of several polypeptide drugs are now produced by bacteria. Unlike the bacteria used in agriculture, which contain genes from other prokaryotes, drug-producing bacteria contain genes from eukaryotes. These genes make pharmaceutically important proteins, including insulin used to treat diabetes, human growth hormone used to treat dwarfism, and interferons and interleukins used to stimulate the immune system. Before these genetically engineered products were available, all such proteins had to be obtained from animals. Human growth hormone was especially scarce because it had to come from human cadavers.

There Are Some Problems in Making Transgenic Plants.

The main problem in making transgenic plants is that, unlike bacterial plasmids, plant DNA, because of its complex

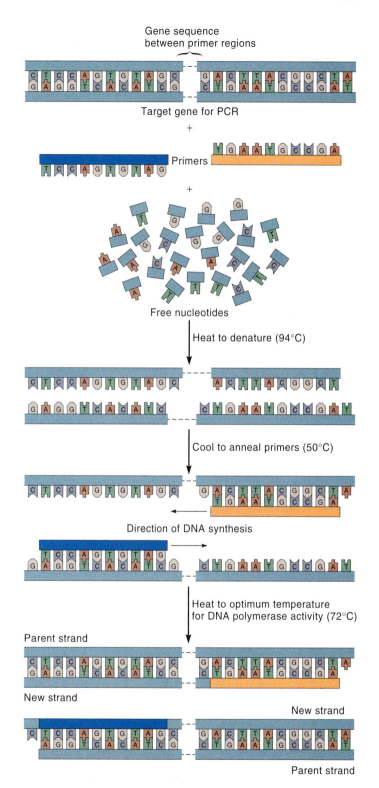

FIGURE A.7

The polymerase chain reaction. Target DNA is amplified by DNA polymerase after primers anneal to their complementary sequences in the target. The cycle is repeated until several million copies of the target gene have been made. Temperatures in each cycle and the number of cycles vary for different target genes and primers.

FIGURE A.8

An industrial fermenter used to grow genetically engineered bacteria. Such large-scale bacterial cultures produce large amounts of the foreign polypeptide at relatively low cost.

structure, cannot be recombined with foreign DNA by the direct use of restriction enzymes, nor can plant cells be transformed easily by absorbing DNA because of difficulties getting the DNA to the plant nucleus. Nevertheless, nature has its own genetic engineers that can overcome these difficulties: parasites that have special genes for inserting their DNA into plant genomes. One such parasite is the bacterium *Agrobacterium tumefaciens,* which causes plants to grow galls (tumors) (fig. A.9). *Agrobacterium* has a plasmid, called the *Ti* plasmid, that is transferred into the host plant's DNA (fig. A.10). In doing so, *Agrobacterium* can genetically transform the plant cells.

The most common way to make a transgenic plant is to use bacteria whose *Ti* plasmid contains an inserted gene of interest. When such bacteria infect plant cells, recombined *Ti* plasmids combine with the plant's chromosomes. The foreign gene is then replicated in each cell cycle as if it were a normal part of the plant genome. Ideally, the newly inserted gene is also expressed in all of the cells that contain it.

When these methods are used on whole plants or tissues, only some cells are infected with recombined *Ti* plasmids and become transgenic. To make an entire plant trans-

FIGURE A.9

The tumorous growths on these *Gladiolus* bulbs arose from cells that were induced to become meristematic by the bacterium *Agrobacterium tumefaciens.* Such growth is known as *crown gall disease.*

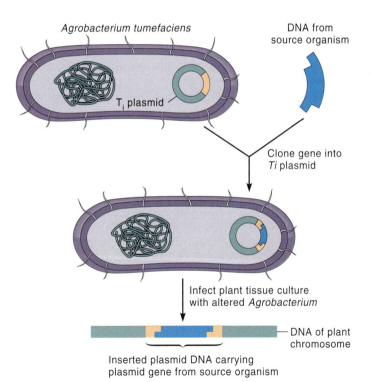

FIGURE A.10

Inserting foreign genes by using the *Agrobacterium Ti* plasmid. First, the bacterium is transformed by the insertion of a foreign gene into its *Ti* plasmid, which carries genes for inserting itself into plant chromosomes. Then a plant cell that is infected with the plasmid receives the foreign gene into its genome.

A.1 Perspective Perspective

Shotgun DNA

Biological vectors are not the only carriers for inserting foreign genes into plant cells. Molecular geneticists have several nonbiological tools for this purpose, the most interesting of which is the gene gun. With this device, DNA is placed on tiny metal beads that are loaded into a shot shell in the barrel of a .22-caliber gun and literally shot into cells. The cells that are hit just right by the metal beads integrate the foreign DNA into their chromosomes about as efficiently as cells that are infected with biological vectors.

---

genic, however, all of its cells must be genetically altered. For many plants, this is a minor problem that is solved by genetically engineering their cells in culture and growing the cultured cells back into whole plants (fig. A.11). The genes of this plasmid are then replicated and expressed as if they belonged to the plant's genome. Such genes reactivate the cell cycle in cells that had stopped in the G_1 phase (see chapter 5). These cells then divide rapidly and form plant tumors.

Antisense Technology Is a New "Trick" in the Game.

Transposons are a segment of DNA that is capable of moving to a new position within the same or another chromosome, plasmid, or cell and thereby transferring genetic properties such as resistance to antibiotics. Although particular genes can be inactivated by inserting transposons or by mutagenic chemicals, these processes are random and inefficient and can alter genes that we do not want to change. We can avoid this problem by using antisense technology, which inactivates specific genes. To appreciate this technique, recall that active genes are codes for polypeptides. When one copy of a strand of DNA is transcribed, the sequence of bases in the DNA is preserved, except that the sequence is exactly complementary to the original code. The mRNA, now containing instructions from DNA in its own language, is then translated into a polypeptide. If done correctly, the message is said to be in correct sense. However, if an exactly complementary (i.e., antisense) RNA were present to combine with the sense RNA, the message of the sense RNA would be nullified.

To make an RNA complement, an antisense message can be made from the "other" strand of DNA and introduced into a cell, or it can be transcribed by the cell after an antisense gene is introduced. When the sense RNA couples with its antisense RNA, it cannot combine with a ribosome and is inactivated. Thus, the original gene is blocked by its own antisense RNA, without affecting other genes.

Antisense technology has been used to control a variety of plant processes, including fruit ripening. Because of its specificity and simplicity, antisense technology has tremendous potential for use in agriculture.

There Are Several Examples of Research in Genetic Technology.

Herbicide Resistance

Most plants are killed by glyphosate (Roundup®), a herbicide that inhibits an enzyme called EPSP synthetase that is required for making aromatic amino acids. (EPSP stands for the product made by EPSP synthetase, which is 5-EnolPyruvyl-Shikimic acid-3-Phosphate.) When plants cannot make aromatic amino acids, their metabolism stops and they die.

Glyphosate-resistant petunias have been made by inserting extra copies of the EPSP synthetase genes into them. These petunias make enough enzyme to overcome inhibition by glyphosate. Plants that contain a bacterial form of EPSP synthetase also resist glyphosate (fig. A.12). Ideally, crops that resist glyphosate do not have to be weeded. Treating a field with the herbicide kills the weeds but does not affect the genetically engineered crop.

Disease Resistance

Botanists have discovered a variety of genes involved in disease resistance by plants. For example, the RPS2 gene in *Arabidopsis thaliana* confers resistance to bacteria such as *Pseudomas syringae*. Similarly, tobacco mosaic virus (TMV) and causes a mosaic pattern of infection on tobacco leaves (fig. A.13). When the gene for the protein coat of TMV is transferred to tobacco, the plants become resistant to the virus. The TMV responds to the genetically engineered tobacco cells as if they were already infected, and since the virus cannot infect cells that are already infected, the plant is immune to infection.

Insect Resistance

Crops can often be protected against insects by being sprayed with bacteria that contain the toxin gene from

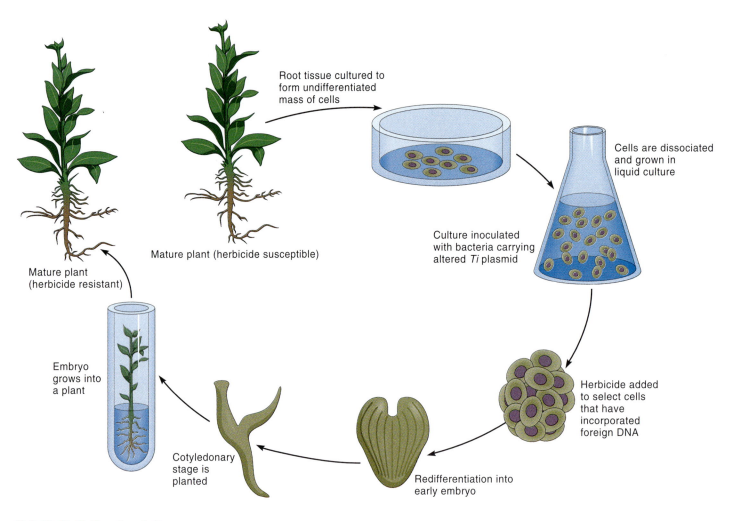

Root tissue cultured to form undifferentiated mass of cells

Cells are dissociated and grown in liquid culture

Culture inoculated with bacteria carrying altered *Ti* plasmid

Mature plant (herbicide susceptible)

Mature plant (herbicide resistant)

Embryo grows into a plant

Herbicide added to select cells that have incorporated foreign DNA

Cotyledonary stage is planted

Redifferentiation into early embryo

FIGURE A.11

Steps in making transgenic plants. Differentiated cells from a plant tissue are put into cell culture, where they grow into an undifferentiated mass of cells. The culture is transferred to a liquid medium and infected with *Ti* plasmids carrying a gene for herbicide resistance. Cells that survive herbicide treatment are transgenic. Plants that are regenerated from these cells are also transgenic, making them resistant to the herbicide in the garden.

Bacillus thuringiensis. This gene has also been inserted into tomato, potato, and tobacco plants. Caterpillars that normally eat the leaves of these species do not eat plants that contain the toxin made by this gene.

Perhaps the most important crop for genetically engineering resistance to insects is cotton; indeed, most of the chemical pesticides used in the United States are used on cotton. Experimental cotton plants that contain the toxin gene from *B. thuringiensis* typically have better resistance to insects and require fewer chemical pesticides.

Drought Resistance

Genes from desert petunias have been transferred into cultivated petunias. The transgenic petunias require much less water than the normal plants. If crop plants likewise can be altered genetically to require less water, there will be less need for irrigation. Such crops will become more important as the demand for water increases in semiarid agricultural regions of the southwestern United States.

Changing Storage Proteins

Grains from cereal grasses are deficient in lysine, an essential amino acid in our diet. However, induced mutations in cultured rice cells have increased their production of lysine and protein. Plants from these mutant cell cultures produce grains containing more lysine and protein than do normal rice plants. Botanists hope to transfer the gene for the high-lysine mutation to other cereal grasses.

Improvement in Fruit Storage

One of the main problems with fruits is that when they are left on the plant too long, they get mushy during shipping and storage. To overcome this problem, fruits are picked and shipped while they are still immature; they are ripened

FIGURE A.12

Normal tomato plants (*right*) and transgenic plants containing a gene for herbicide resistance (*middle*) were treated with the herbicide glyphosate. All of the normal plants died. Plants in the left row were not treated with herbicide.

later by exposure to the hormone ethylene. Unfortunately, such artificially ripened fruits often have little flavor.

Genetic engineering was used to solve this problem for tomatoes, which become mushy because a carbohydrate-digesting enzyme breaks down cell walls in the fruit. Specifically, botanists have introduced an altered gene to slow the enzyme system that affects the mushiness of the fruit. Consequently, these plants produce fruits that do not become mushy. Such genetically engineered tomatoes that are left to ripen on the parent plant maintain their firmness (and flavor) during storage. The first such genetically engineered tomato, called the Flavr Savr® tomato, was sold in grocery stores in the United States in the mid 1990s.

In countries such as Brazil, farmers often lose 20% to 40% of their crops to storage pests such as weevils. This problem is being addressed by engineering plants whose fruits are pest resistant. For example, botanists have produced bean plants having fruit that produces an inhibitor of a-amylase. This inhibitor blocks the a-amylase secreted by attacking weevils, thereby reducing the harvest's susceptibility to the pest.

Production of Polymers

Sales of nonfood products from genetically transformed plants will grow from about $21 million per year in 1997 to $320 million or more by 2005. The first nonfood commercial product of plant genetic engineering is lauric acid, a 12-carbon fatty acid that is used in making soaps and detergents. The transgenic plant that makes this chemical is rapeseed (*Brassica napus*). Normally, fatty acids in this plant are 18 carbons long, but scientists at Calgene, Inc. have inserted into rapeseed a gene from California bay tree (*Umbellularia californica*) that shuts off fatty acid synthesis after 12 carbons. Oil

FIGURE A.13

This tobacco leaf is infected with tobacco mosaic virus (TMV). Resistance to this virus can be engineered into tobacco by using a gene from the virus itself.

from a few thousand acres of transgenic rapeseed, which is easy to grow and a good oil producer, was marketed in 1995. Other transgenic rapeseed lines, in different stages of development, will be producing lubricants, nylon, and fatty acids for making margarine and shortening.

Besides fatty acids, other polymers from transgenic plants include a biodegradable plastic that is made by bacterial genes inserted into *Arabidopsis thaliana* and polyester-like fibers from similar genes inserted into cotton (*Gossypium hirsutum*).

SUGGESTED READINGS

Articles

Barton, J. H. 1991. Patenting life. *Scientific American* 264(March): 40–46.

Boulter, D. 1995. Plant biotechnology: Facts and public perceptions. *Phytochemistry* 40:1–9.

Cai, D. 1997. Positional cloning of a gene for nematode resistence in sugar beet. *Science* 275:832–34.

Holmberg, N., and L. Bulow. 1998. Improving stress tolerance in plants by gene transfer. *Trends in Plant Science:* 3:61–66.

Moffat, A. S. 1995. Exploring transgenic plants as a new vaccine source. *Science* 268:658–60.

Mullis, K. B. 1990. The unusual origin of the polymerase chain reaction. *Scientific American* 262(April):56–65.

Sarhan, F., and J. Danyluk. 1998. Engineering cold-tolerant crops—throwing the master switch. *Trends in Plant Science* 3:287–88.

Books

Henry, R. 1997. *Practical Applications of Plant Molecular Biology.* London: Chapman & Hall.

Jones, P. G., and J. M. Sutton, eds. 1997. *Plant Molecular Biology: Essential Techniques.* New York: John Wiley and Sons.

Wilson, T. M. A., and J. W. Davies, eds. 1992. *Genetic Engineering with Plant Viruses.* Boca Raton, FL: CRC Press.

APPENDIX B

Appendix B

FUNDAMENTALS OF CHEMISTRY FOR BOTANY STUDENTS

Some basic knowledge of chemistry will help you understand several of the important concepts presented in this book. We place this information in an appendix, not because it is less important, but because many students take introductory chemistry in high school or college before they take a botany class. Thus, this appendix is intended as a review for those fortunate enough to have experience with chemistry and as an introduction for those who have not had such a class. This appendix can also serve as a reference if you encounter chemical terms and concepts new to you.

MOLECULES ARE MADE OF ATOMS AND ELEMENTS.

An **element** is a substance that cannot be broken down into a simpler substance by ordinary chemical means. By this definition, there are 92 naturally occurring elements in the universe. Each element is represented by a symbol, usually the first one or two letters of its name. The important elements in plants are listed in table B.1. (A few of the symbols represent Latin names of elements, such as Au for *aurum,* meaning "gold," and Na for *natrium,* meaning "sodium.") About 22 elements are essential to life, but only 6—carbon, hydrogen, oxygen, nitrogen, sulphur, and phosphorus—constitute most of what we call living matter. The rest are trace elements, which are those used by plants only in extremely small quantities.

Atomic Weight Is the Average Mass of an Atom of an Element.

The smallest possible amount of an element is an **atom.** Atoms consist of even smaller particles, the most stable of which are the subatomic **protons, neutrons,** and **electrons.** Atoms with different numbers of protons are different elements; that is, an **element** is a substance composed of atoms having an identical number of protons in each nucleus. For example, an atom with one proton and one elec-

tron is elemental hydrogen. A helium atom has two protons, two neutrons, and two electrons.

Each proton or neutron has a mass of about 1.7×10^{-24} gram. For convenience, this mass is defined as 1 atomic mass unit. It is also called 1 *dalton,* in honor of John Dalton, who helped develop the atomic theory in the early 1800s. The mass of an electron is about $1/1,836$ that of a proton, so it is often disregarded when considering atomic mass. A summary of the features of several common elements is presented in table B.1.

The Structure and Activity of Atoms Help Us Understand Their Importance in Cells.

Atomic structure depends on how many subatomic particles (protons, neutrons, and electrons) an atom possesses. The smallest atom is hydrogen, which has one proton, one electron, and no neutrons. The electron, which has a negative ($-$) electric charge, is attracted to the positive ($+$) electric charge of the proton. (Neutrons, which occur in all elements except hydrogen, have no electric charge.) The opposing electric charges attract each other, but the particles do not touch. This is primarily because the much smaller electron moves at nearly the speed of light, flying around the proton nucleus but never slowing down enough to fall into it. The electron of a hydrogen atom orbits the proton in the nucleus as though it were a tiny planet orbiting its sun (fig. B.1). Although this planetary model is a useful way to envision the structure of an atom, the actual structure is unknown. Nevertheless, such models help us understand atomic theory and explain the physical properties of atoms.

Although the exact location of an electron cannot be pinpointed, there is a region around the nucleus where it is most likely to be. This region is called its orbital. For the electron of hydrogen, the shape of the orbital is approximately spherical (fig. B.2). Helium has two electrons that also occur in a spherical orbital. Atoms larger than hydrogen and helium have more than two electrons. In these atoms, additional orbitals exist further away from the nucleus to accommodate the added electrons. These outer orbitals are larger, and they occur in different shapes and sizes, in contrast to

TABLE B.1

The Symbols, Atomic Numbers, and Atomic Mass of Some Common Elements

ELEMENT	SYMBOL	ATOMIC NUMBER	ATOMIC MASS
Hydrogen	H	1	1
Boron	B	5	10.8
Carbon	C	6	12
Nitrogen	N	7	14
Oxygen	O	8	16
Sodium	Na	11	23
Magnesium	Mg	12	24.3
Phosphorus	P	15	31
Sulfur	S	16	32
Chlorine	Cl	17	35.4
Potassium	K	19	39.1
Calcium	Ca	20	40.1
Manganese	Mn	25	54.9
Iron	Fe	26	55.8
Cobalt	Co	27	58.9
Copper	Cu	29	63.5
Zinc	Zn	30	65.4
Molybdenum	Mo	42	95.9

These elements are essential for plant life; each has one or more vital roles. If any one is missing, a plant cannot survive. *Atomic number* corresponds to the number of protons in the atomic nucleus; *atomic mass* is the number of protons plus neutrons in each nucleus. Protons are stable, positively charged subatomic particles having a mass 1,836 times that of the electron. Neutrons are electrically neutral subatomic particles, having a mass 1,839 times that of the electron, and are stable when bound in an atomic nucleus. It and the proton form nearly the entire mass of atomic nuclei. A nucleus is a positively charged central region of an atom, composed of protons and neutrons and containing almost all of the mass of the atom. An electron is a stable subatomic particle having a negative electric charge. An *element* is a substance composed of atoms having an identical number of protons in each nucleus. Elements cannot be reduced to simpler substances by normal chemical means.

the relatively small, more or less spherical orbitals of hydrogen and helium.

CHEMICAL BONDS HOLD MOLECULES TOGETHER.

One of the most important features of an electron orbital is that it has enough space for only two electrons. An or-

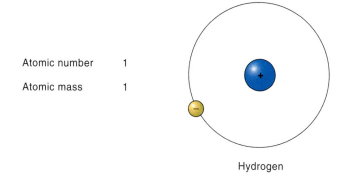

Atomic number   1
Atomic mass   1

Hydrogen

FIGURE B.1
An atom of hydrogen has one proton (+), one electron (−), and no neutrons.

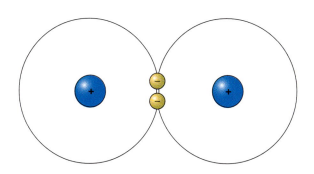

FIGURE B.2
The two atoms of hydrogen gas (H_2) are held together by a covalent bond. In covalent bonds, atoms share valence electrons.

bital with only one electron can attract another electron to fill the available space. In the case of a hydrogen atom, which has just one electron, the orbital is filled by hydrogen attracting an electron from another atom. In this way, two hydrogen atoms can share their single electrons to make a combined orbital with two electrons. The combined orbital of two hydrogen atoms makes a molecule of hydrogen gas, which is designated by the molecular formula H_2. A molecule of hydrogen gas is held together by the chemical bond between the electrons of two hydrogen atoms. This is one of several types of bonds that are based on the behavior of electrons. These bonds are discussed in the next few paragraphs.

A Covalent Bond Is Formed by the Sharing of One or More Electrons Between Atoms.

The chemical bond just described for hydrogen gas is called a **covalent bond** (see fig. B.2), which is a chemical bond formed by the sharing of one or more electrons, especially pairs of electrons, between atoms. A covalent bond is written as a single line that represents a pair of shared electrons. Using this notation, the structural formula for hydrogen gas

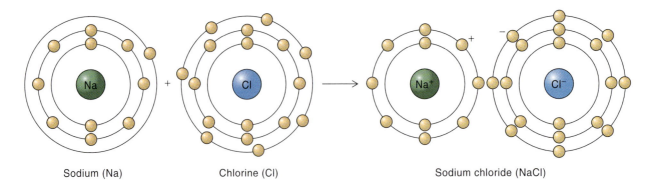

Sodium (Na)   Chlorine (Cl)   Sodium chloride (NaCl)

F I G U R E   B . 3

Sodium and chlorine are bonded ionically to form sodium chloride, or table salt. In ionic bonds, atoms are held together by opposite charges that result from the loss or gain of valence electrons.

is H-H (H_2). You will see this model for a covalent bond often in this book.

Elements with more than one orbital (i.e., all but hydrogen and helium) have more than one electron shell. Unlike the innermost shell, outer shells can contain more than one orbital, up to a maximum of four. For example, a carbon atom has six electrons, two in the innermost orbital and one in each of four outer orbitals in the valence shell. Based on this structure, carbon can share four electrons by covalent bonding. A carbon atom can, for example, bond to four hydrogen atoms. The structural formula of this molecule, methane, is

$$\begin{array}{c} H \qquad H \\ \diagdown \; \diagup \\ C \\ \diagup \; \diagdown \\ H \qquad H \end{array}$$

This molecule can be written more simply as CH_4, which is the molecular formula for methane. Methane is called a compound molecule, or compound, because it consists of more than one kind of element. Except for the simple gases—hydrogen (H_2), oxygen (O_2), and nitrogen (N_2)—virtually all biologically important molecules are compounds.

Carbon can also form double covalent bonds with other carbon atoms or even other atoms such as oxygen. A **double bond** is a covalent bond in which two electron pairs are shared between two atoms. Ethylene (C_2H_4), which provides an example, is a colorless, odorless gas that occurs naturally and is a plant hormone (see chapter 6). The structural formula for ethylene, with a double carbon bond, is

$$\begin{array}{c} H \qquad H \\ | \qquad\quad | \\ C = C \\ | \qquad\quad | \\ H \qquad H \end{array}$$

Double carbon bonds are important in many biological molecules, including fatty acids, nucleic acids, and amino acids.

Interestingly, a carbon atom can also form a triple covalent bond with a nitrogen atom or even with another carbon atom. When bonded to nitrogen, a carbon atom still has one free valence electron that must be shared with another atom. If the other atom is hydrogen, the combination of hydrogen, carbon, and nitrogen makes a molecule of a poisonous gas called hydrogen cyanide. Similarly, when a carbon atom forms a triple bond with another carbon atom, each atom still requires one more covalent bond. When both bonds are made with hydrogen, the compound is a molecule of acetylene (C_3H_2), the highly flammable gas used in blowtorches.

Ionic Bonds Occur Between Two Ions Having Opposite Charges.

In extreme cases, valence electrons are not shared but instead are removed from one atom and transferred to another. This happens between an element that can easily give up an electron, such as sodium, and one that strongly attracts an electron, such as chlorine. The result is that sodium, with 11 protons and 10 electrons, has an extra positive charge; that is, it becomes a positively charged ion (cation). Chlorine then has 17 protons and 18 electrons, which makes it a negatively charged ion (anion). The opposite charges of Na^+ and Cl^- ions attract each other to form the ionic bond that makes sodium chloride (NaCl), or table salt (fig. B.3).

Some ions have more than one charge. Magnesium, for example, gives up two valence electrons and is therefore a doubly charged (divalent) cation, Mg^{+2}. Magnesium can form ionic bonds to two chlorine atoms, forming magnesium chloride ($MgCl_2$).

Besides ionic atoms, entire molecules can become ionic by losing or gaining one or more electrons. Nitrogen and hydrogen, for example, combine to form an ammonium cation (NH_4^+); sulphur and oxygen make a divalent sulfate anion (SO_4^{-2}). Ammonium sulfate, an important ingredient in nitrogen fertilizers, is a compound of these two ions with the molecular formula $(NH_4)_2SO_4$. In this compound, the doubly charged sulfate ion is bonded to two singly charged ammonium ions.

Hydrogen Bonding and the Special Properties of Water Are Critical to Cell Structure and Function.

Nitrogen and oxygen have especially strong attractions for electrons from other atoms. Because of this stronger attraction, they are said to be more *electronegative* than elements such as carbon. Electronegative means having a negative electric charge, or tending to attract electrons when a chemical bond is formed. When hydrogen is covalently bonded to oxygen, the electronegativity of oxygen forces the hydrogen to become only partially positively charged. The partial charge is strong enough to make hydrogen attractive to other electronegative atoms. This attraction is a relatively weak bond, which is neither covalent nor ionic. Rather, this type of bond is called a **hydrogen bond,** which can be defined as a chemical bond in which a hydrogen atom of one molecule is attracted to an electronegative atom, usually of another molecule.

Hydrogen bonds play several significant roles in biological processes. One of these roles is to maintain the shape of large molecules such as enzymes and other proteins that control cellular processes. This happens when a bond between the hydrogen of one part of a molecule and an electronegative atom in another part of the molecule holds the two parts together. Molecular shapes are important when different compounds must fit together in a precise way to complete the chemical reactions of cellular processes.

The properties of water, which makes up over 90% of the mass of most plants, are based on hydrogen bonding. These properties include water's cohesiveness, its high heat of vaporization, and its versatility as a solvent. Each water molecule has the chemical formula H_2O, but the oxygen attracts the hydrogen of neighboring water molecules, and the hydrogen is attracted to other oxygen atoms. Actually, many hydrogen bonds are formed and broken and then re-formed in a fraction of a second, but the overall effect is to give water a cohesive structure. Because of this cohesion, plants can pull water upward against gravity, from roots to the tops of the tallest trees.

Because water molecules are so much more electronegative at the oxygen end than they are at their hydrogen end, they are said to be polar. The "ends" are the poles. When salt is added to water, Na^+ ions are attracted to the oxygen pole, and Cl^- ions are attracted to the hydrogen pole. These *polar attractions* are so strong that the cations and anions are separated and surrounded by water molecules, thereby causing the salt crystals to dissolve in the water. Living cells contain many different kinds of salts that are dissolved in water because of its polarity. The ability of water to dissolve salts also extends to other polar compounds, even

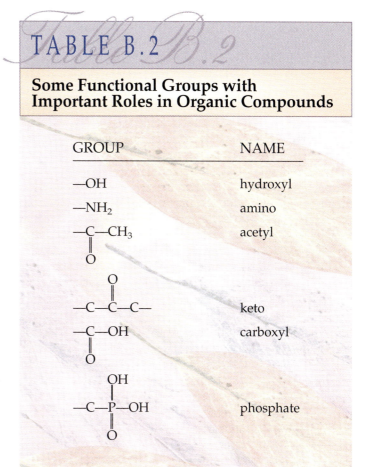

TABLE B.2

Some Functional Groups with Important Roles in Organic Compounds

GROUP	NAME
—OH	hydroxyl
—NH$_2$	amino
—C—CH$_3$ (with =O)	acetyl
—C—C—C— (with =O on middle C)	keto
—C—OH (with =O)	carboxyl
—C—P—OH (with OH and O)	phosphate

though they are not ionic. This property of water contributes to the enormous variety of compounds dissolved in cellular fluids.

Carbon Chains and Rings Form Important Biological Molecules.

The versatility of carbon, as well as its central role in the chemistry of biological processes, comes from the ability of each carbon atom to share four covalent bonds. A carbon atom can be thought of as an intersection of covalent branches in four directions. Covalent bonding occurs in each direction, either with a different element or with another carbon. Because a carbon atom can form a covalent bond with another carbon atom, there is a seemingly infinite variety of possible compounds that have multiple carbons. Known compounds range from the smallest, with one or two carbons (methane, acetylene, ethyl alcohol), to the largest, with up to 30,000 carbons (natural rubber).

Carbon skeletons occur in straight or branched chains or even in closed rings. In each compound, every carbon is fully "accommodated" with four covalent bonds. The simplest compounds are hydrocarbons (i.e., those with only carbon and hydrogen). The complexity of compounds increases dramatically when these functional groups are attached to other kinds of elements. A functional group is an atom or group of atoms that defines the structure of a family of compounds and determines the chemical and biological properties of the family (table B.2).

Compounds with carbon and oxygen can form carboxylic acids, or organic acids. The most common organic acid in plants is acetic acid, which is the basic building block of fats and other lipids. It is this acid that gives vinegar its tangy flavor. Organic acids occur in many metabolic pathways in plants. Some of the most biologically important groups of biomolecules contain nitrogen as the amino group (NH_2). This group is important because it is one of the two functional groups that make up an amino acid (the other is a carboxyl, or acid group; the formula is COOH, which is the functional group characteristic of all organic acids). Amino acids are used for building enzymes and other proteins that are the foundation for almost all biochemical reactions and for a variety of cellular structures. Organic acids are part of the Krebs cycle (see chapter 11).

Sulphur, like oxygen, has two valence electrons and forms a functional group with a hydrogen atom, which is called a sulfhydryl group (-SH). Compounds with sulfhydryl groups are called thiols and are notable for their foul odors. Butanethiol is the major ingredient of skunk spray. Sulphur can also form disulfide bonds (S-S). These bonds form between two cysteine (amino acid) monomers in proteins and are important for stabilizing a protein's three-dimensional structure.

Of all the biologically important functional groups, the most electronegative is the phosphate group. Each phosphate group (PO_4^{-3}) consists of one phosphorous atom and four oxygen atoms. It is derived from a phosphoric acid molecule (H_3PO_4) that, due to the electronegative strength of the oxygens, has released all of its hydrogen atoms as ions. Phosphate groups can form short-lived, unstable covalent bonds with other phosphate groups. The ease with which these bonds can be broken and new ones formed means that the energy stored in the molecule is easily transferred to other molecules during chemical reactions. ATP is an excellent example (see chapters 3, 10, and 11). As phosphate bonds break and new bonds are formed, energy differences in the reactants and products can be used to drive cellular processes. These processes include such functions as building new molecules, moving substances from one place in the cell to another, and controlling what goes through membranes. The phosphate group is also important in the sugar-phosphate backbone of the nucleic acids, DNA and RNA, and in the phospholipids of membranes.

CHEMICAL REACTIONS CONVERT ENERGY AND FORM NEW MOLECULES IN PLANTS.

A chemical reaction is the process of making and breaking chemical bonds and is written as an equation that has the reactants on the left side and the products on the right side. For example, methane gas and oxygen (reactants) can react to make carbon dioxide and water (products) as energy is released:

$$CH_4 + 2O_2 \rightarrow CO_2 + 2H_2O + \text{energy}$$

Notice that all atoms are accounted for in a chemical reaction; that is, the equation is balanced. Reactions cannot create or destroy matter but can only rearrange it. Thus, in this reaction, each methane molecule requires two molecules of oxygen (O_2) to maintain the appropriate proportion of elements in the products. This is a common reaction that occurs at the burners of gas stoves every day. Here, the energy is given off as heat. In cells, the difference in energy between reactants and products can be used to do the work of the cell.

The physical behavior of energy is the driving force underlying all biological processes. Chapters 10 and 11 are specifically devoted to energy and its use by plants. The following paragraphs outline some of the most common chemical reactions that are driven by the energy of plant metabolism. **Metabolism** is the complex of physical and chemical processes occurring within a living cell or organism that are necessary for the maintenance of life. During metabolism, compounds such as sugars are broken down, yielding energy for vital processes simultaneously while other compounds, such as proteins necessary for life, are synthesized.

Oxidation and Reduction Reactions Are Important in Energy Conversions.

The chemical reaction of methane and oxygen presented in the previous section is a combustion reaction; that is, methane and oxygen burn together. In this reaction, the covalent electrons of carbon, which are about evenly shared with the electrons of hydrogen in methane, are pulled further away from the carbon by the strongly electronegative oxygen, forming carbon dioxide (CO_2). This shift of electrons away from an element is called **oxidation.** Each hydrogen is also oxidized, since its electron goes from being equally shared with carbon to being pulled away by the oxygen in water. Oxidation may be defined as (1) the combination of a substance with oxygen and (2) a reaction in which the atoms in an element lose electrons and the valence of the element is correspondingly increased.

When one reactant is oxidized, another reactant undergoes reduction—that is, the reactant molecule is reduced. Reduction may be defined as (1) a decrease in positive valence or an increase in negative valence by the gaining of electrons, (2) a reaction in which hydrogen is combined with a compound, and (3) a reaction in which oxygen is removed from a compound. In the previous reduction, oxygen is reduced; that is, it partially gains electrons from the carbon in CO_2 and from the hydrogen in H_2O. Although the example reaction involves oxygen, oxidation-reduction (redox) reactions do not always require oxygen. Any partial or complete transfer of electrons is a redox reaction. For example, the ionization of sodium and chlorine (NaCl, table salt) fits this description (see fig. B.3):

$$NaCl \rightarrow Na^+ + Cl^-$$

In this reaction, sodium is oxidized (loses an electron), and chlorine is reduced (gains an electron).

The energy that cells need for growth and metabolism is derived from a redox process called cellular respiration (or just respiration). **Cellular respiration** is a series of redox reactions that usually begins with a carbohydrate (such as starch or sucrose) and oxygen and ends with carbon dioxide and water. The following equation summarizes the overall process:

$$C_6H_{12}O_6 + 6\,O_2 \rightarrow 6\,CO_2 + 6\,H_2O + energy$$

The many reactions that occur during respiration, the parts of cells where the reactions occur, and other details about this process are the subject of chapter 11.

Dehydration and Hydrolysis Reactions Synthesize and Break Molecules.

Although oxidation and reduction occur during almost all chemical reactions, many reactions are referred to by the products they yield or by the functional groups that are involved. For example, most biological polymers are built by reactions that yield water as a by-product. These reactions are called **dehydration reactions.** The peptide bonds of proteins, the glycosidic bonds of polysaccharides, the sugar-phosphate bonds of nucleic acids, and the glycerol-fatty acid bonds of lipids are all made by dehydration. For example, where aa = one amino acid:

aa-aa-aa-aa-aa-aa-aa- + aa $\leftrightarrow$ aa-aa-aa-aa-aa-aa-aa-aa + H_2O
(elongating protein)   (protein plus the amino acid)

Dehydration reactions are reversible. The covalent bonds between glucose molecules in starch, for example, can be broken to yield glucose monomers (see chapter 3). This reverse reaction is called **hydrolysis** because it uses water to break the chemical bond between glucose molecules. All bonds formed by dehydration can be broken by hydrolysis, as is shown in the preceding example for a protein. In cells, chemical synthesis by dehydration is catalyzed by one set of enzymes, whereas chemical degradation by hydrolysis is usually catalyzed by another set of enzymes. The metabolic needs of a cell determine which direction a reaction takes. Both kinds of reactions can occur at the same time but in different parts of the cell.

Decarboxylation and Carboxylation Reactions Involve Carbon Dioxide.

Some covalent bonds, such as those in carbon dioxide (O=C=O or CO_2), are stable and difficult to break. The formation of carbon dioxide, with its low potential energy, is the strong driving force that removes CO_2 from organic acids (where R = a carbon skeleton):

$$R\text{-}COOH \rightarrow R\text{-}H + CO_2$$

This **decarboxylation reaction** is common in many cellular processes (e.g., the release of CO_2 during respiration). In fact, several decarboxylations occur for each molecule of glucose that is oxidized during respiration. This means that organic acids play a role in the intermediate steps of this process. Organic acids and their decarboxylation are discussed in chapter 10 ("Energy and Respiration").

Although the removal of carbon dioxide has many metabolic roles, the addition of CO_2, or **carboxylation,** is characteristic of photosynthesis. This reaction is the foundation for making organic molecules from CO_2 in the atmosphere. Energy is required to break the double covalent bonds between carbon and oxygen and form covalent bonds between carbons. Only green plants, algae, and a few bacteria can transform light energy and use it to carboxylate organic compounds. This process is the topic of chapter 10.

Phosphorylation Reactions Involve Phosphate Ions.

Reactions that form or break phosphate bonds are called **phosphorylation reactions.** Perhaps the most common phosphorylation reaction in any cell is the addition of a phosphate group (P_i is the symbol for inorganic phosphate or a phosphate group) to the nucleotide, adenosine diphosphate **(ADP).** The product of this reaction is adenosine triphosphate **(ATP).** Phosphorylation is accompanied by dehydration (loss of H_2O). Thus, phosphorylation can be abbreviated as follows:

$$\text{A-P-P} + P_i + energy \xrightarrow{\text{phosphorylation}} \text{A-P-P-P} + H_2O$$

The covalent bond between the second and third phosphate groups of ATP is easily hydrolyzed, thereby releasing energy to another molecule in a chemical reaction. Thus, *the reverse of this reaction,* where ATP is split and reacts with H_2O, is *hydrolysis:*

$$\text{A-P-P} + P_i + energy \xleftarrow[\text{hydrolysis}]{} \text{A-P-P-P} + H_2O$$

Because of the instability of the phosphate bonds and the demand for this energy to do cell work, each ATP molecule may be recycled to ADP plus P_i in a fraction of a second. ATP is regenerated continuously by the energy derived from respiration and photosynthesis (see chapters 10 and 11).

The structures of glucose, fructose, and galactose. Although these sugars are $C_6H_{12}O_6$, they have different chemical properties.

MOLECULAR SHAPES AND CHEMICAL FORMULA ARE USED TO UNDERSTAND BIOLOGY.

Many organic compounds can have different shapes but the same molecular formula. The simple sugars glucose and fructose, for example, are both $C_6H_{12}O_6$, but they have different chemical properties. Their structural differences are easy to see when their rings are broken open by hydrolysis (fig. B.4). This seemingly small difference is significant. For example, fructose is one of the sweetest sugars, whereas glucose has almost no sweetness at all. Glucose and fructose are structural isomers; that is, they have the same functional groups bonded to different carbon atoms. Galactose has the same functional groups on the same carbon atoms as glucose. However, the hydroxyl group on the fourth carbon of galactose is in a different orientation than it is on glucose (see fig. B.4). This means that glucose and galactose are stereoisomers of each other: they are seemingly identical but, like your left and right hands, are really nonidentical mirror images of each other. Note the number of the carbon atoms in glucose.

SUGGESTED READINGS

Books

Becker, W., J. Reece, and M. Poenie. 1995. *The World of the Cell*, 3rd ed. Menlo Park, CA: Benjamin/Cummings.

Sackheim, G. 1995. *Chemistry for Biology Students*. Menlo Park, CA: Benjamin/Cummings.

ABA (*See* Abscisic Acid)

Abiotic Something that is nonliving and never has been alive

Abscisic Acid (ABA) A plant hormone (growth regulator) associated with water stress and the inhibition of growth; also induces stomatal closing and seed dormancy in many plants

Abscission Zone An area of plants where leaves, flowers, or fruits detach

Accessory Fruit A fruit whose flesh is not derived from ovary tissue; for example, an expanded receptacle

Acclimation The changes in a plant in preparation for new environmental conditions

Acid Rain The combination of water in the atmosphere with sulphur and nitrogenous compounds released by the burning of fossil fuels

Active Transport Movement of solutes across a membrane against their concentration gradient; active transport requires energy from cellular metabolism

Acylglyceride Linkage The covalent bond between the organic acid group, such as in a fatty acid, and one of the three hydroxyl groups of glycerol

Adaptation A heritable characteristic that allows an organism to live in a particular environment

Adaptive Radiation Evolution of divergent forms of a trait in several species that developed from an unspecialized or primitive common ancestor

Adaxial Meristem Tissue that thickens the leaf of a plant

Adenosine Triphosphate (ATP) A nucleotide consisting of adenine, ribose, and three phosphate groups; the major source of usable chemical energy in metabolism; when hydrolyzed, ATP loses a phosphate to become adenosine diphosphate (ADP) and releases usable energy

Adventitious Root A root that arises from a leaf or stem (i.e., not from another root)

Aerenchyma A tissue containing large amounts of intercellular spaces

Aerobic Respiration The process by which organic compounds are oxidized with the release of energy; respiration is aerobic if oxygen is required as the terminal electron acceptor; it is anaerobic if oxygen is not used as an electron acceptor

Aggregate Fruit A fruit with tiny clusters, such as the tiny drupes of blackberry and the follicles of magnolia

Alkaloid A nitrogen-containing base in which at least one nitrogen is part of a ring; examples include nicotine, caffeine, cocaine, and strychnine; alkaloids are often bitter and affect the physiology of vertebrates and other animals

Allele One of the alternative forms of a gene; a gene may have two or more alleles

Allozyme Enzyme that is coded for by different alleles of the same locus; each form is encoded by different alleles

Alternate Leaf Arrangement Type of phyllotaxis, arrangement of leaves on the stem, in which there is one leaf per node

Alternation of Generations A life cycle that alternates between multicellular haploid and diploid generations

Alternative Oxidase The oxidase at the end of a branching chain of electron carriers in plant mitochondria

Alternative Respiratory Pathway (or Chain) The alternative pathway taken by plant mitochondria when cytochrome oxidase is poisoned

Amino Acid The 20 different kinds of monomers that make up most plant proteins

Anabolic Reaction (*See* Anabolism)

Anabolism Biosynthesis; the constructive part of metabolism

Anaerobic Respiration The process by which organic compounds are oxidized with the release of energy; respiration is aerobic if oxygen is required as the terminal electron acceptor; it is anaerobic if oxygen is not used as an electron acceptor

Anaphase The period of mitosis during which centromeres split and sister chromatids become separate chromosomes that begin to move toward opposite poles of the spindle apparatus

Anaphase I The first anaphase of meiosis; in anaphase I, homologous chromosomes move to opposite poles of the meiotic spindle apparatus, resulting in a halving of the number of chromosomes going to each daughter nucleus

Anaphase II The second anaphase of meiosis; in anaphase II, the centromeres divide, thereby allowing the separation of sister chromatids into independent chromosomes

Androecium (plural, **androecia**) Collectively, all of the stamens of a single flower

Angiosperm A plant whose seeds are born in a fruit; an informal name for flowering plant (division Anthophyta or Magnoliphyta)

Animalia One of six kingdoms in the modern system of classification of organisms; others are Fungi, Plantae, Protista, Eubacteria, and Archaebacteria

Annulus An incomplete ring of cells encircling a fern sporangia; unequal drying of the annulus produces a snapping motion that disperses spores

Antennae Complexes Networks of the chloroplast pigments and associated molecules arranged in thylakoids; consist of proteins, about 300 molecules of chlorophyll *a*, and about 50 molecules of carotenoids and other accessory pigments that gather light

Anther The pollen-bearing part of a stamen

Antheridium (plural, **Antheridia**) A unicellular or multicellular structure in which sperm are produced; may be multicellular or unicellular

Anticodon A sequence of three nucleotides in a molecule of transfer RNA, which is complementary to a codon sequence

Apical Meristem The meristem at the tip of a root or shoot in a vascular plant

Archaea (*See* Archaebacteria)

Archaebacteria Primitive prokaryotes with distinctive chemical and structural features

Archegonium A multicellular organ that produces an egg; found in bryophytes and some vascular plants

Artificial Classification A system of classification that does not reflect natural, evolutionary relationships between organisms

Artificial Selection Selection by humans of specific traits in organisms being bred to produce desired characteristics

Ascomycetes A large group of true fungi with septate hyphae; they produce conidiospores asexually and ascospores sexually within asci

Asexual Reproduction (*See* Vegetative Reproduction)

ATP (*See* Adenosine Triphosphate)

Attractant Characteristic that causes an animal to visit a plant, such as color and odor of petals or the movement of flower parts in the wind

Autotrophic An organism that produces its own food, usually by photosynthesis; virtually all plants are autotrophic

Auxin A plant hormone (growth regulator) that influences cellular elongation, among other things; also referred to as indole-3-acetic acid, or IAA

Axil The upper angle between a twig or leaf and the stem from which it grows

Axillary Bud Bud that occurs in the axil of a leaf

Bacteria (*See* Eubacteria)

Bark The part of the stem or trunk exterior to the vascular cambium

Basidiomycetes A large and diverse group of true fungi with septate hyphae; they produce basidiospores externally on basidia

Bilaterally Symmetrical The symmetry that occurs in the plant corolla when petals do not develop equally

Binomial System of Nomenclature A system of applying two-part scientific names to organisms, each name consisting of the genus (generic) name and a species (specific) epithet

Bioenergetics The energy relationships of living organisms

Biogeography The study of geographic distributions of organisms past and present and the mechanisms that caused these distributions

Biological Species Concept A species consists of groups of actually or potentially interbreeding natural populations that produce viable offspring

Biome Similar terrestrial ecosystem in different parts of the world

Biotechnology A way of manipulating a plant's genome to make commercial products

Biotic Pertaining to living organisms

Bivalent A pair of synapsed homologous chromosomes in prophase I

Bloom For algae, an extensive and conspicuous mass of algae resulting from rapid population growth

Blowout A barren area in arctic tundra caused by wind ripping out part of a vegetation mat whose edges became exposed when pulled up by grazing animals

Body Cell One of two cells produced when the generative cell of a gymnosperm male gametophyte divides; the body cell itself later divides, producing two sperm cells

Bordered Pit A pit in which the secondary wall arches over the pit membrane

Botany The scientific study of plants and plantlike organisms

Boundary Layer The thin, moist layer of air adjacent to a transpiring leaf

Bract A structure that is usually leaflike and modified in size, shape, or color

Branch Root A root that arises from another, older root

Brassinosteroid A plant steroid hormone that is an important internal chemical signal involved in structural changes that often occur after exposing a plant to light

Brown Algae One of three main groups of algae consistently recognized by botanists; others are green algae and red algae

Bryophyta The division of nonvascular plants

Bryophyte Member of a division of nonvascular plants; the mosses, hornworts, and liverworts

Bud Scale Tough, overlapping, waterproof leaves that cover and protect buds from low temperatures, desiccation, and pathogens

Bulb A short underground stem covered by fleshy leaf-bases that store food

Bundle Sheath A layer or layers of cells surrounding the vascular bundle; in C_4 plants, the bundle sheath is photosynthetic and prominent

C_4 Pathway The pathway of photosynthesis that occurs in C_4 plants

C_3 Photosynthesis The photosynthetic pathway used by most plants and all algae

Calorie 1,000 calories; the amount of heat required to raise the temperature of 1 liter of water $1°C$; a slice of apple pie contains about 365 Calories

calorie A unit of heat; 1 calorie is the amount of heat required to raise the temperature of 1 gram of water $1°C$; 1 calorie = 4.12 joules

Calvin Cycle Series of enzymatic reactions in which CO_2 is reduced to 3-phosphoglyceraldehyde (a three-carbon compound) and the CO_2 acceptor (ribulose, 1,5-bisphosphate) is regenerated; also referred to as the reductive pentose cycle

Calyptra The covering that partially or entirely covers the capsule of some species of mosses

Calyx Collectively, all of the sepals of a single flower

CAM (*See* Crassulacean Acid Metabolism)

Capsule The sporangium of a bryophyte

Carbohydrate An organic compound consisting of a chain of carbons with hydrogen and oxygen attached, usually in a ratio of 2:1; glucose, sucrose, and starch are carbohydrates

Carbon Fixation Conversion of CO_2 to organic molecules during photosynthesis

Carnivore Animals that may eat herbivores (plant eaters) to obtain energy and building blocks for their bodies

Carotenoid Any compound in a class of yellow, orange, or red fat-soluble accessory pigments that are derived from eight isoprene units linked together; the most widespread carotenoid in plants is beta-carotene

Carpel The ovule-bearing organ of a flower; the flower of many species has more than one carpel, collectively called the gynoecium

Carrion Flower A type of flower that is foul smelling (carrion odor) and attracts flies or beetles as pollinators

Carrying Capacity The maximum number of individuals in any population of an ecosystem that can survive and reproduce

Casparian Strip The suberized layer covering the radial and transverse walls of endodermal walls

Catabolic Reaction (*See* Catabolism)

Catabolism The chemical reactions that break down complex materials

Catkin A pendulous inflorescence, usually made up of flowers of unisexual flowers; typical of such plants as willows and poplars

Cell Cycle Collectively, the repeating processes of cellular growth and division, including mitosis; the complete cell cycle occurs only in cells that divide, other cells being arrested in development at one of the phases of the cycle

Cell Division Cycle (cdc) Gene A gene that regulates transitions from the G_1 phase to the S phase or from the G_2 phase to mitosis

Cell Membrane The semipermeable membrane that surrounds the cytoplasm and is next to the cell wall; also called the plasma membrane or the plasmalemma

Cell Signaling The process of the cell wall binding molecules, such as hormones, released from other cells

Cellular Respiration The energy transformation by which cells extract energy from sugars

Cellulose A polysaccaride of glucose units that constitutes the main part of cell walls of plants; the most abundant structural polysaccharide in plants and the most abundant polymer on earth

Cellulose Fibril The intertwining of several microfibrils

Cell Wall The rigid wall that surrounds each plant cell and provides protection to the cell and gives it shape

Centromere A constricted region of a chromosome where sister chromatids are held together

Chaparral Dense Mediterranean scrub of lower elevations of the North American Pacific Coast. The unique vegetation, which is either evergreen or deciduous in summer, is well adapted to fires

Character State The form or value of a characteristic of an organism

Chlorophyll The pigment responsible for trapping light energy in the primary events of photosynthesis

Chlorophyll *a* The primary photosynthetic pigment; this grass-green pigment occurs in all photosynthetic organisms except photosynthetic bacteria and absorbs light maximally

Chlorophyll *b* A bluish green pigment that absorbs light maximally at 453 and 642 nanometers; occurs in all plants, green algae, and some prokaryotes

Chloroplast Organelle specialized for photosynthesis; chloroplasts occur in cells of aboveground parts of plants

Chromatid One of the two threads of a chromosome that has been duplicated in the S phase of the cell cycle; sister chromatids are held together at their centromere

Chromatin A DNA-protein complex that forms chromosomes

Chromosome Threadlike structure within the cell's DNA that contains genes

Chromosome Theory of Heredity A set of postulates that accounts for the association of genes with chromosomes

Circadian Rhythm Endogenous rhythmic changes occurring in an organism on a daily cycle

Circinate Vernation Irregular growth pattern common in fern fronds in which cells on the lower surface of a developing leaf grow faster than those on the upper surface; a "fiddlehead" shape results

Citric Acid Cycle (*See* Krebs Cycle)

Cladistics A method of classifying and reflecting phylogenetic relationships among organisms, based on an analysis of shared features

Cladogram A line diagram portrayal of a branching pattern of evolution, using the concepts and methods of cladistics

Cladophyll A stem or branch that resembles a leaf

Class A taxonomic category ranking between division and order

Classical Species Concept (*See* Morphological Species Concept)

Climax Community A self-perpetuating community that may become established during succession; its composition is strongly influenced by local climate and soils

Cline Gradual differences in characteristics within a population across a geographic region

Clone An individual or group of individuals that develop vegetatively from cells or tissues of a single parent individual

Codominance A condition that occurs when both alleles of a heterozygous gene are expressed equally

Codon A sequence of three nucleotides in a gene or molecule of mRNA that corresponds to a specific amino acid or to a stop signal at the end of a gene; of the 64 possible codons, 61 are codes for amino acids and 3 are stop codons

Co-evolution The close evolution of plants and animals that results from their interdependency

Coleoptile The protective sheath around the embryonic shoot in grass seeds

Coleorhiza The protective sheath around the embryonic root in grass seeds

Collenchyma A type of cell that is elongated and has an unevenly thickened primary cell wall; these cells support growing regions of shoots

Community Populations of different species living and interacting in the same location

Companion Cell A small cell adjacent to a sieve tube; thought to control the function of sieve tube member

Compartmentalization A method a tree uses to defend itself that involves walling off infections, fortifying cell walls near wounds, and making antimicrobial compounds

Competition An ecological interaction between two organisms to acquire a resource that both need and that is in limited supply

Complete Dominance A condition that occurs when the phenotype of one allele completely masks the phenotype of another allele for a heterozygous gene

Complete Flower A flower that has all four of the main parts: calyx, corolla, androecium, and gynoecium

Complex Carbohydrate The polysaccharide sugars

Composting Using partially decayed organic matter in farming and gardening to enrich the soil and increase its water-holding capacity

Compound Leaf A leaf consisting of two or more independent blades called leaflets

Cone (*See* Strobilus)

Compound Umbel An inflorescence in which each stalk of the primary umbel is terminated by another umbel, characteristic of the carrot family

Conjugation A form of sexual reproduction in protozoa

Consumer The herbivores (plant eaters) and carnivores (animal eaters) in a community of organisms

Convergence The independent evolution of similar structures in organisms that are not closely related

Cork Cambium A meristem that produces the periderm, which protects the underlying tissues of a plant

Corkscrew Shaped (Spirilla) One of three typical shapes of bacteria

Corm An elongate, upright, underground stem

Corolla Collectively, all of the petals of a single flower

Cortex Ground tissue located between the epidermis and vascular bundles of stems and roots

Corymb A flat-topped inflorescence in which the outer flowers open first

Cotyledon Seed leaf; the first leaf formed in a seed; monocots have one cotyledon, and dicots have two

Crassulacean Acid Metabolism (CAM) A type of photosynthesis in which CO_2 is fixed at night into four-carbon acids; during the day, the stomata close, and the carbon is fixed via the Calvin cycle; CAM helps plants conserve water and is often characteristic of xerophytic plants

Crossing-Over The exchange of genetic material between the chromatids of homologous chromosomes during prophase I of meiosis

Cross-Pollination The transfer of pollen from the anthers of one plant to the stigma of a flower on another plant

Cryptomonads One of three divisions of algae; others are euglenoids and dinoflagellates

Cultivar A variety of plant that is selected for cultivation through hybridization and not found in nature

Cutin The main waxy substance in a cuticle; it consists of hydroxylated fatty acids that are linked together in a complex array

Cyanide-Resistant Respiration The continued respiration in plant tissues despite the addition of cyanide

Cyanobacteria A photosynthetic group of bacteria

Cyclin Proteins that activate cell cycle enzymes and cell division cycle (cdc) genes

Cytokinesis The division of cytoplasm into distinct cells; together with mitosis, cytokinesis comprises the phases of the cell cycle involved in cell division

Cytokinin Group of hormones (growth regulators) that promote growth by stimulating cellular division

Cytoplasm All of the material inside the cell membrane but outside the nucleus

Cytoskeleton A network of microscopic filaments that form a mechanical support system in the cell

Cytosol The semifluid matrix that surrounds and bathes the organelles outside the nucleus

Day-Neutral Plant Plant whose flowering is not affected by the length of day

Deciduous Plant Plant that loses all of its leaves during autumn or the dry season

Decomposer An organism, such as bacterium or a fungus, that facilitates recycling of nutrients in an ecosystem through the breakdown of complex molecules to simpler ones

Dendrochronology The study of growth rings of trees to determine past conditions

Density The specific gravity of wood, which is the ratio of the density of the wood to that of water

Desert A biome characterized by low annual precipitation and/or porous soil, low humidity, wide daily fluctuations in temperature, high radiation, and living organisms adapted to these conditions

Deuteromycetes Fungi that have no known sexual reproduction; most reproduce by conidia, and most otherwise have characteristics of ascomycetes; deuteromycetes are also called imperfect fungi

Development A genetically programmed progression from a simpler to a more complex form

Diatom An alga with distinctive silica walls

Dichotomous Key A classification system based on plant features

Dicot A type of angiosperm that belongs to a class whose members are characterized by having two cotyledons (seed leaves) per seed; class Magnoliopsida

Dictyosome A stack of flattened, membranous vesicles that are often branched; dictyosomes are the sites where precursors of cell-wall materials and other cellular components are assembled and prepared for secretion from the cell; dictyosomes are also called Golgi bodies

Differentiation Physical and chemical changes associated with the development and/or specialization of an organism or cell

Diffuse-Porous Wood Wood in which the vessels are distributed uniformly throughout the growth layers

Diffusion The net movement of particles, either dissolved or suspended, from a region of higher concentration to a region of lower concentration; the energy of diffusion is derived from the random motion of particles that is caused by molecular motion; diffusion tends to cause the distribution of particles to become homogenous throughout a medium

Dihybrid Cross A hybridization experiment that follows the inheritance of phenotypes that are controlled by two different genes

Dinoflagellates One of three divisions of algae; others are euglenoids and cryptomonads

Dioecious Having the pollen-producing and the ovule-producing organs on different individuals of the same species; mulberry is an example of a dioecious species

Diploid The condition of having two sets of chromosomes in a nucleus

Directional Selection Selection for a phenotype that is either higher or lower in frequency than the most abundant phenotype

Disaccharide A carbohydrate composed of two monosaccharides that are linked by a covalent bond; sucrose and maltose are examples of disaccharides

Diversifying Selection Selection for the low-frequency (extreme) phenotypes above and below the norm of the population, or selection against the high-frequency phenotype (norm)

Division A taxonomic category between kingdom and class; some botanists use "phylum" instead

DNA (Deoxyribonucleic Acid) The nucleic acid containing four different nucleotides whose simple sugar is deoxyribose; genes are made of DNA; DNA exists as a double helix that can be unwound to replicate itself or to make RNA

Domain A structural and functional portion of a polypeptide, which may be encoded separately by a specific exon; a portion of a protein that has a globular tertiary structure

Dominant A trait that masks an alternative (recessive) trait when the gene for these traits is heterozygous

Dormancy/Dormant A condition in which plant parts such as buds and seeds are temporarily arrested in their development; dormancy is typically seasonal and is broken as environmental conditions change during the year

Double Fertilization In angiosperms, the process by which one sperm cell fertilizes the egg to form a zygote and another sperm cell fertilizes the polar nuclei to form a primary endosperm nucleus

Double Helix The spiral shape of a double strand of DNA

Drupe A fruit with a fleshy mesocarp

Durability The ability of wood to resist weathering and decay

Ecological Trade-off A negative aspect of a characteristic for every positive aspect

Ecological Variation Species show physiological and or morphological variation in different parts of the species' range

Ecologist A scientist who studies organisms and their relationships with the environment

Ecology The study of the interactions of organisms with one another and with their environment

EcoRI An example restriction enzyme that comes from the bacterium *Escherichia coli;* this restriction recognizes the DNA sequence GAATC and then cleaves it between the guanine and the adenine

Ecosystem A major system of organisms that are interacting with one another and with their physical environment

Egg The female sex cell

Electron Transport Chain A sequence of electron carriers that use the energy from electron flow to transport protons against a concentration gradient across the inner mitochondrial membrane

Element A substance that cannot be broken down into a simpler substance by ordinary chemical means

Embryo In plants, the part of the seed that will form the growing seedling after germination; includes a radicle, apical meristem, and embryonic leaf or leaves

Embryo Sac The common name for the female gametophyte of flowering plants

Enation Flap of tissue extending from a stem; possibly the origin of microphylls

Endemic An organism that lives in one geographic location and nowhere else

Endocarp The innermost layer of simple fleshy fruits; the endocarp can be soft, as in tomatoes, or hard and stony, as in peaches

Endodermis Layer of cells inside the cortex and outside the pericycle of roots; radial walls of the endodermis are suberized by the casparian strip

Endoplasmic Reticulum (ER) An extensive network of sheetlike membranes distributed throughout the cytosol of eukaryotic cells; portions that are densely coated with ribosomes are the rough ER; other regions, with fewer ribosomes, are the smooth ER

Endosperm The nutritive storage tissue that grows from the fusion of a sperm cell with polar nuclei in the embryo sac

Endosymbiotic Hypothesis An explanation for the origin of chloroplasts and mitochondria from the descendants of prokaryotes that lived symbiotically in larger prokaryotic hosts

Energy The ability to do work or cause change

Entropy The degree of orderliness in a system

Environmental Biology The study of the effects of population growth on the environment

Enzyme A biological catalyst, usually a protein, that can speed up a chemical reaction by lowering its energy of activation; amylase is an example of an enzyme

Epicotyl The region of an embryo above the attachment point of cotyledons

Epidermis The outermost layer of cells that covers a plant

Epiphyte An organism that is attached to another organism without parasitizing it

Essential Element An element that is required for normal growth and development, that no other element can replace and correct the deficiency, and that has a direct or indirect action in plant metabolism

Ethylene A gaseous plant hormone (growth regulator) that promotes fruit ripening and other physiological responses

Etiolation The abnormal elongation of stems caused by insufficient light; etiolated stems usually lack chlorophyll

Eubacteria The majority of all bacteria; their cell walls contain muramic acid, certain lipids, and other features that distinguish them from archaebacteria

Euglenoids One of three divisions of algae; others are dinoflagellates and cryptomonads

Eukarya (Eucarya) The classification of all eukaryotes; one of three domains in nature

Eukaryote An organism composed of one or more cells containing visibly evident nuclei and organelles

Evergreen A tree such as a pine or oleander whose leaves live 3 to 5 years and can be shed anytime during the year, but not all at once

Exocarp The outermost layer (usually the skin) of simple fleshy fruits

Evolutionary Species Concept The method of defining a species as a group of individuals with a common evolutionary lineage and a common evolutionary fate through time

Exocytosis The process of expelling the contents of a vesicle from a cell by fusing the vesicle membrane with the cell membrane and opening the inside of the vesicle to the outside of the cell

Exon A sequence of DNA within a gene that codes for an amino acid sequence

F₁ Generation The first generation of offspring

Facilitated Diffusion Passive transport through a transport protein

Fascicle A cluster of pine leaves (needles) or other needlelike leaves of gymnosperms

Fatty Acid A long, mostly hydrocarbon chain that has an organic acid group at one end; the most common fatty acids in plants are oleic acid, linoleic acid, and linolenic acid

Female Gametophyte The embryo sac in a plant

Fermentation A process by which energy is obtained from organic compounds without the use of oxygen as an electron acceptor

Fertilization The fusion of sperm and egg in sexual reproduction

Fiber An elongated, thick-walled sclerenchyma cell; helps support or protect the plant

Fibrous Root System Consists of an extensive mass of similar-sized roots

Fiddlehead The curled fern frond prior to unrolling and elongation; also known as a crozier

Filament The stalk of a stamen; or, the vegetative body of filamentous algae and fungi

First Law of Thermodynamics The law of conservation of energy; this law states that energy cannot be created or destroyed but only converted to other forms

Fitness A measure of an individual's evolutionary success; number of its surviving offspring relative to the number of surviving offspring of other individual's within the population

Flagellum (plural, **Flagella**) A hairlike locomotor organelle that protrudes from the cell into the medium surrounding it; flagella enable cells to swim, but the only swimming cells of plants are the sperm cells of some plant groups; flagella also occur in algae, fungi, bacteria, and animals

Flavonoid Any phenylpropanoid-derived compound that is linked to three acetate units and condensed into a multiple-ringed structure; the most common flavonoid is rutin; flavonoids also include naringin, which is a bitter substance in grapefruits

Flora The plants or organisms (other than animals) of a particular region; also a publication devoted to the taxonomy of plants of a particular region

Floriculture The business of growing and selling ornamental potted plants, flowers, fruits, and seeds

Flowerpot Leaves Type of hollow, modified leaf in which falling debris and water may collect

Fluid Mosaic Model A model for the structure of membranes as a fluid phospholipid bilayer through which proteins float in a continually shifting mosaic pattern

Food Chain An interlocking flow of energy involving producers and consumers in an ecosystem

Food Web (*See* Food Chain)

Foot Basal part of a moss sporophyte; the foot is embedded in the gametophyte

Form and Function The idea that the form, or shape, of a structure often reveals, and is directly related to, its function

Frond Photosynthetic leaf blade of a fern

Functional Megaspore The megaspore that, in some types of embryo sac development, is the only one of the four meiotic products to grow into a female gametophyte; the other three spores disintegrate

Fungi One of six kingdoms in the modern system of classification of organisms; others are Animalia, Plantae, Protista, Eubacteria, and Archaebacteria

G_1 Phase During interphase, the portion of the cell cycle that occurs between the end of mitosis and the onset of DNA synthesis; G_1 refers to first gap

G_2 Phase During interphase, the portion of the cell cycle that begins at the end of the S phase and lasts until the beginning of mitosis; G_2 refers to the second gap

Gametangium (plural, **Gametangia**) A cell or structure in which gametes are produced

Gamete A haploid reproductive cell that fuses with another gamete to form a zygote; the female gamete is an egg, and the male gamete is a sperm; in certain kinds of algae and fungi, however, the gametes are neither male nor female

Gametophyte The phase of the plant life cycle that produces gametes

Gel Electrophoresis A technique by which nucleic acids or proteins are separated in a gel that is placed in an electric field

Gemmae (singular, **Gemma**) Asexual plantlets in certain liverworts that can form new gametophytes; often form in gemmae cups

Gemmules An erroneous concept of inheritance; described as packets of heritable information produced throughout a mature organism, transported to the reproductive organs, and packaged into gametes before fertilization

Gene A sequence of DNA that codes for a molecule of mRNA, tRNA, or rRNA or that regulates the transcription of such codes; a gene is the basic unit of heredity

Gene Flow Introduction of genetic material into the gene pool of one population from another population

Gene Pool All of the alleles within a population that are available to future generations

Generative Cell The cell in the pollen grains of angiosperms that divides to form two sperm cells, or the cell in the pollen grains of gymnosperms that divides to form a sterile cell and another cell that divides to form sperm cells

Genetic Code A system of codons (nucleotide triplets) in DNA or RNA that together code for a sequence of amino acids in a polypeptide; 61 of the possible codons are codes for amino acids, the remaining 3 being stop codons that are not translated

Genetic Drift Random changes in gene frequencies within the gene pool of a population

Genetic Engineering The artificial manipulation of genes, or the transfer of genes from one organism to another; synonymous with recombinant DNA technology

Genetic Linkage Linked genes

Genetic Recombination Producing offspring that are genetically different from the parents

Genetic Theory of Evolution A collection of many of Darwin's views on evolution as well as modern ideas about inheritance and genetic change within populations

Genotype An organism's genes, either individually or collectively

Geographic Isolation The separation of populations due to geography

Gibberellin (GA) A type of plant hormone (growth regulator) that affects, for example, stem elongation and seed germination

Girdled The state of a tree when a strip of bark is removed around its entire circumference; girdling usually kills a tree

Glycolysis The anaerobic metabolic pathway by which glucose is broken down into two molecules of pyruvic acid in the cytosol; the substrate-level phosphorylation of two molecules of ADP to ATP and the reduction of two molecules of NAD^+ to NADH occur for the breakdown of each molecule of glucose

Glycoside A molecule combining a secondary metabolite with one or more sugars

Glyoxysome A type of microbody that is common in germinating oilseeds and seedlings that arise from them; glyoxysomes contain enzymes that catalyze the breakdown of fatty acids into acetyl CoA

Golgi (*See* Dictyosome)

Grain In wood, the direction of axial cells relative to the longitudinal axis of the tree; a single plant fruit, such as each kernel on a cob of corn

Grana (singular, **Granum**) Stacks of thylakoids where the photochemical (i.e., "light") reactions of photosynthesis occur

Grassland A biome characterized by the predominance of grasses

Gravitropism The curvature of roots or stems in response to gravity

Green Algae One of three main groups of algae consistently recognized by botanists; others are brown algae and red algae

Green Revolution The dramatic increase in food production since 1950

Ground Tissue Tissue that occurs throughout a plant and whose functions include storage, metabolism, and support

Growth Any irreversible increase in size of an organism and its parts

Growth Ring A growth layer in secondary xylem or secondary phloem, as seen in cross section

Gum A hemicellulose that is secreted by plants, which consists of several kinds of monosaccharides; an example is gum arabic, which is a mixture of the monosaccharides arabinose, galactose, glucose, and rhamnose

Gum Arabic A gum produced by the plant species *Acacia senegal;* this gum is a hemicellulose, which in this case is a complex branched chain consisting of arabinose, galactose, glucose, and rhamnose

Guttation The exudation of liquid water from leaves; caused by root pressure

Gymnosperm A plant also known as an evergreen, such as a pine or fir

Gynoecium (plural, **Gynoecia**) Collectively, all of the carpels of a single flower

Habitat The location, with its own specific set of environmental conditions, where an organism naturally occurs

Half-Life Time required in a chemical reaction for half the original reactant material to decay or be consumed

Haploid The condition of having only one set of chromosomes in a nucleus

Hardwood A woody dicot

Head A short, dense inflorescence, usually made up of many, small flowers, typical of such plants as sunflowers

Heartwood Wood in the center of a tree trunk; usually darker due to the presence of resins, oils, and gums; does not transport water and solutes

Heme A complex organic ring structure, called a protoporphyrin, to which an iron atom is bound; heme occurs in the cytochromes of all organisms and in the hemoglobin of animals

Hemicellulose Primarily a cell-wall polysaccharide of variable composition and structure; hemicellulose that is secreted by plants is also called a gum; (*See* Gum and Gum Arabic)

Herbaceous Nonwoody

Herbarium A systematically arranged collection of dried, pressed, and mounted plant specimens

Herbivore Animals that depend entirely on plants for their energy and building blocks for their bodies

Heterocyst A relatively large, unpigmented, thick-walled, nitrogen-fixing cell that is produced within the filaments of certain cyanobacteria

Heterosis A condition in which crossbred organisms are more fit than inbred organisms because they have more heterozygotic loci

Heterotroph/Heterotrophic An organism that obtains its food from other organisms

Heterozygote Superiority A condition in which individuals heterozygous at one or more loci have higher fitness than an individual with fewer heterozygous loci

Heterozygous A condition in which a gene has two different alleles in a diploid individual

Histone A type of protein that comprises the protein component of chromatin

Homologous Features that have the same evolutionary origin and come from the same ancestor

Homologous Chromosomes Chromosome pairs that have alleles for the same genes

Homozygous A condition in which both alleles of a gene are the same in a diploid individual

Horizon A major layer of soil visible in vertical profile

Host The living organism from which another living organism, known as a parasite, obtains its energy and nutrients

Humus The organic portion of soil; derived from partially decayed plant and animal material

Hybrid The offspring of different parents with different traits

Hybrid Vigor (*See* Heterosis)

Hydrolysis Any chemical reaction that proceeds by the addition of water to break down a molecule; the breakdown of starch by amylase and the breakdown of sucrose by invertase are examples of enzyme-catalyzed hydrolyses

Hydrophilic Refers to chemicals that are freely soluble in water; sugars are examples of hydrophilic compounds

Hydrophobic Refers to chemicals that are not soluble in water but are soluble in nonpolar solvents; lipids and hydrocarbons are generally hydrophobic

Hydrophyte A plant that is adapted to submersion in water or an aquatic environment for at least part of its growing season

Hydroponically A method of growing plants in a nutrient-rich solution rather than in soil to make economic use of root protein secretion

Hyperaccumulator A plant that concentrates certain chemicals in its body at levels 100 times greater than normal

Hypha (plural, **Hyphae**) A single tubular thread of the mycelium of a fungus or similar organism

Hypocotyl The region of an embryo that is between the radicle and the attachment point of the cotyledons

Hypodermis One or more layers of cells just beneath the epidermis that are distinct from the underlying cortical or mesophyll cells

Hypothesis A proposed solution to a scientific problem that must be tested by experimentation; a working explanation based on evidence and suggesting some principle; if disproved, a hypothesis is discarded

Imperfect Flower A flower that lacks either an androecium or a gynoecium

Imperfect Fungi (*See* Deuteromycetes)

Inbreeding Mating within the same plant or between the offspring of an inbred parent

Inbreeding Depression The less vigorous and less productive characteristics of offspring that result from inbreeding

Incomplete Dominance A condition that occurs when the phenotype of one allele only partly masks the phenotype of another allele for a heterozygous gene

Incomplete Flower A flower that has one or more of the parts absent (calyx, corolla, androecium, or gynoecium)

Indusium An outgrowth covering a fern sorus of a fern leaf

Inferior Ovary An ovary located below the other flower parts on a floral axis

Inflorescence A cluster of flowers that are arranged on their axis in a specific pattern

Inheritance of Acquired Characteristics A theory that states traits acquired by an individual during its life were passed to its offspring

Initial Cell Special cells in the meristems that complete the cell cycle

Intercalary Meristem Meristem at the base of a blade and/or sheath of many monocots

Intermediate-Day Plant A plant that flowers only when exposed to days of intermediate length; these plants grow vegetatively if exposed to days that are either too long or too short

Internode Part of the stem between two successive nodes

Interphase Collectively, all of the phases of cell growth apart from cell division

Intron A sequence of DNA within a gene that does not code for an amino acid sequence

Inulin A polymer of fructose having beta-2,1 linkages; an alternative to starch as a storage polysaccharide

Island Biogeography A theory explaining the relationship between a defined habitat area (such as an island) available for organisms and the number and diversity of species in that area

Isozymes Enzymes that have the same function but are encoded from different genes

Joule The amount of energy needed to move 1 kilogram through 1 meter with an acceleration of 1 meter per second per second; 10^7 ergs; 1 watt-second; a slice of apple pie contains about 1.5×10^6 joules

Juvenile Hormone A plant substance that regulates insect development; when the insect eats the plant containing this hormone, its life is altered, usually to benefit the plant

Kinetic Energy The energy of motion; a solute that moves down its concentration gradient has kinetic energy

Kingdom The highest taxonomic category (unless domains are used)

Knot In wood, the base of a branch that has been covered by lateral growth of the main stem

Kranz Anatomy Specialized leaf anatomy characteristic of C_4 plants; characterized by having vascular bundles surrounded by a photosynthetic bundle sheath

Krebs Cycle The metabolic pathway by which acetyl CoA is oxidized in mitochondria to carbon dioxide; each turn of the Krebs cycle also forms one ATP by substrate-level phosphorylation, reduces one NAD^+ to HADH, and reduces one ubiquinone to ubiquinol; the Krebs cycle is also called the citric acid cycle or the tricarboxylic cycle

Landing Platform A large structure on the corolla of a flower that sticks out like a bottom lip and allows an insect to land on the flower

Lateral Meristem Meristem that produces secondary tissue; the vascular cambium and cork cambium are examples of lateral meristems

Laws of Thermodynamics The laws that regulate energy transformations; these laws involve a system and its surroundings; (*See* First Law of Thermodynamics, Second Law of Thermodynamics)

Leaflet An individual part of the dissected blade of a compound leaf

Leaf Primordium A lateral outgrowth from the apical meristem that will eventually form a leaf

Leaf Scar A scar left on a twig when a leaf falls from a stem

Lenticel Spongy areas in the cock surfaces of stems and roots of vascular plants; allows gas exchange to occur across the periderm

Liana A woody vine that is supported by other plants

Life Cycle A series of genetically programmed developmental changes in a plant; in flowering plants, this includes growth of the embryo in a seed into a mature flowering plant, formation of sperm and egg, and sexual reproduction

Light Reactions The photochemical reactions in a plant

Lignin A complex phenylpropanoid polymer that makes cell walls stronger, more waterproof, and more resistant to pests, herbivores, and disease organisms

Limiting factor An environmental factor that negatively affects the growth or behavior of an organism

Linkage The condition of having genes on the same chromosome (linked); alleles of genes that are linked tend to be inherited together

Lipid A polymer that forms water-repellent cell membranes; lipids serve as long-term storage forms of carbon and energy

Loam A soil type consisting of a mixture of sand, silt, and clay

Locus (plural, **Loci**) The position of a gene on a chromosome

Long-Day Plant Plant that flowers when the length of dark is shorter than some critical value; long-day plants flower in spring and summer

Macroevolution Evolutionary changes that refer to the development of new species

Macronutrient Inorganic element required in a large amount for plant growth (e.g., nitrogen, calcium, sulphur)

Marginal Meristem Plant tissue that forms the flattened blade and the stalklike petiole that attaches the blade to the stem

Megaspore A spore that will grow into a female gametophyte

Meiosis Nuclear division in which chromosomes are doubled and then divided twice; the daughter nuclei from meiosis have half the number of chromosomes of the parent nucleus; in plants, meiosis forms spores

Meiosis I The first of two nuclear divisions that, in plants, form spores; in meiosis I, homologous chromosomes synapse, cross over, and move to opposite poles of the meiotic spindle apparatus; the separation of homologous chromosomes in meiosis results in a reduction in chromosome number by one-half in daughter nuclei

Meiosis II The second of two nuclear divisions that, in plants, form spores; in meiosis II, centromeres divide and sister chromatids become independent chromosomes that move to opposite poles of the spindle apparatus

Mendelian Inheritance Refers to patterns of inheritance that were discovered by Gregor Mendel

Meristem Regions of specialized tissue whose cells undergo cell division

Mesocarp The middle layer (often fleshy) of simple fleshy fruits; the mesocarp occurs between the exocarp and the endocarp

Mesophyll Parenchyma tissue between the epidermal layers of a leaf; is usually photosynthetic

Messenger RNA (mRNA) A class of RNA that carries the genetic message of genes to ribosomes, where the message is translated into the amino acid sequence of a polypeptide

Metabolic Pathway The step-by-step sequence of chemical reactions in a cell; the product of one reaction becomes the starting point for another

Metabolism The sum of all chemical reactions occurring in a cell or organism

Metaphase The period of mitosis during which chromosomes become attached to spindle fibers, which align the chromosomes in a circular plane that is perpendicular to the microtubules of the spindle apparatus

Metaphase I The first metaphase of meiosis; in metaphase I, pairs of homologous chromosomes align along an equatorial plane that is perpendicular to the axis of the spindle apparatus

Metaphase II The second metaphase of meiosis; in metaphase II, chromosomes align along an equatorial plane that is perpendicular to the axis of the spindle apparatus

Microevolution Evolutionary changes that occur within a population; may eventually lead to the formation of a new species, but not as a one-time event

Microfibril A complex of cellulose molecules that are twisted together into a strong, threadlike component of cell walls

Micronutrient Inorganic element required in a small amount for plant growth (e.g., boron, copper, zinc)

Micropyle The opening in a ovule through which the pollen tube will enter in angiosperms or through which the pollen grains will enter in gymnosperms

Microspore A spore that will grow into a male gametophyte

Microspore Mother Cell A cell that will undergo meiosis and cytokinesis to produce microspores

Middle Lamella The pectin-containing layer between cells that probably acts as the glue to hold cells together

Mineral Inorganic chemical compound usually made with two or more elements

Mitochondria The organelles in which the energy stored as sugars produced during photosynthesis is converted to ATP; called the powerhouses of the cell

Mitosis The process of nuclear division in which chromosomes are first duplicated, followed by the separation of daughter chromosomes into two genetically identical nuclei; the division of nuclei; together with cytokinesis, mitosis comprises the phases of the cell cycle involved in cell division

Molecular Phylogeny A phylogeny based on molecular data

Monocot A type of angiosperm that belongs to a class whose members are characterized by having one cotyledon (seed leaf) per seed; class Liliopsida

Monoecious Having the pollen-producing and the ovule-producing organs on the same individuals

Monomer The smallest subunit that is a building block of a polymer

Monosaccharide A simple sugar that cannot be broken down by hydrolysis; glucose is an example of a monosaccharide

Morphological Species Concept Traditional concept of taxonomic species surmising that two species are considered distinct if they are sufficiently different morphologically

M Phase The events of mitosis and cytokinesis

mRNA (*See* Messenger RNA)

Mucigel Slimy material secreted by root tips to facilitate growth of the root through soil

Multicellularity Having a body consisting of many cells

Multiple Fruit A fruit such as a cob of corn

Mutation A genetic change; mutations include changes in DNA sequences of genes, rearrangements of chromosomes, and the movements of transposable elements

Mutualism/Mutualistic An interaction that is beneficial to all the organisms involved

Mycelium (plural, **Mycelia**) Collective term for the hyphae of a fungus

Mycorrhiza (plural, **Mycorrhizae**) A mutualistic association between a fungus and the roots of a plant

Mycorrhizal Associations (*See* Mycorrhiza)

NAD⁺ A coenzyme that functions as an electron and hydrogen ion carrier in cellular reactions

NADH A complex of enzymes whose function is to transport protons from NADH across the inner mitochondrial membrane

NADPH An energy-carrying molecule necessary in plant chloroplasts to form sugars produced by photosynthesis

Nastic Movement A movement that occurs in response to a stimulus but whose direction is independent of the direction of the stimulus

Natural Selection Differential reproduction of phenotypes; genotypes and phenotypes vary among organisms, and some of these phenotypes promote reproduction more than other phenotypes

Nectar A sweet exudate secreted by plants to attract insects (e.g., for pollination)

Nectar Guide Targets such as lines on brightly colored flowers that lead insects into a plant; made by UV-absorbing pigments

Nectary A structure in angiosperms that secretes nectar; usually (but not always) associated with flowers

Net Productivity The energy produced in an ecosystem by photosynthesis minus the energy lost through respiration

Netted Venation One or a few prominent midveins in a leaf from which smaller minor veins branch into a meshed network

Nitrogenase A complex of enzymes that convert atmospheric nitrogen gas into ammonia

Nitrogen Fixation Incorporation of atmospheric nitrogen into nitrogenous compounds; done by certain free-living and symbiotic bacteria

Nitrogen-Fixing Bacteria Bacteria that convert gaseous nitrogen to nitrates or nitrites

Node Point where one or more leaves attach to a stem

Nodule Tumorlike swelling on roots of certain higher plants (e.g., legumes) that houses nitrogen-fixing bacteria

Nuclear Envelope The double membrane that surrounds the nucleus

Nucleic Acid An organic acid that is a polymer of mostly four different nucleotides; deoxyribonucleic acid (DNA) and ribonucleic acid (RNA) are the two kinds of nucleic acids

Nucleosome The basic beadlike unit of chromatin in eukaryotes, consisting of DNA that is wound around a core of histone proteins

Nucleotide The subunit of a nucleic acid, consisting of a phosphate group, a simple sugar (either ribose or deoxyribose), and a nitrogen-containing base that is either a purine or a pyrimidine

Operculum The lid of the sporangium in mosses

Opposite Leave Arrangement Type of phyllotaxis, arrangement of leaves on the stem, in which there are two leaves per node and these leaves are positioned on opposite sides of the stem

Organelle A specialized part of a eukaryotic cell that carries out specific functions

Osmosis The diffusion of water or other solvent through a differentially permeable membrane

Osmotic Pressure The water potential of pure water across a membrane; osmotic pressure is an indicator of how concentrated a solution is on the other side of a membrane from pure water

Outbreeding Mating with unrelated individuals

Outcrossing Mating between different individual plants

Ovary The enlarged, ovule-bearing portion of a carpel or of a cluster of fused carpels; after fertilization, an ovary matures into a fruit

Ovule The structure that contains the female gametophyte in seed plants; the female gametophyte is surrounded by a nucellus (megasporangium tissue), which is covered by one or two integuments; when mature, an ovule is called a seed

Oxidation The loss of electrons from an atom or molecule that is involved in an oxidation-reduction (redox) reaction; oxidation removes energy from one substance, which is coupled with the simultaneous addition of energy to another substance by reduction

Oxidative Phosphorylation Phosphorylation of ADP to ATP that uses energy from a proton pump fueled by the electron transport system

Ozone A form of oxygen (O_3) in the stratosphere that, when compared with ordinary oxygen (O_2), more effectively shields living organisms from intense ultraviolet radiation

Paleobotanist Scientists who study fossil plants

Palisade Mesophyll The vertical photosynthetic cells below the upper epidermis of a leaf

Palmately Compound Leaf Type of leaf whose leaflets attach at the same point, much as fingers are attached to the palm of the hand

Panicle A type of inflorescence consisting of a branched main axis with side branches bearing loose clusters of flowers

Parallel Venation Several parallel and prominent veins in a leaf interconnect with smaller, inconspicuous veins

Parasite An organism that obtains energy and nutrients directly from another living organism known as the host

Parenchyma The tissue type characterized by relatively simple, living cells having only primary walls

Parsimony In cladistics, the shortest hypothetical pathway that provides the most likely explanation of an evolutionary event

Passive Transport The unrestricted movement of a substance through a biological membrane; the energy for passive transport is the kinetic energy of movement down a concentration gradient; it is passive because it does not require energy from cellular metabolism

Pectin A gluey polysaccharide that holds cellulose fibrils together; pectins are mostly polymers of galacturonic acid monomers with alpha-1,4 linkages

Pedicel The stalk of a flower in an inflorescence

Peduncle The stalk of a flower or of an inflorescence

Peltate Leaf A simple leaf with a petiole attached to the middle of the blade

PEP Carboxylase An enzyme involved in C_4 photosynthesis

Perfect Flower A flower that has an androecium and a gynoecium

Perfoliate Leaf A simple, sessile leaf that surrounds stems

Pericarp Refers collectively to the layers of ovary tissue in a fruit; pericarp is the preferred term for fruits whose layers cannot be easily distinguished from one another

Pericycle The layer of cells surrounding the xylem and phloem of roots; produces branch roots

Periderm The protective tissue that replaces epidermis; includes cork (phellem), cork cambium (phellogen), and phelloderm

Permafrost Permanently frozen soil

Petal One of the parts of the flower that are attached immediately inside the calyx; collectively, the petals of a single flower are called the corolla; the corolla is usually the part of the flower that is conspicuously colored

Petiole The stalklike part of a leaf that connects the blade to the stem

Pfr A form of phytochrome

Phenotype An organism's observable features, either individually or collectively; the phenotype results from the interaction of the genotype of an organism with its environment

Pheromone A chemical emitted by a plant that attracts insects

Phloem Vascular tissue that transports water and organic solutes

Phospholipid A lipid that has two fatty acids and a phosphate group bound to a molecule of glycerol; phospholipids are important components of membranes

Photon The elementary particle of light

Photoperiod Response to the duration and timing of day and night; the system within plants that measures seasons and coordinates seasonal events such as flowering

Photophosphorylation Production of ATP in an illuminated chloroplast

Photorespiration The light-dependent formation of glycolic acid in chloroplasts and its subsequent oxidation in peroxisomes

Photosynthesis The production of carbohydrates by combining CO_2 and water in the presence of light energy; occurs in chloroplasts and releases oxygen

Photosynthetic Autotroph Organism that uses light energy to make organic compounds from inorganic compounds

Photosystem I The complex that contains P700, the chlorophyll that absorbs maximally at 700 nanometers

Photosystem II The complex that contains P680, the chlorophyll that absorbs maximally at 680 nanometers

Phototropism Growth of a stem or root toward or away from light

Phycocyanin A blue photosynthetic pigment in cyanobacteria and red algae

Phycoerythrin A red phycobilin

Phyllotaxis The arrangement of leaves on a stem

Phylogenetic Reflecting evolutionary relationships

Phytochrome A group of proteinaceous pigments involved in phenomena such as photoperiodism, the germination of seeds, and leaf formation; absorbs red and far-red light

Phytoremediation A process of removing toxic substances from the soil that involves, in part, the use of hyperaccumulators

Pigment Molecule that reflects and absorbs light at particular wavelengths

Pinnately Compound Leaf Type of leaf whose leaflets form pairs along a central, stalklike rachis

Pioneer Plant The first plant to become established on new soil

Pistil A female reproductive structure of a flower; composed of one or more carpels and consisting of ovary, style, and stigma

Pistillate Flower A flower whose reproductive parts consist only of carpels; the kernel-bearing flowers on corn cobs are examples of carpellate flowers

Pith Parenchyma tissue in the center of a stem; located interior to the vascular bundles

Plankton Microscopic organisms that drift in large populations in water

Plantae One of six kingdoms in the modern system of classification of organisms; others are Animalia, Fungi, Protista, Eubacteria, and Archaebacteria

Plant Growth Regulator Plant hormone

Plant Hormone Small organic molecule produced in plants that serves as a highly specific chemical signal between cells

Plant Mineral Nutrition The study of how plants use essential elements

Plasmid A small, circular fragment of DNA in bacteria; a plasmid can be integrated into and replicated with the rest of the bacterial genome; because of their ability to take up foreign DNA, bacterial plasmids are used as vectors for genetic engineering and research

Plasmodesma (plural, **Plasmodesmata**) A tiny, membrane-lined channel between adjacent cells

Plasmolysis Shrinkage of cytoplasm away from the cell wall due to the loss of water by osmosis

Polar Nuclei Nuclei that come from opposite poles of the embryo sac and fuse with a sperm cell to form the primary endosperm nucleus

Pollen Grain A male gametophyte that is surrounded by a microspore wall in seed plants

Pollen Tube The germination tube of a pollen grain, which grows from the stigma, through the style, and into the micropyle of the ovule; the pollen tube carries the sperm cells to the embryo sac

Pollination The transfer of pollen from anthers to the stigma in angiosperms or from male cones directly to the ovule in gymnosperms

Pollination Droplet A sticky exudate at the mouth of the micropyle of a gymnosperm ovule; pollen grains catching in it are slowly withdrawn to the interior (pollen chamber) as the droplet recedes

Pollinivory Pollen consumption

Polymer A molecule consisting of many identical or similar monomers linked together by covalent bonds

Polymerase Chain Reaction (PCR) A procedure by which free nucleotides are assembled into a nucleic chain in a test tube by enabling the activity of a bacterial DNA polymerase to bind them together; the PCR is cycled 30 or more times to produce a millionfold amplification of the target DNA sequence

Polyploid/Polyploidy A condition in which a nucleus has more than two complete sets of chromosomes

Polysaccharide A carbohydrate polymer composed of many monosaccharides that are linked covalently into a chain; polysaccharides include starch, glycogen, and cellulose

Pome An accessory fruit

Population A group of interbreeding individuals of the same species usually occupying the same territory at the same time

Population Genetics The application of genetic laws and principles to entire populations; assumes that evolution is the result of progressive change in the genetic composition of a population rather than individuals

Positively Gravitropic The growth of primary roots as primarily down

Potential Energy The energy stored by matter because of its location or configuration; regarding a solute, the higher its concentration, the steeper is its concentration gradient and the greater is its potential energy; energy available to do work

Pr A form of phytochrome

Primary Cell Wall The usually thin cell wall that forms during cell division; it is part of all but some sperm cells in plants

Primary Growth Growth resulting from the activity of apical meristems

Primary Structure The sequence of amino acids in a protein

Primary Succession Succession that is initiated on bare rock or in water after a disturbance has occurred

Primary Thickening In some monocots, the meristem that increases the thickness of the shoot axis

Primary Xylem The supporting and water-conducting tissue of vascular plants, consisting primarily of tracheids and vessels that derive from primary growth of primary meristems

Procambium A type of undifferentiated plant tissue that gives rise to vascular tissue

Producer An autotrophic organism; producers form the base of food chains in an ecosystem

Prokaryote A cellular organism that does not have a distinct nucleus

Prophase The period of mitosis during which chromosomes condense, first appearing as a mass of elongated threads and later as individual chromosomes

Prophase I The first prophase of meiosis; in prophase I, homologous chromosomes condense, synapse, cross over, and desynapse; chiasmata move to the ends of chromosomes by the end of prophase I

Prophase II The second prophase of meiosis; in prophase II, chromosomes condense, the nuclear envelope disintegrates, and a spindle apparatus is assembled; in many organisms, prophase II is bypassed if telophase I is also bypassed, in which case the meiotic nuclei go directly from anaphase I to metaphase II

Prop Root Adventitious root that forms on a stem above the ground; helps support the plant (as in corn)

Protein A polymer consisting of one or more long chains of monomers linked by chemical bonds between adjacent carbon and nitrogen atoms

Protoctista One of five kingdoms in the modern system of classification of organisms; others are Animalia, Fungi, Monera, and Plantae

Protonema (plural, **Protonemata**) The early, filamentous growth of the gametophyte of bryophytes and ferns

Protoplast The entire plant cell, from the cell membrane inward

Pseudocopulation The transfer of pollen from one flower to another by an insect, which gains nothing from this type of pollination

Pulvinus Jointlike thickening at the base of a petiole; involved in movements of a leaf (or leaflet)

Punctuated Equilibrium A model stating that long geologic time periods with little or no evolutionary change are punctuated by periods of rapid evolution

Punnett Square A gametic grid that is used to show the expected genotypic and phenotypic ratios resulting from a hybridization experiment

Pyrenoid A spherical protein body imbedded in algal chloroplast; surrounded by starch granules and containing RuBP carboxylase; oxygenase involved in photosynthetic starch production

Raceme A simple, elongated inflorescence that contines to grow and produce new flowers at the apex

Rachis The stalklike structure on a leaf along which pairs of pinnately compound leaves attach

Radially Symmetrical The symmetry in the plant corolla that occurs when all petals develop equally

Radicle The root of an embryo

Ray Cell Cell that transports water and dissolved minerals radially in the stem

Reaction Center A chemically unique and special energy-collecting molecule of chlorophyll *a* and associated proteins

Receptacle The region of the floral shoot where the parts of the flower are attached

Recessive A trait that is masked by an alternative (dominant) trait when the gene for these traits is heterozygous

Recombinant DNA Technology (*See* Genetic Engineering)

Red Algae One of three main groups of algae consistently recognized by botanists; others are green algae and brown algae

Reduction The gain of electrons by an atom or molecule that is involved in an oxidation-reduction (redox) reaction; reduction involves the addition of energy to one substance, which is coupled with the simultaneous removal of energy from another substance by oxidation

Reduction Division A synonym for meiosis, specifically for meiosis I

Reproductive Barriers Various mechanisms that prevent reproduction between individuals, usually from different species

Reproductive Isolation The restriction of gene flow into a population

Resin A thick, translucent, combustible organic fluid usually secreted into resin ducts in pines and many other seed plants

Resin Canal A space in a tree that is filled with resin secreted by living parenchyma cells that line the canals

Reward A characteristic that causes an animal to visit flowers of the same kind and move from plant to plant, such as nectar and pollen

Rhizoid Hairlike extension of individual epidermal cells anchoring a thallus and increasing surface area for water absorption

Rhizome A fleshy, horizontal underground stem

Rhizosphere The narrow zone of soil surrounding a root and subject to its influence

Ribosomal RNA (rRNA) The type of RNA that is a component of ribosomes

Ribosome An organelle that is responsible for protein synthesis; ribosomes consist of ribosomal RNA (rRNA) and proteins that are arranged into two subunits, one large and one small

Ribozyme A sequence of RNA that has enzymatic properties; first named from a self-splicing intron

Ring-Porous Wood Wood having larger vessels in early wood than in late wood, thereby producing a ring when viewed in a cross section of wood

RNA (ribonucleic acid) The nucleic acid containing four different nucleotides whose simple sugar is ribose; molecules of RNA, which are made as complements of DNA segments called genes, function in protein synthesis

Rod shaped (bacilli) One of three typical shapes of bacteria

Root Cap A structure that covers the root meristem; helps the growing root penetrate the soil

Root Hairs Epidermal cells just behind the zone of elongation in roots; increase the absorptive surface area of the root

Root Pressure Positive pressure of roots that forces water up the stem; caused by active movement of minerals in root cells that draws water into the vascular system

Rosette Plant A plant whose stem does not elongate; these plants have very short internodes with tightly packed leaves

Rosin The hard, brittle component of resin remaining after volatile parts have been removed

Rough ER A major area of protein synthesis in a cell

rRNA (*See* Ribosomal RNA)

Rubisco A large, complex enzyme that catalyzes the carbon-fixation reaction

RuBP A five-carbon sugar; ribulose-1,5-bisphosphate

Runner (*See* Stolon)

Saprophytic Obtaining food directly from nonliving organic matter

Sapwood Wood found between the vascular cambium and the heartwood; transports water and solutes

Scarification The cutting, abrading, or otherwise softening of the seed coat to induce the seed to germinate

Sclereids Sclerenchyma cells found in tissues varying from pear fruits to the hard shells of some nuts

Sclerenchyma A type of cell that is rigid and produces thick, nonstretchable secondary cell walls; these cells support and strengthen nongrowing, nonextending regions of plants, such as mature stems

Secondary Cell Wall The cell wall that forms interior to the primary cell wall only after cell division is completed; it is restricted to certain cells and often contains lignin

Secondary Growth Growth derived from lateral meristems (e.g., the vascular cambium and cork cambium)

Secondary Metabolite Chemicals that occur irregularly or rarely among plants and that have no known role in plant cells

Secondary Phloem Phloem derived from the outer vascular cambium

Secondary Succession Succession in habitats where the climax community has been disturbed or removed

Secondary Xylem Xylem formed by the vascular cambium; wood

Second Law of Thermodynamics The law of entropy, or disorder; this law states that all energy transformations are inefficient; that is, the amount of concentrated, useful energy decreases in all energy transformations

Secretory Structure Areas in a plant that secrete a variety of substances; some secretory structures are epidermal trichomes and others are parts of the ground and vascular tissues

Seed A mature ovule consisting of a seed coat that surrounds the embryo and associated tissues

Seed Bank The ungerminated but still viable seeds that occur in natural storage in soil

Seed Coat The outer layer of a seed; the seed coat develops from the integument of the ovule

Seed Ferns An extinct group of plants that were characterized by frondlike leaves and seed-bearing structures; classified together in the division Pteridospermophyta

Selection Pressure An environmental factor that promotes or retards reproductive success of a phenotype

Selectively Permeable Refers to a membrane that restricts the passage of some solutes through it

Self-Compatible Refers to the potential for successful reproduction between flowers of the same plant or between stamens and carpels of the same flower

Self-Incompatible Incapable of successful reproduction between flowers of the same plant or between stamens and carpels of the same flower

Self-Pollination The transfer of pollen from the anthers of a flower to the stigma of the same flower or to the stigmas of flowers on the same plant

Semiconservative Replication Refers to the replication of a DNA molecule wherein half of each new double strand consists of one newly synthesized strand and one strand from the parent double helix

Senescence The collective process of aging, decline, and eventual death of plant or plant part

Sepal One of the outermost parts of a flower; collectively, the sepals of a single flower are called the calyx

Separation Layer A layer that forms along the abscission zone of a leaf; cells along the layer form suberin and block water from entering the leaf

Sessile Leaf Leaf lacking a petiole; blades of sessile leaves attach directly to the stem

Seta The stalk that supports the capsule of a moss sporophyte

Sexual Reproduction The process of forming offspring, which involves meiosis and fertilization

Shade Leaves Leaves that grow in the dim light of the forest floor

Short-Day Plant Plant that flowers when the length of dark is longer than some critical value; short-day plants usually flower in autumn

Sieve Area Part of the wall of a sieve element containing many pores through which the protoplasts of adjacent sieve elements are connected

Sieve Elements Cells in the phloem that transport organic solutes; sieve cells and sieve tube members are examples of sieve elements

Sieve Plate The part of a wall of a sieve tube member that has one or more sieve areas

Sieve Pore The 1- to 2-millimeter in diameter spaces that occupy more than 50% of the area of a sieve plate

Sieve Tube A series of sieve tube members arranged end to end and connected by sieve plates

Simple Leaf A leaf having one blade, which may be lobed

Simple Sugar Single (monosaccharides) and double sugars (disaccharides)

Simple Umbel An flat-topped, umbrellalike inflorescence in which the flower stalks of the individual flowers arise from the same point

Sink Where organic solutes are being transported by the phloem; where metabolites such as sugar are used or stored

Smooth ER Endoplasmic reticulum membrane without attached ribosomes

Softwood Coniferous gymnosperm

Soil Particles The smallest pieces of soil

Solute A substance dissolved in a solution

Sori (singular, **Sorus**) Clusters of sporangia; found especially among the algae and lower vascular plants

Source Where organic compounds such as sugar are being made and loaded into the phloem

Speciation Evolutionary formation of a new species

Species (plural, **Species**) A group of similar organisms capable of, or potentially capable of, freely interbreeding and derived from a common ancestor; the scientific names of species are binomials consisting of a genus (generic) name and a specific epithet

Species Diversity The number of species and the number of individuals per species in an ecosystem

Sperm The male sex cell

Sperm Cells The male gametes and a tube cell

S Phase During interphase, the portion of the cell cycle in which DNA synthesis occurs; S refers to the synthesis of DNA

Spherical (Cocci) One of three typical shapes of bacteria

Spike A type of inflorescence consisting of an unbranched, elongated main axis whose flowers have very short or no pedicels

Spindle Apparatus Refers to the elliptically shaped collection of spindle fibers in a cell

Spindle Fibers Nearly parallel microtubules that form between the poles of dividing cells; some spindle fibers attach to chromosomes, but fibers from opposite poles mostly interact with each other; spindle fibers are believed to move chromosomes both by pulling homologous chromosomes in opposite directions and by pushing poles apart

Spine A plant leaf structure modified for protection from herbivores

Spongy Mesophyll Leaf tissue consisting of loosely arranged photosynthetic cells

Sporangium A structure within which the protoplasm becomes converted into an indefinite number of spores

Spore A small reproductive structure, usually consisting of a single cell, that is capable of developing independently into a much larger, mature, and often multicellular body

Sporogenous Tissue Spore-producing tissue

Sporophyte The phase of the plant life cycle that produces spores

Spring Wood Wood produced in the spring; usually characterized by relatively large cells

Stabilizing Selection Selection for a phenotype within the norm of a population, or selection against extreme phenotypes

Stalk Cell One of two cells produced when the generative cell of a gymnosperm male gametophyte divides; immediately before fertilization, the body cell divides, becoming two sperms

Stamen The pollen-producing part of a flower, usually consisting of an anther and a filament; collectively, the stamens of a single flower are called the androecium

Staminate Flower A flower whose reproductive parts consist only of stamens; the tassels at the tops of corn plants are examples of staminate flowers

Starch The most common storage polysaccharide in plants; a long polymer made of repeating glucose monomers

Stele The central vascular cylinder of roots and stems

Sterol A compound derived from six isoprene units linked together in a multiple-ringed structure; beta-sitosterol is an example of a plant sterol; cholesterol is a widely known example of an animal sterol

Stigma The surface of a carpel that is receptive to pollen grains and upon which the pollen grains germinate; a photosensitive eyespot found in certain kinds of algae

Stipule A leaflike appendage that occurs on either side of the base of a leaf (or encircles the stem) in several kinds of flowering plants

Stolon A stem that grows horizontally along the ground

Stoma (plural, **Stomata**) The epidermal structure consisting of two guard cells and the pore between them

Storage Leaves Fleshy, concentric leaves modified to store food

Stratification The exposure of seeds to extended cold periods before they will germinate at warm temperatures

Strobilus (plural, **Strobili**) A stem tip with several closely packed sporangia

Stroma The matrix between the grana in chloroplasts; the site of the biochemical (i.e., "dark") reactions of photosynthesis

Stump Sprouting—the ability of some shrubs to regenerate branches after a fire has destroyed the above-ground parts of the plant

Style A column of carpel tissue arising from the top of an ovary and upon which is the stigma; the style raises the stigma to a receptive position for pollen grains whose pollen tubes must grow through it

Suberin A waxy substance that occurs in cork cells and in the cells of underground plant parts; it consists of hydroxylated fatty acids that are linked together in a complex array

Succession A progression of population replacements in a specific geographic area; it is initiated with pioneer species and may continue to a relatively stable climax community

Succulent A plant having thick, fleshy leaves or stems; succulence is usually an adaptation to water or salt stress

Succulent Stem A stem with a low surface-to-volume ratio that stores large amounts of water and is common in plants growing in deserts

Sucker A sprout on the roots of some plants that forms a new plant

Sucrose The most common simple sugar; a disaccharide consisting of glucose and fructose

Summer Wood Wood produced in the summer; characterized by relatively small cells

Sun Leaves Leaves in a plant canopy that are bathed in intense light

Superior Ovary An ovary located above the other flower parts on a floral axis

Surface Area The part of a plant in contact with its environment

Surface Litter The top layer covering the soil, typically only a few centimeters thick and consisting of fallen leaves

Surroundings The external circumstances, conditions, and objects that affect existence and development; the environment

Survival of the Fittest The concept that those organisms best adapted to the environment produce the most offspring

Synergid A type of cell that occurs next to the egg in an embryo sac; sperm cells entering the embryo sac first pass through one of the synergids

System A naturally occurring group of objects or phenomena; the solar system is an example; a set of objects or phenomena grouped together for classification or analysis

Systematics The classification of organisms into a hierarchy of categories (taxa) based on evolutionary interrelationships

Systematist A scientist who studies the phylogenetic and taxonomic relationships of organisms

Taiga A coniferous forest biome adjacent to arctic tundra in large areas of North America and Eurasia

Taproot A relatively large primary root that produces secondary roots

Taproot System A type of root system consisting of a large taproot and smaller branch roots; common in cone-bearing trees and dicots

Taxon (plural, **Taxa**) Any taxonomic category, such as species or family

Taxonomy The classification, description, and naming of organisms

Telophase The period of mitosis during which chromosomes seem to mimic prophase in reverse; chromosomes steadily elongate and decondense back into diffuse chromatin, and each new daughter nucleus becomes surrounded by a nuclear envelope

Telophase I The first telophase of meiosis; in telophase I, chromosomes decondense, the spindle apparatus disintegrates, and a new nuclear envelope forms around each daughter nucleus; in many organisms, telophase I is bypassed, and the meiotic nuclei go directly from anaphase I to metaphase II

Telophase II The second telophase of meiosis; in telophase II, chromosomes decondense, the spindle apparatus disintegrates, and a new nuclear envelope forms around each of the four new daughter nuclei

Temperate Deciduous Forest A biome dominated by deciduous hardwood trees and located in areas with temperate climates

Tendril A modified leaf or stem in which only a slender strand of tissue constitutes the entire structure

Tepal When sepals of a plant are indistinguishable from the petals except for their location

Texture The arrangement and size of the pores in the wood of a tree

Thallose A plant body undifferentiated into stem, root, or leaf

Thallus (plural, **Thalli**) A body not differentiated into roots, stems, or leaves, as seen in lichens and liverworts

Theory of Biological Classification The main postulates of this theory are that (1) organisms exhibit different degrees of similarities and differences among individual and groups; (2) those organisms that are similar in nearly all respects are a species; (3) species that share some of their features comprise a genus; (4) on the basis of their shared features, similar genera can be organized into a family, and likewise families and larger groups can be organized into successively higher levels of taxonomic hierarchy; (5) the greater the similarity among organisms and among groups, the closer their evolutionary relationship

Theory of Descent with Modification One of three theories of evolution proposed by Darwin that explains the pattern of biological diversity via evolution; (*See* Theory of Natural Selection, Theory of Pangenesis)

Theory of Evolution The most comprehensive theme in biology, which explains how the first forms of life diversified into the organisms of today and how organisms are related to one another

Theory of Inheritance Gregor Mendel's theory of how genetic characteristics are passed from one generation to the next

Theory of Natural Selection One of three theories of evolution proposed by Darwin that explains the primary mechanism by which evolution occurs; (*See* Theory of Descent with Modification, Theory of Pangenesis)

Theory of Pangenesis One of three theories of evolution proposed by Darwin that was an incorrect explanation of inheritance; (*See* Theory of Descent with Modification, Theory of Natural Selection)

Thigmotropism A response to contact with a solid object

Thorn A modified woody stem that terminates in a sharp point

3-PGA A three-carbon acid; 3-phosphoglyceric acid

Thylakoid A saclike membranous structure of the grana of chloroplasts; thylakoids house chlorophylls

Tonoplast The membrane that surrounds a vacuole; also called a vacuolar membrane

Topsoil The layer of soil just beneath the surface litter, which usually extends 10 to 30 centimeters below the surface; this layer has a pH near 7 and contains 10% to 15% organic matter

Totipotent Refers to the notion that every cell has the same genes and therefore the same genetic potential to make all other cell types

Tracheid Elongated, spindle-shaped cell that transports water in the xylem; also helps support the plant; occurs in virtually all vascular plants

Trait A genetically determined characteristic or condition

Transcription The synthesis of a molecule of RNA as a complement to a specific sequence of DNA; transcription occurs in the nucleus

Transfer RNA (tRNA) A type of small RNA molecule that binds to a specific amino acid and to a codon on messenger RNA; it is called transfer RNA because it is associated with the transfer of amino acids to mRNA in ribosomes; more than 40 different tRNA molecules have been found, at least one for each protein amino acid

Transgenic Refers to cells or organisms that contain genes that were inserted into them from other organisms by genetic engineering

Translation The synthesis of an amino acid sequence from a specific sequence of codons along a molecule of mRNA; translation occurs in ribosomes

Transpiration Evaporation of water from leaves and stems; occurs mostly through stomata

Transpiration-Cohesion Hypothesis A hypothesis formulated more than a century ago about how water is transported up a plant; the plant functions like a wick—water evaporating from the leaves draws water up from the roots and the soil

Transport Protein A type of membrane protein that enables the transport of specific solutes across the membrane

Transposable Element A fragment of DNA that is able to multiply and move spontaneously among an organism's chromosomes

Triacylglyceride A combination of a molecule of glycerol with three long-chained organic fatty acids linked to the glycerol by acylglyceride linkages

Trichome An epidermal outgrowth (e.g., a hair or scale)

Triglyceride A naturally occurring molecule of three fatty acids and glycerol that is the chief constituent of fats and oils

Triose-P Triose-phosphate

Triploid Cell The cell formed by the fusion of sperm and two polar nuclei

Trophic Level The grouping of organisms in a community according to their use of energy and nutrients, such as producer, herbivore, carnivore, and decomposer levels

Tropical Rain Forest An endangered tropical biome with exceptional diversity of species

Tropism A response to an external stimulus in which the direction of the response is determined by the direction from which the stimulus comes; tropisms such as phototropism and gravitropism are produced by differential growth

True-Breeding Refers to purebred strains for a given trait, which means the gene for that trait is homozygous

Tube Cell The cell in the pollen grains of seed plants that develops into the pollen tube

Tuber A fleshy, underground stem having an enlarged tip (e.g., potato tuber)

Tundra A vast biome primarily above the Arctic Circle and above the timberlines of mountain ranges further south, whose vegetation includes no typical trees

Turgor Pressure The pressure on a cell wall that is created from within the cell by the movement of water into it

Turpentine The volatile, combustible component of resin

Twining Shoot A plant structure that coils around an object and helps support that plant

Vacuole A membrane-bounded organelle that is filled mostly with water but also may contain water-soluble pigments and other substances; the vacuolar membrane is called the tonoplast

Vascular Bundle Strand of tissue containing primary phloem and primary xylem (and possibly procambrium); often enclosed by a bundle sheath

Vascular Cambium A meristem sandwiched between the xylem and the phloem within each vascular bundle of a dicot stem; this meristem produces new vascular tissue

Vascular Tissue Tissue specialized for long-distance transport of water and minerals; xylem and phloem

Vector In genetics, a virus or bacterial plasmid that can take up a foreign gene and integrate it into the genome of a target organism; in plant reproduction, an animal that carries pollen (a pollinator) from a pollen sac to a stigma or to an ovule

Vegetative Reproduction Reproduction that occurs when tissues or cells from a plant are removed and induced to grow into a new plant that is a clone of the old one; asexual reproduction, which does not involve sperm and egg

Vein Vascular bundle that forms part of the connecting and supporting tissue of a leaf or other expanded organ

Venter The swollen base of an archegonium containing the egg

Vesicle A package of the proteins and other molecules that is the product of the rough ER or smooth ER

Vessel A tubelike structure in the xylem that consists of vessel elements placed end to end and connected by perforations; vessel elements conduct water and minerals; found in nearly all angiosperms and a few other vascular plants

Water Content The water in wood, which accounts for as much as 75% of the weight of a living tree

Water Potential The potential energy of water to move down its concentration gradient; water potential is expressed in units of pressure instead of units of energy because pressure is simpler to measure

Waxes Complex mixtures of fatty acids linked to a long-chain alcohol; also contain free fatty acids with hydroxyl groups and long-chain hydrocarbons

Whorled Leaf Arrangement Type of phyllotaxis, arrangement of leaves on the stem, in which there are more that two leaves per node

Wilt A limp condition of a plant caused by insufficient water in the cells

Window Leaves Leaves shaped like tiny ice-cream cones that grow mostly underground, with only a small, transparent "window" tip protruding just above soil level

Wood Secondary xylem

Xanthophyll A yellow carotenoid; one xanthophyll, zeaxanthin, is the blue-light photoreceptor in shoot phototropism

Xerophyte A plant adapted for growth in arid conditions

Xylem Vascular system specialized for transporting water and dissolved minerals upward in the plant; characterized by the presence of tracheary elements

Yeast Artificial Chromosome (YAC) A yeast chromosome into which large fragments of foreign DNA (millions of base pairs) have been inserted; YACs can be replicated like native chromosomes in yeast cells, thereby cloning large amounts foreign DNA as well

Zone of Cell Division Part of the root tip that includes the apical meristem; where new root cells are produced

Zone of Cell Elongation Part of the root, adjacent to the apical meristem, in which newly formed cells elongate

Zone of Cell Maturation Part of the root in which immature, elongated root cells begin to take on specific functions

Zygomycetes A large group of fungi with primarily coenocytic mycelia; they reproduce asexually by spores produced within sporangia; sexual reproduction includes the formation of zygosporangia

Zygote The diploid cell that is formed by the fusion of two gametes

CREDITS

Credits

LINE ART

Chapter 1

Fig. 1.8: From Ricki Lewis, *Life.* Copyright © 1992 McGraw-Hill Company, Inc. All Rights Reserved. Reprinted by permission.

Chapter 3

Fig. 3.2: From Postlethwait and Hopson, *Nature of Life,* 2nd ed., 1992 McGraw-Hill Company, Inc. Reprinted by permission of the author.; **Fig. 3.4:** From *Biology: The Science of Life,* 3rd ed. by Wallace, Sanders and Ferl. Copyright © 1991 by Harpercollins Publishers, Inc. Reprinted by permission of Addison-Wesley Educational Publishers, Inc.; **Fig. 3.6:** From Purves, Orians, and Heller, *Life: The Science of Biology,* 3rd ed. Copyright © 1992 Sinauer Associates, Inc., Sunderland, MA. Reprinted by permission.; **Fig. 3.7 a, b:** From Postlethwait and Hopson, *Nature of Life,* 2nd ed., 1992 McGraw-Hill Company, Inc. Reprinted by permission of the author.; **Fig. 3.8a:** From Purves, Orians, and Heller, *Life: The Science of Biology,* 3rd ed. Copyright © 1992 Sinauer Associates, Inc., Sunderland, MA. Reprinted by permission.; **Fig. 3.9c:** From Postlethwait and Hopson, *Nature of Life,* 2nd ed., 1992 McGraw-Hill Company, Inc. Reprinted by permission of the author.

Chapter 4

Fig. 4.1b: From Kingsley R. Stern, *Introductory Plant Biology,* 5th ed. Copyright © 1991 McGraw-Hill Company, Inc. All Rights Reserved. Reprinted by permission.; **Fig. 4.2:** From Kingsley R. Stern, *Introductory Plant Biology,* 7th ed. Copyright © 1997 McGraw-Hill Company, Inc. All Rights Reserved. Reprinted by permission.

Chapter 8

Fig. 8.24: From *Botany* by Peter M. Ray, Taylor A. Steeves, and Sara A. Fultz. Copyright © 1983 by Saunders College Publishing, reproduced by permission of the publisher. May not print out or reproduce without permission of the publisher.; **Fig. 8.27:** From *Biology of Plants* by Raven, Evert, and Eichhorn. © 1971, 1976, 1981, 1986, 1992 by Worth Publishers. Used with permission.

Chapter 9

Fig. 9.22 (tree): From Ruth Bernstein and Stephen Bernstein, *Biology,* 1996, The McGraw-Hill Company, Inc. Reprinted by permission of the author.

Chapter 10

Fig. 10.10: From *Botany,* 2nd ed., by W. Jensen and F. Salisbury. © 1984. Reprinted with permission of Brooks/Cole.

Publishing, a division of Thomson Learning. Fax: 800-730-2215; **Perspective fig. 10.5:** From Eldon D. Enger and Bradley F. Smith, *Environmental Science,* 6th ed. Copyright © 1998 McGraw-Hill Company, Inc. All Rights Reserved. Reprinted by permission.

Chapter 12

Fig. 12.8: Courtesy of James E. Canright, *American Journal of Botany,* 39:488, 1952.; **Fig. 12.10b:** Courtesy of the Peabody Museum of Natural History, Yale University; **Fig. 12.16:** From Kingsley R. Stern, *Introductory Plant Biology,* 7th ed. Copyright © 1997 McGraw-Hill Company, Inc. All Rights Reserved. Reprinted by permission.

Chapter 19

Fig. 19.13: From A.R.E. Sinclair and J.M. Fryxell, "The Sahel of Africa: Ecology of a Disaster" in *Canadian Journal of Zoology,* 63:987-994, 1984. Reprinted by permission of the National Research Council of Canada.

Design

Artville/Don Bishop. This art was used as a basis for front matter page heads, end matter page heads, perspective box background texture, Chapter opener background texture and table background texture.

PHOTO

Chapter 1

Opener: © Mark Edwards/Still Pictures; **1.1:** © John Kaprielian/Photo Researchers, Inc.; **1.2a:** © Noboru Komine/Photo Researchers, Inc.; **1.2b:** Courtesy © Norfolk Lavender Ltd.; **1.2c:** © Benjamin H. Kaestner III; **1.2d:** © D. Cavagnaro/VU; **1.3a:** © L.L.T. Rhodes/Earth Scenes; **1.3b:** © Scott Camazine/Photo Researchers, Inc.; **1.4:** © Ivan Polunin/Bruce Coleman, Inc.; **1.5, 1.6a:** © Grant Heilman Photography; **1.6b:** © Cecile Brunswick/Peter Arnold, Inc.; **1.7:** © Edward Parker/OSF/Earth Scenes; **p. 9:** © Fritz Prenzel/Peter Arnold; **1.9:** © James P. Blair/National Geographic Image Collection; **1.10a:** © Michael Holford; **1.10b:** © E.S. Ross; **1.12:** © Verna R. Johnston/Photo Researchers, Inc.; **1.13a:** © Heather Angel; **1.13b:** © Brian Moser/Hutchison Library, London; **1.13c:** © Arthur N. Orans/Horticultural Photography; **1.14:** © Walter E. Harvey/National Audubon Society/Photo Researchers, Inc.; **1.17:** © Dr. Jeremy Burgess/SPL/Photo Researchers, Inc.; **1.20:** © McGraw-Hill Higher Education/Phillip Crawford, University of Oklahoma, photographer; **1.21a:** © Bettmann Archive/Corbis Images; **1.21b:** from Fedoroff, Nina. *Scientific American* 86(June 1984) p. 250. Photos: Fritz W. Goro, Courtesy Nina Fedoroff; **1.22(both):** Courtesy of Monsanto Company; **1.23:** © George Holton/Photo Researchers, Inc.

Chapter 2

Opener: © Wally Eberhart/VU; **2.1:** © Gary W. Carter/VU; **2.2:** © McGraw-Hill Higher Education/Phillip Crawford, University of Oklahoma, photographer; **2.3:** © McGraw-Hill Higher Education/Wayne Elisens, University of Oklahoma, photographer; **2.5:** © Michael Sacca/Earth Scenes; **2.6:** © Heather Angel; **2.8:** © William E. Ferguson; **2.9:** © Harry Engles/Earth Scenes; **2.10:** © Robert & Linda Mitchell; **2.11:** © Glenn Oliver/VU; **2.12:** © McGraw-Hill Higher Education/John Fletcher, University of Oklahoma, photographer; **2.13:** © Patti Murray/Earth Scenes; **2.14:** © McGraw-Hill Higher Education/Gary Hannan, Eastern Michigan University, photographer; **2.15:** © Michael Gadomski/Earth Scenes; **2.18:** © D. Cavagnaro/Peter Arnold, Inc.; **2.19:** © E. Webber/VU; **2.20:** © Biophoto Associates/Photo Researchers, Inc.; **2.21:** © William E. Ferguson; **2.22:** © Barbara J. Miller/Biological Photo Service; **2.23:** © Kingsley Stern; **2.24:** © McGraw-Hill Higher Education/Wayne Elisens, University of Oklahoma, photographer; **2.25:** © McGraw-Hill Higher Education/Sylvester Allred, photographer; **2.26:** © Bill Beatty/VU; **2.27:** © McGraw-Hill Higher Education/Gary Hannan, Eastern Michigan University, photographer.

Chapter 3

Opener: © Donald Specker/Earth Scenes; **3.1:** © Ray Coleman/Photo Researchers, Inc.; **3.11:** © Biophoto Associates/Photo Researchers, Inc.; **3.14d:** © Scott Camazine/Photo Researchers, Inc.; **3.17b:** © Mark Gibson/VU; **3.17c:** © Betsy Fuchs; **3.17d:** © Alan Pitcairn/Grant Heilman Photography; **3.19:** courtesy of Louisville Slugger Museum/Hillerich & Bradsby Co., Inc.

Chapter 4

Opener: © Dr. Jeremy Burgess/SPL/Photo Researchers, Inc.; **4.5:** Courtesy B. Galatis and K. Mitrakos (provided by James Mauseth); **4.6:** © BioPhot; **4.7 a:** © E.H. Newcomb and W.P. Wergin/Biological Photo Service; **4.9:** Micrograph first published in "Papermaking Fibers" figure 11, p. xxxii. Courtesy W.A. Coté, Center for Ultrastructure Studies, College of Environmental Science & Forestry, SUNY; **4.10:** © Biophoto Associates/Photo Researchers, Inc.; **4.12:** © Runk/Schoenberger/Grant Heilman Photography; **4.20:** © Hugh Spencer/Photo Researchers, Inc.; **4.22 a, b:** © BioPhot; **4.22 c:** © Bruce Iverson Photomicrography; **4.23:** © BioPhot; **4.24:** © Dwight R. Kuhn; **4.25:** © Bruno P. Zehnder/Peter Arnold. Inc.; **4.26:** © BioPhot; **4.27:** © Dr. Jeremy Burgess/ SPL/Photo Researchers, Inc.; **4.28:** from Heslop-Harrison, Yolanda. SEM of fresh leaves of *Pinguicula*. *Science* 167(9 Jan 1970:173, fig 1a. © 1970 AAAS; **4.29, 4.30:** © BioPhot; **4.31:** from Anderson, R, and Cronshaw, J. Sieve plate pores in tobacco and bean. *Planta* 91:173-180, 1970. © Springer-Verlag.

Chapter 5

Opener: © K.G. Murti/VU; **5.1g, h:** NIH-Science VU/VU; **5.1i:** from Paulsen, J.R. and Laemmli, U. K. *Cell* 12:817 (1977). © Cell Press; **5.1j:** courtesy Barbara Hamkalo; **5.1k:** © A.L. Olins/Biological Photo Service; **5.1l:** courtesy M.M. Schwesinger. From Novick, Richard P. "Plasmids" *Scientific American* Dec. 1980, p. 104; **5.2:** © Edwin Reschke; **5.15:** courtesy of Calgene, Inc. and Bill Santos Photography; **5.16:** © Runk.Schoenberger/Grant Heilman Photography; **5.20:** © Edwin Reschke.

Chapter 6

Opener: © Runk/Schoenberger/Grant Heilman Photography; p. 141: © Larry Simpson/Photo Researchers, Inc.; **6.3:** © Runk/Schoenberger/Grant Heilman Photography; **6.4a–c:** © Professor Malcolm B. Wilkins, Botany Dept., Glasgow University; **6.5:** © John D. Cunningham/VU; **6.6a:** © John Kaprielian/Photo Researchers, Inc.; **6.6b:** © Richard H. Gross; **6.6c:** © John Kaprielian/Photo Researchers, Inc.; **6.7a, b:** © Runk/Schoenberger/Grant Heilman Photography; **6.10a, b:** © Professor Malcolm B. Wilkins, Botany Dept., Glasgow University; **6.11:** © Willard Clay.

Chapter 7

Opener: © John D. Cunningham/VU; **7.1:** © Lynwood M. Chace/Photo Researchers, Inc.; **7.2:** © John D. Cunningham/VU; **7.3:** © William E. Ferguson; **7.4:** © Dwight R. Kuhn; **7.5, 7.6:** © BioPhot; **7.7 a:** © John D. Cunnngham/VU; **7.7b:** Courtesy J.H. Troughton and L. Donaldson and Industrial Research, Ltd.; **7.8a:** © Edwin Reschke; **7.8b:** © BioPhot; **7.11:** © Omikron/Photo Researchers, Inc.; **7.12:** © Runk/Schoenberger/Grant Heilman Photography; **7.14:** © Michael Kienitz; **7.15:** © Runk/Schoenberger/Grant Heilman Photography; **7.16:** Courtesy © Darrell Vodopich; **7.17:** © Robert & Linda Mitchell; **7.18:** © William E. Ferguson; **7.19:** © Ed Reschke/Peter Arnold, Inc.; **7.20:** © McGraw-Hill Higher Education/Stephen L. Timme, Pittsburg State University, Pittsburg, KS, photographer; **7.21:** courtesy © John G. Mexal, Edwin L. Burke, and C.P.P. Reid; **7.22:** © Runk/Schoenberger/Grant Heilman Photography; **7.24:** © Biophoto Associates/Photo Researchers, Inc.; **7.25a:** © William E. Ferguson; **7.25b:** © Kjell Sandved/Butterfly Alphabet; p. 169: © Walt Anderson/VU **7.26, 7.28:** © William E. Ferguson; **7.30a, b:** © 1991 Regents, University of California/Integrated Pest Management Education; **7.31:** © Jack Wilburn/Earth Scenes.

Chapter 8

Opener: © Eugene Fisher; **8.1a, b:** © Edwin Reschke; **8.2:** © Bruce Iverson Photomicrography; **8.3:** © Bruce Berg/VU; **8.5:** from M.M. Zimmerman. Movement of organic substances in trees. *Science* 133(13 Jan 1961):73–79, fig. 3-4. © AAAS; **8.6:** © John Lemker/Earth Scenes; **8.7:** © Gilbert S. Grant/Photo Researchers, Inc.; **8.8:** © Victor Englebert/Photo Researchers, Inc.; **8.9:** © Charles Gurche; **8.10:** © McGraw-Hill Higher Education/Phillip Crawford, University of Oklahoma, photographer; **8.11:** © Franz Krenn/OKAPIA/Photo Researchers, Inc.; **8.12:** © McGraw-Hill Higher Education/Wayne Elisens,

University of Oklahoma, photographer; **8.13:** © Bernard Photo Productions/Earth Scenes; **8.14:** © Dwight R. Kuhn; **8.16:** © BioPhot; **8.17a:** © Runk/Schoenberger/Grant Heilman Photography; **8.18:** © Jeff Gnass; **8.19:** © Grant Heilman Photography; **8.20:** © Dwight R. Kuhn; **8.21:** © BioPhot; **8.22:** From *Botany* by Peter M. Ray, Taylor A. Steeves and Sara A. Fultz. © 1983 Saunders College Publishing, reprinted by permission of the publisher; p. 195(left): © Kirtley-Perkins/VU; p.195(right): © Porterfield-Chickering/Photo Researchers, Inc.; **8.23a:** © BioPhot; **8.23 b:** © James Bell/Photo Researchers, Inc.; **8.25:** © Walter H. Hodge/Peter Arnold, Inc.; **8.26:** Earth Data Analysis Center, University of New Mexico/NASA; **8.28:** © Dwight R. Kuhn.

Chapter 9

Opener: © Oliver Meckes/Ottawa/Photo Researchers, Inc.; **9.1:** © Bill Beatty/Earth Scenes; **9.2:** © William E. Ferguson; **9.3:** © Nigel Cattlin/Holt Studios International/Photo Researchers, Inc.; **9.4:** © Alford W. Cooper/Photo Researchers, Inc.; **9.6:** © BioPhot; **9.7 a:** © Runk/Schoenberger/Grant Heilman Photography; **9.7b:** © Carolina Biological Supply/Phototake; **9.8a:** © Kjell Sandved/Butterfly Alphabet; **9.8b:** © Stephen P. Parker/Photo Researchers, Inc.; **9.9:** © Science VU/VU; **9.10:** © Richard D. Poe/VU; **9.11:** © Bruce Iverson Photomicrography; **9.12a:** © William E. Ferguson; **9.12b:** © Dwight R. Kuhn; **9.13:** courtesy J.H. Troughton and L. Donaldson and Industrial Research, Ltd.; **9.14:** © Edwin Reschke; **9.15:** © Carl W. May/Biological Photo Service; **9.16a:** © J.N.A. Lott/Biological Photo Service; **9.16b:** © J.N.A/ Lott/Biological Photo Service; **9.17:** © Edwin Reschke; **9.18:** © John D. Cunningham/VU; **9.19:** © George Wilder/VU; **9.20:** © B Meylan and B.G. Butterfield; **9.21:** © John D. Cunningham/VU; **9.24:** © Kingsley Stern; **9.25:** © BioPhot; **9.26:** © McGraw-Hill Higher Education/Phillip Crawford, University of Oklahoma, photographer; **9.27a:** © Brian Rogers/VU; **9.27b:** © Kingsley Stern; **9.28:** © Ghillean Prance/Butterfly Alphabet; **9.29:** © Dwight R. Kuhn; **9.30:** © Kingsley Stern; **9.31:** © Michael P. Gadomski/Photo Researchers, Inc.; **9.32 a, b:** © Walter H. Hodge/Peter Arnold, Inc.

Chapter 10

Opener: © Bruce Iverson Photomicrography; **10.6:** © Dr. Jeremy Burgess/SPL/Photo Researchers, Inc.; **10.12:** © E.H. Newcomb & S.E. Frederick/Biological Photo Service; **10.13a** p. 245: © BioPhot; **10.13b:** © Bruce Iverson Photomicrography.

Chapter 11

Opener: © Professor Malcolm B. Wilkins, Glasgow University; **Box 11.4** p. 267: © Angelina Lax/Photo Researchers, Inc.

Chapter 12

Opener: © Tom Edwards/VU; **12.2a:** © Dwight R. Kuhn; **12.3:** © Charles Steinmetz; **12.4:** © John J. Smith; **12.5:** © Michael T. Stubben/VU; **12.6:** © Jane Burton/Bruce Coleman; **12.7:** © John Gerlach/VU; **12.10:** Courtesy Peabody Museum of Natural History, Yale University; **12.13:** © Stephen Dalton/OSF/Earth Scenes; **12.14a:** © Professor Malcolm B. Wilkins, Botany Dept., Glasgow University; **12.14b:** © M.P.L. Fogden/Bruce Coleman, Inc.; **12.14c, 12.15a:** © William E. Ferguson; **12.15b:** © Professor Malcolm B. Wilkins, Botany Dept., Glasgow University; **12.17a:** © John Shaw/Tom Stack & Associates; **12.18:** © Doug Sokell/VU; **12.19:** © Wolfgang Kaehler; **12.20a:** © E.F. Anderson/VU; **12.20b:** © R.E. Litchfield/SPL/Photo Researchers, Inc.; **12.21:** © Sean Morris/OSF/Earth Scenes; **12.22a:** © Greg Crisci; **12.22b:** © M.J. Coe/OSF/Earth Scenes; **12.22c:** © G.I. Bernard/OSF/Earth Scenes; **12.22d:** © Sean Morris/OSF/Earth Scenes; **12.23a, b:** © Heather Angel; **12.24:** © McGraw-Hill Higher Education/Gary Hannan, Eastern Michigan University, photographer; **12.25:** © John Gerlach/Animals Animals; p. 291: © J.A.L. Cooke/OSF/ Earth Scenes; **12.26:** © J. Alcock/VU; **12.27a:** © Merlin D. Tuttle/Bat Conservation International/Photo Researchers, Inc.; **12.27b:** © Gerald Thompson/OSF/Animals Animals; **12.28a:** © Anthony Mercieca/Photo Researchers, Inc.; **12.28b:** © John R. Brownlie/Bruce Coleman, Inc.; **12.29:** © Scott Camazine/Photo Researchers, Inc.; **12.30a:** © PhotoDisk (World Commerce and Travel 5-202); **12.30b:** © Arthur N. Orans/Horticultural Photography; **12.30c:** © Walter H. Hodge/Peter Arnold, Inc.; **12.31:** © Jane Thomas/VU.

Chapter 13

Opener: © Dwight R. Kuhn; **13.1:** Bettmann Archive/Corbis; **13.7:** Courtesy C.R. Parks (photo by Susan Whitfield); **13.9:** © Evelyne Cudel; **13.10b:** © R.F. Evert; **13.10c:** © Edwin Reschke; **13.15:** © C.A. Hasenkampf/Biological Photo Service; **13.19:** © Kim Todd.

Chapter 14

Opener: © Runk/Schoenberger/Grant Heilman Photography, Inc.; **14.1:** © SPL/Photo Researchers, Inc.; **14.3a:** © Bill Kamin/VU; **14.3b:** © William E. Ferguson; **14.5a, b:** © McGraw-Hill Higher Education/Wayne Elisens, University of Oklahoma, photographer; **14.8:** © Science VU-USM/VU; **14.9:** © William Omerod/VU; **14.10a:** © Heather Angel; **14.10b:** © John M. Thager/VU; **14.12a:** © Bruce Berg/VU; **14.12b:** © Heather Angel; **14.12c:** © William E. Ferguson; **14.14:** © McGraw-Hill Higher Education/Wayne Elisens, University of Oklahoma, photographer; **14.17:** © Robert & Linda Mitchell; **14.19:** © Scott Camazine/Photo Researchers, Inc.; **14.21:** © Runk/Schoenberger/Grant Heilman Photography; **14.22:** © Heather Angel; **14.24:** Smithsonian Tropical Research Institute/Antonio Montaner, photographer.

Chapter 15

Opener: © Dwight R. Kuhn; **15.1:** © R. Konig/Jacana/Photo Researchers, Inc.; **15.2a:** © Robert E. Lyons; **15.2b:** © Jon Bertsch/VU; **15.2c:** © Hans Reinhard/Bruce Coleman, Inc.; **15.3:** © McGraw-Hill Higher Education/Wayne Elisens, University of Oklahoma, photographer; **15.9:** © Runk/Schoenberger/Grant Heilman Photography; **15.12:** © Michael T. Stubben/VU; **15.13:** © Runk/Schoenberger/Grant Heilman Photography; **15.17a:** © Thomas Hovland/Grant Heilman Photography, Inc.; **15.17b:** © Jeff Foott/Bruce Coleman, Inc.; **15.19:** © K.G.

Vock/Okapia/Photo Researchers, Inc.; **15.22:** © Grant Heilman/Grant Heilman Photography.

Chapter 16

Opener: © Randy Morse/Tom Stack & Associates; **16.1a:** © David Phillips/VU; **16.1b:** © M.I. Walker/Photo Researchers, Inc.; **16.2a:** © David Phillips/VU; **16.2b:** © R. Kessel-G. Shih/VU; **16.2c:** © Runk/Schoenberger/Grant Heilman Photography; **16.3:** © G. Musli/VU; **16.6:** © Peter Gregg/Imagery; **16.7a:** © Philip Sze/VU; **16.7b:** © Sinclair Stammers/SPL/Photo Researchers, Inc.; **16.7c:** © Runk/Schoenberger/Grant Heilman Photography; **16.8a:** © Ray Coleman/Photo Researchers, Inc.; **16.8b:** © E.R. Degginger/Bruce Coleman, Inc.; **16.8c:** © Richard Walters/VU; **16.8d:** © Heather Angel; **16.8e:** © Bill Keogh/VU; **16.8f:** © Hans Reinhard/Bruce Coleman, Inc.; p. 391: © E.R. Degginger/Bruce Coleman, Inc.; **16.9a:** © John D. Cunningham/VU; **16.9b:** © Doug Sherman/Geofile; **16.10:** © Robert & Linda Mitchell; **16.11a:** © Bruce Iverson Photomicrography; **16.11b:** © M.I. Walker/Photo Researchers, Inc.; **16.12a:** © Heather Angel; **16.12b:** © Gary Retherford/Photo Researchers, Inc.; **16.13a:** © M.I. Walker/Photo Researchers, Inc.; **16.13b:** © John D. Cunningham/VU; **16.13c:** © Philip Sze; **16.15:** © M. Eichelberger/VU; **16.16a:** © M.I. Walker/Photo Researchers, Inc.; **16.16b:** © Manfred Kage/Peter Arnold, Inc.; **16.17a:** © Heather Angel; **16.17b:** © Tammy Peluso/Tom Stack & Associates; **16.19a:** © Manfred Kage/Peter Arnold. Inc.; **16.19b:** © Veronika Burmeister/VU; **16.20a:** © R.G. Kessel-C.Y. Shih/VU; **16.20b:** © Biophoto Associates/Photo Researchers, Inc.; **16.21:** © John D. Cunningham/VU; **16.22:** © Roland Birke/Peter Arnold, Inc.; **16.23:** © Heather Angel.

Chapter 17

Opener: © Dwight R. Kuhn; **17.1a:** © Robert & Linda Mitchell; **17.1b:** © E. Webber/VU; **17.1c:** © Robert & Linda Mitchell; **17.1d:** © David Cavagnaro/Peter Arnold, Inc.; **17.2a:** © Robert & Linda Mitchell; **17.2b:** © George J. Wilder/VU; **17.4a:** © Jeff Gnass; **17.4b:** © Heather Angel; **17.4c:** © George Holton/Photo Researchers, Inc.; **17.4d:** © William E. Ferguson; **17.4e:** © Kingsley Stern; **17.5:** © Bruce Iverson Photomicrography; **17.6a:** © Kjell Sandved/VU; **17.6b, c:** Courtesy Howard Crum, University of Michigan Herbarium; **17.6d:** © David M. Dennis/Tom Stack & Associates; **17.6e:** Courtesy Howard Crum, University of Michigan Herbarium; **17.7a:** © John D. Cunningham/VU; **17.7b:** © Robert & Linda Mitchell; **17.9b:** Courtesy Ray F. Evert; **17.9c:** Damian S. Neuberger (Courtesy R. Evert); **17.10a:** © M.P. Gadomski/Bruce Coleman, Inc.; **17.10b:** © W. Omerod/VU; **17.11a:** © BIOS/Peter Arnold; **17.11b:** © Fred Bavendam; **17.11c:** © Dan Budnik/Woodfin Camp & Associates; p. 422: © Ira Block; **17.12:** © Grant Heilman/Grant Heilman Photography; **17.14a:** © John D. Cunningham/VU; **17.14b:** © Triarch/VU; **17.15:** © Edwin Reschke; **17.16a, b:** © Runk/Schoenberger/Grant Heilman Photography; **17.16c:** © James W. Richardson/VU; **17.17:** © Cabisco/VU; **17.18a:** © Barry Runk/Grant Heilman, Inc.; **17.18c:** © Patti Murray/Earth Scenes; **17.19a:** © Walter H. Hodge/Peter Arnold, Inc.; **17.19b:** © William E. Ferguson;

17.19c: © R.F. Ashley/VU; **17.20:** © Kjell Sandved/Butterfly Alphabet; **17.21a:** © Runk/Schoenberger/Grant Heilman Photography; **17.21b:** © Kjell Sandved/Butterfly Alphabet; **17.21c:** © Kjell Sandved/VU; **17.23a, b:** Courtesy Warren H. Wagner; **17.24a, b:** courtesy Spencer Barrett, University of Toronto; **p. 433a, b:** © William E. Ferguson; **p. 433c:** © Stephen J. Krasemann/Peter Arnold, Inc.; **17.25a:** The Field Museum, Chicago, (#GEO5637C); **17.25b:** The Field Museum, Chicago, (#CK-49T) Painting by Charles Knight; **17.25c:** Courtesy of Michele E. Kotyk, Patricia G. Gensel and James F. Basinger. From Kotyk, M.E. & J.F. Basinger. The Early Devonian (Pragian zosterophyll) *Bathurstia denticulata.* Hueber, *Canadian J. Bot.* (2000).

Chapter 18

Opener: © Gary Meszaros/Bruce Coleman, Inc.; **18.1:** Field Museum, Chicago (#GEO85723C), photo by Diane Alexander White; **18.3a:** © William E. Ferguson; **18.3b:** © Robert & Linda Mitchell; **18.3c, d:** © William E. Ferguson; **18.4:** © Robert & Linda Mitchell; **18.6a:** © Runk/Schoenberger/Grant Heilman Photography; **18.6b:** © Grant Heilman/Grant Heilman Photography; **18.6c:** © Doug Sokell/VU; **18.6d:** © McGraw-Hill Higher Education/Scott Russell, University of Oklahoma, photographer; **18.7a:** © Robert & Linda Mitchell; **18.7b:** © Stan W. Elams/VU; **18.9:** © Knut Norstog; **18.10a, b:** © Robert & Linda Mitchell; **18.11:** © William E. Ferguson; **18.12:** © Gregory G. Dimijian/Photo Researchers, Inc.; **18.13a:** © Robert & Linda Mitchell; **18.13b:** © Science VU/VU; **18.14a:** © Michael S. Thompson/Comstock; **18.14b:** © Rosamond Purcell; **18.17a** © Fritz Prenzel/Earth Scenes; **18.17b:** © Ed Reschke/Peter Arnold, Inc.; **18.17c:** © Hal Horwitz; **18.18a:** © Grant Heilman/Grant Heilman Photography; **18.18b:** © William H. Allen, Jr.; **18.19:** © Jim Steinberg/Photo Researchers, Inc.; **18.21:** © Walter H. Hodge/Peter Arnold, Inc.; **18.23:** © 1989 Silvester/Black Star; **18.24:** © Ghillean Prance; **18.25:** © Douglas Kirkland/Corbis/Sygma.

Chapter 19

Opener: © Jeff Gnass; **19.1:** © Bruce Iverson Photomicrography; **19.7:** © Don Johnston; **19.8:** © William E. Ferguson; **19.9:** © Runk/Schoenberger/Grant Heilman Photography; **19.10** © William E. Ferguson; **19.11:** © J. Lichter/Photo Researchers, Inc.; **19.12:** © Kim Todd; **19.14:** © McGraw-Hill Higher Education/Wayne Elisens, University of Oklahoma, photographer; **19.15:** © John D. Cunningham/VU; **19.16:** © B.J. O'Donnell/Biological Photo Service; **19.17:** © Brian Parker/Tom Stack & Associates; **19.18:** © Jeff Gnass; **19.19:** © John Cunningham/VU; **19.20:** © D. Newman/VU; **19.21:** © Milton H. Tierney/VU; **19.22:** © Jeff Gnass; **p.486(left):** © Steve Kaufman/Peter Arnold, Inc.; **p.486(right):** © Gregory K. Scott; **19.23:** © Nada Pecnik/VU; **19.24:** © Ron Spomer/VU; **19.25:** © Jane Thomas/VU; **19.26:** © Richard Kolar/Earth Scenes; **19.27, 19.28:** © Jeff Gnass; **19.29:** © Tom McHugh/Photo Researchers, Inc.; **19.30:** © John D. Cunningham/VU; **19.32:** © Kjell Sandved/VU; **19.33:** © Breck P. Kent/Earth Scenes; **19.34:** © Kevin Shafer/Tom Stack & Associates.

Note: Page numbers in boldface indicate term can be found in the illustration.